SOCIAL PSYCHOLOGY

FOURTH EDITION

SOCIAL PSYCHOLOGY

FOURTH EDITION

H. Andrew Michener

UNIVERSITY OF WISCONSIN—MADISON

John D. DeLamater

UNIVERSITY OF WISCONSIN—MADISON

Harcourt Brace College Publishers

Fort Worth   Philadelphia   San Diego   New York   Orlando   Austin   San Antonio
Toronto   Montreal   London   Sydney   Tokyo

Publisher	Earl McPeek
Acquisitions Editor	Linda Marshall
Market Strategist	Kathleen Sharp
Developmental Editor	Janie Pierce-Bratcher
Project Editor	Angela Williams Urquhart / Michele Tomiak
Art Director	David Day
Production Manager	Andrea A. Johnson

Cover illustration: Lisa Henderling/© The Stock Illustration Source, Inc.
ISBN: 0-15-504128-2
Library of Congress Catalog Card Number: 98–87316

Address for Orders
Harcourt Brace College Publishers, 6277 Sea Harbor Drive, Orlando, FL 32887-6777
1-800-782-4479

Address for Editorial Correspondence
Harcourt Brace College Publishers, 301 Commerce Street, Suite 3700, Fort Worth, TX 76102

Web Site Address
http://www.hbcollege.com

Harcourt Brace College Publishers will provide complimentary supplements or supplement packages to those adopters qualified under our adoption policy. Please contact your sales representative to learn how you qualify. If as an adopter or potential user you receive supplements you do not need, please return them to your sales representative or send them to: Attn: Returns Department, Troy Warehouse, 465 South Lincoln Drive, Troy, MO 63379.

Printed in the United States of America

9 0 1 2 3 4 5 6 7   039   9 8 7 6 5 4 3 2

Harcourt Brace College Publishers

Preface

About This Book.   The fourth edition of *Social Psychology* builds on the strengths of prior editions. Most importantly, the book covers the full range of phenomena of interest to social psychologists. While treating intrapsychic processes in detail, it provides strong coverage of social interaction and group processes and of larger-scale phenomena such as intergroup conflict and social movements.

Our goal in writing this book is, at it has always been, to describe contemporary social psychology and to present the theoretical concepts and research findings that make up this broad field. We have drawn on work by a wide array of social psychologists, including both those with sociological and those with psychological perspectives. This book stresses the impact of social structure and group membership on the social behavior of individuals, but it also covers the intrapsychic processes of cognition, attribution, and learning that underlie social behavior. Throughout the book we have used the results of empirical research—surveys, experiments, and observational studies—to illustrate these processes.

New to This Edition.   In developing this edition, we sought not only to keep the reader abreast of changes within the field of social psychology but also to strengthen our presentation of various topics and add research on diversity and differences among racial and ethnic groups. All of the chapters in this edition have been revised and updated. The more important of these changes follow:

- **Chapter 3** (Socialization) has been brought up to date—primarily in its discussion of the effects of day care and divorce. We have included a new section on the interpretive perspective on socialization and supplied new research on the father's involvement in socialization.

- **Chapter 7** (Symbolic Communication and Language) now presents three contemporary perspectives on communication.

- **Chapter 9** (Self-Presentation and Impression Management) covers many of the same topics as before, but we have improved the organization of the material. We have substantially updated **Chapter 10** (Helping and Altruism) and added many new references. We also included a new section on gender differences in helping and improved the coverage of good and bad moods and guilt. We revised **Chapter 11** (Aggression) to reflect the latest research on the effects of violence on television.

- The chapters on groups have been updated and restructured: In **Chapter 13** (Group Cohesion and Conformity), we updated the coverage of cohesion as well as revised and reconceptualized the section on group goals; we also expanded and updated the treatment of minority influence. **Chapter 15** (Group Productivity and Task Performance) contains substantially updated references as well as a new section on idea generation and brainstorming. In **Chapter 16** (Intergroup Conflict), we improved the coverage of stereotypes and biased perception of the out-group as well as the treatment of the impact of intergroup conflict on intragroup processes.

- We revised **Chapter 17** (Life Course and Gender Roles) to improve the content and organization of material. In **Chapter 18** (Social Structure and Personality), the discussion of the effects of roles on physical and mental health has been heavily revised to reflect the latest research.

Content and Organization.   We begin this book with a chapter on theoretical perspectives in social psychology followed by a chapter on research methods. These first two chapters provide the groundwork for all that follows. The remainder of the book is divided into four substantive sections. Section One focuses on individual social behavior. It includes chapters on socialization, self and identity, social perception and cognition, and attitudes. Section Two is concerned with social interaction—the core of social psychology. Each of the chapters in this section discusses how persons interact with others and how they are affected by this interaction. These chapters cover such topics as communication, social influence and persuasion, self-presentation and impression management, helping and altruism, aggression, and interpersonal attraction. Section Three provides extensive coverage of groups. It includes chapters on group cohesion and conformity, group structure and interaction, group productivity and task performance, and intergroup conflict. Section Four considers the relations between individuals and the wider society. These chapters treat the influence of life course and gender roles, the impact of social structure on the individual, deviant behavior, and collective behavior and social movements.

Ease of Use.   Because there are many different ways in which an instructor can organize an introductory course in social psychology, each chapter in this book has been written as a self-contained unit. Later chapters do not presume that the student has read earlier ones. This compartmentation enables instructors to assign chapters in whatever sequence they wish.

Chapters share a standard format. To make the material interesting and accessible to students, each chapter's introductory section poses four to six focal questions. These questions establish the issues discussed in the chapter. The remainder of the chapter consists of four to six major sections, each addressing one of these issues. A summary at the end of each chapter reviews the key points. Thus, each chapter poses several key questions about a topic and then considers these questions in a framework that enables students to easily learn the major ideas.

In addition, the text includes several learning aids. Tables emphasize the results of important studies. Figures illustrate important social psychological processes. Photographs dramatize essential ideas from the text. Boxes in each chapter highlight interesting or controversial issues and studies as well as discuss the applications of social psychological concepts in daily life. Key terms appear in boldface type and are listed alphabetically at the end of each chapter. A glossary of key terms appears at the end of the book.

Acknowledgments.   We extend thanks to reviewers for the fourth edition, including: Peter Burke, Washington State University; Donna Eder, Indiana University; Nancy Eisenberg, Arizona State University; Doug Maynard, Indiana University; Clark McPhail, University of Illinois; Norman Miller, University of Southern California; Diane Shinberg, University of Wisconsin—Madison; Richard Tessler, University of Massachusetts; Steve Wray, Averett College.

Throughout the writing of the various editions of this book, many colleagues have reviewed chapters and provided useful comments and criticisms. We express sincere appreciation to these reviewers of the previous editions: Robert F. Bales, Harvard University; Philip W. Blumstein, University of Washington; Marilyn B. Brewer, University of California at Los Angeles; Peter L. Callero, Western Oregon State College; Bella DePaulo, University of Virginia; Glen Elder, Jr., University of North Carolina at Chapel Hill; Gregory Elliott, Brown University; Richard B. Felson, State University of New York—Albany; John H. Fleming, University of Minnesota; Jim Fultz, Northern Illinois University; Viktor Gecas, Washington State University; Russell G. Geen, University of Missouri; Christine Grella, University of California at Los Angeles; Allen Grimshaw, Indiana University; Elaine Hatfield, University of Hawaii—Manoa; George Homans, Harvard University; Judy Howard, University of Washington; Michael Inbar, Hebrew University of Jerusalem; Dale Jaffe, University of Wisconsin—Milwaukee; Edward Jones, Princeton University; Lewis Killian, University of Massachusetts; Melvin Kohn, National Institute of Mental Health and Johns Hopkins University; Robert Krauss, Columbia University; Marianne LaFrance, Boston College; Robert H. Lee, University of Wisconsin—Madison; David Lundgren, University of Cincinnati; Steven Lybrand, University of Wisconsin-Madison; Patricia MacCorquodale, University of Arizona;

Armand Mauss, Washington State University; Douglas Maynard, University of Wisconsin—Madison; William McBroom, University of Montana; John McCarthy, Catholic University of America; Kathleen McKinney, Illinois State University; Howard Nixon II, University of Vermont; Pamela Oliver, University of Wisconsin—Madison; James Orcutt, Florida State University; Daniel Perlman, University of Manitoba; Jane Allyn Piliavin, University of Wisconsin—Madison; Michael Ross, University of Waterloo, Ontario; David A. Schroeder, University of Arkansas; Melvin Seeman, University of California at Los Angeles; Roberta Simmons, University of Minnesota; Sheldon Stryker, Indiana University; Robert Suchner, Northern Illinois University; James Tedeschi, State University of New York—Albany; Elizabeth Thomson, University of Wisconsin—Madison; Henry Walker, Cornell University; Mark P. Zanna, University of Waterloo, Ontario; Morris Zelditch, Jr., Stanford University; Louis Zurcher, University of Texas.

We express appreciation and thanks to our colleague, Shalom H. Schwartz, co-author of the first and second editions of this book, for his many contributions. We also thank the many students who used the previous editions and who provided us with feedback about the book; we have used this feedback to improve the presentation, pace, and style of the new edition.

Finally, we express thanks to the many professionals at Harcourt Brace in Fort Worth, Texas, who contributed to the process of turning the manuscript into a book. John Matthews, Senior Developmental Editor, provided information and advice early in the process. Janie Pierce-Bratcher, Developmental Editor, worked directly with us throughout the process of preparing the fourth edition; her helpfulness and upbeat attitude are gratefully acknowledged. Angela Urquhart and Michele Tomiak, Senior Project Editors, oversaw the transformation of manuscript into printed pages. Copyeditor Anne Lesser significantly improved the text's lucidity and conciseness. Susan G. Holtz, Photo Editor, worked diligently to find illustrative photographs. Senior Art Director David Day developed the book's design, format, and artwork. Our appreciation to them all.

While this book benefits greatly from feedback and criticisms, the authors accept responsibility for any errors that may remain.

H. Andrew Michener
John D. DeLamater

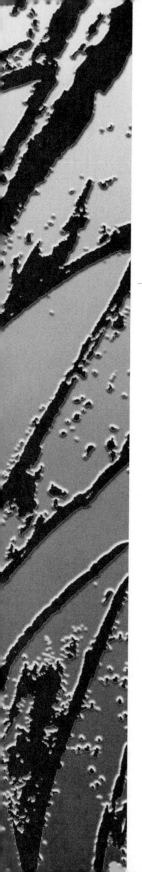

Brief Contents

Contents

SOCIAL PSYCHOLOGY

FOURTH EDITION

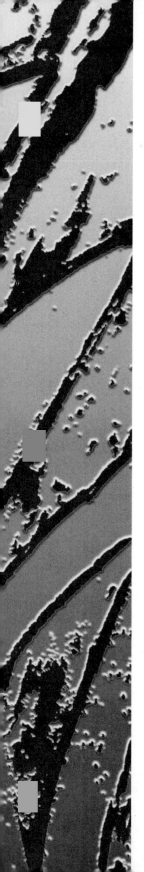

CHAPTER 1
Introduction to Social Psychology

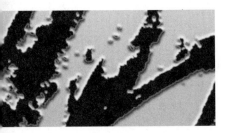

Introduction

- Why are some people effective leaders and others not?

- What makes people fall in love? What makes them fall out of love?

- Why can people cooperate so easily in some situations but not others?

- What effects do major life events like getting married, having a child, or losing a job have on physical health, mental health, and self-esteem?

- What causes conflict between groups? Why do some conflicts persist far beyond the point where participants can expect to achieve any real gains?

- Why do some people conform to norms and laws and others violate them?

- Why do people present different images of themselves in various social situations? What determines the particular images they present?

- What causes harmful or aggressive behavior? What causes helpful or altruistic behavior?

- Why are some groups so much better at performing their tasks than others?

- Why are some people more persuasive and influential than others? What techniques do they use?

- Why do stereotypes of out-groups persist even in the face of information that obviously contradicts them?

Perhaps questions such as these have puzzled you, just as they have perplexed others down through the ages. You might wonder about these issues simply because you want to understand better the social world around you. Or you might want answers for practical reasons, such as increasing your effectiveness in day-to-day relations with others.

Answers to questions such as these come from various sources. One such source is personal experience—things we learn by noting what happens when we interact with others in everyday life. Answers obtained by this means are often insightful, but they are usually limited in scope and generality, and occasionally they are even misleading. Another source is informal knowledge or advice from others who describe their own experiences to us. Answers obtained by this means are sometimes reliable, sometimes not. A third source is the conclusions reached by various thinkers—philosophers, novelists, poets, and men and women of practical affairs—who, over the centuries, have written about these issues. Often their answers have filtered down and taken the form of sayings, or aphorisms, that make up commonsense knowledge. We are told, for instance, that punishment is essential to successful child rearing ("Spare the rod and spoil the child") and that joint effort is an effective way to accomplish large jobs ("Many hands make light work"). Principles such as these reflect certain truths, and they appear to provide guidelines for action in some cases.

Although commonsense knowledge may have some merit, it also has certain drawbacks, not the least of which is that it often contradicts itself. For example, we hear that people who are similar will like one another ("Birds of a feather flock together") but also that persons who are dissimilar will like each other ("Opposites attract"). We learn that groups are wiser and smarter than individuals ("Two heads are better than one") but also that problem solving by groups entails many compromises and inevitably produces mediocre results ("A camel is a racehorse designed by a committee"). Each of these contradictory statements may hold true under particular conditions, but without a clear statement of when they apply and when they do not, aphorisms provide little insight regarding relations among people. They provide even less guidance in situations where we must make decisions. For example, when facing a choice that entails risk, which guideline should we use—"Nothing ventured, nothing gained" or "Better safe than sorry"?

If sources such as personal experience and commonsense knowledge have only limited value, how are we to attain an understanding of social interaction and relations among people? Are we forever restricted to intuition and speculation, or is there a better alternative?

One resolution to this problem—the one pursued by social psychologists—is to obtain accurate knowledge about social behavior by applying the methods of science. That is, by taking systematic observations of behavior and formulating theories that are subject to test and potential disconfirmation, we can attain a valid and comprehensive understanding of human social relations.

One goal of this book is to present some of the major findings from systematic research by social psychologists. In this chapter, we lay the foundation for this effort by addressing the following issues:

1. What exactly is "social psychology"? What are the core concerns of the field of social psychology?
2. What broad theoretical perspectives prevail within social psychology today? What are the strengths and weaknesses of each theory?
3. Is social psychology a science? That is, does social psychology have those properties that are the hallmarks of any scientific field?

What Is Social Psychology?

There are various ways to answer the question "What is social psychology?" One is to offer a formal definition of the field. Another is to list in detail the topics investigated by social psychologists. Yet another is to compare and contrast social psychology with its allied fields, psychology and sociology. In this section, we do all of these.

A Formal Definition

We define **social psychology** as the systematic study of the nature and causes of human social behavior. Note certain features of this definition. First, it states that the main concern of social psychology is human

social behavior. This includes many things—the activities of individuals in the presence of others, the processes of social interaction between two or more persons, and the relationships between individuals and the groups to which they belong.

Second, the definition states that social psychology addresses not only the nature of social behavior but also the causes of such behavior. Social psychologists seek to discover the preconditions that cause various social behaviors. Causal relations among variables are important building blocks of theory; and in turn, theory is crucial for the prediction and control of social behavior.

Third, the definition indicates that social psychologists study social behavior in a systematic fashion. In fact, they rely explicitly on research methodologies, including such formal procedures as experimentation, structured observation, and sample surveys. A description of the research methods used by social psychologists appears in Chapter 2.

Core Concerns of Social Psychology

Another way to answer the question "What is social psychology?" is to describe the topics that social psychologists actually study. Social psychologists investigate human behavior, of course, but their primary concern is human behavior in a social context. There are four *core concerns,* or major themes, within social psychology: (1) the impact that one individual has on another; (2) the impact that a group has on its individual members; (3) the impact that individual members have on the groups to which they belong; and (4) the impact that one group has on another group. The four core concerns are shown schematically in Figure 1.1.

Impact of Individuals on Individuals Individuals are affected by others in many ways. In everyday life, *communication* from others may significantly influence a person's understanding of the social world. Attempts by others at *persuasion* may change an individual's *beliefs* about the world and his or her *attitudes* toward persons, groups, or other objects. Suppose, for example, that Carol tries to persuade Debbie that all nuclear power plants are dangerous and undesirable,

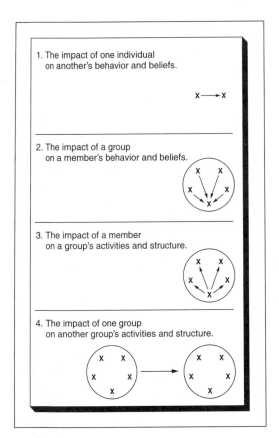

1. The impact of one individual on another's behavior and beliefs.

2. The impact of a group on a member's behavior and beliefs.

3. The impact of a member on a group's activities and structure.

4. The impact of one group on another group's activities and structure.

Figure 1.1   The Core Concerns of Social Psychology

and therefore should be closed. If successful, Carol's persuasion attempt would probably change Debbie's beliefs and perhaps affect her future actions (picketing nuclear power plants, advocating nonnuclear sources of power, and the like).

Beyond influence and persuasion, the outcomes obtained by individuals in everyday life are often affected by the actions of others. A person caught in an emergency situation, for instance, may be helped by an altruistic bystander. In another situation, one person may be damaged or wounded by another's aggressive acts. Social psychologists have investigated the nature and origins of both *altruism* and *aggression,* as well as other interpersonal motivations such as cooperation and competition.

Also relevant here are various *interpersonal sentiments.* One individual may develop strong attitudes toward another (liking, disliking, loving, hating) based on who the other is and what he or she does. Debbie, for instance, may like Louis but dislike David and Will. Social psychologists investigate these issues to discover why individuals develop positive attitudes toward some but negative attitudes toward others.

Impact of Groups on Individuals   A second concern of social psychology is the impact of a group on the behavior of its individual members. Because individuals belong to many different groups—families, work groups, seminars, and clubs—they spend many hours each week interacting with others. Groups influence and regulate the behavior of their members, typically by establishing norms or rules. One result of this is *conformity,* the process by which a group member adjusts his or her behavior to bring it into line with group norms. For example, college fraternities and sororities have norms—some formal and some informal—that stipulate how members should dress, what meetings they should attend, who they can date and who they should avoid, how they should behave at parties, and the like.

Groups also exert substantial long-term influence on their members through *socialization,* a process that enables groups to regulate what their members learn. Socialization assumes that the members will be adequately trained to enact the roles they play in the group and the larger society. It shapes the knowledge, values, and skills of group members. One product of socialization is language skills; another product is political and religious beliefs and attitudes; yet another is our conception of *self.*

Impact of Individuals on Groups   A third concern of social psychology is the impact of individuals on group processes and products. Just as any group influences the behavior of its members, these persons, in turn, may influence the group itself. Individuals contribute to *group productivity* and *group decision making.* In addition, they provide *leadership,* which involves the enactment of various functions (planning, organizing, controlling) necessary for successful group performance. Without effective leadership, coordination among members will falter and the group drift or fail. In addition, individuals and minority coalitions often *innovate change* in group structure and procedures. Both leadership and innovation, of course, depend on the initiative, insight, and risk-taking ability of individuals.

Impact of Groups on Groups

Impact of Groups on Groups A fourth concern of social psychology is the impact of one group on the activities and structure of another group. Relations between two groups may be friendly or hostile, cooperative or competitive. These relationships, which are based in part on members' identities and may entail group *stereotypes,* can affect the structure and activities of each. Of special interest is *intergroup conflict,* with its accompanying tension and hostility. Violence may flare up, for instance, between two teenage street gangs disputing territorial rights or between racial groups competing for scarce jobs. Conflicts of this type affect the interpersonal relations between groups as well as within each group. Social psychologists have long studied the emergence, persistence, and resolution of intergroup conflict.

Relation to Other Fields

Social psychology bears a close relationship to several other fields, especially sociology and psychology. To understand this relationship, first consider these other fields.

Sociology is the scientific study of human society. It addresses such topics as social institutions (family, religion, and politics), stratification within society (class structure, race and ethnicity, sex roles), basic social processes (socialization, deviance, social control), and the structure of social units (groups, networks, formal organizations, bureaucracies).

In contrast, *psychology* is the scientific study of the individual and individual behavior. Although this behavior may be social in character, it need not be. Psychology addresses such topics as human learning, perception, memory, intelligence, emotion, motivation, and personality.

Social psychology bridges the gap between sociology and psychology. In fact, some view it as an interdisciplinary field. Both sociologists and psychologists have contributed to social psychological knowledge. Social psychologists working in the sociological tradition rely primarily on sample surveys and observational techniques to gather data. These investigators are most interested in the relationship between individuals and the groups to which they belong. They emphasize such processes as socialization, conformity and deviation, social interaction, self-presentation, leadership, recruitment to membership, cooperation and competition, and the like. Social psychologists working in the psychological tradition rely heavily on laboratory experimental methodology. Their primary concern is how an individual's behavior and internal states are affected by social stimuli (often other persons). They emphasize such topics as the self, person perception and attribution, attitudes and attitude change, personality differences in social behavior, social learning and modeling, altruism and aggression, interpersonal attraction, and so on.

Thus sociologically oriented and psychologically oriented social psychologists differ in their outlook and emphasis. As we might expect, this leads them to formulate different theories and to conduct different programs of research. Yet these differences are best viewed as complementary rather than as conflicting. Social psychology as a field is the richer for them.

Theoretical Perspectives in Social Psychology

Yesterday at work, Warren reported to his boss that he would not be able to complete an important project on schedule. To Warren's surprise, the boss became enraged and told him to complete the task by the following Monday—or else! Warren was not entirely sure what to make of this behavior—the boss had shouted at him—but he decided to take the threat seriously. That evening, talking with his girlfriend Alice, Warren announced that he would have to work overtime at the office, so he could not take her to a party on Friday evening as originally planned. Alice immediately got mad at Warren—she definitely wanted to go and he had promised several times to take her—and threw a paperweight at him. By now, Warren was very distressed and also a little perplexed.

Reflecting on these two events, Warren noticed that they had some characteristics in common. To explain the behavior of his boss and his girlfriend, he formed a general proposition: "If you fail to deliver on promises and thereby block someone's goals, he or she will get mad at you." He was happy with this simple formulation until the next day when he read an unusual newspaper story: "MAN IS FIRED FROM JOB, THEN SHOOTS HIS DOG IN ANGER." Warren wondered about this event and then concluded that his own theory needed revision. The new version included several propositions: "If someone's goals are blocked, he or she will become frustrated. If someone is frustrated, he or she will become aggressive. If someone is aggressive, he or she will attack either the source of the frustration or a convenient surrogate."

In his own way, Warren is starting to do informally the same thing that social psychologists do more elaborately and systematically. Starting from some observations regarding social behavior, Warren is attempting to formulate a theory to explain the observed facts. As this term is used here, a **theory** is a set of interrelated propositions that organizes and explains a set of observed phenomena. Theories usually pertain not just to some particular event but to whole classes of events. Moreover, as Warren's example indicates, a theory goes beyond mere observable facts in that it postulates causal relations among variables. If a theory is valid, it enables its user to explain the phenomena under consideration and to make predictions about events not yet observed.

In social psychology, no single theory explains all phenomena of interest; rather, the field includes many different theories. It is useful to distinguish between middle-range theories and theoretical perspectives. **Middle-range theories** are narrow, focused frameworks that identify the conditions that produce a specific social behavior. They are usually scientific-causal in nature; that is, they are formulated in terms of cause and effect. For example, one middle-range theory tries to explain the processes by which persuasion produces attitude change (Petty & Cacioppo, 1986a, 1986b). Another middle-range theory tries to specify how majorities and minorities within groups differ qualitatively in the ways they influence their targets (Moscovici, 1985; Nemeth, 1986). Yet another middle-range theory specifies the conditions under which contact between members of different racial and ethnic groups will cause stereotypes to change or disappear (Rothbart & John, 1985). Throughout this book, we describe many middle-range theories.

In addition to middle-range theories, social psychology includes **theoretical perspectives.** Broader in scope than middle-range theories, theoretical perspectives offer general explanations for a wide array of social behaviors in a variety of situations. These general explanations are rooted in explicit assumptions about human nature. Theoretical perspectives serve an important function for the field of social psychology. By making certain assumptions regarding human nature, a theoretical perspective establishes a vantage from which we can examine a range of social behaviors. Because any perspective highlights certain features and downplays others, it enables us to more clearly "see" certain aspects or features of social behavior. The fundamental value of any theoretical perspective lies in its applicability across many situations; it provides a frame of reference for interpreting and comparing a wide range of social situations and behaviors.

Social psychology includes several distinct theoretical perspectives. Four of the more important ones are (1) role theory, (2) reinforcement theory, (3) cognitive theory, and (4) symbolic interaction theory. In the following sections, we discuss each of these perspectives.

Role Theory

Several months ago, Barbara was invited to participate in a stage production of Molière's comedy *The Learned Woman.* She was offered the role of Martine, a kitchen servant dismissed from her job for using poor grammar. Barbara enthusiastically accepted the role and learned her part well. The theater group presented the play six times over a period of 3 weeks. Barbara played the role of Martine in the first four shows, but then she got sick. Fortunately, Barbara's understudy was able to substitute as Martine during the final two shows. Barbara's performance was very good, but so was the understudy's. In fact, one reviewer wrote that it was difficult to tell them apart.

Barbara's friend Craig is more interested in football than in theater. A member of the college football team, Craig plays the position of fullback. Although very large and strong, he is a third-string player because he has the unfortunate habit of fumbling the ball, sometimes at the worst possible moment. But Craig believes that with another year's experience and some improvements in his technique, he could perform better than the other fullbacks and win a place in the team's starting lineup.

Although active in different arenas, Craig and Barbara have something in common: They are both performing roles. When Barbara appears on stage, she performs the role of kitchen servant. When Craig appears on the football field, he performs the role of fullback. In both cases, their behavior is guided by role expectations held by other people. Roles consist of a set of rules (that is, expectations held by others) that function as plans or blueprints and guide behavior.

Barbara's role is defined by very specific expectations. Her part calls for her to say certain things and perform certain actions at specified points in the plot. There is virtually no room for her to improvise or deviate from her lines. Craig's role is also fairly specific.

He has to carry out given assignments—running and blocking—on each of the plays by his team. There is some latitude in exactly how he does these things, but not a great deal. Whenever he misses a block, all the coaches and players know it.

In everyday life, we all perform roles. Anyone who holds a job is performing a role. For instance, unlike Barbara's play-acting role as a kitchen servant, an advertising executive's work role does not dictate exactly what lines are to be spoken. But it will certainly specify what goals should be pursued, what tasks must be accomplished, and what performances are required.

The theoretical perspective that best addresses behavior of this type is **role theory** (Biddle, 1979, 1986; Heiss, 1981; Turner, 1990). Role theory holds that a substantial proportion of observable, day-to-day social behavior is simply persons carrying out their roles, much as actors carry out their roles on the stage or ballplayers theirs on the field.

Propositions in Role Theory The following propositions are central to the role theory perspective:

1. People spend much of their lives participating as members of groups and organizations.

To run smoothly, a production like this newscast requires that all the participants perform tasks and enact roles specified by their work group.

2. Within these groups, people occupy distinct positions (fullback, advertising executive, police sergeant, and the like).

3. Each of these positions entails a **role,** which is a set of functions performed by the person for the group. A person's role is defined by expectations (held by other group members) that specify how he or she should perform.

4. Groups often formalize these expectations as **norms,** which are rules specifying how a person should behave, what rewards will result for performance, and what punishments will result for non-performance.

5. Individuals usually carry out their roles and perform in accordance with prevailing norms. In other words, people are primarily conformists; they try to meet the expectations held by others.

6. Group members check each individual's performance to determine whether it conforms with the norms. If an individual meets the role expectations held by others, then he or she will receive rewards in some form (acceptance, approval, money, and so on). If he or she fails to perform as expected, however, then group members may embarrass, punish, or even expel that individual from the group. The anticipation that others will apply sanctions ensures performance as expected.

Impact of Roles   Role theory implies that if we (as analysts) have information about the role expectations for a specified position, we can then predict a significant portion of the behavior of the person occupying that position. According to role theory, to change a person's behavior, it is necessary to change or redefine his or her role. This might be done by changing the role expectations held by others with respect to that person or by shifting that person into an entirely different role (Allen & Van de Vliert, 1982). For example, if the football coach shifted Craig from fullback to tight end, Craig's behavior would change to match the role demands of his new position. Craig himself may experience some strain while adjusting to the new role, but his behavior will change.

Role theory maintains that a person's role determines not only behavior but also beliefs and attitudes. In other words, individuals bring their attitudes into congruence with the expectations that define their roles. A change in role should lead to a change in attitude. One illustration of this effect appears in a classic study of factory workers by Lieberman (1965). In the initial stage of this study, researchers measured the attitudes of workers toward union and management policies in a midwestern home appliance factory. During the following year, a number of these workers changed roles. Some were promoted to the position of foreman, a managerial role; others were elected to the position of shop steward, a union role.

About a year after the initial measurement, workers' attitudes were reassessed. The attitudes of workers who had become foremen or shop stewards were compared to those of workers who had not changed roles. The recently promoted foremen expressed more positive attitudes than the nonchangers toward the company's management and the company's incentive system, which paid workers in proportion to what they produced. In contrast, recently elected shop stewards expressed more positive attitudes than the nonchangers toward the union and favored an incentive system based on seniority, not productivity. The most efficient explanation of these results is that the workers' attitudes shifted to fit their new roles, as predicted by role theory.

In general, the roles that people occupy not only channel their behavior but also shape their attitudes. Roles can influence the values that people hold and affect the direction of their personal growth and development. We discuss these topics in more depth in Chapters 3, 14, and 18.

Limitations of Role Theory   Despite its usefulness, role theory has difficulty explaining certain kinds of social behavior. Foremost among these is *deviant behavior,* which is any behavior that violates or contravenes the norms defining a given role. Most forms of deviant behavior, whether simply a refusal to perform as expected or something more serious like commission of a crime, disrupt interpersonal relations. Deviant behavior poses a challenge to role theory because it contravenes the assumption that people are essentially conformist. Of course, a certain amount of deviant behavior can be explained by the fact that people are sometimes ignorant of the norms. Deviance may also result whenever people face conflicting and/or incompatible expectations from several other people (Miles, 1977). In general, however, deviant behavior is an unexplained and problematic exception from the standpoint of role theory. In Chapters 13 and 19, we discuss the conditions that cause deviant behavior and the reactions of others to such behavior.

Even critics of role theory acknowledge that a substantial portion of all social behavior can be explained as conformity to established role expectations. But role theory does not and cannot explain how role expectations came to be what they are in the first place. Nor does it explain when and how role expectations change. Without accomplishing these tasks, role theory can provide no more than an incomplete explanation of social behavior.

Reinforcement Theory

Reinforcement theory, another major perspective on social behavior, begins with the premise that social behavior is governed by external events. Its central proposition is that people will more likely perform a specific behavior if it is followed directly by the occurrence of something pleasurable or by the removal of something aversive; likewise, they will more likely refrain from performing a particular behavior if it is followed by the occurrence of something aversive or by the removal of something pleasant.

The use of reinforcement is illustrated by an early study by Verplanck (1955). The study's point was to show that one person can alter the course of a conversation by the selective use of social approval (a reinforcer). Students conducting the study sought out situations in which each could be alone with another person and conduct a conversation. During the first 10 minutes, the student engaged the other in polite but neutral chitchat; the student was careful neither to support nor to reject opinions expressed by the other. During this period, the student privately noted the number of opinions expressed by the other and unobtrusively recorded this information by doodling on a piece of paper.

After this initial period, the student shifted behavior and expressed approval whenever the other ventured an opinion. The student indicated approval with reinforcers like "I agree," "That's so," and "You're right," and by smiling and nodding in agreement. The student continued this pattern of reinforcement

for 10 minutes, all the while noting the number of opinions expressed by the other.

Next, the student shifted behavior again and suspended reinforcement. Any opinions expressed by the other were met with noncommittal remarks or subtle disagreement. As before, the student noted the number of opinions expressed.

The results of the study show that during the "reward period" (when the student expressed approval), the subjects expressed opinions at a higher rate than they had during the initial baseline period. Moreover, during the "extinction period" (when the student suspended approval), about 90% expressed opinions at a lower rate than they had during the "reward period." Overall, the subjects' behavior during the conversation was substantially influenced by social approval.

Some Concepts of Reinforcement Theory

Reinforcement theory has a long tradition within psychology. It began at the turn of the century with research by Pavlov and by Thorndike, and evolved through the work of Allport (1924), Hull (1943), and Skinner (1953, 1971). The reinforcement perspective holds that behavior is determined primarily by external events, not by internal states. Thus the central concepts of reinforcement theory refer to events that are directly observable. Any event that leads to an alteration or change in behavior is called a *stimulus.* For example, a traffic light that changes to red is a stimulus, as is a wailing tornado siren. The change in behavior induced by a stimulus is called a *response.* Drivers respond to red lights by stopping; families respond to tornado sirens by rushing for shelter. A **reinforcement** is any favorable outcome that results from a response; reinforcement strengthens the response—that is, it increases the probability it will be repeated. In Verplanck's study, the students' social approval was a positive reinforcer that strengthened the subjects' response of expressing opinions. Responses that are not reinforced tend to disappear and not be repeated.

Reinforcement is important in some forms of learning, most notably through conditioning (Mazur, 1998). In **conditioning,** a contingency is established between emitting a response and subsequently receiving a reinforcement. If a person emits a particular response and this response is then reinforced, the connection between these is strengthened; that is, the person will more probably emit the same response in the future in hopes of again receiving reinforcement.

A related process, *stimulus discrimination,* occurs when a person learns the exact conditions under which a response will be reinforced. For example, Karl, a young child, has learned that if his mother rings the dinner bell (a stimulus), he should respond by coming indoors, washing his hands, and sitting in the appropriate place at the table. His mother then puts food on his plate (a reinforcer). He has also learned, however, that if he performs the same response (washing his hands and sitting down at the table) without first hearing the stimulus (dinner bell), his mother merely tells him that he's too early and cannot have food until later. Thus Karl has learned to discriminate among stimulus conditions (bell vs. no bell), and he knows that reinforcement (food) is obtained only by making the response in the presence of a specific stimulus (bell).

Social Learning Theory   Although learning based on reinforcement and conditioning is important, it is not the only form of learning. A central proposition of **social learning theory** (Bandura, 1977) is that one person (the learner) can acquire new responses simply by observing the behavior of another person (the *model*). This observational learning process, called **imitation,** is distinguished by the fact that the learner neither performs a response nor receives any reinforcement. Many social responses are learned through imitation. For instance, children learn ethnic and regional speech patterns by imitating adult speakers around them.

In imitation, the learner watches the model's behavior and thereby comes to understand how to behave in a similar manner. Learning of this type can occur without any external reinforcement. But the issue of whether the learner will actually perform the behaviors learned through observation may hinge on the consequences that performance has for the learner—that is, on whether or not the learner receives reinforcement for performance. A young girl, for example, might observe that her older sister puts on makeup before going out with friends; in fact, if she watches closely enough, she might learn precisely

how to apply makeup the right way. But whether or not the little girl actually puts makeup on herself and wears it around the house may depend heavily on reinforcements she receives for doing so. If she knows, for instance, that her mother strongly disapproves of little girls wearing cosmetics, she may hesitate to use what she has learned from her big sister.

In sum, learning theory holds that individuals acquire new responses through conditioning and imitation. Both conditioning and imitation are important processes in socialization, and they help to explain how persons acquire complex social behaviors. Chapter 3 discusses these processes in more detail.

Social Exchange Theory   Another important process based on the principle of reinforcement is social exchange. **Social exchange theory** (Cook, 1987; Homans, 1974; Kelley & Thibaut, 1978) uses the concept of reinforcement to explain stability and change in relations between individuals. This theory assumes that individuals have freedom of choice and often face social situations in which they must choose among alternative actions. Any action provides some rewards and entails some costs. There are many kinds of socially mediated rewards—money, goods, services, prestige or status, approval by others, and the like. The theory posits that individuals are hedonistic—they try to maximize rewards and minimize costs. Consequently, they choose actions that produce good profits (profits = rewards – costs) and avoid actions that produce poor profits.

As its name indicates, social exchange theory views social relationships primarily as exchanges of goods and services among persons. People participate in relationships only if they find that these provide profitable outcomes. An individual judges the attractiveness of a relationship by comparing the profits it provides against those available in other, alternative relationships. If a person is participating in a social relationship and receiving certain outcomes, then the level of outcomes available in the best alternative relationship is termed that person's *comparison level for alternatives.* More concretely, suppose an executive is employed by a food products manufacturer when she unexpectedly is offered an attractive job by a competing firm. The new job entails some additional

An exchange taking place—a scalper offers tickets to a football fan for a sold-out game. In transactions of this type, the price is often determined through negotiation between buyer and seller.

ments with attractive outsiders are not readily available. In other words, they are more likely to stay when the rewards are high, the costs are low, and the comparison level for alternatives is low. Effects of this type are predicted by social exchange theory.

Exchange theory also predicts the conditions under which people try to change or restructure their relationships. Central to this is the concept of **equity** (Adams, 1963; Walster, Walster, & Berscheid, 1978). A state of equity exists in a relationship when participants feel that the rewards they receive are proportional to the costs they bear. For example, a supervisor may earn more money than a line worker and receive better benefits on the job. But the line worker may nevertheless feel the relationship is equitable because the supervisor bears more responsibility and has a higher level of education.

If, for some reason, a participant feels the allocation of rewards and costs in a relationship is inequitable, then the relationship is potentially unstable. People find inequity difficult to tolerate—they may feel cheated or exploited and become angry. Social exchange theory predicts that people will try to modify an inequitable relationship. Most likely, they will attempt to reallocate costs and rewards so that equity is established.

Limitations of Reinforcement Theory

Despite its usefulness in illuminating why relationships change and how people learn, reinforcement theory has been criticized on various grounds. One criticism is that reinforcement theory portrays individuals primarily as reacting to environmental stimuli, rather than as initiating behavior based on imaginative or creative thought. The theory does not account easily for creativity, innovation, or invention. A second criticism is that reinforcement theory largely ignores or downplays other motivations. It characterizes social behavior as hedonistic, with individuals striving to maximize profits from outcomes. Thus it cannot easily explain selfless behavior such as altruism and martyrdom. Despite its limitations, reinforcement theory has enjoyed substantial success in explaining why individuals persist in emitting certain behaviors, how they learn new behaviors, and how they influence the behavior of others through

responsibilities, but it also pays a considerably higher salary and provides more benefits. This job offer has the effect of substantially increasing the executive's comparison level for alternatives. In this case, exchange theory predicts that she will leave her job for the new one or possibly use the outside offer as a bargaining chip when dealing with her current employer. She likely will stay with her current employer only if he promotes her to a new position with greater rewards.

Concepts of this type apply not only to work relations but also to personal relations. For instance, a study of heterosexual couples in long-term dating relationships shows that rewards and costs can explain whether persons stay in or exit from such relationships (Rusbult, 1983; Rusbult, Johnson, & Morrow, 1986). Results of this study indicate that individuals are more likely to stay when the partner is physically and personally attractive, when the relationship does not entail undue hassle (high monetary costs, broken promises, arguments), and when romantic involve-

exchange. Ideas from reinforcement and exchange theory are discussed throughout this book, especially in Chapters 3, 8, 10, 11, 12, and 14.

Cognitive Theory

Another theoretical perspective within social psychology is **cognitive theory,** the basic premise of which is that the mental activities of the individual are important determinants of social behavior. These mental activities, called **cognitive processes,** include perception, memory, and judgment, as well as problem solving and decision making. Cognitive theory does not deny the importance of external stimuli, but it maintains that the link between stimulus and response is not mechanical or hardwired. Rather, the individual's cognitive processes intervene between external stimuli and behavioral responses. Individuals not only actively interpret the meaning of stimuli but also select the actions to be made in response to stimuli.

Historically, the cognitive approach to social psychology has been influenced by the ideas of Koffka, Kohler, and other theorists in the *gestalt* movement within psychology. Central to gestalt psychology is the principle that people respond to configurations of stimuli rather than to a single, discrete stimulus. In other words, people understand the meaning of a stimulus only by viewing it in the context of an entire system of elements in which it is embedded. A chess master, for example, would not assess the importance of a chess piece on the board without considering its location and strategic capabilities vis-à-vis all the other pieces currently on the board. To comprehend the meaning of any element, we must look at the whole of which it is a part.

Modern cognitive theorists (Fiske & Taylor, 1991; Markus & Zajonc, 1985; Wyer & Srull, 1984) depict humans as active in selecting and interpreting stimuli. According to this view, people do more than react to their environment; they actively structure their world cognitively. First, because they cannot possibly attend to all the complex stimuli that surround them, they select only those stimuli that are important or useful to them and ignore the others. Second, they actively control what categories or concepts they use to interpret the stimuli in the environment. One

implication of this, of course, is that several individuals can form dramatically different impressions of a complex stimulus in the environment.

Consider, for example, what happens when several people view a vacant house displaying a bright "FOR RENT" sign. When a building contractor passes the house, he pays primary attention to the quality of the house's construction. He sees lumber, bricks, shingles, and glass, as well as some repairs that need to be made. Another person, a potential renter, sees the house very differently. She notes that it is located close to her job and wonders whether the neighborhood is safe and whether the house is expensive to heat in winter. The realtor trying to rent the house construes it in still different terms—cash flow, occupancy rate, depreciation, mortgage, and amortization. One of the preschool kids living in the neighborhood has yet another view; observing that no person has lived in the house for several months, she is convinced the house is haunted.

Cognitive Structure and Schemas Central to this perspective is the concept of **cognitive structure,** which refers broadly to any form of organization among cognitions (concepts and beliefs). Because a person's cognitions are interrelated, cognitive theory gives special emphasis to exactly how they are structured and organized in memory, as well as to how they affect a person's judgments.

Social psychologists have proposed that individuals use specific cognitive structures called **schemas** to make sense of complex information about other persons, groups, and situations. The term "schema" is derived from the Greek word for "form," and it refers to the form or basic sketch of what we know about people and things. For example, our schema for "law student" might be a set of traits thought to be characteristic of such persons: intelligent, analytic and logical, argumentative (perhaps even combative), thorough and workmanlike with an eagle eye for details, strategically skillful in interpersonal relations, and (occasionally) committed to seeing justice done. Our schema, no doubt, reflects our own experience with lawyers and law students, as well as our conception of what traits are necessary for success in the legal profession. That we hold this schema does not

Individuals in this crowd at a rock concert may be viewing the same battle of the bands, but they surely are perceiving it in different ways. What one perceives depends heavily on one's schemas.

mean we believe everyone with this set of characteristics is a law student or that every law student will have all of these characteristics. We might be surprised, however, if we met someone who impressed us as unmethodical, illogical, withdrawn, inarticulate, inattentive, sloppy, and not very intelligent, and then later discovered that she was a law student.

Schemas are important in social relations because they help us interpret the environment efficiently. Whenever we encounter a person for the first time, we usually form an impression of what he or she is like. In doing this, we not only observe the person's behavior but also rely on knowledge of similar persons we have met from the past—that is, we use our schema regarding this type of person. Schemas help us process information by enabling us to recognize which personal characteristics are important in the interaction and which are not. They structure and organize information about the person, and they help us remember information better and process it more quickly. Sometimes they fill gaps in knowledge and enable us to make inferences and judgments about others.

To illustrate further, consider a law school admissions officer who faces the task of deciding which candidates to admit as students. To assist in processing applications, he uses a schema for "strong law student candidate" that is based on traits believed to predict success in law school and beyond. The admis-

sions officer doubtless pays close attention to information regarding candidates that is relevant to his schema for law students, and he most likely ignores or downplays other information. LSAT scores do matter, whereas eye color does not; undergraduate GPA does matter, whereas ability to throw a football does not; and so on.

Schemas are rarely perfect as predictive devices, and the admissions officer probably will make mistakes, admitting some candidates who fail to complete law school and turning down some candidates who would have succeeded. Then, too, another admissions officer with a different schema might admit a different set of students to law school. Despite their drawbacks, schemas are often quite superior to using no systematic framework at all. Schemas are discussed in more detail in Chapter 5.

Cognitive Consistency   One way to study cognitive structure is to observe changes that occur in a person's cognitions when these are under challenge or attack. The changes will reveal facts about the underlying structure or organization of cognitions. An important idea emerging from this approach is the **principle of consistency** (Heider, 1958; Newcomb, 1968), which maintains that individuals strive to hold ideas that are consistent or congruous with one another, rather than ideas that are inconsistent or incongruous. If a person holds several ideas that are incongruous or inconsistent, then he or she will experience internal conflict. In reaction, he or she will likely change one or more ideas, thereby making them consistent and resolving this conflict.

As an illustration, suppose you hold the following cognitions about your friend Jeff: (1) Jeff has been a good friend for 6 years; (2) you dislike hard drugs and the people who use them; and (3) Jeff has recently started using hard drugs. These cognitions are obviously interrelated, and they are also incongruous with one another. The principle of consistency predicts that a change in cognitions will occur. That is, you will change either your negative attitude toward drugs or your positive attitude toward Jeff, or possibly you will intervene and try to change Jeff's behavior.

Social psychologists have developed several useful theories based on the general notion of consistency.

Among these are *balance theory* and the *theory of cognitive dissonance* (see Chapter 6).

Cognitive theory has made many important contributions to social psychology. It treats such diverse phenomena as self-concept (Chapter 4), perception of persons and attribution of causes (Chapter 5), attitude change (Chapter 6), impression management (Chapter 9), and group stereotypes (Chapters 5 and 16). In these contexts, cognitive theory has produced many insights and striking predictions regarding individual and social behavior.

Limitations of Cognitive Theory   One drawback of cognitive theory is that it simplifies—and sometimes oversimplifies—the ways in which people process information, an inherently complex phenomenon. Another drawback is that cognitive phenomena are not directly observable; they must be inferred from what people say and do. This means that compelling and definitive tests of theoretical predictions from cognitive theory are sometimes difficult to conduct. Overall, however, the cognitive perspective is among the more popular and productive approaches within social psychology.

Symbolic Interaction Theory

A fourth perspective in social psychology is **symbolic interaction theory** (Charon, 1995; Stryker, 1980, 1987). Important early contributions to this perspective include works by Mead (1934) and Blumer (1969b). Like the cognitive perspective, symbolic interactionism stresses cognitive process (thinking and reasoning), but it places more emphasis on the interaction between the individual and society. The basic premise of symbolic interactionism is that human nature and social order are products of symbolic communication among people. In this perspective, a person's behavior is constructed through give-and-take during interaction with others. Behavior is not merely a response to stimuli, nor is it merely an expression of inner biological drives, profit maximization, or conformity to roles or norms. Rather, a person's behavior emerges through communication and interaction with others.

People can communicate successfully with one another only to the extent that they ascribe similar *meanings* to objects. An object's meaning for a person depends not so much on the properties of the object itself but on what the person might do with the object. In other words, an object takes on meaning only in relation to a person's plans. A wine merchant, for example, might see a glass bottle as a container for her product; an interior decorator might see it as an attractive vase for some silk flowers; a man in a drunken brawl might see it as a weapon with which to hit his opponent.

Symbolic interaction theory views humans as proactive and goal seeking. People formulate plans of action to achieve their goals. Many plans, of course, can be brought to realization only through cooperation with other people. To establish cooperation with others, meanings of things must be shared and consensual. If the meaning of something is unclear or contested, an agreement must be developed through give-and-take before cooperative action is possible. For example, if a man and woman have begun to date one another, and he invites her up to his apartment, exactly what meaning does this proposed visit have? One way or another, they will have to achieve some agreement about the purpose of the visit before joint action is possible. In symbolic interaction terms, they would need to develop a consensual *definition of the situation*. The man and woman might achieve this through explicit negotiation or perhaps through tacit, nonverbal communication. But without some agreement regarding the definition of the situation, the woman may have difficulty in deciding whether to accept the invitation, the man may find himself behaving in an atypically awkward manner, and cooperative action will be difficult.

Symbolic interactionism portrays social interaction as having a tentative, developing quality. To fit their actions together and achieve consensus, people interacting with one another must continually negotiate new meanings or reaffirm old meanings. Each person formulates plans for action, tries them out, and then adjusts them in light of responses by others. Thus, social interaction always has some degree of unpredictability and indeterminacy.

Central to social interaction is the process of *role taking,* in which an individual imagines how he or she looks from the other person's standpoint. For example, if an employee is seeking an increase in salary, he might first imagine how his boss would react to one type of request or another. To do this, he might use knowledge gained in past interactions with her, as well as recall what he has heard from others about her reactions to salary requests. By viewing his action from her standpoint, he may be able to anticipate what type of request would produce the desired effect. If he then actually makes a request of this type and she reacts as expected, his role taking has succeeded. Through the role-taking process, cooperative interaction is established.

Symbolic interaction theory emphasizes that a person can act not only toward others but also toward his or her **self.** That is, an individual can engage in self-perception, self-evaluation, and self-control, just as he or she might perceive, evaluate, and control others. One important component of self is identity, the person's understanding as to who he or she is. For interaction among persons to proceed smoothly, there must be some consensus with respect to the identity of each. In other words, for each person there must be an answer to the question "Who am I in this situation, and who are these other people?" Only by answering this question in some detail can each person understand the implications (meanings) that others have for his or her plan of action.

Sometimes a person's identity is very unusual, and in consequence interaction becomes awkward, difficult, or even impossible. Consider the old tale by Cervantes of a man, temporarily deranged, who thought he was made of glass (Shibutani, 1961). This man's conception of himself created problems both for him and for others. Whenever people came near, he screamed and implored them to keep away for fear they would shatter him. He refused to eat anything hard and insisted on sleeping only in beds of soft straw. Concerned that loose tiles might fall on him from the rooftops, he walked in the middle of the street. When a wasp stung him in the neck, he did not swat it away because he was afraid of smashing himself to bits. Because glass is transparent and skin is

not, he claimed that his body's unusual construction enabled his soul to perceive things more clearly than others, and he offered to assist people perplexed by difficult problems. He gradually developed a reputation for astonishing insight, and many persons came to him seeking advice. In the end, a wealthy patron hired a bodyguard to protect him from outlaws and the mischievous boys who threw stones at him.

In daily life, of course, we are not likely to meet someone who believes he or she is made of glass. But we might encounter people who believe they are unusually fragile or remarkably strong or superhumanly intelligent or in contact with the supernatural. Persons with unusual identities can create problems in social interaction, and they make it difficult to achieve consensus. Cooperative action is difficult without such consensus, for people simply do not know how to relate to individuals who insist they are Superman, Napoleon, Goldilocks, or Jesus Christ.

The self occupies a central place in symbolic interaction theory because social order is hypothesized to rest in part on self-control. The individual strives to maintain self-respect in her own eyes, but because she is continually engaging in role taking, she sees herself from the standpoint of the others with whom she interacts. To maintain self-respect she thus must meet the standards of others, at least to some degree.

Of course, the individual will care more about the opinions and standards of some persons than about those of others. The persons whose opinions she cares most about are called **significant others.** Typically, these are people who control important rewards or who occupy central positions in groups to which the individual belongs. Because their positive opinions are highly valued, significant others have relatively more influence over the individual's behavior.

In sum, the symbolic interactionist perspective has several strong points. It recognizes the importance of the self in social interaction. It stresses the central role of symbolic communication and language in personality and society. It addresses the processes involved in achieving consensus and cooperation. And it illuminates why people try to maintain face and avoid embarrassment. Many of these topics are discussed in detail in later chapters. The self is discussed in Chapter 4, symbolic communication and language are taken up in Chapter 7, and self-presentation and impression management are treated in Chapter 9.

Limitations of Symbolic Interaction Theory

Critics of symbolic interactionism have pointed to various shortcomings. One criticism concerns the balance between rationality and emotion. Some critics argue that this perspective overemphasizes rational, self-conscious thought and deemphasizes unconscious or emotional states. A second criticism concerns the model of the individual implicit in symbolic interaction theory. The individual is depicted as a specific personality type—an "other-directed" person who is concerned primarily with maintaining self-respect by meeting others' standards. A third criticism of symbolic interactionism is that it places too much emphasis on consensus and cooperation and therefore neglects or downplays the importance of conflict. The perspective does recognize, however, that interacting people may fail to reach consensus, despite their efforts to achieve it. The symbolic interactionist perspective is at its best when analyzing fluid, developing encounters with significant others; it is less useful when analyzing self-interested behavior or principled action.

A Comparison of Perspectives

The four theoretical perspectives discussed here—role theory, reinforcement theory, cognitive theory, and symbolic interaction theory—differ with respect to the issues they address. They also differ with respect to the variables they treat as important and those they treat as irrelevant or incidental. In effect, each perspective makes different assumptions about social behavior and focuses on different aspects of such behavior.

In this section, we compare the various perspectives in terms of four dimensions: (1) the theory's central concepts or focus; (2) the primary social behaviors explained by the theory; (3) the theory's basic assumptions regarding human nature; and (4) the factors that, according to the theory, produce change in

a person's behavior. Table 1.1 summarizes this comparison by showing the position of each perspective on each of these dimensions.

Central Concepts   Each of the theoretical perspectives places primary emphasis on different concepts. Role theory emphasizes roles and norms, defined by group members' expectations regarding performance. Reinforcement theory explains observable social behavior in terms of the relationship between stimulus and response and the application of reinforcement. Cognitive theory stresses the importance of schemas and cognitive structure in determining judgments and behavior. Symbolic interaction theory emphasizes the self and role taking as crucial to the process of social interaction.

Behaviors Explained   Although overlapping to some degree, the four theoretical perspectives differ with respect to the behaviors or outcomes they try to explain. Role theory emphasizes role behavior and attitude change that results from occupying roles. Reinforcement theory focuses on learning and on the impact of rewards and punishments on social interaction. Cognitive theory centers on the mediating effects of a person's beliefs and attitudes on his or her overt response to social stimuli, and it also focuses on factors that produce change in beliefs and attitudes. Symbolic interactionism stresses the sequences of behaviors occurring in interaction among people.

Assumptions About Human Nature   The four theoretical perspectives differ in their fundamental assumptions regarding human nature. Role theory, for instance, assumes that people are largely conformist. It views people as acting in accord with role expectations held by group members. In contrast, reinforcement theory views people's acts—what they learn and how they perform—as determined primarily by patterns of reinforcement. Cognitive theory stresses that people perceive, interpret, and make decisions about the world. They formulate concepts and develop beliefs, and they act on the basis of these structured cognitions. Symbolic interaction theory assumes that people are conscious, self-monitoring

Table 1.1 Comparison of Theoretical Perspectives in Social Psychology

Dimension	Theoretical Perspective			
	Role Theory	**Reinforcement Theory**	**Cognitive Theory**	**Symbolic Interaction Theory**
Central Concepts	Role	Stimulus-response; reinforcement	Cognitions; cognitive structure	Self; role taking
Primary Behaviors Explained	Behavior in role	Learning of new responses; exchange processes	Formation and change of beliefs and attitudes	Sequences of acts occurring in interaction
Assumptions About Human Nature	People are conformist and behave in accordance with role expectations	People are hedonistic; their acts are determined by patterns of reinforcement	People are cognitive beings who act on the basis of their cognitions	People are self-monitoring actors who use role taking in interaction
Factors Producing Change in Behavior	Shift in role expectations	Change in amount, type, or frequency of reinforcement	State of cognitive inconsistency	Shift in others' standards, in terms of which self-respect is established

beings who use role taking to achieve goals through interactions with others.

Change in Behavior The four theoretical perspectives differ in their conception of what produces change in behavior. Role theory maintains that to change someone's behavior, it is necessary to change the role he or she occupies. Different behavior results when the person shifts roles because the new role entails different expectations and demands. Reinforcement theory, in contrast, holds that change in behavior results from changes in the type, amount, and frequency of reinforcement received. Cognitive theory maintains that change in behavior results from changes in beliefs and attitudes; it further postulates that changes in beliefs and attitudes often result from efforts to resolve inconsistency among cognitions. Symbolic interactionism holds that people try to maintain self-respect by meeting the standards of significant others; the issue of which standards are relevant is usually resolved through negotiation. For behavior to change, the standards held by others and accepted as relevant must shift first. A person will detect this shift in standards (by role taking) and consequently change his or her behavior.

Is Social Psychology a Science?

As we have noted, social psychology is the systematic study of the nature and causes of human social behavior. The field of social psychology includes not only theoretical perspectives and middle-range theories, but also a large body of facts and empirical generalizations obtained through research. Given this, we can ask whether social psychology is a science. That is, can we consider social psychology to be a scientific field in the same sense that we consider physics or biology to be scientific?

Characteristics of Science

Any scientific field rests on several basic assumptions. First, scientists assume that a real, external world exists independently of ourselves. This world is subject to investigation by observers. Second, scientists assume that relations in this world are organized in terms of cause and effect. This assumption—which in practice is a working hypothesis—is termed the **principle of determinism.** In its starkest form, this principle holds that there are discoverable causes for all events in a science's domain of interest. Third, scientists assume that knowledge concerning this external world is objective. Facts discovered by one scientist can be checked and verified by others.

In addition to these assumptions, any field that is a science has certain critical characteristics, or hallmarks. These include:

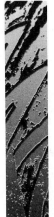

1.1 Milestones in the History of Social Psychology

Social psychology can trace its historical roots to the philosophers of ancient Greece and even more firmly to the psychological and social theorists of 19th-century Europe. However, the discipline as we know it today is largely a product of efforts by 20th-century researchers and theorists, many of them American (Allport, 1985; Jones, 1985).

The following list summarizes some important milestones in social psychology's development from the beginning of the 20th century up to 1960.

1898 Norman Triplett publishes the first social psychological experiment ("The dynamogenic factors in pacemaking and competition"). It investigates social facilitation, a process whereby a person's performance on a familiar task improves in the presence of others performing the same task.

1902 Charles Horton Cooley publishes an influential book, *Human Nature and the Social Order,* which presents the idea that the self and society are ultimately the same thing, although viewed at different levels of abstraction.

1908 William McDougall and E. H. Ross independently publish the first textbooks in the field. Although both books are titled *Social Psychology,* they differ greatly in content. The book by McDougall (a psychologist) stresses the importance of instincts and innate drives in determining behavior; the book by Ross (a sociologist) discusses groups, crowds, and crazes and emphasizes interpersonal processes such as suggestion and imitation.

1918 W. I. Thomas and F. Znaniecki begin their field study of attitudes within immigrant populations (Polish peasants) in Chicago.

1922 Morton Prince establishes the first major social psychology journal, the *Journal of Abnormal and Social Psychology* (which in 1965 becomes the *Journal of Personality and Social Psychology*).

1924 Floyd Allport, writing from a stimulus-response (behaviorist) perspective, publishes a social psychology textbook that is one of the first systematic treatments of the field. This book advocates the use of the experimental method in social psychology and sets forth a research agenda for the next decade.

1928 L. L. Thurstone publishes a path-breaking paper showing how attitudes can be measured.

1934 George Herbert Mead, a symbolic interactionist, publishes his seminal work on the self.

1934 J. L. Moreno develops sociometry, a system for measuring patterns of social interaction based on individuals' choices regarding who they would prefer to associate with.

1934 Richard T. LaPiere investigates inconsistencies between attitudes (racial prejudice) and related behaviors (discrimination) in a field setting.

1936 Muzafer Sherif, by creating social norms in a controlled setting, demonstrates that complex and

continued on next page

1. Any science is based on *observation of facts*. No field can be considered a science unless it includes observation. Thus so-called "armchair" disciplines without practitioners who make observations cannot be scientific fields.

2. Any science uses an explicit, formal *methodology*. This methodology is a set of procedures that must be followed by an investigator when establishing something as a known fact. Because any investigator in the field can use this methodology, one

scientist's findings can, in principle, be verified by others.

3. Any science involves the *cumulation of facts and generalizations*. Once relationships are observed to exist in the world, this knowledge is never lost. Of course, facts sometimes undergo reinterpretation of meaning, but the essential information is still available.

4. Any science includes a body of *theory*. This consists of at least one (but often many) theories that

---

continued from previous page

realistic social situations can be studied experimentally in a laboratory.

1936 George Gallup develops methods for conducting public opinion polls and surveys.

1937 J. L. Moreno founds *Sociometry,* a journal devoted to research on structure and process in groups and networks. (The journal is later renamed *Social Psychology Quarterly.*)

1939 Kurt Lewin, Ronald Lippitt, and Ralph White, using Lewin's field theory, study group members' reactions to various styles of leadership (autocratic, democratic, and laissez-faire).

1943 Theodore Newcomb investigates the effects of social pressures on attitudes held by students at Bennington College.

1943 William Foote White uses the technique of participant observation to study the activities of teenage street gangs.

1946 Solomon Asch demonstrates that cognitive set can influence the impressions people form of others.

1950 Robert Freed Bales develops a categoric framework for systematically observing communication and role differentiation in task groups.

1950 George Homans publishes *The Human Group,* a seminal theoretical treatise on group structure and process.

1951 Solomon Asch demonstrates some conditions under which individuals in a group will conform to the position of a majority when their beliefs are questioned.

1953 Carl Hovland and co-workers at Yale University publish the results of a programmatic study of persuasion and attitude change.

1954 Gordon Allport publishes *The Nature of Prejudice,* an important analysis of intergroup prejudice and stereotyping.

1957 Leon Festinger proposes the theory of cognitive dissonance, an approach to attitude change based on the idea that people strive for consistency between behavior and attitudes.

1958 Fritz Heider publishes *The Psychology of Interpersonal Relations,* which lays the foundation for attribution theory and research.

1959 John Thibaut and Harold Kelley publish *The Social Psychology of Groups,* a general theory of social exchange and interpersonal relations.

Since 1960, social psychology has continued to expand rapidly and develop as a field. The 1960s saw a large number of laboratory studies of cognitive dissonance, as well as an increased concern with such phenomena as altruism, aggression, and interpersonal attraction. The 1970s witnessed the growth of attribution theory and expanding interest in interpersonal relations. The 1980s saw a renewed concern with the social self and greater emphasis on the cognitive aspects of social behavior. The field grew rapidly, and many new social psychology journals began publication.

In the 1990s, social psychologists are continuing to investigate social cognition, emotions and the self, language and communication, interpersonal relationships, group performance and decision making, prejudice and intergroup relations, and numerous other topics. In a fundamental sense, the chapters of this book constitute a summary and distillation of the central concerns of social psychology as the field stands today.

serve to systematize and organize empirical observations. Theory also serves to guide new empirical investigation.

5. After it attains a reasonable level of development, any science provides at least some degree of *prediction and control* over selected aspects of the environment.

When we hold up a well-developed natural science such as physics or biology against this list, we see immediately that it has all these hallmarks. These sciences are based on observation of the world, have a formal methodology to guide research, have an accumulation of established facts, possess a body of well-developed theory, and provide at least a moderate degree of prediction and/or control regarding selected aspects of the world.

Social Psychology as a Science

Can social psychology be considered a science? That is, if we hold it up against the criteria listed above, does it measure up? Social psychology certainly has some of the hallmarks of science. Consider the first hallmark—reliance on empirical observation. Social psychology clearly is based on empirical observation and classification of facts. The field consists of many thousands of empirical studies. Social psychology also meets the second hallmark, for it relies on widely shared methodological procedures for conducting empirical investigation. Among the most widely employed methods within social psychology are experimentation and systematic sample surveys. (Chapter 2 of this book discusses methodology in social psychology.)

To a fair degree, social psychology also meets the third hallmark, the cumulation of observed facts. Social psychologists continue to gather facts regarding the conditions under which specific behaviors occur. Of course, more is known about some types of social behavior than others, but increasingly sophisticated studies have continued to expand the frontiers of knowledge.

Social psychology also measures up fairly well against the fourth criterion, reliance on theory. Although social psychology has no single unified theory

covering all phenomena in the field, it does have several theoretical perspectives, such as those reviewed here. It also has numerous middle-range theories that make predictions regarding specific types of social behavior under restricted conditions. A large number of empirical studies in social psychology are attempts to test predictions from middle-range theories. We will encounter many of these middle-range theories in subsequent chapters of this book.

If any problem arises for social psychology as a science, it is primarily with respect to the fifth hallmark, which holds that after attaining a reasonable level of development, any science provides some degree of prediction and/or control with respect to the phenomena investigated by the field. Whether social psychology can accomplish this feat today is unclear.

Although social psychology does fairly well in "explaining" social behavior (that is, identifying the conditions under which various forms of social behavior occur), it does less well in predicting future events or providing a basis for control of behavior. Of course, we can point to some successes in prediction and control. For example, political election forecasts based on sample surveys have been fairly accurate in recent years, and programs for modification of interpersonal behavior based on reinforcement principles have proved effective. Nevertheless, social psychology does not excel in prediction and control. A large percentage of practicing social psychologists believe that, in the coming years, the field's capacity to predict interpersonal behavior will improve to some degree (Lewicki, 1982). At the present time, however, social psychology is no match for the mature physical sciences in this respect.

Part of the problem stems from the nature of social psychological theory. Few theories that make explicit predictions have much generality. They cover only limited ranges of phenomena or apply only under very restrictive (and sometimes artificial) conditions. Theories often fail to predict accurately when attempts are made to apply them to new settings.

If the problem ran no deeper than this, we might be optimistic that social psychology will soon predict social behavior with great accuracy. We might conclude, for instance, that social psychology merely needs better, more refined theories. To some degree this is true, but unfortunately the problem is more complex. As noted above, any science is based on the principle of determinism; it assumes the world is organized in terms of cause and effect. Science tries to develop laws based on the notion that if X causes Y on one occasion, then X will again cause Y on some similar occasion in the future. Although the assumption of determinism works well for physical phenomena, it may not work as well for human social behavior. Because human beings are conscious and self-aware, they can exercise some control over their own behavior. Unlike atoms or rocks, they are capable of making decisions and suddenly changing their behavior. They can exercise free will, at least to some degree. For this reason, some critics have argued there will never be anything approaching true "universal laws" describing social behavior. Certainly, it is difficult to reconcile the scientific assumption of determinism with the concept of human free will. Of course, this concern is not unique to social psychology. It besets all the social sciences.

As we have shown, social psychology displays many of the characteristics of the more mature physical sciences. It is based on observation of the social world and relies on a formal methodology to guide research. It has accumulated many descriptive facts regarding human social behavior, and it possesses bodies of formal theory. Nevertheless, it has not yet achieved the same degree of accuracy in prediction as the mature physical sciences. Although it does offer compelling explanations for many types of observed social behavior, social psychology has provided only a modest degree of predictability and control of social behavior.

Summary

This chapter considered the fundamental characteristics of social psychology and important theoretical perspectives in the field.

What Is Social Psychology?   There are several ways to characterize social psychology. (1) By definition, social psychology is the systematic study of the nature and causes of human social behavior. (2) Social psychology has several core concerns. These concerns include the impact of one individual on another's behavior and beliefs, the impact of a group on a member's behavior and beliefs, the impact of a member on the group's activities and structure, and the impact of one group on another group's activities and structure. (3) Social psychology has a close relationship with other social sciences, especially sociology and psychology. Although they emphasize different issues and often use different research methods, both psychologists and sociologists have contributed importantly to social psychology.

Theoretical Perspectives in Social Psychology   A theoretical perspective is a broad theory based on particular assumptions about human nature that offers explanations for a wide range of social behaviors. This chapter discussed four theoretical perspectives: role theory, reinforcement theory, cognitive theory, and symbolic interaction theory. (1) Role theory is based on the premise that people conform to norms defined by the expectations of others. It is most useful in explaining the regular and recurring patterns apparent in day-to-day activity. (2) Reinforcement theory assumes that social behavior is governed by external events, especially rewards and punishments. Reinforcement theory helps to explain not only how people learn but also when social relationships will change. (3) Cognitive theory holds that such processes as perception, memory, and judgment are significant determinants of social behavior. The theory treats ideas and beliefs as organized into structures (schemas) and relies on various principles (such as the principle of consistency) to explain change in attitudes and beliefs. Differences in cognitions help to illuminate why individuals may behave differently from one another in a given situation. (4) Symbolic interaction theory holds that human nature and social order are products of communication among people. It stresses the importance of the self, of role taking, and of consensus in social interaction. It is most useful in explaining fluid, contingent encounters among people.

Is Social Psychology a Science?   (1) To ascertain whether social psychology is a science, we must first identify the five hallmarks that characterize any science. These are: scientists engage in empirical observation of the world, use a formal research methodology, cumulate knowledge of facts, develop formal theories to explain facts, and employ these theories to provide some degree of prediction and control. (2) Social psychology meets the first four of these hallmarks, but falls short of meeting the fifth. To date, social psychology has provided only a modest degree of predictability and control over human social behavior.

Key Terms

cognitive processes (p. 12)
cognitive structure (p. 12)
cognitive theory (p. 12)
conditioning (p. 9)

equity (p. 11)
imitation (p. 10)
middle-range theory (p. 6)
norm (p. 8)
principle of consistency (p. 13)
principle of determinism (p. 18)
reinforcement (p. 9)
reinforcement theory (p. 9)
role (p. 8)
role theory (p. 7)
schema (p. 12)
self (p. 15)
significant others (p. 15)
social exchange theory (p. 10)
social learning theory (p. 10)
social psychology (p. 3)
symbolic interaction theory (p. 14)
theoretical perspective (p. 6)
theory (p. 6)

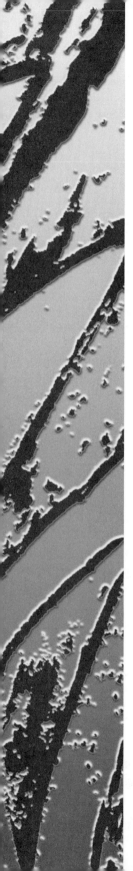

CHAPTER 2

Research Methods in Social Psychology

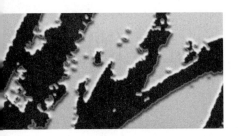

Introduction

The field of social psychology relies heavily on *empirical research,* the systematic investigation of observable phenomena (behavior, events) in the world. Researchers try to collect information about behavior and events in an accurate and unbiased form. This information, which may be either quantitative or qualitative, enables social psychologists to describe reality in detail and to develop theories about social behavior.

When conducting empirical research, investigators usually employ a **methodology,** a set of systematic procedures that guides the collection and analysis of data. In a typical study, investigators first develop a research design. Then they go into a laboratory or field setting and collect the data. Next they code and analyze the data to test hypotheses and arrive at various conclusions about the behaviors or events under investigation. Throughout this process, investigators follow specific procedures to assure the validity of the findings.

When investigators report their research to the wider community of social psychologists, they describe not only the results but also the methodology used to obtain the results. By reporting their methods, they make it possible for other investigators to independently verify their findings.

Independent verification of research findings is one of the hallmarks of any science. Suppose, for instance, that an investigator were to report some unanticipated empirical findings that ran contrary to established theory. Other investigators might wish to replicate the study to see whether they could obtain the same findings in other settings with different human subjects. Through this process, investigators

with differing perspectives can identify and eliminate biases in the original study. If the results are replicable, they stand a better chance of being accepted by other social psychologists as reliable, general findings.

Questions About Research Methods

In this chapter we review the research methods used in contemporary social psychology. This will provide a foundation for understanding and evaluating the empirical studies discussed throughout the text. We address the following questions:

1. What are the basic goals that underlie social psychological research? What form do research hypotheses assume? What steps can researchers take to assure the validity of their findings?
2. What are the defining characteristics of research methods such as surveys, laboratory and field experiments, naturalistic observation, and archival research? What are the strengths and weaknesses of each?
3. What ethical issues are important in the conduct of social psychological research? What safeguards are available to protect the rights of human subjects?

Characteristics of Empirical Research

The major research methods used by social psychologists include surveys, experiments, field observation, and archival studies. Before looking at these in detail, we review some issues common to all forms of empirical research. Among these are the objectives typically underlying empirical research, the nature of the hypotheses guiding research, and the factors affecting the validity of research findings.

Objectives of Research

Investigators conduct social psychological studies for a variety of reasons. Their objectives usually include one or more of the following: describing reality, identifying correlations among variables, testing causal hypotheses, and developing and testing theories.

In some studies, the central objective is simply to describe reality in accurate and precise terms. An investigator may wish to characterize some behavior or describe the features of a social process. Description is often the paramount goal when a researcher investigates a phenomenon about which little or nothing is known. Even when investigating more familiar phenomena, a researcher may wish to ascertain the frequency with which a particular attitude or behavior occurs in a specified group or population. For instance, during election years researchers routinely conduct public opinion polls to learn how Americans feel about political candidates, issues, and parties. Their goal is to describe public sentiment with great accuracy and precision.

A second objective of research is to ascertain whether a correlation exists between two or more behaviors or attributes. Researchers might conduct a survey, for example, to find out whether frequency of sexual intercourse before marriage is related to religious identity (DeLamater & MacCorquodale, 1979) or whether attitudes on abortion are related to sex-role ideology (Barnartt & Harris, 1982). Investigators usually want to know not only whether two variables are correlated but also how strongly they are related. Although a correlation between variables may reflect an underlying causal relation, two variables can be correlated without one causing the other; this will happen, for instance, if both are caused by a third variable. Correlation alone is not sufficient evidence for causation.

A third objective of research, then, is to discover the causes of some behavior or event. When pursuing this goal, the researcher first develops a *causal hypothesis,* which is a statement that a difference (or change) in one behavior or event produces a difference (or change) in another behavior or event. For instance, an investigator might hypothesize that studying for an exam in groups will produce higher grades than studying for the exam individually. After specifying the hypothesis, the investigator collects data to test the hypothesis. To support the hypothesis of causality, this test must show that differences (or changes) in one

variable produce differences (or changes) in another. In addition, the design of the test must preclude or eliminate plausible alternative (noncausal) interpretations of the data. Frequently, the best way to test a causal hypothesis is by an experiment, a topic discussed in greater detail below.

A fourth objective of social psychological research is to test existing theories and to develop new ones. As used here, a *theory* is a set of interrelated hypotheses that explains some observable behavior(s) or event(s). Frequently, a theory serves as a basis for predicting future events. Tests of theories resemble tests of hypotheses, except that several interrelated hypotheses are assessed at once. In some cases, investigators juxtapose theories that make different predictions, and the results of the test may enable them to reject one theory in favor of another.

Research Hypotheses

In broad terms, a **hypothesis** is a conjectural statement of the relation between two or more variables. Although a few social psychological studies are exploratory and originate more from wide-ranging curiosity than from explicit hypotheses, most studies involve one or more hypotheses at the start. To test whether a hypothesis is correct or not, investigators first ask what observations would be expected if the hypothesis is true; then they take some observations or measures of reality and compare these with what is expected under the hypothesis. If a discrepancy is noted, it constitutes evidence against the hypothesis and may lead to rejection.

There are various types of hypotheses. Some hypotheses are noncausal in nature. For example, "Variables X and Y are correlated, such that high levels of X occur with low levels of Y" (negative correlation). Noncausal hypotheses make statements about observed relations among variables, but they stop short of indicating whether X causes Y, Y causes X, both X and Y cause one another, or neither X nor Y causes the other.

Other hypotheses are explicitly causal in nature. For instance, a causal hypothesis relating two variables might take the form "X causes Y" or "Higher levels of X produce lower levels of Y" or "An increase in X will produce a decrease in Y." Sometimes, of course, causal hypotheses are more explicit and qualified in scope. For example, "If conditions A and B are present, then an increase of 1 unit in X will cause a decrease of 6 units in Y."

Causal hypotheses always include at least two variables—an independent variable and a dependent variable. An **independent variable** is any variable considered to cause or have an effect on some other variable(s). A **dependent variable** is any variable caused by some other. The dependent variable changes in response to changes in the independent variable. In the example above where X causes Y, variable X is the independent variable and Y is the dependent variable.

Another important type—the **extraneous variable**—is any variable that is not expressly included in the hypothesis but nevertheless has a causal impact on the dependent variable. Extraneous variables are widespread in social psychology because most dependent variables of interest have more than one cause. In some studies, researchers have an explicit idea regarding most of the extraneous variables (besides the independent variable X) that cause Y; more typically, however, they do not know many of the extraneous variables that impact Y.

Validity of Findings

We cannot take for granted that the findings of any given study have validity. To elaborate this point, consider a situation in which an investigator is studying deviant behavior. In particular, she is investigating the extent to which cheating occurs on exams by college students. Reasoning that it is more difficult for persons monitoring an exam to keep students under surveillance in large classes than in smaller ones, she hypothesizes that a higher rate of cheating will occur on exams in large classes than small. To test this hypothesis, she collects data on cheating in both large classes and small ones and then analyzes the data. Results show that more cheating per student occurs in the larger classes. Thus the data apparently support the investigator's research hypothesis.

A few days later, however, a critic points out that all the large classes in the investigator's study used multiple-choice exams, whereas all the small classes used short-answer and essay exams. The investigator immediately realizes that an extraneous variable (exam format) is confounded with the independent variable (class size) and may be operating as a cause in her data. The apparent support for her research hypothesis (more cheating in large classes) may be nothing more than an artifact. Perhaps the true effect is that more cheating occurs on multiple-choice exams than on essay exams, irrespective of class size.

We say that the findings of a study have **internal validity** if they are free from contamination by extraneous variables. Internal validity is a matter of degree; findings may have high or low internal validity. The investigator's findings about the effect of class size on cheating obviously have low internal validity because of the possibly confounding effect of exam format. Internal validity is very important. Without internal validity, a study cannot provide clear, interpretable results.

Extraneous variables, if they are present, can reduce internal validity in either of two basic ways. First, an extraneous variable can mask the true effects of the independent variable (X) on the dependent variable (Y). This can happen if the hypothesis (X causes Y) is true but the extraneous variable influences the dependent variable in a direction opposite to that of the independent variable. Thus, even if X actually causes Y, the data would not reveal this fact. Second, an extraneous variable can create an apparent causal effect of X on Y when none actually exists. This may have happened in the study of class size and cheating. The results initially appeared to show an effect of class size on cheating, but later it became clear that this effect may have been due to the exam format, an extraneous variable (Campbell & Stanley, 1963; Cherulnik, 1983).

To achieve results with higher internal validity, an investigator might repeat the study with an improved design. For instance, our investigator might repeat her study with only one exam format (say, multiple-choice) in both large and small classes. Then she could test whether class size affects the rate of cheating on multiple-choice exams. By holding constant the extraneous variable (exam format), her new design will have greater internal validity. Better still, she might use a more complex design that includes all four logical possibilities (that is, small class: multiple choice; small class: essay; large class: multiple choice; and large class: essay). She could analyze the data from this design to estimate separately the impacts of these variables (class size, exam format) on cheating. In effect, this design converts an extraneous variable (exam format) into a second independent variable. Although better, it is not a perfect design because other extraneous variables could still be operating as causes of cheating—and they may be confounded with class size and exam format.

Although very important, internal validity is not the only concern of the investigator. Another concern, **external validity,** is the extent to which a causal relationship, once identified in a particular setting with a particular population of subjects, can be generalized to other populations, settings, or time periods. Even if an investigator's results have internal validity, they may lack external validity; that is, they may hold only for the specific group and setting studied, and not generalize to others. For instance, if the investigator studying cheating and class size conducted her study in a two-year college, there is no assurance that the findings (whatever they turned out to be) would also apply to students in other settings, such as high schools or four-year colleges or universities. In general, external validity is important and desirable because the results of a study often have practical importance only if they generalize beyond the particular setting in which they appeared.

Research Methods

Although there are many ways of collecting data about social behavior, most social psychological studies use one or another of four main methods: surveys, experiments, naturalistic observation, and archival research based on content analysis. We discuss each of these methods in turn.

Surveys

A **survey** is a procedure for collecting information by asking members of some population a set of questions and recording their responses. The survey technique is very useful for identifying the average or typical response to a question as well as the distribution of responses within the population. It is also useful for identifying how groups of respondents differ from one another. For instance, Young (1984) used a survey to test the hypothesis that adolescent females pursuing innovative nontraditional careers hold different values from those pursuing traditional careers. In this study, a sample of adolescent females were asked to rank the importance of a series of instrumental values (ambitious, independent, capable, forgiving, obedient, and so on). Results showed that the values *clean, forgiving,* and *obedient* were ranked more highly by females aspiring to traditional careers, whereas the values *courageous, imaginative,* and *independent* were ranked more highly by females aspiring to innovative careers. These value differences provide one way of representing differences among females in sex-role beliefs.

Purpose of a Survey   Investigators often conduct surveys to obtain self-reports from individuals about their own attributes—that is, their attitudes, behavior, and experiences. Information of this type enables investigators to discover the distribution of attributes in the population and to determine whether a relationship exists between two or more attributes of interest.

One form of survey—the public opinion poll—has become very common in the United States. Several organizations specialize in conducting surveys that measure the frequency and strength of favorable or unfavorable attitudes toward public issues, political figures, and candidates for office. These polls play a significant role in American politics, for their findings increasingly influence public policy and the positions taken by political figures (Halberstam, 1979; Ratzan, 1989). All of the presidential candidates used results from such polls to guide their decisions during the 1996 campaign, and they will undoubtedly do so again in 2000.

In addition, investigators often use surveys to obtain data about various social problems. For instance, government agencies and individual researchers have conducted surveys on pregnancy and contraception use among teenagers (Beck & Davies, 1987; Zelnik & Kantner, 1981) and on alcohol and drug use by teenagers (Pascale & Sylvester, 1988; Skager & Fisher, 1989). Information about the extent of such activities and the people involved in them is requisite to developing effective social policies.

Finally, investigators often conduct surveys with the primary objective of making basic theoretical contributions to social psychology. For instance, many studies of socialization processes and outcomes, psychological well-being, discrimination and prejudice, attitude-behavior relationships, and collective behavior have utilized survey methods.

Types of Surveys   There are two basic types of surveys—those based on interviews and those based on questionnaires. In an **interview survey,** a person serving as an interviewer asks a series of questions and records the answers from the respondents. To assure that each respondent in the study receives the same questions, an interviewer usually works from an *interview schedule.* This schedule indicates the exact order and wording of questions. In certain studies, however, the interviewer has flexibility in determining the exact order and wording of questions but must make sure that certain topics are covered. One positive feature of an interview is that the interviewer can adjust the questioning to the respondent. That is, he or she can look for verbal and/or nonverbal signs that the respondent does not understand a question and then repeat or clarify as needed.

Working from a schedule of questions, this survey interviewer carefully records the answers given by a respondent at a mall.

In a **questionnaire survey,** the questions appear on paper, and the respondents read and answer them at their own pace. No interviewer is present. One advantage of questionnaires over interviews is that questionnaires cost less to administer. The cost of a national survey using trained personnel to conduct face-to-face interviews is rather large; it can run as much as $200 to $225 or more per completed interview depending on the length of the interview and other factors. In contrast, the same survey using questionnaires mailed to respondents would cost considerably less, maybe as little as $15 per completed form. The major disadvantage of questionnaires lies in the **response rate,** the percentage of those contacted who complete the survey. An interview study can obtain response rates as high as 75% or better, but mailed questionnaires rarely attain more than a 50% response rate. The lower rate is a serious disadvantage for mailed questionnaires.

A compromise between interviews and questionnaires is the *telephone interview,* which is now the standard method used by public opinion polling organizations such as Gallup and Roper. Investigators are using it increasingly in basic research as well. The telephone interview uses a trained interviewer to ask the questions, but it sacrifices the visual feedback available in a face-to-face interview. It is cheaper (about $60 per completed interview, depending on length) than the face-to-face interview, although it typically involves a somewhat lower response rate (about 65%). Many surveys now use computer-assisted telephone interviewing (CATI). With CATI, the computer randomly selects and dials telephone numbers. Once a potential respondent is on the line, the interviewer takes over and conducts the interview. He or she reads some questions and enters the answers directly into the computer when the respondent gives them. In listing questions to ask, the computer may alter later questions in light of earlier answers by the respondent.

Measurement Reliability and Validity   In surveys, as in any form of research, the quality of measurement is an important consideration. Of primary concern are the reliability and the validity of the instruments. **Reliability** is the extent to which an instrument produces the same results each time it is employed to measure a particular construct under given conditions. A reliable instrument produces consistent results across independent measurements of the same phenomenon. Reliability is a matter of degree; some instruments are highly reliable, but others are less so. Investigators obviously prefer instruments with high reliability and try to avoid those with low.

There are several ways to assess the reliability of an instrument. The first of these is to see if people's responses to an instrument are consistent across time. In this approach, called the *test-retest* method, an investigator applies the measuring instrument to the same respondents on two different occasions, and then he or she compares the first responses with the second responses. If the correlation between the first and second responses is high, the instrument has high reliability; if the correlation is low, the instrument has only low reliability.

A second way to assess the reliability of an instrument is to see if people's responses are consistent across items. This approach is called the *split-half* method. To illustrate, suppose we have a scale of 20 questions measuring psychological well-being. These questions ask respondents about psychological states, such as how often they are sad, nervous, depressed, tense, or irritable and how often they have trouble concentrating, working, or sleeping. Assume that we administer all 20 questions to each of 300 male respondents. To use the split-half method, we would randomly divide the 20 questions into two groups of 10, calculate a score for each respondent on each group of 10, and compute a correlation between the two scores. A high correlation (if it occurs) provides confirmation that the scale is reliable.

Given that a measure is reliable, the next concern is its **validity**—that is, does the instrument actually measure the (theoretical) concept we intend to measure? There are several types of validity, including face validity, criterion validity, and construct validity. First, an instrument has *face validity* if its content is manifestly similar to the behavior or process of interest. If a researcher wishes to measure the frequency of sexual intercourse, for example, the question "How often do you engage in sexual intercourse?" has face validity.

Second, an instrument has *criterion validity* if we can use it to predict respondents' standing on some other variable of theoretical or practical interest. Suppose, for example, that an investigator is concerned with traffic safety on the roads and that she develops an instrument to distinguish good drivers from bad drivers. To establish the instrument's predictive validity, she first administers the instrument to young persons getting their driver's license and then, several years later, checks their driving records for moving violations. If the drivers' scores on the instrument correlate highly with their level of subsequent violations, the instrument has some criterion validity.

Third, an instrument has *construct validity* if it provides a good measure of a theoretical concept being investigated by the research. In general, an instrument has construct validity if it measures what people understand the concept to mean and if it relates to other variables as predicted by the theory under consideration. Establishing construct validity of an instrument can be difficult, especially if the underlying theoretical construct is highly abstract in nature. Suppose, for example, that an investigator's theory includes an abstraction like "intellectual development." Measurement of this concept is somewhat problematic, for there is no readily observable referent, no single behavior or occurrence that the investigator can point to as indicative of "intellectual development." The usual method of establishing construct validity of an instrument is to show that the pattern of correlations between respondents' scores on the instrument and their scores on other variables is what would be expected if the underlying theory holds true.

The Questions   The phrasing of questions used in surveys requires close attention by investigators. Subtle differences in the form, wording, and context of survey questions can produce differences in responses (Krosnick & Schuman, 1988; Schuman & Kalton, 1985). Creating good survey questions is as much art as science, but certain guidelines can help. First, the more precise and focused a question, the greater its reliability and validity. If a question is expressed in vague, ambiguous, abstract, or global terms, respondents may interpret it in different ways, which in turn will produce uncontrolled variation in responses. A second consideration in formulating survey questions is the exact choice of terms used. It is best to avoid jargon or specialized terminology unless one is interviewing a sample of specialists. Likewise, it is important to adjust questions to the educational and reading level of the respondents. A third consideration is the length of questions. Several studies have shown that questions of moderate length elicit more complete answers than very short ones (Anderson & Silver, 1987; Sudman & Bradburn, 1974). A fourth consideration is whether the topic under investigation is potentially a threatening or embarrassing one (sex, alcohol, drugs, money, and so on). In general, threatening questions requiring quantified answers are best expressed in wording and terminology that is very familiar to the respondents (Blair et al., 1977).

The Sample   Suppose a survey researcher wants to ascertain the extent of prejudice toward blacks among white adults in the United States. These white adults constitute the **population** of interest, that is, the set of all people whose attitudes are of interest to the researcher. It would be virtually impossible (and enormously expensive) to interview all people in the population of white adults, so the researcher instead selects a **sample,** or representative subset, from that population to interview.

Selection of the sample is one of the most important aspects of any survey. In some cases, investigators may use a particular sample simply because it is readily available; samples of this type are known as *convenience samples.* A sample consisting of students taking a class, occasionally used in social science research, is a convenience sample. Convenience samples have a major drawback: They usually lack external validity and do not enable the investigator to generalize the findings to any larger population. For this reason, it is better research practice to select some other type of sample, one that is representative of an underlying population. Only when the sample is representative can the results obtained from it (for example, information regarding racial prejudice obtained from survey respondents) be generalized to the entire population. The nature of the sample, therefore, has a major impact on the external validity of the survey.

Two types of systematic samples are commonly used in social psychological surveys. One is the **simple random sample,** in which the researcher selects units, usually individuals, from the population such that every unit has an equal probability of being included. To use this technique, the researcher needs a complete list of members of the population. At a university, for example, she might obtain a list of all students from the registrar. At the city or county level, she might utilize voter registration lists. A frequent problem, especially when the population being studied is large, is the absence of a complete list. Under these circumstances, researchers usually fall back on some substitute, such as a telephone directory. Of course, this limits the population to which one can generalize, because people who are poor or who move frequently may not have telephones, and others may choose not to list their numbers in the directory.

Working from a complete list of the population, the researcher draws a random sample. A common way to do this is to number the people on the list consecutively and then use a table of random numbers to choose persons for the sample. Once the researcher has drawn a random sample, she must take steps to assure that all the members of the sample are interviewed; in other words, the researcher must strive for a high response rate. Without a high response rate, the results of the survey will not be generalizable to the whole population. Bias may result if the people who participate in the study differ in some significant way from those who refuse to participate.

If the population is very large, the investigator may not be able to list all its members and draw a random sample. Under these conditions, researchers frequently employ a **stratified sample.** That is, they divide the population into groups according to important characteristics, select a random sample of groups, and then draw a sample of individuals within each selected group. One illustration is provided by public opinion polls designed to represent the entire adult population of the United States. In surveys such as these, the population is first stratified on the basis of region (Northeast, Midwest, South, Southwest, and West). Next, the population within each region is stratified into urban versus rural. Within urban areas, there may be still further stratification by size of urban area. The result will be numerous *sampling* *units*—population subgroups of known regional and residential type. Some units are then selected for study in proportion to their frequency in the entire population. Thus one would sample more urban units from the Northeast than from the South or Midwest; conversely, one would select more rural units in the latter regions. Finally, within each sampling unit, persons are selected randomly to serve as respondents. Using this technique, one can represent the adult population of the United States with a sample of 1,500 persons and obtain responses accurate within plus or minus 3%.

Causal Analysis of Survey Data

Social psychologist have long used computers to aid in the descriptive analysis of survey data, and they have often relied on such techniques as cross-tabulations and correlations. In recent years, however, some social psychologists have begun to use more sophisticated computer-based techniques to aid in the causal interpretation of survey data. Analysis techniques of this type (such as LISREL and path analysis) require the investigator to postulate a pattern of cause-and-effect relations among a set of variables (Bollen, 1989; Joreskog & Sorbom, 1979). The computer then estimates coefficients of effect from the data. These coefficients indicate the strength of the relationships among the variables, and they provide a test of whether the causal linkages postulated by the theory are indeed present in the data. Using this approach, an analyst can test many alternative theories. Typically, some theories turn out to be inconsistent with the data, and the analyst can reject these in favor of alternative theories that survive the test. One difficulty with this approach is that for problems involving many variables (say, a dozen or more), there are numerous plausible alternative theories. Although this process will eliminate many theories, more than one may survive as tenable.

Panel Studies

One useful extension of the survey technique is the longitudinal survey, or **panel study,** in which a given sample of respondents is surveyed at one point in time and then resurveyed at a later point. For instance, in a panel study, a sample of respondents would be surveyed today by telephone interview or mail questionnaire (this is called the "first wave" of the panel). Then, at some future time (say, 1 year from now), the same respondents would be surveyed again (the "second wave"); the questionnaire items in the second wave are similar to, or an extension of, those used in the first wave. If desired, the same respondents could be surveyed again at a still later point in time (the "third wave"), and so on. In principle, no upper limit exists on the number of waves that might be included in a panel study, although there are practical constraints such as the expense of running the panel and the difficulties in tracking down members of the sample at various times. The waves in a panel study can be spaced either closely together in time or far apart, depending on the study's purpose.

The usual objective of a panel study is to determine whether various outcomes experienced by respondents at later points in time are related to (or determined by) their experiences, attitudes, and relationships at the earlier points in time. For instance, Udry and Billy (1987) used a panel study with two waves to investigate the sexual activity of adolescent boys and girls. The first interview took place when the respondents were in grades 7, 8, or 9 (mean age 13.6 years), and the second interview took place 2 years later, when the respondents were in grades 9, 10, or 11. The objective of the study was to discover which biological and social factors in place at the outset affected the probability of having one's first sexual intercourse during the 2-year period. The results indicated that although social factors (i.e., family situation) had relatively little impact on the probability of making the transition from virgin to nonvirgin for adolescent males, they did have a substantial impact on the probability of transition for females. In particular, adolescent girls who had highly educated mothers, whose mothers were married to and living with their fathers, and whose mothers were less sexually active as adolescents had a lower probability of transition to coitus than other girls.

In general, data from a panel study lend themselves somewhat more readily to causal interpretation than data from a simple cross-sectional survey. The waves in the panel study provide a natural temporal ordering among the variables, which usually provides increased clarity when interpreting the results causally.

Strengths of Surveys　Surveys can provide, at moderate cost, an accurate and precise description of characteristics of a specific population. When a social psychological researcher uses measures that are reliable and valid, employs a sampling design that guarantees representativeness, and takes steps to assure a high response rate from respondents, the survey can produce a clear portrait of the attitudes and social characteristics of a population.

Beyond this, surveys also provide an effective means to study the incidence of various social behaviors. A survey asking people to report their behavior is usually more efficient and cost-effective than observational studies of actual behavior. This is especially true for behavior that occurs only infrequently or in private settings.

Weaknesses of Surveys　As with any methodology, there are certain drawbacks to the survey technique. Both questionnaires and interviews rely on self-reports by respondents. Under certain conditions, self-reports can be invalid sources of information. First, some people may not respond truthfully to questions about themselves. This is not usually a major problem, but it can become troublesome if the survey deals with activities that are highly personal, illegal, or otherwise embarrassing to reveal. Second, even when respondents want to report honestly, they may give wrong information because of imperfect recall or poor memory. This can be a nettlesome problem, especially in surveys investigating the past (for example, historical events or childhood). As an illustration, consider the question "When were you last vaccinated?" This may seem simple and straightforward, but it often produces incorrect responses because many people cannot remember dates. Third, some respondents answering self-report questions have a tendency to fall into a response set. That is, they answer all questions the same way (for example, always agree or disagree) and/or they give extreme answers too frequently. If many respondents adopt a response set, this introduces bias into the survey's results.

Experiments

The **experiment** is the most highly controlled of the research methodologies available to social psychol-

ogists, and it is a powerful method for establishing causality between variables. For a study to be a true experiment, it must have two specific characteristics. First, the researcher must manipulate one or more of the independent variables that are hypothesized to have a causal impact on the dependent variable of concern. Second, the researcher must assign the subjects randomly to the various *treatments*—that is, to the different levels of each of the independent variables.

The term **random assignment** denotes the placement of subjects in experimental treatments on the basis of chance, like flipping a coin or using a table of random numbers. Random assignment is desirable because it handles the effects of extraneous variables. By using random assignment, the researcher creates groups of subjects that are equivalent in all respects except their exposure to different levels of the independent variables. This removes the possibility that these groups will differ systematically on extraneous variables, such as intelligence, personality, or motivation. Thus random assignment enables the investigator to infer that any differences between groups on the dependent variable are due only to the effects of the independent variable(s), not to extraneous variables (Kirk, 1982).

Whereas researchers manipulate the independent variables in an experiment, they simply measure the dependent variable(s). Experimenters can measure dependent variables in many ways. For example, they can monitor subjects' physiological arousal, administer short questionnaires that assess subjects' attitudes, record the interactions that occur between subjects, or score the subjects' performance on tasks. The exact type of measurement used in the experiment depends on the nature of the dependent variable(s) of interest.

Laboratory and Field Experiments　It is useful to distinguish between laboratory experiments and field experiments. *Laboratory experiments* are those conducted in a laboratory setting, where the investigator can control much of the subjects' physical surroundings. In the laboratory, the investigator can determine what stimuli, tasks, information, or situations the subjects will face. This control enables the experimenter to manipulate the independent variables, to measure the dependent variables, to hold constant some known extraneous variables, and to implement random assignment of subjects to treatments. For instance, if an investigator is studying the impact of verbal communication on group productivity in a laboratory setting, he may wish to restrict the interaction among subjects. To do this, he might limit communication to written notes or verbal messages sent by electronic equipment. This practice would not only eliminate the (possibly contaminating) influence of nonverbal communication, but it also would permit the content of any messages to be analyzed later by the experimenter.

Field experiments, in contrast with laboratory experiments, are studies in which investigators manipulate variables in natural, nonlaboratory settings. Usually these settings are already familiar to the subjects. Investigators have used field experiments to study topics ranging from pay inequity in large bureaucratic organizations to altruistic behavior on street corners and subway cars. As compared with laboratory experiments, field experiments have the advantage of high external validity. When conducted in natural and uncontrived settings, they usually have greater mundane realism than laboratory experiments. Moreover, participants in field experiments may not be particularly self-conscious about their status as experimental subjects, a fact that reduces subjects' reactivity. The primary weakness of field experiments is that in natural settings, experimenters sometimes have difficulty manipulating independent variables exactly as they would wish and often have little control over extraneous variables. Consequently the internal validity of field experiments is sometimes much lower than in comparable laboratory experiments.

Conduct of Experiments To illustrate how investigators conduct experiments, consider the following laboratory study, which sought to determine the impact of certain independent variables on whether one person will help another in an emergency (Darley & Latané, 1968). The investigators conducted the study at a university in New York City. Men and women students, serving as subjects, came to the laboratory to participate in a discussion of problems they had encountered in adjusting to the university. The experimenters placed each subject in a separate room within the laboratory and instructed them to communicate with other subjects via an intercom. The rationale given was that this procedure would permit them to remain anonymous while discussing personal problems.

The independent variable was the number of other persons who the subject believed were participating in the discussion (and who would, therefore, later witness an emergency). Depending on experimental treatment, subjects were told there were 1, 2, or 5 other participants. Subjects were randomly assigned to the various levels of this independent variable.

The discussion proceeded with each participant speaking in turn for 2 minutes over the intercom. Thus, depending on experimental treatment, the subject heard the voices of 1, 2, or 5 others. In reality, the subject was hearing a tape recording of other people, not the voices of actual research subjects. (This was the real reason for putting subjects in separate rooms and having them communicate via intercom.) One of these recorded voices admitted somewhat hesitantly that he was subject to nervous seizures. In his second turn, he started to speak normally, but his speech suddenly became disorganized. Soon he lapsed into gibberish and choking sounds and then into silence. Evidently an emergency was occurring. The subject realized that all participants could hear it, although the intercom prevented them from talking to one another.

The dependent variables were whether the subject would leave the room to offer help and how quickly he or she would do so. Those subjects electing to help the victim typically came out of their room looking for the victim. The experimenter timed the speed of the subject's response from the beginning of the victim's speech. The results verified the research hypothesis that the greater the number of witnesses, the less likely a subject was to offer help to the victim.

This carefully controlled experiment allowed a straightforward test of the hypothesis. The manipulated independent variable (number of witnesses) and the measured dependent variable (speed of helping response) were unambiguous. Confounds from

Laboratory research enables the investigator to manipulate independent variables and measure behavior in various ways. In this study of aggressive behavior in children, trained observers collect data by viewing the subject through a one-way mirror.

extraneous variables can be ruled out due to the random assignment of subjects to treatments. From these results, we can conclude that the number of witnesses has a causal effect on the speed of helping response.

Note, however, that although the experiment showed the causal effect to hold, it demonstated this only under those conditions prevailing in the laboratory. The causal effect may or may not hold under other conditions. This can be problematic if the conditions that existed in the laboratory setting are uncommon in daily life. (When, for instance, was the last time you discussed personal issues over an intercom with five strangers in other rooms?) Thus, from this study alone, it is not clear whether we can generalize the cause-and-effect findings from the laboratory

to everyday face-to-face situations. The relationship between the number of others present and a person's reaction to an emergency might be different in other situations.

Although this experiment provides some answers regarding intervention in emergencies, it also raises further questions. Why, for instance, should the number of witnesses present affect a person's willingness to help in an emergency? The researchers conducting this study were aware of this issue and, based on data from a brief questionnaire administered after the experiment, they proposed that subjects in larger groups were slower to help because the responsibility for helping was more diffuse and less focused than in smaller groups. Although this "diffusion of responsibility" explanation is interesting, we must note that this experiment did not demonstrate it to be either true or false. The experiment showed only that under the conditions in the laboratory, the number of witnesses present affected the subjects' helping behavior.

Strengths of Experiments

The strength of experimental studies lies in their high level of internal validity. This makes experiments especially well suited for testing causal hypotheses. Experiments excel over other methods (surveys, field observation, and so on) in this respect.

Experiments have high internal validity precisely because they control or offset all factors other than the independent variable that might affect the dependent variable. Techniques to accomplish this include (1) randomly assigning subjects to treatments, (2) holding constant known extraneous variables, and (3) incorporating extraneous variables as factors in the research design—that is, manipulating them as independent variables, so that they are not confounded with the main independent variables of interest. Another technique is (4) measuring extraneous variables and including them in the data analysis as covariates of the independent variables.

In principle, investigators can design both laboratory experiments and field experiments to have high internal validity. In practice, however, laboratory experiments often have higher internal validity than comparable field experiments. This happens because researchers have more control over extraneous variables in the laboratory than in the field. Field experiments, however, often surpass laboratory experiments with respect to external validity.

Weaknesses of Experiments

One weakness of experimentation is that many social phenomena cannot be studied by this method. Investigators often lack the capacity to manipulate the independent variables of interest or to implement random assignment. Numerous moral, financial, and practical considerations in everyday life restrict what investigators can manipulate experimentally.

Even when the independent variable(s) can be manipulated, experiments face several threats to internal validity. First is the possibility that the experi-mental manipulation may fail. This might occur, for example, if the subjects interpret the manipulation as meaning something other than what the researcher intended. The usual remedy for this problem is to use *manipulation checks*—measures taken after the manipulation that show whether the subjects perceived the manipulation as intended. Use of manipulation checks is routine and widespread in social psychological experiments.

Another threat to internal validity of experiments is the existence of *demand characteristics* (also called *subject effects*). This refers to the possibility that subjects may interpret certain subtle cues in the experimental setting as requiring particular responses (Berkowitz & Troccoli, 1986; Orne, 1969; Thayer & Saarni, 1975). A subject effect occurs, for instance, when subjects bring a stereotyped role expectation or mental set to the experiment and then something in the experimental situation activates that expectation, causing the subjects to emit the role-defined behavior. The most extreme form of subject effect occurs when subjects somehow discover (or think they discover) the hypothesis under investigation by the researcher and then—trying to be "good" subjects—behave so as to make the hypothesis come "true." To reduce subject effects, many studies give standardized instructions (for example, tape-recorded messages) to subjects in all experimental treatments. Then, too, some designs disguise the nature of the research (and the research hypothesis) by providing a *cover story,* a plausible albeit false description of its purpose.

Another threat to internal validity is *experimenter effects,* the possibility that an experimenter may expect subjects to behave in a particular manner (aggressively, cooperatively, and so on) and unwittingly telegraph these expectations to the subjects (Rosenthal, 1966, 1980). The expectations communicated to subjects will likely influence their behavior. This can be a serious problem, especially if the expectancies conveyed by the experimenter change as a function of experimental treatment. Should this occur, the expectations communicated may have some implication for the hypothesis under study (such as speciously making the hypothesis appear to be "true.") Persons designing an experiment can use several techniques to minimize or eliminate experimenter effects. First, they can restrict the experimenters' contact with subjects and standardize their behavior in the experimental setting. This limits the opportunities to transmit expectations. Second, they can keep the research personnel "blind" regarding the hypotheses under study and the treatment to which each subject is assigned. If the research personnel do not hold strong expectations, there obviously will be none to communicate to subjects. Third, they can use a research design with two or more groups of experimenters, each with a different hypothesis concerning the variables of the study. Analysis of data from such a design will show whether experimenter effects are present or absent.

Beyond internal validity, experiments also face problems with external validity. Some experiments take place in settings that seem artificial to subjects and have low apparent realism. This is often true of laboratory experiments, although less true of field experiments. One useful distinction is that between *mundane realism* and *experimental realism* (Aronson et al., 1990). Mundane realism is the extent to which the experimental setting appears similar to natural, everyday situations. Experimental realism, in contrast, is the amount of impact the experimental situation creates—that is, the degree to which the subjects feel involved in the situation.

Low mundane realism need not imply low experimental realism. A laboratory study can have low mundane realism but high experimental realism. Subjects were highly involved, for example, in the study discussed earlier in which the experimenters staged an emergency in the laboratory. Many subjects were nervous and expressed concern when they came out of their room looking for the victim. Most expressed surprise when they later learned that the seizure was simulated, not genuine.

No single solution exists to the problem of establishing high experimental realism. Some investigators use a combination of both laboratory experiments and field experiments when investigating a phenomenon. This approach is often successful, for the field experiments provide the mundane realism that laboratory experiments lack. Other investigators simply note they are more concerned with experimental realism than with mundane realism. If the situation is

real and involving to the subjects, they maintain, then the behavior of the subjects is real and worthy of study.

Field Studies and Naturalistic Observation

Observational research—often termed a **field study**—involves making systematic observations about behavior as it occurs naturally in everyday settings. The data are typically collected by one or more researchers who directly observe the activity of people and record information about it. Field studies have been used to investigate many forms of social behavior in their natural settings. For instance, researchers have observed and recorded data about social interaction between judges and attorneys in the courtroom (Maynard, 1983), between teachers and students in the classroom (Galton, 1987), between couples in informal settings (Zimmerman & West, 1975), between working-class boys and girls in grade school (Thorne, 1993), and between police and juveniles on city streets (Piliavin & Briar, 1964). One study (Suttles, 1968) observed the use of stores, churches, and parks by members of different ethnic groups in one neighborhood, and the results showed that ethnicity was a major determinant of utilization patterns.

Because field studies investigate social behavior in its natural setting, researchers usually make efforts to minimize or limit the extent to which they intrude on that behavior. In fact, field studies are usually less intrusive than surveys or experiments. Whereas a survey often intrudes on people by asking for self-reports and an experiment involves manipulation of the independent variable(s) and random assignment to treatment, a field study involves nothing more intrusive than recording an observation about the behavior of interest.

Field studies differ in how the observers collect and record information. In some studies, observers watch carefully while the phenomenon of interest is occurring and then make notes about their observations at a later time. The advantage of recording afterward is that the observer is less likely to arouse curiosity, suspicion, or antagonism in the participants. In other studies, the observers may record field

Observational research involves careful scrutiny of activities in natural settings. Here, a market researcher observes who is eating what in a food court.

notes at the same time that they observe behavior. For instance, in research on police-citizen encounters (Black, 1980; Lundman, Sykes, & Clark, 1978), trained observers coded the interaction as it occurred. Although taking notes in this manner could potentially be intrusive, it permits more details to be recorded and minimizes any distortion from selective memory on the part of the investigator.

In still other field studies, researchers make audio or video recordings of interactions, and then analyze the tapes later (Whalen & Zimmerman, 1987). Tape recordings may seem a superior alternative to the use of human observers (who may have selective perception), but this is not always the case. Use of recordings maximizes the information obtained, but it can also inadvertently influence behavior if the participants discover they are being taped.

Participant Observation When the behavior of interest occurs in public settings such as restaurants, courtrooms, or retail stores, researchers can simply go to the setting and observe the action directly. The researchers do not need to interact with the persons being observed or to reveal their identities as investigators. However, when the behavior of interest is private or restricted in nature (such as intimate sexual activity, use of illegal drugs, or recruiting new members for a cult), observation is usually more difficult. To investigate activities of this type, research-

ers occasionally use the technique of *participant observation.* In participant observation, members of the research team not only make systematic observations of others' behavior but also interact with these others and play an active role in the ongoing events. The fact of being active participants frequently enables investigators to approach and observe behavior that otherwise would be inaccessible. In participant observation, researchers usually do not engage in overt coding or any other activity that would disrupt the normal flow of interaction. In some instances, they may even need to use an assumed identity, lest their true identity as investigators disrupt the interaction.

One study (Eder, with Evans & Parker, 1995) used observational techniques combined with participation to investigate adolescent school culture in a midwestern community. To observe interaction patterns and topics of conversation among junior high students, the investigators participated over an extended period of time in students' lunchroom groups. They identified themselves (truthfully) as being from a nearby university, and they adopted the role of "quiet friend." They did not affiliate with teachers and avoided appearing to be authority figures of any kind. This approach enabled them to establish sufficient rapport and trust with students that they could ask questions about the students' beliefs regarding gender differences and observe how students' behavior patterns fostered gender inequality.

Unobtrusive Measures Field studies sometimes use *unobtrusive measures,* which are measurement techniques that do not intrude on the behavior under study and that avoid causing a reaction from the persons whose behavior is being measured (Webb et al., 1981). For example, some unobtrusive measures rely on physical evidence left behind by people after they have exited from a situation. One investigator discovered that the rate at which vinyl floor tiles needed replacement in the Chicago Museum of Science and Industry was a good indicator of the popularity of exhibits. Another illustration is the analysis of inventory records and bar bills to unobtrusively measure the alcohol consumption patterns at various nightclubs and bars (Lex, 1986).

Strengths and Weaknesses of Field Studies Like any research method, field studies have both strengths and weaknesses. A primary strength is that observational techniques allow researchers to study activity in real-world settings. Careful observation can provide a wealth of information about behavior as it actually occurs in natural settings. In addition, because these techniques are relatively unintrusive, investigators can use them to investigate sensitive or private behaviors such as drug use or sexual activity that would be difficult to address through intrusive methods like surveys or experiments.

Weaknesses of field studies include their sensitivity to the specific recording methods used. Observations recorded after the fact are often less reliable and valid than those recorded on the spot or those based on audio- or videotaping. In addition, the validity of the observations may depend in part on what identities the investigators project publicly while making their observations; validity may be destroyed if the researchers have been operating covertly and subjects suddenly discover they are under observation. Then, too, the external validity of field observational studies can be problematic because research of this type frequently focuses on only one group or organization or on a sample of interactions selected for convenience.

In some cases, field investigators do not obtain informed consent from the persons being observed prior to the collection of data. Permission is sought for use of the data only after behavior has been observed or conversations tape recorded. Some persons construe this as a serious drawback and object to participant observation on ethical grounds. Of course, this concern has to be judged against the fact that if permission were sought in advance, the behavior under investigation might never occur or might take a different form.

Archival Research and Content Analysis

Although social psychological researchers often prefer to collect original data, it is sometimes possible to test hypotheses and theories by using data that already exist. The term **archival research** denotes the acquisition and analysis (or reanalysis) of information collected previously by others. When archival data of suitable quality exist, a researcher may decide that to analyze them is preferable to collecting and analyzing new data. Archival research usually costs less than alternative methods.

Sources There are many sources of archival data. In the United States, one important source is government agencies. The Census Bureau makes available much of the data it has collected over the years. Census data are a rich source of information about the U.S. population; they often include repeated measures taken at different points in time, which allow an investigator to assess historical trends. The Bureau of Labor Statistics, the Federal Bureau of Investigation, and other agencies also release data to investigators. A second important source of U.S. archival data is the data banks maintained at various large universities. These archives serve as locations where researchers can deposit data they have collected so others can use them. They include, among others, the Interuniversity Consortium for Political and Social Research and the Data Archive on Adolescent Pregnancy and Pregnancy Prevention. A third source of archival data—less used by social psychologists but still important—is formal organizations such as insurance companies and banks. These typically entail over-time data with respect to various measures of financial and economic performance. A fourth source of archival information for research is newspaper articles, a rich source of information about past events. For instance, an investigator wishing to study the reactions of U.S. civilians during a historical event like the 1991 Gulf War might use newspapers as a data source. Other types of printed material (for example, corporate annual reports) can also provide archival data usable in research.

Content Analysis In some cases, an investigator relying on newspaper articles, government documents, or annual reports as archival sources can use the information directly as it appears. All the investigator has to do is extract the information and analyze it, usually via computer. In other cases, however, the investigator faces the problem of how to interpret

Archival research, which involves the analysis of previously collected data, is facilitated by such tools as microfilm readers and computers to access databases.

and code the information from the source. Under these circumstances, he or she may use the technique of **content analysis,** which involves undertaking a systematic scrutiny of documents or messages to identify specific characteristics and then making inferences based on their occurrence. For example, if newspapers serve as the source, the investigator could use content analysis to code the reportage from newspaper articles into a form suitable for systematic statistical analysis.

Researchers have used content analysis to investigate a wide variety of topics. Some studies, for instance, have analyzed the content of lonely-hearts advertisements placed in newspapers by males and females seeking mates (Koestner & Wheeler, 1988). Other studies have addressed such issues as whether the depiction of people with disabilities is distorted in American newspapers (Keller et al., 1990), and

whether the quality press carries more information on health-related issues than the popular press, thereby causing a gap in health knowledge between social classes (Kristiansen & Harding, 1988).

When a researcher conducts a content analysis, the first step is to identify the informational unit to be studied—is it the word, the sentence, the paragraph, or the article? The next step is to define the categories into which the units will be sorted. A further step is to code the units in each document into the categories, and the final step is to look for relations within the categorized data.

As an example of content analysis, consider a study of the relationship between rhetorical forms of speech and applause from the audience (Heritage & Greatbatch, 1986). The investigators hypothesized that political speakers will use certain rhetorical forms—for example, a three-element list—to signal

Table 2.1   Strengths and Weaknesses of Research Methods

		Method			
	Survey	**Laboratory Experiment**	**Field Experiment**	**Observational Study**	**Archival Research**
Internal Validity	Moderate	High	Moderate	Low	Low
External Validity	Moderate	Moderate	High	Moderate	Moderate
Investigator Control	Moderate	High	Moderate	Moderate	Low
Intrusiveness of Measures	Moderate	Moderate	Low	Moderate	Low
Difficulty of Conducting Study	Moderate	Moderate	High	Moderate	Low
Ethical Problems	Few	Some	Some	Many	Few

Note: Entries in the table indicate the strength of the research methods with respect to the various concerns (validity, control, intrusiveness, and the like).

the audience when to applaud. The raw data in this study were the texts of 476 speeches delivered by British political leaders at party meetings. The researchers carefully defined the rhetorical devices and identified their use in the speeches. Then they counted the number of times that speakers used each device and noted whether the audience responded immediately to each use with applause. Results showed that applause was much more likely to occur immediately after the use of certain rhetorical devices (such as a three-element list) than at other points in the speech.

Strengths and Weaknesses of Archival Research
One significant advantage of archival research is its comparatively low cost. By reusing existing information, the investigator avoids the cost of collecting new data in the field or the laboratory. A second advantage is that by using information already on hand, an investigator may complete a study more quickly than otherwise. A third advantage is that an investigator can test hypotheses about phenomena that occur over extended periods of time. In some cases, authorities have kept records for decades or even centuries (such as marriage licenses), and these

can serve as a basis for investigating various issues (such as who marries whom).

One major disadvantage of archival research is the lack of control over the type and quality of information. An investigator must work with whatever others have collected. This may or may not include data on all the variables an investigator wishes to study. There may be doubts, too, regarding the quality of the original research design or the procedures used for collecting data. A second disadvantage of archival research is that to create a reliable and valid content analytic scheme for use with records can be difficult, especially if the records are complex. A third disadvantage is that some sets of records contain large amounts of inconsistent and/or missing information, which will obviously hinder the study and limit the validity of any findings.

Comparison of Research Methods

We have discussed a variety of research methods—surveys, laboratory and field experiments, naturalistic observation, and archival research. Table 2.1 summarizes the strengths and weaknesses of each research method. As the table indicates, no one method of empirical investigation is best for all purposes. A method's appropriateness depends on the phenomenon under study and on the research characteristics most important to the investigator.

Surveys, which provide a useful way of obtaining an accurate description of the attributes of some population, usually have at least moderate internal and external validity, and they pose few ethical problems. Laboratory experiments, which can be especially useful in testing causal hypotheses, are generally high in internal validity, but they may pose some ethical problems (especially if deception is used). Field studies relying on observational techniques tend to have comparatively low internal validity and may confront a variety of ethical issues, but they may still be the best way to investigate previously unexplored social phenomena in their natural settings.

Ethical Issues in Social Psychological Research

In the past 25 years, the public has become concerned about the ethical issues involved in research on humans. A consensus has developed among investigators and others affiliated with the scientific community that subjects who participate in research have certain rights that must be respected. In some cases, protecting those rights requires investigators to limit or modify their research practices.

In the following discussion of ethical issues, we focus first on potential sources of harm to subjects. Then we discuss various safeguards, such as risk-benefit analysis and informed consent, to protect subjects' rights.

Potential Sources of Harm

Harm to subjects from research can take a variety of forms. These include physical harm, psychological harm, and harm from breach of confidentiality.

Physical Harm These days, exposure to physical harm in social psychological research is uncommon. Electric shock, for instance, is used in very few

studies. Investigations to measure the effects of stress do sometimes employ an exercise treadmill or tasks that require subjects to immerse one hand in ice water. As a precaution, investigators usually screen prospective subjects and exclude those with relevant medical conditions. At the onset of a study, investigators are expected to inform the subjects about any risks so they can decide whether they might be harmed by participating. In studies involving physical stress, investigators can and typically do monitor participants for adverse effects throughout the research.

Psychological Harm A more common risk in social psychological research is psychological harm to subjects. This risk is present in studies in which subjects receive negative information about themselves. For example, a not uncommon experimental manipulation is to give subjects false feedback about their physical attractiveness, about others' reactions to them, or about their performance on various tests or tasks. Investigators can use such feedback to raise or lower subjects' self-esteem, to induce feelings of acceptance or rejection by others, or to create perceptions of success or failure on important tasks. These manipulations are effective precisely because they do influence the subject's self-perception.

Negative feedback may cause psychological stress or harm, at least temporarily. For this reason, some investigators believe such techniques should not be employed in research. Others find them acceptable, however, and believe they may be used if alternative, less harmful manipulations are not available. When false feedback is used, an investigator can limit any long-term effects by giving the subjects a thorough *debriefing* after the study. In a debriefing, the investigator provides subjects with a full description of the study and emphasizes the falsity of the feedback. Debriefing should be done immediately after the study to minimize the time that subjects labor under false impressions.

Breach of Confidentiality Confidentiality is another important issue in survey and observational research. Interviewers and observers are frequently able to identify participants, and they may recall

2.1   Ethical Considerations in Research Design

Before conducting a given study, investigators and members of institutional review boards regularly ask certain ethical questions about the proposed research design and its impact on participants. Among the most commonly asked ethical questions are the following:

1. Is it possible that participants in the study might be harmed physically, for example, by strenuous exercise?
2. Does the study give participants false information about themselves or use any other form of deception?
3. Does the study induce participants to engage in behavior that might threaten their self-respect?

4. If the investigators make audio- or videotapes of the participants, will they obtain permission from the participants to use the tapes as a source of data?
5. What steps will the investigators take to preserve the confidentiality of information obtained about the participants?
6. Will the investigators tell potential subjects in advance about foreseeable risks that their participation may entail?
7. Will subjects have a chance to ask questions about the study before they consent to participate?
8. Will the investigators inform the subjects that they have the right to terminate their participation at any time?
9. At the end of the study, will the investigators fully debrief the subjects and tell them about the real nature of the study and its procedures?

details regarding participants' behavior or responses to questions. Were confidentiality to be breached, the effects might be damaging to the participants. This concern arises especially in surveys inquiring about sexual behaviors or sensitive personal matters. It also arises in observational studies of deviant or criminal activities.

One important precaution against breach of confidentiality is to avoid including on the research team any persons who are apt to have social contacts with respondents in other settings. In addition, some investigators refuse to attach any identifying information such as names and addresses to data after it has been collected. Another approach is to keep identifying information separate from questionnaires or behavioral records to prevent breaches of confidentiality.

Observational research often deals with a specific group or organization. During their investigation, researchers may gather information both about the organization itself and about various members. When the findings are published, the investigators typically refer to the organization by a pseudonym and to members by role only. This practice usually suffices to prevent outsiders from identifying the unit and

its members, although it may not prevent members from identifying each other. There are obvious risks to members' positions, reputations, or jobs within the organization if compromising information becomes known to other members.

Box 2.1, "Ethical Considerations in Research Design," lists some major ethical questions that apply to many studies.

Institutional Safeguards

As noted above, researchers can take various steps to prevent harm to participants. Although many people believe voluntary self-regulation by researchers is sufficient, others feel some agency other than the researcher should review research designs to protect the rights and interests of the participants. Accordingly, some institutions have developed safeguards and put them into place. The two most important safeguards are conducting a risk-benefit analysis and obtaining informed consent from all participants.

Risk-Benefit Analysis   The federal government is a major provider of funds for research in the social and biomedical sciences. Many federal departments and agencies have adopted common criteria for the review of research involving human subjects. Under these regulations, both investigators and institutions are responsible for minimizing the risks, of whatever type, to participants in research. The rules encourage researchers to develop designs that expose subjects to no more than "minimal risk," meaning risk no greater than that ordinarily encountered in daily life or during the performance of routine physical or psychological examinations or tests (Code of Federal Regulations, 1992).

In addition, the regulations require each institution that receives funds from federal agencies to establish an institutional review board responsible for reviewing proposed research involving human subjects. The review board (sometimes called a human subjects committee or research ethics committee) assesses the extent to which participants in each proposed study will be placed at risk. As noted above, many social psychological studies involve no foreseeable risks to participants, but if the members of the review board believe participants might be harmed—physically, psychologically, or due to breach of confidentiality—a detailed assessment must be made. That is, the review board conducts a **risk-benefit analysis,** which weighs potential risks to the subjects against anticipated benefits to subjects and the importance of the knowledge that may result from the research. The review board will not approve research involving risk to participants unless it concludes the risk is reasonable in relation to benefits.

Informed Consent   The other major safeguard against risk is the requirement that investigators obtain informed consent from all individuals, groups, or organizations who participate in research studies. **Informed consent** exists when potential subjects or respondents, on being informed (by the investigators) what their participation will involve, agree willingly to participate in the research. Specifically, six elements are essential to informed consent. The researchers should (1) give potential subjects a brief description of the procedures to be employed, although they need not (and usually do not) tell the subjects the purpose or hypothesis of the research; (2) inform subjects

about foreseeable risks from participation; (3) inform subjects about the extent to which the information they provide will be kept confidential and, if possible, give subjects a guarantee that confidentiality or anonymity will be maintained; (4) provide information about what medical or psychological resources, if any, are available to subjects who are adversely affected by participation; (5) offer to answer questions about the study, whenever possible; and (6) inform potential subjects that they have the right to terminate their participation at any time.

In many survey and observational settings, investigators implement informed consent by giving this information to respondents orally. In experiments, especially those involving some risk to subjects, investigators usually obtain written consent from each participant.

Summary

This chapter discussed the research methods used by social psychologists to investigate social behavior, activity, and events.

Characteristics of Empirical Research
(1) Objectives of research include describing reality, identifying correlations among variables, testing causal hypotheses, and testing theory. (2) Research is usually guided by a hypothesis, which specifies a (causal) relationship between two or more variables. (3) Ideally, the findings of empirical research should be high in both internal validity and external validity.

Research Methods   Social psychologists rely heavily on four methods: surveys, experiments, naturalistic observation, and archival research based on content analysis. (1) A survey involves systematically asking questions and recording the answers from respondents. Investigators use surveys to gather self-reported information about attitudes and activities. The quality of the data obtained in a survey depends on the reliability and validity of the measures used. (2) An experiment involves the manipulation of independent variables and the random assignment of

subjects to experimental conditions. Some experiments are conducted in a laboratory, where the investigator has a high degree of control; others are conducted in natural settings, where control may be lower. (3) Naturalistic observation involves collecting data about naturally occurring events. In a field study, observers view an event or activity as it occurs and then record their observations. (4) Archival research involves the analysis of existing information collected by others. Sources of archival data include the Census Bureau and other federal agencies, data archives, and newspapers. Investigators use content analysis to study textual material such as speeches or reports. (5) Each method has particular strengths and weaknesses. What method is best depends on the investigator's aims and concerns.

Ethical Issues in Social Psychological Research

(1) Potential sources of harm to participants in research include physical harm, psychological harm, and breach of confidentiality. Individual investigators can take various steps to prevent or minimize such harm. (2) Institutional safeguards against harm require investigators to minimize risks to participants and to obtain informed consent from participants. Institutional review boards monitor research designs to assure that these conditions are met by investigators.

Key Terms

archival research (p. 38)
content analysis (p. 39)
dependent variable (p. 26)
experiment (p. 32)
external validity (p. 27)
extraneous variable (p. 26)
field study (p. 36)
hypothesis (p. 25)
independent variable (p. 26)
informed consent (p. 43)
internal validity (p. 26)
interview survey (p. 28)
methodology (p. 24)
panel study (p. 31)
population (p. 30)
questionnaire survey (p. 28)
random assignment (p. 32)
reliability (p. 28)
response rate (p. 28)
risk-benefit analysis (p. 43)
sample (p. 30)
simple random sample (p. 30)
stratified sample (p. 30)
survey (p. 27)
validity (p. 29)

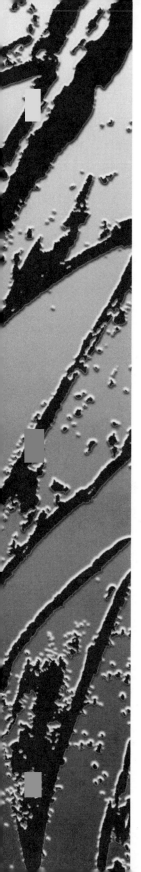

CHAPTER 3
Socialization

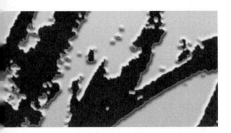

Introduction

A change has come over me. I do things differently from the way I used to. The change has been subtle, slow, but other people have noticed it as well. My wife complains, but it's not my fault. I am suffering from creeping Harry Cohenism.

Harry Cohen is my father. For years, this was nothing more than a statement of fact. He is my father and I am his son and other than that, we had very little in common. He could spell and I could not. He could do math in his head and I could not. He could not sleep well or at all past 6 A.M. and I could sleep soundly and forever and he was always, but always, way older than I was.

Now, however, we are both the same age—middle-aged. He is upper middle-aged, and I am younger middle-aged, but we are essentially the same age.

I sleep like him now. I used to sleep like me—deeply and endlessly. When I was a kid, this would drive my father mad. He would look at me sound asleep in my bed at, say, 1 o'clock in the afternoon, and explode with frustration: "Get up, get up!" He could not understand how anyone could sleep so late. I could not understand how anyone could not. Sleeping, in fact, was one of the very few things I did really well. I excelled at sleeping. But no more. I wake up at 6 in the morning. My eyes open and that is it. I try to go back to sleep, but I cannot. I lie and look at the clock and wait for the sun to come up, but there is no going back to sleep. For a long time, I worried that something was wrong with me. I could not figure out what was happening, and then it occurred to me. I was becoming my father.

I know that by all logic I should be becoming my mother, too. After all, half my genes are hers and, knowing her, they are the dominant ones. But for some reason, I feel that her genes declared themselves early and that from here on out it is my father who is taking over.

After all, it is my father and not my mother who falls asleep in front of the television set. It is my father and not my mother who gets bogged down in detail in the middle of telling a story. And it is my father who will tell you the same story twice—only each time a little bit differently. I am even beginning to like opera.

For a time, I thought that I was my own man, that I could be exactly what I wanted to be. I recognize now that there is such a thing as a genetic manifest destiny—a case of the future really being the past. I don't have any problems with that, though. My father is a grand man—sweet and sensitive and smart. I didn't realize it until recently, but in many ways we're a lot alike—not like my son. Boy can that kid sleep!

I wonder where he gets it from. (Cohen, 1982)

Richard Cohen expresses well one of the striking features of social life. There is great continuity from one generation to the next, continuity in both physical characteristics and in behavior. Genetic inheritance is one source of continuity. But a major contributor to intergenerational similarity is **socialization,** the ways in which individuals learn skills, knowledge, values, motives, and roles appropriate to their position in a group or society (Bush & Simmons, 1990).

How does an infant become "human"—that is, an effective participant in society? The answer is through socialization. As we grew from infancy, we interacted continually with others. We learned to speak English, a prerequisite for participation in our society. We learned basic interaction rituals, such as greeting a stranger with a handshake and a loved one with a kiss. We also learned the socially accepted ways to achieve various goals, both material (food, clothing, shelter) and social (respect, love, help of others). It is obvious that socialization makes us like most other members of society in important ways. It is not so obvious that socialization also produces our individuality. The self and the capacity to engage in self-oriented acts (discussed in Chapter 4) are a result of socialization.

This chapter first examines childhood socialization. By "childhood" we mean the period from birth to adolescence. Childhood is a social concept, shaped by historical, cultural, and political influences (Elkin & Handel, 1989). In contemporary American society, we define children as immature, in need of training at home and of a formal education. At the end of this chapter, we consider the continuing socialization of adults.

The discussion focuses on the following five questions:

1. What are the basic perspectives in the study of socialization?
2. What are the socializing agents in contemporary American society?
3. What are the processes through which socialization occurs?
4. What are the outcomes of socialization in childhood?
5. What is the nature of socialization in adulthood?

Perspectives on Socialization

As we saw at the beginning of this chapter, Richard Cohen attributed the similarity between himself and his father to "genetic manifest destiny." What is the more important influence on behavior—nature or nurture, heredity or environment? This question has been especially important to those who study the child. Although both are influential, one view emphasizes biological development (heredity), whereas the other emphasizes social learning (environment).

The Developmental Perspective

The human child obviously undergoes a process of maturation. He or she grows physically, develops motor skills in a relatively uniform sequence, and begins to engage in various social behaviors at about the same age as most children.

Some theorists view socialization as largely dependent on processes of physical and psychological

Responsiveness to another person develops early in life. By 16 weeks of age, a child smiles in response to a human face. By 28 weeks, a child can distinguish caregivers from strangers.

maturation, which are biologically determined. Gesell and Ilg (1943) documented the sequence in which motor and social skills develop and the ages at which each new ability appears in the average child. They viewed the development of many social behaviors as primarily due to physical and neurological maturation, not social factors. For example, toilet training requires voluntary control over sphincter muscles and the ability to recognize cues of pressure on the bladder or lower intestine. According to developmental theory, when children around age 2½ develop these skills, they learn by themselves without environmental influences.

Table 3.1 lists sequences of development of various abilities that have been identified by observational research. The ages shown are approximate; some children exhibit the behavior at younger ages; others do so later.

As an example, consider the development of responsiveness to other persons. As early as 4 weeks, many infants respond to close physical contact by relaxing. At 16 weeks, babies can discriminate the human face and usually smile in response. They also show signs of recognizing the voice of their usual caregiver. By 28 weeks, the infant clearly differentiates faces and responds to variations in facial expression. At 1 year, the child shows a variety of emotions in

Table 3.1   The Process of Development

	16 Weeks	28 Weeks	1 Year	2 Years	3 Years
Visual Activity	Follows objects with eyes Eyes adjust to objects at varying distances	Watches activity intently Hand-eye coordination	Enjoys watching moving objects (like TV picture)	Responds to stimuli in periphery of visual field Looks intently for long periods	
Interpersonal	Smiles at human face Responds to caregiver's voice Demands social attention	Responds to variation in tone of voice Differentiates people (fears strangers)	Engages in responsive play Shows emotions, anxiety Shows definite preferences for some persons	Prefers solitary play Rudimentary concept of ownership	Can play cooperatively with older child Strong desire to please Sex differences in choice of toys, materials
Vocal Activity	Vocalizes pleasure (coos, gurgles, laughs) Babbles (strings of syllablelike sounds)	Vocalizes vowels and consonants Tries to imitate sounds	Vocalizes syllables Practices two to eight known words	Vocalizes constantly Names actions Repeats words	Uses three-word sentences Likes novel words
Bodily Movement	Can hold head up Can roll over	Can sit up	Can stand Can climb up and down stairs	Can run Likes large-scale motor activity—push, pull, roll	Motion fluid, smooth Good coordination
Manual Dexterity	Touches objects	Can grasp with one hand Manipulates objects	Manipulates objects serially	Good control of hand and arm	Good fine-motor control—uses fingers, thumb, wrist well

SOURCES: Adapted from Caplan, 1973; Gesell and Ilg, 1943.

response to behavior by others. He or she seeks interaction with adults or with siblings by crawling or walking to them and tugging on clothing. Thus recognition of, responsiveness to, and orientation toward adults follow a uniform developmental pattern. The ability to interact with others depends in part on the development of visual and auditory discrimination.

The Social Learning Perspective

Whereas the developmental perspective focuses on the unfolding of the child's own abilities, the social learning perspective emphasizes the child's acquisition of cognitive and behavioral skills from the environment. Successful socialization requires that the child acquire considerable information about the world. First, the child must learn many physical or natural realities (Baumrind, 1980), such as what animals are dangerous and which things are edible. Children also must learn about the social environment. They must learn the language used by people around them to communicate their needs to others. They also need to learn the meanings their caregivers associate with various actions. Children need to learn to identify the kinds of persons encountered in the immediate environment. They need to learn what behaviors they can expect of people, as well as the expectations for their own behavior.

According to the social learning perspective, socialization is primarily a process of children learning the shared meanings of the groups in which they are reared (Shibutani, 1961). Such variation in meanings gives groups and subcultures (and societies) their distinctiveness. Although the content—what is learned—varies from group to group, the processes by which it takes place are universal. This viewpoint emphasizes the adaptive nature of socialization. The infant learns the verbal and interpersonal skills necessary to interact successfully with others. Having acquired these skills, children can perpetuate the meanings that distinguish their social groups and even add to or modify these meanings by introducing innovations of their own.

Recent research on socialization considers both the importance of developmental processes and the influence of social learning. The developmental age of the child obviously determines which acts the child can perform. Infants less than 6 months old cannot walk. All cultures have adapted to these developmental limitations by coordinating the performance expectations placed on children with the maturation of their abilities. However, developmental processes are not sufficient for the emergence of complex social behavior. In addition to developmental readiness, social interaction—learning—is necessary for the development of language. This is illustrated by the case of Isabelle, who lived alone with her deaf-mute mother until the age of 6½. When she was discovered, she was unable to make any sound other than a croak. Yet within 2 years after she entered a systematic educational program, her vocabulary numbered more than 1,500 words and she had the linguistic skills of a 6-year-old (Davis, 1947).

Thus both nature and nurture influence behavior. Developmental processes produce a readiness to perform certain behaviors. The content of these behaviors is determined primarily by social learning, that is, by cultural influences.

The Interpretive Perspective

Socialization occurs primarily through social interaction. Whereas the learning perspective emphasizes the process of learning—for example, the role of reinforcement in the acquisition of behavior—the interpretive perspective (Corsaro, 1992) focuses on the interaction itself. Drawing on symbolic interaction theory, this perspective views the child's task as the discovery of the meanings common to the social group (such as the family). This process of discovery requires communication with parents and other adults and with other children. Especially important is the child's participation in **cultural routines,** which are recurrent and predictable activities that are basic to day-to-day social life (Corsaro & Eder, 1995). Greeting rituals, common games, and mealtime patterns are examples of such cultural routines. These routines provide a sense of security and of belonging to a group.

According to this perspective, development is a *reproductive process.* Children don't simply learn culture. In daily interaction, children use the language and interpretive skills they are learning or discovering.

As they become more proficient in communicating and more knowledgeable about the meanings shared in the family, children attain a deeper understanding of the culture. Children, through interaction, acquire and reproduce the culture.

When children communicate with each other (as in school or at play), they don't simply imitate the acquired culture. They use what they have learned to create their own somewhat unique peer culture. Children in a schoolyard take a traditional game, such as hide-and-seek, and change the rules to fit their needs and the social context in which they are enacting the game. The changed rules become part of a new routine of hide-and-seek. Thus, from an early age, children are not just imitating culture, but creating it.

The Impact of Social Structure

A fourth perspective emphasizes the influence of social structure. Socialization is not a random process. Teaching new members the rules of the game is too important to be left to chance. Socialization is organized according to the sequence of roles that newcomers to the society ordinarily pass through. In American society, these include familial roles, such as son or daughter, and roles in educational institutions, such as preschooler, elementary school student, and high school student. These are age-linked roles; we expect transitions from one to another to occur at certain ages. Distinctive socialization outcomes are sought among those who occupy each of these roles. Thus we expect young children to learn language and basic norms governing such diverse activities as eating, dressing, and bowel and bladder control. Most preschool programs will not enroll a child who has not learned the latter.

Furthermore, social structure designates the persons or organizations responsible for producing desired outcomes. In a complex society such as ours, there is a sequence of roles and a corresponding sequence of agents. From birth until age 5, the family is primarily responsible for socializing children. From ages 6 to 12 children are elementary school students; we expect elementary school teachers to teach them the basics. Next, adolescents become high school students with yet another group of agents to further develop their knowledge and abilities.

This view is sociological; it considers socialization as a product of group life. It calls our attention to the changing content of and responsibility for socialization throughout the individual's life.

Agents of Childhood Socialization

Socialization always involves (1) someone who serves as a source of what is being learned; (2) a learning process; (3) a person who is being socialized; and (4) something that is being learned. We refer to these four components as *agent, process, target,* and *outcome*. This section considers the three primary agents of childhood socialization—family, peers, and school. Later sections focus on the processes and outcomes of childhood socialization.

Family

At birth, infants are primarily aware of their own bodies. Hunger, thirst, or pain create unpleasant and perhaps overwhelming bodily tension. The infant's primary concern is to remove these tensions and satisfy bodily needs. To meet the infant's needs, adult caregivers must learn to read the infant's signals accurately (Ainsworth, 1979). Also, infants begin to perceive their principal caregivers as the source of need satisfaction. These early experiences are truly interactive (Bell, 1979). The adult learns how to care effectively for the infant, and the infant forms a strong emotional attachment to the caregiver.

Is a Mother Necessary?   Does it matter who responds to and establishes a caring relationship with the infant? Must there be a single principal caregiver in infancy and childhood for effective socialization to occur?

Psychoanalytic theory (Freud) asserts that an intimate emotional relationship between infant and caregiver (almost always the mother at the time Freud wrote) is essential to healthy personality development. This was one of the first hypotheses to be studied empirically. To examine the effects of the absence of a single close caregiver on children, researchers have studied infants who are institutionalized. In the earliest reported work, Spitz (1945, 1946) studied an institution in which six nurses cared for 45 infants under 18 months old. The nurses met the infants' basic biological needs. However, they had limited contact with the babies, and there was little evidence of emotional ties between the nurses and infants. Within 1 year, the infants' scores on developmental tests fell dramatically from an average of 124 to an average of 72. Within 2 years, one third had died, 9 had left, and the 21 who remained in the institution were severely retarded. These findings dramatically support the hypothesis that an emotionally responsive caregiver is essential.

The findings of research on institutionalized children led to the conclusion that infants need a secure **attachment**—a warm, close relationship with an adult that produces a sense of security and provides stimulation—to develop the interpersonal and cognitive skills needed for proper growth (Ainsworth, 1979). Moreover, being cared for in such a relationship provides the foundation of the infant's sense of self.

For many years, gender-role definitions in American society made mothers primarily responsible for raising children. Fathers' parental responsibility was to work outside the home and provide the income needed by the family. The division of labor in many families conformed to these definitions. As a result, some concluded that a warm, intimate, continuous relationship between a child and its *mother* is essential to normal child development (Bowlby, 1965). Perhaps only in the mother-infant relation can the child experience the necessary sense of security and emotional warmth. According to this view, other potential caregivers have less emotional interest in the infant and may not be adequate substitutes.

Research on parent-child interaction indicates that if mothers are sensitive to the child's needs and responsive to his or her distress in the first year of life, the child is more likely to develop a secure attachment (Belsky, 1990). This is true in both two-parent and mother-only families. Infants who are securely attached to their mothers in the first 2 years of life evidence less problem behavior and more cooperative behavior from ages 4 to 10. Thus secure mother-infant attachment is associated with positive outcomes (see Box 3.1, "Attachment in Children and Adults").

Since 1960, gender-role definitions have been changing. Married women with children are increasingly working outside the home (see Figure 17.3). What effects do maternal employment and child care have on children?

Effects of Maternal Employment

Studies that compare children whose mothers are employed with similar children whose mothers are not report some differences (Lamb, 1982). Children whose mothers return/start to work within a few weeks or months of the child's birth have an insecure maternal attachment at 12 months of age. Children whose mothers return/start to work in the second or third year following the birth are less cooperative at ages 4 to 6 than children whose mothers remained at home (Belsky & Eggebeen, 1991). Among grade school children from lower-class families, sons of employed mothers show poorer psychological adjustment than sons of nonemployed mothers (Lamb, 1982). Thus maternal employment during childhood has some negative effects. However, a study of 147 employed mothers and their children (MacEwen & Barling, 1991) found it was the amount of conflict between the maternal and work roles that was associated with children's behavior problems and anxiety, not work itself.

Among adolescents, maternal employment is associated with positive outcomes. Youths whose mothers are employed have less stereotyped gender-role attitudes and expectations. A study of the likelihood of graduating from high school, in a sample of 1,258 young people ages 19 to 23, found that maternal employment during adolescence had a positive effect on school completion. Moreover, maternal employment during childhood had no effect on completion (Haveman, Wolfe, & Spaulding, 1991).

What about the effects of day care? It depends on the type of care and the age of the child (Belsky, 1990). Studies of private centers with low child/staff ratios and trained caregivers found that the children were emotionally secure and cooperative. Studies of children in family day care report similar findings. Studies of children in other types of settings, for example centers with high child/staff ratios, report less positive outcomes. Recent studies find that for children in day care during the first year of life, the

3.1   Attachment in Children and Adults

Which of the following best describes your feelings about relationships?

A. I find it relatively easy to get close to others and am comfortable depending on them and having them depend on me. I don't often worry about being abandoned or someone getting too close to me.

B. I am somewhat uncomfortable being close to others; I find it difficult to trust them completely, difficult to allow myself to depend on them. I am nervous when anyone gets too close and often love partners want me to be more intimate than I feel comfortable being.

C. I find that others are reluctant to get as close as I would like. I often worry that my partner doesn't really love me or won't want to stay with me. I want to merge completely with another person and this desire sometimes scares people away.

Each of these statements represents one *attachment style,* an individual's characteristic way of relating to significant others (Hazan & Shaver, 1987). The first describes a secure style, the second an avoidant style, and the third the anxious/ambivalent style.

The roots of the individual's style may be found in childhood. Ainsworth (1979) identified three styles of attachment in caregiver-child interactions. The attachment style of a young child is assessed by observing how the child relates to his or her caregiver when distressed (for example, by a brief separation in a strange environment). The secure child readily approaches the caregiver and seeks comfort. The avoidant child does not approach the caregiver and appears detached. The anxious/ambivalent child approaches the caregiver and expresses anger or hostility toward him or her. Children as young as 2 behave consistently in one of these ways when distressed.

Research suggests that we bring the style we developed as children into our intimate adult relationships (Shaver, Hazan, & Bradshaw, 1988). Surveys of adults (for example, Hazan & Shaver, 1987) find that about 55% describe themselves as secure, 25% as avoidant, and 20% as anxious/ambivalent. Men and women who describe themselves as secure report that their romantic relationships involve interdependence, trust, and commitment (Simpson, 1990). Adults who describe themselves as avoidant say that they do not trust others and are afraid of getting close (Feeney & Noller, 1990). Those who are anxious/ambivalent report intense emotions toward the partner and a desire for deep commitment in a relationship.

Thus the characteristic attachment style we develop in childhood may affect our intimate relationships throughout life.

emotional and behavioral consequences may be good or bad, depending on the quality of care.

Father's Involvement With Children   The broadening of maternal role definitions to include work outside the home has been accompanied by changes in expectations for fathers. This new ideology of fatherhood, promoted by television and film, encourages active involvement of fathers in child care and child rearing (Parke, 1996). Some men have adopted these expectations for themselves; others have not. Research finds that the father's contribution is primarily through play, and that fathers engage in more rough and tumble play with children than mothers. Such play is thought to facilitate the child's development of motor skills. These patterns are found in white, African American, and Hispanic two-parent families (Parke, 1996).

Several variables influence the extent of father's involvement with their children. Maternal attitudes are one important factor; a father is more involved when the mother encourages and supports his participation. Maternal employment is another influence. Husbands of employed women are more involved in child care, and in some cases provide full-time care for the child. Finally, a study found that lower levels of stress on the job and greater support from co-workers for being an active father were associated with greater involvement (Volling & Belsky, 1991). Thus research suggests that work stressors have negative effects on both fathers' and mothers' involvement in child rearing.

Effects of Divorce

Half of all marriages end in divorce; 60% of these divorces involve children. Divorce usually involves several major changes in the life of a child: a change in family structure, a change in residence, a change in the family's financial resources, and perhaps a change of schools. Therefore, it is difficult to isolate the effects of divorce, the change in family structure, independently of these other changes. Research suggests that the effects vary by the age of the child and the length of time since separation. Divorce may affect the children's intellectual, emotional, and social development (Hetherington, Cox, & Cox, 1982). These effects peak about 1 year after the separation, and many of the differences between children living with a single parent and children living with both parents get smaller in the second year.

Children in divorced families have more negative interactions with both mothers and peers, compared with children in intact families (Demo & Acock, 1988). There is often a loss of control over the child by the mother and more oppositional behavior by the children in single-parent families. The effects are especially strong on boys living in single-mother families. These boys have higher rates of behavior problems and interpersonal conflicts both at home and at school (Hetherington, 1991). These outcomes may reflect disrupted or reduced parenting of children during and after a divorce.

Divorce has a strong impact on adolescents. Young people living with a single mother receive less encouragement and help with schoolwork (Astone & McLanahan, 1991). As a result, they are less likely to complete high school than adolescents from intact families (Garfinkel & McLanahan, 1986). In the social arena, adolescents from families disrupted by divorce are more active in dating and more likely to be sexually active at a given age (Demo & Acock, 1988). Women who spend part of their childhood in single-parent families are more likely to have a child while in their teens; they are also more likely to marry early and to have that marriage end in divorce (McLanahan & Bumpass, 1988). Both reduced educational attainment and early parenthood and marriage result in a higher rate of poverty among adults raised in single-parent families (McLanahan & Booth, 1989). Finally, as adults, people who experienced parental divorce are less well adjusted psychologically and are more likely to use alcohol and drugs and commit crimes (Amato & Keith, 1991).

Most single-parent families are headed by a mother. Are these adverse consequences for children following divorce a consequence of the gender of the parent or of the fact that there is only one parent? A careful study of 3,738 parents of children ages 5 to 19 suggests it is the latter. Both single mothers and single fathers report lower rates of supervision and control; families with two adults report higher levels of control (Thomson, McLanahan, & Curtin, 1992). A recent study of the adjustment of young people 4.5 years after a divorce found that most adolescents had adjusted well. The best adjustment was associated with a close relationship between adolescent and parent and the exercise of management and control by the parent (Buchanan, Maccoby, & Dornbusch, 1996).

Social Class and Socialization Techniques

An important influence on socialization is social class. In the United States, the lower and middle classes differ from one another in their discipline techniques. Specifically, there is a negative relationship between social class and the use of physical punishment (Gecas, 1979). Working-class parents are more likely to report using physical punishment, such as pushing, shoving, slapping, or spanking a child (Straus, Gelles, & Steinmetz, 1980). Middle- and upper-class parents are more likely than lower-class parents to rely on reasoning and to use love withdrawal (shame and guilt) in disciplining their children. Such a parent responds to a rule violation by saying "I'm ashamed of you" or "Good children don't do things like that." These techniques are based on the child's emotional attachment to the parent.

Kohn (1969) argues that the goal of discipline varies with social class. Middle-class parents are likely to pay attention to intentions and punish intentional rule violations but not accidental ones. In working-class families, children are more likely to be punished for acts with harmful consequences, regardless of intentions. Research supports these hypotheses (Gecas,

1979). In a study of parents of third graders (Gecas & Nye, 1974), both fathers and mothers were asked, "If your child is playing and accidentally breaks something of value, what would you do?" and "If your child intentionally disobeys after you have told him to do something, what would you do?" The indicator for social class was based on the male's occupational group, either white-collar or blue-collar. White-collar parents frequently reported they would respond differently to the two situations, whereas blue-collar parents were more likely to report similar responses to the two situations.

Peers

As the child grows, peers become increasingly important as socializing agents (see Box 3.2, "The Peer Group"). The peer group differs from the family on several dimensions. These differences influence the type of interaction and thus the kinds of socialization that occur. The family consists of persons who differ in status or power, whereas the peer group is composed of status equals. From an early age the child is taught to treat parents with respect and deference. Failure to do so will probably result in discipline, and the adult will use the incident as an opportunity to instruct the child about the importance of deference (Cahill, 1987; Denzin, 1977). Interaction with peers is more open and spontaneous; the child does not need to be deferential or tactful. Thus children at the age of 4 bluntly refuse to let children they dislike join their games. With peers, they may make comments that adults consider insulting, such as "You're ugly," to another child. Daily exposure to other children facilitates the development of social skills (Roopnarine, 1985).

Membership in a particular family is ascribed, whereas peer interactions are voluntary (Gecas, 1990). Thus peer groups offer children their first experiences in exercising choice over whom they relate to. The opportunity to make such choices contributes to the child's sense of social competence and allows interaction with other children who complement the developing identity.

Unlike the child's family, peer groups in early and especially middle childhood (ages 6 to 10) are usually homogeneous in sex and age. A survey of 2,299 children in third through twelfth grade measured the extent to which they belonged to tightly knit peer groups, the size of such groups, and whether they were homogeneous by race and gender (Shrum & Cheek, 1987). The results are displayed in Table 3.2. The proportion belonging to a group peaked in sixth grade and then declined. The size of peer groups declined steadily from third through twelfth grade, as did heterogeneity by race and gender.

Peer associations make a major contribution to the development of the child's identity. Children learn the role of friend in interactions with peers, contributing to greater differentiation of the self (Corsaro & Rizzo, 1988). Peer and other relationships outside the family provide a basis for establishing independence; the child ceases to be exclusively involved in the roles of offspring, sibling, grandchild, and cousin. These

Table 3.2   Characteristics of Peer Groups During the School Years

Grade Level	Percentage Belonging to a Group	Average Group Size	Percentage Heterogeneous Groups
3	34.9	8.0	46
4–6	46.3	7.9	27
7–8	32.8	6.5	10
9–12	23.2	6.3	19

SOURCE: Adapted from Shrum and Cheek, 1987, tables 1 and 2.

3.2 The Peer Group

American society is highly segregated by age. Most of us spend most of our time with people of about the same age. This is especially true in childhood and adolescence because age segregation is the fundamental organizing principle of our schools. Several recent studies provide important insights into the nature of peer groups and their significance for socialization.

Among preschool-age children, a major concern is social participation. Kids in American society learn about the role of *friend* and the expectations associated with that role. Their understanding of this role provides a basis for evaluating their relationships with other children. As children begin to play in groups, maintaining access to the group becomes an issue. Children become concerned with issues of inclusion and exclusion—who is in the group and who is not. These issues remain important ones throughout childhood and into adolescence (Adler & Adler, 1995).

Peer groups reflect the desire of children to gain some control over the social environment and to utilize that control in concert with other children (Corsaro & Eder, 1995). Children become concerned with gaining control over adult authority, and they learn that a request or plea by several children is more likely to be granted. In elementary school, children develop a strong group identity, which is strengthened by minor rebellions against adult authority. Thorne (1993) observed that in one combined fourth- and fifth-grade classroom, most of the students had contraband, small objects such as toy cars and trucks, nail polish and stuffed animals, which were prohibited by school rules. By keeping these items in their desks, and by displaying or exchanging them at key moments during class, the kids were attempting to cope with their relative lack of power. Both children and adolescents assert themselves by making fun of and mocking teachers and administrators.

Peer groups play a major role in socializing young persons to gender-role norms. As children move through elementary school, they increasingly form groups that are homogeneous by gender (see Table 3.2). For instance, in one study, Thorne (1993) observed a "geography of gender" in the schoolyard. Boys generally were found on the playing fields, whereas girls were concentrated in the areas closer to the building and the on playground equipment. Children who violated these gender boundaries risked being teased or even ridiculed. Thorne (1993) identifies several varieties of **borderwork,** which is "interaction across—yet interaction based on and even strengthening—gender boundaries" (p. 64). One form of borderwork was the *chase,* which almost always involved a boy chasing a girl or vice versa. Another form was *Cooties,* or treating an individual or group as contaminated, which also was often cross-gender; girls were often identified as the ultimate source of contamination, whereas boys typically were not. Finally, *invasion* occurred when a group of boys physically occupied the space that girls were using for some activity; Thorne never observed girls invading a boys' game. All of these activities involve the themes of sex and aggression, themes common to heterosexual relationships in American society. The implicit message is that boys and their activities are more important than girls and their activities.

In another study, Eder (1995) and her colleagues observed peer relationships in a middle school for 3 years. During the sixth, seventh, and eighth grades, young adolescents shift their focus from gender-role norms to norms governing male-female relationships. Males learn from other males the "proper" view of females; in some (but not all) groups, the prescribed view was that females were objects of sexual conquest. Girls learn to view boys as potential participants in romantic relationships. Public teasing and ridicule of those who violate norms, common in elementary school, are replaced by gossip and exclusion from the group as sanctions for violations of group norms in middle school.

Eder also observed that the status hierarchy in the school generally reproduced the class structure of the wider community. Status was accorded to students based on popularity. One became popular by being visible. The most visible students were those who were on athletic teams and the cheerleading squad. Participating in these activities required money because they were not funded by the school. Further, the teams and cheerleaders relied on parents to transport them to games, giving an advantage to students who had one parent who did not work, or parents whose jobs allowed them to take time off for such activities. Not surprisingly, the popular, visible students were those from middle-class families.

alternate, nonfamilial identities may provide a basis for actively resisting parental socialization efforts (Stryker & Serpe, 1982). For example, a parent's attempt to enforce certain rules may be resisted by the child whose friends make fun of children who behave that way.

School

Unlike the peer group, school is intentionally designed to socialize children. In the classroom, there is typically one adult and a group of children of similar age. There is a sharp status distinction between teacher and student. The teacher determines what skills he or she teaches and relies heavily on instrumental learning techniques (with such reinforcers as praise, blame, and privileges) to shape student behavior (Gecas, 1990). School is the child's first experience with formal and public evaluation of performance. Every child's behavior and work is evaluated by the same standards, and the judgments are made public to others in the class, as well as to parents.

We expect schools to teach reading, writing, and arithmetic. But they do much more than that. Teachers use the rewards at their disposal to reinforce certain personality traits, such as punctuality, perseverance, and tact. Schools teach children which selves are desirable and which are not. Thus children learn a vocabulary that they are expected to use in evaluating themselves and others (Denzin, 1977). The traits chosen are those thought to facilitate social interaction throughout life in a particular society. In this sense, schools civilize children.

Social comparison has an important influence on the behavior of schoolchildren. Because teachers make public evaluations of the children's work, each child can judge his or her performance relative to others. These comparisons are especially important to the child because of the homogeneity of the classroom group. Even if the teacher deemphasizes a child's low score on a spelling test, the child interprets the performance as a poor one relative to those of classmates. A consistent performance affects a child's image of self as a student.

An observational study of children in kindergarten, first, second, and fourth grades documented the development of social comparison in the classroom (Frey & Ruble, 1985). In kindergarten, comparisons were to personal characteristics, for example, liking ice cream. Comparisons of performance increased sharply in first grade; at first these were blatant but became increasingly subtle in second and fourth grades.

Processes of Socialization

How does socialization occur? We examine three processes that are especially important: instrumental conditioning, observational learning, and internalization.

Instrumental Conditioning

When you got dressed this morning, chances are you put on a shirt or blouse, pants, a dress, or a skirt that had buttons, hooks, and/or zippers. When you were younger, learning how to master buttons, hooks, zippers, and shoelaces undoubtedly took considerable time, trial and error, and slow progress accompanied by praise from adults. You acquired these skills through **instrumental conditioning,** a process in which a person learns what response to make in a situation in order to obtain a positive reinforcement or avoid a negative reinforcement. The person's behavior is "instrumental" in the sense that it determines whether he or she is rewarded or punished.

The most important process in the acquisition of many skills is a type of instrumental learning called shaping (Skinner, 1953, 1957). **Shaping** refers to learning in which an agent initially reinforces any behavior that remotely resembles the desired response and later requires increasing correspondence between the learner's behavior and the desired response before providing reinforcement. Shaping thus involves a series of successive approximations, in which the learner's behavior comes closer and closer to resembling the specific response desired by the reinforcing agent.

In socialization, the degree of similarity between desired and observed responses required by the agent depends in part on the learner's past performance. In this sense, shaping is interactive in character. In teaching children to clean their rooms, parents initially reward them for picking up their toys. When children show they can do this consistently, parents may require that the toys be placed on certain shelves as the condition for a reward. Shaping is more likely to succeed if the level of performance required is consistent with the child's abilities. Thus a 2-year-old may be praised for drawing lines with crayons, whereas a 5-year-old may be expected to draw recognizable objects or figures.

Reinforcement Schedules

When shaping behavior, a socializing agent can use either positive reinforcement or negative reinforcement. Positive reinforcers are stimuli whose presentation strengthens the learner's response; positive reinforcers include food, candy, money, or high grades. Negative reinforcers are stimuli whose withdrawal strengthens the response, such as the removal of pain.

In everyday practice, it is rare for a learner to be reinforced each time the desired behavior is performed. Instead, reinforcement is given only some of the time. In fact, it is possible to structure when reinforcements are presented to the learner, using a reinforcement schedule.

There are several possible reinforcement schedules. The *fixed-interval* schedule involves reinforcing the first correct response after a specified period has elapsed. This schedule produces the fewest correct responses per unit time; if people are aware of the length of the interval, they will respond only at the beginning of the interval. It is interesting that many schools give examinations at fixed intervals, such as the middle and end of the semester; perhaps that is why many students only study just before an exam. The *variable-interval* schedule involves reinforcing the first correct response after a variable period. In this case, individuals cannot predict when reinforcement will occur, so they respond at a regular rate. Grading a course based on several surprise, or "pop," quizzes utilizes this schedule.

The *fixed-ratio* schedule provides a reinforcement following a specified number of correct, nonreinforced responses. Paying a worker on a piece rate, such as $5 for every three items produced, utilizes this pattern. If the reward is sufficient, the rate of behavior may be high. Finally, the *variable-ratio* schedule provides reinforcement after several nonrewarded responses, with the number of responses between reinforcements varying. This schedule typically produces

Shaping is a process through which many complex behaviors, such as playing the cello, are learned. Initially the socializer (teacher or parent) rewards behavior that resembles the desired response. As learning progresses, greater correspondence between the behavior and the desired response is required to earn a reward, such as praise.

the highest and most stable rates of response. An excellent illustration is the gambler, who inserts quarters in a slot machine for hours, receiving only occasional random payoffs.

Punishment

By definition, **punishment** is the presentation of a painful or discomforting stimulus (by a socializing agent) that decreases the probability the preceding behavior (by the learner) will occur. Punishment is one of the major child-rearing practices used by parents. Up to 90% of parents questioned in surveys reported that they punished their children occasionally (Sears, Maccoby, & Levin,

1957). Punishment is frequently physical; in a survey of 1,146 families, 71% of the parents reported slapping or spanking children (Gelles, 1980). Because punishment is so widely used, it is important to ask whether it actually works as a socialization technique.

Research indicates that punishment is effective in some circumstances but not others. One influence on its effectiveness is timing. In a study of fourth- and fifth-grade boys, each was presented with a pair of toys and told to select one of the pair. The punishment consisted of a verbal reprimand: "No. That's for older boys." The time between the act and the occurrence of punishment was systematically varied. In one group, the boys were reprimanded just as they touched the toys. In another, the reprimand was given after the boys picked up the toy. In a control group, the boys were warned before their choice not to touch one of the pair. The researchers then measured the boys' behavior when they were later left alone with the two toys. Boys who had been reprimanded just as they picked up the toy were least likely to play with the toy. The results suggest that the longer the delay between act and punishment, the less effective the punishment will be (Aronfreed & Reber, 1965).

The effectiveness of punishment may be limited to the situation in which it is given. Because punishment is usually administered by a particular person, it may be effective only when that person is present. This probably accounts for the fact that when their parents are absent children may engage in activities their parents earlier had punished (Parke, 1969, 1970).

Another factor is whether punishments are accompanied by a reason (Parke, 1969). Providing a reason allows the child to generalize the prohibition to a class of acts and situations. Yelling "No!" as a child reaches out to touch the stove may suppress that behavior. Telling the child not to touch it because it is "hot" enables him or her to learn to avoid hot objects as a group. Finally, consistency between the reprimands given by parents and their own behavior makes punishment more effective than if parents do not "practice what they preach" (Mischel & Liebert, 1966).

Self-Reinforcement and Self-Efficacy

Children learn hundreds, if not thousands, of behaviors via instrumental learning. The performance of some of these behaviors remain **extrinsically motivated**—that is, they depend on whether someone else rewards appropriate behaviors or punishes inappropriate ones. However, the performance of other activities becomes **intrinsically motivated**—that is, performed in order to achieve an internal state the individual finds rewarding (Deci, 1975). Research has demonstrated that external reward does not always improve performance. Providing a reward for a behavior that is intrinsically motivated, such as drawing, may actually reduce the frequency or quality of the activity (Lepper, Greene, & Nisbett, 1973).

Closely related to the concept of intrinsic motivation is self-reinforcement. As children are socialized, they learn not only specific behaviors but also performance standards. Children learn not only to write but to write neatly. These standards become part of the self; having learned them, children use them to judge their own behavior and thus become capable of **self-reinforcement** (Bandura, 1982b). The child who has drawn a house and comes running up to her father with a big smile exclaiming, "Look what I drew!" has already judged the drawing as a good one. If her father agrees, standards and self-evaluation are confirmed.

Successful experiences with an activity over time creates a sense of competence at the activity, or *self-efficacy* (Bandura, 1982c). That, in turn, makes us more likely to seek opportunities to engage in that behavior. The greater our sense of efficacy, the more effort we expend at a task and the greater our persistence in the face of difficulty. For instance, a young girl who perceives herself as a good basketball player is more likely to try out for a team. Conversely, experiences of failure to perform a task properly, or of the failure of the performance to produce the expected results, creates the perception that we are not efficacious. Perceived lack of efficacy is likely to lead to avoidance of the task. A boy who perceives himself as poor at spelling will probably not enter the school spelling bee.

Observational Learning

Children love to play dress-up. Girls put on skirts, step into high-heeled shoes, and totter around the room; boys put on sport coats and drape ties around their necks. Through observing adults, children have learned the patterns of appropriate dress in their society. Similarly, children often learn interactive rituals, such as shaking hands or waving goodbye, by watching others perform the behavior and then doing it on their own.

Observational learning, or modeling, refers to the acquisition of behavior based on observation of another person's behavior and of its consequences for that person (Shaw & Costanzo, 1982). Many behaviors and skills are learned this way. By watching other people (the models) perform skilled actions, children can increase their own skills. The major advantage of modeling is its greater efficiency compared with trial and error learning.

Does observational learning lead directly to performance of the learned behavior? No; research has shown that there is a difference between learning a behavior and performing it. People can learn how to perform a behavior by observing another person, but they may not perform the act until the appropriate opportunity arises. Considerable time may elapse before the observer is in the presence of the eliciting stimulus. A parent in the habit of muttering "damn" when he spills something may, much to his chagrin, hear his 3-year-old say "damn" the first time she spills milk. Children may learn through observation many associations between situational characteristics and adult behavior, but they may not perform these behaviors until they occupy adult roles and find themselves in such situations.

Even if the appropriate stimulus occurs, people may not perform behaviors learned through observation. An important influence is the consequences experienced by the model following the model's performance of the behavior. For instance, in one study (Bandura, 1965), nursery school children watched a film in which an adult (model) punched, kicked, and threw balls at a large inflated rubber Bobo doll. Three versions of the film were shown to three groups of children. In the first, the model was rewarded for his acts: A second adult appeared and gave the model soft drinks and candy. In the second version, the model was punished: The other adult spanked the model with a magazine. In the third version, there were no rewards or punishments. Later, each child

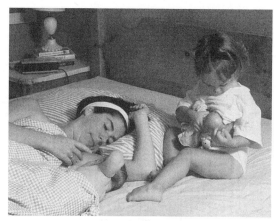

Observational learning is an important process through which children learn the appropriate behaviors.

was left alone with various toys, including a Bobo doll. The child's behavior was observed through a one-way mirror. Children who observed the model who was punished were much less likely to punch and kick the doll than the other children.

Did these children not learn the aggressive behaviors, or did they learn them by observation but not perform them? To answer this question, the experimenter returned to the room and offered a reward for each act of the model that the child could reproduce. Following this offer, children in all three groups were equally able to reproduce the acts performed by the model. Thus a child is less likely to perform an act learned by observation if the model experienced negative consequences.

Whether children learn from observing a model also depends on the characteristics of the model. Children are more likely to imitate high-status and nurturant models than models who are low in status and nurturance (Bandura, 1969). Preschool children given dolls representing peers, older children, and adults consistently chose adult dolls as people they would go to for help and older children as people they would go to for teaching (Lewis & Brooks-Gunn, 1979). Children also are more likely to model themselves after nurturant persons than cold and impersonal ones. Thus socialization is much more likely to be effective when the child has a nurturant, loving primary caregiver.

Internalization

Often we feel a sense of moral obligation to perform some behavior. At other times, we experience a strong internal feeling that a particular behavior is wrong. Usually, we experience guilt if these moral prescriptions or prohibitions are violated.

Internalization is the process by which initially external behavioral standards (for example, those held by parents) become internal and subsequently guide the person's behavior. An action is based on internalized standards when the person engages in it without considering possible rewards or punishments. Various explanations have been offered of the process by which internalization occurs, but all of them agree that children are most likely to internalize the standards held by more powerful or nurturant adult caregivers.

Internalization is an important socializing process. It results in the exercise of self-control. People conform to internal standards even when there is no surveillance of their behavior by others and, therefore, no rewards for that conformity. People who are widely admired for taking political or religious actions that are unpopular, for standing up for their beliefs, often do so because those beliefs are internalized.

Outcomes of Socialization

Persons undergoing socialization acquire new skills, knowledge, and behavior. In this section, we discuss some specific outcomes of the process, including gender role, linguistic and cognitive competence, moral development, and orientation toward achievement and work.

Gender Role

"Congratulations, you have a girl!" Such a pronouncement by a birth attendant may be the single most important event in a new person's life. In large part, the gender assigned to the infant—male or female—has a major influence on the socialization of that child.

Every society has differential expectations regarding the characteristics and behavior of males and females. In our society, men traditionally have been expected to be competent—competitive, logical, able to make decisions easily, ambitious. Women have been expected to be high in warmth and expressiveness—gentle, sensitive, tactful (Broverman et al., 1972). Parents employ these as guidelines in socializing their children, and differential treatment begins at birth. Male infants are handled more vigorously and roughly, whereas females are given more cuddling (Lamb, 1979). Boys and girls are dressed differently from infancy and may be given different kinds of toys to play with.

In addition, mothers and fathers differ in the way they interact with infants. Mothers engage in behavior oriented toward fulfilling the child's physical and emotional needs (Baumrind, 1980), whereas fathers engage the child in rough-and-tumble, physically stimulating activity (Walters & Walters, 1980). These differences are found in white, African American, and Hispanic families (Parke, 1996). Thus almost from birth, infants are exposed to models of masculine and feminine behavior. Fathers begin to pay special attention to their sons at about age 1 and reduce the interaction with their daughters (Lamb, 1979). At the same time, the child increasingly prefers the same-sex parent. More frequent interaction with and a preference for the same-sex parent account for Richard Cohen's observation that he is much more like his father than his mother.

By age 2, the child's gender identity—his or her conception of self as male or female—is firmly established (Money & Ehrhardt, 1972). Boys and girls show distinct preferences for different types of play materials and toys by this age. Between the ages of 2 and 3, differences in aggressiveness become evident, with boys displaying more physical and verbal aggression than girls (Hyde, 1984). By age 3, children more frequently choose same-gender peers as playmates; this increases opportunities to learn gender-appropriate behavior via modeling (Lewis & Brooks-Gunn, 1979). By age 4, the games typically played by boys and girls differ; groups of girls play house, enacting familial roles, and groups of boys play trucks.

Parents are an important influence on the learning of **gender role,** the behavioral expectations associated with one's gender. Children learn

Children and adolescents learn gender-role expectations through interaction with adults. Meeting his hero may have a lifelong impact on this youth.

gender-appropriate behaviors by observing their parents' interaction. Children also learn by interacting with parents, who reward behavior consistent with gender roles and punish behavior inconsistent with these standards. The child's earliest experiences relating to members of the other gender occur in interaction with the opposite-gender parent. A female may be more likely to develop the ability to have warm, psychologically intimate relationships with males if her relationship with her father was of this type (Appleton, 1981).

Obviously, boys are not alike in our society, and neither are girls. The specific behaviors and characteristics the child is taught depend partly on the gender-role expectations held by the parents. These in turn depend on the network of extended family—grandparents, aunts and uncles, and other relatives—and friends of the family. The expectations held by these people are influenced by the institutions to which they belong, such as churches and work organizations (Stryker & Serpe, 1982). With regard to religion, research suggests that the differences among denominations in socialization techniques and in outcomes such as gender-role attitudes have declined in the past 30 years (Alwin, 1986). The data suggest that church attendance is more influential than the denomination to which one belongs.

Schools also teach gender roles. Teachers may reward appropriate gender-role behavior; they often reinforce aggressive behavior in boys and dependency in girls (Serbin & O'Leary, 1975). A more subtle influence on socialization is the content of the stories that are read and told in preschool and first-grade classes. Many of these stories portray males and females as different. Typically men are depicted as rulers, adventurers, and explorers; women are wives (Weitzman et al., 1972). A study of award-winning books for children published in the 1980s found that women continue to be portrayed as passive and dependent more often than men (Kortenhaus & Demarest, 1993).

A major influence on gender-role socialization is the mass media. Researchers analyzing the contents of television programs, television advertising, feature films, and other media report that portrayals of men and women and girls and boys reinforce traditional definitions of gender roles. A content analysis of 175 episodes of 41 animated TV series found that male characters were portrayed as independent, athletic, ambitious, and aggressive, whereas females were shown as dependent, emotional, domestic, and romantic (Thompson & Zerbinos, 1995). A study of the fiction in *Seventeen* and *'Teen,* the two largest selling magazines for teenage girls, found that the stories reinforced traditional messages (Peirce, 1993). Half of the conflicts were about relationships, and half the female characters relied on someone else to solve their problems. Adult men in the stories were doctors, lawyers, and bankers; adult women were nurses, clerical workers, and secretaries. Perhaps the most stereotyped portrayals are found on MTV. An analysis of 40 music videos found that men engaged in more dominant, aggressive behavior, and women engaged in subservient behavior; women were frequently the object of explicit, implicit, and aggressive sexual advances (Sommers-Flanagan, Sommers-Flanagan, & Davis, 1993).

Linguistic and Cognitive Competence

Another important outcome of socialization is the ability to interact effectively with others. We discuss two specific competencies: language and the ability to cognitively represent the world.

Language   Using language to communicate with others is a prerequisite for full participation in social groups (Shibutani, 1961). The child's acquisition of speech reflects both the development of the necessary perceptual and motor skills and the impact of social learning (Bates, O'Connell, & Shore, 1987).

Three main components of language are the sound system (phonology), the words and their associated meanings (lexicon), and rules for combining words into meaningful utterances (grammar). Young children appear to acquire these in sequence, first mastering meaningful sounds, then learning words, and finally learning sentences. In reality, acquiring speech is a process that involves all three at the same time and continues throughout childhood.

Language acquisition in the first 3 years passes through four stages (Bates, O'Connell, & Shore, 1987). The *prespeech* stage lasts for about 10 months and involves speech perception, speech production, and early intentional communication. In the first few weeks of life, infants can perceive all of the speech sounds. They begin producing sounds at 2 to 3 months and begin producing sounds specific to their parent's language at 4 to 7 months. Speech production involves imitation of the sounds they hear. With regard to intentional communication, observational data indicate that vocal exchanges involving 4-month-old infants and their mothers are patterned (Stevenson et al., 1986). Vocalization by *either* infant or mother was followed by silence, allowing the other to respond. Vocalization by one was likely to be followed by vocalization by the other, a pattern like that found in adult conversation.

The first intentional use of gestures occurs at about 9 months. At this age, infants orient visually to adults rather than to desired objects, such as a cookie. Further, if an initial gesture is not followed by the adult engaging in the desired behavior, the infant repeats the gesture or tries a different gesture.

The second, or *first-word,* stage occurs at 10 to 14 months, and involves the infant's recognition that things have names. The first words produced are usually nouns that name or request specific objects (Marchman, 1991). Obviously, this ability to use names reflects cognitive as well as linguistic development.

At about 18 months, there is a vocabulary burst, with a doubling in a short time of the number of words used correctly. The suddenness of this increase suggests the maturation of some cognitive abilities. This, in turn, is followed by an increase in the complexity of vocalizations leading to the *first-sentence* stage at 18 to 22 months. Examples of such sentences include "See truck, Mommy" and "There go one." Such speech is *telegraphic*—that is, the numbers of words is greatly reduced relative to adult speech (Brown & Fraser, 1963). At the same time, such utterances are clearly more precise than the single-word utterances of the 1-year-old child.

The fourth stage, *grammaticization,* occurs at 24 to 30 months. The child's use of language now reflects the fundamentals of grammar. Children at this age frequently overgeneralize, applying rules indiscriminately. For example, they add an appropriate ending to a novel word when it is incorrect—"He runned." Such usage indicates the child understands that there are rules. At about the same age, a child puts a series of acts in the conventional sequence, for example, undressing a doll, bathing it, drying it, and dressing it. Perhaps both activities reflect the maturation of an underlying ability to order arbitrary units.

An important process in learning to make grammatically correct sentences is speech expansion. That is, adults often respond to children's speech by repeating it in expanded form. In response to "Eve lunch," the mother might say, "Eve is eating lunch." One study showed that mothers expanded 30% of the utterances of their 2-year-old children (Brown, 1964). Adults probably expand on the child's speech to determine the child's specific meaning. Speech expansion contributes to language acquisition by providing children with a model of how to convey more effectively the meanings they intend.

The next stage of language development is highlighted by the occurrence of *private speech,* in which children talk loudly to themselves, often for extended periods. Private speech begins about age 3, increases in frequency until age 5, and disappears by about age 7. Such talk serves two functions. First, it contributes to the child's developing sense of self. Private speech is addressed to the self as object, and it often includes the application of meanings to the self, such as "I'm a girl." Second, private speech helps the child develop an awareness of the environment. It often consists of naming aspects of the physical and social environment. The repeated use of these names solidifies the child's understanding of the environment. Children also often engage in appropriate actions as they speak, reflecting their developing awareness of the social meanings of objects and persons. Thus a child may label a doll a "baby" and dress it and feed it.

Gradually, the child begins to engage in dialogues, either with others or with the self. These conversations reflect the ability to adopt a second perspective. Thus, at age 6, when one child wants a toy that another child is using, the first child frequently offers to trade. She knows the second child will be upset if she merely takes the toy. This movement away from a self-centered view also may reflect maturational changes. Dialogues require that the child's own speech meshes with that of another.

Cognitive Competence
Children must develop the ability to represent in their own minds the features of the world around them. This capacity to represent reality mentally is closely related to the development of language.

The child's basic tasks are to learn the regularities of the physical and social environment and to store past experience in a form that can be used in current situations. In a complex society, there are so many physical objects, animals, and people that it is not possible for a child (or an adult) to remember each as a distinct entity. Things must be categorized into inclusive groupings, such as dogs, houses, and girls. A category of objects and the cognitions of an individual about members of that category (for example,

dog) make up a *schema.* Collectively, our schemas allow us to make sense out of the world around us.

Young children must learn schemas. Learning language is an essential part of the process because language provides the names around which schemas can develop. It is noteworthy that the first words children produce are usually nouns that name objects in the child's environment. At first the child uses a few very general schema. Some children learn the concept "dog" at 12 to 14 months and then apply it to all animals, to dogs, cats, birds, and cows. Only with maturation and experience does the child develop the abstract schema "animals" and learn to discriminate between dogs and cats.

Researchers can study the ability to utilize schemas by asking children to sort objects, pictures, or words into groups. Young children (ages 6 to 8) rely on visual features, such as color or word length, and sort objects into numerous categories. Older children (ages 10 to 12) increasingly use functional or superordinate categories, such as foods, and sort objects into fewer groups (Olver, 1961; Rigney, 1962). With age, children become increasingly adept at classifying diverse objects and treating them as equivalent.

These skills are very important in social interaction. Only by having the ability to group objects, persons, and situations can we determine how to behave toward them. Person schemas and their associated meanings are especially important to smooth interaction. Even very young children differentiate people by age (Lewis & Brooks-Gunn, 1979). By about 2 years of age, children correctly differentiate babies and adults when shown photographs. By about age 5, children employ four categories: little children, big children, parents (ages 13 to 40), and grandparents (age 40 plus).

As children learn to group objects into meaningful schemas, they learn not only the categories but also how others feel about such persons. Children learn not only that Catholics are people who believe in the Trinity of Father, Son, and Holy Spirit but also that their parents like or dislike Catholics. Thus children acquire positive and negative attitudes toward the wide range of social objects they come to recognize. The particular schemas and evaluations children learn are influenced by the social class, religious, ethnic, and other subcultural groupings to which those who socialize them belong.

Moral Development

In this section we discuss moral development in children and adults. Specifically, we focus on acquiring knowledge of social rules and on how children become capable of making moral judgments.

Knowledge of Social Rules
To interact effectively with others, people also must learn the social rules that govern interaction and adhere to them. Beliefs about which behaviors are acceptable and which are unacceptable for specific persons in specific situations are termed **norms.** Without norms, coordinated activity would be very difficult, and we would find it hard or impossible to achieve our goals. Therefore, each group, organization, and society develops rules governing behavior.

Early in life, an American child learns to say "please," a French child, "s'il vous plaît," and a Serbian child, "molim te." In every case, the child is learning the value of conforming to arbitrary norms governing requests. Learning language trains the child to conform to linguistic norms and serves as a model for the learning of other norms. Gradually, through instrumental as well as observational learning, the child learns the generality of the relationship between conformity to norms and the ability to interact smoothly with others and achieve personal goals.

What influences which norms children will learn? The general culture is one influence. All American children learn to cover most parts of the body with clothing. The position of the family within the society is another influence. Parental expectations reflect social class, religion, and ethnicity. Thus the norms taught vary from one family to another. Interestingly, parents often hold norms that they apply distinctively to their own children. Mothers and fathers expect certain behaviors of their own sons or daughters but may have different expectations for other people's children (Elkin & Handel, 1989). For instance, they may expect their children to be more polite than other children in interaction with adults. Parental expectations are not constant over time; they change as the child grows older. Parents expect greater politeness from a

10-year-old than from a 5-year-old. Finally, parents adjust their expectations to the particular child. They consider level of ability and experiences relative to other children; they expect better performance in school from a child who has done well in the past than from one who has had problems in school. In all of these ways, each child is being socialized to a somewhat different set of norms. The outcome is a young person who is both similar to most others from the same social background and unique in certain ways.

When children begin to engage in cooperative play, about 4 years of age, they begin to experience normative pressure from peers. The expectations of age-mates differ in two important ways from those of parents. First, children bring different norms from their separate families and, therefore, introduce new expectations. Thus, through peers, children first become aware of other ways of behaving. In some cases, peers' expectations conflict with those of parents. For example, many parents do not allow their children to play with toy guns, knives, or swords. Through involvement with their peers, children may become aware that other children routinely play with such toys. As a result, some children experience normative conflict and discover the need to develop strategies for resolving such conflicts.

Another way that peer group norms differ from parental norms is that the former reflect a child's perspective (Elkin & Handel, 1989). Many parental expectations are oriented toward socializing the child for adult roles. Children react to each other as children and are not concerned with long-term outcomes. Thus peers encourage impulsive, spontaneous behavior rather than behavior directed toward long-term goals. Peer-group norms emphasize participation in group activities, whereas parental norms may emphasize homework and other educational activities that may contribute to academic achievement.

When children enter school, they are exposed to a third major socializing agent, the teacher. In school, children are exposed to *universalistic* rules, norms that apply equally to all children. The teacher is much less likely than the parents to make allowances for the unique characteristics of the individual; children must learn to wait their turn, to control impulsive and spontaneous behavior, and to work without a great deal of supervision and support. In this regard, the

At school, children get their first exposure to universal norms—behavioral expectations that are the same for everyone. Although parents and friends treat the child as an individual, teachers are less likely to do so.

school is the first of many settings in which the individual is treated primarily as a member of the group rather than as a unique individual.

Thus school is the setting in which children are first exposed to universalistic norms and the regular use of symbolic rewards such as grades. Such settings become increasingly common in adolescence and adulthood, in contrast with the individualized character of familial settings.

Moral Judgment We not only learn the norms of our social groups, but we also develop the ability to evaluate behavior in specific situations by applying certain standards. The process through which children become capable of making moral judgments is termed **moral development.** It involves two components: (1) the reasons one adheres to social rules and (2) the bases used to evaluate actions by self or others as good or bad.

How do children evaluate acts as good or bad? One of the first people to study this question in detail was Piaget, the famous Swiss developmental psychologist. Piaget's methodology was to read a young child a set of stories. In each story, the central character performed an act that violated social rules. In one story, for example, the central character was a young girl who, contrary to rules, was playing with scissors and made a hole in her dress. Piaget asked the children to evaluate the behaviors of the characters in the stories (that is, to indicate which characters were naughtier) and then to explain their reasons for these judgments. Based on this work, Piaget concluded there were three bases for moral judgments: amount of harm/benefit, actor's intentions, and the application of agreed upon rules or norms (Piaget, 1965).

More recently, Kohlberg has extended Piaget's work by analyzing in greater detail the reasoning by which people reach moral judgments. He uses stories involving conflict between human needs and social norms or laws. Here is an example:

> In Europe, a woman was near death from cancer. One drug might save her, a form of radium that a druggist in the same town had recently discovered. The druggist was charging $2,000, ten times what the drug cost him to make. The sick woman's husband, Heinz, went to everyone he knew to borrow money, but he could only get together about half of what it cost. He told the druggist that his wife was dying and asked him to sell it cheaper or let him pay later. But the druggist said, "No." The husband got desperate and broke into the man's store to steal the drug for his wife. (Kohlberg, 1969)

Respondents are then asked: Should Heinz have done that? Was Heinz right or wrong? What obligations did Heinz and the druggist have? Should Heinz be punished?

Kohlberg proposes a developmental model with three levels of moral reasoning, each level involving two stages. This model is summarized in Table 3.3.

Kohlberg argues that the progression from stage 1 to stage 6 is a standard or universal one, and that all children begin at stage 1 and progress through the stages in order. Practically no one consistently reasons at stage 6, and few regularly use stage 5 considerations. Most people reason at stages 3 or 4. Several studies have shown that such a progression does occur (Kuhn et al., 1977). If the progression is universal, then children from different cultures should pass through the same stages in the same order. Again, data suggest that they do (White, Bushnell, & Regnemer, 1978). On the basis of such evidence, Kohlberg claims that this progression is the natural human pattern of moral development. He also believes that attaining higher levels is better or more desirable.

Although moral development is an interesting topic of study in its own right, some investigators have explored the relationship between moral judgment and behavior. Studies of cheating in schoolwork, one involving sixth graders (Krebs, 1967) and one involving college students (Malinowski & Smith, 1985), found that those whose moral development had reached higher stages were less likely to cheat. Participants in political activities such as demonstrations and sit-ins often claim their behavior is based on moral principles. But is it? A study by Haan, Smith, and Block (1968) explored the relationship between political activities and moral development in young adults. Respondents were asked about participation in such activities as meetings, demonstrations, picketing, and marches. In general, there was a consistent

Table 3.3   Kohlberg's Model of Moral Development

Preconventional Morality:

Moral judgment based on external, physical consequences of acts.

Stage 1: Obedience and punishment orientation. Rules are obeyed in order to avoid punishment, trouble.

Stage 2: Hedonistic orientation. Rules are obeyed in order to obtain rewards for the self.

Conventional Morality:

Moral judgment based on social consequences of acts.

Stage 3: "Good boy/nice girl" orientation. Rules are obeyed to please others, avoid disapproval.

Stage 4: Authority and social-order maintaining orientation. Rules are obeyed to show respect for authorities and maintain social order.

Postconventional Morality:

Moral judgments based on universal moral and ethical principles.

Stage 5: Social-contract orientation. Rules are obeyed because they represent the will of the majority, to avoid violation of rights of others.

Stage 6: Universal ethical principles. Rules are obeyed in order to adhere to one's principles.

SOURCE: Adapted from Kohlberg, 1969, table 6.2.

increase from those classified in stage 3 to those classified in stage 6 in the percentage reporting each activity (Candee & Kohlberg, 1987).

Kohlberg's model is an impressive attempt to specify a universal model of moral development. However, there are limitations to it. First, like Piaget, he locates the determinants of moral judgment within the individual. He does not recognize the influence of the situation. Studies of judgments of aggressive behavior (Berkowitz et al., 1986), of driving while intoxicated (Denton & Krebs, 1990), and decisions about reward allocation (Kurtines, 1986) found that both moral stage and type of situation influenced moral judgment.

Second, Kohlberg's model has been criticized as sexist, not applicable to the processes women use in moral reasoning. Gilligan (1982) identifies two conceptions of morality: a morality of justice and a morality of caring. She argues that the former is characteristic of men and is the basis of Kohlberg's model. She believes the latter is more characteristic of women. A study of the considerations that men and women had used in resolving personal moral dilemmas reports results consistent with Gilligan's thesis (Ford & Lowery, 1986).

Third, Kohlberg shows little interest in the influence of social interaction on moral reasoning. In response to this limitation, Haan (1978) has proposed a model of interpersonal morality. Moral decisions and actions often result from negotiations among people in which the goal is a "moral balance." Participants attempt to balance situational characteristics, such as the options available, with their individual interests to arrive at a decision that allows them to preserve their sense of themselves as moral persons. Haan (1978, 1986) presented moral dilemmas to groups of friends and asked them to decide. In some cases, the decisions were more influenced by individual moral principles; in others, by the group interaction.

Achievement and Work Orientations

One of the persistent questions about human behavior is what guides its direction and explains its intensity. All of us make choices among alternatives, appear

consistent in our behavior across situations, and often persist in attempting to achieve a goal even in the face of adversity. Social psychologists employ the concept of motivation to account for these phenomena. A *motive* is a disposition within the person that produces behavior directed toward goals; social motives are those developed in interaction with other persons.

The Achievement Motive   One very important social motive is the **achievement motive,** a conscious or unconscious desire to reach high standards of excellence (McClelland, 1961). The strength of this motive is positively related to academic performance, even among people who are equal in intellectual ability. It is also positively associated with the tendency to engage in entrepreneurial activity, including taking risks when the outcome is under one's control. People with high levels of this motive are more likely to engage in innovative activity and to try to anticipate future events.

Research indicates that high levels of the achievement motive do not necessarily result in high performance. For performance to occur, one must not only have high achievement motivation, but also hold the expectation that high performance will, in fact, be rewarded (McClelland & Winter, 1969).

Other research suggests that whether mothers are employed influences the achievement motivation of their children (Lamb, 1982). Interestingly, the effect depends on the child's gender. Adolescent daughters of working mothers have higher achievement motivation than daughters whose mothers are at home, whereas adolescent sons of working mothers are less motivated to achieve.

Orientations Toward Work   Work is of central importance in social life. In recognition of this, occupation is a major influence on the distribution of economic and other resources. We identify others by their work; its importance is evidenced by the fact that one of the first questions we ask a new acquaintance is "What do you do?"

Most adults want to work at jobs that provide economic and, perhaps, other rewards. Therefore, it is not surprising that a major part of socialization is the learning of orientations toward work. By the age of 2, the child is aware that adults "go work" and asks why. A common reply is "Mommy goes to work to earn money." A study of 900 elementary school children found that 80% of first graders understood the connection between work and money (Goldstein & Oldham, 1979). The child, in turn, learns that money is needed to obtain food, clothing, and toys. The child of a physician or nurse might be told "Mommy goes to work to help people who are sick." Thus from an early age the child is taught the social meaning of work.

Occupations vary tremendously in character. One dimension on which jobs differ is closeness of supervision: A self-employed auto mechanic has considerable freedom, whereas an assembly-line worker may be closely supervised. The nature of the work varies: Mechanics deal with things, salespeople deal with people, lawyers deal with ideas. Finally, occupations such as lawyer require self-reliance and independent judgment, whereas an assembly-line job does not. So the meaning of work depends on the type of job the individual has.

Adults in different occupations should have different orientations toward work, and these orientations should influence how they socialize their children. Based on this hypothesis, extensive research has been conducted on the differences between social classes in the values transmitted through socialization (Kohn, 1969). Fathers are given a list of traits, including good manners, success, self-control, obedience, and responsibility, and asked to indicate how much they value each for their children. Underlying these specific characteristics, a general dimension—"self-direction versus conformity"—is usually found. Data from fathers of 3- to 15-year-old children indicate that the emphasis on self-direction and reliance on internal standards increases as social class increases. The relationship of values and social class is found not only in samples of American fathers but also in samples of Japanese and Polish fathers (Kohn et al., 1990).

These differences in the evaluations of particular traits reflect differences in the conditions of work. In general, middle-class occupations involve the manipulation of people or symbols, and the work is not closely supervised. Thus these occupational roles require people who are self-directing and who can make judgments based on knowledge and internal standards. Working-class occupations are more routinized and more closely supervised. They require workers with a conformist orientation. Kohn argues that fathers value those traits in their children that the father associates with success in his occupation.

Do differences in the value parents place on self-direction influence the kinds of activities they encourage their children to participate in? A study of 460 adolescents and their mothers (Morgan, Alwin, & Griffin, 1979) examined how maternal emphasis on self-direction affects the young person's grades in school, choice of curriculum, and participation in extracurricular activities. The researchers reasoned that parents who value self-direction encourage their children to take college preparatory courses because a college education is a prerequisite to jobs that provide high levels of autonomy. Similarly, they expected mothers who value self-direction to encourage extracurricular activities because such activities provide opportunities to develop interpersonal skills. The researchers did not expect differences in grades. The results confirmed all three predictions. Thus parents who value particular traits in their children do encourage activities that they believe are likely to produce those traits.

Adult Socialization

The process of socialization occurs not only in childhood but continues throughout life (Bush & Simmons, 1990). Its focus changes, however. In childhood, socializing efforts are directed at such basic outcomes as gender role, the acquisition of language, and the learning of social norms. In adolescence, socialization is focused on the acquisition of traits such as independence, responsibility, and the ability to relate to others. In adulthood, it is concerned with equipping the individual to function effectively in adult roles. In this section, we discuss three processes important in adult socialization: role acquisition, anticipatory socialization, and role discontinuity.

Through anticipatory socialization, individuals acquire skills and knowledge of roles they hope to assume in the future. Highly visible aspects of roles (makeup) are much easier to acquire than hidden aspects (attitudes toward husbands).

Role Acquisition

Throughout our lives, we move out of some roles and into new roles. The major roles we acquire as adults include intimate partner or spouse, parent, work roles, and, later, the roles of grandparent and retiree. Each of these changes involves *role acquisition,* learning the expectations and skills associated with the new role and entry into the role.

In recognition of the need to train people for new roles, many groups and organizations provide socialization opportunities. Certain agents are given the responsibility for teaching potential or new role occupants the necessary information and skills. Often this training occurs on the job. The novice checker in a supermarket works with an experienced clerk at first. Initially, the experienced clerk performs the work, perhaps instructing the novice as she does so. Gradually, the roles are reversed. As the novice becomes more skilled at both the mechanics of using the price

scanner and interacting with shoppers, she does more and more of the work. Eventually, if she demonstrates the requisite competence, she is on her own. This process can be observed in hundreds of occupational settings. In some cases there is a formal role designation for those undergoing socialization, such as "trainee."

Alternatively, there may be a separate period of formal training outside the organization before the person occupies the role. Many educational programs are designed to prepare people for roles they will play in the future. Training programs or schools for beauticians, flight attendants, dental technicians, and truck drivers are only a few examples of this type of socialization.

Anticipatory Socialization

In addition to intentional training before and after a role is acquired, there may be **anticipatory socialization**—activities that provide people with knowledge, skills, and values of a role they have not yet assumed. The teenager learning about sexual activity from the boasts of an older friend or from an X-rated movie is undergoing anticipatory socialization. So is the aspiring diplomat or politician who attends closely to the behavior of the president of the United States in a television interview.

Anticipatory socialization is different from explicit training because it is not intentionally designed as role preparation by socialization agents (Clausen, 1968; Heiss, 1990). Anticipatory socialization can ease the transition into new roles, but it is more effective for some roles than for others (Bush & Simmons, 1990; Thornton & Nardi, 1975). First, anticipatory socialization usually works best for future roles that are highly visible. Socialization of this type usually prepares children more effectively for the parent than for the spouse role, for example. Children see the parent role directly when interacting with their own parents, whereas important aspects of interaction among spouses occur when children are away or asleep.

Second, anticipatory socialization eases the transition if future roles are presented accurately. The interactions between spouses that children do observe are often intentionally laundered to hide negative feelings and conflict, resulting in poor anticipatory socialization and incorrect expectations about role demands.

Third, anticipatory socialization works better if there is certainty or agreement regarding role demands and expectations. Anticipatory socialization for retirement is difficult, for example, because we lack clear, consensual norms for how the elderly should behave after they leave the workforce (Matthews, 1977; Rosow, 1974). Should they ease up and accept dependence gracefully or stay active and insist on their independence?

Successful anticipatory socialization entails goal setting, planning, and preparation for future roles. Only by setting at least tentative occupational and family goals during our teenage years, for example, can we effectively plan our educational and social lives. Preparation occurs through part-time jobs, special courses, reading, talking with informed individuals, and so on. People also prepare for transitions by trying out elements of their anticipated roles. This is what couples planning marriage are doing when they take joint vacations, live together for a trial period, and share purchases.

Role Discontinuity

The acquisition of a new role does not always proceed smoothly, even if there has been anticipatory socialization. Changing roles can be stressful. A new role can involve meeting new expectations, performing new tasks, and interacting with new types of people. It may necessitate moving to a new location. Furthermore, acquiring a role may involve losses as well as gains. The person may have to leave a prior role to move into the new one.

Entering a new role can be especially difficult when there is **role discontinuity**—that is, when the values and identities associated with a new role contradict those of earlier roles (Benedict, 1938). Upon entering a discontinuous role, we must revise our former expectations and aspirations. Retirement, for example, often creates role discontinuity. During their working years, career-oriented adults are expected to strive for autonomy and productivity and to build their identities around their work. Retirement introduces contradictory expectations for such people. They must now assume more dependent and less productive roles and rebuild their identities around these discontinuous expectations (Mortimer & Simmons, 1978).

Role transitions of particular social importance are sometimes marked by *rites of passage,* public ceremonies or rituals in which the individual's new status is affirmed (Glaser & Strauss, 1971; Van Gennep, 1908/1960). Christenings, bar mitzvahs, graduations, marriage ceremonies, and retirement parties are common rites of passage in American society. They signify both to the individual and to others that the person now has a new identity and new behaviors, rights, and duties are appropriate. These ceremonies serve as social occasions for giving emotional support, advice, and material aid to those adopting new roles.

Summary

Socialization is the process through which infants become effective participants in society. It makes us like all other members of society in certain ways (shared language), but distinctive in other ways.

Perspectives on Socialization (1) One approach to the study of socialization emphasizes biological development; it views the emergence of interpersonal responsiveness and the development of speech and of cognitive structure as influenced by maturation. (2) Another approach emphasizes learning and the acquisition of skills from other persons. (3) A third approach emphasizes the child's discovery of cultural routines as he or she participates in them. Society organizes this process by making certain agents responsible for particular types of socialization of specific persons.

Agents of Childhood Socialization There are three major socializing agents in childhood. (1) The family provides the infant with a strong attachment to one or more caregivers. This bond is necessary for the infant to develop interpersonal and cognitive skills. Family composition and social class influence socialization by influencing the amount and kind of interaction between parent and child. (2) Peers provide the child with equal status relationships and are an important influence on the development of self. (3) Schools teach skills—reading, writing, and arithmetic—as well as traits like punctuality and perseverance.

Processes of Socialization Socialization is based on three different processes. (1) Instrumental conditioning, the association of rewards and punishments with an act, is a basis for learning both behaviors and performance standards. Studies of the effectiveness of various child-rearing techniques indicate that rewards do not always make a desirable behavior more likely to occur and punishments do not always eliminate an undesirable act. Through instrumental learning, children develop the ability to judge their own behaviors and to engage in self-reinforcement.

(2) We learn many behaviors and skills by observation of models. We may not perform these behaviors, however, until we are in the appropriate situation. (3) Socialization also involves internalization, the acquisition of behavioral standards and making them part of the self. This process enables the child to engage in self-control.

Outcomes of Socialization (1) The child gradually learns a gender role, the expectations associated with being male or female. Whether the child is independent or dependent, aggressive or passive, depends on the expectations communicated by parents, kin, and peers. (2) Language skill is another outcome of socialization; it involves learning both words and the rules for combining them into meaningful sentences. Related to the learning of language is the development of thought and the ability to group objects and persons into meaningful categories. (3) The learning of social norms involves parents, peers, and teachers as socializing agents. Children learn that conformity to norms facilitates social interaction. Children also develop the ability to make moral judgments. (4) Children acquire motives, dispositions that produce sustained, goal-directed behavior. One such motive is the desire to reach high standards of excellence, the achievement motive. Orientations toward work are influenced primarily by parents; middle-class families emphasize self-direction, whereas working-class families emphasize conformity.

Adult Socialization Socialization continues throughout life. In adulthood, it involves preparing the person to successfully enact major roles such as intimate partner and parent. (1) Role acquisition is often accompanied by on-the-job training. In addition, time spent in the role of trainee may be prerequisite to entry into a new role. (2) Anticipatory socialization involves unintentional role preparation. Its effectiveness depends on the visibility and accuracy of the portrayal of the new role and consensus about the expectations for the role. (3) Role discontinuity makes role acquisition especially difficult. Discontinuity can be eased by anticipatory socialization and by rites of passage.

Key Terms

achievement motive (p. 67)

anticipatory socialization m(p. 69)

attachment (p. 51)

borderwork (p. 55)

cultural routines (p. 49)

extrinsically motivated behavior (p. 58)

gender role (p. 60)

instrumental conditioning (p. 56)

internalization (p. 60)

intrinsically motivated behavior (p. 58)

moral development (p. 65)

norm (p. 64)

observational learning (p. 59)

punishment (p. 57)

role discontinuity (p. 70)

self-reinforcement (p. 58)

shaping (p. 56)

socialization (p. 46)

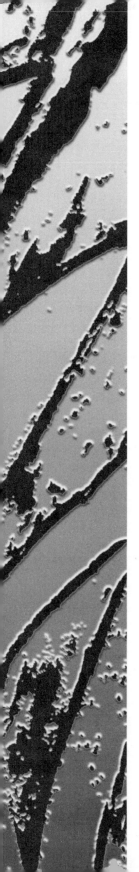

CHAPTER 4
Self and Identity

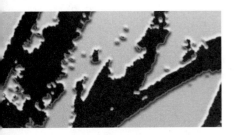

Introduction

An amnesia victim who walked into a pizza parlor six weeks ago and said "Help me, I don't know who I am," now faces a new identity crisis.

"I've gotten lots of calls saying I'm two entirely different people," said the man, who dubbed himself John Jackson after he was taken to a homeless shelter.

Jackson's ordeal began May 3, when he woke up beside railroad tracks and Interstate 64 near where the borders of West Virginia, Ohio and Kentucky meet. "I woke up in a field and the first thing I thought was 'It's cold.' And then I wondered, 'Where am I?' And then, 'Who am I?' Then I realized I had no answers and it got very frightening."

"He was bewildered more than anything else," said mission cook Virginia Berry. "He was dressed OK, had on blue jeans. He has a beard but it's well-trimmed and shaped and been taken care of. His hair's that way, too."

The man asked mission staff to call him Jackson because he remembers roads and a baseball park in Jacksonville, Fla. Mission workers asked Jacksonville newspapers to publish the man's photograph and on Friday he was inundated by calls from people claiming to know him.

Jackson did not recognize any of the callers, although some of the details they provided matched the few pieces he has to his puzzle.

One group of people who saw Jackson's photograph told him they believe he is Roy Moses, 33, who frequently vanished for months at a time "on a whim." Moses' stepmother, stepsister and brother-in-law all talked to Jackson on the telephone "and they seem convinced that's who I am." But Jackson said the family could provide few details, "so I'm not so sure."

Another Jacksonville resident said he recognized Jackson as Michael Shawper, 40, a career Navy man. The caller knew several little known naval bases that Jackson remembers, "and the knowledge of history and geography, that matches. But he said the last time he had seen Shawper was 11 years ago in Guam."

Shawper and Moses both attended the same college in Jacksonville. One has been married two or three times, the other once. "So I'm either an educated responsible man or an educated irresponsible man," Jackson said. "I can't be both, that's for sure. One of them has to be wrong. They both may be wrong." (*Wisconsin State Journal,* June 12, 1988, p. 4A)

"Who am I?" Few human beings in Western societies live out their lives without pondering this question. Some people pursue the search for self-knowledge and for a meaningful identity eagerly; others pursue it desperately. College students in particular are often preoccupied with discovering who they are. Few, however, have experienced the existential uncertainty faced by John Jackson.

Each of us has unique answers to this question, answers that reflect our **self-schema,** or *self-concept,* the organized structure of cognitions or thoughts we have about ourselves. The self-schema is comprised of our perceptions of our social identities and personal qualities, and generalizations about the self based on experience.

The content of self-schema is often assessed by having people answer the question "Who am I?" This "test" is the focus of Box 4.1, "Measuring Self-Concepts." Before you read on, take a few moments and respond to this question yourself in the space provided. For comparison, read the answers of a 9-year-old boy and a female college sophomore to the question "Who am I?" Their responses are listed at the bottom of the box.

4.1 Measuring Self-Concepts

In order to study self-concepts, we need ways to measure them. Many methods have been used. For example, one approach asks people to check those adjectives on a list (intelligent, aggressive, trusting, and so on) that describe themselves (Sarbin & Rosenberg, 1955). In another approach (Osgood, Suci, & Tannenbaum, 1957), people rate themselves on pairs of adjectives (strong-weak, good-bad, active-passive): Are they more like one of the adjectives in the pair or more like its opposite? Another technique, developed by Miyamoto and Dornbusch (1956), asks people whether they have more or less of a characteristic (self-confidence, likeableness) than members of a particular group (such as fraternities, sororities, and so on). In yet another technique, people sort cards containing descriptive phrases (interested in sports, concerned with achievement) into piles according to how accurately they think the phrases describe them (Stephenson, 1953).

Each of these popular methods provides respondents with a single standard set of categories to use in describing themselves. Using the same categories for all respondents makes it easy to compare the self-concepts of different people. These methods have a weakness, however. They do not reveal the unique dimensions that individuals use in spontaneously thinking about themselves. For this purpose, techniques that ask people simply to describe themselves in their own words are especially effective (Kuhn & McPartland, 1954; McGuire & McGuire, 1982).

Instructions for the "Who Am I?" technique for measuring self-concepts (Gordon, 1968) are provided below. You can try this test yourself.

In the 15 numbered blanks write 15 different answers to the simple question "Who Am I?" Answer as if you were giving the answers to yourself, not to somebody else. Write the answers in the order they occur to you. Don't worry about "logic" or "importance."

I Am

1. _____
2. _____
3. _____
4. _____
5. _____

6. _____
7. _____
8. _____
9. _____
10. _____

11. _____
12. _____
13. _____
14. _____
15. _____

The following responses have been obtained from two persons, Josh and Arlene.

Josh: A 9-year-old male

a boy
do what my mother says, mostly
Louis's little brother
Josh
have big ears
can beat up Andy
play soccer
sometimes a good sport
a skater
make a lot of noise
like to eat
talk good
go to third grade
bad at drawing

Arlene: A female college sophomore

a person
member of the human race
daughter and sister
a student
people-lover
people-watcher
creator of written, drawn, and spoken (things)
    creations
music enthusiast
enjoyer of nature
partly the sum of my experiences
always changing
lonely
all the characters in the books I read
a small part of the universe, but I can change it
I'm not sure?! (Gordon, 1968)

Five major questions are addressed in this chapter.

1. What is the self and how does it arise?
2. How do we acquire unique identities, the categories we use to specify who we are? How do we use them to locate ourselves in the world relative to others?
3. How do our identities guide our plans and behavior?
4. How does self-awareness influence the ways we think and feel?
5. Feelings of evaluation inevitably accompany thoughts about ourselves. Where do they come from and how do they affect our behavior? How do we protect our self-esteem against attack?

The Nature and Genesis of Self

The Self as Source and Object of Action

We can behave in a wide variety of ways toward other persons. For example, if Bob is having coffee with Carol, he can perceive her, evaluate her, communicate with her, motivate her to action, attempt to control her, and so on. Note, however, that Bob also can act in the same fashion toward himself—that is, he can engage in self-perception, self-evaluation, self-communication, self-motivation, and self-control. Behavior of this type, in which the individual who acts and the individual toward whom the action is directed are the same, is termed *reflexive behavior*.

For example, if Bob, a student, has an important term paper due Friday, he engages in the reflexive process of self-control when he pushes himself ("Work on that history paper now"). He engages in self-motivation when he makes a promise to himself ("You can go out for pizza and a movie Friday night"). Both processes are part of the self. To have a self is to have the capacity to engage in reflexive actions, to plan, observe, guide, and respond to our own behavior (Bandura, 1982c; Mead, 1934).

Our understanding of reflexive behavior and the self is drawn from the symbolic interaction theory. By definition, the **self** is the individual viewed as both the source and the object of reflexive behavior. Clearly,

the self is both active (the source that initiates reflexive behavior) and passive (the object toward which reflexive behavior is directed). The active aspect of the self is labeled the "I" and the object of self-action is labeled the "Me" (James, 1890; Mead, 1934).

It is useful to think of the self as a continuing process (Gecas & Burke, 1995). Action involving the self begins with the "I," with an impulse to act. For example, Bob wants to see Carol. In the next moment that impulse becomes the object of self-reflection, and, hence, part of the "Me" ("If I don't work on that paper tonight, I won't get it done on time"). Next, Bob responds actively to this self-awareness, again an "I" phase ("But I want to see Carol, so I won't write the paper"). This, in turn, becomes the object to be judged, again a "Me" phase ("That would really hurt my grade"). So Bob exercises self-control and sits down at his desk to write. The "I" and "Me" phases continue to alternate as every new action (I) becomes in the next moment the object of self-scrutiny (Me). Through these alternating phases of self we plan, act, monitor our actions, and evaluate outcomes (Marcus & Wurf, 1987).

Mead portrays action as guided by an internal dialogue. People engage in conversations in their minds as they regulate their behavior. They use words and images to symbolize their ideas about themselves, other persons, their own actions and others' probable responses to them. This description of the internal dialogue suggests three capacities human beings must acquire in order to engage successfully in action. They must (1) develop an ability to differentiate themselves from other persons, (2) learn to see themselves and their own actions as if through others' eyes, and (3) learn to use a symbol system or language for inner thought. In this section, we examine how children come to differentiate themselves and how they learn to view themselves from others' perspectives. We also discuss how language learning is intertwined with acquiring these two capacities.

Self-Differentiation

To take the self as the object of action, we must—at a minimum—be able to recognize ourselves. That is, we must distinguish our own faces and bodies from

To take the self as the object of our action—observing and modifying our own behavior—we must be able to recognize ourselves. Although infants are not born with this ability, they acquire it quickly.

those of others. This may seem elementary, but infants are not born with this ability. At first they do not even discriminate the boundaries between their own bodies and the environment. Cognitive growth and continuing tactile exploration of their bodies contribute to infants' discovery of their physical uniqueness. So does experience with caregivers who treat them as distinct beings. Studies of when children can recognize themselves in a mirror suggest that most children are able to discriminate their own image from others' by about 18 months (Bertenthal & Fischer, 1978). Research indicates that children become capable of representing self-other contingencies (for example, if I do X, she does Y) at 18 to 24 months old (Higgins, 1989).

Children must not only learn to discriminate their physical selves from others, but they also must learn to discriminate themselves as a social object. Mastery of language is critical in children's' efforts to learn the latter (Denzin, 1977). Learning one's own name is one of the earliest and most important steps in acquiring a self. As Allport (1961) put it, "By hearing his name repeatedly the child gradually sees himself as a distinct and recurrent point of reference. The name acquires significance for him in the second year of life. With it comes awareness of independent status in the social group" (p. 115).

A mature sense of self entails recognizing that our thoughts and feelings are our private possessions. Young children often confuse processes that go on in their own minds with external events (Piaget, 1954). They locate their own dreams and nightmares, for example, in the world around them. The distinction between self and nonself sharpens as social experience and cognitive growth bring children to realize that their own private awareness of self is not directly accessible to others. By about age 4, children report that their thinking and knowing goes on inside their heads. Asked further, "Can I see you thinking in there?" they generally answer "No," demonstrating their awareness that self-processes are private (Flavell, Shipstead, & Croft, 1978).

Changes in the way children talk also reveal their dawning realization that the self has access to private information. During their first years of talking, children's speech patterns are the same whether they are talking aloud to themselves or directing their words to others. Gradually, however, they begin to distinguish speech for self from speech for others (Vygotsky, 1962). Speech for self becomes abbreviated until it is virtually incomprehensible to the outside listener, whereas speech for others becomes more elaborated over time. "Cold" suffices for Amy to tell herself she wants to take off her wet socks. But no one else would understand this without access to her private knowledge. When addressing others, Amy would expand her speech to include whatever private information they would need to understand ("Gotta change my wet socks. They're making me cold."). This reflects her growing awareness that each self has its own unique store of knowledge.

Access to private information about the self leads to systematic differences in adults' self-descriptions compared to descriptions of others (McGuire & McGuire, 1986). Descriptions of the self focus on what one does, on physical action and on affective reactions to others. Descriptions of others focus on who the person is, on social interactions and on his or her cognitive reactions. Further, people perceive themselves as more complex than other people (Sande, Goethals, & Radloff, 1988). Did your responses in "Measuring Self-Concepts" reflect these characteristics of self-descriptions?

Role Taking

Recognizing that one is physically and mentally differentiated from others is only one step in the genesis of self. Once we can differentiate ourselves from others we also can recognize that each person sees the world from a different perspective. The second crucial step in the genesis of self is **role taking,** the process of imaginatively occupying the position of another person and viewing the self and the situation from that person's perspective (Hewitt, 1997).

Role taking is crucial to the genesis of self because through it the child learns to respond reflexively. Imagining others' responses to the self, children acquire the capacity to look at themselves as if from the outside. Recognizing that others see them as objects, children can become objects (Me) to themselves (Mead, 1934). They can then act toward themselves to praise ("That's a good girl"), to reprimand ("Stop that!"), and to control their own behavior ("Wait your turn").

Long ago, C. H. Cooley (1908) noted the close tie between role taking and language skills. One of the earliest signs of role-taking skills is the correct use of the pronouns "you" and "I." To master the use of these pronouns requires taking the role of self and of the other simultaneously. Most children firmly grasp the use of "I" and "You" by the middle of their third year (Clark, 1976), which suggests that children are well on their way to effective role taking at this age. Studies indicate that children develop the ability to infer the thoughts and expectations of others between ages 4 and 6 (Higgins, 1989).

The Social Origins of Self

Our self-schema is produced in our social relationships. Throughout life, as we meet new people and enter new groups, our view of self is modified by the feedback we receive from others. This feedback is not an objective reality that we can grasp directly. Rather, we must interpret others' responses in order to figure out how we appear to them. We then incorporate others' imagined views of us into our self-schema.

To dramatize the idea that the origins of self are social, Cooley (1902) coined the term *looking-glass self.* The most important looking glasses for children are their parents and immediate family and, later, their playmates. They are the child's **significant others,** the people whose reflected views have greatest influence on the child's self-concepts. As we grow older, the widening circle of friends and relatives, schoolteachers, clergy, and fellow workers provides our significant others. The changing images of self we acquire through our lives depend on the social relationships we develop (see Table 4.1).

Play and the Game   Mead (1934) identified two sequential stages of social experience leading to the emergence of the self in children. He called these stages *play* and *the game.* Each stage is characterized by its own form of role taking.

In the play stage, young children imitate the activities of people around them. Through such play, children learn to organize different activities into meaningful roles (nurse, doctor, firefighter). For example, using their imaginations, children carry sacks of mail, drop letters into mailboxes, greet homeowners, and learn to label these activities as fitting the role "mail carrier." At this stage, children take the roles of others one at a time. They do not recognize that each role is intertwined with others. Playing mail carrier, for example, the child does not realize that mail carriers also have bosses to whom they must relate. Nor do children in this stage understand that the same person simultaneously holds several roles—that mail carriers are also parents, store customers, and golf partners.

The game stage comes later, when children enter organized activities such as complex games of house, school, and team sports. These activities demand

Table 4.1 Significant Others Mentioned in Self-Descriptions, by Age

	Ratio of the Frequency of Mentioning		
Age	1 **Parents** **versus** **Teachers**	2 **Brothers and Sisters** **versus** **Friends and Fellow Students**	3 **Nonfamily Members** **versus** **Extended Family**
7 years	1.7 to 1	1.7 to 1	4 to 1
9 years	1 to 1.4	1 to 1.4	8 to 1
13 years	1 to 1	1 to 1	13 to 1
17 years	1 to 2.3	1 to 2.3	49 to 1

Note: In this study, 560 boys and girls were asked, "Tell us about yourself." The children's responses suggest that their self-definitions in terms of other people tend to shift away from family members with age—from parents to teachers (column 1), from brothers and sisters to friends and fellow students (column 2), from extended family members (cousins, aunts, uncles) to nonfamily members (column 3). For example, 7-year-olds mentioned parents almost twice as often as teachers, and 17-year-olds mentioned teachers more than twice as often as parents (column 1).

SOURCE: Adapted from McGuire and McGuire, 1982.

interpersonal coordination because the various roles are differentiated. Role taking at the game stage requires children to imagine the viewpoints of several others at the same time. For Ellen to play shortstop effectively, for example, she must adopt the perspectives of the infielders and of the base runners as she fields the ball and decides where to throw. In the game, children also learn that different roles relate to each other in specified ways. Ellen must understand the specialized functions of each position, the ways the players in different positions coordinate their actions, and the rules that regulate baseball.

The Generalized Other Repeated involvement in organized activities lets children see that their own actions are part of a pattern of interdependent group activity. This experience teaches children that organized groups of people share common perspectives and attitudes. With this new knowledge, children construct a **generalized other**—a conception of the attitudes and expectations held in common by the members of the organized groups with whom they interact. When we imagine what "the group" expects of us, we are taking the role of the generalized other. We are also concerned with the generalized other when we wonder what "people" would say, or what "society's standards" demand. As children grow older,

they control their own behavior more and more from the perspective of the generalized other. This helps them resist the influence of impulse, or of specific others who just happen to be present at the moment.

Over time, children internalize the attitudes and expectations of the generalized other, incorporating

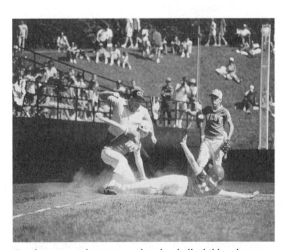

By playing complex games such as baseball, children learn to organize their actions into meaningful roles and to imagine the viewpoints of other players at the same time. Role taking enables the catcher to coordinate effectively with teammates, for example, to tag a runner out at home plate.

them into their self-concepts. But building up self-concepts involves more than accepting the reflected views of others. We may misperceive or misinterpret the responses that others direct to us, for example, because of our less than perfect role-taking skills. Others' responses may themselves be contradictory or inconsistent. We also may resist the reflected views we perceive because they conflict with our prior self-concepts or with our direct experience. A boy may reject his peers' view that he is a sissy, for example, because he previously thought of himself as brave and could still visualize his experience of beating up a bully.

Self-Evaluation   The views of ourselves that we perceive from others usually imply positive or negative evaluations. These evaluations also become part of the self we construct. Actions that others judge favorably produce positive self-concepts. In contrast, when others disapprove or punish our actions, the self-concepts we derive are negative.

Identities: The Self We Know

In "Measuring Self-Concepts" Arlene described herself as a person, daughter, student, people-lover, and creator of things. This is the self she knows, a self that includes specific identities. **Identities** are the meanings attached to the self by self and others (Gecas & Burke, 1995). When we think of our identities, we are actually thinking of various plans of action that we expect to carry out. When Arlene identifies herself as a student, for example, she has in mind that she plans to attend classes, write papers, take exams, and so on. If Arlene does not engage in these behaviors, she will have to relinquish her student identity.

In this section we consider five questions about the self we know: (1) How do our roles influence the identities we include in this self? (2) How do group memberships influence the self we know? (3) What aspects of self do people note in their actual self-descriptions? (4) What evidence is there that the self we know is based on the reactions we perceive from others? (5) How do the aspects of self that people note vary from one situation to another?

Role Identities

Each of us occupies numerous positions in society—student, friend, son or daughter, customer. Each of us, therefore, enacts many different social roles. We construct identities by observing our own behavior and the responses of others to us as we enact these roles. For each role we enact, we develop a somewhat different view of who we are, an identity. Because these identities are concepts of self in specific roles, they are called **role identities.** The role identities we develop depend on the social positions available to us in society. As a result, the self we know is linked to society fundamentally through the roles we play. It reflects the structure of our society and our place in it (McCall & Simmons, 1978; Stryker, 1980).

Do societal role expectations strictly dictate the contents of our role identities? Apparently not. Consider, for example, the role expectations for the college instructor. Some instructors deliver lectures, whereas others lead discussions; some encourage questions, and others discourage them; some assign papers, and others do not. As this example indicates, role expectations usually leave individuals room to improvise their own role performances. It is probably more accurate to think of people as "making" their roles—that is, shaping them—rather than as conforming rigidly to role expectations (Turner, 1978).

Several influences affect the way we make the roles we enact. Conventional role expectations in society set a general framework. In the role of student, for example, you must submit assigned papers. Within this general framework, you can fashion your actual role performances to reflect your personal characteristics and competencies. You can select topics that interest you and highlight your strengths and cover your weaknesses. You also mold your role performances to impress your audience (writing in the style the instructor prefers). Finally, you adjust your different performances to maintain some consistency among them (trying for a level of quality consistent with your other course work). Because each person makes roles in a unique, personal fashion, we each derive somewhat different role identities even if we occupy similar social positions. Consequently, our role identities as student, team player, and so on, differ from the role identities of others who also occupy these positions.

Social Identities

A second source of identities is membership in social categories or groups, such as nationality, race/ethnicity, or political affiliation. A definition of the self in terms of the defining characteristics of a social group is a **social identity** (Hogg, Terry, & White, 1995; Tajfel & Turner, 1979). Each of us associates certain characteristics with members of specific groups. These characterizations—Chicago Bulls fans are loud, women are emotional—define the group. If you define yourself as a member of the group, these become standards for your thoughts, feelings, and actions. If your interactions with other group members confirm the importance of these attributes, they become part of the self you know. Research indicates that cognitive representations of the self and of groups to which the person belongs are closely linked (Smith & Henry, 1996).

Group members usually perceive the group and therefore themselves in positive terms. They rate traits perceived as typical of the group more favorably than they rate other traits. However, this bias does not lead members to violate social reality as defined by nonmembers of the group (Ellemers et al., 1997).

Social groups are often defined in part by reference to other groups. The meaning of being a Young Republican is related to the meaning of being a Young Socialist and a Young Democrat. The meaning of being a male in American society is closely related to the meaning of being female. Thus, when membership in a group becomes a salient basis for self-definition, perceptions of relevant out-groups are also made salient. Often there is an accentuation effect, an emphasis on perceived differences and unfavorable evaluations of the out-group and its members (Hogg, Terry, & White, 1995). Thus negative stereotypes directed at persons of a different gender, race, or religion are often closely related to the self-concept of the person who holds them. Research indicates that both in-group favoritism and out-group hostility are reinforced in conversations between group members (Harasty, 1997).

Actual Self-Descriptions

If the self we know includes identities, people's actual self-descriptions should reflect their major social roles and group memberships. Do they? To answer this question, we examine responses people give to the "Who am I?" questionnaire. These responses reveal three general types of self-description: identities, personal qualities, and self-evaluations (Gordon, 1968).

Identities   In describing themselves, most respondents spontaneously list their role identities as members of occupational, educational, or family groups among their self-descriptions. They also frequently mention identities as members of various social categories such as religion, athletic teams, race or ethnicity. There is usually substantial public agreement about the role identities a person claims (Kuhn & McPartland, 1954).

The other two types of self-description—personal qualities and self-evaluations—refer to conceptions of self that others may dispute or ignore. Our conceptions of how moody we are, how moral, or how competent are not matters of public consensus. These self-concepts express the personal, idiosyncratic variation in our self-descriptions.

Personal Qualities   In describing their personal qualities, people mention most frequently the styles of interpersonal behavior (introverted, cool) that distinguish the way they fashion their unique role performances. People also mention the emotional or psychological styles (optimistic, moody) that characterize these performances. Individual preferences point to specific ways people express their role identities. For example, a person who sees herself as a musician expresses this role identity differently depending on whether she prefers Bach or rock. Body image, the aspect of the self we recognize earliest, remains important throughout life. Beyond this, our self extends to include our material possessions, such as our clothing, house, car, CDs, and so on (James, 1890).

Self-Evaluations   The third type of self-description refers to the ways we evaluate ourselves. We form these self-evaluations when reflecting on the adequacy of our role performances, on the extent to which we live up to the standards to which we aspire. Our evaluations most commonly focus on our competence, self-determination, moral worth, or unity. Self-evaluations also influence the ways we express our role identities. A musician, for example, pursues opportunities to perform in public more persistently if she sees herself as competent than if she thinks she is never quite good enough. Self-evaluations are so important that the concluding section of this chapter will be devoted to them.

Research on Self-Concept Formation

Two of the key theoretical ideas discussed so far are (1) the formation of the self-schema involves the adoption of role identities, and (2) a person's self-concept is shaped by the reactions that he or she receives from significant others during social interaction. Each of these has been the focus of empirical research.

The Adoption of Role Identities   Self-schemas are formed in part by adopting role identities. The identities available depend on the culture. One difference between cultures is whether a culture is individual or collective (Triandis, 1989). Individualistic cultures emphasize individual achievement and associated identities such as president, team captain, idealist, and outstanding player. Collectivist cultures emphasize values that promote the welfare of the group and associated identities such as son (family), Catholic (religion), Italian (ethnicity), and American. According to research, the self-schemas of persons in individualistic cultures (e.g., the United States) include more individual identities, whereas those of persons in collectivist cultures include more group-linked identities (Triandis, McCusker, & Hui, 1990). Studies indicate that whether one identifies with a group in which she can claim membership depends on how easily one can be identified as a member of that group, for example, by name or skin color (Lau, 1989). It also depends on the general salience or visibility of that group in society.

The adoption of a role identity involves a process that occurs over time. First, the individual recognizes that the identity is available. Through contact with others, he or she learns the behavioral expectations and the social evaluation associated with the role. For a time the person may distance himself from others who have the identity. He may be uncertain about whether he can claim the role identity or unsure about how to enact the role. Alternatively, he may quickly embrace the role and seek opportunities to enact it. A study of college athletes found that, over time, the athlete role identity became more salient and other role identities, such as son and student, became more peripheral (Adler & Adler, 1989). These changes were reflected in changes in the amount of time spent enacting each role.

Reflected Appraisals   The idea that the person bases his or her self-schema on the reactions he or she perceives from others during social interaction is captured by the term *reflected appraisals.* Studies of this process (Marsh, Barnes, & Hocevar, 1985; Miyamoto & Dornbusch, 1956) typically compare people's self-ratings on various qualities (intelligence, self-confidence, physical attractiveness) with the views of themselves that they perceive from others. The studies also compare self-ratings with actual views of others. Results of these studies support the hypothesis that it is the perceived reactions of others, rather than their actual reactions, that are crucial for self-concept formation (Felson, 1989).

One study of reflected appraisals analyzed perceptions of leadership in small groups (Riley & Burke, 1995). Groups of four persons met and engaged in discussions on four separate occasions. After each discussion, each member rated self and others on leadership identity and leadership performance scales. Self and others' ratings of self on the identity scale were similar, and both were consistent with scores on the leadership behavior scale. In other words, a shared meaning structure developed among the discussants, and the individual's perception of self was consistent with others' appraisals.

Research has focused on the differential effect of various significant others on one's appraisal of self in particular roles/domains. Felson (1985; Felson & Reed, 1986) has studied the relative influence of parents and peers on the self-perceptions of fourth through eighth graders about their academic ability, athletic ability, and physical attractiveness. The results indicate that parents affect self-appraisals in the areas of academic and athletic ability, whereas peers are an important influence on perceived attractiveness.

One aspect of attractiveness is weight. Although there is an objective measure of weight (pounds, or pounds in relation to height), it is the social judgment ("too fat," "too thin," or "just right") that is incorporated into the self-concept. A study of adolescent health obtained self-appraisals of weight from 6,500 adolescents, as well as appraisals from their parents and a physician (Levinson, Powell, & Steelman, 1986). These young people were generally unhappy with their weight, with males judging themselves to be too thin and females judging themselves to be overweight. For both, parental appraisal was significantly related to the young person's judgment, whereas the physician's rating was not.

Typically, a person's self ratings are related more closely to his or her perceived ratings by others than to the actual ratings by others. Why is this so? Three reasons are especially important. First, others rarely provide full, honest feedback about their reactions to us. Second, the feedback we do receive is often inconsistent and even contradictory. Third, the feedback is frequently ambiguous and difficult to interpret. It may be in the form of gestures (shrugs), facial expressions (smiles), or remarks that can be understood in many different ways. For these reasons, we may know little about others' actual reactions to us. Instead, we must rely on our perceptions of others' reactions to construct our self-concepts (Schrauger & Schoeneman, 1979).

Evidence that self-concepts are related to the perceived reactions of others does not, of itself, demonstrate that self-concepts are actually formed in response to these perceived reactions. However, one study (Mannheim, 1966) does suggest such an impact of the perceived reactions of others on self-concepts. The investigators in this study asked college dormitory residents to describe themselves and to report how they thought others viewed them. Several months later, self-concepts were measured again. In the interim, students' self-concepts had moved closer to the views they had originally thought that others held. Change toward the perceived reactions of others had indeed occurred. Similarly, a longitudinal study of delinquent behavior found that parental appraisals of youth as delinquent were associated with subsequent self-appraisals as delinquent; self-appraisal as delinquent was in turn related to delinquent behavior (Matsueda, 1992).

The Situated Self

If we were to describe ourselves on several different occasions, the identities, personal qualities, and self-evaluations mentioned would not remain the same. This is not due to errors of reporting. Rather, it demonstrates that the aspects of self that enter our awareness and matter to us most depend on the situation. The **situated self** is the subset of self-concepts chosen from our identities, qualities, and self-evaluations that constitutes the self we know in a particular situation (Hewitt, 1997). Markus and Wurf (1987) refer to the current, active, accessible self-representations as the *working self-concept.*

The self-concepts most likely to enter the situated self are those distinctive in the setting and relevant to the ongoing activities. Consider a black woman for whom being black and being a woman are both important self-concepts. When she interacts with black men, she is more likely to think of herself as a woman. When she interacts with white women, she is more likely to be aware she is black. Similarly, whether gender is part of your situated self depends in part on the gender composition of those present (Cota & Dion, 1986). Male and female college students placed in a group with two students of the opposite gender were more likely to list gender in their self-descriptions than members of all-male or all-female groups. Thus self-concepts that are distinctive or peculiar in the social setting tend to enter into the situated self (McGuire & McGuire, 1982) (see Figure 4.1).

Our activities also determine the self-concepts that constitute the situated self. A job interview, for

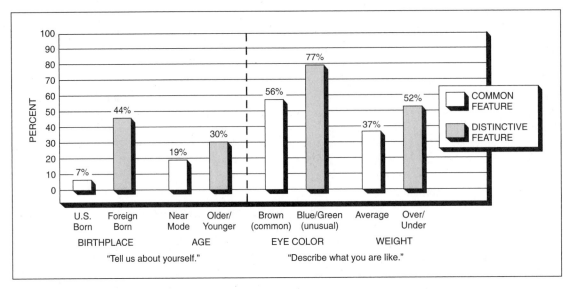

Figure 4.1 Percentage of Students Who Mention a Feature Spontaneously as Part of Their Self-Concept

A group of 252 sixth graders from 10 classrooms were asked to describe themselves. Students mentioned a particular feature (for example, birthplace) more often if that feature distinguished them from their classmates. Because these are characteristics on which we stand out from our social groups, attracting more notice and social comment, we are more likely to build them into our self-concepts.

SOURCE: Adapted from McGuire and Padawer-Singer, 1976.

example, draws attention to your competence; a party makes your body image more salient. The self we experience in our imaginings and in our interactions is always situated because setting characteristics and activity requirements make particular self-concepts distinctive and relevant.

Identities: The Self We Enact

How does the self influence the planning and regulation of social behavior? The general answer to this question is that we are motivated to plan and to perform behaviors that confirm and reinforce the identities we wish to claim for ourselves (Burke & Reitzes, 1981; Markus & Wurf, 1987). In elaborating on this answer, we examine three more specific questions: (1) How are behaviors linked to particular identities? (2) Of the different identities available to us, what determines which ones we choose to enact in a situation?

(3) How do our identities lend unity and consistency to our behavior?

Identities and Behavior

Earlier we noted that self-schemas include both role identities and personal qualities. Some people place greater emphasis on one than on the other. For instance, consider responses to the question "Who am I?" Some people emphasize role identities (daughter, psychology major), whereas others list primarily personal qualities (easygoing, friendly). Leary and colleagues (1986) predicted that the behavioral preferences of these two groups would differ. In the area of recreation, people who emphasize social aspects would prefer team sports (softball, volleyball, basketball), and those who emphasize personal qualities would prefer individual sports (swimming, running, aerobics). Furthermore, those whose self-concept is predominantly social should prefer occupations that offer social rewards, such as status and friendship. Those whose self-concept is predominantly personal should prefer jobs that offer rewards such as opportunities for self-expression and personal growth. The results verified both predictions.

The link between identities and behaviors is through their common meanings (Burke & Reitzes, 1981). If members of a group agree on the meanings of particular identities and behaviors, they can regulate their own behavior effectively. They can plan, initiate, and control behavior to generate the meanings that establish the identities they wish to claim. If members do not agree on these meanings, however, people have difficulty establishing their preferred identities. If Roberta sees no connection between competitiveness and femininity, for example, she will have trouble establishing a feminine identity in the eyes of friends who think being feminine means being noncompetitive.

Choosing an Identity to Enact

Each of us has many different identities. Each identity suggests its own lines of action. These lines of action are not all compatible, however, nor can they be pursued simultaneously in a single situation. If you are at a family reunion in your parents' home, for example, you might wish to claim an identity as a helpful son/daughter, an aspiring poet, or a witty conversationalist. These identities suggest different, even conflicting, ways of relating to the other guests. What influences the decision to enact one rather than another identity? Several factors affect such choices.

The Hierarchy of Identities　The many different role identities we enact do not have equal importance for us. Rather, we organize them into a hierarchy according to their **salience,** their relative importance to the self-schema. This hierarchy exerts a major influence on our decision to enact one or another identity (McCall & Simmons, 1978; Stryker, 1980). First, the more salient an identity is to us, the more frequently we choose to perform activities that express that identity (Stryker & Serpe, 1981). Second, the more salient an identity, the more likely we

are to perceive that situations offer opportunities to enact that identity. Only a person aspiring to the identity of poet, for example, would perceive a family reunion as a chance to recite his or her poems. Third, we are more active in seeking opportunities to enact salient identities (searching for an open-mike poetry reading). Fourth, we conform more with the role expectations attached to the identities that we consider the most important.

What determines whether a particular identity occupies a central or a peripheral position in the salience hierarchy? In general, several factors affect the importance we attach to a role identity: (1) the resources we have invested in constructing the identity (time, effort, and money expended, for example, in learning to be a sculptor); (2) the extrinsic rewards that enacting the identity has brought (purchases by collectors, acclaim by critics); (3) the intrinsic gratifications derived from performing it (the sense of competence and aesthetic pleasure obtained when sculpting a human figure); and (4) the amount of self-esteem staked on enacting the identity well (the extent to which a positive self-evaluation has become tied to being a good sculptor). As we engage in interaction and experience greater or lesser success in performing our different identities, their salience shifts.

Social Networks　Each of us is part of a network of social relationships. These relationships may stand or fall on whether we continue to enact particular role identities. The more numerous and significant the relationships that depend on enacting an identity, the more committed we become to that identity (Callero, 1985). Consider, for example, your role as a student. Chances are that many of your relationships—with roommate(s), friends, instructors, and perhaps a lover—depend on your continued occupancy of the student role. If you left school, you could lose a major part of your life. Given this high level of commitment, it isn't surprising that being forced to leave school is traumatic for many students.

The more commitment we have to a role identity, the more important that identity in our hierarchy. For instance, adults for whom participating in religious activities was crucial for maintaining everyday social relationships ranked their religious identity as relatively important compared with their parent, spouse, and worker identities (Stryker & Serpe, 1981). Similarly, the importance rank that undergraduates gave to various identities (student, friend, son/daughter, athlete, religious person, and dating partner) depended on the importance to them of the social relationships maintained by enacting each identity (Hoelter, 1983).

Need for Identity Support
We are likely to enact those of our identities that most need support because they have recently been challenged. For instance, suppose that someone has recently had difficulty getting a date. That person may now choose actions calculated to elicit responses indicating she is an attractive dating partner. We also tend to enact identities likely to bring intrinsic gratifications (such as a sense of accomplishment) and extrinsic rewards (such as praise) that we especially need or miss at the moment. For example, if, after hours of solitary study, you feel a need for relaxed social contact, you might seek gratification by going to a student lounge or union to find someone to chat with.

Situational Opportunities
Social situations are restrictive; they let us enact only some identities profitably, not others. Thus, in a particular situation, the identity we choose to enact depends partly on whether the situation offers opportunities for profitable enactment. Regardless of the salience of your identity as poet, if no one wants to listen to your poems, there will be no opportunity to enact that identity.

In a series of studies, Kenrick and colleagues (1990) asked students to rate the extent to which various personal qualities could be displayed in each of six different settings. The traits were adjustment, dominance, intellectual ability, likability, social control, and social inclination. The students agreed that one can display intellectual ability in academic settings, but not in recreational ones. Behaviors expressive of dominance can be displayed in athletic and business settings, but not religious ones. Finally, there are opportunities to display adjustment and social inclination in recreational settings, but not in church.

Identities as Sources of Consistency

Although the self includes multiple identities, people usually experience themselves as a unified entity. One reason is the influence of the salience hierarchy. Another reason is that we use several strategies that verify our perceptions of self.

Salience Hierarchy
Our most salient identities provide consistent styles of behavior and priorities that lend continuity and unity to our behavior. In this way, the importance hierarchy helps us construct a unified sense of self from our multiple identities.

The hierarchy of identities influences consistency in three ways. First, the hierarchy provides us with a basis for choosing which situations we should enter and which ones we should avoid. A study of the everyday activities of college students (Emmons, Diener, & Larsen, 1986) found clear patterns of choice and avoidance in each student's interactions; these patterns were consistent with the person's characteristics, such as sociability.

Second, the hierarchy influences the consistency of behavior across different situations. In another study, each person was asked to report the extent to which each of 10 affective states and 10 behavioral responses occurred in various situations, again over a 30-day period (Emmons & Diener, 1986). The results indicated a significant degree of consistency across situations. Third, the hierarchy influences consistency in behavior across time. Serpe (1987) studied a sample of 310 freshmen, collecting data at three points during their first semester in college. The survey measured the salience at each point of five identities: academic ability, athletic/recreational involvement, extracurricular involvement, personal involvement (friendships), and dating. There was a general pattern of stability in salience. Change in salience was more likely for those identities where there was greater opportunity for change, for example, dating.

Although the self-concept exhibits consistency over time, it may change (Demo, 1992). Life transitions may change the roles we play and the situations we encounter. This creates a need to exit from one or more roles, adopt new roles, and change the salience

The more important an identity is to us, the more consistently we act to express it, regardless of others' reactions. Are any of your identities so important that you would express them the way these Hassidic Jews do?

hierarchy. During times such as adolescence and retirement, we are likely to feel a weakened sense of unity and confusion about how to behave. This has been called an *identity crisis* (Erikson, 1968). To overcome such confusion, we must reorganize our identity hierarchy, giving greater importance to identities based on our newly available or remaining social positions. A retiree may successfully reorganize the hierarchy, for example, by upgrading identities based on new hobbies (gardener) and on continuing social ties (witty conversationalist).

Self-Verification Strategies We experience ourselves as consistent across time and situations because we employ several strategies that verify our self-perceptions (Banaji & Prentice, 1995). One set of strategies consists of behaviors that lead to self-confirming feedback from others. First, we engage in selective interaction; we choose as friends, roommates, and intimates those people who share our view of self. Second, we display identity cues that elicit identity-confirming behavior from others. In a hospital setting, most people treat a middle-aged man wearing a white coat as a physician. Third, we behave in ways that enhance our identity claims, especially when those claims are challenged. In one study, white students who viewed themselves as unprejudiced were led to believe they were prejudiced toward blacks. When they were subsequently approached by a black panhandler, they gave him more money than did students whose egalitarian identity had not been threatened (Dutton & Lake, 1973) (see Figure 4.2).

Another set of strategies involve the processing of feedback from others. As noted in the next section, we

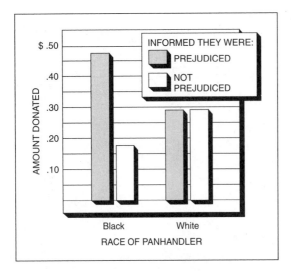

Figure 4.2 Acting to Reaffirm Threatened Identities

White students' self-identities as egalitarians were either threatened by false feedback from a lie detector, indicating that they were racially prejudiced, or confirmed by feedback that they were not prejudiced. Following the lie detector test, students were solicited either by a black or a white panhandler. Those who were approached by the black panhandler after their egalitarian identity had been threatened donated the most money. In this way they "proved" they were not racially prejudiced and reaffirmed their recently threatened egalitarian identity. Donations to the white panhandler were unaffected by the feedback about prejudice. These results demonstrate how the desire to affirm threatened identities influences behavior.

SOURCE: Adapted from Dutton and Lake, 1973.

often do this in ways that make others' responses to us seem to support our self-view.

There are limits to the extent to which we engage in self-verifying strategies. At times we want accurate feedback about our abilities or another person's view of our relationship with him or her. When we want such feedback, and we have the necessary cognitive resources (attention, energy), we evaluate feedback from others by comparing it with our self-representations (Swann & Schroeder, 1995). This evaluation may lead to changes in behavior, for example, to move toward a goal or a desired identity, or to a change in self-representation.

The Self in Thought and Feeling

We are often preoccupied by our own thoughts and feelings, as well as by information especially relevant to us. At a noisy, crowded party you can often barely hear the conversation you are directly involved in. But should someone mention your name, even halfway across the room, you are likely to hear it and immediately to shift your attention in that direction.

In this section we discuss four ways in which the self affects our thoughts and feelings: (1) the impact of information's relevance to the self on the processing of that information; (2) ways that focusing attention on the self influences the relationship between our identities and our behavior; (3) the effect of discrepancies in the self on mood; and (4) ways the self influences the emotions we experience.

Self-Schema

The influence of self on thought occurs through the operation of the self-schema (Greenwald and Pratkanis, 1984). One's self-schema influences cognitive processes in several ways (Markus & Wurf, 1987). The self-schema influences the speed and certainty with which we process information, how we interpret feedback from others, and the storage in and retrieval from memory of information.

The self-schema influences the processing of incoming information. Compare Sara, for whom an athlete identity is important, with Maggie, who does not think about herself as either athletic or nonathletic. Sara judges more quickly and confidently whether traits like agile, clumsy, muscular, and puny apply to her than Maggie does. She also rejects more strongly information purporting to show that she is either more coordinated than she had previously thought or less. In short, people are quicker and more certain when judging and interpreting information related to their important identities or qualities.

The self-schema influences the way we interpret feedback. When discussing reflected appraisals, we noted that the feedback we receive from others is always incomplete, frequently ambiguous, and sometimes inconsistent. The self-schema determines how we receive and process this feedback. We pay more attention to relevant information and selectively focus on information that confirms our self-concepts, especially highly salient identities. It also provides us with a basis for interpreting the responses of others to us. Because of the influence of self-schema, we typically perceive more confirmation of our self-concept than actually exists (Swann, 1987).

The self-schema also influences memory. In one study, subjects were led to believe that either extroversion or introversion was a desirable trait. They were then asked to remember information about themselves relevant to the trait. In both conditions, subjects remembered more information consistent with the trait and remembered it more quickly (Sanitioso, Kunda, & Fong, 1990). Thus memory for events is better the more such events relate to the self. Relating information to the self has also been shown to enhance learning and memory for factual material. We remember a list of personality traits better, for instance, if we think about how each trait applies to us while trying to memorize the list (Rogers, 1977).

Thus the important identities and other interrelated self-concepts constituting our self-schema provide a finely tuned set of mental categories that we use to process information.

Effects of Self-Awareness

While eating with friends, reading a book, or participating in conversation, your attention is usually directed toward the objects, people, and events that surround you. But what happens if—upon looking up—you discover a photographer with his lens focused on you, snapping away? Or what if you suddenly notice your image reflected in a large mirror? In such circumstances most of us become self-conscious. We enter a state of **self-awareness**—that is, we take the self as the object of our attention and focus on our own appearance, actions, and thoughts. This corresponds to the "Me" phase of action (Mead, 1934).

Numerous circumstances cause people to become self-aware. Mirrors, cameras, and recordings of our own voice cause self-awareness because they directly present the self to us as an object. Unfamiliar situations and blundering in public also cause self-awareness because they disrupt the smooth flow of ac-

tion and interaction. When this happens, we must attend to our own behavior more closely, monitoring its appropriateness and bringing it into line with the demands of the situation. In general, anything that reminds us we are objects of others' attention will increase our self-awareness.

How does self-awareness influence behavior? When people are highly self-aware, they are more likely to be honest and to report more accurately on their mood state, psychiatric problems, and hospitalizations (Gibbons et al., 1985). In general, people who are self-aware act in ways more consistent with personal and social standards (Wicklund, 1982; Wicklund & Frey, 1980). Their behavior is controlled more by the self. In the absence of self-awareness, behavior is more automatic or habitual. Society gains control over its members through the self-control individuals exercise when they are self-aware (Shibutani, 1961). This is because the standards to which people conform are largely learned from significant groups in society. Self-awareness is thus often a civilizing influence.

The most widely endorsed theory to explain these effects of self-awareness assumes that attention to self activates the self-schema, which in turn leads to self-evaluation (Gibbons, 1990; Wicklund, 1975). We may evaluate the self at any of three levels: experiential, behavioral, and global.

A focus on experience heightens our current affect or mood and elicits behavior associated with that affect; for example, anger may elicit attack, joy may elicit playfulness. A focus on behavior leads to consideration of behavioral standards. The standards may be internal/personal ones. If there is a discrepancy between behavior and internal standards, the person will attempt to align behavior with her standard.

Alternatively, the salient standard may be external or social. When social standards and internalized standards correspond, an increase in self-awareness brings behavior closer to ideals. But what if social standards conflict with internalized standards, as when groups pressure individuals to change their attitudes or to violate their personal standards? When this happens, the impact of increased self-awareness depends on whether our attention is drawn to the public, social aspects of self or to the private, covert aspects of self

(Scheier & Carver, 1981). If we attend to public aspects of self (our public image, our mannerisms), we may respond to group influence and modify our attitude or behavior; however, we will not abandon our internal standard (Gibbons, 1990). But if we attend to private aspects of self (our personal values, attitudes, and internalized standards), the effect is the opposite: We guide our behavior more by personal standards and yield less to group influence.

The third possibility is that self-focus may elicit a global evaluation of self. The reference point for this evaluation may be the ideal self. The nature and effects of real:ideal self-discrepancies will be discussed below.

These findings suggest that groups enhance their social control over individual behavior when they expose individuals to conditions like an attentive audience, unfamiliar circumstances, and socially awkward tasks that increase awareness of the public self. Interestingly, these are precisely the conditions used so effectively by cults.

Effects of Self-Discrepancies

Research has shown that the relationships between components of the self-schema influence one's emotional state. There are three components of the self-concept: self as one is *(actual),* as one would like to be *(ideal),* and as one ought to be *(ought).* When we evaluate ourselves, we typically use the ideal self or the ought self as the reference point. When the actual self matches the ideal self, we feel satisfaction or pride. However, when there is a *self-discrepancy,* that is, a component of the actual self is the opposite of a component of the ideal self or the ought self, we experience discomfort (Higgins, 1989).

According to **self-discrepancy theory,** the two types of discrepancy produce two different emotional states. Someone who has an actual:ideal discrepancy will experience dejection, sadness, or depression. Someone who perceives an actual:ought discrepancy will experience fear, tension, or restlessness. The theory predicts that the larger the discrepancy, the greater the discomfort.

In a study designed to test these hypotheses (Higgins, Klein, & Strauman, 1985), students were asked to list up to 10 attributes each of the actual self, the ideal self, and the ought self. Discrepancy was measured by comparing two lists, for example, the actual and the ideal; a self-state listed in both was a match, whereas a self-state listed on one list with its antonym (opposite) listed on the other was a mismatch. The self-discrepancy score was the number of mismatches minus the number of matches. Discomfort was measured by several questionnaires. The results showed that as the actual:ideal discrepancy increased, the frequency and intensity of reported dissatisfaction and depression increased. As the actual:ought discrepancy increased, the frequency and intensity of reported fear and irritability increased.

Self-discrepancy scores also are related to various behaviors. A study of satisfaction with one's body and of eating disorders found that a form of ideal:actual discrepancy was associated with bulimic behaviors, whereas an actual:ought discrepancy was associated with anorexic behaviors (Strauman et al., 1991).

Influences of Self on Emotions

At first, knowing how we feel appears to be a straightforward matter. We know whether we are happy or sad because we feel these emotions directly. But the matter is more complex than this. When you feel tightness in your stomach and sweaty palms, how do you know immediately whether you are excited about your date for tonight or fear the exam you have to take tomorrow? Closer examination suggests that the emotions we recognize are often merely plausible explanations we generate for physiological reactions. Whether we interpret our clenched teeth as a sign of fear or of sexual excitement depends on whether we are watching a horror film or an erotic movie.

Cognitive Labeling Theory According to the cognitive labeling theory, we actively construct our emotions (Schachter, 1964). The theory proposes that emotional experience is the result of the following three-step sequence:

1. An event in the environment produces a physiological reaction.

2. We notice the physiological reaction and search for an appropriate explanation.
3. By examining situational cues (What was happening when I reacted?), we find an emotion label (joy, disgust) for the reaction.

The theory further assumes that arousal is a generalized state, so that different emotions are not physiologically distinguishable. This implies that general arousal can signify virtually any emotion, depending on the situation.

In one study of this issue (Schachter & Singer, 1962), researchers gave students an injection of epinephrine, a drug that produces physiological arousal. They informed one group of students that this injection would probably cause them to experience a pounding heart, flushed face, and trembling. They told a second group nothing about the drug's effects. All students then waited with a confederate who, while appearing to be another student, was actually employed by the researchers. Depending on the experimental treatment, the confederate behaved either euphorically (shooting crumpled paper at a wastebasket, flying paper airplanes, playing with a hula hoop) or angrily (reacting with hostility to items on a questionnaire and finally tearing it up).

According to the theory, students in the informed group would not need to seek an explanation for their arousal, because they knew their symptoms were drug induced. Students in the uninformed group, however, lacked an adequate explanation for their symptoms and thus would need to search the environment for cues to help them label their feelings. Results confirmed these predictions. Students in the uninformed group adopted the label for their arousal suggested by the environment. That is, those who waited with the euphoric confederate described themselves as happy, and those who waited with the angry confederate described themselves as angry. The self-descriptions of the informed group, in contrast, were largely unaffected by the confederate's behavior.

Numerous later studies have expanded these findings to additional emotions (Kelley & Michela, 1980). They show that people who are unaware of the true cause of their physiological arousal can be induced to view themselves as anxious, guilty, amused, or sexually excited by placing them in environments that suggest these emotions (Dutton & Aron, 1974; Zillman, 1978). As the theory predicts, environmental conditions influence people's labeling of their physiological arousal only when they are not aware of its true origins.

Later research suggests that the emotional label sometimes precedes the awareness of arousal (Leventhal, 1984; Pennebaker, 1980). We begin with a belief that we are experiencing a particular emotion and, only then, search our bodily sensations for signs to verify our belief. If environmental cues give us reason to believe we are angry, we attend to our flushed face and racing heart and verify our anger. If the cues suggest we are happy, we attend to our feelings of alertness and trembling and confirm our happiness. At any given time, our physiological state may afford evidence to support several emotion labels.

This extension of cognitive labeling theory suggests two ways in which role identities may influence the emotions and sensations we experience. First, particular role identities provide situational cues that we use in interpreting sensations. Hypochondriacs verify their identity as sickly by attending to their stuffed noses, raspy throats, and uneven pulse. People who view themselves as healthy overlook these same physical signs. Second, role identities imply how sensations should be interpreted. The pounding heart that accompanies a drop to earth from an airplane signifies thrill to a veteran skydiver but terror to a novice paratrooper. Each interprets the same physical sign in line with his or her role identity.

Sentiments Most of the emotions we have discussed (such as fear, joy, anger) are a response to immediate situations and fade quickly when arousal subsides. Complex feelings like grief, love, or jealousy, however, arise out of our enduring social relationships. These socially significant feelings are called **sentiments** (Gordon, 1990); each is a pattern of sensations, emotions, actions, and cultural beliefs appropriate to a social relationship. Sentiments such as grief, loyalty, envy, and patriotism develop around our attachments to family, friends, fellow workers, and country.

Sentiments reflect the nature of our social relationships and the changes in them. Grief and nostalgia reflect social losses. Jealousy and envy reflect problems over control of possessions. Anger and resentment reflect betrayal of commitments. We label our feelings with the culturally appropriate sentiment to make sense of our diverse emotional responses. For example, Mark's joy in Laurie's presence, his sorrow in her absence, his anger when she is criticized, and his fear when threatened with her loss make sense if he labels his feeling "love." Like simpler emotions, sentiments are produced by cognitive labeling. In choosing a sentiment label, however, we consider all the information we have about our enduring relationship.

As we develop our role identities, we also learn which sentiments are appropriate to them (Denzin, 1983). By expressing appropriate sentiments, we affirm or modify our identities and the social relationships in which they are embedded. To affirm our identity as a romantic partner, for example, we must express love, jealousy, anxiety, tenderness, and ecstasy, all at the appropriate times and places.

Emotion Work   On occasion, the self takes an especially direct, active role in the control of emotions. At one time or another most of us have psyched ourselves up, forced ourselves to have a good time even when we were tired. Or we have tried to feel grateful in the face of an unwanted gift or displayed a stiff upper lip despite severe disappointment. These are all instances of **emotion work,** attempts to change the intensity or quality of our feelings to bring them into line with the requirements of the occasion (Hochschild, 1983). Emotion work is needed when we find that we are violating **feeling rules**—rules that dictate what people with our role identities ought to feel in a given situation. One study asked students how a person who was experiencing each of 128 events should feel (Heise & Calhan, 1995). The students agreed in naming a specific emotion for one third of the events, evidence of feeling rules.

There are two basic kinds of emotion work: (1) evocation of feelings that are not present but should be; and (2) suppression of feelings that are present but should not be. For example, an airline flight attendant is expected to feel calm and cheerful as she interacts with passengers. But suppose she has been working for 10 hours, serving hundreds of people on three different flights? Fatigue and irritation may be her main feelings. If so, she must then work directly on her own emotions to evoke feelings of cheerfulness and suppress feelings of irritation.

Among methods used to evoke suitable emotions and suppress unsuitable ones are adopting appropriate postures, shaping our facial expressions, breathing quickly or deeply, and imagining a situation that produces the required feeling. To recapture some cheerfulness, the flight attendant may take a deep breath and relax to reduce her irritation, and imagine how good it will feel to be home tonight. Emotion work modifies our actual feelings. An alternative strategy, employed when we are not personally committed to the identity to which the feeling rules apply, is to control only the outward expression of our feelings (Gordon, 1990).

Role Taking and Emotion   Earlier in this chapter we emphasized the importance of role taking in the development of self-schema. It is also a source of two kinds of emotional experience (Shott, 1979). *Reflexive role-taking emotions* are feelings about the self evoked by viewing the self from another's perspective. When we accomplish an important goal—an A on a major paper, a successful recital, or an athletic victory—we may role-take and imagine the positive reactions of family and close friends. This evokes the emotion of pride. Conversely, when we fail at an important task, or engage in behavior that violates group norms, we may imagine others' reactions and feel shame or guilt. Less serious violations, such as eating the salad with the wrong fork, may evoke embarrassment. In each of these cases, viewing one's own behavior from the perspective of others arouses emotion. These emotions play an important role in social control. Our ability to anticipate the reactions of others and the resulting emotion we feel leads to greater effort in the face of difficulty, or self-control that overrides the impulse to violate social norms or shirk a responsibility.

Empathic role-taking emotions are evoked by placing yourself in the position of another. Hearing a survivor recount the devastation caused by a tornado or the injuries and deaths caused by a bomb elicits sadness and perhaps tears in the listener. Reading a tale of great achievement or heroism may elicit feelings of joy and accomplishment. Empathic role taking may be the source of motivation to prevent future suffering or achieve a great feat. The ability to experience an empathic emotion depends on the situation and on characteristics of the observer. In a study of empathic embarrassment, groups of women observed another woman perform a very embarrassing or an ambiguous task (Marcus, Wilson, & Miller, 1996). Most women reported feeling embarrassed as they observed another perform the embarrassing task. Some observers of the innocuous task also reported feeling embarrassed, suggesting that people vary in their readiness to perceive emotion in others.

Self-Esteem

Do you have a positive attitude about yourself, or do you feel you do not have much to be proud of? Overall, how capable, successful, significant, and worthy are you? Answers to these questions reflect **self-esteem,** the evaluative component of the self-concept (Gecas & Burke, 1995).

This section addresses four questions: (1) How is self-esteem assessed? (2) What are the major sources of self-esteem? (3) How is self-esteem related to behavior? (4) What techniques do we employ to protect our self-esteem?

Assessment of Self-Esteem

Our overall self-esteem depends on how we evaluate our specific role identities and personal qualities. We evaluate each as relatively positive or negative. For instance, you may consider yourself a competent athlete and a worthy friend, but an incompetent debater and an unreliable employee. According to theory, our overall level of self-esteem is the product of these individual evaluations, with each identity weighted according to its salience (Rosenberg, 1965; Sherwood, 1965).

Ordinarily, we are unaware of precisely how we combine and weight the evaluations of our specific identities. If we weight our positively evaluated identities as more important, we can maintain a high level of overall self-esteem while still admitting to certain weaknesses. If we weight our negatively evaluated identities heavily, we will have low overall self-esteem even though we have many valuable qualities.

Sources of Self-Esteem

Why do some of us enjoy high self-esteem and others suffer low self-esteem? To help answer this question, consider three major sources of self-esteem—family experience, performance feedback, and social comparisons.

Family Experience As you might expect, parent-child relationships are important for the development of self-esteem. From an extensive study of the family experiences of fifth and sixth graders, Coopersmith (1967) concluded that four types of parental behavior promote higher self-esteem: (1) showing acceptance, affection, interest, and involvement in children's affairs; (2) firmly and consistently enforcing clear limits on children's behavior; (3) allowing children latitude within these limits and respecting initiative (children setting their own bedtime and participating in making family plans); and (4) favoring noncoercive forms of discipline (denying privileges and discussing reasons, rather than punishing physically). Findings from a representative sample of 5,024 New York high school students corroborate these conclusions (Rosenberg, 1965).

Family influences on self-esteem confirm the idea that the self-concepts we develop mirror the view of ourselves communicated by significant others. Children who see that their parents love, accept, care about, trust, and reason with them come to think of themselves as worthy of affection, care, trust, and respect. Conversely, children who see that their parents do not love and accept them may develop low self-esteem. A longitudinal study of adolescents found that excessive parental shaming and criticism were associated with low self-esteem and depression (Robertson & Simons, 1989).

Research also suggests that self-esteem is produced by the reciprocal influence of parents and their children on each other (Felson & Zielinski, 1989). Children with higher self-esteem exhibit more self-confidence, competence, and self-control. Such children are probably easier to love, accept, reason with, and trust. Consequently, they are likely to elicit responses from their parents that further promote self-esteem.

As young people move into adolescence, their overall or global self-esteem becomes linked to the self-evaluations tied to specific role identities. A study of 416 sixth graders found that evaluations of self as athlete, son/daughter, and student were positively related to global self-esteem (Hoelter, 1986). Also, the number of significant others expands to include friends and teachers, in addition to parents. The relative importance of these others appears to vary by gender. A study of 1,367 high school seniors found that the perceived appraisals of friends had the biggest impact on females' self-esteem, whereas perceived appraisals of parents had the biggest impact on males' self-esteem (Hoelter, 1984). For both males and females, teachers' appraisals were second in importance.

Not everyone can win an Olympic gold medal. But, for all of us, an inner sense of self-esteem depends on experiencing ourselves as causal agents who make things happen, overcome obstacles, and attain goals.

Performance Feedback   Everyday feedback about the quality of our performances—our successes and failures—influences our self-esteem. We derive self-esteem from experiencing ourselves as active causal agents who make things happen in the world, who attain goals and overcome obstacles (Franks & Marolla, 1976). In other words, self-esteem is based partly on our sense of efficacy—of competence and power to control events (Bandura, 1982c). People who hold low-power positions (such as clerks, unskilled workers) have fewer opportunities to develop efficacy-based self-esteem because such positions limit their freedom of action. Even so, people seek ways to convert almost any kind of activity into a task against which to test their efficacy and prove their competence (Gecas & Schwalbe, 1983). In this way they obtain performance feedback useful for building self-esteem.

Social Comparison   To interpret whether performances represent success or failure, we must often compare them with our own goals and self-expectations or with the performances of others. Getting a B on a math exam, for example, would raise your sense of math competence if you had hoped for a C at best, but it would shake you if you were counting on an A. The impact of the B on your self-esteem also would vary, depending on whether most of your friends got A's or C's.

Social comparison is crucial to self-esteem because the feelings of competence or worth we derive from a performance depend in large part on whom we are compared with, both by ourselves and others. Even our personal goals are largely derived from our aspirations to succeed in comparison with people we admire. We are most likely to receive evaluative feedback from others in our immediate social context—our family, peers, teachers, and work associates. We are also most likely to compare ourselves with these people and with others who are similar to us (Festinger, 1954; Rosenberg & Simmons, 1972).

A study of job applicants clearly demonstrates the effect of social comparison on self-esteem (Morse & Gergen, 1970). After each applicant had completed a set of forms, including a self-esteem scale, another applicant entered the waiting room. For half the participants, the second applicant wore a dark business suit, carried an attaché case, and communicated an aura of competence. The remaining participants each waited with an applicant who wore a smelly sweatshirt and no socks, and appeared dazed. Several minutes later, while still in the presence of the highly impressive or unimpressive competitor, applicants completed additional forms, including a second self-esteem scale. Applicants exposed to the obviously inferior competitor revealed a large increase in self-esteem from the first to the second self-esteem measurement; among those faced with the impressive competitor, self-esteem dropped substantially.

Losing one's job is generally interpreted as a serious failure in our society. A national survey of American employees reveals that job loss undermined self-esteem, but the size of the drop in self-esteem depended on social comparison (Cohn, 1978). In neighborhoods with little unemployment, persons who lost their jobs suffered a large drop in self-esteem. In neighborhoods where many others were unemployed too, the drop was less. This difference points to the importance of the immediate social context for defining success/failure. The impact of social comparison suggests that members of minorities that are discriminated against should have low self-esteem. However, the research summarized in Box 4.2, "Minority Status and Self-Esteem," indicates that this is not true.

Self-Esteem and Behavior

People with high self-esteem often behave quite differently from those with low self-esteem. Research findings indicate that high self-esteem is associated with active and comfortable social involvement, whereas low self-esteem is a depressing and debilitating state (Coopersmith, 1967; Rosenberg, 1979; Wylie, 1979).

Compared with those having low self-esteem, children, teenagers, and adults with higher self-esteem are socially at ease and popular with their peers. They are more confident about their own opinions and judgments and more certain of their perceptions of self (Campbell, 1990). They are more vigorous and assertive in their social relations, more ambitious, and more academically successful. During their school years, those with higher self-esteem participate more in extracurricular activities, are elected more frequently to leadership roles, show greater interest in public affairs, and have higher occupational aspirations. Persons with high self-esteem achieve higher scores on measures of psychological well-being (Rosenberg et al., 1995). Adults with high self-esteem experience less stress following the death of a spouse and cope with the resulting problems more effectively (Johnson, Lund, & Dimond, 1986).

The picture of people with low self-esteem forms an unhappy contrast. People low in self-esteem tend to be socially anxious and ineffective. They view interpersonal relationships as threatening, feel less positively toward others, and are easily hurt by criticism. Lacking confidence in their own judgments and opinions, they yield more readily in the face of opposition. They expect others to reject them and their ideas and have little faith in their ability to achieve. In school, they set lower goals for themselves, are less successful academically, less active in the classroom and in extracurricular activities, and less popular. People with lower self-esteem appear more depressed and express more feelings of unhappiness and discouragement. They more frequently manifest symptoms of anxiety, poor adjustment, and psychosomatic illness.

Most of these contrasts are drawn from comparisons between naturally occurring groups of people who report high or low self-esteem. It is difficult to determine, therefore, whether self-esteem causes these behavior differences or vice versa. For example, high self-esteem may enable people to assert their opinions more forcefully and thus to convince others. But the

4.2 Minority Status and Self-Esteem

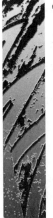

Members of racial, religious, and ethnic minorities may have special problems in developing positive self-esteem. Because of prejudice, minority group members are likely to see a negative image of themselves reflected in appraisals by members of other groups. When they make social comparisons of their own educational, occupational, and economic success with that of the majority, they are likely to compare unfavorably. Therefore, we might assume that members of minority groups interpret their performances and failures to achieve as evidence of their basic lack of worth and competence, that they have low self-esteem.

Is this hypothesis true? Hundreds of studies have sought to determine whether minority status undermines self-esteem in America (Porter & Washington, 1993; Wylie, 1979). The vast majority of studies offer little support for the conclusion that minorities (racial, religious, or ethnic) have significantly lower self-esteem. Further, research suggests that self-esteem among racial and ethnic minorities has two components. One is **group self-esteem,** how the person feels as a member of a racial or ethnic group. The other is *personal self-esteem,* how the person feels about the self (Porter & Washington, 1993).

With regard to personal self-esteem, studies indicate that African Americans have as high self-esteem as whites (Rosenberg & Simmons, 1972; Rotheram-Barus, 1990; Simmons et al., 1978). Reflected appraisals from significant others affect minority group members just as they do whites. The self-esteem of black schoolchildren is strongly related to their perception of what their parents, teachers, and friends think of them. These appraisals are not negative (Rosenberg, 1973, 1990). Living in segregated neighborhoods, minority group children usually see themselves through the unprejudiced eyes of

their own group, not the prejudiced eyes of others. The self-esteem of black adults is related to the quality of their relationships with family and friends and their involvement in religion (Hughes & Demo, 1989). Few studies have been done of the personal esteem of Mexican Americans, Puerto Ricans, and Asian Americans; however, there is no reason to believe that members of these groups have low self-esteem.

Group self-esteem, in contrast, is not associated with reflected appraisals. Among black Americans, group esteem includes black consciousness, identification of self with other blacks, and support for independent black politics. High group esteem among blacks is associated with higher education and more frequent contact with whites, not with relationships with family and friends (Demo & Hughes, 1990). Research indicates that Puerto Ricans, Mexican Americans, and Asian Americans have high levels of group esteem (Porter & Washington, 1993). Other data suggest that when members of these groups receive negative feedback from members of other groups, they attribute it to racial prejudice (Crocker et al., 1991). One study suggests that many persons are reclaiming American Indian group identity because of governmental policies that are making resources available and Indian political activism (Nagel, 1995).

But what about the effects of social comparisons? Many minority persons are disadvantaged in terms of education, occupation, and income. Minority individuals do compare themselves with the majority, but they often do not blame themselves for their disadvantaged position. Minorities can protect their personal self-esteem by blaming the system of discrimination for their lesser accomplishments. Indeed, minority statuses such as race, religion, and ethnicity show virtually no association with self-esteem (Jacques & Chason, 1977; Rotheram-Barus, 1990). Social failure affects self-esteem only when people attribute it to poor individual achievement (Rosenberg & Pearlin, 1978).

experience of influencing others, in turn, may increase self-esteem. Thus reciprocal influence, rather than causality from self-esteem to behavior, is probably most common (Rosenberg, Schooler, & Schoenbach, 1989).

Protecting Self-Esteem

What grade would you like to get on your next exam in social psychology—an A or a C? Your answer depends, in part, on whether your self-esteem is high or low. We often think that everybody wants positive feedback from others, to have others like them, to be successful, in other words, to experience *self-enhancement*. As noted in the last section, people with high self-esteem expect to perform well and usually do. People with low self-esteem, in contrast, expect to perform poorly and usually do. People are motivated to protect their self-esteem, whether it is high or low. Most people have high self-esteem and want self-enhancing feedback. Some people have low self-esteem; to verify their self-evaluation, they want self-derogating feedback.

People use several techniques to maintain their self-esteem. We will examine four of them (McCall & Simmons, 1978).

Manipulating Appraisals

We choose to associate with people who share our view of self and avoid people who do not. For example, a study of interaction in a college sorority revealed that women associated most frequently with those they believed saw them as they saw themselves (Backman & Secord, 1962). People with negative self-views seek people who think poorly of them (Swann & Predmore, 1985).

Another way to maintain our self-esteem is by interpreting others' appraisals as more favorable or unfavorable than they actually are. For instance, college students took an analogies test and subsequently were given positive, negative, or no feedback about their performance (Jussim, Coleman, & Nassau, 1987). Each student then completed a questionnaire. Students who were high in self-esteem perceived the feedback, whether positive or negative, as more positive than students low in self-esteem.

Selective Information Processing

Another way we protect our self-esteem is by attending more to those occurrences that are consistent with our self-evaluation. In one study, subjects high or low in self-esteem performed a task; they were then told either that they succeeded or that they failed at the task. On a later self-rating, all the subjects gave biased ratings. High self-esteem subjects who succeeded increased their ratings, whereas their low self-esteem counterparts did not. Low self-esteem subjects who failed gave themselves lower ratings, whereas high self-esteem subjects who failed did not (Schlenker, Weigold, & Hallam, 1990). Memory also acts to protect self-esteem. People with high esteem recall good, responsible, and successful activities more often, and those with low esteem are more likely to remember bad, irresponsible, and unsuccessful ones.

Selective Social Comparison

When we lack objective standards for evaluating ourselves, we engage in social comparison (Festinger, 1954). By carefully selecting others with whom to compare ourselves, we can further protect our self-esteem. We usually compare ourselves with persons who are similar in age, sex, occupation, economic status, abilities, and attitudes (Suls & Miller, 1977; Walsh & Taylor, 1982). We tend to avoid comparing ourselves with the class valedictorian, homecoming queen, or star athlete, thereby forestalling a negative self-evaluation.

Once people make a social comparison, they tend to overrate their relative standing (Felson, 1981). This is illustrated by self-ratings obtained from a large sample of American adults (Heiss & Owens, 1972). Only 2% rated themselves "below average" as parents, spouses, sons or daughters, or in the qualities of trustworthiness, intelligence, and willingness to work. The latter were probably people with low self-esteem.

Selective Commitment to Identities

Still another technique involves committing ourselves more to those self-concepts that provide feedback consistent with self-evaluation and downgrading those which provide feedback that challenges it. This protects overall self-esteem because self-evaluation is based most heavily on those identities and personal qualities we consider most important.

People tend to enhance self-esteem by assigning more importance to those identities (religious, racial, occupational, family) they consider particularly admirable (Hoelter, 1983). They also increase or decrease identification with a social group when the group becomes a greater or lesser potential source of esteem (Tesser & Campbell, 1983). In one study, students were part of a group that either succeeded or failed at a task (Snyder, Lassegard, & Ford, 1986). On measures of identification with the group, students belonging to a successful group claimed closer association (that is, basked in the reflected glory), whereas those in an unsuccessful group distanced themselves from the group. Similarly, students are more apt to wear clothing that displays university affiliation following a football victory rather than a defeat. They also identify more with their school when describing victories ("We won") than defeats ("They lost"), thereby enhancing or protecting self-esteem (Cialdini et al., 1976).

People who want to verify low self-esteem behave differently. Low self-esteem participants who were members of a successful group downplayed their connection to the group and minimized their contribution to its success. Low self-esteem participants were more likely to link themselves to the successful group when they were *not* members of it (Brown, Collins, & Schmidt, 1988).

All four techniques for protecting self-esteem described here portray human beings as active processors of social events. People do not accept social evaluations passively or allow self-esteem to be buffeted by the cruelties and kindnesses of the social environment. Nor do successes and failures directly affect

The self-esteem of many sports fans is influenced in part by the performance of their team. If the team wins, they can boost their self-esteem by identifying strongly with the team. If the team loses, fans are likely to protect their self-esteem by identifying less with the team.

self-esteem. The techniques described here testify to human ingenuity in selecting and modifying the meanings of events in the service of self-esteem.

Summary

The self is the individual viewed both as the source and the object of reflexive behavior.

The Nature and Genesis of Self   (1) The self is the source of action when we plan, observe, and control our own behavior. The self is the object of action when we think about who we are. (2) Newborns lack a sense of self. Later, they come to recognize that they are physically separate from others. As they acquire language, they learn that their own thoughts and feelings are also separate. (3) Through role taking, children come to see themselves through others' eyes. They can then observe, judge, and regulate their own behavior. (4) Children construct their identities based on how they imagine they appear to others. They also develop self-evaluations based on the perceived judgments of others.

Identities: The Self We Know   The self we know includes multiple identities. (1) Some identities are linked to social roles we enact. (2) Some identities are linked to our membership in social groups or categories. These may be associated with in-group favoritism and out-group stereotyping. (3) Individuals mention role identities very often when describing themselves. They also mention personal qualities and self-evaluations. (4) We form self-concepts primarily through learning and adopting identities. The self we know is primarily influenced by the perceived reactions of others. (5) The self we know varies with the situation. We attend most to those aspects of our selves that are distinctive and relevant to the ongoing activity.

Identities: The Self We Enact   The self we enact expresses our identities. (1) We choose behaviors to evoke responses from others that will confirm particular identities. To confirm identities successfully, we must share with others our understanding of what these behaviors and identities mean. (2) We choose which identity to express based on that identity's salience, need for support, and situational opportunities for enacting it. (3) We gain consistency in our behavior over time by striving to enact important identities. We also employ several strategies that lead to verification of our self-conceptions.

The Self in Thought and Feeling The self affects both thought and feeling. (1) We perceive and process information more effectively if it relates to our important identities. We learn and remember information better if it relates to the self. (2) When attention is drawn to the self, we become self-aware and take greater control over our behavior. We then conform more with our personal standards and with salient social standards. (3) Discrepancies between components of the self may cause sadness, depression, fear, or restlessness. (4) The emotions we experience also depend on the self. Our identities cause us to notice particular sensations and influence the interpretations and emotional labels we apply to them. When our emotions seem socially inappropriate, we try to modify our feelings. Role taking may evoke a variety of emotions, including pride, shame, sadness, and joy.

Self-Esteem Self-esteem is the evaluative component of self. Most people try to maintain positive self-esteem. (1) Overall self-esteem depends on the evaluations of our specific identities. (2) Self-esteem derives from three sources: family experiences of acceptance and discipline, direct feedback on the effectiveness of actions, and comparisons of our own successes and failures with those of others. (3) People with higher self-esteem tend to be more popular, assertive, ambitious, academically successful, better adjusted, and happier. (4) We employ numerous techniques to protect self-esteem. Specifically, we seek reflected appraisals consistent with our self-view, process information selectively, carefully select those with whom we compare ourselves, and attribute greater importance to qualities that provide consistent feedback.

Key Terms

emotion work (p. 92)

feeling rules (p. 92)

generalized other (p. 79)

group self-esteem (p. 96)

identity (p. 80)

role identity (p. 80)

role taking (p. 78)

salience (p. 85)

self (p. 76)

self-awareness (p. 89)

self-discrepancy theory (p. 90)

self-esteem (p. 93)

self-schema (p. 74)

sentiments (p. 91)

significant others (p. 78)

situated self (p. 83)

social identity (p. 81)

CHAPTER 5
Social Perception and Cognition

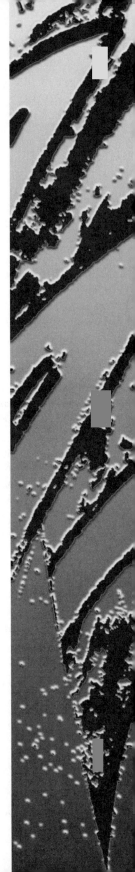

Introduction

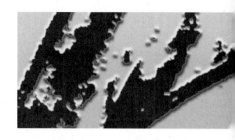

It is 10 P.M., and the admitting physician at the psychiatric hospital is interviewing a respectable-looking man who has asked for treatment. "You see," the patient says, "I keep on hearing voices." After taking a full history, the physician diagnoses the man as a schizophrenic and assigns him to an inpatient unit. The physician is well trained and makes the diagnosis with apparent ease. Yet to diagnose correctly someone's mental condition is a difficult problem in social perception. The differences among paranoid, schizophrenic, depressed, and normal are not always easy to discern.

In one unusual study (Rosenhan, 1973), eight pseudo-patients who were actually research investigators gained entry into mental hospitals by claiming to hear voices. During the intake interviews, the pseudo patients gave true accounts of their backgrounds, life experiences, and present (quite ordinary) psychological condition. They falsified only their names and their complaint of hearing voices. Once in the psychiatric unit, the pseudo patients stopped simulating symptoms of abnormality. They reported that the voices had stopped, talked normally with other patients, and made observations in their notebooks. Although some other patients suspected that the investigators were not really ill, the staff continued to believe they were. Even upon discharge, the pseudo patients were still diagnosed as schizophrenic, although now it was "schizophrenia in remission."

A person who presents himself to a psychiatric hospital for admission may pose a difficult problem in perception for the hospital staff. Is he really "mentally ill" and in need of hospitalization, or is he "healthy"? Is he no longer able to function in the outside world? Or is he merely faking and trying to get a break from his work or his family?

To determine what kind of person the potential patient really is, the admitting physician must gather information about the person and classify it as indicating illness or health. Then the doctor must combine these facts to form a general diagnosis (paranoia, schizophrenia, or depression) and specify what form of treatment the person needs. While performing these actions, the doctor is engaging in social perception. Broadly defined, **social perception** refers to constructing an understanding of the social world from the data we get through our senses. More narrowly defined, social perception refers to the processes by which we form impressions of other people's traits and personalities.

In making his or her diagnosis, the physician not only forms an impression about the traits and characteristics of the new patient, but he or she also tries to understand the causes of that person's behavior. He or she tries, for instance, to figure out whether the patient acts as he does because of some internal dispositions or because of external pressures from the environment. Social psychologists term this process *attribution.* In attribution, we observe others' behavior and then infer backward to causes—intentions, abilities, traits, motives, and situational pressures—that explain why people act as they do.

Social perception and attribution involve more than passively registering the stimuli that impinge on our senses. Our expectations and cognitive structures influence what we notice and how we interpret it. The intake physician at the psychiatric hospital, for example, expects to meet people who are mentally ill, not research investigators. So he or she gathers information and interprets it in ways based on that expectation.

Most of the time, the impressions we form of others are sufficiently accurate to permit smooth interaction. Yet, as our example shows, social perception and attribution can be unreliable. Even an intelligent and skilled observer such as the intake physician can misperceive, misjudge, and reach the wrong conclusions.

This chapter focuses on the processes of social perception and attribution. It addresses the following questions:

1. How do we make sense of the flood of information that surrounds us? What functions do schemas serve in social perception?
2. Why do we use person schemas and group stereotypes? What problems do they solve and what difficulties do they create?
3. How do we form impressions of others? That is, how do we integrate the diverse (or even contradictory) information we receive about someone into a coherent, overall impression?
4. How do we ascertain the causes of other people's behavior and interpret the origins of actions we observe? For instance, when we judge someone's behavior, how do we know whether to attribute the behavior to that person's internal dispositions or to the external situation impacting that person?
5. What sorts of attributional errors do we commonly make in judging the behavior of others, and why do we make such errors?

Schemas

The human mind, among other things, is a sophisticated system for processing information. One of the most basic mental processes is **categorization,** our tendency to perceive stimuli as members of groups or classes rather than as isolated entities. For instance, at the theater, we see a well-groomed woman on stage wearing a short dress and dancing on her toes; rather than viewing her as a novel entity, we immediately categorize her as a "ballerina."

How do we go about assigning people or other entities to categories? For instance, how do we know the woman should be categorized as a "ballerina" and not as an "actress" or a "cheerleader"? To categorize some person, we usually compare that person to the prototype of the category. A **prototype** is an abstraction that represents the "typical" or quintessential instance of a class or group. Usually, prototypes are specified in terms of a set of attributes. For example, the prototype of a "cultured person" is someone who

is knowledgeable about literature, classical music, fine food, and foreign cultures and who indulges these tastes by regularly attending concerts, eating at fine restaurants, and traveling worldwide.

The categories into which we encode persons or entities are not isolated from one another. Rather, they link together in various ways and constitute a structure. For instance, we may think of a person (Jonathan) not only as having various attributes (tall, wealthy) but also as bearing certain relations with other persons or entities (friend of Caroline, stronger than Bill, owner of a Honda). These other persons or entities will themselves have attributes (Caroline: thin, athletic, brunette; Bill: short, fat, mustachioed; Honda: blue, four-door, new). They also have relations with still other persons and entities (Caroline: cousin of Bill, wife of George; Bill: friend of George, owner of a Buick). In this way, we build a cognitive structure consisting of persons, attributes, and relations.

Social psychologists use the term **schema** to denote a well-organized structure of cognitions about some social entity such as a person, group, role, or event. Schemas usually include information about an entity's attributes and about its relations with other entities. To illustrate, suppose that Martha (who is somewhat cynical about politics) has a schema about the role of "member of Congress." In Martha's schema, the occupant of such a role will insist that he (or she) serves the needs of his constituents, yet will vote for the special interests of the PACs from which he takes money, run TV advertisements containing half-truths at election time, spend more time in Washington, D.C., than in his home district, use congressional franking privileges to send out mail supporting his reelection, strive constantly to avoid scandal over ethics, vote large pay raises and retirement benefits for himself, and above all, never do anything that might lessen his power or influence.

Someone else, of course, may hold a less cynical view of politics than Martha and have a different schema about the role of "member of Congress." But, like Martha's, this schema will likely incorporate such elements as the congressman's or congresswoman's typical activities, relations, motives, and tactics. Whatever their exact content, schemas enable us to organize and remember facts, to make inferences that go beyond facts immediately available, and to assess new information (Fiske & Linville, 1980; Wilcox & Williams, 1990).

Types of Schemas

There are several distinct types of schemas, including person schemas, self-schemas, group schemas, role schemas, and event schemas (Taylor & Crocker, 1981). People use all of these in everyday life.

Person schemas are cognitive structures that describe the personalities of others. Person schemas can apply either to specific individuals (Ross Perot, Al Bundy, your mother) or to types of individuals (an introvert, a manic-depressive, a sociopath, and so on). Person schemas organize our conceptions of others' personalities and enable us to develop expectations about others' behavior.

Self-schemas are structures that organize our conception of our own qualities and characteristics (Catrambone & Markus, 1987; Markus, 1977). A self-schema consists of the dimensions that you use to think about yourself. For instance, if you conceive of yourself as independent (as opposed to dependent), you may see yourself as individualistic, unconventional, and assertive. Then if you behave in a manner consistent with your self-schema, you may refuse to accept money from your parents, refuse to ask others for help with schoolwork, take a part-time job, and the like. (Self-schemas are discussed in detail elsewhere in this book, especially in Chapter 4.)

Group schemas (or *stereotypes*) are schemas regarding the members of a particular social group or social category (Hamilton, 1981). Stereotypes indicate the attributes and behaviors considered typical of members of that group or social category. American culture includes a wide variety of stereotypes, including those of different races (blacks, Hispanics, Asians), religious groups (Protestants, Catholics, Jews), and ethnic groups (Germans, Irish, Poles, Greeks, Italians).

Role schemas indicate what attributes and behaviors are typical of persons occupying a particular role in a group. Our example of Martha's conception of the role of member of Congress illustrates a role schema. Role schemas exist for most occupational roles—nurses, cab drivers, store managers, and the like. Observers might use a role schema to understand and to predict the behavior of the role occupant.

Event schemas (or *scripts*) are schemas regarding important, recurring social events (Abelson, 1981; Hue & Erickson, 1991; Schank & Abelson, 1977). In our society, these events include weddings, funerals, graduation ceremonies, job interviews, cocktail parties, first dates, and the like. An event schema specifies the activities that constitute the event, the predetermined order or sequence for these activities, and the persons (or role occupants) participating in the event. Scripts can be revealed by asking people to describe what typically happens during an event. In one study, researchers asked male and female college students to describe the typical sequence of activities on a first date (Rose & Frieze, 1993). There was substantial agreement between males and females, as shown in Table 5.1. Several activities were mentioned by more than half of the participants, including groom and dress, pick up date, and take date home. Several activities were included in both male and female scripts, including worry about appearance, leave, confirm plans, eat, and go home. Reflecting the impact of gender roles, both men and women agreed that the male would take the initiative in picking up the date, taking the date home, and kissing her goodnight. Notice that a host of activities were not mentioned, such as taking a driver's license test or going to the dentist; these are not appropriate for a first date. In addition, note that the script specifies a sequence or expected order for the various activities—the man will not kiss the woman goodnight before they eat dinner.

Schematic Processing

Schemas enable us to process large amounts of information efficiently. They do this in several ways. First, they influence our capacity to recall information by making certain kinds of facts more salient and easier to remember. In addition, they guide our inferences and judgments about people and things. In this section, we discuss the effects of schemas on cognition.

Table 5.1   Core Actions of the First Date Script

Script for Woman	Script for Man
GROOM AND DRESS*	
BE NERVOUS	
Worry about appearance	WORRY ABOUT APPEARANCE
PICK UP DATE (BY MAN)	PICK UP DATE
	MEET PARENTS/ ROOMMATES
Leave	Leave
Confirm plans	Confirm plans
Get to know and evaluate date	Get to know and evaluate date
TALK, JOKE, LAUGH	TALK, JOKE, LAUGH
GO TO MOVIES, SHOW, PARTY	
Eat	EAT
Take date home (by man)	TAKE DATE HOME
Kiss goodnight (by man)	Kiss goodnight
Go home	Go home

*Capital letters indicate the action was mentioned by 50% or more of the participants; lowercase letters indicate the action was mentioned by fewer than 50% of the participants.

SOURCE: Rose and Frieze, 1993.

Schematic Memory   Human memory is largely reconstructive. That is, we do not usually remember all the precise details of what transpired in a given situation or setting as if we were some kind of motion picture camera recording all the images and sounds. Instead, we typically remember some facts of the situation and then rely on schemas to fill in other details. Schemas organize information in memory and therefore affect what we remember and what we forget (Hess & Slaughter, 1990; Sherman, Judd, & Park, 1989). When trying to recall something, people often remember better those facts that are consistent with their schemas. For instance, one study (Cohen, 1981) investigated the impact of an occupational role

schema on recall. Subjects viewed a videotape of a woman celebrating her birthday by having dinner with her husband at home. Half the subjects were told the woman was a librarian; the other half were told she was a waitress. Some characteristics of the woman were consistent with the schema of a librarian. She wore glasses, had spent the day reading, had previously traveled in Europe, and liked classical music. Other characteristics of the woman, however, were consistent with the schema of a waitress. She drank beer, had a bowling ball in the room, ate chocolate birthday cake, and flirted with her husband. Later on, when subjects tried to recall details of the videotape, they recalled most accurately those facts consistent with the woman's occupational label. That is, subjects who thought she was a librarian remembered facts consistent with the librarian schema, whereas those who thought she was a waitress remembered facts consistent with the waitress schema.

What about memory for material inconsistent with schemas? Several studies have tested recall of three types of information: material consistent with schemas, material contradictory to schemas, and material irrelevant to schemas. Results show that people recall both schema-consistent and schema-contradictory material better than schema-irrelevant material. In other words, material consistent with expectations or contrary to expectations is recalled better than material that is simply irrelevant to the schema (Cano, Hopkins, & Islam, 1991; Higgins & Bargh, 1987). Persons recall schema-contradictory material better when the schema itself is concrete rather than abstract (Pryor, McDaniel, & Kott-Russo, 1986); for instance, a person schema formulated in concrete terms (spends money wisely, often tells lies, brags about her accomplishments) provides a better basis for recall than one built in abstract terms (practical, dishonest, egotistical). In addition, persons who are expert on a topic or who hold quickly accessible schemas are more likely to recall schema-contradictory material than are persons who are not expert (Bargh & Thein, 1985; Fiske, Kinder, & Larter, 1983; Higgins & Bargh, 1987).

Schematic Inference   Schemas affect the inferences we make about persons and other social entities (Fiske & Taylor, 1991). That is, they supply missing facts when gaps exist in our knowledge. If we know certain facts about a person but are ignorant about others, we fill in the gaps by inserting suppositions consistent with our schema for that person. For example, knowing your roommate is a nonsmoker, you can infer he will not want to spend time with your new friend who smokes. Of course, use of schemas can lead to erroneous inferences. If the schema is incomplete or does not correctly mirror reality, then some mistakes are likely.

Schemas—especially well-developed schemas—can help us infer new facts. For instance, if a physician diagnoses a patient as having chicken pox, he may be able to make various inferences about the disease. These may include how the patient contracted the disease, the full range of symptoms that will be present, what side effects or complications might arise, what treatment will be effective, and the like. For another person who has no schema regarding this disease, inferences of this type would be virtually impossible.

Schematic Judgment
Schemas can influence our judgments or feelings about persons and other entities. For one thing, the schemas themselves may be organized in terms of evaluative dimensions; this is especially true of person schemas. For another thing, the level of complexity of our schemas affects our evaluations of other persons. Greater schematic complexity leads to less extreme judgments. That is, the greater the complexity of our schemas about groups of people, the less extreme are our evaluations of persons in those groups. This is called the *complexity-extremity effect.*

For instance, in one study (Linville & Jones, 1980), white college students evaluated a person applying for admission to law school. Depending on treatment, the applicant was either white or black and had an academic record that was either strong or weak. Results show an interaction effect between academic record and race. Subjects rated a weak black applicant more negatively than a weak white applicant, but they rated a strong black applicant more positively than a strong white applicant. Judgments about black applicants were more extreme (in both directions) than those about white applicants because the subjects' schema for their own in-group (whites) was more

complex than their schema for the out-group (blacks). Further research (Linville, 1982) shows that the complexity-extremity effect also holds for other attributes such as age. College students have less complex schemas for the elderly than for persons their own age, so they are more extreme in judgments of the elderly.

Drawbacks of Schematic Processing
Although schemas provide certain advantages, they also entail some corresponding disadvantages. These drawbacks include the following: First, people are overly accepting of information that fits consistently with a schema. In fact, some research suggests that perceivers show a *confirmatory bias* when collecting new information relevant to schemas (Higgins & Bargh, 1987; Snyder & Swann, 1978). That is, when given an opportunity to obtain new information to test hypotheses based on schemas, perceivers tend to ask questions that will obtain information supportive of the schemas rather than questions that will obtain information disconfirmatory of the schemas.

Second, when faced with missing information, people fill in gaps in knowledge by adding elements that are consistent with their schemas. Sometimes these added elements turn out to be erroneous or factually incorrect. When this happens, it will, of course, create inaccurate interpretations or inferences about people, groups, or events.

Third, because people are often reluctant to discard or revise their schemas, they occasionally apply schemas to persons or events even when the schemas do not fit the facts very well. Forced misapplication of a schema may lead to incorrect characterization and inferences, and this in turn can produce inappropriate or inflexible responses toward other persons or groups.

Schemas as Cultural Elements

The concept of *schema* is important within social psychology for several reasons. At the individual level, schemas clearly affect memory, inference, and judgment. Beyond this, schemas are cultural elements (Forgas & Bond, 1985; Harris et al., 1988). That is, schemas are socially shared patterns of thought that can be communicated intact from one person to another. Schemas about types of persons, groups, and events are often shared by members of a group. In fact, holding specific schemas may be a characteristic of group membership. Members of another group may have different schemas about the same types of persons or events.

Groups differ in the schemas used by their members. For example, when a professional football player leaves one team and joins another, he has to learn the offensive and defensive scripts (plays) used by his new team. These may or may not resemble those used by his old team. Because schemas shared by members affect their capacity to work together, differences between members in the schemas used may produce differences in team performance.

Differences in schemas can affect the relations between two (or more) groups. If the schemas used by one group differ materially from the schemas used by another, the two groups may have difficulty finding common ground. Under these conditions, schematic processing and stereotyping may lead to conflict or inhibit alliances between groups. In one study, for example, two pro-environment groups were discussing whether to form a coalition. One group usually made its decisions by majority vote, whereas the other made decisions by reaching consensus after thorough discussion. When observers from the first group attended a long meeting held by the second, it appeared inefficient and aimless to them; they went back to their own group and recommended against joining the proposed coalition (Lichterman, 1995).

Person Schemas and Group Stereotypes

In this section, we describe in some detail two important types of schemas—person schemas and group stereotypes. We review the nature, the content, and the origins of these schemas.

Person Schemas

As noted earlier, person schemas are cognitive structures that describe the personalities of other individu-als. There are several distinct types of person schemas. Some person schemas are very specific and pertain to particular people. For example, Carolyn married George 3 years ago and she knew him for 4 years before that. By this time she has an elaborate schema regarding George, and she can usually predict how he will react to new situations, opportunities, or problems. Similarly, we often have individual schemas for public figures (Earvin "Magic" Johnson—outstanding basketball player, Olympic star, black, talk-show host, HIV positive) or for famous historical personages (Abraham Lincoln—political leader during the Civil War, honest, determined, opposed to slavery, committed to holding the Union together).

Other person schemas are very abstract and focus on relations among personality traits. A schema of this type is an **implicit personality theory**—a set of unstated assumptions about which personality traits are correlated with one another (Anderson & Sedikides, 1991; Grant & Holmes, 1981; Sternberg, 1985). These theories often include beliefs about what behaviors are associated with various personality traits (Skowronski & Carlston, 1989). They are considered "implicit" because in everyday life we usually do not subject our person schemas to close examination, nor are we explicitly aware of their contents. Implicit personality theories are important because they influence our judgments and evaluations of other people (Wegner & Vallacher, 1977).

Implicit Personality Theories and Mental Maps   As do all schemas, implicit personality theories enable us to make inferences that go beyond the information given. Instead of withholding judgment, we use them to flesh out our impressions of a person about whom we have little information. For instance, if we learn someone has a warm personality, we might infer she is also likely to be sociable, popular, good-natured, and so on. If we hear that somebody else is pessimistic, we may infer he is humorless, irritable, and unpopular, even though we lack evidence that he actually has these traits.

We can depict an implicit personality theory as a *mental map* indicating the way traits are related to one another. Figure 5.1 displays such a mental map. Based on judgments made by college students,

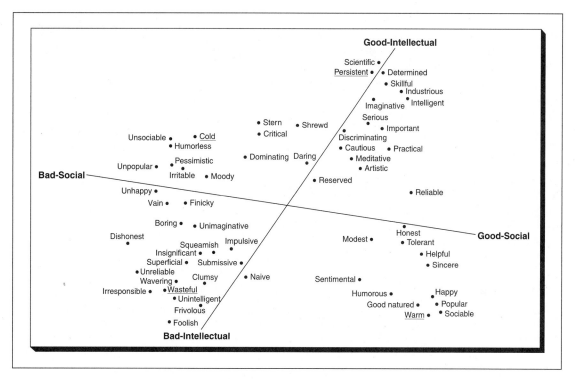

Figure 5.1 Relationships Among Attributes: A Mental Map

Each of us has an implicit theory of personality—a theory about which personality attributes tend to go together and which do not. We can represent our theories of personality in the form of a mental map. The closer attributes are located to each other on our mental map, the more we assume that these attributes will appear in the same person. The mental map shown above is based on the mental maps of many American college students.

SOURCE: Adapted from Rosenberg, Nelson, and Vivekananthan, 1968.

this figure shows how various personality traits stand in relation to one another (Rosenberg, Nelson, & Vivekananthan, 1968). Traits thought to be similar are located close together within our mental map, meaning that people who have one probably have the other. Traits thought to be dissimilar are located far apart, meaning they rarely occur together in one person.

If your mental map resembles the one portrayed in Figure 5.1, you think that people who are wasteful are also likely to be unintelligent and irresponsible (see the lower left part of the map). In addition, you think that people who are persistent are also likely to be determined and skillful (the upper right part of the

map). When you observe that a person has a particular trait, you infer the person possesses traits close to it on the mental map.

As portrayed within some mental maps, personality traits fall along two distinct evaluative dimensions—a social dimension and an intellectual dimension. These dimensions are represented by the lines shown in Figure 5.1. For instance, the traits "warm" and "cold" differ mainly on the social dimension, whereas "frivolous" and "industrious" differ on the intellectual dimension (Rosenberg & Sedlak, 1972). Some traits (such as "important") are good on both the social and the intellectual dimensions; other traits (such as "unreliable") are bad on both. In fact, traits tend to be either good on both dimensions or bad on both dimensions.

That traits are arranged along evaluative dimensions in mental maps explains a common bias in impression formation. We tend to judge persons who have several good traits as generally good and those who have several bad traits as generally bad. Once we have a global impression of someone as, say, generally good, we assume (in the absence of explicit information to the contrary) that other positive traits (located nearby in the mental map) also apply to that person. The tendency of our general or overall liking for a person to influence our subsequent assessment of more specific traits of that person is called the **halo effect** (Lachman & Bass, 1985; Thorndike, 1920). The halo effect produces bias in impression formation; it can lead to inaccuracy in our ratings of others' traits and performances (Cooper, 1981; Fisicaro, 1988).

The particular mental map described above (from Rosenberg et al., 1968) organizes traits in terms of two dimensions. However, many different researchers have constructed mental maps from trait data (Conley, 1985; Goldberg, 1981). Depending on the statistical techniques used, the maps produced sometimes have more than two dimensions. For instance, McCrae and Costa (1987) uncovered five dimensions—neuroticism, extraversion, openness, agreeableness, and conscientiousness. Three of these (neuroticism, extroversion, agreeableness) correspond roughly to Rosenberg and colleagues' social traits, and the others (openness, conscientiousness) correspond to intellectual traits. Viewed another way, McCrae and Costa's five-dimension map is a more elaborated implicit personality theory than the Rosenberg group's two-dimension map.

Origins of Person Schemas   Where do implicit personality theories come from? First, they stem from the distilled wisdom of our socializers, who use expressions like "power corrupts," "ignorance is bliss," and "tall, dark, and (fill in the blank)." These expressions tell us which traits go together. Second, and probably more important, the very meanings of words in our language suggest certain relationships. The word "generous" implies both "helpful" and "sympathetic"; "fickle" implies "unreliable" and "unpredict-

able." As a result, when one word applies to a person, we assume other words that share similar meanings also apply (Shweder, 1977). Third, people we know actually do possess sets of personality characteristics that occur together; we observe their actions, infer the traits underlying these actions, and then formulate our implicit personality theories accordingly (Borkenau & Ostendorf, 1987; Stricker, Jacobs, & Kogan, 1974).

Group Stereotypes

- "The Irish are quick-tempered, drunken, pugnacious louts."

- "Blacks are lazy and unreliable, and can't do anything except sing and dance and run and jump."

- "Feminists are left-wing, militant, bra-burning radicals."

- "Jocks may be strong, but they're stupid."

- "Southerners are rednecked, speech-slurring, barefooted bigots."

- "Right-wing Republicans are heartless, racist, elitist reactionaries."

- "Lawyers are shrewd, contentious, overpriced troublemakers."

We have all heard remarks like these—categoric, overstated, and extreme. Each is an example of a group schema, or stereotype. A **stereotype** is a set of characteristics attributed to all members of some specified group or social category (McCauley, Stitt, & Segal, 1980; Taylor, 1981). Stereotypes simplify the complex social world. Because stereotypes are based on the simple expedient of categorizing people into groups, they enable us to make quick judgments about people when we have only minimal information. They allow us to form impressions of people and to predict their behavior merely by knowing the groups to which they belong.

Stereotypes, however, involve overgeneralization. They imply that all members of a particular group or social category have certain attributes. Although stereotypes may contain a "kernel of truth"—some of the members of the stereotyped group may have some of the imputed characteristics—it is usually not true that all members have all the imputed characteristics.

We can hardly avoid making a snap judgment about the personalities of these individuals, but are we right? Stereotypes enable us to form impressions about people merely by knowing the group to which they belong.

For this reason, stereotypes often lead to inaccurate inferences. Consider, for instance, all the persons you know of Irish descent. Perhaps one of them does, as the stereotype suggests, have a quick temper and maybe another did once get into a fistfight. It is certainly false, however, that all your Irish acquaintances spend most of their time fighting and arguing and drinking and eating potatoes. Stereotypes entail overstatement and overgeneralization.

That stereotypes are overgeneralizations does not prevent us from using them in everyday life. Many people are unaware of the impact of stereotypic beliefs on their judgments of others (Hepburn & Locksley, 1983). And, although there is nothing inherent in stereotypes that requires them to be negative, many stereotypes do contain negative elements. Of course, a few stereotypes are positive ("Asians excel at math"; "Graduate students are hard working"), but many others disparage or diminish the group stereotyped.

Common Stereotypes As the examples above suggest, in American society, some widely known stereotypes pertain to ethnic, racial, and gender groups. Ethnic (national) stereotypes held by Americans might include, for example, the view that Germans are industrious and technically minded; Italians, passionate; Irish, quick tempered; and Americans, materialistic (Karlins, Coffman, & Walters, 1969). Investigators have studied ethnic and racial stereotypes for many years, and the results show that the content of stereotypes changes over time. For in-

stance, few of us now believe, as many once did, that the typical American Indian is a drunk, the typical black is superstitious and musical, or the typical Chinese American is conservative and inscrutable. Stereotypes may not have disappeared over time, but they have changed form (Dovidio & Gaertner, 1996).

Just as stereotypes about ethnic and racial groups are commonly held in our society, so also are stereotypes about gender groups. Usually our first observation upon meeting people is to classify them as male or female. This classification is likely to activate an elaborate—although questionable—stereotype. This stereotype depicts males as more independent, dominant, competent, rational, competitive, assertive, and stable in handling crises. It characterizes females as more emotional, sensitive, expressive, gentle, helpful, and patient (Ashmore, 1981; Martin, 1987; Minnigerode & Lee, 1978). (Research on the nature of these stereotypes of males and females is discussed in Box 5.1, "Gender Stereotypes.") Within gender, stereotypes are linked to titles. For instance, the stereotype elicited by the title Ms. differs from that elicited by the title Mrs. Women labeled Ms. are seen as more achieving, more masculine, and less likable than women labeled Mrs. (Dion & Schuller, 1991). In addition to using ethnic, racial, and gender stereotypes, people also stereotype groups defined by occupation, age, political ideology, mental illness, hobbies, school attended, and so on (Milburn, 1987; Miller, 1982).

Origins of Stereotypes How do various stereotypes originate? Some theorists suggest that stereotypes arise out of direct experience with some members of the stereotyped group (Campbell, 1967). We may once have known Italians who were passionate, blacks who were musical, or Japanese who were polite. We then build a stereotype by generalizing. That is, we infer that all members of a group share the attribute we know to be characteristic of some particular members. Although this inference may be based on some true facts, it goes too far and involves overgeneralization.

Other theorists (Eagly & Steffen, 1984) suggest that stereotypes derive in part from a biased distribution of group members into social roles. Roles have associated characteristics, and eventually those characteristics are imputed to the persons occupying

5.1   Gender Stereotypes

One of the most consistent research findings on stereotypes is that many people believe males and females have different personality traits. What are the traits believed to be typical of each sex? Where do these sex stereotypes come from?

A well-known series of studies on sex stereotyping of personality characteristics began by asking male and female college students to list all the ways they thought men and women differed psychologically (Broverman et al., 1972). The researchers then constructed a measuring instrument by listing each trait the students mentioned at least twice on a scale that looked like this:

Not at all aggressive				Very aggressive		
1	2	3	4	5	6	7

Next, the researchers asked various groups of adults, ages 17 to 60, to indicate how much they thought each item characterized adult men and adult women. In numerous studies, respondents cited consistent differences between males and females for 20 traits. To see how aware you are of these stereotypes, indicate with a check on the table whether you think each of the traits listed is more typical of men or of women. Note also whether you consider each trait a desirable or undesirable one for an adult to have.

Broverman and her associates (1972) found that both men and women agree on the gender stereotypes and on the desirability of each trait. The first five traits listed in the table were seen as more typical of men, whereas the next five were seen as more typical of women. That is, men were seen as more independent, aggressive, ambitious, strong, and blunt; women were seen as more passive, emotional, easily influenced, talkative, and tactful. In general, men were perceived as stronger and more

continued on next page

Trait	Most Typical of		Desirable	
	Men	**Women**	**Yes**	**No**
Independent	___	___	___	___
Aggressive	___	___	___	___
Ambitious	___	___	___	___
Strong	___	___	___	___
Blunt	___	___	___	___
Passive	___	___	___	___
Emotional	___	___	___	___
Easily influenced	___	___	___	___
Talkative	___	___	___	___
Tactful	___	___	___	___
Excitable in minor crises	___	___	___	___
Aware of others' feelings	___	___	___	___
Submissive	___	___	___	___
Strong need for security	___	___	___	___
Feelings easily hurt	___	___	___	___
Self-confident	___	___	___	___
Adventurous	___	___	___	___
Acts as a leader	___	___	___	___
Makes decisions easily	___	___	___	___
Likes math and science	___	___	___	___

continued from previous page

confident than women, and women as weaker and more expressive than men. Recent studies find that these stereotypes have persisted over time (Bergen & Williams, 1991; Deaux & Lewis, 1983).

Broverman and associates (1972) also found that most traits stereotyped as masculine were evaluated as desirable, whereas most traits stereotyped as feminine were evaluated as undesirable. In other words, traits associated with men were usually considered to be better than those associated with women. Did your evaluations of trait desirability favor the male stereotyped traits? If not, you may fit in with a trend among educated respondents toward valuing some traditionally feminine traits (emotional) more positively and some traditionally masculine traits (ambitious) more negatively (Der-Karabetian & Smith, 1977; Pleck, 1976). This trend means that even if sex stereotypes persist, women may be evaluated less negatively than before.

Are we more likely to use stereotypes in some situations than in others? Research indicates that the answer is "Yes." One of the functions of schematic or stereotypic processing is that it "fills in the blanks." Thus the less information we have about someone, the more likely we are to rely on stereotypes. In addition, we are more likely to apply stereotypes when cues make group membership salient. Thus if we meet someone whose clothing or grooming emphasizes masculinity or femininity, we are likely to assume our stereotype fits the person. This has interesting implications for women who work in male-dominated professions. They must decide whether to emphasize their distinctive gender by wearing dresses and tasteful makeup or to empha size their similarity by wearing pantsuits and no makeup. Finally, there is a *priming effect* (Fiske & Taylor, 1991); the more recently a stereotype or schema has been used, the more accessible it is and therefore likely to be used again. Gender schemas are among the most frequently activated and therefore are highly accessible.

the roles. If members of some social group disproportionately occupy roles with associated negative characteristics, an unflattering stereotype of that group may eventually emerge. For instance, if members of a given racial group disproportionately occupy jobs that entail despised work, observers eventually may ascribe the negative characteristics of the job to members of that racial group.

Stereotyping may be a natural outcome of social perception. When people have to process and remember a lot of information about many others, they store this information in terms of group categories rather than in terms of individuals (Taylor et al., 1978). In trying to remember what went on in a classroom discussion, you may recall that several women spoke and a black person expressed a strong opinion, although you cannot remember exactly which women spoke or who the black person was. Because people remember behavior by group category rather than by individual, they are likely to form stereotypes of these groups (Rothbart et al., 1978). Remembering that women spoke and a black person expressed a strong opinion, you might infer that women are talkative in general and blacks are opinionated. You would not form these stereotypes if you recalled these attributes as belonging to individuals.

Errors Caused by Stereotypes Because stereotypes are overgeneralizations, they foster various errors in social perception and judgment. First, stereotypes lead us to assume that all members of a group are alike and possess certain traits. Yet individual members of a group obviously differ in many respects. One person wearing a hard hat may shoulder you into the stairwell on a crowded bus; another may offer you his seat. Second, stereotypes lead us to assume that all the members of one group differ from all the members of other groups. Stereotypes of football players and ballet dancers may suggest, for instance, that these groups have nothing in common. In fact, among football players as well as among ballet dancers there are individuals who are patient, neurotic, hardworking, profligate, intelligent, and so on.

Although stereotypes can produce inaccurate inferences and judgments in simple situations, they are especially likely to do so in complex situations. When the judgment to be made is multifaceted and involves a lot of complex data, reliance on stereotypes can prove particularly misleading. If an observer uses a stereotype as a central theme around which to organize information relevant to a decision, he or she may neglect information that is inconsistent with the stereotype (Bodenhausen & Lichtenstein, 1987). By neglecting information that does not fit the stereotype, the observer may overlook some crucial facts and reach a poor decision or judgment.

Although stereotypes involve overstatement and overgeneralization, they resist change even in the face of concrete evidence that contradicts them. This occurs because people tend to accept information that confirms their stereotypes and to ignore or explain away information that disconfirms them (Lord, Lepper, & Mackie, 1984; Snyder, 1981; Weber & Crocker, 1983). Suppose, for example, that Stan (who is straight) stereotypes homosexual males as effeminate, nonathletic, and artistic. If he stumbles into a gay bar, he is especially likely to notice those males who fit this description, thereby confirming his stereotype. But how does he construe any rough-looking, athletic males who are there? There are several ways he can prevent their presence from forcing a change in his stereotype. He might scrutinize them closely for hidden signs of effeminacy, underestimate their number, consider them the exceptions that prove the rule, or even assume they are straight. Through cognitive strategies like these, people explain away contradictory information and preserve their stereotypes.

Impression Formation

Information about other people comes to us from various sources. We may read facts about someone. We may hear something from a third party. From a distance, we may witness acts by the other. We may interact directly with the other and form an impression of that person based on his or her appearance, dress, speech style, or background. Regardless of how we get

bits of information about the other, we (as perceivers) must find a way to integrate these diverse facts into a coherent picture. This process of organizing diverse information into a unified impression of the other person is called *impression formation*. It is fundamental to person perception.

In this section, we discuss impression formation in some detail. First, we review some classic studies of trait centrality. Next, we discuss various formal models of information integration and the role of first impressions. Finally, we consider how impressions of others can become self-fulfilling prophesies.

Trait Centrality

In a classic experiment, Asch (1946) used a straightforward procedure to show that some traits have more impact than others on the impressions we form. Undergraduates in one group received a list of seven traits describing a hypothetical person. These were intelligent, skillful, industrious, warm, determined, practical, and cautious. Undergraduates in a second group received the same list of traits, but with one critical difference: The trait "warm" was replaced by "cold." All subjects then wrote a brief paragraph indicating their impressions and completed a checklist to rate the stimulus person on such other characteristics as generous, wise, happy, good-natured, humorous, sociable, popular, humane, altruistic, and imaginative.

The findings led to several conclusions. First, the students had no difficulty performing the task. They were able to weave the trait information into a coherent whole and to construct a composite sketch of the stimulus person. Second, substituting the trait "warm" for the trait "cold" produced a large difference in the overall impression formed by the students. When the stimulus person was "warm," the students typically described him as happy, successful, popular, and humorous. But when he was "cold," they described him as self-centered, unsociable, and unhappy. Third, the terms "warm" and "cold" had a larger impact than some other traits on the overall impression formed of the stimulus person. This was demonstrated, for instance, by a variation in which the investigator repeated the basic procedure but substituted the pair "polite" and "blunt" in place of "warm" and "cold." Whereas describing the stimulus person as "warm" (rather than "cold") made a great difference in the impressions formed by the students, describing him as "polite" (rather than "blunt") made little difference.

We say that a trait has a high level of **trait centrality** when information about a person's standing on that trait has a large impact on the overall impression we form of that person. In Asch's study, the warm-cold trait displayed more centrality than the polite-blunt trait because differences in warm-cold produced large differences in subjects' ratings of the stimulus person on other attributes, whereas differences in polite-blunt produced only minor differences in subjects' ratings.

A follow-up study (Kelley, 1950) replicated the warm-cold finding in a more realistic setting. Students in sections of a psychology course read trait descriptions of a guest lecturer before he spoke. These descriptions contained adjectives similar to those used by Asch (that is, industrious, critical, practical, determined), but they differed regarding the warm-cold variable. For half the students, the description contained the trait "warm"; for the other half, it contained "cold." The lecturer subsequently arrived at the classroom and led a discussion for about 20 minutes. Afterward, the students were asked to report their impressions of him. Results showed large differences between impressions formed by those who read he was "warm" and those who read he was "cold." Those who had read he was "cold" rated him as less considerate, sociable, popular, good-natured, humorous, and humane than those who had read he was "warm." Because all students saw the same guest instructor in the classroom, differences in their impressions could only stem from the use of "warm" or "cold" in the profile they had read.

How could a single trait embedded in a profile have such an impact on impressions of someone's behavior? Several theories have been advanced, but one plausible explanation holds that the students used a schema—a mental map—indicating what traits go with being warm and what go with being cold. Figure 5.1 helps make this concrete. Looking again at that figure, we note the locations of the attributes

"warm" and "cold" on the map and the nature of the other attributes close by. If the mental maps used by the subjects in the Asch and Kelley studies resembled Figure 5.1, it becomes immediately clear why the subjects judged the warm person as more sociable, popular, good-natured, and humorous than the cold person. These traits are close to "warm" and remote from "cold" on the mental map.

Integrating Information About Others

In everyday life, we often receive a lot of information at once. New experiences with others typically add information to impressions we already hold. Some information we receive may be inconsistent with information we already have. If your uncle, whom you view as loving and tolerant, criticizes your cousin for her sloppy clothes and stringy hair, how do you make sense of this behavior? Most likely, you would try to understand how this contradictory information squares with your earlier impression of your uncle. You might conclude, for instance, that your uncle's intolerance stemmed from his love for your cousin and his desire to protect her from others' criticism. In general, when forming an impression of a person, perceivers try to integrate information about many seemingly contradictory attributes to create a unified and coherent impression of that individual (Asch & Zukier, 1984).

Models of Information Integration

Many models of how perceivers combine information are based on a key assumption, namely, that the most important aspect of a perceiver's impression of a person is the overall positive or negative *evaluation* of that person. Although restrictive, this assumption is justified by two facts. First, empirical studies show that evaluation is the most important dimension on our mental maps of personality traits. Second, this evaluative dimension is crucial when we make practical judgments and decisions. For instance, we may reject a job applicant because we think he is unproductive (negative evaluation) or invite a new acquaintance to a party because we think she is vivacious (positive evaluation).

Assume a theorist wishes to predict the overall evaluation that a perceiver would make of a stranger who has these traits: sincere, friendly, cautious, and dishonest. To do this, the theorist needs to know several facts. First, the theorist must know how positively or negatively perceivers rate each of these traits. Most college students assign highly positive values to such traits as "sincere" (say, +3), less positive values to "friendly" (+2) and "cautious" (+1), and negative values to "dishonest" (−3) (Anderson, 1968).

Next, the theorist needs to know how perceivers combine the trait values to form an overall evaluation of the stranger. Social psychologists have proposed several different models that depict how perceivers combine information on diverse traits to form an overall evaluation of someone. One of these, termed the **additive model,** postulates that perceivers form an overall evaluation by summing the values of all the single traits. An example of the additive model appears in the top panel of Table 5.2 with four different combinations of traits. A key feature of the additive model is that when we add traits with a positive value, we increase the favorableness of our overall impression (column I vs. column II), whereas when we add traits with a negative value, we decrease favorableness (column III vs. column IV).

Although the additive model is plausible, theorists have developed other models as well. One alternative—termed the **averaging model**—postulates that perceivers form an overall evaluation by averaging the values of all the single traits (see the bottom panel of Table 5.2). In the averaging model, the impact of new information depends on whether this information is more favorable or less favorable than the overall impression we already have. Thus incorporating a new, mildly positive trait into a strongly positive impression makes the resulting impression less positive (column I vs. column II), whereas incorporating a mildly negative trait to a strongly negative impression makes the resulting impression less negative (column III vs. column IV).

Although both the additive and averaging models have some appeal, the bulk of evidence from empirical studies of impression formation supports a third model. This is a refinement of the averaging model called the **weighted averaging model** (Anderson, 1981). According to this model, perceivers average the values of the traits to form an impression, but they also give more weight to some information and less to others. For instance, if a stranger is described as "tall, dark, handsome, talkative, blue-eyed, armed, and dangerous," the last two traits on this list will overshadow the others and receive more weight in the perceiver's evaluation of the stranger.

Several factors influence the weights that perceivers assign to trait information. First, they give greater

Table 5.2 A Comparison of Additive and Averaging Models for Forming Impressions

Model	Trait Combinations			
	I	**II**	**III**	**IV**
Additive	sincere +3	sincere +3	serious +1	serious +1
	friendly +2	friendly +2	irresponsible −3	irresponsible −3
	tolerant +1	tolerant +1	dishonest −3	dishonest −3
		cautious +1		unimaginative −1
Overall impression:	+6	+7	−5	−6
Averaging	sincere +3	sincere +3	serious +1	serious +1
	friendly +2	friendly +2	irresponsible −3	irresponsible −3
	tolerant +1	tolerant +1	dishonest −3	dishonest −3
		cautious +1		unimaginative −1
Overall impression:	+2.00	+1.75	−1.67	−1.50

weight to information from highly credible sources than to information from less credible sources. Second, they weight negative attributes more heavily than positive attributes (Hamilton & Zanna, 1972; Ronis & Lipinski, 1985), perhaps because negative information is distinctive in a world where people usually present socially desirable selves. Third, they attend more to attributes that pertain to the purpose or judgment at hand. Fourth, they discount information that is very inconsistent with previous impressions or is redundant with what they already know. Finally, they weight first impressions more heavily than subsequent impressions. We consider the importance of first impressions in more detail below.

First Impressions You have surely noticed the effort that individuals make to create a good impression when interviewing for a new job, entering a new group, or meeting an attractive potential date. This effort reflects the widely held belief that first impressions are especially important and have enduring impact. In fact, this belief is supported by a body of systematic research. Observers forming an impression of a person give more weight to information received early in a sequence than to information received later. This is called the **primacy effect** (Luchins, 1957).

What accounts for the impact of first impressions? One explanation is that after forming an initial impression of a person, we interpret subsequent information in a way that makes it consistent with our initial impression. Having established that your new roommate is neat and considerate, you interpret the dirty socks on the floor as a sign of temporary forgetfulness rather than as evidence of sloppiness and lack of concern. Thus the schema into which an observer assimilates new information influences the interpretation of that information (Zanna & Hamilton, 1977).

A second explanation for the primacy effect holds that we attend very carefully to the first bits of information we get about a person but we pay less attention once we have enough information to make a judgment. It is not that we interpret later information differently; we simply use it less. This explanation assumes that whatever information we attend to most has the biggest effect on our impressions (Dreben, Fiske, & Hastie, 1979). Both the reinterpretation of

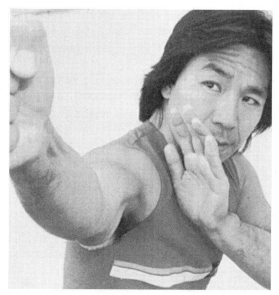

This man makes a first impression as strong, stern, and self-controlled. First impressions are hard to change because people pay less attention to later information. Told the man is fearful, for example, observers are likely to ignore this information or to interpret it as meaning he shows healthy fear in extremely dangerous situations.

later information and a waning of attention produce primacy effects.

Although primacy effects are commonplace, they do not always occur; sometimes the direct opposite happens. Under certain conditions, the most recent information we acquire exerts the strongest influence on our impressions, an occurrence known as the *recency effect* (Jones & Goethals, 1971; Steiner & Rain, 1989). A recency effect is likely to occur when so much time has passed that we have largely forgotten our first impression or when we are judging characteristics which change over time, like moods or attitudes. In laboratory settings, investigators can induce a recency effect by asking perceivers to make a separate evaluation after each new piece of information is received (Stewart, 1965).

Although both primacy effects and recency effects can have an impact on the impressions that people form of one another, primacy effects are especially important in everyday life. In one study investigating the relative impact of primacy and recency effects on impression formation (Jones et al., 1968), subjects observed the performance of a college student on an SAT-type aptitude test. In one condition, the student started successfully on the first few items but then her performance deteriorated steadily. In a second condition, the student started poorly and then gradually improved. In both conditions, the student answered 15 out of 30 test items correctly. After observing one or the other performance, subjects rated the student's intelligence and tried to predict how well she would do on the next 30 items. Although the student's overall performance was the same in both conditions (15 of 30 correct), subjects rated the student as more intelligent when she started well and then tailed off than when she started poorly and improved. They also predicted higher scores for the student on the next series when the student started well than when she started poorly. Clearly, subjects gave more weight to the student's performance on the first few items—a primacy effect.

Impressions as Self-Fulfilling Prophecies

Whether correct or not, the impressions we form of people influence our behavior toward them. Recall, for instance, the study in which students read that their guest instructor was "warm" or "cold" before meeting him (Kelley, 1950). Not only did the students form different impressions of the instructor, they also behaved differently toward him. Those who believed the instructor was "warm" participated more in the class discussion than those who believed he was "cold."

When our behavior toward people reflects our impressions of them, we cause them to react in ways that confirm our original impressions. For example, if we ignore someone because we think she is dull, she will probably withdraw and add nothing interesting to the conversation. Because our own actions evoke appropriate reactions from others, our initial impressions—correct or incorrect—are often confirmed by the reactions of others. When this happens, our impressions become *self-fulfilling prophecies* (Darley & Fazio, 1980).

A study of "getting acquainted" conversations between male and female college students shows vividly how impressions may become self-fulfilling (Snyder, Tanke, & Berscheid, 1977). The investigators provided each male student in the study with a folder of information about a female student; the male student was interested in this because he would subsequently contact the woman by telephone to get acquainted. The folder included a biographical sheet and a photo of the woman. The biographical information was accurate, but (unknown to the male subjects) the photo was actually a snapshot of a different woman. The woman shown in the photo was either very attractive or very unattractive (an experimental manipulation). Each male student was asked to form an impression of the woman's personality based on the biographical sheet and the photo. As expected, men who saw the attractive snapshot rated the woman more positively on personality dimensions than those who saw the unattractive snapshot.

Each man then engaged in a "get acquainted" phone conversation with the woman who had provided the biographical information in his folder. Men who believed their partner was attractive spoke with more animation, sociability, and warmth than those who thought their partner was unattractive. In other words, the impressions the men formed from the snapshots influenced their own behavior. For their part, the women responded over the phone in a more poised, confident, animated, sociable, sexually warm, and outgoing manner when they were speaking with men who thought they were attractive than with men who thought they were unattractive. This occurred even though the women did not realize the men had seen snapshots of any kind. It is easy to understand how the men might interpret the responses of the women as confirmation of their original impressions concerning their partners' sociability. Thus the prophecy was self-fulfilling.

Attribution Theory

When we interact with other people, we observe only their actions and the effects these have. This is fine as far as it goes, but, as perceivers, we want to know why others act as they do. To discern this, we must usually make inferences beyond what we observe. For instance, if a woman performs a favor for us, why is she doing it? Is she doing it because she is fundamentally a generous person? Or is she manipulative and pursuing some ulterior motive? Does her social role require her to do it? Have other people pressured her into doing it? To act effectively toward her and to predict her future behavior, we must first figure out why she behaves as she does.

The term **attribution** refers to the process through which an observer infers the causes of another's behavior. That is, the observer attempts to answer the question "Why did that person act as he or she did?" In attribution, we observe another's behavior and infer backward to its causes—to the intentions, abilities, traits, motives, and situational pressures that explain why people act as they do. Theories of attribution focus on the methods we use to interpret another person's behavior and to infer its sources. They explain how we come to attribute a given behavior to some causes rather than others (Kelley & Michela, 1980; Lipe, 1991; Ross & Fletcher, 1985).

Dispositional Versus Situational Attributions

Fritz Heider (1944, 1958), whose work was an early stimulus to the study of attribution, noted that people in everyday life use commonsense reasoning to understand the causes of others' behavior. They act as "naive scientists" and use something resembling the scientific method in attempting to discern causes of behavior. Heider maintained that whether their interpretations about the causes of behavior are scientifically valid or not, people act on their beliefs. For this reason, social psychologists must study people's commonsense causal explanations of behavior and events. Only by studying these, suggested Heider, can we understand their behavior.

The most crucial decision that observers make is whether to attribute a behavior to the internal state(s) of the person who performed it (this is termed a **dispositional attribution**) or to factors in that person's environment (a **situational attribution**). As an example, consider the attributions an observer might make upon learning that her neighbor is unemployed. She might judge that he is out of work because he is lazy, irresponsible, or lacking in ability. These are dispositional attributions because they attribute the

causes of behavior to his internal states or characteristics. Alternatively, she might attribute his unemployment to the scarcity of jobs in his line of work, to employment discrimination, to the depressed condition of the economy, or to the evils of the capitalist system. These are situational attributions because they attribute his behavior to external causes.

What determines whether an observer attributes an act to a person's disposition or to the situation? One important consideration is the strength of situational pressures on the person. These pressures may include normative role demands, as well as rewards or punishments applied to the person by others in the environment. For example, suppose we see a judge give the death penalty to a criminal. We might infer that the judge is tough (a dispositional attribution). However, suppose we learn that the law in that state requires the death penalty for the criminal's offense. Now we would see the judge not as tough but as responding to role pressures (a situational attribution).

This logic has been formalized as the **subtractive rule,** which states that when making attributions about personal dispositions, the observer subtracts the impact of situational forces from the personal disposition implied by the behavior itself (Trope & Cohen, 1989; Trope, Cohen, & Maoz, 1988). So, considered by itself, the judge's behavior (imposing the death penalty) might imply that she is tough in disposition. The subtractive rule, however, states that the observer must subtract the effect of situational pressures (the state law) from the disposition implied by the behavior itself. When an observer does this, he may conclude the judge is not especially tough or overly inclined to impose the death penalty.

For the judge and the death penalty, using the subtractive rule served to weaken the dispositional attribution and strengthen the situational attribution. In certain other cases, however, applying the subtractive rule strengthens or augments the dispositional attribution. This happens, for instance, whenever a person persists in actions that the environment discourages or punishes. Suppose that while interacting with political conservatives, a woman expressed liberal sentiments on various issues. If she persists in this despite a negative reaction from the others, we attribute the behavior to her personal disposition. That is, applying the subtractive rule, we subtract the impact of situational forces from the personal disposition implied by the behavior itself. This means we subtract a negative quantity (the negative reaction of others) from the disposition implied by the behavior itself. The net effect is to increase or augment the dispositional attribution. Our conclusion as observers: The woman must hold strong liberal sentiments.

Inferring Dispositions From Acts

Although Heider's analysis and the subtractive rule are useful in identifying some conditions under which observers make dispositional attributions, they do not explain which specific dispositions we will ascribe to the person. Suppose, for instance, that you are on a city street during the Christmas season and you see a young, well-dressed man walking with a woman. Suddenly, the man stops and tosses several coins into a Salvation Army pot. From this act, what can you infer about the man's dispositions? Is he generous and altruistic? Or was he trying to impress the woman? Or was he perhaps just trying to clear out some nuisance change from his coat pocket?

When we try to infer a person's dispositions, our perspective is much like that of a detective. We can observe only the *act* (a man gives coins to the Salvation Army) and the *effects* of that act (the Salvation Army receives more resources, the woman smiles at the man, the man's pocket is no longer cluttered with coins). From this observed act and its effects, we must infer the man's dispositions.

According to one prominent theory (Jones, 1979; Jones & Davis, 1965), we perform two major steps when inferring personal dispositions. First, we try to deduce the specific *intentions* that underlie a person's actions. In other words, we try to figure out what the person originally intended to accomplish by performing the act. Second, from these intentions we try to infer what prior personal *disposition* would cause a person to have such intentions. If we think the man intended to benefit the Salvation Army, for example, we infer the disposition "helpful" or "generous." However, if we think the man had some other intention(s), such as impressing his girlfriend, we do not infer he has the disposition "helpful." Thus

we attribute a disposition that reflects the presumed intention.

One problem in inferring dispositions from acts, of course, is that any given act may have multiple effects. In order to make confident attributions, perceivers must decide which effect(s) the person is really pursuing and which effects are merely incidental. This is not always easy to do. When the man donated money to the Salvation Army, for example, was his intention to perform a charitable act or to impress the woman accompanying him? His act accomplished both effects. Before making the inference that the man is "generous" and "helpful" by disposition, an observer must know which effect(s) the man intended the act to produce.

Several factors influence observers' decisions regarding which effect(s) the person is really pursuing and, hence, what dispositional inference is appropriate. These factors include the commonality of effects, the social desirability of effects, and the normativeness of effects (Jones & Davis, 1965).

Commonality   Usually, any act that a person performs will produce more than one effect. This multiplicity of effects makes it difficult for observers to infer dispositions from acts. If any given act produced one, and only one, effect, then inferences of dispositions from acts would always be clear-cut. Because of the multiplicity of effects, however, observers attributing specific intentions and dispositions to a person find it informative to observe the person in situations that involve choice among alternative actions.

Suppose, for example, that a person can engage in either action 1 or in action 2. Action 1, if chosen, will produce effects a, b, and c. Action 2 will produce effects b, c, d, and e. As we can see, two of these effects (b and c) are *common* to actions 1 and 2. The remaining effects (a, d, and e) are unique to a particular alternative; these are *noncommon* effects.

Now suppose the person chooses action 2. What can we infer about his intentions and dispositions? The main inference is that while he may or may not have intended to produce effects b and c, he certainly intended to produce either effect d or effect e (or both) or to avoid effect a. The unique (noncommon) effects of acts enable observers to make inferences regarding intentions and dispositions, but the common effects of two (or more) acts provide little or no basis for inferences (Jones & Davis, 1965).

Thus observers who wish to discern the specific dispositions of a person try to identify effects that are unique to the action chosen. Research shows that the fewer noncommon effects associated with the chosen alternative, the greater the confidence of observers about their attributions (Ajzen & Holmes, 1976).

Social Desirability In many situations, people engage in particular behaviors because they are socially desirable. Yet people who perform a socially desirable act shows us only that they are "normal" and reveal nothing about their distinctive dispositions. Suppose, for instance, that you observe a guest at a party thank the hostess upon leaving. What does this tell you about the guest? Did she really enjoy the party? Or was she merely behaving in a polite, socially desirable fashion? You cannot be sure—either inference could be correct. Now, suppose instead that upon leaving the guest complained loudly to the hostess that she had a miserable time at such a dull party. This would likely tell you more about her because observers interpret acts low in social desirability as indicators of underlying dispositions (Miller, 1976).

Normative Expectations When inferring dispositions from acts, observers consider the normativeness of behavior. *Normativeness* is the extent to which we expect the average person to perform a behavior in a particular setting. This includes conformity to social norms and to role expectations in groups (Jones & McGillis, 1976). Actions that conform to role expectations are uninformative about personal dispositions, whereas actions that violate role expectations lead to dispositional attributions by observers.

A study by Jones, Davis, and Gergen (1961) illustrated the effect of out-of-role behavior on inferences. Subjects listened to a tape-recorded interview of an individual seeking employment either as a sub-

mariner or as an astronaut. The first part of the tape provided a description (by the interviewer) of the ideal job applicant. For the submariner (who had to work long hours in cramped quarters with other people), the ideal characteristics were friendliness, cooperativeness, obedience, and gregariousness (an other-directed person). In contrast, for the astronaut (who at that time traveled alone in space), the ideal characteristics were resourcefulness, thoughtfulness, independence, and a capacity to perform without help or the company of others (an inner-directed person).

Next, subjects heard a tape of the applicant presenting himself for the job. Depending on experimental condition, he sought employment either as a submariner or as an astronaut, and he presented himself either as an other-directed person or as an inner-directed person. The important point is that two of these combinations are role appropriate—the other-directed person applying for submariner job and the inner-directed person applying for the astronaut job. The other two combinations are role inappropriate—the other-directed person applying for the astronaut job and the inner-directed person applying for the submariner job. After listening to the taped interview, the subjects rated the applicant on various trait measures. They also indicated how much confidence they had in their ratings.

Table 5.3 displays the ratings. Subjects rated the two applicants whose behavior was role appropriate (the other-directed submariner and the inner-directed astronaut) as conforming and affiliative. Subjects had low confidence in their ratings of these applicants. These applicants knew the job requirements and presented themselves accordingly during the interview, so subjects could not infer much about them as persons. In particular, subjects could not tell whether the applicants were truly what they claimed to be or merely posing to get the job.

However, when the subjects rated the two applicants whose behavior was role inappropriate (the other-directed astronaut and the inner-directed submariner), the results were different. Subjects rated the inner-directed submariner as independent (nonconforming) and nonaffiliative. Because his behavior was contrary to role requirements, the subjects made strong attributions and reported high confidence in

Table 5.3   Mean Ratings by Subjects of Interviewees

Trait Rated	Role			
	Astronaut		Submariner	
	Inner-Directed	Other-Directed	Inner-Directed	Other-Directed
Conformity	13.09	15.91	9.41	12.58
Affiliation	11.12	15.27	8.64	12.00
Candor	9.68	12.42	12.08	10.09

SOURCE: Adapted from Jones, Davis, and Gergen, 1961.

their ratings. In contrast, subjects rated the other-directed astronaut as very conforming and affiliative. This candidate's behavior was also contrary to role requirements, and subjects again were confident of their ratings.

Note that the subjects in this study followed the subtractive rule, discussed earlier. In their attributions subjects augmented (increased) personal dispositions as causes for the applicants who were role inappropriate (the other-directed astronaut and the inner-directed submariner). Likewise, they discounted (diminished) personal dispositions as causes for the applicants who were role appropriate (the other-directed submariner and the inner-directed astronaut). Both effects are consistent with the subtractive rule.

Covariation Model of Attribution

Up to this point we have examined how observers make attributions regarding a person's behavior in a single situation. Sometimes, however, we have multiple observations of a person's behavior. That is, we have information about a person's behavior in a variety of situations or in a given situation vis-à-vis different partners. Multiple observations enable us to make many comparisons, and these, in turn, facilitate causal attribution.

How do perceivers use multiple observations to arrive at a conclusion about the cause(s) of a behavior? Extending Heider's ideas, Kelley (1967, 1973) suggests that when we have multiple observations of be-

havior, we analyze information essentially the same way a scientist would. That is, we try to figure out whether the behavior occurs in the presence or absence of various factors (actors, objects, contexts) that are possible causes. Then to identify the cause(s) of behavior, we apply the **principle of covariation:** We attribute the behavior to that factor which is both present when the behavior occurs and absent when the behavior fails to occur—the cause that "covaries" with the behavior.

To illustrate, suppose you are working at your part-time job one afternoon when you hear your boss loudly criticizing another worker, Michael. To what would you attribute your boss's behavior? There are at least three potential causes: the *actor* (the boss), the *object* of the behavior (Michael), and the *context* or setting in which the behavior occurs. For example, you might attribute the loud criticism to your boss's aberrant personality (a characteristic of the actor), to Michael's slothful performance (a characteristic of the object), or to some particular feature of the context.

Kelley (1967) suggests that when using the principle of covariation to determine whether a behavior is caused by the actor, object, or context, we rely on three types of information: consensus, consistency, and distinctiveness information.

Consensus refers to whether all actors or only a few perform the same behavior. For example, do all the other employees at work criticize Michael (high consensus), or is your boss the only person who does so (low consensus)?

Consistency refers to whether the actor behaves the same way at different times and in different settings. If your boss criticizes Michael on many different occasions, his behavior is high in consistency. If he has never before criticized Michael, his behavior is low in consistency.

Distinctiveness refers to whether the actor behaves differently toward a particular object than toward other objects. If your boss criticizes only Michael and none of the other workers, his behavior is high in distinctiveness. If he criticizes all workers, his behavior toward Michael is low in distinctiveness.

The causal attribution that observers make for a behavior depends on the particular combination of consensus, consistency, and distinctiveness information that people associate with that behavior. To illustrate, Table 5.4 reviews the scenario in which your boss criticizes Michael. The table displays three combinations of information that might be present in this situation. These combinations of information are interesting because studies have shown they reliably produce different attributions regarding the cause of the behavior.

As Table 5.4 indicates, observers usually attribute the cause of a behavior to the actor (the boss) when the behavior is low in consensus, low in distinctiveness, and high in consistency. In contrast, observers usually attribute a behavior to the object (Michael) when the behavior is high in consensus, high in distinctiveness, and high in consistency. Finally, observers usually attribute a behavior to the context when consistency is low.

Several studies show that, at least in general terms, people use consensus, consistency, and distinctiveness information in the way Kelley theorized (Hewstone & Jaspars, 1987; McArthur, 1972; Pruitt & Insko, 1980). Of course, in any given situation, the combination of available information may differ from the three shown in Table 5.4. In such cases, attributions are more complicated, more ambiguous, and less certain. We usually assign less weight to a given cause if other plausible causes are also present.

Table 5.4 Why Did the Boss Criticize Michael?

Situation: At work today, you observe your boss criticizing and yelling at another employee, Michael.
Question: Why did the boss criticize Michael?

1. Kelley's (1973) model indicates that attributions are made to the actor (boss) when consensus is low, distinctiveness is low, and consistency is high.

  > *Example:* Suppose no other persons criticize Michael (low consensus). The boss criticizes all the other employees (low distinctiveness). The boss criticized Michael last month, last week, and yesterday (high consistency).
  > *Attribution:* The perceiver will likely attribute the behavior (criticism) to the boss. ("The boss is a very critical person.")

2. The model indicates that attributions are made to the stimulus object (Michael) when consensus is high, distinctiveness is high, and consistency is high.

  > *Example:* Suppose everyone at work criticizes Michael (high consensus). The boss does not criticize anyone else at work, only Michael (high distinctiveness). The boss criticized Michael last month, last week, and yesterday (high consistency).
  > *Attribution:* The perceiver will likely attribute the behavior (criticism) to Michael. ("Michael is a lazy, careless worker.")

3. The model indicates that attributions are made to the context or situation when consistency is low.

  > *Example:* Suppose the boss has never criticized Michael before (low consistency).
  > *Attribution:* The perceiver will likely attribute the behavior (criticism) to a particular set of contextual circumstances, rather than to Michael or the boss per se. ("Michael made a remark this morning that the boss misinterpreted.")

Attributions for Success and Failure

For students, football coaches, elected officials, and anyone else whose fate rides on evaluations of their performance, attributions for success and failure are vital. As observers realize, however, attributions of this type are problematic. Whenever someone succeeds at a task, a variety of explanations can be advanced for the outcome. For example, a student who passes a test could credit her own intrinsic ability ("I have a lot of intelligence"), her effort ("I really studied for that exam"), the easiness of the task ("The exam could have been much more difficult"), or even luck ("They just happened to test us on the few articles I read").

These four factors—ability, effort, task difficulty, and luck—are general and apply in many settings. How do observers decide which of these is the "real" cause of an outcome (success or failure)? When observers look at an event and try to figure out the cause for success or failure, they must consider two things. First, they must decide whether the outcome is due to sources within the actor (an internal or dispositional attribution) or due to forces in the environment (an external or situational attribution). Second, they must decide whether the outcome is a stable or unstable occurrence. That is, they must determine whether the cause is a permanent feature (of the actor or the environment) or whether it is labile and changing. Only after observers make judgments regarding internality-externality and stability-instability can they reach conclusions regarding the cause(s) of the success or failure.

As various theorists (Heider, 1958; Weiner, 1986; Weiner et al., 1971) have pointed out, the four factors mentioned above—ability, effort, task difficulty, and luck—can be grouped according to internality-externality and stability-instability. Ability, for instance, is usually considered internal and stable. That is, observers usually construe ability, or aptitude, as a property of the person (not the environment), and they consider it stable because it does not change from moment to moment. In contrast, effort is internal and unstable. Effort, or temporary exertion, is a property of the person that changes depending on how hard he or she tries. Task difficulty depends on objective task characteristics, so it is external and stable.

Luck, or chance, is external and unstable (fickle). Table 5.5 displays these relations.

Determinants of Attributed Causes

Whether observers attribute a performance to internal or external causes depends on how the actor's performance compares with those of others. We usually attribute extreme or unusual performances to internal causes. For example, we would judge a tennis player who wins a major tournament as extraordinarily able and/or as highly motivated. Similarly, we would view a player who turns in an unusually poor performance (for example, loses 6–0, 6–0 in the first round to an unseeded competitor) as weak in ability and/or unmotivated. In contrast, we usually attribute average or common performances to external causes. If defeat comes to a player halfway through the tournament, we are likely to attribute it to tough competition or perhaps bad luck.

Whether observers attribute a performance to stable or unstable causes depends on how consistent the actor's performance is over time. When performances are very consistent, we attribute the outcome to stable causes. Thus, if a tennis player wins tournaments consistently, we would attribute this success to her great talent (ability) or perhaps to the uniformly low level of her opponents (task difficulty). When performances are very inconsistent, however, we attribute the outcomes to unstable causes, rather than stable ones. Suppose, for example, that our tennis player is unbeatable one day and a pushover the next. In this case, we would attribute the outcomes to fluctuations in motivation (effort) or to random

Table 5.5 Perceived Causes of Success and Failure

Degree of Stability	Locus of Control	
	Internal	**External**
Stable	Ability	Task difficulty
Unstable	Effort	Luck

SOURCE: Adapted from Weiner, Heckhausen, Meyer, and Cook, 1972.

external factors such as wind speed, court condition, and so on (luck).

An experimental study (Frieze & Weiner, 1971) clearly illustrates these effects. Subjects were given information about an individual's performance (success or failure) at a given task. They also received information about that person's past success rate on the same and similar tasks, as well as information about others' success rate on that task. These data influenced whether the subjects viewed the actor's performance as consistent (or inconsistent) with his past performance and as similar to (or different from) the performance of others on the same task. Subjects then reported their judgments about the impact of internal factors (ability, effort) and external factors (task difficulty, luck) in causing the actor's performance outcome (success or failure) on the immediate task. The results showed that, first, success was more likely to be attributed to internal factors (ability, effort) than was failure. Second, performance similar to that of others was attributed to external factors (task difficulty), whereas performance different from that of others (such as success while others fail, or failure while others succeed) was attributed to internal factors (ability, effort). Third, performance consistent with one's own past record (such as success when one has succeeded in the past, or failure when one has failed in the past) was attributed to stable factors (ability, task difficulty), whereas performance inconsistent with one's past was attributed to unstable factors (luck, effort).

Consequences of Attributions Attributions for performance are important because they influence both our emotional reactions to success and failure and our future expectations and aspirations. For instance, if we attribute a poor exam performance to lack of ability, we may despair of future success and give up studying; this is especially likely if we view ability as given and not controllable by us. Alternatively, if we attribute the poor exam performance to lack of effort, we may feel shame or guilt, but we are likely to study harder and expect improvement. If we attribute the poor exam performance to bad luck, we may experience feelings of surprise or bewilderment, but we are not likely to change our study habits because the situation will not seem controllable; despite

this lack of change, we might nevertheless expect improved grades in the future. Finally, if we attribute our poor performance to the difficulty of the exam, we may become angry, but we do not strive for improvement (McFarland & Ross, 1982; Valle & Frieze, 1976; Weiner, 1985, 1986).

Cultural Basis of Attributions

Attributions are influenced by schemas. To the extent that the schemas we employ to process information are cultural elements, culture influences the attribution process. These shared representations can influence the causal attributions we make. In a culture with person and group schemas based on race, members may be more likely to attribute the causes of behavior to race than to environmental influences (Hewstone & Jaspars, 1984). Thus whites may ascribe black unemployment to the laziness of black persons (effort) rather than to the lack of appropriate jobs in the central city (luck).

An important difference among cultures is whether a culture is individualistic or collectivistic (Triandis, 1995). Individualistic cultures emphasize the individual and value individual achievement; collectivistic cultures emphasize the welfare of the family and ethnic group over the interests of individuals. We would expect this difference in values to be reflected in the attributions people make regarding behavior. In an individualistic society, one that places a high value on the individual and his or her goals, perceivers are more likely to attribute behavior to individual dispositions. In a collectivist society, one that emphasizes loyalty to family and ethnic group, perceivers are more likely to attribute behavior to environmental or situational influences on behavior.

In one study, researchers compared attributions made by students from an individualistic society (the United States) with those made by students from a collectivist society (Saudi Arabia). Participants in the study were 163 students recruited from U.S. universities and 162 students from a university in Saudi Arabia (Al-Zahrani & Kaplowitz, 1993). Each student was presented with vignettes describing eight situations, four involving achievement and four involving morality. Students were asked to assign responsibility for the outcome to each of several factors. Consistent with the hypothesis, results showed that across the eight situations, U.S. students assigned greater responsibility to internal dispositional factors than Saudi students.

Bias and Error in Attribution

According to the picture we have drawn so far, observers scrutinize their environment, gather information, form impressions, and interpret behavior in rational, if sometimes unconscious, ways. In actuality, however, observers often deviate from the logical methods described by attribution theory and fall prey to biases. These biases may lead observers to misinterpret events and to make erroneous judgments. In this section we consider several major biases and errors in attribution.

Overattributing to Dispositions

Years ago, at the time of the Cuban missile crisis, the Cuban leader Fidel Castro was generally unpopular (even feared) in the United States. In an interesting study done shortly after the crisis, Jones and Harris (1967) asked subjects to read an essay written by another person (a student). Depending on treatment, the essay either strongly supported the Cuban leader or strongly opposed him. Crossed with this, subjects received information about the conditions under which the student wrote the essay. Depending on treatment, they learned either that the essay was written by a student who was assigned by the instructor to take a pro-Castro or anti-Castro stand (no-choice condition) or that the essay was written by a student who was free to choose whichever position he wanted to present (choice condition). The subjects' task was to infer the writer's true underlying attitude about Castro. In the conditions where the writer had free choice, subjects inferred that the content of the essay reflected the writer's true attitude about Castro. That is, they saw the pro-Castro essay as indicating pro-Castro attitudes and the anti-Castro essay as indicating anti-Castro attitudes. In the conditions where the writer was assigned the topic and had

In this portrait of failure, a basketball player experiences depression after losing the championship game at the buzzer. Although failing is never pleasant, its impact may vary. If we attribute our failure to lack of effort, we are likely to feel guilt or shame and to renew our efforts in the future. If we attribute it to lack of ability, we may despair and quit trying.

no choice, subjects still thought the content of the essay reflected the writer's true attitude about Castro, although they were less sure this was so. Subjects made these internal attributions even though it was possible the writer held an opinion directly opposite of that expressed in the essay. In effect, subjects overestimated the importance of internal dispositions (attitudes about Castro) and underestimated the importance of situational forces (role obligations) in shaping the essay.

The tendency to overestimate the importance of personal (dispositional) factors and to underestimate situational influences as causes of behavior is so common that it is called the **fundamental attribution error** (Higgins & Bryant, 1982; Ross, 1977; Small & Peterson, 1981). This error results from a failure by the observer to fully apply the subtractive rule. This tendency was first identified by Heider, who noted that most observers ignore or minimize the impact of role pressures and situational constraints on others and interpret behavior as caused by people's intentions, motives, or attitudes. This bias is especially dangerous when it causes us to overlook the advantages of power built into social roles. For instance, we may incorrectly attribute the successes of the powerful to their superior personal capabilities, or we may incorrectly attribute the failures of persons without power to their personal weaknesses.

Focus of Attention Bias

Another common tendency is to overestimate the causal impact of whomever or whatever we focus our attention on; this is called the **focus of attention bias.** A striking demonstration of this bias appears in a study by Taylor and Fiske (1978). The study involved six subjects who observed a conversation between two persons (Speaker 1 and Speaker 2). Although all six subjects heard the same dialogue, they differed in the focus of their visual attention. Two observers sat behind Speaker 1, facing Speaker 2; two sat behind Speaker 2, facing Speaker 1; and two sat on the sides, equally focused on the two speakers (see Figure 5.2). Measures taken after the conversation showed that observers thought the speaker they faced not only had more influence on the tone and content of the conversation but also had a greater causal impact on the other speaker's behavior. Observers who sat on the sides and were able to focus equally on both speakers attributed equal influence to them.

We perceive the stimuli that are most salient in the environment—those that attract our attention—as most causally influential. Thus, we attribute most causal influence to people who are noisy, colorful, vivid, or in motion. We credit the person who talks the most with exercising the most influence; we blame the person who runs past us when we hear a rock shatter a window. Although salient stimuli may be causally important in some cases, we overestimate their importance (McArthur & Post, 1977).

The focus of attention bias provides an explanation for the fundamental attribution error. The person behaving is the active entity in the environment;

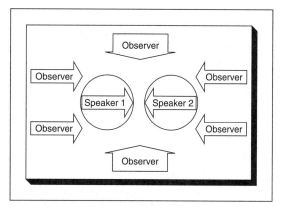

Figure 5.2 The Focus of Attention Bias

This diagram depicts the seating arrangement for speakers and observers in a study investigating the effect of the focus of visual attention on observers' attributions. Arrows indicate visual focus of attention. Following a conversation between both speakers, observers attributed more influence to the speaker they faced than to the other speaker. Observers on the side, however, attributed equal influence to both speakers. This illustrates our tendency to attribute more causal impact to the object of our attention.

SOURCE: Adopted from Taylor and Fiske, 1978.

therefore, that person is likely to capture our attention. Because we direct our attention more to people who act than to the surrounding situation, we attribute more causal importance to people than to their situations.

Actor-Observer Difference

Actors and observers make different attributions for behavior. Observers tend to attribute actors' behavior to the actors' internal characteristics, whereas actors see their own behavior as due more to characteristics of the external situation (Jones & Nisbett, 1972; Watson, 1982). This tendency is known as the **actor-observer difference.** Thus, although other customers in a market may attribute the mix of items in your grocery cart (beer, vegetables, candy bars) to your personal characteristics (hard drinking, vegetarian, chocolate addict), you will probably attribute it to the requirements of your situation (preparing for a party) or the qualities of the items (nutritional value or special treat).

In one demonstration of the actor-observer difference (Nisbett et al., 1973), male students wrote descriptions explaining why they liked their girlfriends and why they chose their majors. Then, as observers, they explained why their best friend liked his girlfriend and chose his major. When explaining their own actions, students emphasized external characteristics, like the attractive qualities of their girlfriends and the interesting aspects of their majors. However, when explaining their friends' behavior, they downplayed external characteristics and emphasized their friends' internal dispositions (preferences and personalities).

Two explanations for the actor-observer difference in attribution are that actors and observers have different visual perspectives and different information. We discuss each of these next.

Visual Perspectives

Visual Perspectives   The actor's natural visual perspective is to look at the situation, whereas the observer's natural perspective is to look at the actor. Thus, the actor-observer difference reflects a difference in the focus of attention. Both the actor and observer attribute more causal influence to that which they focus on.

Storms (1973) reasoned that if the actor-observer difference in attributions was due simply to a difference in perspective, it might be possible to reverse the actor-observer difference by making the actor see behavior from the observer's viewpoint and the observer see the same behavior from the actor's viewpoint. To give each the other's point of view, Storms videotaped a conversation between two people, using two separate cameras. One camera recorded the interaction from the visual perspective of the actor, the other from the perspective of the observer. Storms then showed actors the videotape made from the observer's perspective, and he showed observers the videotape made from the actor's perspective. As predicted, reversing visual perspectives reversed the actor-observer difference in attribution.

Information

Information   A second explanation for the actor-observer difference is that actors have information about their own past behavior that observers lack. Consequently, observers may assume that certain behaviors are typical of an actor, when in fact they are not. This would cause observers to make incorrect dispositional attributions. For example, observers who see a clerk return an overpayment to a customer may assume the clerk always behaves this way. They may then use this behavior as the basis for a dispositional attribution of honesty. However, if the clerk knows he has often cheated customers in the past, he would probably not interpret his current behavior as evidence of his honest nature. Consistent with this, research shows that observers who have a low level of acquaintance with the actor tend to form more dispositional attributions and fewer situational attributions than those who have a high level of acquaintance with the actor (Prager & Cutler, 1990).

Even when observers have some information about an actor's past behavior, they often do not know how changes in context influence the actor's behavior. This is because observers usually see an actor only in limited contexts. Suppose that students observe a professor deliver witty, entertaining lectures in class week after week. The professor knows that in other social situations she is shy and withdrawn, but the students do not have an opportunity to see this. As a result, observers (students) may infer dispositions from apparently consistent behavior that the actor (the professor) knows to be inconsistent across a wider range of contexts.

Motivational Biases

Up to this point, we have considered attribution biases based on cognitive factors. That is, we have traced biases to the types of information that observers have available, acquire, and process. Motivational factors—a person's needs, interests, and goals—are a second source of bias in attributions. When events affect a person's self-interests, biased attribution is likely. Specific motives that influence attribution include the desire to defend deep-seated beliefs, to enhance one's self-esteem, to increase one's sense of control over the environment, and to strengthen the favorable impression of oneself that others have.

The desire to defend cherished beliefs and stereotypes may lead observers to engage in biased attribution. Observers may interpret actions that correspond

What is this man doing? We often try to discern the reasons why people behave as they do.

with their stereotypes as caused by the actor's personal dispositions. For instance, they may attribute a woman executive's outburst of tears during a crisis to her emotional instability because that corresponds to their stereotype about females. At the same time, people attribute actions that contradict stereotypes to situational causes. If the woman executive manages the crisis smoothly, the same people may credit this to the effectiveness of her male assistant. When observers selectively attribute behaviors that contradict stereotypes to situational influences, these behaviors reveal nothing new about the persons who perform them. As a result, the stereotypes persist even in the face of contradictory evidence (Hamilton, 1979).

Motivational biases may also influence attributions for success and failure. People tend to take credit for acts that yield positive outcomes, whereas they deflect blame for bad outcomes and attribute them to external causes (Bradley, 1978; Ross & Fletcher, 1985). This phenomenon, referred to as the **self-serving bias,** is illustrated clearly in a study in which college students were asked to explain the grades they received on three examinations (Bernstein, Stephan, & Davis, 1979). Students who received A's and B's attributed their grades much more to their own effort and ability than to good luck or easy tests. However, students who received C's, D's, and F's attributed their grades largely to bad luck and the difficulty of the tests. Other studies show similar effects (Reifenberg, 1986)

The self-serving bias also appears when athletes report the results of competitions (Lau & Russell, 1980; Ross & Lumsden, 1982). Whereas members of winning teams take credit for winning ("We won"), members of losing teams are more likely to attribute the outcome to an external cause—their opponent ("They won," not "We lost").

Various motives may contribute to this self-serving bias in attributions of performance. For instance, attributing success to personal qualities and failure to external factors enables people to enhance or protect their self-esteem. Regardless of the outcome, they can continue to see themselves as competent and worthy. Then, too, by avoiding the attribution of failure to personal qualities they maximize their sense of control. This in turn supports the belief that they can master challenges successfully if they choose to apply themselves because they possess the necessary ability. Finally, biased attributions enable people to present a favorable public image and to make a good impression on others.

Summary

Social perception is the process of using information to construct understandings of the social world and form impressions of people.

Schemas   A schema is a well-organized structure of cognitions about some social entity. (1) There are several distinct types of schemas: person schemas, self-schemas, group schemas (stereotypes), role schemas, and event schemas (scripts). (2) Schemas organize information in memory and therefore affect what we remember and what we forget. In addition, they guide our inferences and judgments about people and objects. (3) Schemas are transmissible cultural elements. They can be taught by one person and learned by another, and therefore can be shared among members of a social group.

Person Schemas and Group Stereotypes

(1) One important type of person schema is an implicit personality theory—a set of assumptions about which personality traits go together with what other traits. These schemas enable us to make inferences about other people's traits. We can depict an implicit personality theory as a "mental map." (2) A stereotype is a fixed set of characteristics attributed to all members of a given group. American culture includes stereotypes for ethnic, racial, gender, and occupational groups. Because stereotypes are overgeneralizations, they cause errors in inference; this is especially true in complex situations.

Impression Formation

(1) Research on trait centrality using the "warm-cold" variable illustrates how variations in a single trait can produce a large difference in the impression formed by observers of a stimulus person. (2) Perceivers try to combine the bits of information they receive to create a unified impression of others. They average these bits of information after weighting certain types of information more heavily than other types. Information received early usually has a larger impact on impressions than information received later. (3) Impressions become self-fulfilling prophecies when we behave toward others according to our impressions and evoke corresponding reactions from them.

Attribution Theory

Through attribution, people infer an action's causes from its effects. (1) One important issue in attribution is locus of causality—dispositional (internal) versus situational (external) attributions. Observers follow the subtractive rule when making attributions to dispositions or situations. (2) To attribute specific dispositions to an actor, observers first observe an act and its effects and then try to infer the actor's intention with respect to that act. Observers then attribute the disposition that

corresponds best with the actor's inferred intention. (3) Observers who have information about an actor's behaviors in many situations make attributions to the actor, object, or context. The attribution made depends on which of these causes covaries with the behavior in question. They assess covariation by considering consensus, consistency, and distinctiveness information. (4) Observers attribute success or failure to four basic causes—ability, effort, task difficulty, and luck. They attribute consistent performances to stable rather than unstable causes, and they attribute average performances to external rather than internal causes. (5) Cultural factors can affect attribution. In comparison to those in an individualist society, perceivers in a collectivist society may overestimate the importance of situational or environmental influences on behavior.

Bias and Error in Attribution

(1) Observers frequently overestimate personal dispositions as causes of behavior and underestimate situational pressures; this bias is called the fundamental attribution error. (2) Observers also overestimate the causal impact of whatever their attention is focused on. (3) Actors and observers have different attribution tendencies. Actors attribute their own behavior to external forces in the situation, whereas observers attribute the same behavior to the actor's personal dispositions. (4) Motivations—needs, interests, and goals—lead people to make self-serving, biased attributions. People defend deep-seated beliefs by attributing behavior that contradicts their beliefs to situational influences. People defend their self-esteem and sense of control by attributing their failures to external causes and taking personal credit for their successes.

Key Terms

actor-observer difference (p. 125)
additive model (p. 114)
attribution (p. 117)
averaging model (p. 114)
categorization (p. 102)
dispositional attribution (p. 117)

focus of attention bias (p. 125)
fundamental attribution error (p. 124)
halo effect (p. 108)
implicit personality theory (p. 106)
primacy effect (p. 115)
principle of covariation (p. 120)
prototype (p. 102)
schema (p. 102)

self-serving bias (p. 127)
situational attribution (p. 117)
social perception (p. 101)
stereotype (p. 108)
subtractive rule (p. 117)
trait centrality (p. 113)
weighted averaging model (p. 114)

CHAPTER 6
Attitudes

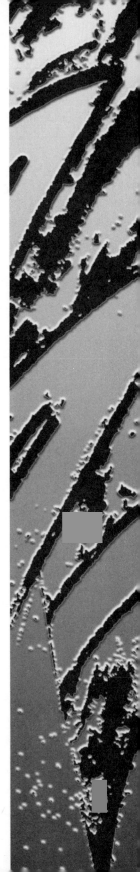

Introduction

The Nature of Attitudes

The Measurement of Attitudes

Attitude Organization and Change

The Relationship Between Attitudes and Behavior

The Reasoned Action Model

Introduction

- "Genesis's music is great!"
- "My human sexuality class is really boring."
- "I like my job."
- "Government spending causes inflation."
- "The law requiring 18-year-olds to register is a lousy law."
- "Guns don't kill people; people kill people."

What do these statements have in common? Each represents an **attitude,** a predisposition to respond to a particular object in a generally favorable or unfavorable way (Ajzen, 1982). A person's attitudes influence the way in which he or she perceives and responds to the world (Allport, 1935; Thomas & Znaniecki, 1918). Attitudes influence attention: The person who likes Genesis's music is more likely to notice news stories about its activities. Attitudes influence behavior: The young man who opposes the draft is more likely to participate in a demonstration against the draft law.

Because attitudes are an important influence on people, they occupy a central place in social psychology. But what exactly is an attitude? What do we mean by a "predisposition to respond"? Further, how do we measure a person's attitudes? We cannot study predispositions directly; instead, we must rely on various measures that reflect a person's attitudes. Moreover, a particular attitude does not exist in isolation. The person who believes that government spending causes inflation has a whole set of beliefs about the role of government in the economy, and this attitude about spending is related to those other beliefs. If attitudes influence behavior, perhaps we can change behavior by changing attitudes. This leads us to ask, "How do attitudes change?" Politicians, lobbyists, auto manufacturers, and brewers spend billions of dollars every year trying to create favorable attitudes. Even if they succeed, do these attitudes affect our behavior?

In this chapter we consider these four questions:

1. What is an attitude? Where do attitudes come from and how are they formed?

2. How do we find out someone's attitudes? How are attitudes measured?
3. How are attitudes linked to other attitudes? How does this organization influence attitude change?
4. What is the relationship between attitudes and behavior?

The Nature of Attitudes

An attitude exists in a person's mind; it is a mental state. Every attitude is about something, an object. In this section we consider the components of an attitude, the sources of attitudes, and their functions.

The Components of an Attitude

Consider the following statement: "My human sexuality class is really boring." This attitude has three components: (1) beliefs or cognitions, (2) an evaluation, and (3) a behavioral predisposition.

Cognition An attitude includes an object label, rules for applying the label, and a set of cognitions or knowledge structure associated with that label (Pratkanis & Greenwald, 1989). The person who doesn't like his or her human sexuality class perceives it as involving certain content, taught by a particular person. Often we cannot prove whether particular beliefs are true or false. For example, economists and government officials disagree on whether government spending causes inflation, with both sides equally convinced they are right.

Evaluation An attitude also has an evaluative (or affective) component. "It's boring" indicates that the course arouses a mildly unpleasant emotion in the speaker. Stronger negative emotions include dislike, hatred, or even loathing: "I can't stand punk rock." Of course, the evaluation may be positive: "I like Genesis's music," or "This food is terrific!" The evaluative component has both a direction (either positive or negative) and an intensity (ranging from very weak to very strong). The evaluation component distinguishes an attitude from other types of cognitive elements.

Behavioral Predisposition An attitude involves a predisposition to respond or a behavioral tendency toward the object. "It's boring" implies a tendency to avoid the class. "I like my job" suggests an intention to go to work. Persons having a specific attitude are inclined to behave in certain ways toward an object and not in other ways.

Relationships Between the Components

Cognitive, evaluative, and behavioral components all have the same object, so we would expect them to form a single, relatively consistent whole. However, these three components are distinct; if they were identical we would not need to distinguish among them. Thus we should be able to measure each component, and we should be able to find a relationship between them. A survey of women's attitudes toward contraceptives found that beliefs, feelings, and actions are distinct and somewhat related (Kothandapani, 1971).

The degree of consistency between components is related to other characteristics of the attitude. Greater consistency between the cognitive and affective components is associated with greater attitude stability and resistance to persuasion (Chaiken & Yates, 1985). Greater consistency is also associated with a stronger relationship between attitude and behavior, as we will discuss later in this chapter.

Attitude Formation

"I like my job."

"Blacks are lazy."

"Asians are good at math."

"Nuclear power plants are dangerous."

"The law requiring 18-year-olds to register is a lousy law."

"Guns don't kill people; people kill people."

Where do these attitudes come from? How are they formed? The answer lies in the processes of social learning, or socialization (discussed in Chapter 3). Attitudes may be formed through reinforcement (instrumental conditioning), through associations of stimuli and responses (classical conditioning), or by observing others (observational learning).

We acquire an attitude toward our classes and jobs through *instrumental conditioning*—that is, learning based on direct experience with the object. If you experience rewards related to some object, your attitude will be favorable. Thus, if your work provides you with good pay, a sense of accomplishment, and compliments from your co-workers, your attitude toward it will be quite positive. Conversely, if you associate negative emotions or unpleasant outcomes with some object, you will dislike it. For example, repeated exposure to bland, overcooked food leads many students to have a very negative attitude toward cafeteria food.

Only a small portion of our attitudes are based on direct contact with the object, however. We have attitudes about many political figures we have never met. We have attitudes toward members of certain ethnic or religious groups, although we have never been face-to-face with a member of those groups. Attitudes of this type are learned through our interactions with third parties. We learn some attitudes from our parents as part of the socialization process. Research shows that children's attitudes toward male-female relations (gender roles), divorce, and politics frequently are similar to those held by their parents (Glass, Bengston, & Dunham, 1986; Thornton, 1984). This influence also involves instrumental learning; parents typically reward their children for adopting the same or similar attitudes.

Friends are another important source of our attitudes. The attitude that the law requiring selective service registration is bad, for example, may be learned through interaction with peers. A classic study of Bennington College women by Newcomb (1943) demonstrated the impact of peers on the political attitudes of college students. Although most of these women grew up in wealthy, politically conservative families, the faculty of Bennington had very liberal political attitudes. The study demonstrated that first-year students who maintained close ties with their families and did not become involved in campus activities remained conservative. Women who became active in the college community and who interacted more frequently with other students gradually became more liberal. Presumably, the students at Bennington rewarded the liberal attitudes of their peers.

We acquire attitudes and prejudice toward a particular group through *classical conditioning,* in which a neutral stimulus gradually acquires the ability to elicit a response through repeated association with other stimuli that elicit that response. Children learn at an early age that "lazy," "dirty," "stupid," and many other characteristics are undesirable. Children themselves are often punished for being dirty, or hear adults say, "Don't be stupid!" If they hear their parents (or others) refer to members of a particular group as lazy or stupid, children increasingly associate the group name with the negative reactions initially elicited by these terms. Several experiments have shown that classical conditioning can produce negative attitudes toward groups (Lohr & Staats, 1973; Staats & Staats, 1958).

Another source of attitudes is the media, especially television and films. Here, the mechanism may be *observational learning.* The attitude that "Nuclear power plants are dangerous" may result from exposure to TV news and newsmagazines. The media provide interpretive packages or frames about an object that may influence the attitudes of viewers and readers. A content analysis of media coverage of nuclear power found that, from 1970 to 1979, a secondary frame for such coverage was "runaway"—if nuclear reactors get out of control, they cannot be stopped. After the accident at Three Mile Island nuclear plant in 1979, "runaway" became the primary frame (Gamson & Modigliani, 1989).

The Functions of Attitudes

We acquire attitudes through learning. But why do we retain them? That is, why do attitudes stay with us for months or years? One answer is that they serve

important functions for us (Katz, 1960; Pratkanis & Greenwald, 1989). One way of measuring the function(s) of an attitude is to have people write essays about an attitude object and analyze the essays for themes or patterns (Herek, 1987).

Each attitude serves at least one of three functions. The first is the heuristic or instrumental function. We develop favorable attitudes toward objects that aid or reward us and unfavorable attitudes toward objects that thwart or punish us. Once they are developed, attitudes provide a simple and efficient means of evaluating objects. Businesspeople learn that Republican politicians frequently propose and vote for legislation that benefits business. Learning that a new candidate is a Republican, the businessperson immediately has a favorable attitude.

Second, attitudes serve a schematic or knowledge function—that is, they provide us with a meaningful environment and guide behavior. The world is too complex for us to understand. We group people, objects, and events into categories, or schema, and develop simplified (stereotyped) attitudes that allow us to treat individuals as members of a category. Our attitudes about that category (object) provide us with meaning, with a basis for making inferences about the person (Bodenhausen & Wyer, 1985). The belief that blacks are untrustworthy leads some whites to be guarded in their interaction with blacks. Reacting to every member of the group in the same way is more efficient, if less satisfying, than trying to learn about each as an individual.

Stereotypes of groups are often associated with intense emotions. A strong like or a strong dislike for members of a specific group is called **prejudice.** Prejudice and stereotyping go together, with people using their stereotyped beliefs to justify prejudice toward members of the group. (Stereotypes are discussed in Chapter 5.) The emotional component of prejudice can lead to intergroup conflict (see Chapter 16).

Third, attitudes both define the self and maintain self-worth. Some attitudes express the individual's basic values and reinforce self-image. Many conservatives in our society have negative attitudes toward abortion, racial integration, and equal rights for women. Thus a person whose self-concept includes conservatism may adopt these attitudes because they express that self-image. Some attitudes symbolize a person's identification with or membership in particular groups or subcultures. The attitude "Guns don't kill people; people kill people" is widespread among members of the National Rifle Association. Holding this attitude may be both a prerequisite to acceptance by other group members and a symbol of loyalty to the group.

Finally, some attitudes protect the person from recognizing certain thoughts or feelings that threaten his or her self-image or adjustment. For instance, an individual (say, Tom) may have feelings that he cannot fully acknowledge or accept, such as hostility toward his father. If he recognized this hostility, he would feel very guilty because such sentiments are contrary to his upbringing. So instead of acknowledging that he hates his father, Tom may direct anger and hatred toward members of a minority group or authority figures such as police officers or teachers. Research indicates that experiences which threaten a person's self-esteem, such as failing a test, lead to a more negative evaluation of other groups (Crocker et al., 1987), particularly among persons whose self-esteem was initially high.

The Measurement of Attitudes

When you meet someone, you need information about the person to interact smoothly. You need to find out his or her attitudes about objects that are relevant to your interaction—the class or workplace where you meet, the persons you both know, or current local and national events. Because attitudes are mental states, they cannot be directly observed. Sometimes we can infer someone's attitudes from some form of display associated with the person. A button that says "No Nukes" indicates that the person wearing it is opposed to nuclear power plants. Slogans printed on T-shirts are another source of information about attitudes, as are bumper stickers (Wrightsman, 1969). But we cannot rely on these sources because most people do not put their attitudes on display. To find out someone's attitude, we usually ask the person.

Social psychologists, advertisers, and politicians also are interested in finding out people's attitudes. Often they want to assess the attitudes held by the members of some population. Social psychologists have developed a variety of methods for measuring attitudes, some direct and others indirect.

Direct Methods

The most direct way of finding out someone's attitude is to ask a direct question and record the person's answer. This is the way most of us "study" the attitudes of persons with whom we interact. It is also the technique used by newspaper and television reporters. To make the process more systematic, social psychologists employ several methods, including the single-item measure, Likert scales, and semantic differential techniques (see Box 6.1, "The Measurement of Attitudes"). These measures are obtained by surveys.

It can be difficult to find out another person's attitudes. But some people help us by displaying what they believe.

6.1   The Measurement of Attitudes

Suppose you want to assess attitudes toward premarital sexual behavior. Here are three techniques you could employ.

Single Item

The single item is probably the most common measure of attitudes. An example of this type is:

> I think people should wait until they are married to have sex.
>
> _____ Yes
> _____ No
> _____ Not sure

Likert Scale

The Likert scale consists of a series of statements about the object of interest. The statements may be positive or negative. The respondent indicates how much he or she agrees with each statement. For example:

1. I think people should wait until they are married to have sex.

   ___ Strongly agree   (+ 2)
   ___ Agree   (+ 1)
   ___ Undecided   (0)
   ___ Disagree   (− 1)
   ___ Strongly disagree   (− 2)

2. I think having sex before marriage strengthens the marriage.

   ___ Strongly agree   (− 2)
   ___ Agree   (− 1)
   ___ Undecided   (0)
   ___ Disagree   (+ 1)
   ___ Strongly disagree   (+ 2)

Semantic Differential Scale

The semantic differential scale consists of a number of dimensions on which the respondent rates the attitude object. For example:

> Rate how you feel about premarital sexual intercourse on each of the following dimensions.

good	____	____	____	____	____	____	____	bad
	(+ 3)	(+ 2)	(+ 1)	(0)	(− 1)	(− 2)	(− 3)	
weak	____	____	____	____	____	____	____	strong
	(− 3)	(− 2)	(− 1)	(0)	(+ 1)	(+ 2)	(+ 3)	
fast	____	____	____	____	____	____	____	slow
	(+ 3)	(+ 2)	(+ 1)	(0)	(− 1)	(− 2)	(− 3)	
negative	____	____	____	____	____	____	____	positive
	(− 3)	(− 2)	(− 1)	(0)	(+ 1)	(+ 2)	(+ 3)	
light	____	____	____	____	____	____	____	heavy
	(− 3)	(− 2)	(− 1)	(0)	(+ 1)	(+ 2)	(+ 3)	
exciting	____	____	____	____	____	____	____	boring
	(+ 3)	(+ 2)	(+ 1)	(0)	(− 1)	(− 2)	(− 3)	

Single Items   The use of one question to assess attitudes is very common. The *single-item scale* usually consists of a direct positive or negative statement about the object, and the respondent indicates whether he or she agrees, disagrees, or is unsure. Such a measure is economical; it takes a minimum of time and space to present. It is also easy to score. But the single item is not very precise. Of necessity, it must be general and detects only gross differences in attitude. Using the single-item measure in Box 6.1, we could only separate people into two groups: those who favor premarital abstinence and everybody else.

Likert Scales   We often want to know not only how each person feels about an object but also how each respondent's attitude compares with the attitudes of others. The **Likert scale,** a technique based on summated ratings, provides such information (Likert, 1932).

Box 6.1 includes a two-item Likert scale. Each possible response is given a numerical score, indicated in parentheses. We would assess the respondent's attitude by adding his or her scores for both items. For example, suppose you strongly agree with item 1 (+2) and strongly disagree with item 2 (+2). Your score would be +4, indicating strong opposition to premarital intercourse. Your roommate might strongly disagree with the statement that people should wait until they marry (−2) and might also disagree that premarital sex strengthens a marriage (+1). The resulting score of −1 indicates a slightly positive view of premarital intercourse. Finally, someone who strongly disagrees with item 1 (−2) and agrees with item 2 (−1) would get a score of −3 and could be differentiated from a person who received a −4.

Typically, a Likert scale includes at least four items. The items should be counterbalanced—that is, some should be positive statements, and others should be negative ones. Our two-item scale in the feature has this property; one item is positive, and the other is negative. The Likert scale allows us to order respondents fairly precisely; items of this type are commonly used in public opinion polls. Such a scale takes more time to administer, however, and involves a scoring stage as well.

Semantic Differential Scales

Like most attitude scales, the single-item and Likert scales measure the *denotative* or dictionary meanings of the object to the respondent. However, objects also have a *connotative* meaning, a set of psychological meanings that vary from one respondent to another. For instance, one person may have had very positive experiences with sexual intercourse, whereas another person's experiences may have been very frustrating.

The **semantic differential scale** (Osgood, Suci, & Tannenbaum, 1957) is a technique for measuring connotative meaning. An investigator presents the respondents with a series of bipolar adjective scales.

Each of these is a scale whose ends are two adjectives having opposite meanings. The respondent rates the attitude object on each scale. After the data are collected, the researcher can analyze them by various statistical techniques. Analyses of such ratings frequently identify three aspects of connotative meaning: evaluation, potency, and activity. Evaluation is measured by adjective pairs such as good-bad and positive-negative; potency, by weak-strong and light-heavy; and activity, by fast-slow and exciting-boring.

The example in Box 6.1 includes two bipolar scales measuring each of the three dimensions. Scores are assigned to each scale from +3 to −3; they are then summed across scales of each type to arrive at evaluation, potency, and activity scores. In the example shown, scores on each dimension could range from −6 (bad, weak, and slow) to +6 (good, strong, and fast).

One advantage of the semantic differential technique is that researchers can compare an individual's attitudes on three dimensions, allowing more complex differentiation among persons. Another advantage is that because the meaning it measures is connotative, it can be used with any object, from a specific person to an entire nation. This technique is also used to assess the meaning of role identities (mother, doctor) and role behaviors (hug, cure) (Heiss, 1979; Smith-Lovin, 1990). But a major disadvantage of the semantic differential technique is that it requires more time to administer and to score.

Indirect Methods

All of the methods discussed so far involve asking direct questions. They assume people will honestly report their attitudes toward the object of interest. But is this assumption valid? Some persons might be uneasy if asked questions about their sexual behavior. Also, many people with strong prejudices toward blacks or Asians might be unwilling to express those attitudes to a stranger, the interviewer. Furthermore, these methods assume that every member of the sample has an attitude; some may not, and yet they may answer direct questions as if they did. If either of these assumptions is false, direct methods will yield erroneous results.

To avoid such error, we can measure attitudes indirectly by observing overt behavior. Several researchers have obtained behavioral measures from representative samples.

Suppose your phone rings about 9 P.M. and the caller says, "Hello . . . Ralph's Garage? This is George Williams. Listen, I'm stuck out here. I'm wondering if you'd be able to come out here and take a look at my car?"

You would probably say, "Sorry, this isn't Ralph's Garage." The caller would reply, "This isn't Ralph's Garage? Listen, I'm terribly sorry to have disturbed you, but listen . . . I'm stuck out here on the highway, and that was the last change I had. Now I'm really stuck out here. What am I going to do now? Listen . . . do you think you could do me the favor of calling the garage and letting them know where I am? I'll give you the number. They know me over there."

Would you call? Would it make any difference if the caller sounded white or sounded like a southern black?

Although this phone call appears to have come from a person in need, in actuality it was an instance of the *wrong-number technique,* a procedure for measuring attitudes. Because white persons might not report prejudice toward blacks if they are asked direct questions, the wrong-number technique was developed to provide an indirect method of studying attitudes toward blacks. In one study conducted in Brooklyn, New York (Gaertner & Bickman, 1971), researchers selected subjects from the telephone directory. They called more than 500 whites and 500 blacks and recorded how many of these subjects contacted the garage (actually a confederate of the experimenter, waiting to receive calls). The results showed that whites were more likely to help other whites (65%) than blacks (53%), whereas blacks were equally likely to help both blacks and whites (63%).

A second indirect technique involves littering. As you walk across campus, you are often handed a leaflet about a candidate for campus or political office. If it is for a candidate you support, you probably read and perhaps keep the flyer. But if you dislike the candidate, you will probably get rid of the flyer quickly, perhaps by simply dropping it on the ground. In one study, handbills were placed on windshields of parked cars. As the drivers returned, observers noted whether they kept the flyer or dropped it. As each car left the lot, the interviewer stopped the car and then measured the driver's attitude by asking direct questions. Results showed a strong relationship between dislike of the candidate and littering (Cialdini & Baumann, 1981).

Researchers have studied not only what people throw away but also what they pick up, using the *lost-letter technique* (Schwartz & Ames, 1977). The researcher prepares letters and places them in stamped envelopes addressed to organizations with known positions on major issues. The researcher then drops them in areas of high pedestrian traffic so they appear to have been lost before they were mailed. When addressing the letters, the researcher can vary the organization named on the envelope; for example he can address half of the letters to the National Abortion Rights Action League and the other half to Parents Opposed to Abortion. Or the researcher can address all letters to the same organization and vary the location where they are dropped. The behavioral measure is the percentage of letters returned (the letters are addressed to a post office box controlled by the researcher). Mailing a lost letter to an organization presumably indicates a favorable attitude toward that organization. Variation in the return rate is assumed to reflect variation in the attitudes of a particular neighborhood, campus, or other area where the letters were "lost."

Few studies have compared these indirect methods with direct methods. Evidence indicates, however, that indirect methods are less reliable and less valid than the direct ones (Lemon, 1973). In addition, direct techniques are more sensitive to the differences in attitudes among individuals (Petty & Cacioppo, 1981). Finally, the available evidence suggests that most people do respond honestly to direct questions, even when the questions involve sensitive topics (DeLamater & McKinney, 1982). Unobtrusive methods, such as determining whether "lost" letters are mailed, are a useful approach when it is impossible to ask direct questions or when the researcher believes some persons do not have an attitude on the issue under study.

Attitude Organization and Change

Attitude Structure

Have you ever tried to change another person's attitude toward an object (such as a political candidate or a racial group) or a behavior (such as premarital sex)? If you have, you probably discovered the person had a counterargument for most every argument you put forth. He seemed to have several reasons why he felt his attitude was correct. An individual's attitude toward some object usually is not an isolated unit. It is embedded in a cognitive structure, linked with a variety of other attitudes. We can often find out what other cognitive elements are related to a particular attitude by asking the person why he or she holds that attitude. Consider the following interview.

INTERVIEWER: Why do you think premarital sexual intercourse is bad?

BILL: Because sex outside of marriage is wrong; it is against the teachings of God in the Bible.

INTERVIEWER: Are there any other reasons?

BILL: Well, I think people who have sex before marriage are usually promiscuous, and they could spread AIDS.

INTERVIEWER: Any other reasons?

BILL: Um . . . yeah. They may get pregnant, and teenage pregnancy causes a lot of problems.

This exchange indicates Bill's reasons for his attitude. More than that, it illustrates the two basic dimensions of attitude organization, vertical and horizontal structure (Bem, 1970).

Vertical Structure   Bill is opposed to premarital sex because it violates his religious beliefs. Specifically, it violates the biblical injunction against

intercourse outside of marriage. Bill accepts this injunction because he views the Bible as a statement of God's teachings. Bill's attitude toward premarital intercourse ultimately rests on his belief in God. The unquestioning acceptance of the credibility of some authority, such as God, is termed a *primitive belief* (Bem, 1970).

Attitudes are organized hierarchically. Some attitudes (primitive beliefs) are more fundamental than others. The linkages between fundamental beliefs and minor beliefs in cognitive structure are termed *vertical*. Vertical linkages signify that a minor belief is derived from or dependent on a primitive belief. Such a structure is portrayed in the center of Figure 6.1.

A fundamental or primitive belief, such as a belief in God, is often the basis for a large number of specific or minor attitudes (Bem, 1970). For example, Bill probably is opposed to murder, adultery, and other sins mentioned in the Bible. Changing a primitive belief may result in widespread changes in the person's attitudes. If Bill meets members of the Unification Church ("Moonies"), they may attempt to persuade him that the Reverend Moon is the only legitimate religious authority. If Bill is converted, the resulting change in primitive beliefs will lead to changed attitudes toward many objects, including family and friends.

Horizontal Structure   When the interviewer asked Bill why he was opposed to sex before marriage, Bill gave two other reasons. One was his belief that people who engage in premarital sexual intercourse are promiscuous and that promiscuity spreads AIDS. The other reason was his belief that premarital sex leads to teenage pregnancy and such undesirable consequences as birth defects. These belief structures are portrayed in the right-hand and left-hand columns of Figure 6.1. When an attitude is linked to more than one set of underlying beliefs—that is, when there are two or more different justifications for it—the linkages are termed *horizontal.*

An attitude with two or more horizontal linkages, or justifications, is more difficult to change than one based on a single primitive belief. Even if you show Bill statistical evidence that AIDS is not associated with premarital intercourse, his religious beliefs and

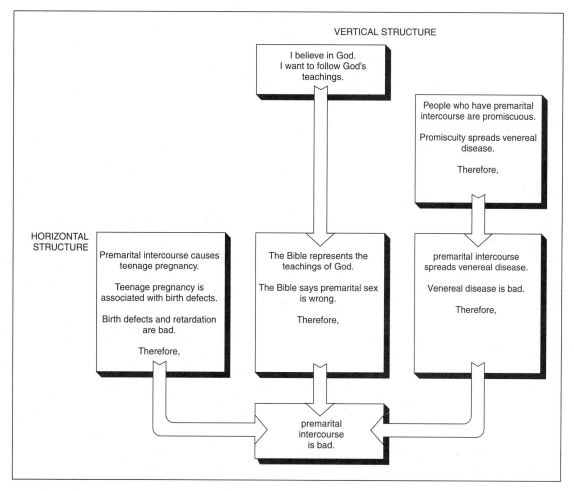

Figure 6.1 The Structure of Attitudes

his concern about teenage pregnancy make it unlikely his attitude will change.

One way to study attitude structure is to interview people and identify vertical and horizontal linkages. A different approach is to study response latency, how long it takes a person to reply to an attitude question (Judd et al., 1991). What is your attitude toward extramarital sex? What is your attitude toward vegetarians? Chances are it took you longer to retrieve from memory your attitude toward vegetarians. Your attitudes about various types of sexual behavior were primed or activated by our discussion of premarital sex and should be associated with short latencies. The

shorter the latency, the closer two attitudes are in a person's attitude structure.

Balance Theory

An important theory about the linkages between attitudes is balance theory. This theory assumes we prefer consistency and that cognitive systems generally are consistent.

The Drive Toward Consistency
The elements of cognitive structure are called **cognitions.** A cognition is an individual's perception of personal

attitudes, beliefs, and behaviors. Bill perceives himself as someone who believes in God and follows God's teachings. These two cognitions go together; we are not surprised that Bill perceives both as applying to him. Many of his attitudes are consistent with what he perceives as God's teachings. For example, he has very negative attitudes toward adultery and murder. Given his attitude toward premarital sex, we would expect Bill to abstain from intercourse until he marries. Indeed, he has never engaged in that behavior. Thus Bill's behavior is consistent with his attitudes.

Consistency among a person's cognitions—that is, beliefs and attitudes—is widespread. If you have liberal political values, you probably favor medical assistance programs for the poor. If you value equal rights for all persons, you probably support affirmative action plans. Also, you may try hard to behave in nonsexist ways when you interact with members of the opposite sex. The observation that most people's cognitions are consistent with one another implies that individuals are motivated to maintain that consistency. Several theories of attitude organization are based on this principle. In general, these *consistency theories* hypothesize that if inconsistency develops between cognitive elements, people are motivated to restore harmony between elements.

Balanced Cognitive Systems

One important consistency theory is **balance theory,** which was formulated by Heider (1958) and elaborated by Rosenberg and Abelson (1960). To see how balance theory works, consider the following statement: "I'm going to vote for Steve Smith; he's in favor of reducing taxes." Balance theory is concerned with cognitive systems like this one. This system contains three elements—the speaker, P; another person (candidate Steve Smith), O; and an impersonal object (taxes), X. According to balance theory, two types of relationships may exist between elements. **Sentiment relations** refer to sentiments or evaluations directed toward objects and people; a sentiment may be either positive (liking, endorsing) or negative (disliking, opposing), symbolized as + or −. **Unit relations** refer to the extent of perceived association between elements. For example, a positive unit relation may result from ownership, a social relationship (such as friendship or marriage), or causality. A negative relation indicates

dissociation, like that between ex-spouses or members of groups with opposing interests. A null relation exists when there is no association between elements.

Using these terms, let's analyze our example. We can depict this system as a triangle (see Figure 6.2A). Balance theory is concerned with the elements and their interrelations from P's viewpoint. In our first example (Figure 6.2A), the speaker favors reduced taxes, perceives Steve Smith as favoring reduced taxes, and intends to vote for Steve. This system is balanced. By definition, a **balanced state** is one in which all three sentiment relations are positive or in which one is positive and the other two are negative. Consider another example (Figure 6.2B). Suppose you favor legalizing possession of marijuana, and candidate Mary Smith wants mandatory prison sentences for its possession. Your cognitions would be balanced if you disliked Mary Smith.

Imbalance and Change

According to balance theory, an **imbalanced state** is one in which two of the relationships between elements are positive and one is negative or in which all three are negative. Consider Judy and Mike, who are seniors in college. They have been going together for 3 years and are in love. Mike is thinking about going to law school. Judy doesn't want him to stay in school after he gets his bachelor's degree. She doesn't want to have to work full-time while he goes to school for 3 more years. Figure 6.2C illustrates the situation from Mike's viewpoint. It is imbalanced; there are two positive relations and one negative one.

In general, an imbalanced situation like this is unpleasant. When subjects are presented hypothetical triads like those shown in Figure 6.2C and asked to rate each triad, imbalanced triads are rated less pleasant than balanced ones (Price, Harburg, & Newcomb, 1966). Balance theory assumes people will try to restore balance, that is, they will eliminate perceived imbalance among cognitions by changing one or another of their attitudes.

There are three basic ways to restore balance. First, Mike may change his attitudes so the sign of one of the relations is reversed (Tyler & Sears, 1977). For instance, Mike may decide he does not want to attend law school (Figure 6.2D). Alternatively, Mike may decide he does not love Judy, or he may persuade Judy

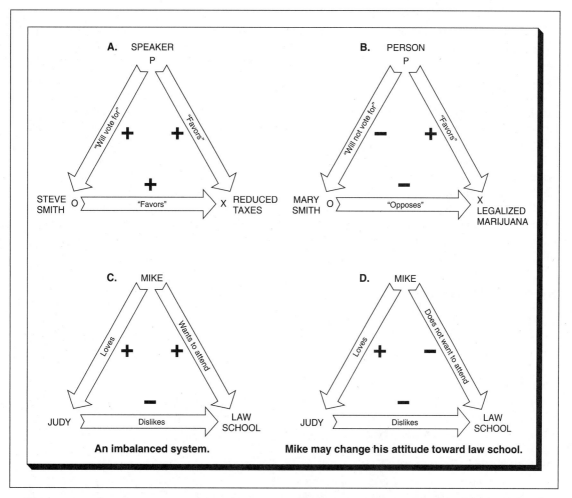

Figure 6.2 Balanced Cognitive Systems and Resolution of Imbalanced Systems

When the relationships among all three cognitive elements are positive (A), or when one relationship is positive and the other two are negative (B), the cognitive system is balanced. When two relationships are positive and one negative, the cognitive system is imbalanced. In (C), Judy's negative attitude toward law school creates an unpleasant psychological state for Mike. He can resolve the imbalance by deciding he does not want to go to law school (D), by deciding he does not love Judy, or by persuading Judy to like law school.

it is a good idea for him to go to law school. Each of these involves changing one relationship so the system of beliefs contains either zero or two negative relationships.

Second, Mike can restore balance by changing a positive or negative relation to a null relation (Steiner & Rogers, 1963). Mike may decide that Judy doesn't know anything about law school and her attitude toward it is irrelevant. Third, Mike can restore balance by differentiating the attributes of the other person or object (Stroebe et al., 1970). For instance, Mike may distinguish between major law schools, which require all the time and energy of their students, and less prestigious ones, which require less work. Judy is correct in her belief that they would have to postpone marriage if he goes to Yale Law School. However, Mike believes he can go to a local school part-time and also work.

Which technique will a person use to remove imbalance? Balance is usually restored in whichever way is easiest (Rosenberg & Abelson, 1960). If one relationship is weaker than the other two, the easiest mode of restoring balance is to change the weaker relationship (Feather, 1967). Because Mike and Judy have been seeing each other for 3 years, it would be very difficult for Mike to change his sentiments toward Judy. It would be easier for him to change his attitude toward law school than to get a new fiancée. However, Mike would prefer to maintain their relationship and go to law school. In this case, he may attempt to change Judy's attitude, perhaps by differentiating the object (law schools). If this influence attempt fails, Mike will probably change his own attitude toward law school.

Theory of Cognitive Dissonance

Another major consistency theory is the **theory of cognitive dissonance.** Whereas balance theory deals with the relationship among three cognitions, dissonance theory deals with consistency between two or more elements (behaviors and attitudes). There are two situations in which dissonance commonly occurs: (1) after a decision, or (2) when one acts in a way that is inconsistent with his or her beliefs.

Postdecisional Dissonance   Susan will begin her junior year in college next week. She needs to work part-time to pay for school. After 2 weeks of searching for work, she receives two offers. One is a part-time job doing library research for a faculty member she likes and it pays $5.50 per hour with flexible working hours. The other is a job in a restaurant as a cashier that pays $8 per hour but has working hours from 5 P.M. to 9 P.M., Thursdays, Fridays, and Saturdays. Susan has a hard time choosing between jobs. Both are located near campus, and she thinks she would like either one. Whereas the research job offers flexible hours and easier work, the cashier's job pays more and offers her the opportunity to meet interesting people. In the end, Susan chooses the cashier's job, but she is experiencing dissonance.

Dissonance theory (Festinger, 1957) assumes there are three possible relationships between any two cognitions. Cognitions are consistent, or *consonant,* if

one naturally or logically follows from the other. They are *dissonant* when one implies the opposite of the other. The logic involved is *psycho logic* (Rosenberg & Abelson, 1960)—logic as it appears to the individual, not logic in a formal sense. Two cognitive elements also may be irrelevant; one may have nothing to do with the other. In Susan's case, the decision to take the cashier's position is consonant with its convenient location, the higher pay, and the opportunities to meet people, but it is dissonant with the fact that she will be responsible for hundreds of dollars and has to work weekend nights (see Figure 6.3).

Having made the choice, Susan is experiencing **cognitive dissonance,** a state of psychological tension induced by dissonant relationships between cognitive elements. Although dissonance is a cognitive concept, it has physiological correlates. A study used the galvanic skin response (GSR)—a widely utilized measure of physiological arousal—as a measure of the

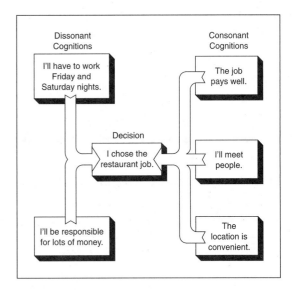

Figure 6.3   Postdecisional Dissonance

Whenever we make a decision, there are some cognitions—attitudes, beliefs, knowledge—that are consonant with that decision, and other cognitions that are dissonant with it. Dissonant cognitions create an unpleasant psychological state that we are motivated to reduce or eliminate. In this example, Susan has chosen a job and is experiencing dissonance. Although three cognitions are consistent with her decision, two other dissonant cognitions are creating psychological tension.

tension induced by dissonance (Elkin & Leippe, 1986). As expected, subjects who were placed in a dissonance-arousing situation showed an elevated GSR.

Some decisions produce a large amount of cognitive dissonance, others very little. The magnitude of dissonance experienced depends on the proportion of elements that are dissonant with a person's decision. In Susan's case, there are three consonant and only two dissonant cognitions, so she will experience moderate dissonance. The magnitude is also influenced by the importance of the elements. She will experience less dissonance if it is not important that she has to work every Friday and Saturday, more dissonance if an active social life on weekends is important to her.

Dissonance is an uncomfortable state. To reduce dissonance, the theory predicts that Susan will change her attitudes. She can either change the cognitive elements themselves or change the importance associated with the elements. It is hard to change cognitions. She chose the restaurant job, and she made a commitment to work weekend nights and to be responsible for large sums of money. Alternatively, Susan can change the relative importance of her cognitions. She can emphasize the importance of one (or more) of the consonant cognitions and deemphasize one (or more) of the dissonant cognitions. Although she has to work to earn money, she can emphasize the fact that the cashier's job pays well. Although she would prefer to be able to go out on weekends, this is less important than otherwise because the cashier's job will allow her to meet people.

In a laboratory study of postdecision dissonance, undergraduate women were given a choice between two products, such as a toaster and a coffeemaker. Subjects rated the attractiveness of each item before and after their choice. Researchers predicted that to reduce dissonance, the women would minimize the importance of cognitions dissonant with the decision. That is, after the choice is made, the attractiveness of the item chosen would increase and the attractiveness of the item not chosen would decrease. Results verified these hypotheses (Brehm, 1956).

Counterattitudinal Behavior

A second circumstance that produces dissonance occurs when a person behaves in a way that is inconsistent with his or her attitudes. Such situations may involve *forced compliance,* that is, pressures on a person to comply with a request to engage in counterattitudinal behavior.

Imagine you have volunteered to serve in a psychology experiment. You arrive at the lab and are told you are participating in a study of performance. You are given a pegboard and told to turn each peg exactly one-quarter turn. After you have turned the last peg, you are told to start over, to turn each peg another one-quarter turn. Later you are told to remove each peg from the pegboard and then to put each peg back. After an hour of such activity, the experimenter indicates you are finished. The experimenter says, "We are comparing the performance of subjects who are briefed in advance with that of others who are not briefed. You did not receive a briefing. The next subject is supposed to be briefed, but my assistant who usually does this couldn't come to work today." He then asks you to help out by telling a waiting subject that the tasks you have just completed were fun and exciting. For your help, he offers you either $1 or $20.

In effect, you are being asked to lie, to say the boring and monotonous tasks you performed are enjoyable. If you actually tell the next subject the tasks are fun, you may experience cognitive dissonance afterward. Your behavior is inconsistent with your cognition that the tasks are boring. In addition, lying to the next subject is dissonant with your beliefs about yourself—that you are moral and honest. To reduce dissonance, you can change one of the cognitions. Which will you change? You cannot change your awareness that you told the next subject the task is fun. The only cognition open to change is your attitude toward the tasks, which can change in the direction of greater liking for the tasks.

The theory of cognitive dissonance predicts, first, that you will change your attitudes toward the tasks (like them better) and, second, that the amount of change will depend on the incentive you were paid to tell the lie. Specifically, the theory predicts that greater attitude change will occur when the incentive to tell the lie is low ($1) rather than high ($20), because you will experience greater dissonance under low incentive than you would under high incentive.

These predictions were tested in a classic experiment by Festinger and Carlsmith (1959). In this

People use various strategies for handling the dissonance aroused by messages that are inconsistent with their behavior. Faced with these two ads, a nonsmoker resolves the inconsistencies by emphasizing the importance of health and denying that smokers enjoy life more than nonsmokers. A smoker resolves the inconsistencies by denying the risk of cancer and emphasizing the link between smoking and a life of leisure.

study, most of the subjects agreed to brief the next subject. They told him that the tasks were interesting and that they had fun doing them. A secretary then asked each subject to rate the experiment and the tasks. These ratings provide the measures of the dependent variable. As expected, control subjects who did not brief anyone and were not offered money rated the tasks as very unenjoyable and did not want to participate in the experiment again.

What about the experimental subjects who were paid money to tell a lie? For those receiving $20, the situation was not very dissonant. The money provided ample justification for engaging in counterattitudinal behavior (lying). In the $1 condition, however, the subjects experienced more dissonance because they did not have the justification for lying provided by a large amount of money. These subjects could not deny they lied, so they reduced dissonance by changing their attitude—that is, by increasing their liking for the task and the experiment. Results of this study confirmed the predictions from dissonance theory. Subjects in the high-incentive ($20) condition experienced little dissonance and rated the task and experiment negatively, whereas those in the low-incentive ($1) condition experienced more dissonance and rated the task and experiment positively.

These results reflect the **dissonance effect:** the greater the reward or incentive for engaging in counterattitudinal behavior, the less the resulting attitude change. The opposite of this is the **incentive effect:** the greater the incentive for engaging in counterattitudinal behavior, the greater the resulting attitude change.

Under what conditions does each of these effects occur? Research suggests that the dissonance effect is more likely when subjects choose (or have the illusion of choosing) whether or not to engage in the behavior. In one study (Sherman, 1970), subjects were asked to write essays taking a position on current issues that contradicted their own attitudes. They were paid either 50 cents or $2.50 for the essay. In one condition, subjects were allowed to choose whether or not to write the essay; in the second condition, students were not given a choice. For those subjects given a choice, the results showed a dissonance effect: Subjects given little incentive (50 cents) wrote longer, more persuasive essays and showed more attitude change than those given larger incentive ($2.50). For those subjects given no choice, the results showed an incentive effect: Subjects given more incentive ($2.50) wrote longer, more persuasive essays and showed more attitude change than those given a small incentive (50 cents).

Dissonance occurs only in some situations (Wicklund & Brehm, 1976). To experience dissonance, a person must be committed to a belief or course of action (Brehm & Cohen, 1962). In addition, the person must believe he or she chose to act voluntarily and is thus responsible for the outcome of the decision (Linder, Cooper, & Jones, 1967). This is shown in the case of Susan, who chose the cashier's job. If the owner of the restaurant were Susan's father and he demanded she work for him, she would have had little or no postdecision dissonance.

Is Consistency Inevitable?

If our beliefs and behavior were always consistent, all of our cognitions would be in harmony. Obviously, that is not the case. Practically every adult in the United States knows that cigarette smoking causes lung cancer, yet millions continue to smoke. Most of us overindulge in a favorite food (pizza, chocolate) or beverage (soda, beer) at least occasionally, although we know it is not healthy to do so. When we do, our behavior is inconsistent with the belief that overindulgence is unhealthy. How is it that people can hold mutually inconsistent cognitions?

For one thing, many of our cognitions never come into contact with each other. In order for people to experience a strain toward consistency, they must perceive the *implicational relationship* between two in-

consistent cognitions or attitudes (Lavine, Thomsen, & Gonzales, 1997). Research indicates that thought about attitude objects, or about the relationship between an attitude and one's values, may lead to recognition of this relationship. The strength of the perceived implicational relationship determines the strength of the consistency pressure. A related reason for inconsistency is that some of our behavior is mindless. Because we do not think about our actions, we are unaware that they are inconsistent with our beliefs (Triandis, 1980). The cigarette smoker often lights up without consciously thinking about the act; sometimes he is surprised to find a lit cigarette in the ashtray. Thus the relationship between the act and one's knowledge is often not salient to the person.

A third reason inconsistency occurs is that each belief, attitude, or self-perception is embedded in a larger structure of consistent, related attitudes, beliefs, and self-perceptions. For example, Bill's attitude toward premarital sexual intercourse, discussed earlier, was embedded in a structure of other attitudes and values. Although two attitudes may be inconsistent, each may be related to several other consonant attitudes. To change one or the other would create new inconsistencies. In effect, people tolerate some inconsistencies to avoid others.

Finally, people vary in the strength of the preference for consistency. In one set of experiments, half of the participants did not show this preference (Cialdini, Trost, & Newsom, 1995). It has also been suggested that the motivation toward consistency is stronger in Western—United States, Western Europe—than in Eastern cultures—Japan, India (Markus & Kitayama, 1991).

The Relationship Between Attitudes and Behavior
Do Attitudes Predict Behavior?

We have seen how behavior can affect our attitudes and how people sometimes change their attitudes when their behavior appears to contradict them. However, most people think of attitudes as the source of behavior. For example, we often assume that when we know a person's attitude toward an object (another person, volleyball, Genesis's music), we can predict how that person will behave toward the object. If you know someone enjoys volleyball, you would expect her to accept your invitation to play volleyball with friends. When we are able to predict another person's responses, we can decide how to behave toward that person in order to achieve our own goals. But can we truly predict someone's behavior if we know his or her attitudes?

In 1930 the social scientist Richard LaPiere traveled around the United States by car with a Chinese couple. At that time, there was considerable prejudice against the Chinese, particularly in the West. The three travelers stopped at more than 60 hotels, auto camps, and tourist homes and more than 180 restaurants. They kept careful notes about how they were treated. In only one place were they denied service. Later, LaPiere sent a questionnaire to each place, asking whether they would accept Chinese guests. He received responses from 128 establishments; 92% of them indicated that they would *not* serve Chinese guests (LaPiere, 1934). Evidently there can be a great discrepancy between what people do and what they say.

Many studies on the topic have found only a modest correlation between attitude and behavior (Wicker, 1969). The correlation (*r*) is a measure of the relationship between two variables and may range from −1.00 to +1.00. If there is no relationship between variables, the correlation is zero. If two variables increase together (or decrease together), the correlation is positive. A survey of 33 studies of attitudes and behavior found that the average correlation between these two variables is +0.30 or less, a modest correlation. Several reasons why the relationship is not stronger have been suggested. In this section, we consider four variables that influence the relationship between attitudes and behavior: (1) the activation of the attitude, (2) the characteristics of the attitude, (3) the correspondence between attitude and behavior, and (4) situational constraints on behavior.

Activation of the Attitude

Each of us has thousands of attitudes. Most of the time a particular attitude is not within our conscious aware-ness. Moreover, much of our behavior is mindless or spontaneous (Fazio, 1990). We act without thinking, that is, without considering our attitudes. For an attitude to influence behavior, it must be **activated,** that is, brought from memory into conscious awareness (Zanna & Fazio, 1982).

An attitude is usually activated by exposure of the person to the object, particularly if the attitude was originally formed through direct experience with the object (Fazio, Powell, & Herr, 1983). Earlier sections of this chapter may have activated your attitudes toward many objects, such as Genesis's music, blacks, premarital sexual activity, and cigarette smoking. Thus one way to activate attitudes is to arrange situations in which persons are exposed to relevant objects. Soft lighting, a cozy fire, and glasses of wine are all associated with seduction; we often set up these cues in the hope of activating our partner's positive attitudes toward romantic and sexual activity.

Attitudes differ in the ease with which they can be activated—that is, they differ in *accessibility.* Some attitudes are highly accessible and are activated automatically by the presentation of the object, such as stereotypes (Devine, 1989). One measure of the degree to which an attitude is accessible is the speed of activation. Attitudes activated instantaneously are, by definition, highly accessible. Other attitudes are activated more slowly—that is, they are less accessible (Fazio et al., 1986). The more accessible an attitude, the greater its influence on how we categorize an object and our judgments about that object (Smith, Fazio, & Cejka, 1996).

Evidence also indicates that the more accessible an attitude, the more it is likely to guide future behavior. This was shown, for example, in a study of the impact of accessibility on voting in the 1984 presidential election (Fazio & Williams, 1986). In June and July 1984, 245 people were questioned about their attitudes toward presidential candidates Ronald Reagan and Walter Mondale. The latency of the answer—how quickly the person replied to the question about each candidate—was used as a measure of accessibility. After the election, each person was asked whom he or she voted for. The more accessible the attitude—that is, the more quickly the person replied to the question about the candidate—the more likely the person was to vote for that candidate in November.

Characteristics of the Attitude

The relationship between attitude and behavior is also affected by the nature of the attitude itself. Four characteristics of attitudes that may influence the relationship are (1) the degree of consistency between the affective (evaluative) and the cognitive components, (2) whether the attitude is grounded in personal experience, (3) the strength of the attitude, and (4) whether it is stable over time.

Affective-Cognitive Consistency At the beginning of the chapter, we identified three components of an attitude: cognition, evaluation (affect), and behavioral predisposition. When we consider the relation between attitude and behavior, we are looking at the relationship between the first two components and the third. Not surprisingly, research has shown that the degree of consistency between the affective and cognitive components affects the attitude-behavior relationship. That is, the greater the consistency between cognition and evaluation, the greater the strength of the attitude-behavior relation.

Recall that the cognitive component is a belief about the attitude object (for example, "Capital punishment is necessary to protect society"). The affective component is the emotion associated with the object ("I am strongly in favor of capital punishment"). In this case, there is a high degree of affective-cognitive consistency. Now suppose another person endorses the belief but is opposed to capital punishment. Whose behavior could you confidently predict? The first person is much more likely to write letters to legislators supporting the death penalty and to vote for candidates who advocate its use.

Although most Americans have attitudes about abortion, only a minority act on their beliefs like these demonstrators. People with strong attitudes, whether pro or con, are more likely to engage in such behavior.

In one experiment, subjects' beliefs and evaluations regarding capital punishment were assessed by questionnaires (Chaiken & Yates, 1985). Next, subjects who were either high or low in consistency were asked to write two essays, one on the death penalty and one on an unrelated topic. The death penalty essays written by high-consistency subjects were much more internally consistent; that is, their attitudes were part of an internally consistent structure. Furthermore, high-consistency subjects dealt with discrepant information by discrediting it or minimizing its importance, making their attitudes more resistant to change.

Direct Experience

Suppose you have a positive attitude toward an activity based on having done it once, and your roommate has a positive attitude based on hearing you rave about it. Which of you is more likely to accept an invitation to engage in it again?

One study (Regan & Fazio, 1977) provides an answer to this question. The behavior of interest was the proportion of time spent playing with several kinds of puzzles. Subjects in the direct-experience condition played with sample puzzles; those in the indirect-experience condition were given only descriptions of the puzzles. Researchers then asked subjects to respond to some attitude measures and later gave them an opportunity to play with the puzzles. They discovered that the average correlation between attitude and behavior was much higher for subjects who had direct experience than for those who did not.

Attitudes based on direct experience are more predictive of subsequent behavior for several reasons (Fazio & Zanna, 1981). The best predictor of behavior is past behavior; the more frequently you have played tennis in the past, the more likely you are to play it in the future (Fredricks & Dossett, 1983). An attitude is a summary of a person's past experience; thus an attitude grounded in direct experience predicts future behavior more accurately. In addition, direct experience makes more information available about the object itself (Kelman, 1974). In a test of the hypothesis that the attitude-behavior relation will increase as the amount of information increases, researchers studied three different behaviors, including voting for specific candidates in an election (Davidson et al., 1985). The results indicated that both the amount of information and direct experience increase the relationships.

Strength

Suppose you ask two friends which candidate they like in the upcoming presidential election. One replies, "I'm voting for X"; the other hedges a bit, saying, "Well, maybe I'll vote for Y." Which person's behavior do you think you could predict? In general, the greater the strength of an attitude, the more likely it is to influence behavior.

Studies of voting behavior find that many of the errors in predictions occur among those who report indifference to the election—that is, people who have weak or uncertain attitudes (Schuman & Johnson, 1976). In one study, researchers measured people's attitudes before an upcoming election. Subjects completed a 15-item Likert scale and indicated how certain they were of each response (Sample & Warland, 1973). For each subject, two measures were constructed: one of attitude and one of certainty. The 243 subjects were divided into high- and low-certainty groups. Researchers noted how subjects voted in the election 15 days after they had completed the questionnaire. Among subjects who were more certain of their attitudinal responses, the attitude-behavior correlation was 0.47, whereas among those who were less certain the correlation was 0.06. The average for all subjects was in the usual range (0.29).

Attitudes based on direct experience with the object may be held with greater certainty. Certainty is also influenced by whether affect or cognition was involved in the creation of the attitude. Attitudes formed based on affect (for example, fear of snakes) are more certain than attitudes based on cognition (for example, a preference for Hondas based on reading analyses of auto quality) (Edwards, 1990).

The **relevance** of an attitude, the extent to which the issue or object directly affects the person, is an important influence on its strength. Framing an issue in relevant terms (e.g., tuition increases on your college campus) brings to mind important consequences for you, such as the need for greater income. Framing it in irrelevant terms (e.g., tuition increases on campuses in Russia) may elicit no thought of personal consequences (Lieberman & Chaiken, 1996). A study of the reactions of 1,300 adults in the Boston area to busing students to achieve racial integration included several measures of relevance: Was busing occurring in the respondent's neighborhood, did she or he have a child in the public schools, and was his or her neighborhood integrated racially? The survey also measured racial attitudes, including tolerance, and who the respondent voted for in 1972, Nixon or McGovern. A much stronger relationship between racial attitude and voting behavior was found among adults for whom busing was a relevant issue (Crano, 1997).

Temporal Stability   Most studies attempting to predict behavior from attitudes measure people's attitude first and their behavior weeks or months later. A modest or small correlation may mean a weak attitude-behavior relationship. Or it could mean people's attitudes have changed in the interim period. If the attitude changes after it is measured, the person's behavior may be consistent with his or her present attitude, although it appears inconsistent with our measure of the attitude. Thus, to predict behavior from attitudes, the attitudes must be stable over time.

In general, we would expect that the longer the time between the measurement of attitude and of behavior, the more likely the attitude will change and the smaller the attitude-behavior relationship will be. In a study designed to test this possibility (Schwartz, 1978), an appeal was mailed to almost 300 students to volunteer as tutors for blind children. Earlier, students had filled out a questionnaire measuring general attitudes toward helping others, as well as questions about tutoring blind children. Some students had filled out the questionnaire 6 months earlier; some, 3 months earlier; some, both 3 and 6 months earlier; and still others had not seen the questionnaire. The correlation between attitude toward tutoring and actually volunteering was greater over the 3-month period than over the 6-month period. Thus, to avoid problems of temporal instability, the amount of time between the measurement of attitudes and of behavior should be brief.

Some attitudes evidence a remarkable degree of stability, however. Thornton (1984) studied the attitudes of 458 women toward divorce; their attitudes were measured in 1962, 1977, and 1980. He found substantial stability over the 18-year period, particularly among women who attended church regularly. Marwell, Aiken, and Demerath (1987) studied the political attitudes of 220 white young people who spent the summer of 1964 organizing blacks in the South to vote. They measured the same attitudes of two thirds of these activists in 1984, two decades later. The extreme radical attitudes these people held in 1965 softened in the intervening 20 years; but, in general, these people remained liberal and committed to the needs of disadvantaged groups (see Box 6.2, "Where Are They Now? Persistence and Change in Political Attitudes").

What is the relationship between age and attitude stability? One hypothesis is that attitudes are very unstable in adolescence and young adulthood and very stable thereafter. Alternatively, attitudes may become more stable gradually throughout life. Analyses of data on sociopolitical attitudes from three longitudinal studies indicate that attitudes stabilize in young adulthood and remain relatively unchanging thereafter (Alwin & Krosnick, 1991; Krosnick & Alwin, 1989).

Correspondence

Attitudes are more likely to predict behavior when the two are at the same level of specificity (Schuman & Johnson, 1976). For example, suppose you have invited a casual acquaintance to dinner, and you want to plan the menu. You know she is Italian, so she probably likes Italian food. But can you predict with confidence that she will eat green noodles with red clam sauce? Probably not. A favorable attitude toward a type of cuisine does not mean the person will eat every dish of that type.

Many studies have attempted to predict from general attitudes to specific behaviors. For instance, some studies of racial prejudice have tried to predict from people's general attitudes toward blacks to specific behaviors, such as willingness to have one's photograph taken with particular blacks in particular settings (Green, 1972). Not surprisingly, the relationship between attitude and behavior was weak. A

6.2   Where Are They Now? Persistence and Change in Political Attitudes

The decade from 1961 to 1971 was a period of political activism, when tens and perhaps hundreds of thousands of Americans engaged in activities such as demonstrations, sit-ins, voter registration drives, and community organizing in an attempt to bring about social and political change. College campuses were frequently the setting for these activities, and college students were some of the most active and visible participants. These people were often called "radicals" because they were calling for large-scale changes: civil rights for blacks, the withdrawal of American troops from Vietnam, and the restructuring of colleges and universities to give students more power. In the early 1970s, the student movement subsided, and the faces of its leaders faded from front pages and television screens.

Where are they now? What happened to those student activists? What happened to their sometimes radical attitudes as they grew older and conditions in American society changed? One possibility is that their radical attitudes reflected youthful idealism and as they became involved in family and work their attitudes mellowed. In fact, we might find that yesterday's activists are today's establishment, that their education gave them access to positions as lawyers, bankers, and stockbrokers in the 1980s. Others argue that the activists of the 1960s

were the product of unique historical experiences—the presidency of John F. Kennedy, the war in Vietnam—that created radical attitudes which will last throughout their lives (DeMartini, 1983). According to this view, we may find the former activists working in social service and community action settings.

An unusual longitudinal study provides us with some answers. In early 1965, 300 young people—95% of them students—volunteered to spend 10 weeks in the South, working to get blacks registered to vote. During their training, 231 of the activists completed a questionnaire that measured various characteristics and attitudes. The volunteers were young, from metropolitan and urban areas, and from politically liberal families (Demerath, Marwell, & Aiken, 1971). Their attitudes were very liberal, if not radical: 90% favored federal intervention to secure civil rights for blacks in the South, and 71% felt a large portion of the federal budget should be devoted to social programs. One third expressed pessimism regarding American institutions, saying they had grave doubts about democracy as a political system and that the system had proven itself incapable of coping with racial discrimination.

In 1984 and 1985 the researchers attempted to locate those who participated in the original study. The researchers were able to find and obtain data from 145

continued on next page

---

general attitude is a summary of many feelings about an object under a variety of conditions or about a whole class of objects. Logically, it should not predict behavior in any particular single situation. But it might predict a composite measure of several relevant behaviors (Weigel & Newman, 1976).

What about predicting a specific behavior, such as whether your Italian guest will eat green noodles and red clam sauce? Just as general attitudes best predict a composite index of behavior, we need a specific measure of attitude to predict a specific behavior. We can think of an attitude and a behavior as having four elements: an action (eating), object or target (green

noodles and red clam sauce), context (in your home), and time (tomorrow night). The greater the degree of **correspondence**—that is, the number of elements that are the same in the two measures—the better we can predict behavior from attitudes (Ajzen & Fishbein, 1977).

A study of birth control use by 244 women (Davidson & Jaccard, 1979) demonstrated that attitudinal measures which exhibit correspondence with the behavioral measure are better predictors of behavior. In this study, the behavior of interest was whether women used birth control pills during a particular 2-year period. Attitude was measured in four

continued from previous page

of the 231 (63%). They compared those in the follow-up sample with the original group on 13 background characteristics. The only significant difference was on gender; there were proportionately more males in the 1984–1985 group because the researchers had greater difficulty locating the women who had been members of the original group.

The follow-up study yields an interesting picture of student activists 19 years later (Marwell, Aiken, & Demerath, 1987). In 1984–1985 they were all in their mid-40s and often embedded in work and family roles. What happened to their attitudes? First, the researchers compared selected political attitudes of the former activists with those of a national sample surveyed in 1982. The activists differed on all nine items, generally picking the most liberal response possible. Next, the researchers compared each individual respondent's attitudes in 1984–1985 with his or her attitudes in 1963. Changes were noted in three areas. First, the activists' commitment to nonviolence as a strategy was significantly lower in 1984–1985. Second, their attitudes toward the South had become less hostile in the intervening years; this may in part reflect the fact that conditions in the South have changed. Finally, they had less trust in federal political institutions than they had in 1965. These results suggest a softening of the extreme attitudes held by these people in the 1960s. At the same time, the former activists were still very liberal.

Another study compared 212 activists who spent the summer of 1964 in Mississippi with 118 persons who were accepted by the same program but did not participate (McAdam, 1989). All of these people were interviewed in 1984. Volunteers were more politically active in the 1980s than the nonparticipants. Also, volunteers had significantly lower incomes in 1984, suggesting that some of them were working in social service and community settings.

A study of the relationship between activism between 1967 and 1973 and later political participation among women also reports a positive relationship (Cole & Stewart, 1996). The focus of activism in the late 1960s differed. Questioned in 1992, black women were more likely to report participation in the civil rights movement, whereas white women reported participation in the women's movement. Black women reported significantly more activism during their student years, and reported more frequent participation in 17 political behaviors in the 2 years prior to the interview. Women who were more politically active at midlife reported a greater sense of their own efficacy in changing the political system and a stronger desire to improve the world for future generations.

The data from all of these studies indicate more persistence than change in political attitudes and rates of political participation among those who were young adults in the years 1965 to 1973, at least among people who were politically active as youth.

---

ways. The measure of the women's general attitude toward birth control had only one element in common with the behavior (object). The correlation between this general measure and behavior was a modest 0.323, as shown in Figure 6.4. When the attitude measure had two elements in common with the behavior (object and action), the correlation rose to 0.525. Finally, an attitude measure that included three elements (object, action, and time)— "Do you plan to use birth control pills in the next 2 years?"—was most highly correlated with the behavioral measure. Thus attitudinal measures having high correspondence with the behavioral measure are

better predictors of behavior than attitudinal measures having low correspondence.

Earlier in this chapter, we mentioned that in LaPiere's study most establishments that served the Chinese couple later said they would not. The lack of a relationship between attitude and behavior in LaPiere's study may be due to lack of correspondence. The behavioral measure was whether a particular Chinese couple (object) was served (action) in a particular restaurant or hotel (context) on a particular day (time). However, LaPiere's questionnaire measuring attitudes simply asked whether Chinese guests would be served. Thus there was correspondence

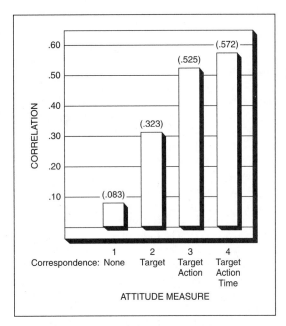

Figure 6.4 Correlations of Attitude Measures That Vary in Correspondence With Behavior

Every behavior involves a target, action, context, and time. In order to predict behavior from attitude, the measures of attitude and behavior should correspond—that is, involve the same elements. The larger the number of elements in common, the greater the correlation between attitude and behavior. Researchers obtained four measures of attitudes toward birth control from 244 women: (1) general attitude toward birth control; (2) attitude toward birth control pills; (3) attitude toward using pills; and (4) attitude toward using pills in the next 2 years. The behavioral measure was actual use of pills during the 2-year period. Note that as correspondence increased from zero to three elements, the correlation between attitude and behavior also increased.

SOURCE: Adapted from Davidson and Jaccard, 1979.

between the measures on only one element—the action—which may account for the discrepancy LaPiere found between attitude and behavior.

Situational Constraints

If you believe tuition increases at your college are necessary to maintain the quality of your education—

retain the best faculty, provide ready access to books, journals, and computers, and so on—and you attend a meeting of Students for Educational Quality, your behavior reflects your attitude. If you are opposed to tuition increases and you participate in a protest march, your behavior is consistent with your attitude.

Suppose, however, that you oppose tuition increases but find yourself in a conversation with your partner's parents in which they express support for the increases. Would you voice your opposition—that is, behave in a manner consistent with your attitudes—or not? Your reaction would probably depend partly on the strength and certainty of your attitude (Pratkanis & Greenwald, 1989). If you are strongly opposed to tuition increases, you may speak your mind. But if you are moderately opposed, you may decide to avoid an argument and behave in a way that is inconsistent with your attitude. In LaPiere's study, for instance, hotel and restaurant employees confronted by a white man and a Chinese couple may have felt compelled to serve them rather than run the risk of creating a scene by refusing to do so.

Situational constraint refers to an influence on behavior due to the likelihood that other persons will learn about a behavior and respond positively or negatively to it. Situational constraints often determine whether our behavior is consistent with our attitudes. In fact, how we behave is frequently a result of the interaction between our attitudes and constraints present in the situation (Warner & DeFleur, 1969). This relationship is summarized in Figure 6.5, using attitudes toward college tuition increases as an example. A conversation between someone weakly opposed to increases and your partner's parents (weak pressures) would be a situation of conflict for the individual, whereas someone in the same situation who is strongly opposed to the increases is more likely to voice that opposition.

Sometimes we feel constrained by the possibility that others may learn of our behavior. At other times those around us exert direct social influence; they communicate specific expectations about how we should behave. The greater the agreement among others about how we should behave, the greater the situational constraint on persons whose attitudes are inconsistent with the situational norms (Schutte,

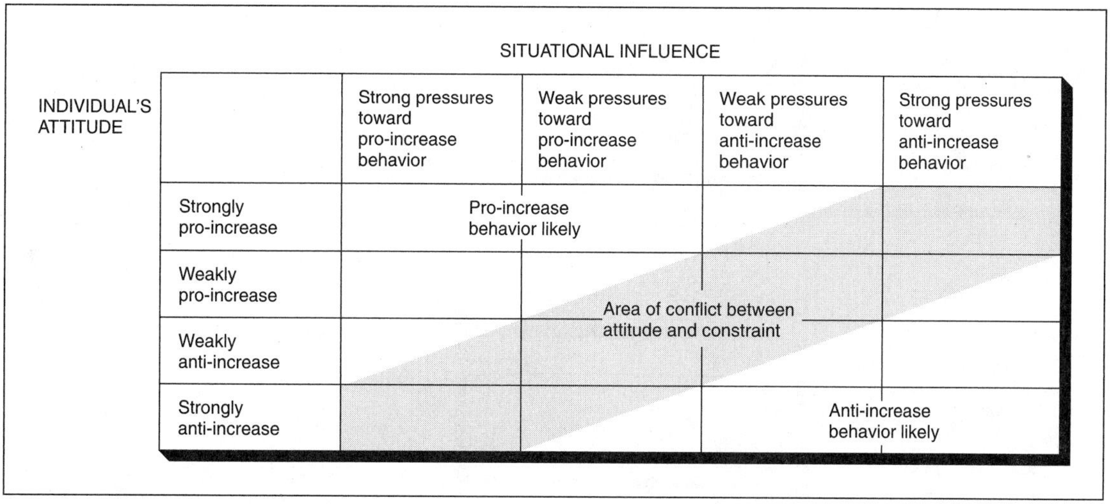

SITUATIONAL INFLUENCE				
INDIVIDUAL'S ATTITUDE	Strong pressures toward pro-increase behavior	Weak pressures toward pro-increase behavior	Weak pressures toward anti-increase behavior	Strong pressures toward anti-increase behavior
Strongly pro-increase	Pro-increase behavior likely			
Weakly pro-increase		Area of conflict between attitude and constraint		
Weakly anti-increase				
Strongly anti-increase			Anti-increase behavior likely	

Figure 6.5 **The Influence of Attitude and Situational Constraints on Behavior**

Our behavior is influenced not only by our attitudes but also by situational constraints, the behavior of others, or the likelihood that others will find out what we do. When the individual has a strongly held attitude and situational influences encourage behavior consistent with the attitude, there will be a strong relationship between attitude and behavior. But when situational influences produce pressure to behave in ways inconsistent with one's attitude or when the attitude is weak, behavior and attitude are less likely to be consistent.

SOURCE: Adapted from Warner and DeFleur, 1969, figure 3.

Kendrick, & Sadalla, 1985). Under these conditions, there is a weaker relationship between attitudes and behavior. Consequently, the less visible our behavior is to others, the more likely it is our behavior and attitudes will be consistent (Acock & Scott, 1980).

But what if persons whose opinions we value are not actually present? Several studies have assessed the impact of reference groups on the attitude-behavior relationship. Such research involves measuring the subject's attitudes toward some object and then asking him or her to indicate the position of various social groups about that object. One survey assessed adults' attitudes toward drinking alcoholic beverages and the degree to which their friends approved of drinking (Rabow, Newman, & Hernandez, 1987). When attitudes and social support were congruent—that is, when the respondent's and friends' views were the same—there was a much stronger relation between attitudes and behavior than when attitudes and social support were not congruent. Another study found that the perceived norms of their friends influenced

whether women engaged in regular exercise, but only for those who identified strongly with the reference group (Terry & Hogg, 1996).

The Reasoned Action Model

In the preceding section, we identified several influences on the relationship between a single attitude and behavior. At times, an object or situation may elicit multiple attitudes. In these cases, predicting behavior is more difficult. When several attitudes are invoked, the individual often engages in deliberative processing of information (Fazio, 1990). He or she considers the attributes of the object or situation, the relevant attitudes, and the costs and benefits of potential behaviors. One important attempt to specify this process is the **theory of reasoned action,** developed by Fishbein and Ajzen (1975; Ajzen & Fishbein, 1980). This model is based on the assumption that behavior is rational, and it incorporates several factors

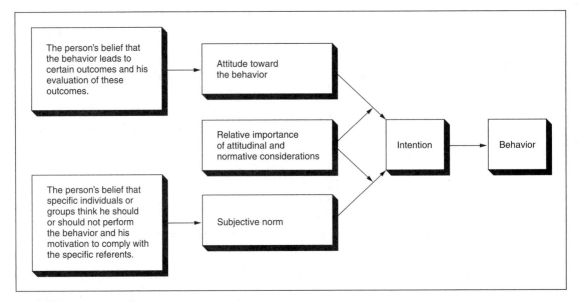

Figure 6.6   The Reasoned Action Model

Note: Arrows indicate the direction of influence.

SOURCE: Ajzen and Fishbein, 1980, figure 1-1.

that have been shown to affect the consistency between attitudes and behavior (see Figure 6.6).

According to the reasoned action model, behavior is determined by behavioral intention. Behavioral intention is primarily influenced by two factors: attitude (AB) and subjective norm (SN). Attitude refers to positive or negative feelings about engaging in a behavior. **Subjective norm** is the individual's perception of others' beliefs about whether or not a behavior is appropriate. (In other words, subjective norm is one form of situational constraint.) The reasoned action model also specifies the determinants of attitude and of subjective norm. Attitude is influenced by a person's beliefs about the likely consequences of the behavior and a person's evaluation—positive or negative—of each of those outcomes. Subjective norm is influenced by a person's beliefs about the reactions of other persons or groups to the behavior and his or her motivation to comply with their expectations.

Formal Model

The model can be summarized using the following expression:

Behavior = Behavioral intention
= Attitude + Subjective norm

Suppose Bill has two good friends who are followers of Reverend Moon. They have been giving him literature and encouraging him to join the Unification Church. To predict what Bill will do, we need to know his attitude (A_β) and his perception of how others will react to his joining this cult. Attitude is the sum (Σ) of beliefs (b) about the likelihood of various consequences of the act and evaluation (e)—positive or negative—of each consequence. That is,

$$A_\beta = \Sigma(b \times e)$$

Bill has several beliefs about joining the Moonies. If he joins, he will gain a greater sense of purpose in his life, and his physical needs will be met. Also, he will have to end his relationship with Cindy, his girlfriend. Finally, Bill has read that former cult members claim they had to relinquish their personal freedom when they joined. These beliefs and their evaluations are shown in Table 6.1. Bill is certain consequences (1) and (2) would occur; hence their value (b) is +3. His

Table 6.1 Determining Attitude *(A$_\beta$)* From Beliefs *(b)* and Evaluations *(e)*

Consequences of Joining the Unification Church	Belief (b)	Evaluation (e)	Product (b × e)
(1) Gain a sense of purpose	+ 3	+ 3	+ 9
(2) Have one's physical needs provided for	+ 3	+ 1	+ 3
(3) Loss of relationship with Cindy	+ 2	− 2	− 4
(4) Loss of some personal freedom	+ 1	− 3	− 3
		Attitude = Σ *b* × *e* =	+ 5

Table 6.2 Determining Subjective Norm From Normative Beliefs *(NB)* and Motivation to Comply *(MC)*

Significant Others	Normative Beliefs (NB)	Motivation to Comply (MC)	NB × MC
Parents	− 3	+ 2	− 6
Friends	+ 3	+ 2	+ 6
Cindy	− 2	+ 3	− 6
			S$_N$ = Σ (NB)(MC) = − 6

evaluation of consequence (1) is very positive (+3), whereas his evaluation of having his physical needs cared for is less positive (+1). He believes it is likely that he will have to give up Cindy (+2), which would be unpleasant (−2). He is skeptical about the claim that he will lose his freedom (+1), although he would be very upset if that occurred (−3). Bill's attitude is +5, the value of *b* × *e*.

Subjective norm *(S$_N$)* is the product of normative beliefs *(NB)*—expectations about how significant others will react—and motivation to comply *(MC)* with each; that is,

$$S_N = \Sigma \ (NB \times MC)$$

For Bill, the significant others are his parents, his peers, and his girlfriend, Cindy. His parents are strongly opposed to his joining a cult (−3), and he is moderately motivated to comply with their views (+2). He is equally motivated to comply with his friends' views (+2), who strongly favor his joining (+3). And he is highly motivated to comply with Cindy, who opposes his becoming a Moonie (see summary in Table 6.2).

Thus the value of *SN* is the product of *NB* × *MC*, or −6. Behavioral intention is the (weighted) sum of attitude and subjective norm. Assuming equal weights, the model predicts that Bill will not join the Unification Church; that is, (+5) + (−6) = −1. Although his attitude is positive, the subjective norm is negative.

Assessment of the Model

The reasoned action model combines several elements discussed earlier in this chapter. It has been used to predict behaviors such as whether a pregnant woman will breast-feed her baby (Manstead, Proffitt, & Smart, 1983). When combined with quantitative measures of the components of attitudes, this model can predict a specific behavior under specific circumstances. For instance, one study attempted to predict weight loss among college women (Schifter & Ajzen, 1985). The participants' subjective intention, attitude, and subjective norm about losing weight were measured. Several other variables were also assessed, including whether or not the woman had a detailed plan regarding weight loss. Six weeks later, the amount

This Hari Krishna member is passing out literature and seeking new recruits. A person who considers joining this cult will be influenced not only by his or her attitudes, but also by subjective norms—the anticipated reaction of family and friends.

of weight actually lost was measured. The amount of weight lost was associated with intention and with having a detailed plan; intention to lose weight was determined by attitude and subjective norm.

This model has been criticized (Liska, 1984) because it assumes our behavior is determined largely by our intentions. This assumption is not always correct; in some situations, our past behavior may be even more influential than our intentions. For example, whether one has donated blood in the past is a much better predictor of whether a person will donate blood in the next 4 months than his statement about whether he intends to do so (Bagozzi, 1981). In effect, much of our behavior is habitual and may not match our conscious intentions. The effect of prior behavior is particularly strong when the stated intention is not compatible with the individual's self-identity (Granberg & Holmberg, 1990). Conversely, a significant relationship between intention and weight loss over an 8-week period was noted among women whose self-schema was consistent with their intention (Kendzierski & Whitaker, 1997).

Also, research suggests our behavior may be affected not only by intentions but also by whether we have the resources or the ability needed to carry out the intention. As a result, it has been suggested an additional variable should be added to the model, **perceived behavioral control** (Ajzen, 1985). A study of intentions to engage in safer sex among 403 undergraduates found that attitude and subjective norm explained substantial variation in intention to use condoms in the next 3 months. Even more variance was explained when perception of one's ability to use condoms was added to the analysis (Wulfert & Wan, 1995). The revised model is referred to as the *theory of planned behavior*.

Summary

The Nature of Attitudes Attitudes have three characteristics. (1) Every attitude has three components: cognition, evaluation, and a behavioral predisposition toward some object. (2) We learn attitudes through reinforcement, through repeated associations of stimuli and responses, and by observing others. (3) Attitudes are useful; they may serve instrumental and knowledge functions, and they define and maintain self.

The Measurement of Attitudes

There are two types of attitude measures. (1) Direct methods involve asking a direct question and recording the answer. Such methods include the use of single items, Likert scales, and semantic differential techniques. (2) Indirect methods involve observing overt behavior. Such methods are useful when a direct question might elicit a false response. Examples include the wrong-number and lost-letter techniques.

Attitude Organization and Change

An attitude is usually embedded in a larger cognitive structure and is based on one or more fundamental or primitive beliefs. Consistency theories assume that when cognitive elements are inconsistent, individuals will be motivated to change their attitudes or behavior to restore harmony. Balance theory assesses the relationship among three cognitive elements and suggests ways to resolve imbalance. Dissonance theory identifies two situations in which inconsistency often occurs: after a choice between alternatives or when people engage in behavior that is inconsistent with their attitudes. The theory also cites two ways to reduce dissonance: by changing one of the elements or by changing the importance of the cognitions involved.

The Relationship Between Attitudes and Behavior

The attitude-behavior relationship is influenced by four variables: activation of the attitude, characteristics of the attitude, correspondence, and situational constraints. (1) For an attitude to influence behavior, it must be activated, and the person must use it as a guide for behavior. (2) The relationship is stronger if affective-cognitive consistency is high and if the attitude is based on direct experience, is strong (relevant), and is stable over time. (3) The relationship is stronger when the measures of attitude and behavior correspond in action, object, context, and time. (4) Situational constraints may facilitate or prevent the expression of attitudes in behavior.

The Reasoned Action Model

This model suggests that behavior is determined by behavioral intention. In turn, intention is determined by a person's attitudes and perceptions of social norms. This model allows precise predictions of behavior, and some studies report results consistent with such predictions. However, it may not apply to some types of behavior, such as behavior determined by habit.

Key Terms

activation of an attitude (p. 146)
attitude (p. 131)
balance theory (p. 140)
balanced state (p. 140)
cognitions (p. 139)
cognitive dissonance (p. 142)
correspondence (p. 150)
dissonance effect (p. 144)
imbalanced state (p. 140)
incentive effect (p. 144)
Likert scale (p. 136)
perceived behavioral control (p. 156)
prejudice (p. 133)
relevance (p. 148)
semantic differential scale (p. 136)
sentiment relations (p. 140)
situational constraint (p. 152)
subjective norm (p. 154)
theory of cognitive dissonance (p. 142)
theory of reasoned action (p. 153)
unit relations (p. 140)

CHAPTER 7
Symbolic Communication and Language

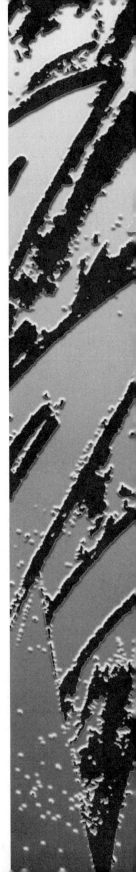

Introduction

Communication is a basic ingredient of every social situation. Imagine playing a game of basketball or buying a pair of shoes without some form of verbal or nonverbal communication. Without communication, interaction breaks down and the goals of any social encounter are foiled. Indeed, it would be simply impossible to arrange commercial transactions, courtroom trials, birthday parties, or any other social occasion without communication.

Communication is the process whereby people transmit information about their ideas, feelings, and intentions to one another. We communicate through spoken and written words, through voice qualities and physical closeness, through gestures and posture. Often communication is deliberate: We smile, clasp our beloved in our arms and whisper, "I love you." Other times, we communicate meanings that are unintentional. A Freudian slip, for instance, may tell our listeners more than we want them to know.

Because people do not share each other's experiences directly, they must convey their ideas and feelings to each other in ways that others will notice and understand. We often do this by means of symbols. **Symbols** are (arbitrary) forms that are used to refer to ideas, feelings, intentions, or any other object.

Symbols represent our experiences in a way that others can perceive with their sensory organs—through sounds, gestures, pictures, even fragrances. But if we are to interpret symbols as others intend, their meanings must be socially shared. To communicate successfully we must master the ways for expressing ideas and feelings that are accepted in our community.

Symbols are arbitrary stand-ins for what they represent. A green light could as reasonably stand for "stop" as for "go," the sound *luv* as reasonably for negative as for positive feelings. The arbitrariness of symbols becomes painfully obvious when we travel in foreign countries. We are then likely to discover that the words and even the gestures we take for granted fail to communicate accurately. A North American who makes a circle with thumb and index finger to express satisfaction to a waiter may be in for a rude surprise if he is eating at a restaurant in Ghana, where the waiter may interpret his gesture as a sexual invitation. In Venezuela, it may be interpreted as a sexual insult! The traveler may then have serious difficulties straightening out these misunderstandings because he and the waiter lack a shared language of verbal symbols to discuss them.

Language and nonverbal forms of communication are amazingly complicated. They must be understood and used with flexibility and creativity. Most of us fail on occasion to communicate our ideas and feelings with accuracy or to understand others' communications as well as we might wish. Yet considering the problems a communicator must solve, most people do surprisingly well. This chapter begins with an examination of language, moves on to nonverbal communication, then analyzes the impacts of communication and social relationships on each other. Finally, this chapter considers the delicate coordination involved in our most common social activity—conversation. The chapter addresses the following questions:

1. What is the nature of language, and how is it used to grasp meanings and intentions?
2. What are the major types of nonverbal communication, and how do they combine with language to convey emotions and ideas?
3. How do social relationships shape communication, and how does it in turn express or modify those relationships?
4. What rules and skills do people employ to maintain a smooth flow of conversation and to avoid disruptive blunders?

Language and Verbal Communication

Although people have created numerous symbol systems (mathematics, music, painting), language is the main vehicle of human communication. All people possess a spoken language. There are thousands of different languages in the world. This section addresses several crucial topics regarding the role of language in communication. These include the nature of language, three perspectives on how people attain understanding through language use, and how language and thought are related to one another.

Linguistic Communication

Little is known about the origins of language, but humans have possessed complex spoken languages since earliest times (Kiparsky, 1976; Lieberman, 1975). **Spoken language** is a socially acquired system of sound patterns with meanings agreed on by the members of a group. We will examine the basic components of spoken language and some of the advantages of language use.

Basic Components Spoken languages include sounds, words, meanings, and grammatical rules. Consider the following statement of one roommate to another: "Wherewereyoulastnight?" What the listener hears is a string of sounds much like this, rather than the sentence, "Where were you last night?" To understand a string of sounds and to produce an appropriate response, people must recognize the following components: (1) the distinct sounds of which the language is composed (the *phonetic* component); (2) the combination of sounds into words (the *morphologic* component); (3) the common meaning of the words (the *semantic* component); and (4) the conventions for putting words together built into the language (the *syntactic* component, or grammar). We are rarely conscious of manipulating all these components during conversation, although we do so regularly and with impressive speed.

Unspoken languages, such as Morse code, computer languages, and the sign languages of the deaf,

Signing by the interpreter parallels the oral message by the speaker. Although sign language lacks the phonetic component, it possesses the morphologic, semantic, and syntactic components of language.

lack a phonetic component, although they do possess the remaining components of spoken language. People who use sign languages, for example, use upper body movements to signal words (morphology) with shared meanings (semantics), and they combine these words into sentences according to rules of order (syntax). For a communication system to be considered a language, morphology, semantics, and syntax are all essential. Linguists study these components, seeking to uncover the rules that give structure to language. Social psychologists are more interested in how language fits into social interaction and influences it, and in how language expresses and modifies social relationships (Giles, Hewstone, & St. Clair, 1981).

Advantages of Language Use Words—the symbols around which languages are constructed—provide abundant resources with which to represent ideas and feelings. The average adult native speaker of English knows the meanings of some 35,000 words, and actively uses close to 5,000. Because it is a symbol system, language enhances our capacity for social action in several ways.

First, language frees us from the constraints of the here and now. Using words to symbolize objects, events, or relationships, we can communicate about things that happened last week or last year, and we can discuss things that may happen in the future. The ability to do the latter allows us to coordinate our behavior with the activities of others.

Second, language allows us to communicate with others about experiences we do not share directly. You cannot know directly the joy and hope your friend feels at bearing a child, nor her grief and despair at her mother's death. Yet she can convey a good sense of her emotions and concerns to you through words, even in writing, because these shared symbols elicit the same meanings for you both.

Third, language enables us to transmit, preserve, and create culture. Through the spoken and written word, vast quantities of information pass from person to person and from generation to generation. Language also enhances our ability to go beyond what is already known and to add to the store of cultural ideas and objects. Working with linguistic symbols, people generate theories, design and build new products, and invent social institutions.

We turn now to three models of communication: the encoder-decoder model, the intentionalist model, and the perspective-taking model. We consider how each model views the communication process and we discuss communication accuracy.

The Encoder-Decoder Model

Language is often thought of as a medium of communication that one person uses to transmit information to another. The **encoder-decoder model** views communication as a process in which an idea or feeling is encoded into symbols by a source, transmitted to a receiver, and decoded into the original idea or feeling (Krauss & Fussell, 1996). This process is portrayed in Figure 7.1.

Communication Process   According to this model, the basic unit is the message, which has its origin in the desire of the speaker to communicate. A message is constructed when the speaker encodes the information he or she wishes to communicate into a combination of verbal and nonverbal symbols. The message is sent via a channel, whether face-to-face interaction, telephone, electronic communication, or in writing. The listener must decode the message in order to arrive at the information he or she believes the speaker wanted to communicate.

Communication Accuracy   The goal of communication is to transfer the message content accurately from speaker to listener. The speaker hopes to create in the listener the mental image or feeling that the speaker intends to convey. The listener is also

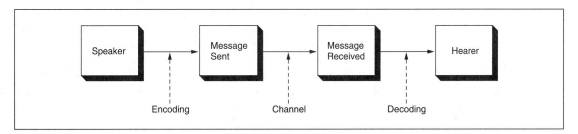

Figure 7.1   The Encoder-Decoder Model

According to the encoder-decoder model, communication originates in the speaker's desire to convey an idea or feeling. She encodes the message into a set of symbols, and transmits it to the hearer. The hearer decodes the message. The more codable the idea or feeling in the language, the more accurate the communication.

motivated to achieve accuracy, in order to coordinate his or her behavior with that of the speaker. **Communication accuracy** refers to the extent to which the message inferred by the listener matches the message intended by the speaker. According to this model, the primary influence on accuracy is *codability,* which is the extent of interpersonal agreement about what something is called. Codability is partly a function of language. Early research focused on the codability of colors (Lantz & Stefflre, 1964). In this research, one person (the encoder) was shown a color and asked to describe that color in words. This verbal message was then sent to a second person (the decoder), who tried to use the verbal description to identify the color intended by the encoder. Some colors are much more easily coded in the English language (fire engine red) than others (the reddish color of a sunset). By extension, some ideas and feelings are easily expressed in English, whereas others are much more difficult to put into words. In general, messages that are easily coded are more accurately transmitted.

Codability involves agreement about what something is to be called. It also depends on the extent to which speaker and listener define symbols (words, gestures) in the same way. This in turn depends on the language to which each was socialized. Thus a common cause of miscommunication is differences in language between speaker and listener. This is obvious when we try to converse with someone from a different country. It is less obvious but perhaps just as important when we converse with someone of a different race, class, or gender (Maynard & Whalen, 1995).

At times, the processes of encoding and decoding are very deliberate or mindful (Giles & Coupland, 1991). If we are preparing a speech, we may consciously consider alternative ways to phrase a message and alternative gestures to use when communicating. Listening to a speaker, we may pay careful attention to the words used, the speed and volume of the spoken message, the gestures, and the posture of the speaker in order to decide which message is the correct one. We are often mindful of the encoding and decoding process in novel situations or in communicating about novel topics.

Communication is not always a process of consciously translating ideas and feelings into symbols and then transmitting these symbols deliberately in hopes that the listener will interpret them correctly. Much communication occurs without any self-conscious planning. In familiar or routine situations, we often rely on a *conversational script,* a sequentially organized series of utterances that occur with little or no conscious thought. Thus, when you enter a restaurant, you can interact with the waitperson without much mental effort because you both follow a conversational script that specifies what each of you should say and in what sequence. Communication accuracy is typically high in situations governed by conversational scripts. When you order food in a fast-food outlet, you usually get exactly what you want with minimum effort.

If conversation is scripted, listeners probably do not pay careful attention to the idiosyncratic features of a message. They tend to remember the generic content of a message, but not its unusual characteristics. In a field experiment testing this prediction, students were approached by a stranger, who asked for a piece of paper. Prior to the request, one half of the students were asked to pay attention to it; the other half were not forewarned. Later, students in the first condition were more likely to remember the specific words used in the request than students in the second condition (Kitayama & Burnstein, 1988).

The Intentionalist Model

The encoder-decoder model emphasizes messages consisting of symbols whose meaning is widely understood. It directs our attention to the literal meaning of verbal messages. Often, however, messages are not interpreted literally. For example, in most theaters, the feature film is preceded by the message, "Please, silence during the show." But are members of the audience expected to be silent? No. They can laugh if the film is a comedy, boo at the villain, and applaud when the bad guy/gal gets what he or she deserves. Most of us understand this message in terms of its intention; we should not whisper or talk to those seated near us. For this type of communication, we need a different model.

According to the **intentionalist model,** communication involves the exchange of *communicative intentions,* and messages are merely the means to this end (Krauss & Fussell, 1996). The speaker selects the message she believes is most likely to accomplish her intent. "Please, silence during the show" is intended to keep us from disturbing other members of the audience, and we understand it to mean that.

Communication Process

The origin of communication is the speaker's desire to achieve some goal or to have some effect on the hearer. But there is not a fixed or one-to-one relation between words and intended effects, so the speaker can use a variety of messages or utterances to achieve the intended effect. For example, imagine you are studying in your living room, and you want your roommate to bring you something to drink. Table 7.1 lists some of the utterances you might use to make the request. Which one would you choose?

According to the intentionalist model, decoding the literal meaning of a message is only part of the process of communication. The hearer must also infer the speaker's underlying intention in order to respond appropriately. Asked, "How is that lemonade we bought?" a satisfactory response to the literal message is "Good." If the communication is to be successful, however, your roommate needs to infer your intention, that he should bring you a glass of lemonade. Both the process of selecting a message to convey your intention and inferring another's intention from their utterances is carried out according to social conventions.

Communication Accuracy

According to this model, accuracy in communication is accuracy in understanding the intentions of the speaker. To achieve accuracy requires a more sophisticated processing than merely interpreting the literal meaning of the message. When inferring the speaker's intention, the listener needs to take into account the context, especially (1) the status or role relationship between speaker and listener, and (2) the social context in which the communication occurs. If you and your roommate are lovers, you might choose a less polite form of the request, such as option 2 in Table 7.1, and you would expect a less polite response than if the two of you are simply sharing the residence. If your parents are visiting at the time, your request to them is likely to take a different form, such as option 3.

Table 7.1

Utterances that can be used to achieve the same effect as the request "Get me a drink of lemonade."

1. Get me a glass of lemonade.
2. Can you get me some lemonade?
3. Would you get me some lemonade?
4. Would you get me something to drink?
5. Would you mind if I asked you to get me some lemonade?
6. I'm thirsty.
7. Did you buy some lemonade at the store?
8. How is that lemonade we bought?

According to **speech act theory,** utterances both state something and do something (Searle, 1979). In Table 7.1, utterances 1 to 6 state the speaker's desire for a drink (or specifically for lemonade), whereas utterances 7 and 8 do not. But all eight of the utterances perform an action; each has the force of a request. The significance of an utterance is not its literal meaning, but what it contributes to the work of the interaction within which it occurs (Geis, 1995). The use of language to perform actions is rule governed; these rules influence both the creation and interpretation of speech acts. To achieve accurate communication, both speaker and hearer must be aware of these rules. Miscommunication is caused not only by the lack of shared meaning of symbols, but also by the lack of shared understanding of the rules governing the use of speech to perform actions.

To determine whether the message has achieved the intended effect, the speaker relies on the feedback provided by the listener's reaction. If the reaction indicates that the listener interpreted the message accurately, the speaker may elaborate, change the topic, or end the interaction. However, if the reaction suggests the listener inferred a meaning different from the intended one, the speaker often attempts to send the same message, perhaps using different words and gestures. For example, when Jim asked Susan, a co-worker, for a date, Susan replied that she liked him as a friend and she was busy Saturday night. Her intended message was that she was not interested in developing a relationship with Jim. Several days later, Jim tried to give Susan six red roses. Inferring that Jim had not received her intended message, Susan refused the roses and told Jim directly that she was not interested in dating him.

The Cooperative Principle   Mutual understanding is a cooperative enterprise. Because language does not convey thoughts and feelings in an unambiguous manner, people must work together to attain a shared understanding of each other's utterances (Goffman, 1983). A speaker must cooperate with a listener by formulating the content of speech acts in a manner that reflects the listener's way of thinking about objects, events, and relationships.

In turn, the listener must cooperate by actively trying to understand. He or she must go beyond literal meanings of words to infer what the speaker is really saying. A listener must make a creative effort to cope with a speaker's tendency to formulate speech acts indirectly. Without such an effort, a listener would not understand speech acts that leave out words ("Paper come?"), abbreviate familiar terms ("See ya in calc."), and include vague references ("He told him he would come later.").

According to Grice (1975), listeners assume that much talk is based on the **cooperative principle.** That is, listeners ordinarily assume the speaker is behaving cooperatively by trying to be (1) informative (giving as much information as is necessary and no more), (2) truthful, (3) relevant to the aims of the ongoing conversation, and (4) clear (avoiding ambiguity and wordiness).

The cooperative principle is more than a code of conversational etiquette. It is crucial to the accurate transmission of meaning. Often a listener can reach a correct understanding of otherwise ambiguous talk only by assuming the speaker is trying to satisfy this principle. Consider, for example, how the relevance assumption (3) enables the conversationalists to understand each other in the following exchange:

TONY: I'm exhausted.

CAROLYN: Fred will be back next Monday.

On the surface, Carolyn's statement seems unrelated to Tony's declaration. In some contexts, we might infer she has changed the subject, indirectly sending the message that she does not care about Tony's physical state. In fact, however, Carolyn is stating that she and Tony won't have to work as hard after their colleague Fred returns to the office next week. But why does she expect Tony to understand? Because she expects him to assume she is adhering to the relevance maxim, that her comment relates to what he said.

The cooperative principle is also crucial for speech forms like sarcasm or understatement to succeed. In sarcasm or understatement, speakers want listeners to recognize that their words mean something quite different from what they seem to convey. One way we signal listeners that we intend our words to imply something different is by obviously violating one or two maxims of the cooperative principle while holding to the rest. Consider Carrie's sarcastic reply when asked what she thought of the lecturer: "He was so exciting that he came close to keeping most of us awake the first half hour." By flouting the maxim of clarity (responding in an unclear, wordy way) while still being informative, truthful, and relevant, Carrie implies the lecturer was in fact a bore.

The Perspective-Taking Model

A third model views the process of communication as both creating and reflecting a shared context between speaker and listener. This approach maintains that symbols do not have an invariant meaning across situations. According to the **perspective-taking model,** communication involves the exchange of messages using symbols whose meaning grows out of the interaction itself.

The Communication Process   Communication involves the use of verbal and nonverbal symbols whose meaning depends on the shared context created by the participants. The development of this shared context requires reciprocal *role taking,* in which each participant places himself or herself in the role of the other in an attempt to view the situation from the other's perspective. The context created by the ongoing interaction changes from minute to minute; each actor must be attentive to these changes in order to communicate successfully, as both speaker and listener.

7.1 The Linguistic Relativity Hypothesis

Does the language we speak influence the way we think about and experience the world? The most famous theory on this question—the Sapir-Whorf **linguistic relativity hypothesis**—holds that language "is not merely a reproducing instrument for voicing ideas, but is itself a shaper of ideas, the program and guide for the individual's mental activity" (Whorf, 1956). Two forms of this hypothesis—strong and weak—have been proposed.

According to the strong form of the linguistic relativity hypothesis, language determines our perceptions of reality, so we cannot perceive or comprehend distinctions that don't exist in our own language. Orwell's description of "Newspeak," the language developed by the totalitarian rulers in his novel *1984* (1949), portrays in frightening terms how language restricts thought:

> Don't you see that the whole aim of Newspeak is to narrow the range of thought? In the end we shall make thoughtcrime literally impossible because there will be no words in which to express it. . . . Every year fewer and fewer words, and the range of consciousness always a little smaller. . . . The revolution will be complete when the language is perfect. (pp. 46–47)

Orwell's description suggests that language determines thought through the words it makes available to people. We cannot talk about objects or ideas for which we lack words. The ways we think about the world are determined by the way our language slices up reality.

The strong form of the linguistic relativity hypothesis has not fared well in research. Consider some of the facts. Some languages have only two basic words (dark and white) to cover the whole spectrum of colors. Yet people from these and all other known language groups can discriminate and communicate about whatever large numbers of colors they are shown (Heider & Olivier, 1972). Most likely any concept can be expressed in any language, although not with the same degree of ease and efficiency. Before either TV or the word "television"

existed, for example, someone undoubtedly referred to the concept of "a device that can transmit pictures and sounds over a distance." When new concepts are encountered, people invent words (laser) or borrow them from other languages ("sabotage" from French, "goulash" from Hungarian).

Thus the strict linguistic relativity hypothesis that language determines thought has found little support. But there is considerable evidence for a weaker form of this hypothesis. The weaker form says that each language facilitates particular forms of thinking because it makes some events and objects more easily codable or symbolizable. In fact, the availability of linguistic symbols for objects or events has been shown to have two clear effects. First, it improves the efficiency of communication about these objects and events. Second, it enhances success in remembering them.

Communication efficiency is often improved when the language includes labels that distinguish among similar objects. For example, one tribal group (the Hanunoo, for whom rice is a staple food) has 92 names for rice (Brown, 1965). Each name conveys the shape, color, texture, state, and so on, of a different type of rice. This makes communication easy and precise. To convey the same information in English would be possible, but less efficient.

The availability of linguistic symbols also affects memory for objects and persons. This was shown in a study that involved subjects who spoke English and/or Chinese (Hoffman, Lau, & Johnson, 1986). This study used English- and Chinese-language descriptions of two persons whose traits could be easily labeled in English but not in Chinese, and of two persons whose traits could be easily labeled in Chinese but not in English. Three groups of subjects read the descriptions: English monolinguals, Chinese-English bilinguals who read in Chinese, and Chinese-English bilinguals who read in English. Subjects' memory of the descriptions was assessed. Results showed that memory was much better when the information about the target conformed to labels in the subject's language of processing. These results lend support to the weak form of the linguistic relativity hypothesis.

Communication Accuracy In the perspective-taking model, communication is much more than transmitting and receiving words with fixed, shared meanings. Conversationalists must select and discover the meanings of words through their context. In ordinary social interaction, the meanings of whole sentences and conversations may be ambiguous. Speakers and listeners must jointly work out these meanings as they go along.

Successful communication depends on *intersubjectivity*; each participant needs information about the other's status, view of the situation, and plans or intentions. Strangers rely on social conventions, rules about interpersonal communication. They categorize other participants and utilize stereotypes as a basis for making inferences about the plans and intentions of the other persons who are present in the setting. Persons who know each other can draw on their past experience with each other as a basis for effective communication.

Interpersonal Context According to this model, both the production and interpretation of communication are heavily influenced by the interpersonal context in which each occurs (Giles & Coupland, 1991). Three ways in which this context influences communication are via norms, cognitive representations of prior similar situations, and emotional arousal.

Every social situation includes norms regarding communicative behavior. These norms specify what topics are appropriate and inappropriate for discussion, what language is to be used, and how persons of varying statuses should be addressed. Depending on these norms, we use one or another of various *speech repertoires*, ways of communicating the same literal message that vary in words, tone, and so on (Giles & Coupland, 1991). Imagine a man who wishes another person to close a door. To his son, in his home, he might say, "Close the door." To his son at work, he might say, "Please close the door, Tom." To an employee, he could ask, "Would you close the door?" These different ways of making a request reflect differences in speech rules, which depend on the relationship between speaker and listener and the setting.

Each new situation evokes representations of prior similar situations and the language one has used or heard in them (Chapman et al., 1992). These conversational histories provide us with the contents of our speech repertoires. Each of us has a set of things we say when we meet a stranger our own age at a party; these are opening lines that in the past have been effective in facilitating conversation with strangers at parties. If instead you met the same stranger on a plane, you might use different speech acts.

The processing of messages by listeners is also influenced by these contextual factors. Listeners interpret messages in light of the rules operating in situations, their past experience, and the emotions elicited in them. When speaker and listener have the same understanding of the normative demands, communication should be quite accurate. Similarly, if a situation evokes the same representations and emotions in both, it is likely the listener will accurately interpret the speaker's message. Communication across group and cultural boundaries is often difficult precisely because speaker and listener differ in their assumptions and experiences, even though they may speak the same language.

The accuracy of indirect or covert communication depends heavily on shared knowledge. In a series of experiments, subjects were asked to compose messages, either in writing or on videotape, taking a position they did not believe in. They were also instructed to try to covertly inform the reader/viewer that they did not hold that position. Most of the subjects used the device of including false information about themselves in the message. Friends of the subjects who read/viewed the message detected the deception, whereas strangers did not (Fleming et al., 1991).

Sociolinguistic Competence To attain mutual understanding, language performance must be appropriate to the social and cultural context. Otherwise, even grammatically acceptable sentences will not make sense. "My mother eats raw termites" is grammatically correct and meaningful; it reflects linguistic

Successful communication is a complex process. These two kids are combining language, interpersonal spacing, and body language to accomplish the sharing of a secret.

competence. But as a serious assertion by a North American, this utterance would draw amazed looks. It expresses an idea that is incongruous with American culture, and listeners would have difficulty interpreting it. In a termite-eating culture, however, the same utterance would be quite sensible. This example shows that successful communication requires **sociolinguistic competence**—knowledge of the implicit rules for generating socially appropriate sentences. Such sentences make sense to listeners because they fit with the listeners' social knowledge (Hymes, 1974).

Speech that clashes with what is known about the social relationship to which it refers suggests a speaker is not sociolinguistically competent (Grimshaw, 1990). Speakers are expected to use language appropriate to the status of the individuals they are discussing and to their relationship of intimacy. For example, competent speakers would not state seriously: "The janitor ordered the president to turn off the lights in the Oval Office." They know that low-status persons do not "order" those of much higher status; at most they "hint" or "suggest." Referring to a relationship of true intimacy, sociolinguistically competent speakers would not say, "The lover bullied her beloved." Rather, they would select such socially appropriate verbs as "coaxed" or "persuaded." In short, competent speakers recognize that social and cultural constraints make some statements interpretable in a situation and others uninterpretable.

Thus successful communication is a complex undertaking. A speaker must produce a message that has not only an appropriate literal meaning, but also an intention or goal appropriate to the relationship and setting. The message must reflect the present degree of intersubjectivity between speaker and hearer, consistent with the interactional context. And the message must also signify the statuses of the participants (Geis, 1995). It is remarkable that each of us communicates successfully many times each day.

Nonverbal Communication

Have you ever been in a situation in which you tried to communicate without using words? Perhaps you were interacting with someone who was deaf or someone who was too far away for your words to be heard. Imagine you are looking out of a window of your third-floor dorm room or apartment. You notice a man on the sidewalk below, dressed immaculately in a three-piece suit, pacing back and forth. He looks up and sees you and immediately begins to gesture. He points to you, then to some other window and then to his watch. His movements are quick and sharp. His face is tense. What is he trying to communicate to you?

Even without the use of words, most of us can make some inferences about the man's message and emotional state. We do so by interpreting his nonverbal communication. This section examines three questions concerning nonverbal communication: (1) What are the major types of nonverbal communication? (2) How is emotion communicated through facial expressions? (3) What is gained and what problems arise because nonverbal and verbal communication are combined in ordinary interaction?

Types of Nonverbal Communication

By one estimate, the human face can make some 250,000 different expressions (Birdwhistell, 1970). In addition to facial expressions, nonverbal communication utilizes many other bodily and gestural cues. Four major types of nonverbal cues are described here and summarized in Table 7.2.

Paralanguage   Speaking involves a great deal more than the production of words. Vocal behavior includes loudness, pitch, speed, emphasis, inflection, breathiness, stretching or clipping of words, pauses, and so on. All the vocal aspects of speech other than words are called **paralanguage.** This includes such highly communicative vocalizations as moaning, sighing, laughing, and even crying. Shrillness of voice and rapid delivery communicate tension and excitement in most situations (Scherer, 1979). Various uses and interpretations of paralinguistic and other nonverbal cues will be examined later in this chapter. For now,

Table 7.2   Types of Nonverbal Communication

Type of Cue	Definition	Examples	Channel
Paralanguage	Vocal (but nonverbal) behavior involved in speaking	Loudness, speed, pauses in speech	Auditory
Body language (kinesics)	Silent motions of the body	Gestures, facial expressions, eye gaze	Visual
Interpersonal spacing (proxemics)	Positioning of body at varying distances and angles from others	Intimate closeness, facing head-on, looking away, turning one's back	Primarily visual; also touch, smell, and auditory
Choice of personal effects	Selecting and displaying objects that others will associate with you	Clothing, makeup, room decorations	Primarily visual; also auditory and smell

Nonverbal cues suffuse words with life, emphasizing them and clarifying their meaning. Nonverbal cues such as the posture and direct gaze of these two persons carry their own message.

see how many distinct meanings you can give to the sentence, "George is on the phone again" by varying the paralinguistic cues you use.

Body Language
The silent movement of body parts—scowls, smiles, nods, gazing, gestures, leg movements, postural shifts, caressing, slapping, and so on—all constitute **body language.** Because body language entails movement, it is known as *kinesics* (from the Greek *kinein* meaning "to move"). Although paralinguistic cues are auditory, we perceive kinesic cues visually. The body movements of the man in our example were probably particularly useful to you in interpreting his feelings and intentions.

Interpersonal Spacing
We also communicate nonverbally by using **interpersonal spacing** cues—

positioning ourselves at varying distances and angles from others (standing close or far away, facing head-on or to one side, adopting various postures, and creating barriers with books or other objects). Because proximity is a major means of communication between people, this type of cue is also called *proxemics.* When there is very close positioning, proxemics can convey information through smell and touch as well.

Choice of Personal Effects
Although we usually think of communication as expressed through our bodies, people also communicate nonverbally through the personal effects they select: their choices of clothing, hairstyle, makeup, eyewear (glasses or contact lenses), and the like. A uniform, for example, may communicate social status, political opinion, lifestyle, and occupation, revealing a great deal about how its

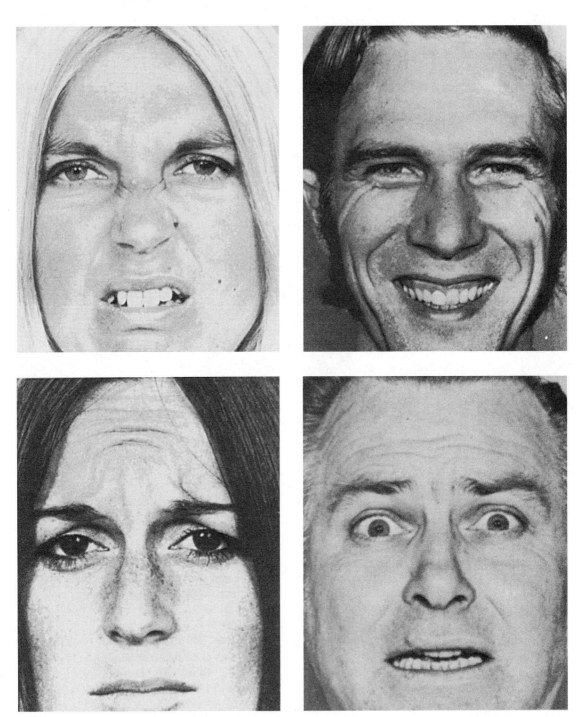

People from many cultures use the same distinctive facial expressions to express six primary emotions: happiness, sadness, surprise, fear, anger, and disgust. Can you identify the emotion expressed by each of these faces? Cross-cultural agreement in the expression and recognition of emotions suggests that these facial expressions are innate.

wearer is likely to behave (Joseph & Alex, 1972). You may have made assumptions about the status and lifestyle of the man in our sketch based on the fact that he wore a three-piece suit. The deliberate use of personal effects to communicate impressions is discussed in Chapter 9.

For the most part, nonverbal cues—like language—are learned rather than innate. As a result, the meanings of particular nonverbal cues may vary from culture to culture. Other features of nonverbal communication may have universal meanings, however. These universals are based in our biological nature. The nonverbal communication of emotion reveals an interesting combination of learned and innate features.

Facial Communication of Emotion

More than 2,000 years ago, the Roman scholar Pliny the Elder said, "The face of man is the index to joy and mirth, to severity and sadness." When we want to hide our feelings, we look away or cover our face. If we must show our face, we try carefully to compose our expressions (Goffman, 1959). That we shield our natural facial expressions to conceal our true emotions from others suggests we share two beliefs: (1) people express their emotions in distinctive ways on their faces, and (2) observers can accurately recognize the emotions others are experiencing. Are these two beliefs correct? This section addresses this question.

Emotions Expressed by the Face Research indicates that people communicate six different emotions by distinctive facial expressions: happiness, sadness, surprise, fear, anger, and disgust (Ekman & Friesen, 1975; Harper, Wiens, & Matarazzo, 1978). Certain facial features are crucial for the expression of each emotion. When shown only the lower face, for instance, nearly everyone in one study was able to identify happiness but was unsuccessful in identifying fear (Ekman, Friesen, & Tomkins, 1971). Both fear and sadness are judged best from the eyes/eyelid area (Boucher & Ekman, 1975).

Learned or Innate? Social psychologists are especially interested in the communication of emo-

tion because it reflects a meeting among biological, social, and cultural influences. A century ago, Darwin first proposed that facial expressions of emotion are an innately determined part of our biological heritage. If this is so, all peoples of the world should exhibit very similar expressions when experiencing the same emotion. Individuals should also be able to recognize emotions expressed by members of other cultures.

To test the universality of expression, photos of individuals from many different cultural groups were shown to members of other groups (Ekman & Friesen, 1975). Members of each group successfully recognized the facial emotions expressed by members of other cultural groups. This suggests that all were responding to a common set of facial expressions which represents the primary emotions across cultures. Observations of children who were born blind provide convincing added support for the universality of emotional expression. These blind children, who cannot learn how to express emotions from seeing others, still smile, laugh, and frown much like sighted children (Eibl-Eibesfeldt, 1979).

One study provides an impressive demonstration of the universality of recognition of the six basic emotions (Ekman et al., 1987). Three photographs were selected that depicted each of the six emotions. The photos were black-and-white pictures of the head and shoulders of a Caucasian man or woman. The photographs were shown to college student observers in 10 countries, including Western and non-Western nations. These subjects were asked to identify the emotion of the person in each photo. The results are displayed in Table 7.3. In every country, from 65% to 98% of the observers correctly identified the emotion portrayed in the pictures.

Cultural Influences What is universal in facial expressions is the particular combination of facial muscles that move when we experience a given emotion. There are strong cultural influences, however, on the actual expression of emotion in everyday interaction. These influences take two main forms. First, through learning, culture helps determine which stimuli evoke particular emotions. Thus cultural groups differ in the emotions that various odors, sounds,

Table 7.3   Single-Emotion Judgment Task: Percentage of Subjects Within Each Culture Who Chose the Predicted Emotion

Nation	Happiness	Surprise	Sadness	Fear	Disgust	Anger
Estonia	90	94	86	91	71	67
Germany	93	87	83	86	61	71
Greece	93	91	80	74	77	77
Hong Kong	92	91	91	84	65	73
Italy	97	92	81	82	89	72
Japan	90	94	87	65	60	67
Scotland	98	88	86	86	79	84
Sumatra	69	78	91	70	70	70
Turkey	87	90	76	76	74	79
United States	95	92	92	84	86	81

SOURCE: Adapted from Ekman et al., 1987.

events, and so on, evoke in their members. For example, cultural learning influences whether the event of death elicits sadness or happiness for the deceased.

Second, culture influences the expression of emotion through display rules. **Display rules** are culture-specific norms for modifying facial expressions of emotion to make them fit with the social situation (Ekman, 1972). Display rules are typically learned in childhood. They become habits that automatically control facial muscles. The impact of situation on emotional display has been documented in several studies. A study of the facial expressions of Olympic gold medalists focused on displays of happiness during the medal awards ceremony (Fernandez-Dols & Ruiz-Belda, 1995). Although medalists were judged by observers of the videotaped ceremonies to be happy throughout, they only smiled when they were interacting with others. Display rules may require modifying facial expressions of emotion in one of several ways. They may require (1) greater intensity in the expression of an emotion, (2) less intensity in the expression of an emotion, (3) complete neutralization of the emotional expression, or (4) masking one emotion with a different one. If cultures vary in the intensity of emotional displays considered appropriate, we would expect observers to have a hard time assessing the intensity of an emotional display by someone from another culture. In the study by Ekman and his colleagues (1987) using observers from 10 cultures,

each observer was asked to rate how intensely the person in the photograph was experiencing happiness, anger, and so on. Whereas judgments of the emotion showed high levels of agreement across cultures, the judgments of intensity showed lower levels of agreement from one culture to another. In an experiment with women as participants, facial displays were found to be influenced not only by the intensity of the emotional experience and the situation, but also by the relationship between the woman and the "audience" (Hess, Banse, & Kappas, 1995). The most intense display occurred when the emotion was intense, the display norm allowed display, and the other participant was a friend.

In response to our earlier question, we can conclude that people do indeed express primary emotions in distinct ways on their faces, and that others can accurately recognize these emotions. These universal features of the facial expression of emotion are innate. In order to communicate emotions effectively in everyday interaction, however, people must learn and employ the display rules of their own culture.

Combining Nonverbal and Verbal Communication

When we speak on the telephone or shout to a friend in another room we are limited to communicating

7.2 Flirting

A distinctive class of communicative behaviors is flirting, or courtship signaling (Birdwhistell, 1970). It is widely believed that males initiate contact, and therefore relationships, with females. Often it is the male who physically approaches and initiates verbal interaction. But a small body of research indicates that females take the initiative, using nonverbal signals in encouraging the male to initiate interaction.

Flirting refers to a class of nonverbal facial expressions and behavior, exhibited by women, which serves to attract the attention and elicit the approach of a man (Moore, 1985). Monica Moore has studied flirting for 20 years. She began by observing women and recording their nonverbal behavior. Observation of 200 white women, judged to be ages 18 to 35, yielded a catalog of flirtatious behavior. A woman not accompanied by a man was selected at random and observed for at least 30 minutes, in settings where there were at least 20 people present, both men and women. Observers recorded every behavior of the focal subject and its consequences. Flirting, or a *nonverbal solicitation behavior,* was defined as a behavior that resulted in males' attention within 15 seconds. The resulting catalog includes 52 behaviors, including facial and head movements (room-encompassing glance, fixed gaze, head toss), gestures (palming—palm facing a male; primping—touching hair or clothing), and posture (lean toward male, touch).

Subsequent research (Moore & Butler, 1989) describes behaviors that attract male attention and those which maintain his attention after interaction begins. Male attention is likely to follow a room-encompassing glance, a smile while looking at him, patting or smoothing the hair, the "lip lick," or a head toss. Male attention is maintained by frequent head nods while he talks, leaning close to him, and touching or brushing part of the body against him.

Moore provides contextual evidence for the assertion that these behaviors are courtship signals. If these behaviors are intended to attract male attention, we should observe them in contexts where such solicitations are likely, such as singles bars, but not in settings where no males are present, such as women's group meetings.

Flirting is a complex behavior that conveys interest in being approached by another person. These young women are using posture, smiles, and direct gaze to attract the attention of the young men.

She studied 10 women in each of four social settings: singles bars, university snack bar, university library, and women's center meeting. Again, focal sampling was employed; a woman was observed only if at least 25 people were present and she was not accompanied by a man. The display of courtship signals was clearly context specific (see table on next page). Women were much more likely to engage in these behaviors in the singles bar, and least likely to engage in them at women's center meetings.

A recent study (Moore, 1995) looks at the display of nonverbal solicitation behaviors by adolescent girls. Observations were made at school events, swimming pools, and shopping malls, when there were at least 20 males and females present. One hundred girls judged to be between 13 and 16 were randomly selected. Each girl was observed for at least 30 minutes. Overall, girls exhibited the same behaviors, but much less frequently than adult women in similar settings. Also, the behaviors often had

continued on next page

continued from previous page

an exaggerated and playful quality. These observations are consistent with the conclusion that young women are learning to flirt, and rehearsing these behaviors in appropriate contexts.

Moore's research program clearly documents female influence on the initiation of relationships. Flirting is an important component of the development of relation-

ships. The selection of a superior mate, one with desirable qualities, is important to both men and women. Traditional gender-role norms specify the male as the initiator of interaction. Flirting allows a female to influence who approaches her. It also signals the male that his overture will be welcome, reducing his anxiety about making an approach.

The Impact of Social Context on Display Frequency and Number of Approaches

	Singles Bar	Snack Bar	Library	Women's Meetings
Number of subjects	10	10	10	10
Total number of displays	706	186	96	47
Mean number of displays	70.6	18.6	9.6	4.7
Mean number of categories utilized	12.8	7.5	4.0	2.1
Number of approaches to the subject by a male	38	4	4	0
Number of approaches to a male by the subject	11	4	1	0

Note: The tabulated data are for a 60-minute observation interval. Asymmetry in display frequency: $x^2 = 25.079$, df = 3, $p < 0.001$; asymmetry in number of categories utilized: $x^2 = 23.099$, df = 3, $p < 0.001$.
SOURCE: Adapted from Moore, 1985.

through verbal and paralinguistic channels. When we wave to arriving or departing passengers at the airport we use only the visual channel. Ordinarily, however, communication is multichanneled. Information is conveyed simultaneously through verbal, paralinguistic, kinesic, and proxemic cues.

What is gained and what problems are caused when different communication channels are combined? If they appear to convey consistent information, they reinforce each other and communication becomes more accurate. But if different channels convey information that is inconsistent, the message may produce confusion or even arouse suspicion of deception. In this section we examine some outcomes of apparent consistency and inconsistency among channels.

Reinforcement and Increased Accuracy

The multiple cues we receive often seem redundant, each carrying the same message. A smile accompanies a compliment delivered in a warm tone of voice; a scowl accompanies a vehemently shouted threat. But multiple cues are seldom entirely redundant, and they are better viewed as complementary (Poyatos, 1983). The smile and warm tone convey that the compliment is sincere; the scowl and vehement shout imply the threat will be carried out. Thus multiple cues convey added information, reduce ambiguity, and increase the accuracy of communication (Krauss, Morrel-Samuels, & Colasante, 1991).

Taken alone, each channel lacks the capacity to carry the entire weight of the messages exchanged in the course of a conversation. By themselves, the verbal aspects of language are insufficient for accurate communication. Paralinguistic and kinesic cues supplement verbal cues by supporting and emphasizing them. The importance of paralinguistic cues is illustrated in a study of students from a Nigerian secondary school and teachers' college (Grayshon, 1980). Although these students took courses in English and knew the verbal language well, they did not know the paralinguistic cues of British native speakers. The students listened to two British recordings with identical verbal content. In one recording, paralinguistic cues indicated that the speaker was giving the listener a brush-off. In the other recording, paralinguistic cues indicated that the speaker was apologizing. Of 251 students, 97% failed to perceive any difference in the meanings the speaker was conveying. Failure to distinguish a brush-off from an apology could be disastrous in everyday communication. Accurate understanding requires paralinguistic as well as verbal knowledge.

Our accuracy in interpreting events is greatly enhanced if we have multiple communication cues, rather than verbal information alone. The value of a full set of cues was demonstrated in a study of students' interpretations of various scenes (Archer & Akert, 1977). Students observed scenes of social interaction that were either displayed in a video broadcast or described verbally in a transcript of the video broadcast. Thus students received either full, multichannel communication or verbal cues alone. Afterward students were asked to answer questions about what was going on in each scene, questions that required going beyond the obvious facts. Observers who received the full set of verbal and nonverbal cues were substantially more accurate in interpreting social interactions. For instance, of those receiving multichannel cues, 56% correctly identified which of three women engaged in a conversation had no children; this compared with only 17% of those limited to verbal cues. These findings convincingly demonstrate the gain in accuracy from multichannel communication.

Resolving Inconsistency At times the messages conveyed by different channels appear inconsistent with one another. This makes communication and interaction problematic. What would you do, for example, if your instructor welcomed you during office hours with warm words, a frowning face, and an annoyed tone of voice? You might well react with uncertainty and caution, puzzled by the apparent inconsistency among the verbal and nonverbal cues you were receiving. You would certainly try to figure out the instructor's "true" feelings and desires, and you might also try to guess why the instructor was sending such confusing cues.

The strategies people use to resolve apparently inconsistent cues depend on their inferences about the reasons for the apparent inconsistency (Zuckerman, DePaulo, & Rosenthal, 1981). Inconsistency could be attributed to the communicator's ambivalent feelings, to poor communication skills, or to an intention to deceive. A large body of research has compared the relative weight we give to messages in different channels when we do not suspect deception.

In one set of studies, people judged the emotion expressed by actors who posed contradictory verbal, paralinguistic, and facial signals (Mehrabian, 1972). These studies showed that facial cues were most important in determining which feelings are interpreted as true. Paralinguistic cues were second, and verbal cues were a distant third. Later research, exposing receivers to more complete combinations of visual and auditory cues, replicated the finding that people rely more on facial than on paralinguistic cues when the two conflict. This preference for facial cues increases with age from childhood to adulthood, indicating it is a learned strategy (DePaulo et al., 1978).

People also use social context to help them judge which channel is more credible (Bugenthal, 1974). They consider whether the facial expression, tone of voice, or verbal content are appropriate to the particular social situation. If people recognize a situation as highly stressful, for example, they rely more on the cues that seem consistent with a stressful context (a strained tone of voice) and less on cues that seem to contradict it (a happy face or verbal assertion of calmness). If the emotional expression is ambiguous, situational cues determine the emotion that observers attribute to the person (Carroll & Russell, 1996). For example, a person in a frightening situation displaying an expression of moderate anger was judged to be afraid. In short, people tend to resolve apparent inconsistencies between channels in favor of the channels whose message seems most appropriate to the social context.

Social Structure and Communication

So far, this chapter has examined the nature of verbal and nonverbal communication and some consequences of the fact that everyday communication usually combines the two. But how do social relationships shape communication? And how does communication express, maintain, or modify social relationships? These questions pinpoint social psychology's concern with the reciprocal impacts of social structure and communication on each other. This section investigates three aspects of these impacts. First, it discusses the links between styles of speech and position in the stratification system. Second, it analyzes ways that communication creates and expresses the two central dimensions of relationships—status and intimacy. Third, it examines social norms that regulate interaction distances and some of the outcomes when these norms are violated.

Social Stratification and Speech Style

The way we speak both reflects and recreates our social relationships (Giles & Coupland, 1991). Every sociolinguistic community recognizes variation in the way its members talk. One style is usually the preferred or standard style. In addition, there is often one or more other, nonpreferred styles.

Consider an example of each style. As you enter a theater, a young man approaches you. He asks, "Would you please fill out this short survey for me?" Depending on your mood, you might comply with his request. But what if he asked, "Wud ja anser sum questions?" Many people would be less likely to comply with this request.

The first request employs standard American English. **Standard speech** is characterized by diverse vocabulary, proper pronunciation, correct grammar, and abstract content. It takes into account the listener's

perspective. Note the inclusion of "please" in the first request, which indicates the speaker's recognition that he is asking for a favor. **Nonstandard speech** is characterized by limited vocabulary, improper pronunciation, incorrect grammar, and is direct. It is egocentric; the absence of "please" and "for me" in the second request makes it sound like an order.

In the United States, as in many other countries, speech style is associated with social status (Giles & Coupland, 1991). The use of standard speech is associated with high socioeconomic status and with power. People in positions of economic and political power are usually very articulate and grammatically correct in their public statements. In contrast, the use of nonstandard speech is associated with low socioeconomic status and low power.

Speech style is also influenced by the interpersonal context. In informal conversations with others of equal status, such as at some parties or in bars, we often use nonstandard speech, regardless of our socioeconomic status. In more formal settings, especially public ones, we usually shift to standard speech. Thus our choice of standard or nonstandard speech gives listeners information about how we perceive the situation.

Studies in a variety of cultures have found systematic differences in how people evaluate speakers using standard and nonstandard speech. In one study, students in Kentucky listened to tape recordings of young men and women describing themselves. Four of the recordings, two by men and two by women, were of speakers with "standard" American accents. Four others, identical in content, were of speakers with Kentucky accents. On the average, students gave the standard speakers high ratings on status and the nonstandard speakers low ratings on status (Luhman, 1990).

Nonstandard speech involves limited vocabulary, is rooted in the present, and does not allow for elaboration and qualification of ideas. As a result, some analysts advocate so-called *deficit theories,* which claim that people who use nonstandard speech are less capable of abstract and complex thought. These theories also claim that nonstandard speech styles are typical of lower-class, black, and other culturally disadvantaged groups in America, Great Britain, and other societies. Combining these two claims, deficit theorists argue that the children from disadvantaged groups perform poorly in school because their restricted language makes them cognitively inferior. Their poor academic performance in turn leads to unemployment and poverty in later life.

The strongest criticism of deficit theories has come from Labov (1972). Based on interviews in natural environments, he demonstrates that "black English," which has been described as nonstandard speech, is every bit as rich and subtle as standard English. Black English differs from standard English mainly in surface details like pronunciation ("ax" = ask) and grammatical forms ("He be busy" = He's always busy). Nonstandard speech may appear impoverished because nonstandard speakers feel less relaxed in the social contexts where they are typically observed (schools, interviews), and so they limit their speech. Social researchers or other "outsiders" who observe nonstandard speakers may also inhibit their language (Grimshaw, 1973). When interviewed by a member of their own race, for instance, black job applicants used longer sentences and richer vocabularies, and employed words more creatively (Ledvinka, 1971). Overall, speech differences between groups have not been shown to reflect differences in cognitive ability (Thorlundsson, 1987), and deficit theories have not received much support empirically.

In 1996 a somewhat different issue arose when the Oakland, California, school board decided that black English is a distinct language, *ebonics* (*ebony phonics*). The board decided to make classes available in ebonics in the hope it would improve black students' comprehension of schoolwork. Several federal agencies and African American leaders including Jesse Jackson reject the idea that ebonics is a "second language."

Communicating Status and Intimacy

The two central dimensions of social relationships are status and intimacy. Status is concerned with the exercise of power and control. Intimacy is concerned with the expression of affiliation and affection that creates social solidarity (Kemper, 1973). Verbal and nonverbal communication express and maintain particular levels of intimacy and relative status in relationships. Moreover, through communication we may challenge existing levels of intimacy and relative status and negotiate new ones (Scotton, 1983).

Communication can signal our view of a relationship only if we recognize which communication behaviors are appropriate for an expected level of intimacy or status and which are inappropriate. The following examples suggest that we easily recognize when communication behaviors are inappropriate. What if you: Repeatedly addressed your mother as Mrs. _____? Used vulgar slang during a job interview? Draped your arm on your professor's shoulder as she explained how to improve your test answers? Looked away each time your beloved gazed into your eyes? Each of these communication behaviors would probably make you uncomfortable, and they would doubtlessly cause others to think you inept, disturbed, or hostile. Each behavior expresses levels of intimacy or relative status easily recognized as inappropriate to the relationship. In the following, we survey systematically how specific communication behaviors express, maintain, and change status and intimacy in relationships.

Status Forms of address clearly communicate relative status in relationships. Inferiors use formal address (title and last name) for their superiors ("When is the exam, Professor Levine?"), whereas superiors address inferiors with familiar forms (first name or nickname: "On Friday, Daphne"). Status equals use the same form of address with one another. Both use either formal (Ms. / Mr. / Mrs.) or familiar forms (Carol / Bill), depending on the degree of intimacy between them (Brown, 1965). When status differences are ambiguous, individuals may even avoid addressing each other directly. They shy away from choosing an address form because it might grant too much or too little status.

A shift in forms of address signals a change in social relationships, or at least an attempted change. During the French Revolution, in order to promote equality and fraternity, the revolutionaries demanded that everyone use only the familiar *(tu)* and not the formal *(vous)* form of the second-person pronoun, regardless of past status differences. Presidential candidates try to reduce their differences with voters by inviting the use of familiar names (Bob, Bill). In cases where there is a clear status difference between people, the right to initiate the use of the more familiar or equal forms of address belongs to the superior ("Why don't you drop that 'Doctor' stuff?"). This principle also applies to other communication behaviors. It is the higher status person who usually initiates changes toward more familiar behaviors such as greater eye contact, physical proximity, touch, or self-disclosure.

We each have a speech repertoire, different pronunciations, dialects, and a varied vocabulary from which to choose when speaking. Our choices of language to use with other people express a view of our relative status and may influence our relationships. People usually make language choices smoothly, easily expressing status differences appropriate to the situation (Gumperz, 1976; Stiles et al., 1984). Teachers in a Norwegian town, for instance, were observed to lecture to their students in the standard language (Blom & Gumperz, 1972). When they wished to encourage student discussion, however, they switched to the local dialect, thereby reducing status differences. Note how your teachers also switch to more informal language when trying to promote student participation.

An experiment involving groups comprised of a manager and two workers studied the effect of authority and gender composition of the group on verbal and nonverbal communication (Johnson, 1994). The researcher created a simulated retail store; the manager gave instructions to the subordinates and monitored their work for 30 minutes. The interaction was coded as it occurred. Authority affected verbal behavior; subordinates talked less, were less directive, and gave less feedback compared to superiors, regardless of gender. Gender affected the nonverbal behaviors of smiling and laughing; females in all female groups smiled more than males in all male groups.

Language choice by bilinguals also expresses status in relationships. Paraguayans speak Spanish with persons of higher status, but switch to the tribal language, Guarani, among equals (Rubin, 1962). Bilingual Puerto Ricans in New York use English in the formal, status-differentiated settings of school and work. They typically switch to Spanish with their friends who are status equals (Greenfield, 1972).

Paralinguistic cues also communicate and reinforce status in relationships. People of higher status interrupt their partners more during conversations and talk more themselves. Inferiors grant status to others by not interrupting, and by responding with "M-hmn" more frequently at appropriate intervals to indicate they are following the conversation. Among status equals, these paralinguistic behaviors are distributed more equally (LaFrance & Mayo, 1978; Leffler, Gillespie, & Conaty, 1982).

An experimental study of influence in small three-person groups varied systematically the paralanguage of one member (Ridgeway, 1987). This member, a confederate, was most influential when she spoke rapidly, in a confident tone, and gave quick responses. She was less influential when she behaved dominantly (spoke loudly, gave orders) or submissively (spoke softly, in a pleading tone). A subsequent study found that a person who spoke in a task-oriented style (rapid speech, upright posture, eye contact) or a social style (moderate volume, relaxed posture) was more influential (Carli, LaFleur, & Loeber, 1995); persons who spoke using dominant or submissive paralanguage were less influential. Thus engaging in the paralinguistic behaviors appropriate to the statuses of group members enhances one's influence; engaging in behaviors inappropriate to one's status, for example, like a superior toward equals, reduces one's influence.

Body language also serves to express status. When status is unequal, people of higher status tend to adopt relatively relaxed postures with their arms and legs in asymmetrical positions. Those of lower status stand or sit in more tense and symmetrical positions. The amount of time we spend looking at our partner, and the timing, also indicate status. Higher status persons look more when speaking than when listening, whereas lower status persons look more when listening than when speaking. Overall, inferiors look more at their partners, but they are also first to break the gaze between partners. Finally, superiors are much more likely to intrude physically on inferiors by touching or pointing at them (Dovidio & Ellyson, 1982; LaFrance & Mayo, 1978; Leffler, Gillespie, & Conaty, 1982).

Recent research has focused on the impact of facial maturity on the status and intimacy of one's interactions. In one study, judges rated 114 people (50 males, 64 females) on the extent to which each had a "baby face" (Berry & Landy, 1997). These 114 people kept diaries of their social encounters for 1 week, yielding records for 5,106 interactions. Men who were judged to have babyish faces reported less influence on (i.e., lower status) and greater intimacy of their interactions with women, compared to men with mature faces. Facial maturity was not related to variation in the interactions reported by women. Another study found a "baby-faced" overgeneralization effect (Zebrowitz, Voinescu, & Collins, 1996). Both men and women with babyish faces were perceived as more honest by observers.

Intimacy   Communication also expresses another central dimension of relationships, intimacy. One way we signal intimacy or solidarity is by addressing each other with first names. The exchange of title and last names is common for strangers. In other languages, speakers express intimacy by their choice of familiar versus formal second-person pronouns. As noted earlier, the French can choose between the familiar *tu* or formal *vous;* the Spanish have *tú* or *usted.*

Our choice of language is another way to express intimacy. For example, Paraguayan men usually court women in Spanish; after marriage, they converse with their wives in the tribal language as a sign of increased intimacy (Rubin, 1962). Similarly, residents of a Norwegian town were found to use the formal version of their language with strangers and the local dialect with friends. They spoke the formal version when transacting official business in government offices and then switched to dialect for a personal chat with the clerk after completing their business (Blom & Gumperz, 1972). The use of slang gives strong expression to in-group intimacy and solidarity. Through slang, group members assert their own shared social identity and express their alienation from and rejection of the out-group of slang illiterates.

The intimacy of a relationship is clearly reflected in and reinforced by the content of conversation. As a relationship becomes more intimate, we disclose more personal information about ourselves. Intimacy is also conveyed by conversational style. In one study (Hornstein, 1985), telephone conversations were recorded and later analyzed; the conversations were between strangers, acquaintances, and friends. Compared to strangers, friends used more implicit openings ("Hi," or "Hi. It's me."), raised more topics, and were more responsive to the other conversationalist (for example, asked more questions). Friends also used more complex closings, for example, making concrete arrangements for the next contact. Conversations of acquaintances were more like those of strangers.

The **theory of speech accommodation** (Beebe & Giles, 1984; Giles, 1980) illustrates an important way that people use verbal and paralinguistic behavior to express intimacy or liking. According to this theory, people express or reject intimacy by adjusting their speech behavior during interaction to converge with or diverge from their partner's. To express liking or evoke approval, they make their own speech behavior more similar to their partner's. To reject intimacy or communicate disapproval, they accentuate the differences between their own speech and their partner's.

Adjustments of paralinguistic behavior demonstrate speech accommodation during conversations (Taylor & Royer, 1980; Thakerar, Giles, & Cheshire, 1982). Individuals who wish to express liking tend to shift their own pronunciation, speech rate, vocal intensity, pause lengths, and utterance lengths during conversation to match those of their partner. Individuals who wish to communicate disapproval modify these vocal behaviors in ways that make them diverge more from their partner's. Among bilinguals, speech accommodation may also determine the choice of language (Bourhis et al., 1979). To increase intimacy, bilinguals choose the language they believe their partner would prefer to speak. To reject intimacy, they choose their partner's less preferred language.

Accommodation is evident even in very subtle paralinguistic cues. Using audiotapes of interviews by talk-show host Larry King of 25 guests (stars, athletes, politicians), analyses indicated voice convergence between partners (Gregory & Webster, 1996). Lower status persons accommodated their voices to higher status persons. Moreover, student ratings of the status of Larry King and of his guests were correlated with the voice characteristics that showed convergence.

The ways we express intimacy through body language and interpersonal spacing are well recognized. For instance, research supports the folklore that lovers gaze more into each other's eyes (Rubin, 1970). In fact, we tend to interpret a high level of eye contact from others as a sign of intimacy. We communicate liking by assuming moderately relaxed postures, moving closer and leaning toward others, orienting ourselves face-to-face, and touching them (Mehrabian, 1972). Increasing emotional intimacy is often accompanied by increasing body engagement, from an arm around the shoulders to a full embrace (Gurevitch, 1990). There is an important qualification to these generalizations, however. Mutual gaze, close distance, and touch reflect intimacy and promote it only when the interaction has a positive cast. If the interaction is generally negative—if the setting is competitive, the verbal content unpleasant, or the past relationship antagonistic—these same nonverbal behaviors intensify negative feelings (Schiffenbauer & Schiavo, 1976).

Normative Distances for Interaction

American and Northern European tourists in Cairo are often surprised to see men touching and staring intently into each other's eyes as they converse in public. Surprise may turn to discomfort if the tourist engages an Arab male in conversation. Bathed in the warmth of his breath, the tourist may feel sexually threatened. In our own communities, in contrast, we are rarely made uncomfortable by the overly close approach of another. People apparently know the norms for interaction distances in their own cultures and they conform to them. What are these norms, and what happens when they are violated?

Normative Distances Edward Hall (1966) described four spatial zones that are normatively prescribed for interaction among middle-class Americans. Each zone is considered appropriate for particular types of activities and relationships. *Public distance* (12–25 feet) is prescribed for interaction in formal encounters, lectures, trials, and other public events. At this distance, communication is often one-way, sensory stimulation is very weak, people speak loudly, and they choose language carefully. *Social distance* (4–12 feet) is prescribed for many casual social and business transactions. Here, sensory stimulation is low. People speak at normal volume, do not touch one another, and use frequent eye contact to maintain smooth communication. *Personal distance* (1–4 feet) is prescribed for interaction among friends and relatives. Here, people speak softly, touch one another, and receive substantial sensory stimulation by sight, sound, and smell. *Intimate distance* (0–18 inches) is prescribed for giving comfort, making love, and aggressing physically. This distance provides intense stimulation from touch, smell, breath, and body heat. It signals unmistakable involvement.

Many studies support the idea that people know and conform to the normatively prescribed distances for particular kinds of encounters (LaFrance & Mayo, 1978). When we compare different cultural and social groups, both similarities and differences in distance norms emerge. All cultures prescribe closer distances for friends than for strangers, for example. The specific distances for preferred interactions vary widely, however. Latin Americans, Arabs, Greeks, and French use shorter interaction distances than Americans, British, Swiss, and Swedes (Sommer, 1969). Females tend to interact with one another at closer distances than males do in various cultures (Sussman & Rosenfeld, 1982). Social class may also influence interpersonal spacing. In Canadian school yards, lower-class primary school children were observed to interact at closer distances than middle-class children, regardless of race (Scherer, 1974).

Differences in distance norms may cause discomfort in cross-cultural interaction. People from different countries or social classes may have difficulty in interpreting the amount of intimacy implied by each other's interpersonal spacing and in finding mutually comfortable interaction distances. Cross-cultural training in nonverbal communication can reduce such discomfort. For instance, Englishmen were liked more by Arabs with whom they interacted when the Englishmen had been trained to behave nonverbally like Arabs—to stand closer, smile more, look more, and touch more (Collett, 1971).

Despite the crowded circumstances, the people in this airport waiting area are maintaining some privacy. Strangers feel uncomfortable when they must intrude on each other's personal space. To overcome this discomfort, they studiously ignore each other, avoiding touch, eye contact, and verbal exchanges.

Two aspects of interpersonal spacing that clearly influence and reflect status are physical distance and the amount of space each person occupies. Equal status individuals jointly determine comfortable interaction distances and tend to occupy approximately equal amounts of space with their bodies and with the possessions that surround them. When status is unequal, superiors tend to control interaction distances, keeping greater physical distance than equals would choose. Superiors also claim more direct space with their bodies and possessions than inferiors (Gifford, 1982; Hayduk, 1978; Leffler, Gillespie, & Conaty, 1982).

Violations of Personal Space What happens when people violate distance norms by coming too close? In particular, what do we do when strangers intrude on our personal space?

The earliest systematic examination of this question included two parallel studies (Felipe & Sommer, 1966). In one, strangers approached lone male patients in mental hospitals to a point only 6 inches away. In the other, strangers sat down 12 inches away

from lone female students in a university library. The mental patients and the female students who were approached left the scene much more quickly than the other patients and students who were not approached. After only 2 minutes, 30% of the patients who were intruded on had fled, compared with none of the others. Among the students, 70% of those whose space was violated had fled by the end of 30 minutes, compared with only 13% of the others. Results of this study are shown in Figure 7.2.

Flight is not the only response to space violation. In addition, people protect their privacy by turning their backs on intruders, leaning away, and placing

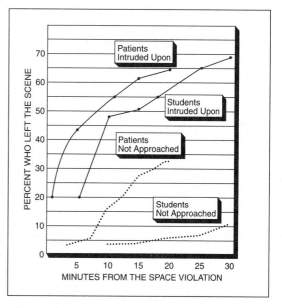

Figure 7.2 Reactions to Violations of Personal Space

How do people react when strangers violate norms of interpersonal distance and intrude on their personal space? A common reaction is illustrated here. Strangers sat down 12 inches away from lone female students in a library or approached lone male patients in a mental hospital to within 6 inches. Those who were approached left the scene much more quickly than control subjects who were not approached. Violations of personal space often produce flight.

SOURCE: Adapted from Felipe and Sommer, 1966.

barriers such as books, purses, or elbows in the intervening space. Only rarely do people react verbally to space violations (Patterson, Mullens, & Romano, 1971). The reaction to violations depends in part on the setting in which they occur. Whereas violations of space norms at library tables lead to flight, violations in library aisles lead to the person spending more time in the aisle (Ruback, 1987). Similarly, intrusion into the space of someone using a public telephone is associated with the caller spending more time on the phone (Ruback, Pape, & Doriot, 1989). It is possible that when you are looking for a book or talking on the phone, a violation of distance norms is distracting so it takes you longer to complete your task.

Violating personal space is uncomfortable for intruders too. Individuals required to pass through the personal space of others who are engaged in conversation feel more awkward and display more unpleasant facial expressions than individuals who merely pass nearby (Efran & Cheyne, 1974). Because of their discomfort, people avoid intruding. They will drink from a water fountain much less frequently, for instance, if someone is standing within a foot of it (Baum, Riess, & O'Hara, 1974).

Staring is a powerful way to violate another's privacy without direct physical intrusion. Staring by strangers elicits avoidance responses, indicating it is experienced as an intense negative stimulus. When stared at by strangers, for instance, pedestrians cross the street faster, and drivers speed away from intersections more quickly (Ellsworth, Carlsmith, & Henson, 1972; Greenbaum & Rosenfeld, 1978).

Conversational Analysis

Although conversation is a regular daily activity, we all have trouble communicating at times. The list of what can go wrong is long and painful: inability to get started, irritating interruptions, awkward silences, failure to give others a chance to talk, failure to notice that listeners are bored or have lost interest, changing topics inappropriately, assuming incorrectly that others understand, and so on. This section examines the ways people avoid these embarrassing and annoying blunders. To maintain smooth-flowing conversation requires knowledge of certain rules and communica-

tion skills that are often taken for granted. We discuss some of the rules and skills that are crucial for initiating conversations, regulating turn taking, and coordinating conversation through verbal and nonverbal feedback.

Initiating Conversations

Conversations must be initiated with an attention-getting device, a summons to interaction. Greetings, questions, or the ringing of a telephone can serve as the summons. But conversations do not get under way until potential partners signal they are attending and willing to converse. Eye contact is the crucial nonverbal signal of availability for face-to-face interaction. Goffman (1963) suggests that eye contact places a person under an obligation to interact: When a waitress permits eye contact, she places herself under the power of the eye-catcher.

The most common verbal lead into conversation is a **summons-answer sequence** (Schegloff, 1968). Response to a summons ("Jack, you home?" "Yeah.") indicates availability. More importantly, this response initiates the mutual obligation to speak and to listen that produces conversational turn taking. The summoner is expected to provide the first topic—a conversational rule little children exasperatingly overlook. Our reactions when people violate the summons-answer sequence demonstrate its widespread acceptance as an obligatory rule. When people ignore a summons, we conclude either that they are intentionally insulting us, socially incompetent, or psychologically absent (sleeping, drunk, crazy).

Telephone conversations exhibit a common sequential organization. Consider the following conversation between a caller and a recipient:

0. (ring)
1. Recipient: Hello?
2. Caller: This is John.
3. R: Hi.
4. C: How are you?
5. R: Fine. How are you?
6. C: Good. Listen, I'm calling about . . .

The conversation begins with a summons-answer sequence (lines 0, 1). This is followed by an identification-recognition sequence (lines 2, 3); in this example, the recipient knows the caller, John, recognizes his voice, so he does not state his name. Next, there is a trading of "how are you" sequences (lines 4–6). Finally, at line 6, John states the reason for the call. This organization is found in many types of telephone calls. However, in an emergency, when seconds count, the organization is quite different (Whalen & Zimmerman, 1987). Consider the following example:

0. (ring)
1. R: Mid-City Emergency.
2. C: Um yeah. There's a fire in my garage!
3. R: What's your address?

Notice that the opening sequence is shortened; both the greeting and the "how are you" sequences are omitted. In emergency calls, the reason for the call is stated sooner. Note also that the recognition element of the identification-recognition sequence is moved forward, to line 1. Both of these changes facilitate communication in an anonymous, urgent situation. However, if the dispatcher answers a call and the caller says, "This is John," that signals an ordinary call. Thus the organization of conversation clearly reflects situational contingencies.

Regulating Turn Taking

A pervasive rule of conversation is to avoid bumping into someone verbally. To regulate turn taking, people use many verbal and nonverbal cues, singly and together, with varying degrees of success (Duncan & Fiske, 1977; Kendon, Harris, & Key, 1975; Sacks, Schegloff, & Jefferson, 1978).

Signaling Turns Speakers indicate their willingness to yield the floor by looking directly at a listener with a sustained gaze toward the end of an utterance. People also signal readiness to give over the speaking role by pausing, and by stretching the final syllable of their speech in a drawl, terminating hand gestures, dropping voice volume, and tacking relatively meaningless expressions ("you know") on to the end of their utterances. Listeners indicate their desire to talk by inhaling audibly as if preparing to speak. They also tense and move their hands, shift their head away from the speaker, and emit especially loud vocal signs of interest ("Yeah," "M-hmn").

Speakers retain their turn by avoiding eye contact with listeners, tensing their hands and gesticulating, and increasing voice volume to overpower others when simultaneous speech occurs. People who persist in these behaviors are soon viewed by others as egocentric and domineering. They have violated an implicit social rule: "It's all right to hold a conversation, but you should let go of it now and then" (Richard Armour).

Verbal content and grammatical form of speech also provide important cues for turn taking. People usually exchange turns at the end of a meaningful speech act, after an idea has been completed. First priority for the next turn goes to any person explicitly addressed by the current speaker with a question, complaint, or other invitation to talk. People expect turn changes to occur after almost every question, but not necessarily after other pauses in conversation (Hanni, 1980). It is difficult to exchange turns without using questions. When speakers in one study were permitted to use all methods except questions for signaling their desire to gain or relinquish the floor, the length of each speaking turn virtually doubled (Kent, Davis, & Shapiro, 1978).

Turn Allocation Much of our conversation takes place in settings where turn taking is more organized than in spontaneous conversations. In class discussions, meetings, interviews, and therapy sessions, for example, responsibility for allocating turns tends to be controlled by one person, and turns are often allocated in advance. Prior allocation of turns reduces strains that arise from people either competing for speaking time or avoiding their responsibilities to speak. Allocation of turns also increases the efficiency of talk. It can arrange a distribution of turns that best fits the task or situation—a precisely equal distribution (as in a formal debate) or just one speaker (as in a football huddle).

Feedback and Coordination

We engage in conversation to attain interpersonal goals—to inform, persuade, impress, control, and so on. To do this effectively, we must assess how what we say is affecting our partner's interest and understanding as we go along. Both verbal and nonverbal feedback help conversationalists in making this assessment. Through feedback, conversationalists coordinate what they are saying to each other from moment to moment. Responses called **back-channel feedback** are especially important for regulating speech as it is happening. These are the small vocal and visual comments a listener makes while a speaker is talking, without taking over the speaking turn. They include such responses as "Yeah," "M-hmn," short clarifying questions ("What?" "Huh?"), brief repetitions of the speaker's words or completions of his or her utterances, head nods, and brief smiles. When conversations are proceeding smoothly, the fine rhythmic body movements of listeners (swaying, rocking, blinking) are precisely synchronized with the speech sounds of speakers who address them (Condon & Ogston, 1967). These automatic listener movements are another source of feedback that indicates to speakers whether they are being properly "tracked" and understood (Kendon, 1970).

Both the presence or absence and the timing of back-channel feedback influence speakers. In smooth conversation, listeners time their signs of interest, agreement, or understanding to occur at the end of long utterances, or when the speaker turns his or her head toward them. When speakers are denied feedback, the quality of their speech deteriorates. They become less coherent and communicate less accurately. Their speech becomes more wordy, less efficient, and more poorly fitted to the specific information needs of their partner (Kraut, Lewis, & Swezey, 1982). Lack of feedback causes such deterioration because it prevents speakers from learning several things about their partners. They cannot discern whether their partners (1) have relevant prior knowledge they need not repeat; (2) understand already so they can wrap up the point or abbreviate; (3) have misinformation they should correct; (4) feel confused so they should backtrack and clarify; or (5) feel bored so they should stop talking or change topics.

Alerted to the possible loss of listener attention and involvement by the absence of feedback, speakers employ attention-getting devices to evoke feedback. One such attention-getting device is the phrase "You know." Speakers frequently insert "You know" into long speaking turns immediately prior to or following pauses if their partner seems to be ignoring their invitation to provide feedback or to accept a speaking turn (Fishman, 1980).

Another device a speaker can use to regain the attention of another participant is to ask a question. Such displays of uncertainty—for example, "What was the name of that guy on the Oprah Winfrey show?"—restructures the interaction by getting listeners more involved (Goodwin, 1987). If the speaker shifts his or her gaze to a specific person as the speaker asks the question, it will draw that person into the conversation.

The fact that feedback influences the quality of speech has another interesting consequence. Listeners who frequently provide their conversational partners with feedback also understand their partner's communication more fully and accurately. Through their feedback, active listeners help shape the conversation to fit their own information needs. This finding reinforces a central theme of this chapter: Accurate communication is a shared social accomplishment.

Feedback is important not only in conversations, but also in formal lectures. Lecturers usually monitor members of the audience for feedback. If listeners are looking at the speaker attentively, and nodding their heads in agreement, the lecturer infers his or her message is understood. Quizzical or "out of focus" expressions, however, suggest failure to understand. Similarly, members of the audience use feedback from the lecturer to regulate their own behavior; a penetrating look from the speaker may be sufficient to end a whispered conversation between listeners.

An important form of feedback in many lectures is applause. Speakers may want applause for a variety of reasons, not just ego gratification. Sometimes lecturers subtly signal the audience when to applaud; audiences watch for such signals in order to maintain their involvement. As an illustration, analysis of 42 hours of recorded political speeches suggests there is a narrow range of message content that stimulates applause (Heritage & Greatbatch, 1986). Attacks on political opponents, foreign persons, and collectivities; statements of support for one's own positions, record, or party; and commendations of individuals or groups generate applause. When these messages are framed within particular rhetorical devices, applause is from two to eight times more likely. For example:

> Speaker: "Governments will argue [pause] that resources are not available [short pause] to help disabled people. [long pause] The fact is that too much is spent on the munitions of war, [long pause] and too little is spent [applause begins] on the munitions of peace."

In this example, the speaker uses the rhetorical device of contrast or antithesis. Using this device, the speaker's point is made twice. Audiences can anticipate the completion point of the statement by mentally matching the second half with the first. This rhetorical device is an "invitation to applaud," and in the example the audience begins to applaud even before the speaker completes the second half.

Summary

Communication is the process whereby people transmit information about their ideas and feelings to one another.

Language and Verbal Communication
Language is the main vehicle of human communication. (1) All spoken languages consist of sounds that are combined into words with arbitrary meanings and put together according to grammatical rules. (2) According to the encoder-decoder model, communication involves the encoding and sending of a message by a speaker, and the decoding of the message by a listener. Accuracy depends on the codability of the idea or feeling being communicated. (3) In contrast, the intentionalist model argues that communication involves the speaker's desire to affect the listener, or the transmission of an intention. The context of the communication influences how messages are sent and interpreted. (4) The perspective-taking model argues that communication requires intersubjectivity, the shared context created by speaker and listener. Thus communication is a complex undertaking; to attain mutual understanding, conversationalists must express their message in ways listeners can interpret,

take account of others' current knowledge, and actively work to decipher meanings.

Nonverbal Communication
A great deal of information is communicated nonverbally during interaction. (1) Four major types of nonverbal communication are paralanguage, body language, interpersonal spacing, and choice of personal effects. (2) People from all cultures express and recognize the same six primary emotions in distinctive facial expressions. During interaction, however, people modify facial expressions of emotion according to cultural display rules. (3) Information is usually conveyed simultaneously through nonverbal and verbal channels. Multiple cues may add information to each other, reduce ambiguity, and increase accuracy. But if cues appear inconsistent, people must determine which cues reveal the speaker's true intentions.

Social Structure and Communication
The ways we communicate with others reflect and influence our relationships with them. (1) In every society, speech that adheres to rules governing vocabulary, pronunciation, and grammar is preferred or standard. Its use is associated with high status or power and is evaluated favorably by listeners. Nonstandard speech is used by lower status persons and evaluated negatively. (2) We express, maintain, or challenge the levels of relative status and intimacy in our relationships through our verbal and nonverbal behavior. Status and intimacy influence and are influenced by forms of address, choice of dialect or language, interruptions, matching of speech styles, gestures, eye contact, posture, and interaction distances. (3) The appropriate interaction distances for particular types of activities and relationships are normatively prescribed. These distances vary from one culture to another. When strangers violate distance norms, people flee the scene or use other devices to protect their privacy.

Conversational Analysis
Smooth conversation depends on conversational rules and communication skills that are often taken for granted.

(1) Conversations are initiated by a summons to interaction. They get under way only if potential

partners signal availability, usually through eye contact or verbal response. (2) Conversationalists avoid verbal collisions by taking turns. They signal either a willingness to yield the floor or a desire to talk through verbal and nonverbal cues. In some situations turns are allocated in advance. (3) Effective conversationalists assess their partner's understanding and interest as they go along through vocal and visual feedback. If feedback is absent or poorly timed, the quality of communication deteriorates. An effective speech also involves coordination between speaker and audience; the timing of applause is a joint accomplishment.

Key Terms

back-channel feedback (p. 184)
body language (p. 169)
communication (p. 159)

communication accuracy (p. 162)
cooperative principle (p. 164)
display rules (p. 172)
encoder-decoder model (p. 161)
flirting (p. 173)
intentionalist model (p. 162)
interpersonal spacing (p. 169)
linguistic relativity hypothesis (p. 165)
nonstandard speech (p. 176)
paralanguage (p. 168)
perspective-taking model (p. 164)
sociolinguistic competence (p. 167)
speech act theory (p. 163)
spoken language (p. 160)
standard speech (p. 176)
summons-answer sequence (p. 182)
symbols (p. 159)
theory of speech accommodation (p. 179)

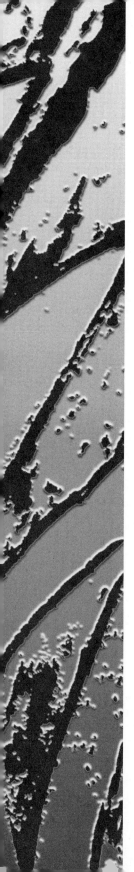

CHAPTER 8
Social Influence and Persuasion

Introduction

Consider some examples of social influence:

- In front of her house, Carol is met by Bernice, a neighbor. Bernice has heard that a waste-management company plans to open a new landfill only ⅛ mile from their neighborhood. Very opposed to this prospect, Bernice argues that the landfill would pose dangers to health and diminish land values. She asks Carol to attend a meeting and sign a petition against the landfill. Somewhat alarmed by developments, Carol finds Bernice's view persuasive and agrees to sign.

- One evening, the owner in a 24-hour convenience store is suddenly confronted by a bandit wearing a ski mask and brandishing a pistol. The man issues a threat: "Hand over your money or I'll blow you away!" Facing a choice between two undesirable alternatives—losing his money or his life—the hapless victim opens the cash register and hands over the money.

- During the Gulf War, a U.S. infantry commander orders a platoon of men to attack an enemy bunker. The danger involved is great. Night has fallen, the entire area is covered with antipersonnel mines, and defenders have been returning fire from the bunker. Despite these obstacles, the troops move out as ordered.

These stories illustrate various forms of social influence. By definition, **social influence** occurs when one person's behavior causes another person to change an opinion or to perform an action that he or she would not otherwise perform. In social influence, one person (the **source**) engages in some behavior (such as persuading, threatening or promising, or issuing

orders) that causes another person (the **target**) to behave differently from how he or she would otherwise behave. In the preceding illustrations, the sources were Bernice (persuading Carol), the midnight bandit (threatening his victim), and the infantry commander (ordering his troops to attack).

Various outcomes can result when social influence is attempted. In some cases, the influencing source may produce **attitude change**—a change in the target's beliefs and attitudes about some issue, person, or situation. Attitude change is a fairly common result of social influence. In other cases, however, a source may not really care about changing the target's attitudes but only about securing compliance. **Compliance** occurs when the target's behavior conforms to the source's requests or demands. Some influence attempts, of course, produce both attitude change and compliance.

Moreover, we must recognize that many influence attempts simply fail. Attempts at persuasion often prove ineffectual, producing little or no change in the target. Although orders issued by authorities usually obtain compliance, targets sometimes respond with defiance or open revolt. Because influence attempts vary in their degree of success, one concern in this chapter is to discern the conditions under which influence attempts are effective in producing compliance and attitude change.

Forms of Social Influence

Influence attempts can be either open or manipulative (Tedeschi, Schlenker, & Lindskold, 1972). In open influence, the attempt is readily apparent to the target. The target understands that someone is trying to change his or her attitudes or behavior. In manipulative influence, the attempt is hidden from the target. Examples of manipulative influence include ingratiation and tactical self-presentation. We discuss manipulative influence in Chapter 9. This chapter focuses on open influence.

We can distinguish many forms of open influence. Among the more important forms are (1) the use of persuasive communication to change the target's attitudes or beliefs, (2) the use of threats and/or promises to gain compliance, and (3) the use of orders based on legitimate authority to gain compliance.

These forms differ with respect to the resources or *bases of power* utilized by the source during an influence attempt (Raven, 1993).

Consider, first, the resources involved in persuasion. When attempting to persuade, the source uses information to change the target's attitudes and beliefs about some issue, person, or situation. Certain types of information are more useful than others in bringing about persuasion. For instance, a persuasion attempt is more likely to succeed if the source can introduce facts not already known to the target; likewise, success is more likely if the source can advance compelling and valid arguments not previously considered by the target. Having the right type of information is crucial when attempting persuasion.

Influence attempted by means of threats and/or promises is based on control of punishments and rewards, rather than information. If a threat is to be effective in gaining compliance, the target must believe the source controls whether the threatened punishment will be imposed or not. The same is true for influence based on promises, except that this involves control of rewards rather than punishments. If the target believes the source has no real control over the punishments or rewards involved, then the threat or promise is unlikely to succeed in influencing the target.

Influence through the use of orders from an authority or officeholder is based on acceptance of that authority's legitimacy by the target person. Influence of this type is especially common within formal groups or organizations. When attempting influence by invoking legitimate authority, the source makes demands of the target that is vested in his or her role within the group. Such an attempt will succeed only if the target believes the source actually holds a position of authority and has the right to issue orders of the kind involved in the influence attempt.

Because influence attempts can vary greatly in their degree of success, and we all use social influence in our relationships with others, it is important to understand the conditions under which influence attempts are effective. Specifically, in this chapter we address the following issues:

1. What factors determine whether a communication will succeed in persuading a target to change beliefs and/or attitudes? In what ways, for instance, do characteristics of the source and the target, as well as properties of the message itself, determine whether the persuasion attempt will be effective?

2. Under what conditions do threats and promises prove successful in gaining compliance from the target? When two interacting persons can threaten and punish one another, under what conditions will threats intensify and conflict escalate?

3. When a person in authority issues an order, under what conditions are targets most likely to obey it?

Attitude Change via Persuasion

Day in and day out, others bombard us with messages that seek to persuade. As an example, consider what happens to Steve Maxwell on a typical day. Early in the morning, Steve's clock radio comes on. Before Steve can get out of bed, a cheerful announcer is trying to sell him a new mouthwash. On the way into work, one of the persons in his car pool attempts to persuade Steve to vote for a particular candidate in the upcoming election. At lunch, a friend mentions her plans to attend a concert the following weekend and urges him to come along. In mid-afternoon, Steve listens to an argument from a co-worker who wants to change some paperwork procedures in the office. When he arrives home in the evening, Steve opens his mail. One letter is a carefully worded appeal from a charitable organization asking him to volunteer his time. Other letters are junk-mail flyers asking for money. Later that night, when Steve is watching television, advertisers bombard him endlessly with ads for their products—laundry soaps, lite beers, anti-dandruff shampoos, and imported automobiles.

All these messages received by Steve have something in common: They seek to persuade. **Persuasion** may be defined as changing the beliefs or attitudes of a target through the use of information or argument. Persuasion is widespread in social interaction and assumes many different forms (McGuire, 1985). In this section we consider various facets of message-based persuasion. First we examine how targets process messages. Next we discuss the communication-persuasion paradigm. Finally we consider

Political lobbyists target a congressman in an effort to persuade him to vote favorably on an upcoming bill.

characteristics of sources, messages, and targets that affect the persuasiveness of a message.

Processing Persuasive Messages

When a target receives a message advocating a position different from what he or she believes, the target can respond in various ways. Of course, the target may accept the message and change attitudes or beliefs. But messages intended to persuade do not always achieve their objective, and other reactions are possible.

Reactions to Persuasive Messages

Rather than accept the argument in a message, the target might simply *reject the message* or refuse to listen further. For instance, rather than heed the automobile advertisement or the beer commercial, the television viewer can switch channels. Another possible response to a persuasive communication is to *derogate the source.* Instead of changing attitudes, the target might dismiss the communicator as poorly informed, excessively partisan, or simply illogical. A

third possible response is to listen to the message but to *suspend judgment* on the issue. If the target knows some contrary facts not discussed in the message, he or she might decide to seek additional information before concluding that one viewpoint or another is correct. A fourth possible response is to *distort the message.* In this case, the target misperceives or misconstrues the content of the message (bringing it, perhaps, into line with his or her own cognitive schemas). The message might produce attitude change, although not necessarily what the source intended. A fifth possible response is to *attempt counterpersuasion.* That is, the target might react by arguing back and trying to change the source's own beliefs and attitudes. Obviously, counterpersuasion is not possible with messages transmitted via TV or other mass media, but people frequently use it in face-to-face situations.

Central and Peripheral Routes

The **elaboration likelihood model,** a theory advanced by Petty and Cacioppo and their co-workers (Petty & Cacioppo, 1986a, 1986b; Petty et al., 1994), undertakes to explain the processes by which messages produce attitude change. The elaboration likelihood model identifies two basic routes through which a message may alter a target's existing attitudes: the *central route* and the *peripheral route.*

Persuasion via the central route occurs when a target scrutinizes the arguments contained in a persuasive message, interprets and evaluates them (taking into account what he or she already knows), and then integrates them into a coherent position. Petty and Cacioppo use the term *elaboration* to describe the process whereby the target thinks through the implications of the arguments contained in a message. In elaboration, attitude change occurs only when the arguments are strong, internally coherent, and consistent with known facts. Information held by the target before receiving the message, as well as new information in the message itself, comes into play during elaboration.

In contrast, persuasion via the peripheral route occurs when, instead of elaborating the content of the message, the target pays attention primarily to extraneous cues linked to the message. Among these cues are the characteristics of the source (expertise, trustworthiness, likableness), superficial characteristics of the message (message length), or characteristics of the situation (response of other audience members). If a target processes the message via the peripheral route, he or she does not elaborate the message but instead appraises peripheral cues and uses them as a basis for accepting or rejecting the message.

The elaboration likelihood model states that what matters most about a message is not its properties per se, but whether it undergoes elaboration. What factors determine whether a message will undergo elaboration? Elaboration requires that a target have both the motivation to pay attention to the message's arguments and the ability to pay attention to the arguments. Thus the target's involvement with the issue is an important factor. If the target is highly involved with and cares about the specific issue addressed in a message, the message is more likely to be elaborated than if the target is uninvolved with the issue. Another factor is the target's *need for cognition,* a personality trait. Persons high on need for cognition characteristically enjoy thinking issues or problems through in detail. These persons are more likely to elaborate a given message than persons who are low on this trait (Cacioppo et al., 1996). Other factors relevant to elaboration include whether or not the target is distracted by noise or other disturbances in the situation or whether the target is tired. Obviously, a target will be less able to elaborate a message under conditions where he or she is distracted or tired. Elaboration occurs only when a target both wants to consider message arguments and is able to do so.

If a given message undergoes elaboration, attitude change by the target (if it occurs) depends largely on the strength and quality of its arguments (central route factors). If the message does not undergo elaboration, however, attitude change by the target depends on such factors as the identity and credibility of the source or the reaction of other audience members (peripheral route factors). This matter of what route was involved in attitude change is sometimes important. Attitudes established via the central route tend to be more strongly held than those established via the peripheral route. That is, attitudes obtained via the central route tend to be more long lasting and more resistant to change because the target has thought through the issue in more detail (Chaiken, 1980; Petty et al., 1994).

Communication-Persuasion Paradigm

Consider the general question "Who says what to whom with what effect?" Some theorists believe this question provides a useful framework for organizing modern research on persuasion. In this question, the "who" refers to the source of a persuasive message, the "whom" refers to the target, and the "what" refers to the content of the message. The phrase "with what effect" refers to the various responses of the target to the message. These elements (source, message, target, response) are fundamental components of the **communication-persuasion paradigm.** Figure 8.1 displays this paradigm and shows how these components interrelate.

As this figure indicates, each of these components is relevant to persuasion. First, properties of the source can affect how the target audience construes the message. For instance, characteristics such as the expertise and trustworthiness of the source can affect

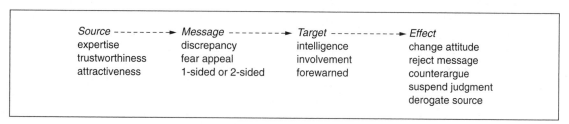

Figure 8.1 The Communication-Persuasion Paradigm

whether or not a target changes attitudes. Second, properties of the message itself can have a significant impact on its persuasiveness. For instance, whether a message carries a fear appeal or presents only one-sided arguments can affect whether a persuasion attempt is successful. Third, characteristics of the target also enter the picture. For instance, what a target already believes about an issue, as well as the extent of that person's involvement and commitment, can affect whether a message leads to attitude change or merely to rejection.

In the discussion that follows, we consider all these factors—properties of the source, the message, and the target—and we examine their impact on attitude change.

The Source

Suppose we ask 25 persons selected at random to read a persuasive communication (such as a newspaper editorial) that advocates a position on a nutrition-related topic. We tell this group that the message came from a Nobel Prize–winning biologist. At the same time, we ask another 25 persons to read the same message, but we tell this group that it came from a cook at a local fast-food establishment. Subsequently, we ask both groups to indicate their attitude toward the position advocated in the message. Which group of persons will be more persuaded by the communication?

Most likely, the persons who received the message from the prize-winning biologist will be more persuaded than those who received it from the fast-food cook. Because we assembled the groups at random, the only systematic difference between them is the source who wrote the message. Any difference in their reaction to the message must therefore be due to that factor.

Why should the source's identity make any difference? The identity of the source provides the target with information above and beyond the content of the message itself. Because some sources are more credible than others, the target may pay attention to the source's identity when assessing whether to believe the message. The term **communicator credibility**

denotes the extent to which the communicator is perceived by the target as a believable source of information. Note that the communicator's credibility is "in the eye of the beholder." Thus, a given source may be credible for some audiences but not for others.

Many factors influence the extent to which a source is credible. Two of these—the source's expertise and the source's trustworthiness—are of special importance. We examine each of these in more detail.

Expertise   Generally, a message from a source having a high level of expertise relevant to the issue brings about greater attitude change than a similar message from a source having a lower level of expertise (Chebat, Filiatrault, & Perrien, 1990; Haas, 1981; Maddux & Rogers, 1980). This occurs because targets may be more accepting and less critical of messages from high-expertise sources.

The impact of source expertise is illustrated by a study in which subjects were exposed to a message advocating the passage of a consumer protection bill by the U.S. Senate (Sternthal, Dholakia, & Leavitt, 1978). For some subjects, the message was ascribed to a lawyer who was educated at Harvard and had extensive experience in consumer issues (a high-credibility source). For other subjects, the message was ascribed to a citizen interested in consumer affairs (a low-credibility source). Afterward, subjects indicated their reaction to the message. In general, subjects who initially opposed the position advocated in the message expressed more unquestioning agreement and less counterargument when the message came from the high-credibility source than when it came from the low-credibility source.

The source's expertise interacts with the target's involvement and knowledge in determining attitude change. When the target has little involvement or prior knowledge on a given issue, messages from highly expert sources produce more attitude change than those from less expert sources. But the more involving the issue or the more knowledge the target has about the issue, the less likely it is that communicator expertise will make much difference in persuasion (Rhine & Severance, 1970). When involvement and knowledge are high, the target is more likely to engage in detailed processing and elaboration, so the content of the message itself becomes the overriding determinant of attitude change (Petty & Cacioppo, 1979; Stiff, 1986).

Trustworthiness Although expertise is an important factor in communicator credibility, it is not the only one. Under some conditions, a source can be highly expert but still not very credible. As an example, suppose your car is running poorly, so you take it into a garage for a tune-up. A mechanic you had never met before inspects your car. He identifies several problems, one of which involves major repair work on the engine. The mechanic offers to complete this work for $680 and claims your car will soon fall apart without it. The mechanic may be an expert, but can you accept his word that the expensive repair is necessary? How much does he stand to gain if you believe his message?

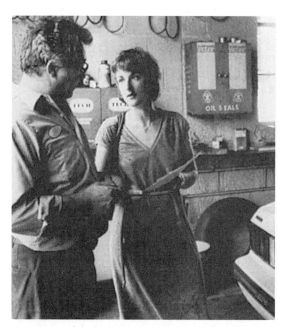

As an automobile owner listens to the message from the garage mechanic, she assesses not only the quality of the argument but also the credibility of the communicator. He may have expertise, but can he be trusted?

As this example shows, the target pays attention not only to a communicator's expertise but also to his or her motives. If the message appears highly self-serving and beneficial to the source, the target may distrust the source and discount the message (Hass, 1981). In contrast, communicators who argue against their own vested interests seem especially candid and trustworthy. For example, suppose an employee of a local business told you that you should not purchase a product made by her company but one made by a Japanese competitor. Her remarks would probably be unexpected, but they would have more impact than if she had argued for purchasing her own American-made model. Even if you normally prefer to buy products made in the United States, you might think twice in this case. A source who violates our initial expectations by arguing against his or her own vested interest appears especially trustworthy and, therefore, is persuasive (Eagly, Wood, & Chaiken, 1978; Walster, Aronson, & Abrahams, 1966).

Trustworthiness also depends on the source's identity because this carries information about the source's goals and values. A source perceived as having goals similar to the audience will be more persuasive than one perceived as having dissimilar goals (Berscheid, 1966; Cantor, Alfonso, & Zillmann, 1976). For example, a given policy proposal will be received differently by conservative Republicans depending on whether it was advanced by Dan Quayle (a Republican) or Jesse Jackson (a Democrat). The political identity of the source reveals much about his underlying goals and values, and these, in turn, affect perceived trustworthiness.

Attractiveness The physical attractiveness of the source can affect the extent to which a message is persuasive. Advertisers regularly select attractive individuals as spokespersons for their products, as we can see in television and magazine advertisements. Because it is rewarding to look at attractive spokespersons, these advertising messages receive more attention than they otherwise would. Higher source attractiveness leads us to give greater attention to the message, and higher levels of attention facilitate greater persuasion (Chaiken, 1986). In addition, because physical attractiveness leads to liking, we like attractive persons more than unattractive ones, and thus are sometimes more positively disposed to accept products or positions they advocate (Eagly & Chaiken, 1975; Horai, Naccari, & Fatoullah, 1974).

A source's attractiveness can have greater effect when combined with other factors. In one study investigating the impact of persuasive advertisements for suntan oil, the subjects received a message that, depending on treatment, contained either strong or weak arguments and came from either an attractive or an unattractive female spokesperson (DeBono & Telesca, 1990). Results showed that, in general, the attractive source was more persuasive than the unattractive one. But the attractive source was especially persuasive when the message arguments were strong, rather than weak. When the arguments were weak, attractiveness made very little difference in persuasion.

Effect of Multiple Sources Factors other than the source's expertise and trustworthiness can affect whether a message is persuasive. **Social impact theory** (Jackson, 1987; Latané, 1981; Sedikides & Jackson, 1990), which is a general framework applicable to both persuasion and obedience, states that the impact of an influence attempt is a direct function of strength (i.e., social status or power), immediacy (i.e., physical or psychological distance), and number of influencing sources. A target is more likely to be influenced when the sources are strong (rather than weak), when the sources are physically close (rather than remote), and when the sources are numerous (rather than few).

Although not all the predictions from social impact theory have yet been fully tested (Jackson, 1986; Mullen, 1985), there is support for one of the theory's more interesting predictions, that regarding the impact of multiple sources. The theory predicts that a message is more persuasive when a target receives it from multiple sources rather than from a single source. Consistent with this prediction, several studies have shown that a message presented by several different sources is more persuasive than the same message presented by a single source (Harkins & Petty, 1981b, 1987; Wolf & Bugaj, 1990; Wolf & Latané, 1983). This is especially true when the arguments

presented in the message are strong rather than weak. Strong messages coming from multiple sources receive greater scrutiny and foster more issue-relevant thinking by the target, which leads to attitude change; however, weak messages from multiple sources may receive added scrutiny but produce no extra attitude change (Harkins & Petty, 1981a).

Certain qualifications apply to this multiple-source effect. First, for multiple sources to have more impact than a single source, the target must perceive the multiple sources to be independent of one another. If the target believes the sources colluded in sending their messages, the impact of multiple sources tends to vanish, and the communication has no more effect than if it came from a single source (Harkins & Petty, 1983).

Second, there is an upper limit to increases in persuasion from the multiple-source effect (Tanford & Penrod, 1984). Adding more and more sources increases persuasion, but only up to a point. For instance, a message from three independent sources is more persuasive than the same message from a single source, but a message coming from, say, 13 sources may not be appreciably more persuasive than the same message coming from 11 sources. Beyond some threshold, more sources do not produce additional attitude change.

The Message

Persuasive communications differ dramatically in their content. Some messages contain arguments that are highly factual and rational, whereas others contain emotional appeals that motivate action by arousing fear or greed. Messages differ in their detail and complexity (simple vs. complex arguments), their strength of presentation (strong vs. weak arguments), and their balance of presentation (one-sided vs. two-sided arguments). These properties affect how a target scrutinizes, interprets, and elaborates a message. For this reason, they pertain more to the central route than to the peripheral route. In this section, we discuss the impact of these properties on persuasion.

Message Discrepancy Suppose a woman told you that Elizabeth II, the queen of England, is 5 feet

4 inches tall. Would you believe her? What if she said 5 feet 10 inches tall—would you believe that? How about 6 feet 3 inches? Or 7 feet 6 inches? You may not know how tall the queen actually is, but you probably have a rough idea. Although you might believe 5 feet 10 inches, you would probably doubt 6 feet 3 inches and certainly doubt 7 feet 6 inches. The message asserting that the queen is 7 feet 6 inches tall is highly discrepant from your beliefs.

By definition, a **discrepant message** is one that advocates a position different from what the target believes. Discrepancy is a matter of degree; some messages are highly discrepant, others less so. To cause a change in beliefs and attitudes, a message must be at least somewhat discrepant from the target's current position; otherwise, it would just reaffirm what the target already believes. Up to a certain point, greater levels of message discrepancy lead to greater change in attitudes (Jaccard, 1981). A moderately discrepant message is likely to be more effective in changing a target's beliefs and attitudes than a message that is only slightly discrepant. Of course, it is possible for a message to be so discrepant that the target simply dismisses it. To say the queen of England is 7 feet 6 inches tall is just not believable.

There is an important interaction between message discrepancy and source credibility. Sources with high credibility produce maximum attitude change at higher levels of discrepancy than do sources with low credibility. Thus a target is more likely to accept a highly discrepant message from a high-credibility source than from a low-credibility source. Highly discrepant messages from a low-credibility source are ineffective because the target may derogate the source. Figure 8.2 summarizes the joint impact of message discrepancy and communicator credibility on attitude change.

Many empirical studies report findings consistent with the relationships shown in Figure 8.2 (Aronson, Turner, & Carlsmith, 1963; Fink, Kaplowitz, & Bauer, 1983; Rhine & Severance, 1970). In one experiment, for instance, subjects were given a written message regarding the number of hours of sleep that people need each night to function effectively (Bochner & Insko, 1966). In some cases, the message was attributed to a Nobel Prize–winning physiologist

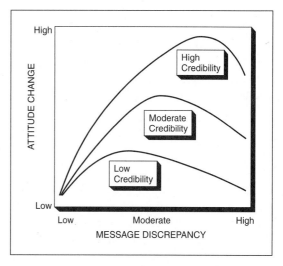

Figure 8.2 **Attitude Change as a Function of Communicator Credibility and Message Discrepancy**

These three curves summarize the relationship between message discrepancy and attitude change conditional on the credibility of a source. Note that messages from low-credibility communicators produce maximum attitude change at moderate levels of discrepancy, whereas messages from high-credibility communicators produce maximum attitude change at high levels of discrepancy.

(high credibility); in other cases, it was attributed to a YMCA director (medium credibility). The arguments contained in the message were identical for all subjects, with one important exception. In some cases the message proposed that people need 8 hours of sleep per night; in others, the message proposed 7 hours; in others, 6 hours; and so on down to zero hours of sleep per night. Most subjects began the experiment believing people need approximately 8 hours of sleep each night. Therefore, these messages differed in level of discrepancy.

Results of this study show that the more discrepant the position advocated by the high-credibility source (the Nobel Prize winner), the greater the amount of attitude change. Only when the source argued for the most extreme position (zero hours of sleep) did the subjects refuse to believe the message.

The same pattern appeared for the medium-credibility source (the YMCA director), except that his effectiveness peaked out at moderate levels of discrepancy (3 hours of sleep per night). For very extreme positions (2 hours of sleep or less), the medium-credibility source was less effective. Thus this study demonstrates that sources with higher credibility produce the greatest amounts of attitude change at higher levels of discrepancy.

Fear Arousal   Most messages intended to persuade use either rational appeals or emotional appeals. Rational appeals are factual in nature; they present specific, verifiable evidence to support claims. Rational appeals frequently address a need already felt by the audience and provide the missing solution; that is, these messages are drive reducing. Emotional appeals, in contrast, try to arouse basic drives and to stimulate a need where none was present. These messages are drive creating.

Perhaps the most common emotional appeals are those involving fear. Fear-arousing messages are especially useful when the source is trying to motivate the target to take some specific action. A political candidate, for example, may warn that if voters elect her opponent to office, the nation will become embroiled in international conflict. Likewise, in an antismoking advertisement on TV, a victim dying of throat cancer and emphysema warns young persons that if they start smoking cigarettes, they may end up as diseased victims themselves. In each of these cases, the source is using a fear-arousing communication. Messages of this type direct the target's attention to some negative or undesired outcome that is likely to occur unless the target takes certain actions advocated by the source (Higbee, 1969).

Some studies have shown that communications arousing high levels of fear produce more change in attitude than communications arousing low levels of fear (Dembroski, Lasater, & Ramires, 1978; Leventhal, 1970). If a message arouses fear and the targets believe attending to the message will show them how to cope with this fear, then they may analyze the message carefully and change their attitudes via the central route (Petty, 1995). Fear-arousing communi-

cations have been effective in persuading people to do many things, including reduce their cigarette smoking, drive more safely, improve their dental hygiene practices, change their attitudes toward communist China, and so on (Insko, Arkoff, & Insko, 1965; Leventhal, 1970; Leventhal & Singer, 1966).

Some studies suggest, however, that fear-arousing messages can fail if they are too strong and create too much fear. If people feel very threatened, they may become defensive and deny the reality or the importance of the threat, rather than think rationally about the issue (Johnson, 1991; Liberman & Chaiken, 1992). In this sense, a message arousing moderate fear may prove more effective than one arousing extremely high fear.

The impact of fear-arousing communications is shown clearly by a study in which college students received messages advocating inoculations against tetanus (Dabbs & Leventhal, 1966). These messages described tetanus as easy to catch and as producing serious, even fatal, consequences. The message also indicated that inoculation against tetanus, which could be obtained easily, provided effective protection against the disease. Depending on experimental treatment, subjects received either high-fear, low-fear, or control communications. In the high-fear condition, the messages described tetanus in extremely vivid terms (thereby creating a high level of fear and apprehension). In the low-fear condition, the messages described tetanus in less detailed terms (thereby creating no more than low to moderate fear). In the control condition, the message provided little detail about the disease (thereby arousing no fear).

To determine the message's effectiveness, the students were asked whether they thought it was important to get a tetanus inoculation and whether they actually intended to get one. Responses showed that students exposed to the high-fear message had stronger intentions to get shots than those exposed to the other messages. Moreover, records kept at the university health service indicated that students receiving the high-fear message were more likely to obtain inoculations during the following month than were students receiving the other messages. Overall, higher levels of fear led to greater attitude and behavior change in this study.

More generally, fear-arousing messages are effective only when certain conditions are met. First, the message must assert that if the target does not change behavior, he or she will suffer serious negative consequences. Second, the message must show convincingly that these negative consequences are highly probable. Third, the message must recommend a specific course of action that, if adopted, will enable the target to avoid the negative consequences. A message that predicts negative consequences but fails to assure the target that he or she can avoid these by taking specific action will produce little attitude change. Instead, it will leave the target feeling that the negative consequences are inevitable regardless of what he or she may do (Job, 1988; Maddux & Rogers, 1983; Patterson & Neufeld, 1987).

One-Sided Versus Two-Sided Messages

When a source uses rational rather than emotional appeals, other message characteristics also come into play. One such characteristic is the number of viewpoints, or "sides," represented in the message. A one-sided message emphasizes only those facts that explicitly support the position advocated by the source. A two-sided message, in contrast, presents not only the position advocated by the source but also opposing viewpoints. For example, if a man used a one-sided message to persuade his wife to spend their vacation at the seashore, he would mention only the reasons for going to the shore. If he used a two-sided message, he would mention both the reasons for going and the reasons for not going. Of course, if he really preferred the seashore, in the two-sided message he might also try to refute or discredit the reasons for not going.

Which is more effective, a one-sided message or a two-sided message? The answer depends heavily on the nature of the target audience. One-sided messages have the advantage of being uncomplicated and easy to grasp. They are more effective when the audience already agrees with the source; they also tend to be effective when the audience does not know much about the issue, for they keep the audience blind to opposing viewpoints. Two-sided messages are more complex, but they have the advantage of making the source appear less biased and more trustworthy. They are more effective when the audience initially opposes the source's viewpoint or knows a lot about the alternative positions (Karlins & Abelson, 1970; Sawyer, 1973).

The Target

To this point, we have discussed how the characteristics of the source and the content of the message affect persuasion. Yet it is also true that characteristics of the target play a role in persuasion. One important target characteristic that affects persuasion is the degree to which the target is involved with the issue. Another important characteristic is whether the target has been immunized against persuasion. In the following, we consider both of these.

Involvement With the Issue

One important attribute of targets is the extent of their involvement with a particular issue (Johnson & Eagly, 1989; Petty & Cacioppo, 1990). Suppose, for example, that someone advocates a fundamental change at your college, such as increasing the number of comprehensive exams required for graduation. The proposed change would take effect in September of next year. Many undergraduates would probably be very involved with this issue because the change would affect their chances of getting a college degree. Now suppose the source advocated that the change take place 10 years in the future rather than next September. Today's students would probably have little interest in this proposal because they will finish college long before any changes take effect.

Involvement with the issue affects the way a target processes a message. When highly involved, a target wants to scrutinize the message closely and think carefully about its content. Strong arguments are likely to produce substantial attitude change, whereas weak arguments are likely to produce little or no attitude change. In contrast, the target who is uninvolved has less motivation to scrutinize the message or think carefully about it. If any change in attitude occurs, it results more from peripheral

This magazine advertisement is part of a media campaign to sell perfume. To define the intended market for the product, the ad uses the technique of reversing traditional gender roles.

factors (such as source expertise or trustworthiness) than from the arguments themselves (Chaiken, 1980; Leippe & Elkin, 1987; Petty, Cacioppo, & Heesacker, 1981).

In one study, a message similar to that just described was presented to a group of college students (Petty, Cacioppo, & Goldman, 1981). The message proposed that college seniors be required to take a comprehensive exam before graduation. Three independent variables were manipulated in this study. The first variable was personal involvement with the issue. Half the subjects were told the new policy would take effect next year at their college (high involvement), whereas the other half were told the policy would take effect 10 years in the future (low involvement). The second variable was the strength of the message's argument. Half the subjects received eight strong and cogent arguments in favor of the proposal; the other subjects received eight weak and spe-

cious arguments. The third variable was the expertise of the source. Half of the subjects were told the source of the message was a professor of education at Princeton University (high-expert source); the other half were told the source was a student at a local high school (low-expert source).

The results of this study appear in Figure 8.3. In the high-involvement condition, the target's attitude toward comprehensive exams was determined primarily by the strength of the arguments. Strong arguments produced significantly more attitude change than weak ones. The expertise of the source had no significant impact on attitude change. In the low-involvement condition, attitudes were determined primarily by the source's expertise. The high-expert source produced more attitude change than the low-expert source. The strength of the arguments had little effect on this group.

Thus the target's involvement with the issue moderated which factor was the primary determinant of attitude change. For subjects with high involvement, the strength of the argument was more important than source expertise (a peripheral factor) because subjects cared about the issue. For those with low involvement, source expertise were more important because subjects had little motivation to scrutinize the arguments. Similar findings have been reported by Chaiken and Maheswaran (1994).

Immunization Against Persuasion   Most people try to defend their attitudes and beliefs—especially those about important issues—from attack. When attitudes are based on a large amount of information, they usually are easy to defend, but when attitudes are based on little information, they may be more vulnerable to attack. One class of beliefs that targets find hard to defend is the *cultural truism*. Truisms are beliefs widely accepted by the members of a given culture. Here are some examples of American truisms: "Everyone should brush their teeth after every meal if at all possible"; "The effects of penicillin have been, almost without exception, of great benefit to humankind"; and "Mental illness is not contagious." Truisms are rarely questioned, so people have little practice in defending them against attack.

Interested in how persons develop resistance to persuasion, McGuire (1964) proposed that a target

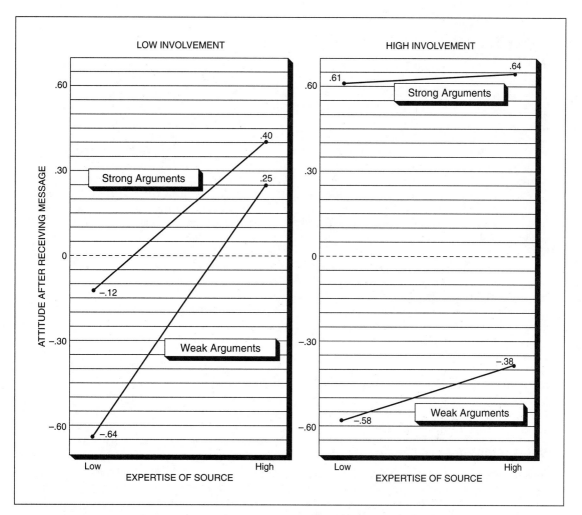

Figure 8.3 The Effects of Personal Involvement on Persuasion

In this study, students received a message advocating that college seniors be required to take a comprehensive exam prior to graduation. Half the students were told that the new policy would take effect next year (high involvement), whereas the others were told the policy would take effect 10 years later (low involvement). Results show that students in the high-involvement condition were affected primarily by the strength of the arguments rather than the expertise of the source, whereas students in the low-involvement condition were affected primarily by the expertise of the source rather than the strength of the arguments.

SOURCE: Adapted from Petty, Cacioppo, and Goldman, 1981.

can be "immunized" against persuasion. He specified various immunization treatments that would enable target persons to defend their beliefs (such as cultural truisms) against an attack. One such treatment, called a *refutational defense,* is analogous to medical immunization in which a patient receives a small dose of a pathogen so he or she can develop antibodies. The refutational defense consists of giving the target (1) information that is discrepant with the truism and (2) arguments that refute this discrepant information and support the truism. By exposing a target to weak attacks and allowing the target to refute them, this

8.1   Persuasion via Mass Media

Although many of the persuasive messages we receive daily come to us through face-to-face interaction with other persons, a significant portion of them arrive via the mass media. The term **mass media** refers to those channels of communication that enable a source to reach a large audience. Whereas face-to-face communications typically can reach only a small audience, the mass media can potentially influence many people. The most influential mass medium in the United States is television, followed by newspapers, radio, and magazines (Atkin, 1981).

Not everyone has equal exposure to the mass media. For example, in the United States women view more television than men. Children and retirees view more television than adolescents and working adults. Viewing is negatively related to the level of education, income, and occupational status (Comstock et al., 1978; Newspaper Advertising Bureau, 1980).

Consideration of the mass media immediately raises a central question: To what extent are communications transmitted by the mass media effective in changing the beliefs and attitudes of large numbers of people? We look at this issue from the standpoint of media campaigns.

Media Campaigns

A **media campaign** is a systematic attempt by a source to use the mass media to change attitudes and beliefs of a select target audience. Media campaigns are common in the industrialized world. They are used by advertisers to sell new products or services and by political parties to sway voters' sentiments. They are also used by public officials to change citizens' behavior—"Only you can prevent forest fires," "Take a bite out of crime," "Friends don't let friends drive drunk," and so on (Cummings et al., 1989; Farhar-Pilgrim & Shoemaker, 1981; Solomon, 1982).

Each year, advertisers spend a lot of money—tens of billions of dollars—on media campaigns (McCombs & Eyal, 1980). Nevertheless, most media campaigns do not produce large amounts of attitude change. In general, messages sent via mass media have only a small impact on their target audience's attitudes (Barber & Grichting, 1990; Bauer, 1964). Consider, for example, what occurs during presidential campaigns. Soon after the political conventions nominate their candidates in midsummer, most Americans know how they intend to vote in the upcoming November election. Although the parties spend millions of dollars on political advertising during the fall campaign, they will not change many voters' attitudes. In most presidential elections, only about 7% to 10% of the voters switch their preferences regarding candidates during the campaign. It is difficult for campaigners to change political attitudes via the mass media.

Of course, from another perspective, a shift of 7% to 10% may be very significant. Many professional advertisers and politicians are satisfied if their media messages shift public opinion a few percentage points in the intended direction. In a close political race, a gain of 1% or 2% might be sufficient to win the election. Thus, even though a media campaign might be a disappointment in terms of producing widespread attitude change, it may simultaneously be a success from the standpoint of getting a candidate elected (Mendelsohn, 1973). Even a small amount of attitude change may be sufficient to justify the cost of the media campaign in a close election.

Why are media campaigns usually able to produce only small amounts of attitude change? There are several reasons. First, there is the phenomenon of *selective exposure.* Many messages do not reach the audience they are intended to influence because audience members attend mostly to those sources with which they agree. Instead

continued on next page

immunization treatment builds up the target's resistance and prepares the target to resist stronger attacks on the truism in the future.

Some research (McGuire & Papageorgis, 1961) has demonstrated the effectiveness of a refutational

defense against persuasion attempts. College students received messages attacking three different cultural truisms. Two days before the attack, the students had received an immunization treatment to foster resistance to persuasion. For one truism, they received

continued from previous page

of reaching persons who disagree with the message (and whose opinions might therefore be changeable), many media communications are heard by persons who already agree with the message (and whose opinions will therefore be reinforced, not changed). In media exposure, persons encounter more messages supporting than not supporting their preexisting attitudes (Klitzner, Gruenewald, & Bamberger, 1991; Sears & Freedman, 1967).

Second, even if the intended targets receive messages from the media, they may reject them or derogate the source. Recipients of media communications are certainly not passive, and the impact of a message depends heavily on the uses and gratifications that the audience can obtain from the information (Dervin, 1981; Swanson, 1979). For example, in selling consumer products, media persuasion is more effective when the target's involvement with the decision is low and when he or she perceives relatively small differences among alternative products. In contrast, the impact of the media is slight when involvement with the decision is high and the differences between products appear clear-cut (Chaffee, 1981; Ray, 1973).

Third, even when a target finds a media message compelling, he or she may be subject to counterpressures that inhibit attitude change (Atkin, 1981). Some of these pressures come from social groups (such as family, friends, and co-workers); these groups may exert influence that nullifies the media's impact. In addition, targets are exposed to conflicting persuasive communications and cross pressures transmitted via the media. For example, beer advertisements would probably be enormously successful if only one manufacturer advertised its product. But because many brands advertise, media messages offset one another.

Other Effects of Media Campaigns

Although media campaigns do not usually cause a massive change in attitudes, they do exert other impacts on audiences. First, they are effective in strengthening preexisting attitudes. In other words, they reinforce and buttress preferences already held by the target audience. Televised debates between presidential candidates, for example, usually strengthen existing attitudes rather than change them (Kraus, 1962; Sears & Whitney, 1973).

In addition to strengthening preexisting attitudes, mass media are successful in creating attitudes toward objects that previously were unknown or unimportant to the audience. Advertisers use media campaigns to cultivate positive attitudes toward newly introduced products (e.g., breakfast cereals, new-design toys for children). Political parties also use media campaigns to create positive attitudes toward new little-known candidates running for office. For example, former President Jimmy Carter is widely known and well liked today by the American public. But when he first ran for the presidency in 1976, the situation was very different. Although Carter had been governor of Georgia, he was almost entirely unknown outside the South. To win the Democratic nomination for president, Carter desperately needed more name recognition among voters. His political handlers solved this problem by launching a media campaign based on the theme "Jimmy Who?" By asking this question, the campaign sought first to pique voters' curiosity and subsequently to create a positive view of this previously unknown candidate (Patterson, 1980). To a lesser degree, the same problem faced Michael Dukakis in the 1988 election and Bill Clinton in the 1992 election. Neither one of these politicians was well known initially outside his home state, and each tried to use the media to establish a positive public image from the outset.

---

a refutational defense. For a second truism, they received a different immunization treatment called a *supportive defense*—information containing elaborate arguments in favor of the truism. For the third truism, they received no defense. Following exposure to attacks on their attitudes, students rated the extent of their agreement with each of the truisms. Results show that the refutational defense provided a high level of resistance to persuasion. The supportive defense provided less resistance than the refutational defense.

And, when no defense was present, there was still less resistance to persuasion. In general, the findings demonstrate the effectiveness of the refutational defense in creating resistance to persuasion.

Compliance With Threats and Promises

Important though attitude change is, it is not the only outcome of social influence. Another important outcome is *compliance*, that is, behavioral conformity by the target to the source's requests or demands. In some situations a source cares only about securing compliance from a target, not about changing the target's beliefs and attitudes. Of course, a source might attempt to obtain compliance indirectly through persuasion. If the source can change what the target believes, he or she also may change how the target behaves and thereby obtain compliance. However, if persuasion fails to produce compliance, the source may still be able to obtain compliance by the direct use of threats and/or promises.

For example, consider how Richard Sorenson uses promises to obtain compliance. Sorenson, a home owner, lives in the upper peninsula of Michigan, where it snows heavily each winter. One cold day in January, a blizzard dumps 8 inches of snow on his driveway and sidewalk. Soon afterward, he notices his neighbor's teenage son is clearing the driveway next door with a snowblower. Seeking to clear his own driveway without lifting a shovel himself, Sorenson approaches the boy and says, "If you use your snowblower to clear my driveway and sidewalk, I will pay you $10." This is an influence attempt in the form of a promise. Sorenson promises to pay the boy $10 in return for a specified performance.

Influence based on promises and threats differs from persuasion in a fundamental way. In using persuasion, the source tries to change the way a target views the situation. Sorenson, for example, might have attempted to persuade the boy that shoveling snow is enormous fun or that clearing out the driveway would make him a good neighbor. These appeals, if successful, would change how the boy looks at the situation, but they would not actually change the situation itself. This contrasts with the use of promises and/or threats, where the source does restructure the situation. By promising to pay money for a clear driveway, Sorenson has added a new reinforcement contingency to the situation—money in return for snow removal. He hopes this will induce the boy to comply.

Effectiveness of Threats and Promises

To discuss the effectiveness of threats and promises, we must define these terms more precisely. A **threat** is a communication from one person (the source) to another (the target) that takes the general form, "If you don't do X (which I want), then I will do Y (which you don't want)" (Boulding, 1981; Tedeschi, Schlenker, & Lindskold, 1972). For example, an employer might say to her employee, "If you don't complete this project before the deadline, I'll reduce your salary." If the employee needs his salary to keep food on the table and has no other job prospects, he will take the threat seriously and do his best to comply with the demand.

When a source issues a threat, the sanction threatened can be virtually anything—a physical beating, the loss of a job, a monetary fine, the loss of love. The important point is that for a threat to be effective, the target must want to avoid the sanction. If the employee threatened by his boss happens to have a new job lined up elsewhere, he will not care whether his boss intends to cut his salary, and the threat will have little impact.

A **promise** is similar to a threat, except that it involves contingent rewards, not punishments. A person using a promise says, "If you do X (which I want), then I will do Y (which you want)." Notice that a promise involves a reward controlled by the source. Richard Sorenson promises a payment of $10 if his teenage neighbor clears the driveway and sidewalk. People frequently use promises in exchanges, both monetary and nonmonetary.

By issuing a promise, the source creates a set of options for the target. Suppose, for example, that the source makes the promise, "If you clear the snow from my driveway with your snowblower, I will pay you $10." In response, the target can (1) comply with the source's request and clear the driveway, (2) refuse to comply and let the matter drop, or (3) make a counteroffer ("How about $15? It's a long driveway, and the snow is very deep"). In similar fashion, a threat also creates a choice for the target. Once a threat is issued, the target can (1) comply with the threat, (2) refuse to comply, or (3) issue a counterthreat (Boulding, 1981).

The range of possible responses to threats and promises raises a fundamental question: Under what conditions will threats and promises be successful in gaining compliance, and under what conditions will they fail? Certain characteristics of threats and promises, such as their magnitude and credibility, affect the probability that the target will comply.

Magnitude of Threats and Promises

In promises, the greater the magnitude of the reward offered by a source, the greater the probability of compliance by the target (Lindskold & Tedeschi, 1971). For example, a factory supervisor might obtain compliance from a worker by offering a large incentive: "If you are willing to work the late shift next month, I'll approve your request for 4 extra days of vacation in September." The worker's reaction might be less accommodating, however, if his supervisor offered only a trivial incentive: "If you work the late shift next month, I'll let you take your coffee break 5 minutes early today."

A similar principle holds for threats. Compliance with threats varies directly with the magnitude of the punishment involved. Other factors being equal, targets dismiss threats that entail trivial consequences, but they are more likely comply with threats that entail large and serious consequences (Miranne & Gray, 1987).

Credibility of Threats and Promises

Suppose you own a little puppy that often runs wild. Your not-so-nice neighbor hates dogs. One day she issues a threat: "If you don't keep your dog off my property, I will call the city dogcatcher." This threat is troublesome because your dog romps on her property frequently. But would your neighbor really do what she says, or is she merely bluffing? Let's say you are willing to comply and tether your dog if the threat is real,

but you do not want to comply if it is merely a bluff. Unfortunately, there is no surefire, risk-free way to find out whether the threat is credible. The only true way to test your neighbor's credibility is to call her bluff—that is, to refuse to comply. Then, if your neighbor was merely bluffing, that fact will quickly become evident. Of course, if she was not bluffing, you will suffer the consequences. Retrieving your puppy from the dogcatcher may cost some money and entail some anxious moments.

Bluffing or not, any threatener wants the target to believe the threat is credible and to comply with his demand. He does not want the target to call his bluff. After all, a successful threat is one that obtains compliance without actually having to be carried out. If the target refuses to comply, the threatener must either admit he is bluffing or incur the costs of carrying out the threat.

To judge the credibility of a threat, targets gauge the cost to the source of carrying out the threat. Threats that cost a lot to carry out are less credible than those costing less. Targets also estimate the credibility of a threat from the social identity of the source. A threat involving physical violence, for example, is more credible if it comes from a karate expert wearing a black belt than if it comes from the proverbial 97-pound weakling.

The SEV Model

A threat's **subjective expected value (SEV)** is a measure of the pressure that the target feels from the threat. The level of SEV depends on several factors. SEV increases as the threat's credibility increases and as the magnitude of punishment threatened increases (Stafford et al., 1986; Tedeschi, Bonoma, & Schlenker, 1972). When both credibility and punishment magnitude are high, the SEV of the threat is high; consequently, the target feels a lot of pressure to comply. When both credibility and punishment magnitude are low, SEV is low; consequently, the target feels little or no pressure to comply. When one is high and the other is low (as, for example, when magnitude is high but credibility is low), the threat has a moderate to low SEV.

The *SEV model of threat effectiveness* predicts a target's compliance with a threat depends on two factors: the SEV of the threat and the cost to the target

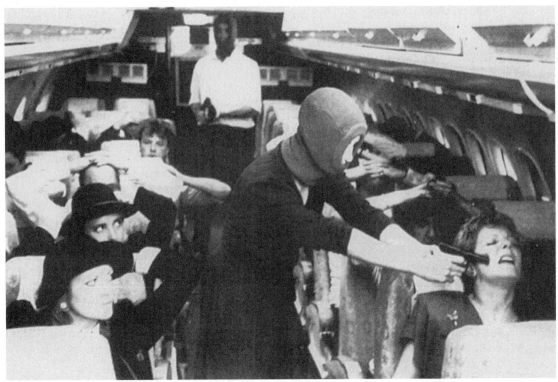

A terrorist issues a threat at gunpoint during an airplane hijacking. Targets are more likely to comply when threats are both large and credible.

of complying with the threat (Tedeschi, Bonoma, & Schlenker, 1972). That is, when deciding whether or not to comply with the threat, a target estimates both the SEV of the threat and the cost of complying. These factors have opposite effects on compliance. The probability of compliance increases directly as a function of SEV, but it decreases as a function of the cost to the target of complying.

Various empirical studies have supported the SEV model of threat effectiveness (Faley & Tedeschi, 1971; Grasmick & Bryjak, 1980; Stafford et al., 1986). For example, in one laboratory study by Horai and Tedeschi (1969), several pairs of persons played a game repeatedly over 150 trials. These pairs were composed such that one member was a naive subject and the other was a confederate of the researchers programmed to follow a specific strategy. Both the subject and the confederate were required to make certain choices that determined how points would be allo-

cated during each trial. The object of the game was to gain as many points as possible.

In this game, threat capability was unilateral; that is, the confederate could issue threats but the subject could not. To influence the subject's choice and gain extra points, the confederate occasionally made threats in this form: "If you do not pick choice 1 on the next trial, I will take [some number of] points away from your counter." The magnitude of the punishment threatened by the confederate was either high (20 points), medium (10 points), or low (5 points), depending on experimental treatment. The credibility of these threats was manipulated by varying the frequency with which the confederate carried them out if the subject did not comply. In the high-credibility condition, the confederate carried out threats 90% of the time; in the medium-credibility condition, 50% of the time; and in the low-credibility condition, 10% of the time.

The results of this study show that both the magnitude of punishment and the credibility of the threat affect the subject's level of compliance. Compliance increased as a direct function of both variables. Compliance was greatest under the highest magnitude of punishment and the highest level of threat credibility. Results of this study are consistent with the SEV model.

Box 8.2, "Your Money or Your Life," discusses compliance to threats in cases of attempted robbery where coercion is used illegitimately.

Although we have described the SEV model as it pertains to threats, it also applies to promises. Of course, the relevant variables in this case are the magnitude of the reward and the credibility of the promise. Results of one study showed that targets were more influenced when the reward promised was large rather than small and when the source had credibility in following through on his promises (Lindskold et al., 1970). Consistent with the SEV view, these conclusions hold true only when the reward promised for compliance is greater than the rewards that might be gained from refusing to comply.

Problems in Using Threats and Promises

Threats and promises pose certain problems for their user. First, the person using threats or promises must maintain surveillance to determine whether the target has complied. This typically entails some cost to the source in the form of time and trouble to monitor the

---

8.2   "Your Money or Your Life"

Perhaps the strongest threat that someone can make is one against your life. Threats of this type frequently occur during robbery or holdup attempts involving a lethal weapon, such as a knife or firearm. How do robbery victims react to threats made against their lives? To answer this question, Luckenbill (1982) studied 201 cases of robbery and attempted robbery that occurred in a Texas city between February 1976 and March 1977. Police records indicated that victims take several factors into account before deciding whether to comply with a threat. First, they consider the source's capacity to inflict serious injury or death. For example, compliance was high (75%) when the source had a lethal weapon (a knife or a gun) and was in a position to use it. As two victims noted,

> "I wasn't going to try anything because he had a piece (firearm). When's he's got a piece, you give him the money. That's all there is to it. If you try anything, he might shoot you."
>
> "I would've killed him if I could've gotten to my pistol. But . . . I couldn't do a thing, you know. He was standing right there watching me. And his pistol was loaded. I could see the steel bullets in the cylinder. And he was shak-

ing so bad he might've shot me if I tried anything. If I had a clear chance, you know, I would've nailed him. But hell, I'm not going to risk my life for a few measly bucks." (Luckenbill, 1982, p. 814)

When the source had no lethal weapon, however, compliance by the targets was low (5%). According to police records, the target opposed the demand for money in almost all cases in which the robber did not have a weapon.

A second factor considered by the victims was the source's intent regarding the use of force—that is, whether the source's threat was contingent or noncontingent. If the threat was contingent ("If you don't give me your money, then I'll shoot you"), victims assumed they would not be hurt if they complied. However, if the threat was noncontingent ("I'm going to shoot you and take the money"), most victims concluded that the robber intended to inflict harm indiscriminately. Luckenbill's data show that in cases where the robber issued a noncontingent threat, victims actively opposed the robber because they feared imminent death.

In general, these findings support the view that compliance with a threat is likely when the subjective expected value of the threat is large, provided that compliance provides an assured means of avoiding punishment or harm.

target's behavior. The problems involved in maintaining surveillance can be especially troublesome if the source uses threats rather than promises. When the source uses threats, the target not only may fail to comply, but also may attempt to conceal his or her noncompliance and thereby avoid punishment (Ring & Kelley, 1963). If the target engages in concealment, then adequate surveillance and detection may entail even greater costs for the source.

Another problem is that threats often cause resentment or hostility by the target toward the influencing source (Zipf, 1960). In some cases, these negative interpersonal sentiments may linger long after the threat is finished. Promises usually do not entail this difficulty because they are based on rewards and not punishments. A target is likely to develop more positive feelings toward someone who uses promises than toward someone who uses threats (Rubin & Lewecki, 1973). Thus, when the source enjoys a good relationship with the target and wants to maintain it, he or she may prefer to use promises, or at least to avoid using threats.

Bilateral Threat and Negotiation

To this point we have considered various means of influence, including persuasion, threats, and promises. We have relied on the simplifying assumption that influence flows in only one direction, from the source to the target. But in many real-life situations, both persons are able to exercise influence. That is, when person A (the source) attempts to influence person B (the target), person B (now the source) may respond by attempting to counterinfluence person A (now the target). Two-way influence can surely involve persuasion or promises, but it is especially interesting when it entails threats.

Negotiation   Consider first a situation where (1) two individuals have strongly opposing interests, and (2) each person has the capacity to issue threats and inflict punishments on the other. This situation carries the potential for open conflict, but, if the individuals involved are circumspect, they may hesitate to use threats and try instead to negotiate an acceptable solution. **Negotiation** is defined as communication

between opposing sides in a conflict, wherein the parties make offers and counteroffers and a solution occurs only when the parties reach an agreement (Pruitt & Carnevale, 1993). At first, the negotiators might attempt to find a straightforward compromise, such as one based on the split-the-difference principle. This compromise would yield a solution midway between person A's most preferred outcome and person B's most preferred outcome. If a solution like this is not acceptable to the negotiators, they may attempt something more creative like an *integrative solution,* which is a compromise whereby the two parties make trade-offs on issues according to their relative importance to each side (Neale & Bazerman, 1991; Pruitt, 1983). Specifically, each side makes its largest concessions on issues that are unimportant to it but important to the other side; the result is a solution that provides greater total benefits to negotiators than a split-the-difference compromise.

Although the notion of an integrative solution is appealing, such solutions are surprisingly difficult to achieve in practice, in part because there are barriers to identifying them (Ross & Ward, 1995; Thompson, 1990). For one thing, participants in a negotiation frequently fail to discover one another's true interests. And the more the participants have at stake in a negotiation, the more biased their perceptions of one another. They may come to view one another as opponents and distrust any proposals made by the other side (Thompson, 1995). If the participants overlook interests held in common or fail to discover one another's true interest, integrative solutions will prove elusive.

Deterrence and Conflict Spiral   If attempts at negotiation fail, the situation (as perceived by the participants) now has a different, more ominous character. The situation has become one in which (1) two individuals have strongly opposing interests, (2) each person has the capacity to issue threats and inflict punishments on the other, and (3) negotiations have shown there is little realistic possibility for compromise between them. Circumstances like these can become quite volatile and erupt into open conflict. If one or both participants believe compromise is impossible, then the situation may change quickly from simple negotiation to bilateral threat. A **bilateral threat** situation is one in which both participants issue threats and inflict damage on one another.

Bilateral threat situations involve a tension between the temptation to use punishment against the other and the fear of having punishment used against oneself. Most bilateral threat situations end up in one of two ways (Lawler, 1986; Lawler & Ford, 1995). On one hand, they can end in a state of mutual **deterrence,** which is a kind of armed stand-off. If both participants especially fear potential damage inflicted by the other, a state of mutual deterrence exists and the participants are likely to refrain from initiating threats and punishments (Bacharach & Lawler, 1981, chapter 5; Michener & Cohen, 1973). Alternatively, bilateral threat situations can devolve into a **conflict spiral,** a form of escalation. In a conflict spiral, the participants insist on compliance to their demands, even though they realize the other can inflict damage. They exchange successively larger threats and counterthreats to bolster their demands and end up inflicting more and more damage on one another (Deutsch & Krauss, 1962; Lawler, Ford, & Blegen, 1988; Smith & Anderson, 1975).

Power Differences and Deterrence One factor of special interest in bilateral threat situations is the relative strength or power of the participants. Several studies have investigated bilateral situations involving *asymmetrical power relations*—that is, situations where both persons have threat and punishment capability, but one person has more than the other. Power differences have an effect on the participants' use of threats, on concessions made, and on deterrence.

First, situations of unequal power are less stable and less likely to result in mutual deterrence than situations of equal power. If the participants have equal coercive power, the situation is less likely to result in a conflict spiral with threats and punishments than if the participants have unequal power (Lawler, Ford, & Blegen, 1988). Moreover, given equal power, participants are less likely to engage in damage tactics when they are both strong rather than weak, because both fear retaliation by the other (Hornstein, 1965; Lawler, 1986; Lawler & Bacharach, 1987).

In situations of unequal power, the high-power negotiators tend to behave exploitatively, whereas those with low power behave more submissively (Rubin & Brown, 1975). Given unequal power, the high-power negotiators usually maintain firm aspirations, adopt tougher tactics, and prefer not to make concessions. Not only do they avoid initiating concessions, they may refuse to match concessions from the low-power negotiator (Michener et al., 1975; Smith & Leginski, 1970). In contrast, low-power negotiators frequently match any concessions made by the other.

In unequal bilateral situations, the low-power negotiator often hesitates to use threats or to behave aggressively because he or she fears retaliation from the high-power opponent (who can, after all, inflict a lot of damage). The deterrence effect is unmistakable: As the punishment magnitude of the high-power person increases, the aggressive behavior of the low-power person decreases (Michener & Cohen, 1973).

Despite this deterrence effect, situations of unequal power are still more likely to result in escalation than situations of equal power. Persons having lower but still significant levels of power often refuse to comply with demands from stronger negotiators. Under these conditions, even a single punitive act may be sufficient to trigger a conflict spiral. Negotiators, in fact, often overmatch or exceed the other's threats at low levels of conflict, an unhappy tendency that can lead to conflict spirals (Youngs, 1986). One threat provokes another of larger size, making an agreement ever more difficult to achieve.

Obedience to Authority

To this point, we have discussed attitude change via persuasion and compliance based on threats and promises. Important though these are, they are not the only forms of influence used in everyday life. We have all witnessed situations in which—without the use of threat, promises, or persuasion—one person issues an order and another person complies. For example, a baseball umpire tosses an unruly manager out of the game and orders him to leave the field; the manager, after showing his resentment by slamming

The lines of authority become salient as a military commander gives orders to his troops prior to a mission in the jungle.

his cap to the ground and kicking first base, grudgingly complies. The umpire does not attempt to persuade the manager to leave voluntarily; he simply issues an order directing the manager to leave. Compliance in this case is based on the fact that both the umpire and the manager are participating in a larger social system (two ball clubs playing a game) in which behavior is regulated by rules and roles. The capacity of the source (umpire) to influence the target (manager) stems from the rights conferred by their roles within the game. Under the rules of the game, the umpire has the right to throw the manager out of the game for disruptive behavior.

When persons occupy roles within a group, organization, or larger social system, they accept certain rights and obligations vis-à-vis other members in that social unit. Typically, these rights and obligations give one person authority over another with respect to certain acts and performances. **Authority** refers to the capacity of one member to issue orders to others— that is, to direct or regulate the behavior of other members by invoking rights vested in his or her role. When the umpire tosses the manager out of the game, the basis of his power is legitimate authority.

Orders by police officers, decisions by judges, directives by corporate executives, and exhortations by members of the clergy—all these entail authority and the invocation of norms. A source can exercise authority only when, by virtue of the role that he or she occupies in a social group, others accept his or her right to prescribe behavior regarding the issue at hand (Kelman & Hamilton, 1989; Raven & Kruglanski, 1970). In exercising authority, the source invokes a norm and thereby obliges the target(s) to comply. The greater the number of persons the source can directly or indirectly influence in this manner, and the wider the range of behaviors over which the source has jurisdiction, the greater his or her authority within the group (Michener & Burt, 1974; Zelditch, 1972).

In this section, we discuss influence based on legitimate authority. First, we consider some interesting experimental studies of destructive obedience. These studies illustrate the extremes to which authorities can push behavior. Second, we consider some factors that determine whether a target will comply with an authority's directives or refuse to comply.

Experimental Study of Obedience

We note that obedience to authority frequently produces beneficial results, for it facilitates coordination among persons in groups or collective settings. Civil order hinges on obedience to orders from police officers and judicial officials, and effective performance in work settings often depends on following the directives of bosses or employers.

Yet if obedience to authority is unquestioning, it can sometimes produce disquieting or undesirable outcomes. For instance, in one study, hospital nurses received orders from doctors to administer a drug to a patient. The order came via telephone, and the nurses involved did not previously know the doctors giving the order. The drug was one not often used in the hospital; hence it was not very familiar to the nurses. The dosage prescribed was heavy and substantially exceeded the maximum listed on the package. Results showed that nearly all the nurses in this study were nevertheless ready to follow orders and administer the drug at the prescribed dosage (Hofling et al., 1966). Of course, conditions in this study were very favorable for obedience; under different conditions, obedience rates would not be so high. Subsequent research has indicated, for instance, that when nurses are more familiar with the medicine involved and when they are able to consult freely with their colleagues, rates of obedience are considerably lower (Rank & Jacobson, 1977).

In some cases, obedience to authority can produce very negative consequences, especially if the orders involve actions that are morally questionable or reprehensible. History provides many examples, such as the My Lai massacre during the Vietnam War when soldiers obeyed Lieutenant Calley's orders to kill innocent villagers. Or consider the activity of the Third Reich of Nazi Germany during the 1930s and 1940s, which produced the Holocaust. In complying with the dictates of Hitler's authoritarian government, some German citizens committed acts that most people consider morally unconscionable—beatings, confiscation of property, torture, and murder of millions of others. This may seem like madness, but Hannah Arendt (1965) has argued that most participants in the Holocaust were not psychotics or sadists who enjoyed committing mass murder but just ordinary individuals exposed to powerful social pressures.

To explore the limits of obedience to legitimate authority, Stanley Milgram carried out a program of experimental research in a laboratory setting (Milgram, 1965a, 1974, 1976; Miller, Collins, & Brief, 1995) . Milgram created a hierarchy in which one person (the experimenter, who assumed the role of authority) directed another person (the subject) to engage in actions that hurt a third person (a confederate, who played the role of victim). The primary goal of this research was to understand the conditions under which subjects would follow morally questionable orders to hurt the confederate.

At the outset, Milgram (1963) recruited 40 adult males to serve as subjects. These men, contacted through newspaper advertisements, were adults (ages 20–50) with diverse occupations (labor, blue-collar, white-collar, and professional). When a subject arrived for the experiment, he found that another person (a gentle 47-year-old male accountant) had also responded to the newspaper advertisement. This person, ostensibly another subject, was actually a confederate of the experimenter. The subjects were told by the experimenter that the purpose of research was to study the effects of punishment (i.e., electric shock) on learning. One of the subjects was to occupy the role of learner, and the other was to occupy the role of teacher. Subjects drew lots to determine their roles; unknown to the subject, the drawing was rigged such that the confederate was selected as the learner. The confederate was then taken into the adjacent room, strapped into an "electric chair," and electrodes were attached to his wrist. He mentioned he had some heart trouble and expressed concern that the shock might prove dangerous. The experimenter, who was dressed in a lab coat, replied that the shock would be painful but would not cause permanent damage. The subjects and the confederate then participated in a paired-associates learning task. The subject, in the role of teacher, read pairs of words over an intercom system to the confederate in the adjacent room, and the confederate was supposed to memorize these. After reciting the entire list of paired words, the subject then tested the amount learned by the confederate. Going through the list again, he read aloud the first word of each pair and four alternatives for the second word of the pair, like a multiple-choice exam. The confederate's task was to select the correct alternative response for each item.

Consistent with the cover story that they were investigating the effects of punishment on learning, the experimenter ordered the subject to shock the learner whenever he made an incorrect response. This shock was to be administered by means of an electric generator that had 30 voltage levels, ranging from 15 to 450 volts. The subject was directed to set the first shock at the lowest level (15 volts) and then, with each successive error, to increase voltage to the next higher level. Thus the subject was to increase the voltages from 15 to 30 to 45, up to the 450-volt maximum. On the shock generator, the lowest voltage level (15 volts) was labeled *slight shock;* a higher level (135 volts) read *strong shock;* higher still (375 volts) read *danger: severe shock;* the highest level (450 volts) was ominously marked *XXX.* In actuality, this equipment was a dummy generator, and the confederate never received any shock, but its appearance was quite convincing to subjects.

Soon after the session began, it became apparent that the confederate was a slow learner. Although he got a few answers right, his responses were incorrect on most trials. Subjects reacted by administering ever-higher levels of shock, as they had been ordered to do. When the shock level reached 75 volts, the confederate (who was still in the adjacent room) grunted loudly. At 120 volts, he shouted that the shocks were becoming painful. At 150 volts, he demanded to be released from the experiment ("Get me out of here! I won't be in the experiment anymore! I refuse to go on!"). At 270 volts, his response to the shock was an agonized scream. (Actually, the shouts and screams that subjects heard from the adjacent room came from tape recordings; this made learner's response uniform for all subjects.)

Whenever a subject expressed concern or dismay about the procedure, the experimenter urged him to persist ("The experiment must continue." "You have no other choice; you must go on."). At the 300-volt level, the confederate shouted in desperation that he wanted to be released from the electric chair and would not provide any further answers to the test. In reaction, the experimenter directed the subject to treat any refusal to answer as an incorrect response. At the 315-volt level, the learner gave out a violent scream. At the 330-volt level, he fell completely silent, and from that point on nothing more was heard from him. Stoically, the experimenter directed the subject to continue toward the 450-volt maximum even though the learner did not respond.

The basic question addressed by this study was this: What percentage of the subjects would continue to administer shock up to the 450-volt maximum? The results showed that of the 40 subjects, 26 (or 65%) continued to the end of the shock series (450 volts). Although they could have refused to proceed, not a single subject stopped before administering 300 volts. These results are displayed in Figure 8.4.

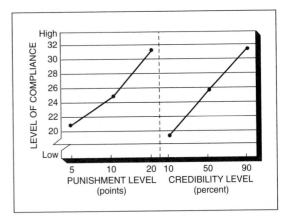

Figure 8.4   Compliance With Threats

Compliance is a function of both the magnitude of the punishment threatened and the credibility of the threat. The greater the punishment threatened, the greater the compliance with the threat. Likewise, the greater the credibility of the threat, the greater the compliance.

SOURCE: Adapted from Horai and Tedeschi, 1969.

Despite the tortured reaction of the confederate, most subjects followed the experimenter's orders.

Understandably, this situation was very stressful for subjects, and many felt some concern for the learner's welfare. As the shock level rose, subjects grew increasingly worried and agitated. Some began to sweat or laugh nervously, and many pleaded with the experimenter to check the learner's condition or to end the study immediately. A few subjects became so distressed that they refused to follow the experimenter's orders. The overall level of compliance in this study, however, was quite high, reflecting the enormous impact of directives from a legitimate authority.

Factors Affecting Obedience to Authority

As Milgram's results show, persons in authority usually obtain compliance with their orders, especially when these are accepted as legitimate and/or backed by potential force. Nevertheless, orders from an authority can set off a complex process, which can lead to various responses (Blass, 1991). Compliance does not always occur, and subordinate members sometimes defy orders from an authority. Although most subjects in Milgram's research obeyed orders, some refused to comply. Other studies have reported similar effects: Obedience is the most common response to authority, but defiance occurs in some cases (Martin & Sell, 1986; Michener & Burt, 1975a). This raises a basic issue: Under what conditions do people comply with authority, and under what conditions do they refuse? What factors affect the probability that group members will comply with authority?

Certain factors affecting compliance are straightforward. For instance, it matters whether the person issuing orders uses an overt display of symbols, such as wearing a uniform with authoritative insignia (Bushman, 1988). Other things equal, a direct display of authority symbols increases compliance. In one study (Sedikides & Jackson, 1990), visitors at the bird exhibit of the Bronx Zoo were approached by a person who told them not to touch the handrail of the exhibit. Results showed they were significantly more likely to obey this directive when it came from a person dressed in a zookeeper uniform than when it came from a person dressed in casual clothes. The use of authoritative symbols may also have played a part in the Milgram studies, where the experimenter wore a white lab coat.

Another factor that matters is whether the person in authority can back up his or her demands with punishment in the event of noncompliance. Although this was not an explicit factor in Milgram's studies, other research has manipulated the magnitude of the potential punishment wielded by the authority, and results support the view that greater punishment magnitude leads to higher levels of compliance (Michener & Burt, 1975a).

Milgram (1974) extended his basic experiment to study some other factors that affect compliance with orders. For instance, one variation manipulated the degree of surveillance by the experimenter over the subject (Milgram 1965a, 1974). In one condition, the experimenter sat a few feet away from the subject during the experiment, maintaining direct surveillance; in another condition, after giving basic instructions, the experimenter departed from the laboratory and issued orders by telephone from a remote location. Results show that the number of obedient subjects was almost three times greater in the face-to-face condition than in the order-by-telephone condition. In other words, obedience was greater when subjects were under direct surveillance than under remote surveillance. During the telephone conversations, some subjects specifically assured the experimenter that they were raising the shock level, when in actuality they were using only the lowest shock and nothing more. This tactic enabled them to ease their conscience while at the same time avoiding a direct confrontation with authority.

In another variation, Milgram (1974) manipulated the subject's physical proximity to the victim. Findings showed that bringing the victim closer to the subject (and therefore increasing the awareness by subjects of the learner's suffering) substantially reduced the subjects' willingness to administer shock. In the extreme case, when the victim was seated right next to the subject, obedience decreased substantially. Tilker (1970) reported similar results and also showed that expressly making subjects totally responsible for their own actions rendered them less likely to administer shocks to the learner.

Obedience to authority is also affected by the subject's position in a larger chain of command. Kilham and Mann (1974) used a Milgram-like situation in which a subject (the executant) actually pushed the buttons to administer shock while another subject (the transmitter) simply conveyed the orders from the experimenter. Results showed that obedience rates were approximately twice as high among transmitters as among executants. In other words, persons positioned closer to the authority but farther from the unhappy task of throwing the switch were more obedient.

Summary

Social influence occurs when behavior by one person (the source) causes another person (the target) to change an opinion or to perform an action he or she would not otherwise perform. Important forms of open influence include persuasion, use of threats and promises, and exercise of legitimate authority.

Attitude Change via Persuasion Persuasion is a widely used form of social influence intended to produce attitude change. (1) There are many possible reactions to persuasive messages. Although some attempts at persuasion succeed, others merely lead to rejection of the message, derogation of the source, or suspension of judgment. Persuasive messages can change attitudes and beliefs through either the central route or the peripheral route. (2) The communication-persuasion paradigm points to many factors—properties of the source, the message, and the target—that affect whether a message will change beliefs and attitudes. (3) Certain attributes of the source affect a message's impact. Sources who are credible (that is, highly expert and trustworthy) are more persuasive than sources who are not. Attractive sources are more persuasive than unattractive ones, especially if message arguments are strong. A message coming from multiple independent sources is likely to have more impact than the same message from a single source. (4) Message characteristics also determine a message's effectiveness. Highly discrepant messages are more persuasive when they come from a source having high credibility. Fear-arousing messages are most effective when they specify a course of action that can avert impending negative consequences. One-sided messages have more impact than two-sided messages when the target already agrees with the speaker's viewpoint or is not well informed. (5) Attributes of the target also determine a message's effectiveness. Targets highly involved with an issue scrutinize messages closely and are more influenced by the strength of the arguments than by peripheral factors, such as communicator credibility. Targets exposed to a refutational defense show a high level of resistance to persuasion, whereas those exposed to a supportive defense show somewhat less resistance.

Compliance With Threats and Promises
Threats and promises are influence techniques used to achieve compliance (not persuasion) from the target. In using threats and promises, the source alters the environment of the target by directly manipulating reward contingencies. (1) The effectiveness of a threat depends on both the magnitude of the punishment involved and the probability it will be carried out. Greater compliance results from high magnitude and high probability. Similar effects hold true for promises, although these naturally involve rewards rather than punishments. (2) Use of threats and promises pose several problems for the source. For threats to be effective, the source must maintain surveillance over the target. In addition, threats often arouse resentment or hostility toward the source and therefore disrupt relationships. (3) When influence is bilateral rather than unilateral, participants may attempt to resolve their differences through negotiation. If successful, negotiation may lead to an integrative solution; solutions of this type, however, are difficult to achieve. If each party has the capacity to inflict punishment, bargaining may escalate into a conflict spiral. If the two persons have very unequal coercive power, the weaker one will probably hesitate to issue threats and will more likely comply with the demands of the stronger one. But bilateral situations are volatile, in part because threats by one party often provoke threats by the other.

Obedience to Authority Authority refers to the capacity of one group member to issue orders or make requests of other members by invoking rights vested in his or her role. (1) Research on obedience to authority shows that subjects will comply with orders to administer extreme levels of electric shock to an innocent victim. (2) Obedience to authority is more likely to occur when the authority is dressed in uniform, when the authority can back up orders with punishments, when subjects are under direct surveillance by the person issuing orders, when subjects are distant from rather than close to the victim, and when subjects are transmitters rather than executants of a command.

Key Terms

attitude change (p. 188)
authority (p. 208)

bilateral threat(p. 207)
communication-persuasion paradigm (p. 191)
communicator credibility (p. 192)
compliance (p. 188)
conflict spiral (p. 207)
deterrence (p. 207)
discrepant message (p. 195)
elaboration likelihood model (p. 190)
mass media (p. 200)
media campaign (p. 200)
negotiation (p. 206)
persuasion (p. 189)
promise (p. 202)
social impact theory (p. 194)
social influence (p. 188)
source (p. 188)
subjective expected value (SEV) (p. 203)
target (p. 188)
threat (p. 202)

CHAPTER 9

Self-Presentation and Impression Management

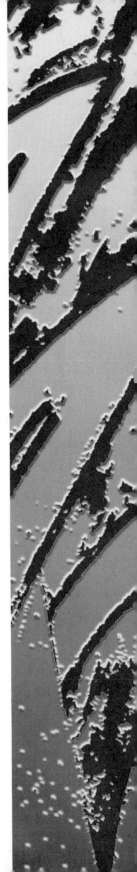

Introduction

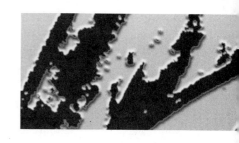

Remember the shabbily dressed street violinist you saw downtown? Your friend remarked on his artistic demeanor and sensitive eyes, and you agreed it was a shame a guy with such talent couldn't afford to study at Juilliard. When people dropped coins into his battered violin case, you both felt pleased.

If you live in New York City, you may have seen Richard Wexler. By day, Richard lives the good life in his fashionable Manhattan apartment. But come nightfall, Wexler takes off his expensive clothes and Italian-made loafers, slips into torn sneakers, tattered jeans, and a frayed shirt, and transforms himself into Richie, the ragged street artist. He hails a cab for Broadway, where he sets up a sign on the sidewalk ("VIOLINIST NEEDS MONEY TO FURTHER STUDIES") and serenades theatergoers with sentimental show tunes. On a typical night, Richie pulls in more than $300 for an hour's fiddling, enough to buy himself expensive watches and Bermuda vacations ("Gold in the Streets," 1978).

Richie is a true artist—an artist at managing impressions. He masterfully creates the perception of himself he desires. Richie capitalizes on an important social psychological principle: The way we perceive others' behavior and the impressions we form of them largely determine how we act toward them. Because we respond to their social identities, others find it advantageous to control the images they present to us (Schlenker, 1980). Thus, by presenting himself as a "struggling violinist," Richie influences the behavior of passersby and achieves his goal of making a lot of money.

Although few of us make our living by creating a false impression, we all present particular images of who we are. When we shout or whisper, dress up or dress down, smile or frown, we actively influence the impressions others form of us. In fact, presenting some image of ourselves to others is an inherent aspect of all social interaction. The term **self-presentation** refers to the processes by which individuals attempt to control the impressions that others form of them in social interaction (Leary, 1995; Leary &

Kowalski, 1990; Schlenker & Weigold, 1992). The individuals involved either may be self-aware of these processes or not.

For certain purposes, it is useful to distinguish among authentic self-presentation, ideal self-presentation, and tactical self-presentation (Baumeister, 1982; Kozielecki, 1984; Swann, 1987). In *authentic self-presentation,* our goal is to create an image of ourselves in the eyes of others that is consistent with the way we view ourselves (our real self). In *ideal self-presentation,* our goal is to establish a public image of ourselves that is consistent with what we wish we were (our ideal self). In *tactical self-presentation,* our concern is to establish a public image of ourselves that is consistent with what others want us to be. We may do this, for instance, by claiming to have some attributes they value, even if we really do not have them.

Persons engaging in tactical self-presentation usually have some ulterior motive(s) in mind. In some cases, they want others to view them positively because it will enable them to get some reward(s) that others control. Richard Wexler, for example, was trying to obtain some easy money to support his comfortable lifestyle. In other cases, they are trying to pass as specific kinds of persons in hopes of gaining access to individuals and situations that are otherwise unavailable. If an undercover narcotics agent is trying to set up a sting, for example, he may first need to infiltrate the drug operation, create the impression that he is an experienced drug runner, and gain the confidence of the bad guys. In tactical self-presentation, a person cares only about the impact of the image he or she is presenting to others, not about whether that image is consistent with his or her real self or ideal self. When a person uses self-presentation tactics calculated to manipulate the impressions formed of him or her by others, we say that he or she is engaging in **tactical impression management.**

Of course, there are hybrid situations where a person uses several forms of self-presentation simultaneously. For instance, a woman might try to remain largely authentic in self-presentation (giving others a correct impression of her) but also try to hide a few little flaws (so that others form an exaggeratedly positive impression of her).

This chapter considers the ways in which people actively determine how others perceive them. It addresses the following questions:

1. What content is conveyed through self-presentation in everyday life? What factors—both personal and situational—affect self-disclosure between persons?
2. What impression management tactics can we use when we want to claim a particular identity such as "overworked employee," "attractive date," or "competent student?" What factors influence our choice to use one impression management tactic rather than another?
3. To what extent can people detect when others are using impression management tactics against them? What cues reveal that an impression manager is trying to deceive them?
4. What are some of the consequences when people try, but ultimately fail, to project the social identities they desire?

Self-Presentation in Everyday Life

In this section, we discuss self-presentation in everyday situations. Our primary concern is authentic self-presentation, although we must recognize that many processes in authentic self-presentation also apply to tactical self-presentation. In everyday settings, people routinely project specific social identities, and they must take care that others understand and accept their identity claims. For example, when a temporarily out-of-work individual meets a potential employer during a job interview, she may naturally strive to create a positive first impression and claim the identity of "productive worker." However, she has to be careful

to create an authentic impression and not to claim too much. If she is hired for the job, it would be quite difficult to maintain a false image for very long when she has to perform on the job.

To project a social identity successfully, an individual must share with others some understandings about the situation in which they are participating. Successful self-presentation involves efforts, first, to establish a workable definition of the situation and, second, to disclose information about the self that is consistent with the claimed identity. We discuss each of these topics in this section.

Definition of the Situation

For interaction to be successful, participants in a situation must share some understandings about their social reality. *Symbolic interactionist theory* (Blumer, 1962; Charon, 1995; Stryker, 1980) holds that for social interaction to proceed smoothly, people must somehow achieve a shared **definition of the situation**—an agreement about who they are, what their goals are, what actions are proper, and what their behaviors mean. Persons can establish a definition of the situation by various means. In some interactions, they can establish a shared definition by actively negotiating the meaning of events (McCall & Simmons, 1978; Stryker & Gottlieb, 1981). In other interactions, people may invoke preexisting *event schemas* to provide a definition of the situation. Event schemas are particularly relevant when the event is of a common or recurring type, such as weddings, job interviews, funerals, first dates, and the like.

To establish a definition of the situation, people must agree on the answers to several questions: (1) What type of social occasion is at hand? That is, what is the frame of the interaction? (2) What identities do the participants claim, and what identities will they grant one another? Consider these issues in turn.

Frames   The first requirement in defining the situation is for people to agree regarding the type of social occasion in which they are participating. Is it a wedding? a family reunion? a job interview? The type of social occasion that people recognize themselves to be in is called the frame of the interaction

(Goffman, 1974; Manning & Cullum-Swan, 1992). More strictly, a **frame** is a set of (widely understood) rules or conventions pertaining to a transient but repetitive social situation that indicate which roles should be enacted and which behaviors are proper. When people recognize a social occasion to be a wedding, for example, they immediately expect a bride, a groom, and someone authorized to perform the ceremony to be present. They also know that the other guests attending are mostly friends and relatives of the couple and that it is acceptable (indeed, appropriate) to kiss the bride.

Participants usually know the frame of interaction in advance or else they discover it quickly once interaction commences. Sometimes, however, there will be conflict and they must negotiate the frame of interaction. When parents send their wayward teenage daughter to a physician for a talk, for example, the discussion may begin with subtle negotiations about whether this is a "psychiatric interview" or merely a "friendly chat." Once established, the frame limits the potential meanings that any particular action can have (Gonos, 1977). If the persons involved define the situation as a "psychiatric interview," for example, any jokes the teenager tells may end up being interpreted as symptoms of illness, not as inconsequential banter.

Identities

Another issue in defining a situation is for people to agree on the identities they will grant one another and, relatedly, on the roles they will enact. That is, people must agree on the type of person they will treat each other as being. The frame of the situation places limits on the identities that any person might claim. For example, a teenager in a psychiatric interview cannot easily claim an identity as a "normal, well-adjusted kid." And employers would find it incongruous and bizarre if a young woman tried to claim the identity of "blushing bride" in a job interview.

Each person participating in an interaction has a **situated identity**—a conception of who he or she is in relation to the other people involved in the situation (Alexander & Rudd, 1984; Alexander & Wiley, 1981). Identities are "situated" in the sense that they pertain to the particular situation. For in-

Even if they have done nothing wrong, these teenagers had best dramatize their innocence by presenting themselves to this police officer with polite deference. True identities may not be self-evident because perceivers are biased by the stereotypes and expectations they bring to a situation.

stance, the identity projected by a person while discussing a film (insightful critic) differs from the identity projected by the same person when asking for a small loan (reliable friend). Situated identities usually facilitate smooth interaction. For this reason, people sometimes adopt particular situated identities in public settings even though they may not accept them privately (Muedeking, 1992). To avoid unpleasant arguments, for example, you might relate to your friend as if she were an insightful or reliable person even though privately you believe she is not.

Much of the time, our identities are not self-evident to others because their perceptions of us filter through the stereotypes and implicit personality theories they bring to a situation. These bias the identities they perceive and grant to us. Thus, even if the identity claimed by us is authentic (in the sense of being consistent with our self-concept), we may need to highlight or dramatize it (Goffman, 1959). For instance, consider some adolescents who are innocent of any wrongdoing. If they display their usual nonchalant, defiant image when stopped by police, they risk being detained or arrested. They are more likely to avoid arrest if they dramatize their innocence by presenting a polite, deferential demeanor (Piliavin & Briar, 1964).

Self-Disclosure

A primary means we use to make authentic identity claims is to reveal certain facts about ourselves. When we first meet someone, we usually discuss only safe or superficial topics and reveal rather little about ourselves. Eventually, however, as we get to know the other better, we disclose more revealing and intimate details about ourselves. This might include information about our needs, attitudes, experiences, aspirations, and fears (Archer, 1980). This process of revealing personal aspects of one's feelings and behavior to others is termed **self-disclosure** (Derlega et al., 1993; Jourard, 1971).

Although self-disclosure can be unilateral, in everyday life it usually is bilateral or reciprocal. There is a widely accepted social norm that one person should respond to another's disclosures with disclosures at a similar level of intimacy (Rotenberg & Mann, 1986; Taylor & Belgrave, 1986). This is termed the norm of *reciprocity in disclosure.* Most people follow it, although strict reciprocity in disclosure is more prevalent in new relationships or developing friendships than in established ones where people already know a lot about one another (Davis, 1976; Won-Doornink, 1979). In addition, we are more likely to reveal more personal information to those we initially like and find attractive (Collins & Miller, 1994; Kleinke & Kahn, 1980).

Males and females differ somewhat in self-disclosure. There is evidence that, overall, females disclose more than males, although this difference is rather small (Dindia & Allen, 1992). This difference is conditioned by certain factors, one of which is the sex of the person to whom disclosures are made. When the partner to whom disclosures are made is female, females disclose more than males; but when the partner to whom disclosures are made is male, females do not disclose more than males. Another

These chess players seem to be building trust and liking through reciprocal self-disclosure, an important process in authentic self-presentation.

conditioning factor is the nature of the relationship between the discloser and the partner. When the person to whom disclosures are made is a stranger, there is rather little difference between males and females in the amount of disclosure. But when the person to whom disclosures are made has an established relationship with the discloser (such as friend, spouse, or parent), females disclose more than males (Dindia & Allen, 1992).

Males tend to find it more difficult than females to reveal information about themselves and their relationships (Davidson & Duberman, 1982; Hacker, 1981). This occurs, in part, because some men view excessive disclosure as a sign of weakness (Cunningham, 1981). Moreover, when men and women engage in self-disclosure about their personal shortcomings, they discuss different things. Men tend to discuss such topics as unwarranted risk taking or excessively aggressive behavior, whereas women discuss such topics as immature behavior or flaws in their appearance. (Derlega et al., 1981).

Self-disclosure usually leads to liking and social approval from others. People who reveal a lot of information about themselves tend to be liked more than people who disclose at lower levels (Collins & Miller, 1994). This holds especially true if the content of the self-disclosure complements what their partner has revealed (Daher & Banikiotes, 1976; Davis & Perkowitz, 1979).

Although self-disclosure usually produces liking, there is such a thing as revealing too much about oneself. Self-disclosure that violates the audience's normative expectations may actually produce disliking. For instance, self-disclosure that is too intimate for the depth of the relationship (such as a new acquaintance describing the details of her latest bladder infection) will not strengthen the friendship and may just create the impression that the discloser is indiscreet or maladjusted (Cozby, 1972). Likewise, self-disclosure that reveals negatively valued attributes (a person discussing his prison record for felonious assault) or profound dissimilarities with the partner (a believer revealing his strong religious commitment to a nonbeliever) may produce disliking (Derlega & Grzelak, 1979).

Perhaps not surprisingly, level of self-disclosure is related to loneliness. Young adults low in self-disclosure feel more lonely and isolated than those high in self-disclosure (Mahon, 1982). Lonely persons tend to have fewer skills in self-presentation and are less effective in making themselves known to others than are nonlonely persons (Solano, Batten, & Parish, 1982). The self-disclosure style of lonely persons may impair the normal development of social relations.

Tactical Impression Management

As we noted above, self-presentation is inherent in social situations. Most people strive to create images of themselves that are authentic or true—that is, consistent with their own self concept. Nevertheless, under certain conditions, individuals may try to present themselves in such a way as to create false, exaggerated, or misleading images in the eyes of others. The process of creating false images is called *tactical impression management.*

There are various reasons we might engage in tactical impression management (Jones & Pittman, 1982; Tetlock & Manstead, 1985). One major reason is to make others like us more than they would otherwise (ingratiation). Other reasons for impression management are to make others fear us (intimidation), respect our abilities (self-promotion), respect our morals (exemplification), or feel sorry for us (supplication). Usually, however, if we examine the underlying reason for engaging in ingratiation or self-promotion or intimidation, we find that the main motive is to increase our control over valued outcomes mediated by other persons (Tedeschi, 1981).

In this section we examine some of the tactics used in impression management. In particular, we discuss managing appearances, ingratiation, aligning actions, and altercasting.

Managing Appearances

People often try to plan and control their appearance. As used here, the term *appearance* refers to everything about a person that others can observe. This includes clothes, grooming, overt habits such as smoking or chewing gum, choice and arrangement of personal

Physical appearance is important in impression management. If impression management is to be successful, one's appearance in the eyes of the audience must be consistent with the identity one claims. If he lacked the make-up and costume of a clown, this performer would have a hard time convincing even young children that he really is a clown.

possessions, verbal communications (accents, vocabulary), and nonverbal communications. Through the appearances we present, we show others the kind of persons we are and the lines of action we intend to pursue (DePaulo, 1992; Stone, 1962).

Physical Appearances and Props

Many everyday decisions regarding appearance—which clothes to wear, how to arrange our hair, whether to shave, and so on—stem from our desire to claim certain identities. In some situations, we arrange our clothing and accessories to achieve certain effects. This would be true, for example, if we were attending a dance or going on a date. It is also true when we go for a job interview, as illustrated in a study of female job applicants (von Baeyer, Sherk, & Zanna, 1981). Some applicants in this study were led to believe that their male interviewer felt the ideal female employee should conform closely to the traditional female stereotype (passive, gentle, and so on); other applicants were led to believe that he felt the ideal female em-

ployee should be nontraditional (independent, assertive, and so on). Results showed that applicants managed their physical appearance to match their interviewer's stereotyped expectations. Those expecting to meet the traditionalist wore more makeup and used more accessories, such as earrings, than those planning to meet the nontraditional interviewer.

The impression an individual makes on others depends not only on clothes, makeup, and grooming, but also on *props* in the environment. The impression a woman makes on her friends and acquaintances, for instance, depends in part on the props she uses—the big pile of books she places on her desk, the music she selects for her CD player, the wine she serves, and the like. Others will make inferences regarding her character and interests from the props that surround her.

Regions

Goffman (1959) draws a parallel between a theater's front and back stages and the "regions" we use in managing appearances. He uses the term **front regions** to denote settings in which people carry out interaction performances and exert efforts to maintain appropriate appearances vis-à-vis others. One example of a front region is a restaurant's dining room, where waiters smile and courteously offer to help customers. **Back regions** are settings inaccessible to outsiders in which people knowingly violate the appearances they present in front regions. Behind the kitchen doors, the same waiters shout, slop food on plates, and even mimic their customers. In general, persons use back regions to prepare, rehearse, and rehash the performances that occur in front regions.

Front and back regions are often separated by physical or locational barriers to perception, like the restaurant's kitchen door. These barriers facilitate impression management because they block access to the violations of images that occur in back regions. Any breakdown in these barriers will undermine the ability of persons to manage appearances. In recent years, for example, one such breakdown has occurred regarding national political figures. Because the mass media are pervasive, they sometimes catch presidents, senators, and other officials off guard. National figures are shown expressing views and performing actions they would strongly prefer to keep hidden from the public. American presidents find it difficult to project a heroic identity when the media show one bumping his head on a door, another nearly collapsing while jogging, and a third nodding off during an audience with the pope. It was easier to be a hero in the days of Jefferson or Lincoln, before the invention of electronic media that penetrate the barriers between regions (Meyrowitz, 1985).

Ingratiation

Most people want to be liked by others. Not only do we find it inherently pleasant, but being liked may gain us a promotion or a better grade, and it may save us from being fired or flunked. How do we persuade others to like us? Although much of the time we are authentic and sincere in our relations with others, occasionally we may resort to ingratiation (Jones, 1964; Jones & Wortman, 1973). **Ingratiation** refers to the

deliberate use of deception to increase a target person's liking for us. Usually, the use of ingratiation stems from an ulterior motive, such as gaining some tangible benefits that the target person controls. To ingratiate ourselves with others, we may try to flatter them or exaggerate our own admirable qualities. When successful, ingratiation usually reduces the power and the decisional freedom of the target person vis-à-vis the ingratiator (Pandey, 1981).

Certain preconditions make ingratiation more likely. Individuals may try to ingratiate themselves when they depend on the target person for certain benefits, and believe (or assume) that the target person is more likely to grant those benefits to someone he or she likes than to someone he or she does not like. In addition, people are more likely to use ingratiation tactics when the target is not constrained by regulations and can therefore exercise his or her

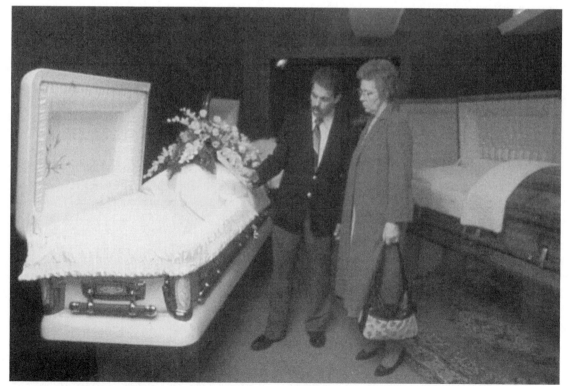

An undertaker employs a range of impression management techniques to create an atmosphere of quiet comfort. The opulent front region, a casket showroom, is carefully separated from the back region where a team of workers prepares the body for burial.

discretion in distributing rewards (Jones et al., 1965). For example, students are more likely to try to ingratiate themselves with instructors whose grading may be influenced by their liking for a student than with instructors who base their grades on objective multiple-choice exams.

There are a number of ingratiation tactics, each involving deception. Among these are opinion conformity (pretending to share the target person's views on important issues), other enhancement (outright flattery or complimenting the target person), and selective self-presentation (either exaggerating one's own admirable qualities or feigning modesty). We discuss these in the following sections.

Opinion Conformity   Faced with a target person who has discretionary power, an ingratiator may try to curry favor by expressing insincere agreement on important issues. This tactic, termed *opinion conformity,* is often successful because people tend to like others who hold opinions similar to their own (Byrne, 1971). Of course, obvious or excessive opinion conformity on issue after issue would quickly arouse a target's suspicion, so a clever ingratiator mixes conformity on important issues with disagreement on unimportant issues.

Opinion conformity sometimes requires us to tailor the content of the opinions we express to match a target person's general values rather than any specific opinions he or she may hold. Such matching of opinions to values was illustrated by an experiment in which subjects were working for a supervisor (Jones et al., 1965). The supervisor was a target for ingratiation because although failing at the task of judging advertising slogans, subjects needed a positive rating from the supervisor to earn money. Some subjects were led to believe the supervisor valued solidarity (friendliness and social compatibility); others were led to believe the supervisor valued productivity (efficiency and independence). When given an opportunity to express opinions publicly, subjects shifted their opinions to reflect their supervisor's values. Subjects in the solidarity condition shifted their expressed opinions on various issues to conform to those of the supervisor (thereby increasing their apparent social compatibility). In contrast, subjects in the productiv-

ity condition shifted their expressed opinions away from those of the supervisor, thereby matching his values (by showing their independence). In both cases, subjects selected a strategy that would increase their supervisor's liking for them.

This experiment investigated ingratiation directed upward in a hierarchy (from a worker to a high-status supervisor). Indeed, some evidence indicates that persons tend to show more ingratiation responses (of all kinds) toward their boss than toward a stranger or a friend (Bohra & Pandey, 1984). Yet recent findings (Gordon, 1996) indicate that ingratiation attempts directed upward (toward persons of higher status) are less likely to be effective in promoting liking than are ingratiation attempts directed laterally (toward persons of equal status) or downward (toward persons of lower status). High-status targets, aware that others may have a motive to ingratiate, may be somewhat more vigilant than equal- or low-status targets.

Other Enhancement   A second ingratiation tactic is *other enhancement*—that is, using flattery on the target person. To be effective, flattery cannot be careless or indiscriminate. More than two centuries ago, Lord Chesterfield (1774) stated that people are best flattered in those areas where they wish to excel but are unsure of themselves. This hypothesis was tested in a study, similar to that discussed above, in which female subjects were told that their supervisor valued either efficiency or sociability (Michener, Plazewski, & Vaske, 1979). The supervisor was a target for ingratiation because the subjects' earnings depended on the evaluations they received from her. Before the supervisor made these evaluations, the subjects had an opportunity to flatter her. The experimenter asked them to rate the supervisor's efficiency and sociability and indicated that the supervisor would see the ratings. The results showed that the supervisor's values channeled the form of flattery the subjects used. Subjects who believed the supervisor valued efficiency publicly rated her higher on efficiency than on sociability, whereas subjects who believed she valued sociability publicly rated her higher on sociability than on efficiency. Thus the subjects were discriminating in their use of praise and matched it to the target's values. In addition, they avoided extreme ratings that might suggest insincerity.

Beyond flattery, other enhancement can take some other forms. One of these is discussed in Box 9.1, "Playing Dumb."

Selective Self-Presentation

A third ingratiation tactic is *selective self-presentation*, which involves the explicit presentation or description of one's own attributes to increase the likelihood of being judged attractive by the target. There are two distinct forms of selective self-presentation: *self-enhancement* and *self-deprecation*. When using self-enhancement, a person advertises his or her strengths, virtues, and admirable qualities. If successful, this tactic creates a positive public identity and gains liking by others. In

---

9.1 Playing Dumb

"Playing dumb" is an ingratiation tactic used with some frequency in interaction. When playing dumb, impression managers pretend to be less intelligent or knowledgeable than they really are. By playing dumb they present themselves as inferior, thereby giving the target person a sense of superiority. Thus playing dumb is a form of other enhancement.

Although popular belief and early research suggest that women are more prone than men to play dumb (Wallin, 1950), a national survey of American adults indicates just the opposite (Gove, Hughes, & Geerken, 1980). Significantly more males than females said that they had pretended, at least once, to be less intelligent or knowledgeable than they really were. As shown in the table, men play dumb more often than women in most of the situations examined, including work.

What leads people to play dumb? The data indicate that people who use this technique tend to be young, highly educated males (Gove et al., 1980). In contemporary American society, these persons are likely to hold lower status positions in competitive occupations where knowledge is valued (junior executives, law clerks, graduate students, and the like). Because of their youth, many of these people are located at the bottom of an occupational ladder they aspire to climb.

These people interact with superiors in settings where intelligence and knowledge are prized. Under these circumstances, a person's relatively low status may require deferring to superiors despite one's own knowledge and ability. People may stand to gain by hiding any intellectual superiority they feel—that is, by playing dumb.

Interestingly, men tend to play dumb more than women with their boss and co-workers, and women tend to play dumb more than men with their spouse. Gove and colleagues (1980) suggest that this may reflect the continuing cultural expectation that a wife should refrain from displaying superior knowledge which might challenge her husband's assumed superiority, at least on certain issues.

Percentage of People Who Reported "Playing Dumb"

	Target Person						
	Overall	**Date**	**Spouse**	**Boss**	**Co-workers**	**Friends**	**Strangers**
Males (1,065)	31.1	8.2	6.0	13.1	14.6	12.6	15.8
Females (1,182)	22.9	9.0	10.0	5.2	6.3	8.7	7.7

SOURCE: Adapted from Gove, Hughes, and Geerken, 1980.

contrast, when using self-deprecation, a person makes only humble or modest claims. Self-deprecation can be an effective way to increase others' approval and liking, especially when it aligns the ingratiator with such important cultural values as honesty and objectivity in self-appraisal.

Of these two variations, self-enhancement is probably the more common. For instance, when subjects in experimental settings are asked to describe themselves in such a way as to make a good impression, the general tendency is for them to emphasize their positive attributes and play down their weaknesses (Jones, 1964). They may mention their honesty, competence, and friendliness directly, or they may imply their admirable qualities indirectly by, for example, publicly attributing their behavior to appropriate motives (Tetlock, 1981).

Although often effective, the tactic of selectively emphasizing our admirable qualities can be risky. This is especially true if the target knows enough about us to suspect we are boasting or if uncontrollable future events could prove our claims invalid. Wise ingratiators, therefore, use self-enhancing descriptions only when these risks are minimal—that is, when the target person does not know them well and has no way to check their future performances (Frey, 1978; Schlenker, 1975).

Because of the risks inherent in self-enhancement, the opposite approach—a self-deprecating or modest self-presentation—is often a safer tactic. To be effective, however, self-deprecation must be used in moderation. Excessively harsh and vigorous public self-criticism may gain expressions of support from others, but these expressions run the risk that others may actually believe them and form a negative private evaluation of the person using them (Powers & Zuroff, 1988). A more effective form of self-deprecation is an assured, matter-of-fact modesty that understates or downplays one's (substantial) achievements. In one study of self-deprecation, subjects read about a teacher who had either succeeded or failed in teaching his pupil and who gave either a self-enhancing or a self-deprecating explanation for the outcome (Tetlock, 1980). Subjects liked the teachers more when teachers responded to their success or failure with modest,

self-deprecating claims. They viewed these teachers as more concerned for others than teachers whose responses were self-enhancing. Thus, in this case, modesty was an effective technique.

In another study, members of a group were asked to evaluate other members following the group's success or failure at a task (Forsyth, Berger, & Mitchell, 1981). Group members reported greater liking for those who took blame for the group's failure or credited others for the group's success (self-deprecation) than for those who blamed others for failure and claimed credit themselves for the group's success (self-enhancement). Taken together, these studies suggest that when observers have evidence about someone's performance—whether favorable or unfavorable—self-deprecation can be an effective tactic of ingratiation.

Aligning Actions

During interaction, occasional failures of impression management are inevitable. Others may sometimes catch us performing actions that violate group norms (missing an appointment) or contravene laws (running a red light). Such actions potentially undermine the social identities we have been claiming and disrupt smooth interaction. When this occurs, people engage in **aligning actions**—attempts to define their apparently questionable conduct as actually in line with cultural norms. Aligning actions repair cherished social identities, restore meaning to the situation, and reestablish smooth interaction (Hunter, 1984; Spencer, 1987). In this section, we discuss two important types of aligning actions: disclaimers and accounts.

Disclaimers   When people anticipate that their impending actions will disrupt smooth social interaction, invite criticism, or threaten their established identity, they often employ disclaimers. A **disclaimer** is a verbal assertion intended ward off any negative implications of impending actions by defining these actions as irrelevant to one's established identity (Bennett, 1990; Hewitt & Stokes, 1975). By using disclaimers, they suggest that although the impending acts ordinarily imply a negative identity, theirs is an extraordinary case. For example, before making a bigoted remark, a person may point to her extraordinary credentials ("My best friend is Hispanic, but . . . "). Disclaimers are also used prior to acts that would normally undermine one's identity as moral ("I know I'm breaking the rules, but . . . ") or as mentally competent ("This may seem crazy to you, but . . . "). These disclaimers emphasize that although the person is aware the act could threaten her identity, she is appealing to a higher morality or to a superior competence.

Still other disclaimers plead for a suspension of judgment until the whole event is clear: "Please hear me out before you jump to conclusions." When individuals are not certain how others will react to new information or suggestions, they are more likely to preface their actions with hedging remarks ("I'm no expert, but . . . " or "I could be wrong, but . . . "). Such remarks proclaim in advance that possible mistakes or failures should not reflect on one's crucial identities.

Although disclaimers can be helpful in smoothing interaction, they tend to lose their effectiveness when they are overused. One study of disclaimers in dyadic interaction found that whereas stimulus persons who used disclaimers to a limited degree were not evaluated negatively by observers, others who used a large number of disclaimers were viewed as unrealistic and were evaluated more negatively (Bell, Zahn, & Hopper, 1984).

Accounts   After individuals engage in disruptive behavior, they may try to repair the damage by using accounts. **Accounts** are the explanations people offer to mitigate responsibility after they have performed acts that threaten their social identities (Harvey, Weber, & Orbuch, 1990; Scott & Lyman, 1968; Semin & Manstead, 1983). There are two main types of accounts: excuses and justifications. *Excuses* reduce or deny one's responsibility for the unsuitable behavior by citing uncontrollable events ("My car broke down"), coercive external pressures ("She made me do it"), or compelling internal pressures ("I suddenly felt dizzy and couldn't concentrate on the exam"). Presenting an excuse reduces the observer's tendency to

hold the individual responsible or to make negative inferences about his character (Riordan, Marlin, & Kellogg, 1983; Weiner et al., 1987). Excuses also preserve the individual's self-image and reduce the stress associated with failure (Snyder, Higgins, & Stucky, 1983). *Justifications* admit responsibility for the unsuitable behavior but also try to define the behavior as appropriate under the circumstances ("Sure I hit him, but he hit me first") or as prompted by praiseworthy motives ("It was for his own good"). Justifications reduce the perceived wrongness of the behavior.

One theorist (Schonbach, 1980) suggests that the account sequence consists of the following: (1) a failure event (for example, arriving late for a meeting), (2) a reproach by another person ("We agreed to start at 8 o'clock sharp. Why are you an hour late?"), (3) an account ("My car wouldn't start"), and (4) an evaluation by the other person of whether the account is acceptable ("Hmmm . . . I wonder if that's the whole story.").

Persons are more likely to accept accounts when the content appears truthful and conforms with the explanations commonly used for such behavior (Riordan et al., 1983). Accounts are honored more readily when the individual who gives them is trustworthy, penitent, and of superior status and when the identity violation is not serious (Blumstein, 1974). Thus we are more likely to accept a psychiatrist's quiet explanation that he struck an elderly mental patient because she kept shouting and would not talk with him than to accept a delinquent's defiant use of the same excuse to explain why he struck an elderly woman.

Altercasting

The tactics discussed so far illustrate how people claim and protect identities. The actions of one person in an encounter will place limits on who the others can claim to be. Therefore, to gain an advantage in the interaction, we might try to impose identities on others that complement those we claim. We might also pressure others to enact roles that mesh with the roles we want for ourselves. **Altercasting** is the use of tactics to impose roles and identities on others. Through altercasting, we place others in situated identities and roles that are to our advantage (Weinstein & Deutsch-berger, 1963).

In general, altercasting involves treating others as if they already have the identities and roles that we wish to impose. Teachers engage in altercasting when they tell a student, "I know you can do better than that." This remark pressures the student to live up to an imposed identity of competence. Altercasting can entail carefully planned duplicity. An employer may invite subordinates to dinner, for example, casting them as close friends in hopes of learning employee secrets.

People frequently use altercasting to put some-one on the defensive. "Explain to the voters why you can't control the runaway national debt," says the challenger, altercasting the incumbent official as in-competent in dealing with the economy. Should the incumbent rise to her own defense, she admits that the charge merits discussion and the negative identity may be correct. Should she remain silent, she implies acceptance of the altercast identity. Putting one's ri-vals on the defensive is an effective technique because a negative identity is difficult to escape.

When bargaining over identities, persons often use altercasting to achieve an edge. A study of iden-tity bargaining between members of dating couples il-lustrates the use of altercasting (Blumstein, 1975). The researchers instructed women to claim an iden-tity of "healthily assertive" by altercasting their dates into a more submissive identity. The women did this by making critical remarks to their dates such as, "Must you insist on making all the decisions?" Some men conceded the assertive identity claimed by their dates by presenting a self consistent with the altercast ("Sorry I've been so pushy. Whatever you say goes."). Other men rejected their date's assertive identity and pressured her to return to a submissive identity ("You always liked me to make the decisions before. What's up?"). The research showed that one of the factors de-termining whether people resist altercasting is domi-nance, a personality variable. Men who had earlier rated dominance as an important part of their self-concept resisted their date's altercasting, whereas men who had rated dominance as unimportant were more prone to accept the submissive identity imposed by the women. In general, the men responded to alter-casting by conceding identities that were unimpor-tant to their overall self-concept but by retaining iden-tities that were more central.

Detecting Deceptive Impression Management

To this point, we have discussed various techniques used by impression managers to project identities and control relationships. Now we shift our focus and look at the person toward whom impression manage-ment tactics are directed. Impression managers try to create a deceitful, false image. This image may or may not be challenged by the person targeted. In some cases, the target accepts the false image because he or she has little to gain by questioning the sincerity of the impression manager. For instance, funeral direc-tors strive to convey an air of sympathy and concern although they did not know the deceased person. Mourning relatives realize the sympathy is superficial, but they ask very few questions because they would only be more upset to discover the mortician's true feelings of boredom and indifference.

In other cases, however, accurate detection of deception is crucial for protecting our own interests. In attempting to win a contract, for example, builders (such as those who put new roofs on old buildings) may claim to be reliable businesspeople and skilled ar-tisans even when they are total frauds. For the home owner about to make a down payment, it is worth thousands of dollars to determine whether the build-er's hearty handshake belongs to a responsible con-tractor or a fly-by-night operator.

How do people go about trying to unmask the impression manager? In general, they attend to two major types of information: the ulterior motives the other person has for an action and the nonverbal cues that accompany the action. In this section, we discuss both of these.

Ulterior Motives

Recognition that another person has a strong ulterior motive for his or her behavior usually colors an inter-action. For example, when a used car salesman tells us that a battered vehicle with sagging springs was driven only on Sundays by his retired aunt, his ulterior motive is transparent, and we are certain to suspect deceit. In such a case, we will probably discount what the salesman says about any car on his lot or even refuse to do business with him.

When ulterior motives become apparent to a target, they undermine the success of tactical impression management. For instance, in one study (Dickoff, 1961), an experimenter expressed praise for the performances of her subjects under two different conditions. In one condition, when the experimenter had no apparent ulterior motive, this flattery worked. Subjects liked her better the more she praised them. In the second condition, however, when the experimenter had an obvious ulterior motive (she wanted them to volunteer for another experiment), subjects discounted her remarks, and the flattery did not increase their liking for her.

Ironically, the very conditions that increase the temptation to use ingratiation tactics also make the target more vigilant. As noted above, ingratiators are especially prone to use such tactics as flattery or opinion conformity when the target person controls important rewards and can use discretion in distributing them. Unhappily for ingratiators, these same conditions alert the target to be vigilant and to expect deception. This state of affairs, termed the *ingratiator's dilemma,* means that ingratiators must be doubly careful to conceal ulterior motives and avoid detection under conditions of high dependency. As documented by Gordon (1996), ingratiation attempts that are transparent tend to be relatively ineffective, sometimes to the point of backfiring. Ingratiators usually understand this, and indeed, some evidence indicates that ingratiators avoid using tactics such as opinion conformity under conditions of blatant power inequality; they are more likely to use them under conditions less likely to alert the target (Kauffman & Steiner, 1968).

Nonverbal Cues of Deception

It is widely believed that nonverbal cues, such as the avoidance of eye contact and nervous gestures, are telltale signs of deception. Trial lawyer Louis Nizer (1973), for example, asserts that certain identifiable cues reveal when witnesses are attempting to deceive jurors. Years earlier, Sigmund Freud wrote, "He who has eyes to see and ears to hear may convince himself that no mortal can keep a secret. If his lips are silent, he chatters with his fingertips; betrayal oozes out of him at every pore" (1905, p. 78). Although these views are a bit overblown and exaggerated, research indicates nonverbal cues do provide a basis for detecting deception at a rate somewhat better than chance (DePaulo, Zuckerman, & Rosenthal, 1980; Kraut, 1980).

Cues Indicating Deception When people interact face-to-face, they send messages through both verbal and nonverbal channels. People transmit meanings not only by words (verbal expressions), but also by facial expressions, bodily gestures, and voice quality. The multichannel nature of communication can pose problems for impression managers because the meanings transmitted through some of these channels are more controllable than those transmitted through others. For instance, if an impression manager is trying to deceive a target, he or she may tell a lie verbally, but then inadvertently reveal true intentions or emotions through nonverbal channels. The term *nonverbal leakage* denotes the inadvertent communication of true intentions or emotions through nonverbal channels (Ekman & Friesen, 1969, 1974).

An impression manager will generally have a high level of control over his or her verbal expression (choice of words) and a fair amount of control over facial expressions (smiles, frowns, and so on). The deceiver may have less control, however, over body movements (arms, hands, legs, feet) and over voice quality and vocal inflections (the pitch and waver of her voice). The nonverbal channels that are least controllable—voice quality and body movements—are the ones that leak the most information (Blanck & Rosenthal, 1982; DePaulo & Rosenthal, 1979).

Several studies have demonstrated that the fundamental pitch of the voice is higher when someone is lying than when telling the truth (DePaulo, Stone, & Lassiter, 1985; Ekman, Friesen, & Scherer, 1976). The difference is fairly small—individuals cannot discriminate just by listening—but electronic analysis of vocal signals can reliably detect when an impression manager is lying. Other vocal cues associated with deception include speech hesitation (liars hesitate more), speech errors (liars stutter and stammer more), and response length (liars give shorter answers) (DePaulo et al., 1985; Zuckerman, DePaulo, & Rosenthal, 1981).

Certain facial and body cues are also associated with deception. Tip-offs regarding deception include eye pupil dilation (liars show more dilation) and blinking (liars blink more); another tip-off is self-directed gestures (liars touch themselves more) (DePaulo et al., 1985). The musculature of a smile is slightly different when people are lying than when they are telling the truth. Lying smiles contain a trace of muscular activity usually associated with expression of disgust, fear, or sadness (Ekman, Friesen, & O'Sullivan, 1988).

Accuracy of Detection   How good are observers at detecting acts of deception? Although some people believe they can always detect deception when it is used by an impression manager, recent research suggests the contrary. Results of most experiments reveal that observers are not especially adept at correctly identifying when others are lying. Rates of detection are generally somewhat better than chance but not especially good in absolute terms (Ekman & O'Sullivan, 1991; Zuckerman, DePaulo, & Rosenthal, 1981). This occurs in part because observers often use the wrong cues or do not rely on the most useful kinds of information in judging whether someone is lying.

Difficulty in detection is illustrated by a study in which airline travelers at an airport in New York were asked to participate in a mock inspection procedure (Kraut & Poe, 1980). Some of these travelers were given "contraband" to smuggle past inspection while others carried only their own legitimate luggage. All subjects were instructed to present themselves as honest persons. (As motivation, the researchers offered travelers prizes up to $100 for appearing honest.) Later on, professional customs inspectors and lay judges watched videotaped playbacks of each of the travelers and tried to decide which ought to be searched. Results showed that both the customs inspectors and the inexperienced judges failed to identify a substantial proportion of the travelers who were smuggling the contraband. The rate of detection, even

by the customs inspectors, was no better than chance. Interestingly, however, the professional inspectors and inexperienced judges agreed on which travelers should be searched. This resulted because the inspectors and judges used the same behavioral cues as indicative of deception. Travelers were more likely to be selected for search if they were young and lower class, appeared nervous, hesitated before answering questions, gave short answers, avoided eye contact, and shifted their posture frequently. Unfortunately for the inspectors, these cues were imperfect indicators of deception.

Why aren't observers better at detecting deception? First, nonverbal behaviors that reveal deception, such as high vocal pitch and short response length, are imperfect indicators. They do arise from deception, but they can also result from conditions unrelated to deception, such as excitement or anxiety. In such circumstances, the innocent will appear guilty, and observers will make mistakes in detection. Second, certain cues commonly believed to reveal deception do not actually do so (DePaulo et al., 1985). These cues include speech rate (people think liars talk slower), smiling (people think liars smile less), gaze (people think liars engage in less eye contact), and postural shifts (people think liars shift more). If observers rely heavily on these cues, they will make mistakes in detection. Third, certain skilled impression managers are able to give near flawless performances when deceiving. One study (Riggio & Friedman, 1983) finds evidence that certain people can give off what seem to be honest emotional cues (facial animation, some exhibitionism, few nervous behaviors) even when they are deceiving. If an impression manager has this capability, he or she will appear innocent, again causing mistakes in detection by observers. Fourth, face-to-face interaction is a two-way street: Impression managers not only exhibit behavior, but also observe the reactions of their audiences. The feedback from audiences in face-to-face situations is fairly rich, and it often provides impression managers with a clear indication whether their attempts at deception are succeeding. If they are not succeeding, they may be able to adjust or fine-tune their deceptive communications to be more convincing.

The picture is not entirely bleak, however. Observers' success in detecting deception can be increased by special discrimination training (Zuckerman, Koestner, & Alton, 1984). Moreover, success in detecting deception can be affected by instructions given to observers. For instance, one study (DePaulo, Lassiter, & Stone, 1982) varied the instructions given to observers in face-to-face interaction. When given instructions to pay particular attention to auditory cues and to downplay visual cues, observers were more successful in discriminating truth from deception than when they were given instructions to pay attention to both visual and auditory cues. By emphasizing auditory and downplaying visual cues, observers more fully attended those cues that are least under an impression manager's control, such as voice quality. In general, lack of attention to verbal content and paralinguistic cues seriously impairs the ability to detect deception (Geller, 1977; Littlepage & Pineault, 1978).

Ineffective Self-Presentation and Spoiled Identities

Social interaction is a perilous undertaking, for it is easily disrupted by challenges to identity. Some of us may recover when a challenge occurs, but others will be permanently saddled with spoiled identities. In this section, we discuss what happens when tactical impression management fails. First, we consider embarrassment, a spontaneous reaction to sudden or transitory challenges to our identities. Second, we examine cooling-out and identity degradation, which are deliberate actions aimed at destroying or debasing the identities of persons who fail repeatedly. Third, we analyze the fate of those afflicted with stigmas—physical, moral, or social handicaps that may spoil their identities permanently.

Embarrassment and Saving Face

Embarrassment is the feeling we experience when the public identity we claim in an encounter is discredited (Edelmann, 1987; Semin & Manstead, 1982, 1983). Many people describe it as an uncomfortable feeling of exposure, mortification, awkwardness, and chagrin (Miller, 1992; Parrott & Smith, 1991). It may entail such physiological symptoms as blushing, in-

creased heart rate, and increased temperature (Edelmann & Iwawaki, 1987).

We experience embarrassment not only when our own identity is discredited but also when the identities of people with whom we are interacting are discredited (Miller, 1987). In this sense, embarrassment is contagious. It may be more acute when our own (adequate) performance serves as a frame of reference that highlights the inadequacy of others' performances (Bennett & Dewberry, 1989). We feel embarrassment at others' spoiled identities because we have been duped about the assumptions on which we built our interaction with them, including our unwarranted acceptance of their identity claims (Edelmann, 1985; Goffman, 1967). For example, someone who claims to be an outstanding ballplayer will experience embarrassment when he drops the first three routine fly balls to center field, but the manager who let him play in a crucial game also will feel embarrassment and chagrin for accepting the ballplayer's claim of competence.

Sources of Embarrassment   In several studies, investigators have analyzed hundreds of cases of embarrassment to ascertain the conditions that produce this feeling (Gross & Stone, 1970; Miller, 1992; Sharkey & Stafford, 1990). Results show that any of several conditions can precipitate embarrassment. First, people feel embarrassed if it becomes publicly apparent that they lack the skills to perform in a manner consistent with the identity they claim. This is the plight, for example, of the math professor who suddenly discovers that he cannot solve the demonstration problem he has written on the chalkboard. Closely related to lack of skill is cognitive shortcoming, such as forgetfulness. A host's inability to remember others' names during introductions at a dinner party can cause embarrassment for all concerned.

Another condition that precipitates embarrassment is violation of privacy norms. If one person barges unaware into a place he or she does not belong (such as a bathroom occupied by another), then both persons are likely to experience embarrassment at the violation of privacy. The sudden and unexpected conversion of a back region into a front region is embarrassing for those whose identities are tarnished or discredited.

We can read the embarrassment on the face of this chief of police as he announces that one of his own trusted officers has been arrested on a drug charge. People experience embarrassment when an important social identity they claim for themselves or accept in others is discredited.

A further condition that often precipitates embarrassment is awkwardness or lack of poise. A person can lose poise if he or she trips, stumbles, spills coffee, or miscoordinates physically with others. Loss of control of equipment (a dentist dropping her drill), of clothing (a speaker splitting his pants), or of one's own body (trembling, burping, and worse) also will destroy poise. In general, poise vanishes and embarrassment increases whenever we lose control over those aspects of our self-presentation that we ordinarily manage routinely.

Responses to Embarrassment   A continuing state of embarrassment is uncomfortable for everyone involved. For this reason, it is usually in everyone's interest to *restore face*—that is, eliminate the conditions causing embarrassment. The major responsibility for restoring face lies with the person whose actions produced the embarrassment, but interaction part-

ners frequently try to help the embarrassed person restore face (Levin & Arluke, 1982). For instance, if a party guest slips and falls while demonstrating his dancing prowess, his partner might help him save face by remarking that the floor tiles seem newly waxed and very slippery. Mutual commitment to supporting each other's social identities is a basic rule of social interaction (Goffman, 1967).

To restore face, the embarrassed person will often apologize, provide an account, or otherwise realign his or her actions with the normative order (Knapp, Stafford, & Daly, 1986; Metts & Cupach, 1989). When providing accounts, people will either make excuses that minimize their responsibility or offer justifications that define their behavior as acceptable under the circumstances. If the interaction partners accept these accounts—and partners have been known to accept the lamest excuses rather than endure continuing embarrassment—a proper identity is restored. If accounts are unavailable or insufficient, the embarrassed individual may offer an *apology* for the discrediting behavior and admit that his or her behavior was wrong. In this way the person reaffirms threatened norms and reassures others that he or she will not violate those norms again.

When our behavior discredits a particular, narrow identity, we can sometimes restore face through an exaggerated reassertion of that identity. A man whose masculine identity is threatened by behavior suggesting he is infantile, for example, might try to reassert his courage and strength. In a test of this hypothesis (Holmes, 1971), some male subjects were asked to suck on a rubber nipple, a pacifier, and a breast shield—all embarrassing experiences. Other subjects were asked to touch surfaces such as sandpaper and cloth. The men were next asked how intense an electric shock they would be willing to endure later in the experiment. Men who faced the embarrassing experiences indicated willingness to endure more intense shocks than men who faced no threat to their masculine identity. By taking the intense shocks, the embarrassed men could present themselves as tough and courageous, thereby reasserting their threatened masculinity.

Sometimes people embarrass others intentionally and make no effort to help them to save face. In such circumstances, the embarrassed persons are likely to react aggressively. They may vigorously attack the judgment or character of those who embarrassed them. Some research indicates that an aggressive response to embarrassment is more likely between status unequals than between status equals (Sueda & Wiseman, 1992). Alternatively, they may assert that the task on which they failed is worthless or absurd (Modigliani, 1971). Finally, they may retaliate against those who embarrassed them intentionally. Retaliation not only reasserts an image of strength and achieves revenge, but it also forestalls future embarrassment by showing resolve to punish those who discredit us.

Cooling-Out and Identity Degradation

When people repeatedly or glaringly fail to meet performance standards or to present appropriate identities, others cease to help them save face. Instead, they may act deliberately to modify the offenders' identities or to remove them from their positions in interaction. Failing students are dropped from school, unreliable employees are let go, tiresome suitors are rebuffed, schizophrenics are institutionalized. Persons will modify an offender's identity either by cooling-out (Goffman, 1952) or by degradation (Garfinkel, 1956), depending on the social conditions surrounding the failure.

The term **cooling-out** refers to gently persuading a person whose performance is unsuitable to accept a less desirable, although still reasonable, alternative identity. A counselor at a community college may cool-out a weak student by advising him to switch from premed to an easier major, for example, or by recommending that he seek employment after completing community college rather than transfer to a university. Persons engaged in cooling-out seek to persuade offenders, not to force them. Cooling-out actions usually protect the privacy of offenders, console them, and try to reduce their distress. Thus the counselor meets privately with the student, emphasizes the attractiveness of the alternative, listens sympathetically to the student's concerns, and leaves the final choice up to him.

The process of destroying the offender's identity and transforming him or her into a "lower" social type

is termed **identity degradation.** Degradation establishes the offender as a nonperson, an individual who cannot be trusted to perform as a normal member of the social group because of reprehensible motives. This is the fate of a political dissident who is fired from her job, declared a threat to society, and relegated to isolation in a prison or work camp.

Identity degradation imposes a severe loss on the offender, so it usually is done forcibly. Identity degradation often involves a dramatic ceremony, such as a criminal trial or sanity hearing or impeachment proceeding, in which a denouncer acts in the name of the larger society or the law (Scheff, 1966). In such ceremonies, persons who had previously been treated as free, competent citizens are brought before a group or individual legally empowered to determine their "true" identity. They are then denounced for serious offenses against the moral order. If the degradation succeeds, offenders are forced to give up their former identities and to take on new ones like "criminal" or "insane" or "dishonorably discharged."

Two social conditions strongly influence the choice between cooling-out and degradation: the offender's prior relationships with others and the availability of alternative identities (Ball, 1976). Cooling-out is more likely when the offender has had prior relations of empathy and solidarity with others and when alternative identity options are available. For example, during a breakup, lovers who have been close in the past can cool-out their partners by offering to remain friends. Identity degradation is more likely when prior relationships entailed little intimacy and when respectable alternative identities are not readily available. Thus strangers found guilty of molesting children are degraded and transformed into immoral, subhuman creatures.

Stigma

A **stigma** is a characteristic widely viewed as an insurmountable handicap that prevents competent or morally trustworthy behavior (Goffman, 1963; Jones et al., 1984). There are several different types of stigma. First are physical challenges and deformities—missing or paralyzed limbs, ugly scars, blindness, deafness. Second are character defects—dishonesty, unnatural passions, psychological derangements, treacherous beliefs. These may be inferred from a known record of crime, imprisonment, sexual abuse of children, mental illness, or radical political activity, for example. Third are characteristics such as race, sex, and religion that, in particular segments of society, are believed to contaminate or morally debilitate all members of a group.

Once recognized during interaction, stigmas spoil the identities of the persons having them. No matter what their other attributes, stigmatized individuals are likely to find that others will not view them as fully competent or moral. As a result, social interaction between "normal" and stigmatized persons is shaky and uncomfortable.

Sources of Discomfort

Discomfort arises during interaction between "normals" and stigmatized individuals because both are uncertain what behavior is appropriate. "Normals" may fear, for example, that if they show direct sympathy or interest in the stigmatized person's condition, they will be intrusive ("Is it difficult to write with your artificial hand?"). Yet, if they ignore the defect, they may make impossible demands ("Would you help me move the refrigerator?"). To avoid being hurt, stigmatized individuals may vacillate between shamefaced withdrawal (avoiding social contact) and aggressive bravado ("I can do anything anyone else can!").

Another source of discomfort for "normals" is fear that associating with a stigmatized person may discredit them ("If I befriend a convicted criminal, people may wonder about my trustworthiness."). In recent times, this problem has arisen frequently with respect to AIDS, a heavily stigmatized condition due in part to its association with drug use and homosexuality (and the lingering fear of transmission). Persons with AIDS experience the stigma, of course; but beyond that, the compassionate confidants who provide care and social support for persons with AIDS often encounter social difficulties as well. The stigma associated with AIDS results in some of them being rejected by friends and family (Jankowski, Videka-Sherman, & Laquidara-Dickinson, 1996).

Effects on Behavior and Perceptions

"Normals" react toward stigmatized persons with an attitude of *ambivalence* (Katz, 1981; Katz, Wackenhut, & Glass, 1986). Toward a quadriplegic, for instance, "normals" have feelings of aversion and revulsion but also feelings of sympathy and compassion. This ambivalence creates a tendency toward behavioral instability, in which extremely positive or extremely negative responses may occur toward the stigmatized person, depending on the specific situation.

When interacting with stigmatized individuals, "normals" alter their usual behavior. They gesture less than usual, refrain from expressing opinions that reflect their actual beliefs, maintain less eye contact, and terminate the encounters sooner (Edelmann et al., 1983). In addition, "normals" speak faster in interactions with stigmatized persons than with other "normals," ask fewer questions, agree less, make more directive remarks, and allow the stigmatized fewer opportunities to speak (Bord, 1976). By limiting the responses of the stigmatized, "normals" reduce uncertainty and diminish their own discomfort.

For their part, stigmatized persons also have difficulty interacting with "normals." Remarkably, the mere belief that we have a stigma—even when we do not—leads us to perceive others as relating to us negatively. In a dramatic demonstration of this principle (Kleck & Strenta, 1980), some female subjects were led to believe that a woman with whom they would interact had learned they had a mild allergy (a nonstigmatizing attribute). Other female subjects believed that the woman would view them as disfigured due to an authentic-looking scar that had been applied to their faces with stage makeup (a stigmatizing attribute). In fact, the interaction partner had no knowledge of either attribute. In the allergy condition, the partner had received no medical information whatever. In the scar condition, there was actually no scar to be seen because the experimenter had surreptitiously removed the scar just before the discussion.

After a 6-minute discussion with the interaction partner, the subjects described their partners' behavior and attitudes. Those subjects who believed they had a facial scar remarked more frequently that their partners had stared at them. They also perceived their partners as more tense, more patronizing, and less attracted to them than the nonstigmatized subjects did. Judges who viewed videotapes of the interaction perceived none of these differences. This is not surprising because the partner knew nothing about either disability. However, these results show that people who believe they are stigmatized perceive others as relating negatively to them. This occurs even if the others are not, in fact, doing anything negative or irregular. These findings are illustrated in Figure 9.1.

When people believe they are stigmatized, they tend not only to perceive the social world differently but also to behave differently. In one study, for instance, one group of mental patients believed the person with whom they were interacting knew their psychiatric history, whereas another group thought their stigma was safely hidden (Farina et al., 1971). Patients in the first group performed more poorly on a cooperative test and found the task more difficult. Moreover, outside observers of the interaction perceived these patients to be more anxious, more tense, and less well adjusted.

Coping Strategies Stigmatized persons adopt various strategies to avoid awkwardness in their interactions with "normals" and to establish the most favorable identities possible (Gramling & Forsyth, 1987). Persons who are handicapped or physically challenged often must choose between engaging in interaction (thereby disclosing their stigma) or withdrawing from interaction (concealing their stigma) (Lennon et al., 1989). A stutterer, for instance, may refrain from introducing himself to strangers; were he to introduce himself, he could do so only at the risk of stumbling over his own name and drawing attention to his stigma (Petrunik & Shearing, 1983). People whose speech reveals their stigmatized foreign origin or lack of education face a similar dilemma when meeting strangers.

In interaction, stigmatized persons often try to induce "normals" to behave tactfully toward them and to build relationships around the aspects of their selves that are not discredited. Their strategies depend on whether their stigma can be defined as temporary,

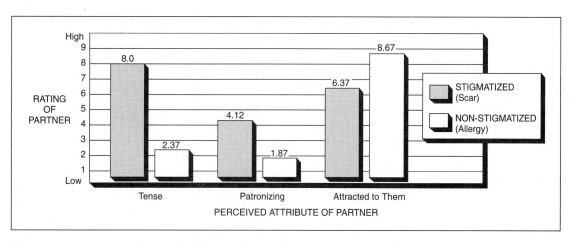

Figure 9.1 **Perceptions of Interaction Partners by Stigmatized and Nonstigmatized Individuals**

In this study, some female students were led to believe that a large facial scar stigmatized them in the eyes of their female interaction partner. Others were led to believe their partner knew they had a mild allergy—a nonstigmatized characteristic. In fact, interaction partners were unaware of either of these characteristics. Nonetheless, students who believed they were stigmatized perceived their partners as substantially more tense and patronizing and as less attracted to them. This suggests that the mere belief that we are stigmatized leads us to perceive others as behaving negatively toward us.

SOURCE: Adapted from Kleck and Strenta, 1980.

such as a broken leg on the mend or a passing bout of depression, or whether it must be accepted as permanent, such as blindness or stigmatized racial identity (Levitin, 1975). Persons who are temporarily stigmatized focus attention on their handicap, recounting how it befell them, detailing their favorable prognosis, and encouraging others to talk about their own past injuries. In contrast, people who are permanently stigmatized often try to focus attention on attributes unrelated to their stigma (Davis, 1961). They often use props to highlight aspects of the self that are unblemished, such as proclaiming their intellectual interests (carrying a heavy book), their political involvements (campaign buttons), or their hobbies (a knitting bag).

In cases where a stigma does not force excessive dependency, the permanently stigmatized often try to strike a deal with "normals": They will behave in a nondemanding and nondisruptive manner in exchange for being treated as trustworthy human beings despite their handicaps. Under this arrangement, they are expected to cultivate a cheerful manner, avoid bitterness and self-pity, and treat their stigma as a minor problem with which they are coping successfully (Hastorf, Wildfogel, & Cassman, 1979).

Everyone gains some benefit from handling stigmas in these ways. Stigmatized persons avoid the constant embarrassment of indelicate questions, inconsiderateness, and awkward offers of help. They gain some acceptance and enjoy relatively satisfying interaction in most encounters. "Normals" gain because this resolution assuages the ambivalence they feel toward the stigmatized and spares them the true pain the stigmatized suffer.

Summary

Self-presentation refers to our attempts, both conscious and unconscious, to control the images we project of ourselves in social interaction. Some self-presentation is authentic, but some may be tactical.

Self-Presentation in Everyday Life   Successful presentation of self requires efforts to control

how others define the interaction situation and accord identities to participants. (1) In defining the situation, people negotiate the type of social occasion considered to be at hand and the identities they will grant each other. (2) Self-disclosure is a process through which we not only make identity claims but also promote friendship and liking. Self-disclosure is usually two sided and gradual, and it follows a norm of reciprocity.

Tactical Impression Management   People employ various tactics to manipulate the impressions others form of them. (1) They manage appearances (clothes, habits, possessions, and so on) to dramatize the kind of person they claim to be. (2) They ingratiate themselves with others through such tactics as opinion conformity, other enhancement, and selective presentation of their admirable qualities. (3) When caught performing socially unacceptable actions, people try to repair their identities through aligning actions, which are attempts to align their questionable conduct with cultural norms. They explain their motives, disclaim the implications of their conduct, or offer accounts that excuse or justify their actions. (4) They altercast others, imposing roles and identities that mesh with the identities they claim for themselves.

Detecting Deceptive Impression Management   Observers attend to two major types of information in detecting deceitful impression management. (1) They assess others' possible ulterior motives. If a large difference in power is present, an impression manager's ulterior motives may become transparent to the target, making tactics like ingratiation difficult. (2) They scrutinize others' nonverbal behavior. Although detection of deceit is difficult, observers are more accurate when they concentrate on leaky cues, such as tone of voice, and discrepancies between messages transmitted through different channels.

Ineffective Self-Presentation and Spoiled Identities   Self-presentational failures lead to several consequences. (1) People experience embarrassment when their identity is discredited. Interaction partners usually help the embarrassed person to restore an acceptable identity. Otherwise, embarrassed persons tend to reassert their identity in an exaggerated manner or to attack those who discredited them. (2) Repeated or glaring failures in self-presentation lead others to modify the offender's identity through deliberate actions. Others may try to cool-out offenders by persuading them to accept less desirable alternative identities, or they may degrade offenders' identities and transform them into lesser social types. (3) Many physical, moral, and social handicaps stigmatize individuals and permanently spoil their identities. Interaction between stigmatized and "normal" persons is marked by ambivalence and is frequently awkward and uncomfortable. In general, "normals" pressure stigmatized individuals to accept inferior identities, and stigmatized individuals seek to build relationships around the aspects of their selves that are not discredited.

Key Terms

accounts (p. 225)
aligning actions (p. 224)
altercasting (p. 225)
back region (p. 220)
cooling-out (p. 231)
definition of the situation (p. 216)
disclaimers (p. 224)
embarrassment (p. 229)
frame (p. 217)
front region (p. 220)
identity degradation (p. 231)
ingratiation (p. 221)
self-disclosure (p. 218)
self-presentation (p. 215)
situated identity (p. 217)
stigma (p. 231)
tactical impression management (p. 216)

CHAPTER 10
Helping and Altruism

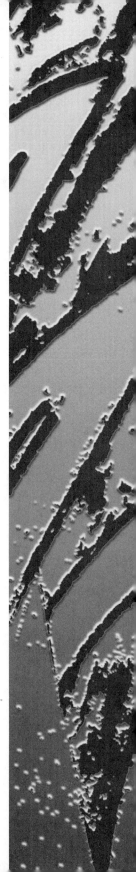

Introduction

Jennifer Beyer, age 22, was driving along Old River Road in Appleton, Wisconsin, on a cold day in February. She was on the way to visit a friend, but when a soaking wet child flagged her down, she pulled over immediately. Shivering and frightened, Jeff Lasrewski hurriedly explained that he and his friend, 9-year old Colin Deeg, had been playing on the frozen Fox River when the ice gave way. Jeff had managed to climb back onto the ice and make it to shore, but Colin was still in the water.

Starting down the riverbank, Jennifer saw Colin splashing in the frigid water. At the point where many others would have stopped because of the great personal risk, she went onto the frozen river to rescue him. Inching her way onto the ice, she tried to use "her scarf to pull Colin out, but the ice cracked and she plunged into the water. At this point, Colin was still conscious but fading fast. In the meantime, Jeff reached another adult, Cyndy Graf, who quickly dialed 911 for help and then ran to the river.

Jennifer grabbed Colin to keep him from going under and tried to get him out of the water. This proved impossible, however, for Colin soon passed out and the weight of his wet clothes made him too heavy to push onto land. Jennifer's limbs were numb with cold by the time police and fire teams reached the river with rescue equipment, but she had kept Colin's head above water and prevented him from drowning. Officers performed CPR on Colin, then rushed the pair to nearby St. Elizabeth's Hospital, where doctors used a bypass machine to warm Colin's blood, which had dropped to 78 degrees. Jennifer was treated for hypothermia. A week later, Colin was doing fine ("Rescuers Respond," 1992; "Wisconsin Town Thanks Hero," 1992).

The story of Jennifer Beyer is extraordinary for its valor and heroism, but everyday life is filled with smaller tales of people helping others in need. Individuals help others in many ways. They may give someone a ride, help change a flat tire, donate blood, make contributions to charity, return lost items to their owners, assist victims of accidents, and so on.

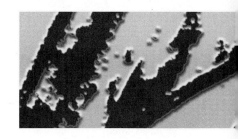

Of course, the mere fact that someone needs help does not mean others will rush to give aid. Humans are capable of vastly different responses to persons in need. Although Jennifer Beyer went onto the ice to rescue Colin Deeg, many others would not have taken that risk. Some will not even stop to help a stranded motorist or make contributions to charitable causes. Thus the challenge for social psychologists is to explain variations in helping behavior. When will people help others, when will they refuse to help, and why? Drawing on research and theory, this chapter addresses the following questions:

1. What motivates one person to help another? How do such factors as cost/reward and empathic concern for others affect helping and altruistic behavior?
2. How do characteristics of the person needing help influence help-giving by others?
3. What impact do cultural factors such as norms and roles have on helping behavior?
4. How do characteristics of the situation influence helping? And how do emotional states of the helper (guilt, mood) affect helping behavior?
5. In emergency situations, what factors determine whether bystanders intervene and offer help?
6. When help is given, what factors determine the recipient's reactions?

To address these questions, we must first define some important terms. Social psychologists use the term **prosocial behavior** to refer to a broad category of actions that are considered by society as beneficial to others and having positive social consequences. A wide variety of specific behaviors can qualify as prosocial, including donation to charity, intervention in emergencies, cooperation, sharing, volunteering, sacrifice, and the like. Prosocial behavior often contrasts with *antisocial behavior,* which refers to activity that is aggressive, violent, destructive, or criminal.

One important type of prosocial behavior is helping. By **helping,** we mean any action that has the consequences of providing some benefit to or improving the well-being of another person. This definition, which is intentionally broad, has certain properties. First, it does not place the intent of the helper at issue. There is no requirement, for instance, that the helper intend to benefit another person with his or her action. Second, the definition does not explicitly address the issue of whether or not the helper can also benefit from giving help. There is no requirement, for instance, that to qualify as helping behavior, an action must benefit another person more than oneself. Under this definition, helping behavior may stem from altruistic concerns, but it may also involve selfish or egoistic motives.

Motivation to Help Others

What motivates one person to help another? There are at least two major views on this issue, rooted in different conceptions of human nature. The first view depicts humans as egoistic or selfish beings concerned primarily with their own gratification. Although this view acknowledges that helping behavior occurs with considerable frequency, it treats helping as always originating from some ulterior, self-serving considerations. For instance, a woman might help another with a shopping chore to get admiration and approval from the other, to avoid feelings of guilt or shame, to obligate the other to her, or to bolster her own self-esteem. Social psychologists often refer to helping behavior motivated by a person's sense of self-gratification as **egoism.**

The second view depicts humans as rather more generous and unselfish beings, capable of real concern for the welfare of others. In this view, at least some helping behaviors originate primarily because humans are able and willing to extend themselves unselfishly to benefit other persons in need. For instance, a bystander may rush to rescue an accident victim in order to relieve the victim's pain and anguish; help like this can be based on altruistic concerns, without any selfish motives. As used here, the term **altruism** refers to an action that is performed voluntarily with the intention of helping someone else and entails no expectation of receiving a reward or benefit in return (except possibly an internal feeling of having done a good deed for someone) (Piliavin & Charng, 1990; Simmons, 1991). Under this definition, whether or not an act is altruistic depends on the intentions of the helper—the helper must intend to benefit the other. Some theorists would prefer to restrict the term *altruism* even further and impose the additional condition that to be considered altruistic, an act must involve at least some cost to the person performing it (Krebs, 1982). Not everyone accepts this self-sacrifice criterion, however, and they point out that a person can be an altruist without being a martyr. But at the very least, this self-sacrifice criterion underscores the widely accepted notion that when an altruist provides help, he or she does not do so to gain personal benefits.

In this section, we look at these two views in more detail. First, we consider the view that humans are primarily egoistic, and we look at the role of reinforcement (rewards and costs) in helping. Then we consider the view that humans can behave altruistically, and we look at the role of empathic concern for others in helping.

Egoism and Cost/Reward Motivation

Although some helping may be altruistic, much of it is egoistic in nature. Some reinforcement theorists maintain that human nature is basically selfish and individuals are motivated to maximize their own net gains (Gelfand & Hartmann, 1982). There seems to be little doubt that considerations of reward and cost often enter decisions to give or withhold help. Some theorists have suggested that individuals do not generally give help unless they perceive that the rewards to themselves for helping outweigh the costs (Lynch & Cohen, 1978; Piliavin et al., 1981).

Rewards for Helping   The rewards that motivate potential helpers are many and varied. They may

A passerby gives a donation to a homeless man in New York City. Altruistic behavior of this type depends on personal attitudes, but it can also result from situational stimuli that make salient the social responsibility norm.

include thanks from the victim, admiration and approval from others, financial rewards and prizes, and recognition for competence. In general, greater anticipated rewards produce more helping (Kerber, 1984). The form of help a potential helper will provide may depend on the specific rewards he or she seeks, and these in turn may depend on his or her own needs. This is illustrated by a study in which undergraduates were invited to volunteer their assistance in projects such as studying unusual states of consciousness (ESP and hypnosis) or counseling troubled high school students (Gergen, Gergen, & Meter, 1972). Most participants chose to help by means of activities that satisfied their personal preferences and needs. Those who enjoyed novelty volunteered more frequently to help with the project on unusual states of consciousness. Those who liked

close social relationships volunteered more frequently to help troubled high school students. Thus the rewards people sought through helping matched their own personal needs.

Costs for Helping and Not Helping   Every helping act imposes costs on the helper. These may include exposure to danger, loss of time, financial costs, expenditure of effort, or exposure to repulsive people and objects. In general, the greater these costs, the less likely persons are to help (Kerber, 1984; Shotland & Stebbins, 1983). Beyond costs for helping, however, there are also some costs to potential helpers for not helping. Among these are public disapproval by others, embarrassment and loss of face, and condemnation by the victim. Students may donate blood, for example, to avoid condemnation for failing to help their fraternity meet its quota. Employees may authorize payroll deductions for the United Way to avoid censure or criticism by their fellows. In some instances, costs are normatively imposed. In certain places, for example, bystanders who fail to help victims not only incur disapproval but may even face criminal prosecution. Some helping is simply compliance with others' demands to avoid socially imposed costs.

Altruism and Empathic Concern

People often react to the distress of others on an emotional level and selflessly offer help in response. The term **empathy** refers to vicariously experiencing an emotion that is congruent with, or possibly identical to, the emotion another person is experiencing (Barnett, 1987; Eisenberg & Miller, 1987). We can distinguish between empathy and *sympathy,* which is a feeling of concern or sorrow for another. Through empathy, another person's pleasure gives us pleasure, and another person's pain causes us pain. The tendency to respond empathically to the perceived emotional states of others is probably biologically based (Hoffman, 1977). People of all ages, including 1-year-olds, respond empathically when seeing others in distress. Empathy is revealed in facial expressions, self-reports of emotion, and measures of physiological arousal (Eisenberg & Fabes, 1990). Empathic responses occur very rapidly, and they are largely automatic.

Empathy is heightened by various situational factors. Believing a victim is similar to oneself—whether in race, sex, personality, or attitudes—promotes empathy. In one study, for example, students observed another student receive painful shocks when he lost at roulette. Students who had been told the victim had a personality similar to their own became more physically aroused than those who thought the victim had a personality different from their own. Those who were more aroused also donated more of their own money to help the victim (Krebs, 1975). Empathy can also be heightened by situational norms or role expectations that expressly induce role taking (Shott, 1979). When observers actively try to imagine how another person feels rather than simply observing how that person reacts, both empathy and helping behavior increase (Harvey et al., 1980; Toi & Batson, 1982).

Considerable evidence indicates that feelings of empathy for a person in need will lead to help. Numerous studies have shown that greater empathy leads to greater help-giving (Batson et al., 1981; Dovidio, Allen, & Schroeder, 1990; Eisenberg & Miller, 1987; Fultz et al, 1986).

Empathy-Altruism Model   The issue of motivational links between empathy and helping has been addressed by the **empathy-altruism model,** developed by Batson and his co-workers (Batson, 1987, 1991; Batson & Oleson, 1991). This model proposes that adults can experience two distinct states of emotional arousal while witnessing another's suffering: distress and empathy. *Distress* involves unpleasant emotions such as shock, alarm, worry, and upset at seeing another person suffer. *Empathy* entails such emotions as compassion, concern, warmth, and tenderness toward the other (Batson & Coke, 1981). These states of emotional arousal give rise to different motivations.

More specifically, suppose a bystander witnesses another person suffering. If the bystander experiences distress at seeing the other suffer, he or she may be motivated to reduce this unpleasant emotion. By helping the person in need, the bystander not only as-

suages the victim's suffering but also reduces his or her own distress. Notice that although this help benefits the victim, the bystander's fundamental goal in helping is to reduce his or her own distress (an egoistic motive, not an altruistic one). This contrasts with the situation in which a bystander experiences empathy when witnessing the suffering of another. Empathy entails other-oriented feelings such as compassion, tenderness, sympathy, and the like. Feelings of this type may cause the bystander to help the victim, but the point to notice is that this help is motivated fundamentally by a desire to reduce the other's distress (an altruistic motive, not an egoistic one). Thus the arousal states of distress and empathy give rise to very different motives, but either one of these can lead to helping.

Thus the empathy-altruism model states that empathic concern for a person in need evokes altruistic helping (that is, motivation directed toward the goal of benefiting that person, not toward some subtle form of egoistic self-benefit). This model does not preclude the possibility that persons can engage in altruistic acts in the absence of empathy. The model is controversial, however, because it flies in the face of reinforcement theory by asserting that helping can occur without reward to self (Cialdini et al., 1987; Wallach & Wallach, 1991).

The empathy-altruism model has received support from various experiments. Typically, subjects in these studies first witness a person in distress and must decide whether or not to offer help. The independent variables in these studies are level of empathy and ease of escape from the situation. For instance, in one experiment (Batson et al., 1983), subjects served as observers who watched another same-sex participant (Elaine for females, Charlie for males) over closed circuit television (actually a videotape). In the role of worker, Elaine (Charlie) performed a sequence of 10 digit-recall trials and received electric shocks at random intervals; when these shocks were applied, Elaine (Charlie) reacted with expressions of pain.

Some subjects were instructed that they needed to watch only the first two trials and were then free to leave if they wanted (easy-escape condition). Other subjects were instructed to stay and watch all 10 trials (difficult-escape condition). After the second trial, subjects completed a brief questionnaire that contained adjective scales measuring their personal feelings regarding the situation of Elaine (Charlie). Some of these adjectives reflected personal distress (*alarmed, grieved, upset, worried, disturbed*), and others reflected empathy (*sympathetic, moved, compassionate*). Each subject's average score on the distress items were subtracted later from his or her average score on the empathy items to create a measure of the subject's predominant emotional response to the situation (ranging from empathy to distress).

After completing the questionnaire, the subjects learned (via videotape) that Elaine (Charlie) had had a traumatic experience with shocks as a child, making them extremely difficult to bear. Naturally, this news was disturbing, and the experimenter gave the subjects an opportunity to help out Elaine (Charlie) by taking the remaining shock trials in her (his) place.

Results showed that subjects' helping response (taking Elaine's or Charlie's place) depended on the subject's predominant emotional response (empathy or distress) and on ease of escape (easy or difficult). When empathy was high, the frequency of helping behavior was high irrespective of whether escape was easy or difficult. However, when empathy was low (and hence distress high), the frequency of helping behavior dropped off substantially when escape was easy; subjects left the situation rather than take the shocks themselves. This pattern of behavior is consistent with the empathy-altruism model. Similar findings supporting the model appear in related studies (Batson, 1991; Batson et al., 1981, 1988; Fultz et al., 1986).

For purposes of contrast, a different perspective on altruism is discussed in Box 10.1, "Sociobiological Theory and Self-Sacrificing Behavior."

Characteristics of the Needy That Foster Helping

When in need, some people have a much better chance of receiving help than others. Our willingness to help needy persons depends on various factors. Important among these are whether we know them and like them, whether they are similar to or different

from us, and whether we consider them truly deserving of help.

Acquaintanceship and Liking

We are especially inclined to help people we know and we feel close to. Studies of reactions following natural disasters, for example, indicate that although people generally become very helpful toward others, they tend to give aid first to needy family members, then to friends and neighbors, and lastly to strangers (Dynes & Quarantelli, 1980; Form & Nosow, 1958). We help family members and friends more than strangers in everyday life (Amato, 1990). Even a brief acquaintanceship is sufficient to make us more likely to help someone (Pearce, 1980). Relationships like these increase helping because they involve relatively stronger normative obligations, more intense emotion and empathy, and greater costs if we fail to help.

We are more likely to help someone we like than to help someone we do not like. This effect occurs whether our positive feelings about the other are based on his or her physical appearance, personal characteristics, or friendly behavior (Kelley & Byrne, 1976; Mallozzi, McDermott, & Kayson, 1990). Moreover, we are more likely to help someone who likes us than to help someone who does not. For instance, in one study (Baron, 1971), subjects were led to believe another person either liked them or did not like them, depending on treatment. Results showed that subjects who believed the other liked them were more willing to comply with a request from that person to return some books to the library than were subjects who believed the other did not like them.

Similarity

In general, we are more likely to help others who are similar to ourselves than to help others who are dissimilar (Dovidio, 1984). That is, we are more likely to help those who resemble us in race, attitudes, political ideologies, and even in mode of dress. For instance, with respect to race, several studies have reported that in situations where refusing to help may be justified without threatening one's egalitarian self-image, whites are more likely to help other whites than

10.1 Sociobiological Theory and Self-Sacrificing Behavior

Evolutionary theories of natural selection rely on the Darwinian principle of survival of the fittest. Classic Darwinian theory proposes that any genetically determined physical attribute or trait that helps an individual survive will be passed on to the next generations. Because the individual having the attribute is more likely to survive than others lacking it, he or she will tend to have more offspring, and these offspring will tend to carry the attribute. The offspring themselves will have a greater tendency to survive, pass on the trait to their own offspring, and so on. Eventually, individuals with the attribute will become more numerous than those without.

The principle of survival of the fittest can be used to explain selfish or aggressive behavior. For instance, in many species (chickens, crows, parrots, moles, lions, chimpanzees, and numerous others), the strongest and most aggressive animals occupy the top positions in the group's social hierarchy (or "pecking order"). To fight for position in this hierarchy is adaptive in a Darwinian sense, for it gives the animal control over food, shelter, and other resources needed to survive as well as access to mating partners with which to reproduce. Another form of aggressive behavior, the killing of the young (infanticide), is common in some animal species. When a male langur monkey takes control of another male's harem, for instance, he usually proceeds to murder all the infants sired by his competitor (Hrdy, 1977, 1982). This act has clear Darwinian implications.

Investigators have observed that many animal species manifest not only aggressive behavior, but also altruistic behavior. In some cases, giving help entails great self-sacrifice. Ground squirrels, for instance, frequently sound alarm calls when a predator approaches. These calls warn other squirrels of the threat, but they also draw attention to the squirrel issuing them. That squirrel, therefore, is more likely to be attacked and killed by the predator (Sherman, 1980). When a termite hive is attacked by intruders, soldier termites place themselves in front of the other termites. Many soldier termites sacrifice themselves so the others may live and the nest survive (Wilson, 1971).

The occurrence of helping behavior among animals appears to defy explanation in terms of the principles of natural selection. In consequence, self-sacrificing altruism among animals poses a paradox for evolutionary theorists. If animals give up their lives so that others of their species can survive and prosper, those who are most helpful are the least likely to survive. This means they are less likely to have offspring and may not have any at all. Therefore, how does a biological predisposition to act altruistically toward other species members persist among animals over generations? Why does it not eventually vanish? The same question can be posed, or course, with respect to humans.

A theoretical perspective, **sociobiology,** seeks to resolve this paradox (Wilson, 1975, 1978). Consistent with modern evolutionary theory, sociobiology focuses on the perpetuation of genes rather than on perpetuation of physical attributes (Dawkins, 1976). In this view, the fittest animal is one that passes its genes to subsequent generations. This can happen in two distinct ways: (1) the animal itself produces offspring, or (2) the animal's close relatives, such as brothers, sisters, and cousins (who have similar genes), produce offspring.

Sociobiologists such as Wilson maintain that the same selection process Darwin described for physical attributes also operates (via genes) with respect to behavioral patterns or dispositions. Applied to humans,

continued on next page

to help blacks (Benson, Karabenick, & Lerner, 1973; Dovidio & Gaertner, 1981). However, in situations where refusing to help cannot be justified on non-racist grounds, whites help blacks as much as they help other whites (Frey & Gaertner, 1986).

A series of field studies demonstrated that similarity of opinions and political ideologies increases helping (Hornstein, 1978). In these studies, New York pedestrians came across "lost" wallets or letters that had been planted by researchers in conspicuous places.

continued from previous page

sociobiological principles have been invoked to explain a range of social behaviors that includes sexuality and mate selection, parental attachment to offspring, homosexuality, aggression, and altruism (Buss, 1988; Cunningham, 1981; Rothstein & Pierotti, 1988).

Altruistic behavior (especially self-sacrificing altruism) does not have survival value for an individual. But altruistic acts can increase the probability that the individual's genes will be passed to later generations, provided those acts are directed toward others who hold the same genes. If an individual helps his or her close relatives, that act increase the chances those relatives will survive and eventually have offspring. Because the relatives share many genes with the altruistic individual, reproduction by relatives passes many of the altruist's own genes to the next generation. Thus altruism directed toward close relatives contributes to the survival of an individual's genes in future generations (Krebs & Miller, 1985; Ridley & Dawkins, 1981).

The sociobiological perspective generates some interesting and testable propositions. Among these are the following: (1) Animals are most altruistic toward those who most closely resemble them genetically; therefore, they help immediate family members more than distant cousins, and help distant cousins more than outsiders or strangers. (2) Parents tend to behave altruistically toward healthy offspring (who are likely to survive and pass on their genes) but less altruistically toward sick or unhealthy offspring (who are likely to die soon). (3) Mothers are more altruistic toward their offspring than fathers. Males have the biological capacity to have many children, whereas females (in many species) can have only a small number of offspring. Thus males can perpetuate their genes without much altruism toward any one child, and females try to assure genetic survival by helping each of their (smaller number of) offspring to a greater extent.

Although interesting, the sociobiological perspective is controversial, especially as applied to humans. Critics have attacked this perspective on various grounds. First, they dispute the hypothesis that altruism is genetically transmitted (Kitcher, 1985). If animals and humans behaved in accord with the sociobiological model, they would help only close relatives and rarely or never help those who are genetically unrelated. Yet we know that humans sometimes help others who are unrelated, even total strangers. Some critics argue that to explain altruism of this type, it is necessary to rely on cultural constructs (for instance, norms that indicate when help is appropriate to give, institutions such as religion that define certain unrelated others as appropriate recipients of help, and so on). Sociobiology is at best an incomplete explanation for altruism.

Second and more broadly, sociobiology itself has become an issue in the long-running debate between those who believe human nature is essentially biologically determined and those who believe it is culturally determined. Some sociobiologists accept that culture can influence behavior but believe it does so within limits or boundaries set by biology (Wilson, 1978); they hold that although culture may determine the specific form and intensity of certain behavioral acts, biology establishes the very tendency to engage in broad patterns of behavior. Some critics, however, point out that sociobiological theory, by maintaining that behavioral patterns are transmitted genetically over generations, provides an excuse for accepting the status quo in society. Differences between racial, gender, or ethnic groups in performance and welfare can be dismissed as originating from genetic causes, and hence, as requiring no change (or as being inherently impossible to modify). Obviously, this view does not suit advocates of social change or social reform.

These objects contained information indicating the original owner's views on the Arab-Israeli conflict, on worthy or unpopular organizations, or on trivial opinion items. The owner's views on these topics either resembled or differed from the views known to characterize the neighborhoods in which the objects were dropped. Persons finding the wallets or letters took steps to return them to their owner much more frequently when the owner's views were similar to their own.

Even similarity in something as superficial as mode of dress can increase helping. In one study, for example, a confederate of the researchers dressed either in jeans and a work shirt or in conventional sports clothes and asked college students for a coin to make a telephone call. Students complied with the requests more often when the confederate's mode of dress was similar to their own than when it was different (Emswiller, Deaux, & Willis, 1971). In another study, persons approached in a shopping mall were less likely to help an individual dressed in punk style than one in ordinary street clothes, especially when the individual making the request approached very closely (Glick, DeMorest, & Hotze, 1988).

Deservingness

Suppose you received a call asking you to help elderly people who had just suffered a sharp reduction in income after losing their jobs. Would it matter whether they lost their jobs because they were caught stealing and lying or because their work program was being phased out? A study of Wisconsin homemakers who received such a call showed that respondents were more likely to help if the elderly people had become dependent because their program was cut than because they had been caught stealing (Schwartz & Fleishman, 1978).

What matters in this situation is the potential helpers' causal attribution regarding the origin of need. Potential helpers respond more when the needy persons' dependency is caused by circumstances beyond their control. Such people are true "innocent victims" who deserve help.

In contrast, needs caused by a person's own actions, misdeeds, or failings elicit little help. Various studies have shown that persons are less likely to offer help in cases where others' dependency and need can be attributed to their own failures or transgressions than in cases where others' dependency is not due to their own failures (Bryan & Davenport, 1968; Frey & Gaertner, 1986). For instance, one study found that students were less sympathetic and less likely to help a person who developed AIDS through promiscuous sexual contact than through a blood transfusion (Weiner, Perry, & Magnusson, 1988). In addition,

need viewed as stemming from illegitimate sources undermines helping in several ways. It inhibits empathic concern, blocks our sense of normative obligation, and increases the possibility of condemnation rather than social approval for helping (Brickman et al., 1982).

Even in emergencies, potential helpers are influenced by whether they consider a victim deserving. Consider responses to an emergency staged by experimenters in the New York subway (Piliavin, Rodin, & Piliavin, 1969). Shortly after the subway train left the station, a young man (a confederate) collapsed to the floor of the car and lay staring at the ceiling during the 7-minute trip to the next station. In one experimental condition, the man carried a cane and appeared crippled. In another condition, he carried a liquor bottle and reeked of whiskey. Bystanders helped the seemingly crippled victim quickly, usually leaping to his aid within seconds, but they often left the apparent drunk lying on the floor for several minutes. Much of this difference in response probably reflects the fact that many people—rightly or wrongly—blame drunks for their own plight.

Normative Factors in Helping

Would you intervene in a heated argument between a man and a woman you believe are married? In one experiment (Shotland & Straw, 1976), subjects unexpectedly witnessed a realistic fight between a man and a woman in an elevator. The man attacked the woman, shaking her violently, while she struggled and resisted. In one treatment, the man and woman were depicted as strangers; the woman screamed, "Get away from me! I don't know you!" In the other treatment, they were depicted as married; the woman screamed, "Get away from me! I don't know why I ever married you!" This simple variation greatly affected the subjects' propensity to help: Although 65% of the subjects intervened in the "stranger" fight, fewer than 20% intervened in the "married" fight.

This difference may have been due, in part, to subjects' perception of a greater likelihood of injury to the woman in the stranger fight than in the married fight. However, it may also have been due in part to normative expectations. The subjects who witnessed the married fight said they hesitated to take any action because they were not sure their help was wanted. Almost all the subjects who did not intervene said they felt the fight was "none of my business."

Clearly, wife and husband are social roles, and the relations between wives and husbands (and outsiders) are regulated by some widely understood norms. One of these is that except in the case of physical abuse, outsiders should basically mind their own business and let married couples resolve disputes as they will. When the woman on the elevator identified herself as the man's wife, this norm suddenly became relevant and changed the meaning of intervention. To intervene in the fight would intrude on the marital relationship and might invite reprisals from the husband, the wife, or both. In fact, subjects who thought the attacker was the woman's husband believed he was more likely to attack them if they intervened than did subjects who believed the attacker was a stranger.

In this section, we consider the impact of normative and cultural factors in helping. As the fight study suggests, whether one helps can be influenced by the desire to comply with cultural norms or role expectations held by others. We first discuss the impact of widely held norms on helping behavior, and we then consider helping behavior as role behavior.

Norms of Responsibility and Reciprocity

Cultural norms mandate helping as appropriate under some conditions, and they define it as inappropriate under others. When mandated as appropriate, helping becomes an approved behavior, supported by social sanctions. Here we discuss the responsibility norm and the reciprocity norm, broad social norms that indicate when helping is appropriate.

Social Responsibility Norm
The **social responsibility norm** states that individuals should help others who are dependent on them. People often mention their sense of what they "ought to do"—their internalized standards—when asked why they offer to help (Berkowitz, 1972). For example, Simmons (1991, p. 14) reports the words of a bone marrow donor prior to giving: "This is a life and death situation and you *must* do anything you can to help that person, whether it is family, friends, or [someone] unknown." The word "must" in this statement suggests a norm is operative.

Applicable in many situations, the social responsibility norm is readily activated. Some research suggests that simply informing individuals that another person, even a stranger, is dependent on them is enough to elicit help (Berkowitz, Klanderman, & Harris, 1964). Recognize, however, that there are stronger and weaker versions of the social responsibility norm. Although the norm that we must help dependent kin or needy friends is widely held, the belief that we must help needy strangers or unknown persons is not so universally accepted. Although the awareness of a stranger's dependency sometimes elicits help, it does not always do so. Speeding passersby, for example, frequently disregard stranded motorists they notice on the roadside. Bystanders watch, apparently fascinated but immobile, during rapes and other assaults. Thousands of people reject charity appeals every day. Thus not all persons follow the social responsibility norm, especially as it applies to strangers.

Some theorists have suggested that the social responsibility norm effectively motivates helping only when people are expressly reminded of it. In a test of this hypothesis (Darley & Batson, 1973), theological students were asked to prepare a talk on the parable of the Good Samaritan. On the way to record their talk, the students passed a man slumped in a doorway. Although these students were presumably thinking about the virtues of altruism, they helped the stranger only slightly more than a similar group of students who had prepared a talk on an unrelated topic (careers). A second variable—being in a hurry—had a much stronger impact on the amount of help offered. Students who were in a hurry offered much less help than those who were not. These findings suggest that the social responsibility norm is a fairly weak source of motivation to help and one easily negated by the costs of helping.

The Norm of Reciprocity
Another cultural standard, the **norm of reciprocity,** states that people should (1) help those who have helped them, and

(2) not help those who have denied them help for no legitimate reason (Schroeder et al., 1995; Trivers, 1983). This norm applies to a person who has previously received some benefit from another. Small kindnesses that create the conditions for reciprocity are a common feature of family, friendship, and work relationships. The reciprocity norm is found in different cultures around the world (Gergen et al., 1975).

People report that the reciprocity norm influences their behavior, and behavioral studies have demonstrated that people are inclined to help those who helped them earlier (Bar-Tal, 1976; Wilke & Lanzetta, 1982). Reciprocity is especially likely when the person expects to see the helper again (Carnevale, Pruitt, & Carrington, 1982). People try to match the amount of help they give to the quantity they received earlier, and they are less likely to ask for help when they believe they will not be able to repay the aid in some form (Fisher, Nadler, & Whitcher-Alagna, 1982; Nadler et al., 1985). By matching benefits, people maintain equity in their relationships and avoid becoming overly indebted to others.

People do not reciprocate every benefit they receive, however. Whether we feel obligated to reciprocate depends in part on the intentions we attribute to the person who helped us. We feel more obligated to reciprocate if we perceive the original help was given voluntarily rather than coerced and was chosen consciously rather than accidentally (Gergen et al., 1975; Greenberg & Frisch, 1972). Reciprocity is also more likely when the original benefit appears to have been tailored specifically to the recipient's preferences and needs.

Personal Norms and Helping

Although broad norms like social responsibility and reciprocity undoubtedly affect helping behavior, they are, by themselves, inadequate bases from which to predict the occurrence of helping behavior with precision. There are several reasons for this. First, given the wide variety of contingencies that people encounter, these norms are simply too general to dictate our behavior with any precision in all cases. Second, these norms are not accepted to the same degree by everyone in society; some persons internalize them to a greater extent than others. Third, the social norms

that apply to any given situation occasionally conflict with one another; the social responsibility norm may obligate us to help an abused wife, for example, but the widely accepted norm against meddling in others' marriages tells us not to intervene.

In response to these criticisms, a different type of normative theory has been developed by social psychologists (Schwartz & Howard, 1981, 1984). This theory explains not only the conditions under which norms are likely to motivate helping but also individual differences in helping in particular situations. Instead of dealing with broad social norms, this theory focuses on **personal norms**—feelings of moral obligation to perform specific actions that stem from an individual's internalized systems of values.

To investigate whether feelings of moral obligation motivate helping, investigators use a survey questionnaire to measure respondents' personal norms. For example, a survey on medical transplants might ask, "If a stranger needed a bone marrow transplant and you were a suitable donor, would you feel a moral obligation to donate bone marrow?" This survey would then be followed by an apparently unrelated encounter with a representative of an organization who would ask these individuals for help. In various studies, individuals' personal norms have predicted differences in their willingness to donate bone marrow or blood, to tutor blind children, to work for increased welfare payments for the needy (Schwartz & Howard, 1982), and to participate in community recycling programs (Hopper & Nielsen, 1991).

Helping is most likely to occur when conditions simultaneously foster the activation of personal norms and suppress defenses that might neutralize personal norms. For example, in one study (Schwartz & Howard, 1980), personal norms predicted quite accurately how much time college students would volunteer to tutor blind children. Among students who accepted responsibility for their actions, hours volunteered depended on the strength of students' personal norms. Among students who tended to deny such responsibility, however, there was no relationship between personal norms and volunteering. The results of this study are displayed in Figure 10.1. Other studies (Schwartz & Howard, 1984) have revealed that the relationship between personal norms and helping depends on conditions that support norm activation

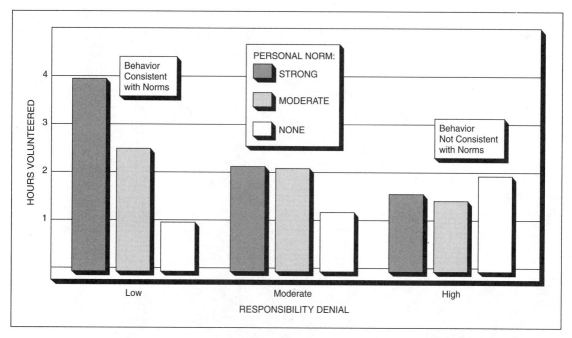

Figure 10.1 Volunteering as a Function of Personal Norms and Responsibility Denial

In a survey on social issues, university students indicated how much of a moral obligation they would feel (personal norm) to read texts to blind children. Three months later, the director of the Institute for the Blind wrote to the students, requesting that they volunteer time for just this purpose. Students who rarely denied responsibility for the consequences of their acts (low responsibility denial) behaved in a manner consistent with their personal norms: The stronger their moral obligation, the more they volunteered. Students moderate in responsibility denial showed weak consistency between personal norms and behavior. Students high in responsibility denial showed no consistency between personal norms and behavior. These findings indicate that the impact of our personal norms on helping behavior depends on whether we accept or deny our own responsibility.

SOURCE: Adapted from Schwartz and Howard, 1980.

(for example, the tendency to notice others' suffering) and suppress defensive denial of norms (for example, the focus of responsibility on the potential helper).

Helping Behavior as Role Behavior

Although some helping behavior is spontaneous bystander intervention or transient assistance, other helping behavior is more properly viewed as role behavior (Gergen & Gergen, 1983). Repetitive or recurring acts of help in the context of groups or organizations are often of this type. Consider, for example, the act of donating blood. Many donors give blood not just on a single occasion but on a recurring basis over time. To explain a pattern of repeated do-

nation, it is useful to think of "blood donor" as a social role. This role is enacted in the context of the medical group or organization that collects the blood. Research by Callero (1985) showed that this role is associated with self-definition, entails expectations from others, induces perception of others in terms of the role, and affects future donation behavior.

The role of blood donor has several interesting characteristics. It is a voluntary role, usually unpaid, involving altruistic behavior toward a beneficiary who is often unknown to the donor. Why would a person continue to enact a role of this type over time? One answer lies in the concept of *role-person merger*—the degree to which the blood donor role reflects the self-concept. Individuals with a high level of

Rescuers carry an injured climber down a mountain to safety. Many people satisfy their own needs, such as the need for approval or for excitement, by helping others.

role-person merger view blood donation as an important component of who they are. Callero, Howard, and Piliavin (1987) reported that role-person merger predicts future blood donation behavior. Higher levels of role-person merger produce higher levels of future blood donation, independent of the effect of personal norms.

Personal and Situational Factors in Helping

Helping behavior is influenced not only by normative factors, but also by situational and personal factors. In this section, we consider how an important situational factor—the presence of models—influences helping. We also consider the impact of mood states in helping. We focus on how various transient emotional states—whether somebody is in a good mood or bad mood or whether he or she feels guilty about something—affect the tendency to give or withhold help.

Modeling Effects

An important factor that promotes helping is the presence of someone else who is helping—that is, a behavioral model. For instance, one study found that the presence of a model increased adults' willingness to donate blood (Rushton & Campbell, 1977). Other studies have shown that the presence of a model increases motorists' tendency to help a stranger change a flat tire and shoppers' tendency to donate coins to the Salvation Army kettle during the Christmas season (Bryan & Test, 1967).

Not surprisingly, modeling effects hold for children as well as for adults. Studies have consistently shown that children behave more generously toward others when they are exposed to generous models than when they are exposed to selfish models (Lipscomb et al., 1982). In one study, for instance, children exposed to models who behaved inconsistently (once generously, once selfishly) donated less than children exposed to a consistently generous model but more than children exposed to a consistently selfish model (Lipscomb et al., 1985).

The presence of a behavioral model tends to increase helping for several reasons. First, a model demonstrates what kinds of actions are possible or effective in the situation. Others who previously did not know how to assist can emulate the model. Second, a helping model conveys the message that to offer help is appropriate in the particular situation. A model may increase the salience of the social responsibility norm; once aware of this norm, others may decide to assist. Third, a model provides information about the costs and risks involved in helping, a consideration that is especially important in situations involving danger. By offering help under conditions of danger or potential damage to self, models demonstrate to others that the risks incurred are tolerable or justified.

Gender Differences in Helping

It is not possible to maintain that males are always more helpful than females, or vice versa, because so much depends on the type of situation and the relevant role demands. Research findings do indicate that men are more likely than women to intervene and offer assistance in emergency situations that entail danger (Eagly & Crowley, 1986). For instance, in one study the investigators interviewed people who had been publicly recognized as heroes by the state of California (these persons had intervened to protect someone during a dangerous criminal act such as a mugging or bank robbery). It turned out that all but one of these persons was a male (Huston et al., 1981). Acting heroically and dealing with risk and danger is often considered part of the traditional male role.

Then, too, men may perceive the costs of dealing with danger as lower than women do because men are physically stronger and more likely to have relevant abilities, such as self-defense training (Huston et al., 1981).

Studies of helping show that people tend most to offer the types of help which are consistent with their gender-role expectations (Eagly & Crowley, 1986; Piliavin & Unger, 1985). Thus, if men are more likely then women to help in heroic or chivalrous situations entailing risk, women are more likely than men to help in situations requiring nurturance, caretaking, and emotional support. For instance, women tend more than men to care for children and aging parents on a day-to-day basis, an important help-giving function (Brody, 1990). Women are also more likely than men to provide their friends with personal favors, emotional support, and informational counseling about personal or psychological problems (Eisenberg & Fabes, 1991; Otten, Penner, & Waugh, 1988).

Good and Bad Moods

A *mood* is a transitory feeling, such as being happy and elated or being frustrated and depressed. A person's mood-state can influence whether he or she gives help to another. In this section, we examine the effects on helping of both good mood and bad mood.

Good Mood and Helping   Substantial evidence shows that when individuals are in a good mood, they are more likely to help others than when they are in a neutral mood (Salovey, Mayer, & Rosenhan, 1991). Good moods promote both spontaneous helping and compliance with requests. But moods can change quickly, and as a person's good mood fades, helping also quickly drops back to normal (Isen, Clark, & Schwartz, 1976).

Almost every experience that puts us in a good mood increases the chance we will help others. For instance, suburban schoolteachers who learned they scored well on a battery of tests donated more to a school library fund than teachers who received no feedback on their performance (Isen, 1970). People

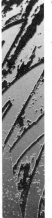

10.2   Urban-Rural Differences in Helping

Are city dwellers less likely than small-town residents to help a stranger? Many believe city dwellers are brusque, untrusting, and unhelpful, whereas small-town residents are friendly, cooperative, and generally helpful. Some research does support the hypothesis that people are less helpful in urban than in rural environments (Hedge & Yousif, 1992; Steblay, 1987). And, despite some exceptions, research generally supports the hypothesis that as population size increases, spontaneous helping of strangers declines (Merrens, 1973; Rushton, 1978).

For instance, Takooshian, Haber, and Lucido (1977) investigated responses of passersby to a request for help from a lost child. Conducted on busy streets in four large cities (Boston, Chicago, New York, and Philadelphia) and in 12 smaller towns in nearby areas, this study utilized children as confederates. A child confederate said to each passerby, "I'm lost. Can you call my house?" Results showed that in the smaller towns, 72% of the subjects offered to help and most of the people approached were sympathetic to the plight of the child. In the cities, however, only 46% offered to help, and many passersby were abrupt and unsympathetic.

Another study (Amato, 1983) investigated urban-rural differences in helping across 55 communities in eastern Australia; these communities ranged from small villages to major cities. Subjects were exposed to requests from strangers for five different forms of help (correcting inaccurate directions, donating money to combat multiple sclerosis, assisting a fallen pedestrian, and so on). Results show that strangers were more likely to receive help in small towns than in bigger towns and cities.

What is it about large cities that leads to lower rates of helping? Research findings suggest that lower rates of helping are due to contextual or environmental conditions, rather than to some kind of special "urban personality." One condition that affects helping in cities is threat to personal security. In the United States, many urban areas have heterogeneous populations and relatively high crime rates (theft, assault, murder). Because individuals are likely to feel less secure in the midst of many dissimilar others, heterogeneity inhibits interaction with strangers and hence lowers the rate of helping (Fischer, 1976; Holahan, 1977). For instance, House and Wolf (1978) found that persons living in cities were less likely to admit an interviewer into their homes than were those living in rural areas. The reason: Urban residents had greater distrust of strangers and fear of possible crime. Persons afraid of interacting with strangers will surely offer less help than those not afraid.

Beyond the issue of danger, city dwellers usually experience more stressful input and stimulus overload than small-town people (Milgram, 1970). Each day, urbanites are bombarded by a plethora of stimuli—flashing lights, loud noises, varied smells, excessive crowding, and the like. To cope with overload, city dwellers tend to ignore stimuli that are nonessential to their personal goals. Stimulus overload reduces people's responsiveness to events in the environment, and thus it indirectly suppresses helping. Several studies have shown that high stimulus overload (such as a very noisy environment) reduces the likelihood of helping a needy stranger (Korte, Ypma, & Toppen, 1975; Sherrod & Downs, 1974). This effect is illustrated by a field experiment in which a man (a confederate) wearing a cast on his arm dropped the books he had been carrying. When ordinary street noises were present, 80% of male passersby helped pick up the books, but when a loud lawn mower was running nearby, only 15% stopped to help (Mathews & Canon, 1975).

Thus many environmental conditions—overload, danger, complexity, crowding, heterogeneity—are aversive and caution inducing. As such, they lessen responsiveness to the environment and diminish helping responses toward strangers. Because these conditions tend to arise more in urban areas than in rural areas, they operate to produce urban-rural differences in helping.

were more likely to help a stranded caller who had dialed the wrong number from a pay phone if they had just received a small gift than if they had received no gift (Isen & Levin, 1972). Other mood-enhancing experiences that have been shown to increase helping include recalling happy experiences, reading statements describing pleasant feelings, hearing good news on the radio, listening to soothing music, and enjoying good weather. Some powerful effects of good mood on helping are shown in Table 10.1.

There are several reasons why being in a good mood increases our propensity to help others (Carlson, Charlin, & Miller, 1988). First, people in a good mood are less preoccupied with themselves and less concerned with their own problems. This allows them to focus more attention on the needs and problems of others, which (through empathy) often leads to helping. Second, people in a good mood often feel relatively fortunate compared to others who are deprived. They recognize their good fortune is out of balance with others' needs; to restore a more just balance, they use their resources to help others (Rosenhan, Salovey, & Hargis, 1981). Third, people in a good mood tend to see the world in a positive light, and they want to maintain the warm glow of happiness. Thus, if by offering help that benefits others they can maintain and even increase their own positive feelings, they will do so. However, people in a good mood may tend to avoid forms of helping that involve unpleasant or embarrassing activities, for these may interrupt or destroy the good mood (Cunningham, Steinberg, & Grev, 1980; Isen & Simmons, 1978).

Bad Mood and Helping Bad mood—feeling sad or depressed—can have rather complex effects on helping. Results of studies show that under some conditions, bad mood (when contrasted with neutral mood) inhibits helping; under other conditions, however, it promotes helping (Carlson & Miller, 1987; Rosenhan et al., 1981).

Bad mood can suppress helping for several reasons. First, it has an impact on the salience of others' needs. People in a bad mood are often concerned about their own problems and therefore are less likely to notice others' needs than people in a neutral mood. When others' needs do not grab the attention of a potential helper because of bad mood, help is less likely to be given (Aderman & Berkowitz, 1983; Rogers et al., 1982). Second, people in a bad mood often see themselves as less fortunate than others. Feeling relatively impoverished and underbenefited, they may resist using their own resources to help others lest they become even more disadvantaged (Rosenhan et al., 1981). The net effect is that the feelings of relative impoverishment associated with bad mood inhibit helping, especially when the costs for helping are large.

Despite these effects, bad mood can sometimes increase helping. One explanation how this can come about is provided by the *negative-state relief hypothesis* (Cialdini, Kendrick, & Baumann, 1982; Cialdini et al., 1987). This hypothesis starts with the assumptions, first, that people experiencing unpleasant feelings (such as sadness) are motivated to reduce them, and second, that many persons have learned from childhood that helping others improves their own

Table 10.1 The Effects of Good Mood on Helping

	Percentage Who Helped		
	Picking up Dropped Papers	**Mailing a Lost Letter**	**Making a Phone Call for a Stranger**
Neutral mood	4%	10%	12%
Good mood	88%	88%	83%

SOURCES: Adapted from Isen, Clark, and Schwartz, 1976; Levin and Isen, 1975; Isen and Levin, 1972.

mood (often through receipt of thanks or praise). The hypothesis predicts that persons of this type, when in a bad mood, will help others primarily as a means to boost their own spirits. This is an egoistic motive for helping, not an altruistic one, because persons are offering help primarily to relieve sadness in themselves rather than to relieve suffering in others.

One implication of the negative-state relief hypothesis is that if people in a bad mood are relieved by some event or act other than giving help, they will not volunteer to help as much as others whose bad mood has not been relieved. Results from several studies are consistent with this prediction (Cialdini, Darby, & Vincent, 1973; Schaller & Cialdini, 1988). Another implication of this hypothesis is that negative moods motivate help-giving only if people believe their moods will be improved by helping. One study on this point manipulated whether subjects did or did not believe their (sad) mood could be changed or improved by helping; results were consistent with expectations from the negative-state relief hypothesis (Manucia, Baumann, & Cialdini, 1984).

Despite some support, we must recognize that the negative-state relief hypothesis is controversial and has been criticized by various writers (Carlson & Miller, 1987; Cialdini & Fultz, 1990; Miller & Carlson, 1990). Not all research evidence is consistent with it, and widespread acceptance will hinge on results of future studies.

Guilt and Helping

Guilt is a negative emotional state aroused when we transgress against others or do something we consider wrong (Tangney, 1992). For instance, we may feel guilty if we accidentally break someone's camera or spill ink on a person's clothes. To study the impact of guilt on helping, investigators in various studies have induced subjects to perform actions that transgress against others (such as killing a laboratory animal, giving painful electric shocks, damaging expensive machinery) and that consequently produce guilt. After the transgression, subjects are presented with an opportunity to help the other; the forms of help in these studies include picking up scattered papers, volunteering to participate in further experiments, donating blood, making telephone calls for an ecology group, and the like (Rosenhan et al., 1981). Studies of this type generally show that individuals are more likely to help others when they believe they have harmed these persons in some way (and feel guilty about it) than when not (Salovey, Mayer, & Rosenhan, 1991).

Guilt from transgressions of this type has led to helping regardless of whether the transgressions have been intentional or unintentional, public or private, in the laboratory or in the field. Such transgressions have also produced help both spontaneously and in response to requests, as well as help that benefits either the victim of the transgression or an unrelated third party.

We can explain these findings in part by assuming that helping allows transgressors to relieve their feelings of guilt. For instance, by giving help to the injured party, a transgressor may be able to directly compensate or make up for the transgression. This would assuage the transgressor's guilt and improve his or her mood.

One problematic finding, however, is that helpfulness induced by guilt is often directed not just toward the injured victim, but toward unrelated third parties who may need help. This raises the possibility that helping under conditions of guilt can involve something more than just direct compensation to victims. In this regard, the *image-reparation hypothesis* (Cunningham, Stineberg, & Grev, 1980) maintains that help given by a transgressor is not necessarily directed at undoing the harm that's been done. Rather, the prosocial response serves to repair the transgressor's tarnished self-image and restore self-esteem. Most of us do not see ourselves as abusive or mean, so giving help to someone—anyone—after a transgression can restore our image as positive, worthwhile, decent persons. Note that this is essentially an egoistic (not an altruistic) motive for giving help. Although the image-reparation hypothesis cannot be taken as a total explanation for helping under conditions of guilt, it identifies an important process and is consistent with a body of research findings (Cunningham et al., 1990).

Bystander Intervention in Emergency Situations

Some of the earliest and most interesting social psychological research on helping was inspired by the tragic murder in 1964 of a young woman named Catherine (Kitty) Genovese. Shortly before 3:20 A.M. on March 13, 1964, in the borough of Queens in New York City, Kitty Genovese parked her red Fiat in the lot of the Long Island Railroad station. Although she lived in Kew Gardens, a quiet middle-class residential area, Kitty must have sensed something wrong. In-stead of taking the short route home, she started running along well-lighted Lefferts Boulevard. She didn't get far before an assailant attacked her.

Milton Hatch awoke at the first scream. Staring from his apartment window, he saw a woman kneeling on the sidewalk directly across the street and a small man standing over her. "Help me! Help me! Oh God, he's stabbed me!" she cried. Leaning out his window, Hatch shouted, "Let that girl alone!" As other windows opened and lights went on, the assailant fled in his car. No one called the police.

With many eyes now following her, Kitty dragged herself along the street, but not fast enough. More

Persons witnessing a train derailment assist in rescuing injured passengers. If bystanders are to intervene, they must interpret the situation as an emergency, decide that they have the responsibility to act, and know how to offer the appropriate form of assistance.

than 10 minutes passed before the neighbors saw her assailant reappear, hunting for her. When he stabbed her a second time, she screamed, "I'm dying! I'm dying!" Still no one called the police. Emil Power would have called, had his wife not insisted someone else must already have done so.

The third, fatal attack occurred in the vestibule of a building a few doors from Kitty's own entrance. Onlookers saw the assailant push open the building door, although few could hear the weak cry that greeted him. Finally, at 3:55, 35 minutes after Kitty's first scream, Harold Klein, who lived at the top of the stairs where Kitty was murdered, called the police. The first patrol car arrived within 2 minutes, but by then it was too late (Seedman & Hellman, 1975).

The tragic story quickly became front-page news in New York and across the country. When it was subsequently discovered that a total of 38 persons had witnessed the stalking and stabbing, the editorials asked why no one had offered help or called the police sooner. A parade of experts attributed the failure to urban apathy, to the depersonalizing environment of New York, to people's unwillingness to get involved.

Although these explanations may be partly true, they don't cut to the heart of the matter. We know that bystanders sometimes help in emergencies, and sometimes not. So for social psychologists, the more fundamental question raised by Kitty's murder is this: Under what conditions will bystanders and witnesses intervene in an emergency and give help? Why do people help in some emergency situations but not in others? In this section, we consider this issue in detail and look at various factors that influence whether a bystander will help a victim.

The Decision to Intervene

The term **bystander intervention** denotes a (quick) response by a person witnessing an emergency to help another who is endangered by events. Whether (and how) to intervene in an emergency is a complex matter, for in responding to an emergency helpers often place themselves in considerable danger.

A theory proposed by Latané and Darley (1970) maintains that bystanders (potential helpers) go through a specific decision-making sequence prior

to actually giving help in emergencies. Specifically, this theory identifies five steps in the process. Before giving help, a potential helper must (1) notice that something is happening (rather than not notice), (2) interpret the situation as an emergency (rather than as a nonemergency), (3) decide that he or she has the responsibility to act (rather than no responsibility to act), (4) know or recognize an appropriate course of action to give help (rather than not know the appropriate course), and (5) implement the chosen course of action (rather than not implement it). The steps in this decision-making process are displayed schematically in Figure 10.2.

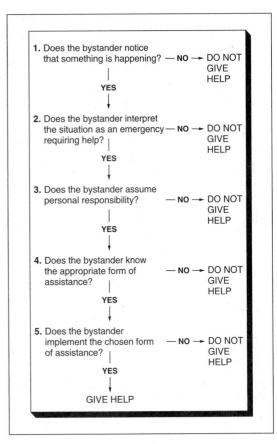

Figure 10.2 Decisions Leading to Intervention in an Emergency

The Bystander Effect

In emergency situations, potential helpers are influenced by their relations with other bystanders (Dovidio, 1984; Latané & Darley, 1970). This influence is apparent at each step in the decision process (Figure 10.2). To investigate the nature of bystander influence, researchers have conducted a variety of laboratory studies that entail simulated emergencies of one kind or another. For instance, in one experiment (Latané & Rodin, 1969), subjects heard a loud crash from the room next door, followed by a woman screaming, "Oh my God, my foot I . . . I . . . can't move it. Oh my ankle. I . . . can't get this . . . thing off me." In another experiment (Darley & Latané, 1968), subjects participating in a discussion over an intercom suddenly heard someone in their group begin to choke, gasp, and call for help, apparently gripped by an epileptic seizure.

In each experiment, the number of people who were supposedly present when the emergency occurred was varied. Subjects believed either that they were alone with the victim or that one or more bystanders were present. Time and again the same finding emerged: As the number of bystanders increased, the likelihood that any one of them would help decreased (Latané & Nida, 1981). Bystanders helped most often and most quickly when they were alone. Knowledge that other potential helpers are present inhibited intervention in an emergency.

Social psychologists use the term **bystander effect** to refer to the finding that as the number of bystanders increases, the likelihood any one bystander will help a victim decreases. This effect is illustrated in Figure 10.3, using data from the epileptic seizure experiment.

Theorists have identified several distinct processes that contribute to the bystander effect. These include social influence regarding the interpretation of the situation, evaluation apprehension, and diffusion of responsibility (Latané et al., 1981; Piliavin et al., 1981). Each of these processes affects specific steps in the decision-making sequence.

Interpreting the Situation   Emergencies and other situations requiring help are often ambiguous,

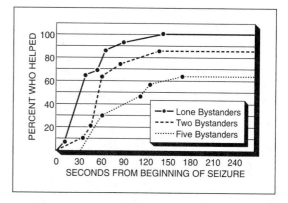

Figure 10.3   The Bystander Effect

Students who were engaged in a discussion via intercom of their adjustment to college life heard one participant begin to choke, then gasp and call for help, as if he were undergoing a serious nervous seizure. Students intervened to help the victim most quickly and most often when they believed they were the lone bystander to witness the emergency. More than 90% of lone bystanders helped within the first 90 seconds after the seizure. Among those who believed other bystanders were present, however, fewer than 90% intervened, even after 4 minutes. The bystander effect refers to the fact that the greater the number of bystanders in an emergency, the less likely any one bystander will help.

SOURCE: Adapted from Darley and Latané, 1968.

at least initially. For instance, is a choking, gasping student in real trouble? When faced with ambiguity, people look to the reaction of others for cues about what is going on. The reactions of others can influence three of the steps in the decision-making sequence leading to helping. If others appear calm, the bystander may decide that nothing special is happening (step 1), or that whatever is happening requires no help (step 2). Likewise, the failure of others to act may influence the bystander to decide there is no appropriate way to help (step 4).

Bystanders often try to appear calm, avoiding overt signs of worry until they see whether others are alarmed or not. Through such cautiousness, they unintentionally encourage each other to define the situation as not problematic. In that way they inhibit each other from helping. The larger the number of apparently unruffled bystanders, the stronger their

High potential costs and diffusion of responsibility inhibit bystander intervention in this knife fight. The man on the ground, wounded and bleeding profusely, clearly needs help, but bystanders flee in the opposite direction. They feel little responsibility for the wounded man and wish to avoid entanglement in the fight, still in progress.

inhibiting influence is on each other. However, consistent with this explanation, increasing the number of bystanders does not inhibit individual helping under certain conditions, such as (1) when observation reveals that others are indeed alarmed (Darley, Teger, & Lewis, 1973), and (2) when the need for help is so unambiguous that others' reactions are unnecessary to define the situation (Clark & Word, 1972).

Evaluation Apprehension Bystanders are not only interested in others' reactions; they also realize that other bystanders are an audience for their own reactions. As a result, bystanders may feel **evaluation apprehension**—concern about what others expect of them and about how others will evaluate their behavior. Evaluation apprehension can either inhibit or promote helping. On one hand, evaluation apprehension inhibits helping when bystanders fear

that others will view their intervention as foolish or wrong. When they see that other witnesses to an emergency are not reacting (as in the Kitty Genovese case), they may infer that the others see no need to intervene or even oppose intervention. In the decision-making sequence, evaluation apprehension mainly affects step 4 (choosing a way to react) and step 5 (deciding whether to implement the chosen course of action).

On the other hand, evaluation apprehension promotes helping if there are no cues to suggest that other witnesses oppose intervention. Bystanders then tend to assume that others approve of intervention. In three laboratory studies demonstrating this effect, bystanders witnessed a convulsive nervous seizure or a violent assault (Schwartz & Gottlieb, 1976, 1980). Knowledge that an audience of other bystanders was present led to increased helping when the audiovisual system prevented each bystander from learning how others were reacting.

Diffusion of Responsibility When one and only one bystander witnesses an emergency, responsibility to intervene is focused wholly on that individual. But when there are multiple bystanders, responsibility to intervene is shared, as is the blame if the victim is not helped. Hence a witness is less likely to intervene when other bystanders are present ("Why should I help? Let someone else do it"). This process wherein a bystander does not take action to help because others persons share responsibility for intervening is called **diffusion of responsibility.**

In the decision sequence, diffusion of responsibility operates primarily at step 3 (bystander decides whether he or she has the responsibility to act). When multiple witnesses are present, bystanders sometimes limit their own responsibility by assuming others have already taken action. Of the 38 witnesses to Kitty Genovese's murder, many claimed they thought someone else must surely have called the police.

Diffusion of responsibility occurs only when a bystander believes the other witnesses are capable of helping. We do not diffuse responsibility to witnesses who are too far away to take effective action or too young to cope with the emergency (Bickman, 1971; Ross, 1971). The tendency to diffuse responsibility is particularly strong if a bystander feels less competent than others who are present. Bystanders helped less, for example, when one of the other witnesses to a seizure was a premed student with experience working in an emergency ward (Pantin & Carver, 1982; Schwartz & Clausen, 1970).

Costs and Emergency Intervention

When deciding whether to offer assistance in an emergency, bystanders typically consider the costs involved. Some interventions involve little or no cost, but others are more costly, and help is presumably easier to give when the cost of doing so is low.

One theory, the **arousal/cost-reward model** of helping (Dovidio et al., 1991; Piliavin et al., 1981),

proposes that bystanders weigh the needs of the victim and their own needs and goals, and then decide whether helping is too costly in the circumstance. Bystanders often take into account several kinds of costs to themselves in emergency situations. First, bystanders consider the *cost of giving direct help.* This includes costs to them if they offer help—lost time, exposure to danger, expenditure of effort, exposure to disgusting experiences, and the like. Second, bystanders consider the *cost of not giving help.* Costs borne by the bystanders if the victim receives no help include the burden of unpleasant emotional arousal while witnessing another's suffering and costs associated with one's personal failure to act in the face of another's need (self-blame, possible blame from others, embarrassment, and the like).

Intervention as a Function of Cost The cost approach to helping hypothesizes that the responses of bystanders in the emergency situation depends on whether the cost to themselves of giving direct help is high or low and on whether the cost to themselves of not giving help is high or low (Piliavin, Piliavin, & Rodin, 1975). Table 10.2 displays the predicted responses of bystanders as a function of cost. The model predicts that direct intervention will occur primarily when the bystander incurs low costs for giving direct help and high costs for not helping the victim. Other modes of response (indirect helping, redefinition of the situation, leaving the scene, and so on) will occur under the other cost combinations (see Table 10.2).

Evidence Regarding Cost Various studies have documented that cost has an impact on help giving. First, some results support the proposition that the greater the cost to self of giving direct help, the less likely one is to help (Darley & Batson, 1973; Shotland & Straw, 1976). This was demonstrated, for instance, in a study conducted in the New York City subway (Allen, 1972). Aboard a subway car, a bewildered-looking man asked the subject (a passenger) whether the train was going uptown or downtown. The man in the neighboring seat, a muscular type who was reading a bodybuilding magazine, responded quickly but

Table 10.2   Predicted Effects of Cost Factors on Helping

Cost of Not Helping the Victim	Cost of Directly Helping the Victim	
	Low	**High**
High	Direct intervention by bystander	Indirect intervention or Redefinition of the situation, disparagement of the victim, etc.*
Low	Variable reactions (largely a function of perceived norms in the situation)	Leaving the situation, ignoring, denial, etc.

This table displays theoretically predicted reactions of an observer to an emergency as a function of costs to the observer and costs to the victim if no help is given.

* The response of redefining the situation or disparaging the victim lowers the cost of giving no help to the victim; this may lead to subsequent responses like leaving the situation, ignoring, denial, etc.

SOURCE: Adapted from Piliavin, Piliavin, and Rodin, 1975.

gave an obviously zwrong answer. (Both the bewildered man and the bodybuilder were confederates.) The subject could help by correcting this misinformation, but only at the risk of challenging the bodybuilder. Whether or not the subjects helped depended on how threatening the bodybuilder appeared to be. Threat was manipulated by varying his reaction to an incident a minute before. When the bodybuilder had previously threatened physical harm to a person who had stumbled over his outstretched feet, only 16% helped. When the bodybuilder had only insulted and embarrassed the stumbler, 28% helped. When the bodybuilder had given no reaction to the stumbler, 52% helped. Thus, the greater the anticipated cost of antagonizing the misinforming bodybuilder, the less likely people were to help the bewildered man.

In addition, studies show that the greater the costs to self of not helping the victim, the more likely one is to help (Berkowitz, 1978; Gottlieb & Carver, 1980). For instance, Staub (1974) conducted a study with a simulated health emergency on a city sidewalk. A male confederate faked a bad knee (he grabbed his knee, fell to the sidewalk, failed to stand up). The study varied the ease with which bystanders could escape from the emergency scene. In one treatment, the emergency happened on the other side of the street, so escape was easy and the cost of not helping was fairly low. In the other treatment, the emergency happened right in front of the bystander on the same side of the street, so escape was difficult and the cost of not helping was higher. Victims received considerably more help when escape was difficult than when it was easy.

Seeking and Receiving Help

To this point, we have focused primarily on giving help rather than receiving it. Yet recipients' reactions to receiving help—and related phenomena such as people's willingness to seek help in the first place—are important topics that also deserve attention. How does it feel to receive help from other persons? Help not only relieves need but also demonstrates to the recipient that someone cares. We might expect, therefore, that recipients will feel gratitude and joy, and of course they frequently do. In other cases, however, help elicits a different emotional reaction from recipients—resentment, hostility, and anxiety. In this section, we look at the various reactions to help and the reasons for them.

Help and Obligation

Although our society has some norms (such as the norms of reciprocity and responsibility) that mandate giving help when needed, it also has norms of adult independence and self-reliance that tell us to avoid asking for help unless really necessary. When help is sought and received, resources (labor, materials) are transferred from one person to another. If the norm of reciprocity is salient in the situation, the person receiving help may feel obligated or indebted to the helper (Greenberg & Westcott, 1983).

In consequence, needy persons (in nonemergency situations) sometimes experience a dilemma. On one hand, they can ask for help and possibly endure some embarrassment and/or social obligation, or they can suffer through the difficulties of trying to solve their problems on their own (Gross & McMullen, 1983). In cases where the recipient has the opportunity and ability to reciprocate, there may be no problem. But in cases where this transfer of resources is strictly one-way (from the helper to the needy), it may create a lingering sense of indebtedness (in the needy toward the helper). Several studies have shown that if a person lacks the capacity to repay or to reciprocate, he or she is less likely to ask for help than otherwise (Nadler, 1991; Wills, 1992); this is especially so if the person needing help has high self-esteem. Persons tend to develop resentment and negative sentiments toward a benefactor they cannot repay (Clark, Gotay, & Mills, 1974; Gross & Latané, 1974).

In addition, help that goes beyond what is absolutely necessary may be resented by the recipient (Schwartz, 1977). We are often chary of people who are overly generous or extend help beyond our ability to reciprocate. Accepting substantial generosity oblig-

ates us to submit to our benefactors' wishes, at least to some degree. Gifts we cannot reciprocate threaten our freedom of action (Greenberg, 1980).

Threats to Self-Esteem

In studying people's reactions to receiving help, theorists have proposed that an important determinant of whether help is appreciated or resented is the extent to which it undermines the recipient's self-esteem (Nadler, 1991; Nadler & Fisher, 1986; Shell & Eisenberg, 1992). As we have noted, to receive help is a mixed blessing. Although it provides relief, it can also impair a recipient's self-esteem and sense of self-reliance. Of course, helpers usually intend just the opposite. The avowed purpose of welfare, for instance, has been to aid impoverished individuals and to help families escape hunger while they establish themselves as self-supporting. The purpose of foreign aid is to enable nations to overcome crises and to develop their own resources. Yet welfare, foreign aid, and other forms of assistance are sometimes given reluctantly or in ways that do not promote these outcomes. Intentionally or otherwise, helpers may communicate the message that those who need and accept help are inferior in status and ability because they fail to display the self-reliance and achievement admired in Western societies (DePaulo & Fisher, 1980; Rosen, 1984). When help is couched in these terms, recipients' acceptance of help may diminish their self-esteem and even change their self-concepts.

Ego Centrality Help is especially threatening to self-esteem—and hence is less likely to be sought or accepted gratefully by persons in need—when it implies inferiority in intelligence, competence, morality, or other qualities central to a recipient's self-concept. Help is less threatening, hence more likely to be sought, when it does not imply any major inadequacy with respect to central personal attributes (Nadler, 1987; Nadler & Fisher, 1984a). For example, help can be nonthreatening if need is explicitly attributed to uncontrollable or chance factors like a drought, epidemic, or unprovoked attack or if the aid is defined as enabling one to overcome a trivial inadequacy in experience or effort.

Similarity of Help Provider

Surveys regarding help seeking for personal and psychological problems indicate that we are most likely to ask our friends, or people who are similar to us, for assistance. Wills (1992) indicates that persons looking for help of this type are several times more likely to seek it from friends, acquaintances, or family members than from professionals or strangers. Generally, help-seeking is more common among close friends and people in communal relationships than among strangers and people in exchange-based relationships.

Nevertheless, the helper's similarity to the recipient is a complex factor in help giving and help seeking. Help that implies an important inadequacy is often more threatening to our self-esteem when it is received from those who are similar to us in attitudes or background than from those who are dissimilar (Nadler, 1987; Nadler & Fisher, 1984a). Similarity can aggravate recipients' self-evaluations because similar helpers are relevant targets for self-comparison ("If we are both alike, why do I need help while you can give it?"). People who accept aid from helpers similar to themselves on a task central to their self-concept report lower self-esteem, less self-confidence, and more personal threat than when they accept aid from dissimilar helpers (DePaulo, Nadler & Fisher, 1983; Nadler, Fisher, & Ben-Itzhak, 1983).

Threat to Self-Esteem as a Motivator

Is the fact that aid threatens recipients' pride and self-esteem wholly undesirable? Not if we consider the long-term consequences (Nadler & Fisher, 1984b). If recipients experience aid as nonthreatening, they feel favorably toward themselves and the helper, but have little motivation to change. Consequently, they will invest little in developing self-reliance and will just continue to seek help in the future. Nations become dependent satellites; individuals become helpless parasites. In contrast, aid that threatens the self may generate negative feelings toward the helper as well as the self, but it can motivate recipients to change. This

motivation promotes self-reliance and drives individuals to reestablish their independence.

Summary

Helping is behavior intended to benefit others. Altruism, a specific kind of helping, is voluntary behavior intended to benefit another with no expectation of external reward.

Motivation to Help Others

There are several motivations for helping. (1) Egoistic considerations of cost and reward often motivate helping. People tend to help more when it leads to rewards, such as social approval, material gain, and satisfaction of personal needs. People often refuse to help when it entails costs, such as danger and effort. They also avoid costs imposed for not helping, such as public disapproval and self-condemnation. (2) Altruistic behavior is often mediated by empathic arousal in response to others' distress. Our similarity to victims increases our empathy. Empathic concern motivates altruism—helping to benefit others with no expectation of reward.

Characteristics of the Needy That Foster Helping

The characteristics of the needy influence whether others will give help. (1) People tend to help those to whom they are related or with whom they are well acquainted. They also tend to give more help to those they like than those they do not like. (2) Needy persons have a better chance of receiving help if they are similar to the potential help giver rather than dissimilar. This holds for similarity in a wide variety of attributes including appearance (race, dress) and attitudes (political opinions). (3) Needy persons have a better chance of receiving help if they are seen as deserving and not the cause of their own plight.

Normative Factors in Helping

(1) Norms define when helping is the appropriate thing to do. The social responsibility norm directs us to help

whoever is dependent on us. The norm of reciprocity tells us to reciprocate intentional benefits. (2) Personal norms, based on internalized values, motivate help when we notice another's need and feel responsible for relieving it. (3) Some forms of helping behavior, such as recurrent donations of blood, are best viewed as role behavior.

Personal and Situational Factors in Helping

(1) The presence of a model (a person actively helping) will increase the probability that others will also offer to help. (2) Consistent with traditional gender roles, men are more likely than women to help in heroic or chivalrous situations entailing risk, whereas women are more likely than men to help in situations requiring nurturance, caretaking, and emotional support. (3) Good moods promote helping because they reduce self-preoccupation, increase the attention paid to others' needs, and increase one's feelings of relative good fortune. Bad moods inhibit helping because they increase self-preoccupation and increase feelings of being relatively underbenefited. However, a bad mood may increase helping when this is perceived as a route to improving one's own mood. (4) Guilt promotes helping because helping boosts transgressors' self-esteem and relieves guilt.

Bystander Intervention in Emergency Situations

(1) Prior to actually giving help in emergencies, bystanders go through a decision sequence. A potential helper must notice that something is happening, interpret the situation as a emergency, decide he or she has the responsibility to act, know or recognize the appropriate form of assistance, and decide to implement the chosen behavior. (2) Bystanders to an emergency influence each other's behavior in three ways. First, the reactions of others affect whether a bystander interprets the situation as one that requires help. Second, the expectations of other bystanders create evaluation apprehension, which may inhibit or enhance helping. Third, the presence of other bystanders fosters diffusion of responsibility,

which reduces helping. (3) Bystanders' responses to the emergency depend on whether the costs to themselves of giving help are high or low and on whether the costs to themselves if the victim receives no help are high or low. Direct intervention occurs primarily when the bystander incurs low costs for helping and high costs for not helping.

Seeking and Receiving Help

(1) From the standpoint of the recipient, help is a mixed blessing. Although providing benefits, it may also cause the recipient to become indebted to the helper, especially if the recipient lacks the capacity or opportunity to reciprocate. In such a case, a recipient may develop resentment or other negative sentiments toward the helper. (2) An important determinant of whether help is appreciated or resented is the extent to which it undermines the recipient's self-esteem. If helpers communicate the message that recipients are inferior in status and competence, the net result may be a reduction in recipients' self-esteem. Help can increase recipients' dependency and weaken their future self-reliance.

Key Terms

altruism (p. 238)
arousal/cost-reward model (p. 257)
bystander effect (p. 255)
bystander intervention (p. 254)
diffusion of responsibility (p. 257)
egoism (p. 238)
empathy (p. 239)
empathy-altruism model (p. 240)
evaluation apprehension (p. 256)
helping (p. 238)
norm of reciprocity (p. 245)
personal norms (p. 246)
prosocial behavior (p. 237)
social responsibility norm(p. 245)
sociobiology (p. 242)

CHAPTER 11
Aggression

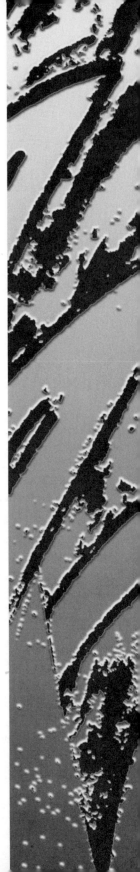

Introduction

■ On Sunday, November 26, 1995, Harvey Kaufman was at work in a subway token booth in New York City. Two men ran up to the booth, squirted a flammable liquid through the coin-change slot in the window of the booth, and set fire to the liquid. The booth exploded, and Kaufman suffered third-degree burns over 80% of his body.

■ Denise Farmer, a 40-year-old mother of two living in Chicago, got up and dressed for work. At 7 A.M., she left her apartment and walked down the stairs. According to police, one or more attackers were waiting at the foot of the stairs. The assailant(s) stabbed her more than 20 times; four of the thrusts penetrated her heart and killed her. Another resident of the building found Farmer, her pockets turned out and empty.

■ On Tuesday, December 7, 1993, Colin Ferguson, 35, boarded a Long Island Railroad commuter train in New York City. As the rush-hour train sped toward the Long Island suburbs, Ferguson stood up, walked down the aisle of the crowded car, and repeatedly fired a semiautomatic pistol at passengers. By the time 3 passengers wrestled him to the floor, he had killed 4 persons and wounded 19 others.

These incidents portray in stark relief a person's ability to inflict pain and death on other human beings. How can we account for such incidents and for the much more common and less extreme forms of aggression—harassment, abuse, assault—that occur several times each minute in the United States? These phenomena are the focus of this chapter.

What Is Aggression?

Defining aggression seems a simple enough task: Aggression is any behavior that hurts another. But this definition considers only the observable consequences of behavior, and ignores the actor's intentions. Hence it often leads to absurd conclusions. Under this definition, for instance, we would consider a surgeon an aggressor if a heart transplant patient died on the operating table despite heroic efforts to preserve the patient's life.

Because intentions are clearly important in defining an act as aggression (Krebs, 1982), we use the following definition: **Aggression** is any behavior intended to harm another person, which that person wants to avoid. According to this definition, a bungled assassination is an act of aggression; it involves intended harm that the target surely would wish to avoid. Heart surgery—approved by the patient and intended to improve his or her health—is clearly not aggression, even if the patient dies. Intended harm may be physical, psychological, or social (for example, harm to the target's reputation).

Drawing on research and theory, this chapter addresses the following questions:

1. What motivates people to aggress against others?
2. How do characteristics of the target influence aggression?
3. How do characteristics of the situation influence aggression?
4. How can we reduce the frequency of aggressive behavior?
5. What influences the incidence of interpersonal aggression—abuse, assault, sexual assault, and murder—in our society?

Aggression and the Motivation to Harm

As the examples in the introduction show, human beings have a remarkable capacity to harm others. Our first question concerns the motivation for human aggression: Why do people turn against others? There are at least four possible answers: (1) People are instinctively aggressive. (2) People become aggressive in response to events that are frustrating. (3) People aggress against others as a result of aversive emotion. (4) People learn to use aggression as an effective means of obtaining what they want. We consider each of these answers in turn.

Aggression as Instinct

The best known proponent of the theory that aggression is an instinct was Sigmund Freud (1930, 1950). In Freud's view, from the moment of conception we carry within us both an urge to create and an urge to destroy. The innate urge to destroy, or **death instinct,** is as natural as our need to breathe. This instinct constantly generates hostile impulses that demand release. We release these hostile impulses by aggressing against others, by turning violently against ourselves (suicide), or by suffering internal distress (physical or mental illness).

Many studies of animal behavior provide evidence that aggression is instinctive. According to Lorenz (1966, 1974), the aggressive instinct has evolved because it contributed to an animal's survival. Animals motivated to fight succeed better in protecting their territory, obtaining desirable mates, and defending their young. Through evolution, animals have also developed an instinct to inhibit their aggression once their opponents signal submission. Humans have no such instinct, however, so in this sense, humans are more dangerous and destructive than animals.

Instinct theories postulate that the urge to harm others is part of our genetic inheritance. As a result, proponents of such theories are pessimistic about the possibility of controlling human aggression. At best, they believe, aggression can be partly channeled into approved competitive activities such as athletics, academics, or business. Social rules that govern the expression of aggression are designed to prevent competition from degenerating into destructiveness. Quite often, however, socially approved competition stimulates aggression: Football players start throwing punches, soccer fans riot violently, and businesspeople destroy competitors through ruthless practices. If aggression is instinctive, we should not be surprised that it is always with us.

Despite the popularity of instinct theories of aggression, most social psychologists find them neither persuasive nor particularly useful. Generalizing findings about animal behavior to human behavior is hazardous. Moreover, cross-cultural studies suggest that human aggression lacks two characteristics typical of instinctive behavior in animals—universality and periodicity. The need to eat and breathe, for example, are universal to all members of a species. They are also periodic, for they rise after deprivation and fall when satisfied. Aggression, in contrast, is not universal in humans. It pervades some individuals and societies but is virtually absent in others. Moreover, human aggression is not periodic. The occurrence of human aggression is largely governed by specific social circumstances. Aggressive behavior does not increase when people have not aggressed for a long time or decrease after they have recently aggressed. Thus our biological makeup provides only the capacity for aggression, not an inevitable urge to aggress. We must look elsewhere to explain why particular people harm others in particular circumstances.

Frustration-Aggression Hypothesis

The second possible explanation of aggressive behavior is that aggression is an internal state elicited by certain events. The most famous view of aggression as an elicited drive is the **frustration-aggression hypothesis** (Dollard et al., 1939). This hypothesis makes two bold assertions. First, every frustration leads to some form of aggression. Second, every aggressive act is due to some prior frustration. In contrast to instinct theories, this hypothesis states that aggression is instigated by external environmental events.

In an early experiment (Barker, Dembo, & Lewin, 1941), researchers showed children a room full of attractive toys. They allowed some children to play with the toys immediately. They made other children wait about 20 minutes, looking at the toys, before they allowed them into the room. The children who waited behaved much more destructively when given a chance to play, smashing the toys on the floor and against the walls. Here, aggression is a direct response to **frustration,** that is, to the blocking of goal-directed activity. By blocking the children's access to the tempting toys, the researchers frustrated them. This in turn elicited an aggressive drive that the children expressed by destroying the researchers' toys.

Several decades of research have led to modifications of the original hypothesis (Berkowitz, 1978). First, studies have shown that frustration does not always produce aggressive responses (Zillman, 1979). Although motivated to behave aggressively, individuals may restrain themselves because of fear of punishment. Being laid off is a frustrating experience. Researchers predicted that small increases in layoffs would lead to violence by those laid off. Large increases, however, would lead to reduced violence because those still working are afraid of being laid off (Catalano, Novaco, & McConnell, 1997). Data from San Francisco supported the predictions. Also, frustration sometimes leads to different responses, such as despair, depression, or withdrawal. Second, research indicates that aggression can occur without prior frustration (Berkowitz, 1989). Even though competitors have not blocked his or her goal-directed activity, the ruthless businessperson or scientist may attempt to destroy rivals out of a desire for wealth and fame.

The frustration-aggression hypothesis implies that the nature of the frustration influences the intensity of the resulting aggression. Two factors in a situation that intensify aggression are the strength of frustration and the arbitrariness of frustration.

Strength of Frustration   The more we desire a goal and the closer we are to achieving it, the more frustrated and aroused we become if blocked. If someone cuts ahead of us as we reach the front of a long line, for example, our frustration will be especially strong. According to theory, this strong frustration should lead to aggressiveness.

A field experiment based on this idea demonstrated that stronger frustration elicits more aggression (Harris, 1974). Researchers directed a confederate to cut ahead of people in lines at theaters, restaurants, and grocery checkout counters. The confederate cut in front of either the 2nd or the 12th person in line. Observers recorded the reactions of the person. As predicted, people at the front of the line responded far more aggressively. They made more than twice as many abusive remarks to the intruder than people at the back of the line.

Arbitrariness of Frustration   People's perceptions of the reasons for frustration markedly influence the degree of hostility they feel. People are apt to feel more hostile when they believe the frustration is arbitrary, unprovoked, or illegitimate than when they

attribute it to a reasonable, accidental, or legitimate cause. As a result, arbitrary or illegitimate frustration elicits more aggression.

In a study demonstrating this principle, researchers asked students to make appeals for a charity over the telephone (Kulik & Brown, 1979). The students were frustrated by refusals from all the potential donors (in reality, confederates). In the legitimate frustration condition, potential donors offered good reasons for refusing (such as "I just lost my job"). In the illegitimate frustration condition, they offered weak, arbitrary reasons (such as "charities are a rip-off"). As shown in Figure 11.1, individuals exposed to illegitimate frustration were more aroused than those

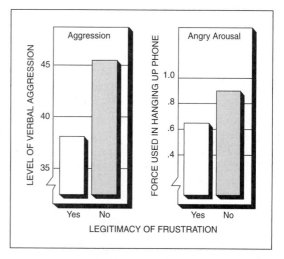

Figure 11.1   Effects of Legitimacy of Frustration on Aggressive Responses

Students were frustrated when potential donors whom they telephoned all refused their appeal for a charity. Half the students heard legitimate reasons for refusal, whereas the others heard illegitimate and arbitrary reasons. Those exposed to illegitimate frustration showed a higher level of angry arousal and aggression than those exposed to legitimate frustration. They slammed the receiver down harder when hanging up and expressed more verbal aggression toward potential donors during their conversations. These findings demonstrate that the perceived reasons for frustration influence angry arousal and aggression.

SOURCE: Adapted from Kulik and Brown, 1979.

exposed to legitimate frustration. They also directed more verbal aggression against the potential donors.

Aversive Emotional Arousal

In the six decades since the original statement of the frustration-aggression hypothesis, research has identified several other causes of aggression. In one study, community residents and university students were asked what events upset or angered them (Averill, 1982). Some replied that legitimate actions by others and unavoidable accidents triggered aggressive reactions. What makes you aggressive? Chances are that insults, especially those involving traits you value, perhaps your intelligence, honesty, ethnicity, or attractiveness, would be on your list. Physical pain also can produce aggression.

Direct attacks often provoke aggressive reactions. We may react sharply to the impatient beeping of another driver's horn. On occasion, drivers have shot and killed other drivers who honked at them. Verbal and physical attacks may arouse us and elicit an aggressive response.

Accidents, insults, and attacks all arouse **aversive affect,** negative affect that people seek to reduce or eliminate (Berkowitz, 1989). Often this affect is in the form of anger, but it can be pain or other types of

A young boy reacts aggressively to a physical attack by another. An attack typically produces anger, which in turn leads to an aggressive response.

discomfort. (For example, later in the chapter we will discuss the evidence that high temperatures and loud noise produce discomfort and aggression.) The resulting aggression is often instrumental, that is, intended to reduce or eliminate the cause of the affect. Turning on an air conditioner, slapping someone who insults you, or shooting an attacker are instrumental actions.

Aggression resulting from aversive affect is called *affective aggression,* in contrast to aggression due to hostile thought/cognition. In one experiment, participants either experienced extreme temperatures or viewed pictures of weapons (Anderson, Anderson, & Denser, 1996). The former increased anger and hostile attitudes; the latter did not.

Social Learning and Aggression

Social learning theories provide a fourth explanation for aggressive behavior. Two processes by which aggression can be learned are modeling and reinforcement.

Modeling   Some people, perhaps many, learn aggressive behavior by observing others commit aggressive acts and then imitating them. An experiment conducted by Bandura and his colleagues (1961) illustrates such learning. Children observed an adult playing with toys. In one condition, the man played with Tinkertoys for 1 minute. Then he played with a 5-foot-tall inflated rubber Bobo doll. He engaged in aggressive behavior toward the doll, including punching and kicking it and sitting on it. These actions, accompanied by shouted aggressive words and phrases, continued for 9 minutes. In the other condition, the man played with the Tinkertoys for the entire 10 minutes. Later, each child was intentionally frustrated. Then the child was left alone in a room with various toys, including a smaller Bobo doll. The children who had observed the aggressive model were much more aggressive toward the doll than those who had observed the nonaggressive model. They engaged in aggressive behavior such as kicking the doll and made comments similar to those they had observed.

Aggressive behavior within the family—child abuse, spouse abuse, or sibling abuse—can be explained by social learning theory. People who abuse their spouses or children often themselves grew up in families in which they either witnessed or were the targets of abuse (Gelles & Cornell, 1990). Growing up in a family in which some members abuse others teaches the child it is acceptable to engage in physical aggression. It also teaches that occupants of certain roles such as husbands or children are appropriate targets for aggression.

Reinforcement   Often people behave aggressively because they anticipate that the aggressive act will be rewarding to them. Muggers may attack a woman to take her money. One child knocks down another to obtain a desired toy. Students destroy library materials to improve their own chances—and worsen others' chances—of doing well on exams. These and other aggressive acts provide rewards to their perpetrators. According to social learning theory, the expectation of reward is a major motive for aggression (Bandura, 1973). Social learning theory holds that aggressive responses are acquired and maintained, like any other social behavior, through experiences of reward.

If the expectation of a reward motivates a person to aggress, which aggressive responses, if any, will he or she perform in a particular situation? The answer depends on two factors: the range of aggressive responses the person has acquired and the cost/reward consequences the person anticipates for performing these responses. A person may be skilled, for example, in using a switchblade knife, a Molotov cocktail, or a sarcastic comment to harm others. People also consider the likely consequences of enacting particular aggressive behaviors in a particular situation. They try to calculate which actions will produce the rewards they seek, and at what cost. These considerations largely determine which aggressive act(s), if any, people perform.

Characteristics of Targets That Affect Aggression

In the preceding section we discussed four potential sources of the motivation to aggress. Once aroused, such motives incline us toward aggressive behavior. Whether aggression occurs, however, depends on char-

acteristics of the **target,** the person toward whom the aggressive behavior is directed. In this section, we discuss three target characteristics: (1) gender and race, (2) attributions for an aggressor's attack, and (3) retaliatory capacity.

Gender and Race

Aggression does not occur at random. If it did, we would observe aggressive behavior by all kinds of people directed at targets of both genders, all ethnic groups, and all ages. In fact, aggression is patterned. First, aggressive behavior usually involves two people of the same race or ethnicity. This is obviously true of aggression within the family, whether it involves child, sibling, spouse, or elder abuse, because most families are ethnically homogeneous. It is also true of violent crime, that is, assault, sexual assault, and murder.

The relationship between aggression and gender depends on the type of aggressive behavior. In cases of abuse within the family, males and females are about equally likely to be the targets. Boys and girls are equally likely to be abused by a parent. Wives abuse their husbands as often as husbands abuse their wives (see Table 11.1) (Gelles & Cornell, 1990).

Violence outside the family, in contrast, involves primarily targets of one gender or the other. More than 95% of reported cases of rape or sexual assault involve a male offender and a female victim. Sexual assaults comprise about 6% of the violent crime reported in the United States. More than 80% of such crime involves **aggravated assault,** an attack by one person on another with the intent of causing bodily injury. These assaults overwhelmingly involve males as both offender and victim. Most murders also involve two males.

These patterns indicate that the display of aggression is channeled by social beliefs and norms. Observing violence within one's family teaches a child that violence directed at children or spouses is acceptable. Similarly, certain beliefs and norms in American society encourage men to direct sexual aggression toward women. Males in our society frequently compete with each other for various rewards, such as influence over each other, status in a group, the companionship of a woman, or other symbols of success. These

Table 11.1 Marital Violence by Husbands and Wives in 1985

Violent Behavior	Incidence Rate (percent reporting each act)		Frequency (number of incidents in 1985)			
			Mean		Median	
	Husband	Wife	H	W	H	W
1. Threw something at spouse	2.9	4.6	3.7	2.7	1.5	1.0
2. Pushed, grabbed, or shoved spouse	9.6	9.1	2.9	3.1	2.0	2.0
3. Slapped spouse	3.1	4.4	2.8	2.7	1.0	1.0
4. Kicked, bit, or hit with fist	1.5	2.5	3.9	2.9	1.5	1.0
5. Hit or tried to hit spouse with something	1.9	3.1	3.6	3.3	1.2	1.1
6. Beat up spouse	.8	.5	4.2	5.7	2.0	2.0
7. Choked spouse	.7	.4	1.9	2.9	1.0	1.0
8. Threatened spouse with knife or gun	.4	.6	4.3	2.0	1.8	1.1
9. Used a knife or gun	.2	.2	18.6	12.9	1.5	4.0
Overall violence (1–9)	21.3	12.4	5.4	6.1	1.5	2.5
Wife-beating/husband-beating (4–9)	3.4	4.8	5.2	5.4	1.5	1.5

SOURCE: Gelles and Cornell, 1990.

competitions often lead to insults that provoke anger or direct physical challenges. Norms in some groups require men to defend themselves in such situations. Observers have often described the American South as having norms that require men to defend themselves against insults. An experiment was designed to examine whether there is a "culture of honor" in the South. Men from either the North or the South were intentionally bumped by another man (a confederate of the experimenters), who then called him an "asshole." Men who grew up in the North generally did not react. Men raised in the South were more upset, more primed for aggression (as shown by elevated testosterone levels), and more likely to engage in aggressive behavior (Cohen et al., 1996). It is likely that the incident caused *aversive affect* in men from the South.

Attribution for Attack

Direct attacks, both verbal and physical, typically produce an aggressive reaction (Geen, 1968; White & Gruber, 1982). Nevertheless, we withhold retaliation when we perceive that an attack was not intended to harm us. We are unlikely to respond aggressively, for example, if we see that a man who has smashed his grocery cart into us was trying to save a child from falling. Aggression following an attack is both more probable and stronger when we attribute the attack to the actor's intentions rather than to accidental or legitimate external pressures (Dyck & Rule, 1978).

Research indicates that attributions for physical abuse by one's spouse play a key role in determining the victim's response. In one study of 70 women, those living with the violent partner sometimes blamed themselves for the abuse. They attributed it to their incompetence, unattractiveness, or talking back to the partner. Others blamed situational factors such as stress. Women who had left their abusive partner, in contrast, blamed him for the abuse (Andrews & Brewin, 1990).

An important influence on attributions is whether the attacker apologizes. An apology often states or implies that the harm done to us by another was unintentional. In one study, an experimenter made mistakes that caused the subject to fail at a task. When the experimenter apologized, subjects refrained from aggression against her. However, as the severity of the harm increased, the effect of the apology decreased (Ohbuchi, Kameda, & Agarie, 1989).

Retaliatory Capacity

Earlier, we identified four potential sources of the motivation to harm someone. Two involve emotions, frustration, and aversive affect. When we experience intense frustration or anger, we usually are not concerned with the possible consequences of aggressive behavior. When we experience less intense emotions or are motivated by the expectation of reward, we may assess the likely consequences of an aggressive act. A consequence of particular importance is retaliation by the target.

Research suggests that the threat of retaliation reduces aggressive behavior. In one experiment, male and female subjects were told to deliver electric shocks to another person; they could select the intensity of the shock. In one condition, researchers told the subjects that after they had delivered shocks, the experiment would be over. In another condition, researchers told the subjects that after they had delivered shocks, they would change places with the other person; that is, the other person could retaliate. Subjects in the latter condition delivered significantly less intense shocks than subjects in the former condition (Rogers, 1980). The finding that expected retaliation reduces aggressive behavior is consistent with social reinforcement theory; anticipated punishment leads to inhibition of a behavior.

Situational Impacts on Aggression

We turn now to characteristics of the situation in which the motive to aggress is aroused. We discuss five features: potential rewards (reinforcements), modeling, norms, stress, and aggressive cues.

Reinforcements

Three rewards that promote aggression are direct material benefits, social approval, and attention. The material benefits that armed robbers, mafiosi, and young bullies obtain by using violence support their aggression. If material benefits are reduced by vigorous law enforcement or by training the bullied children in karate, this aggressive violence will decline.

Despite the general condemnation of aggression, social approval is a second common reward for specific aggressive acts. Virtually every society has norms approving aggression against particular targets in particular circumstances. We honor soldiers for shooting the enemy in war. We praise children for defending their siblings in a fight. Most of us, on occasion, urge friends to respond aggressively to insults or exploitation.

Attention is a third type of reward for aggressive acts. The teenager who aggressively breaks school rules basks in the spotlight of attention from peers even as he is reproached by school authorities. If we ignored aggressive behavior and rewarded cooperation with attention and praise, would this reduce aggressive acts? This is, in fact, what researchers found in a study of aggression among 27 male nursery school children (Brown & Elliott, 1965). In this study, teachers selectively ignored aggressive acts and rewarded cooperation with attention and praise. After 2 weeks, the frequency of verbal and physical aggression declined substantially. Three weeks after these reinforcement schedules were discontinued, however, physical aggression rebounded while verbal aggression continued to decline. The researchers then asked the teachers again to ignore aggression and reward cooperation. After this second treatment, acts of both physical and verbal aggression declined dramatically. (Even two of the most violent boys in the class became friendly and cooperative.) These findings demonstrate that the careful use of rewarding attention and approval can have powerful effects on aggression. Results of this study are shown in Table 11.2.

Modeling

A second situational factor that promotes aggression is the presence of behavioral models. We noted earlier that aggressive behavior is often learned by observing and then imitating a model, as demonstrated by Bandura's research. In specific situations, a model's aggressive behavior may encourage others to behave in similar ways.

Table 11.2 Reducing Aggression in Children Through Selective Reinforcement

Timing of Observations of Children's Behavior	Average Number of Aggressive Acts Observed	
	Verbal	Physical
Before selective reinforcement treatment	22.8	41.2
After first 2-week treatment	17.4	26.0
After 3 weeks of no treatment	13.8	37.8
After second 2-week treatment	4.6	21.0

SOURCE: Adapted from Brown and Elliott, 1965.

In 1992, four Los Angeles police officers were on trial, charged with abusing a black suspect, Rodney King. On May 6, 1992, a California jury returned verdicts of not guilty on all but one of the charges. Within 2 hours, a group of black youth assembled at the intersection of Normandie and Florence avenues; some began throwing rocks and bottles at passing cars. The crowd grew; some participants stopped cars and trucks, pulled the drivers out of their vehicles, and beat them. The aggressive acts of these participants served as a model for others.

Aggressive models provide three types of information that influence observers. First, through their actions, models demonstrate specific aggressive acts that are possible in a situation. The idea of stopping a car, for example, never crossed the minds of most participants until they saw a model do it. The model's acts identified available opportunities for aggressive action.

Second, models provide information about the appropriateness of aggression, about whether it is normatively acceptable in the setting. The behavior of the initial participants in the Los Angeles riot signaled that violence was appropriate. The live television coverage of the riot provided by several local stations unwittingly transmitted this message to tens of thousands of other Angelenos. The result of this process is termed an *emergent norm* (see Chapter 20). Later, after the looting started, television announcers identified the exact locations, undoubtedly attracting additional looters. Research on the location and timing of riots by African Americans between 1961 and 1968 demonstrates that riots in one city tended to spread (diffuse) to other cities in response to mass media coverage (Myers, 1997). Furthermore, prior riots in one city are associated with subsequent riots in the same city (Olzak, Shanahan, & McEneaney, 1996).

Third, models provide information about the consequences of acting aggressively. Observers see whether the model succeeds in attaining goals, whether the behavior is punished or rewarded. Observers are more likely to imitate aggressive behaviors that yield reward and avoid punishment. Hesitation by the police in the first 4 hours of the riot gave participants the impression that smashing cars and beating motorists would go unpunished. Had the police immediately surrounded the initial rioters at Florence and Normandie, observers would probably not have joined in, although they might have erupted in another way.

A successful aggressive model conveys the message that aggression is acceptable in a particular situation. This message matters little when observers are not motivated to do harm. But people who feel provoked and are suppressing their inclination to aggress, like thousands in Los Angeles that afternoon, often lose their inhibitions after observing an aggressive model. They are the most likely to imitate aggression. The aggressive motivation experienced by those men and women is a consequence of decades of racial discrimination and years of harsh treatment of blacks by a white-dominated police force.

Five years later, in 1997, 85% of the 1,000 businesses damaged or destroyed in the riot had been rebuilt. But the underlying tensions remain, making future outbreaks likely (Purdum, 1997).

Norms

Just as there is a positive norm of reciprocity (see Chapter 10), there is a negative norm of reciprocity. This norm—"an eye for an eye, a tooth for a tooth"— justifies retaliation for attacks. In a national survey,

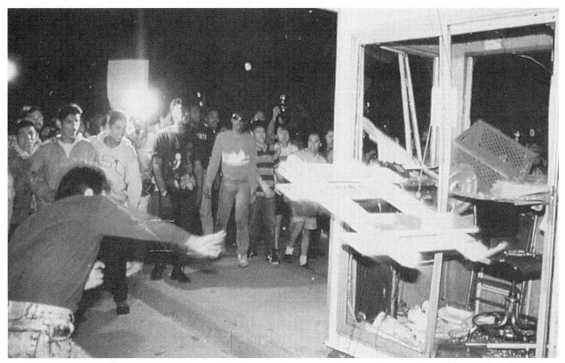

An angry crowd in Los Angeles watches as a young man destroys property following the acquittal of four police officers for beating Rodney King. Aggressive acts such as this can spread rapidly through modeling, especially when large numbers of observers feel frustrated.

more than 60% of American males considered it proper to respond to an attack on one's family, property, or self by killing the attacker (Blumenthal et al., 1972). Milder attacks call for milder retaliation.

The negative reciprocity norm requires the retaliation to be proportionate to the provocation. Numerous experiments indicate that people match the level of their retaliation to the level of attack (Taylor, 1967). In the heat of anger, however, we are likely to overestimate the strength of another's provocation and to underestimate the intensity of our own response. When angry, we are also more likely to misinterpret responses that have no aggressive intent as intentional provocation. Thus even when people strive to match retaliation to provocation, aggression may escalate.

A study of 444 assaults against police officers revealed that escalation of retaliation because of mutual misunderstanding was the most common factor leading to violence (Toch, 1969). Typically the police officer began with a routine request for information. The person confronted interpreted the officer's request as threatening, arbitrary, and unfair, and refused to comply. The officer interpreted this noncompliance as an attack on his or her own authority and reacted by declaring the suspect under arrest. Angered further by the officer's seemingly illegitimate assertion of power, the suspect retaliated with verbal insults and obscenities. From there the incident escalated quickly. The officer angrily grabbed the suspect, who retaliated by attacking physically. This sequence illustrates how a confrontation can spiral into violent aggression even when the angry participants feel they are merely matching their opponents' level of attack.

Stress

Stress increases the likelihood of aggressive behavior. Social stressors such as chronic unemployment and the experience of discrimination are related to aggression because of their effects on frustration and anger. In addition, situational stressors such as temperature and noise also can produce high levels of aggression.

Several studies have shown that temperature is related to the occurrence of violent crime. Analysis of data from 260 cities for the year 1980 found that the higher the average temperature, the higher the rate of assault, sexual assault, and murder (Anderson, 1987). A study of 102 incidents of collective violence or riots found that these events were much more likely to occur on days when the air temperature was between 71 and 90 degrees than when it was cooler (Baron & Ransberger, 1978). When it is hot, people are more irritable and thus more likely to respond to provocation with aggression. Interestingly, the increase in rates of violence peaks at temperatures around 75 degrees; higher temperatures lead to reduced violence, perhaps because people want to escape the heat and break off interaction with others (Cohn & Rotton, 1997).

Research also indicates that loud noise is associated with aggressive behavior. In one laboratory study, all the subjects were provoked and then put in a situation that allowed them to aggress against another person. Researchers exposed the subjects in one condition to uncomfortably loud noise; in the other condition, the noise was at everyday levels. Subjects in the first condition delivered twice as many shocks as subjects in the second condition (Konecni, 1975). Other research suggests that it may not be noise per se that causes aggression but stress that results from the inability to control the noise (Geen, 1978).

Aggressive Cues

Whether the motivation to harm actually leads to aggressive acts depends in part on the presence of suitable aggressive cues in the environment (Berkowitz, 1989). These cues may intensify the aggressive motivation or lower inhibitions.

Aggressive cues can intensify the arousal an angry person experiences. Seeing a gun or a violent film, for example, may further arouse a frustrated person and make him or her more aggressive. People who have been frustrated respond more aggressively when in the presence of a gun than in the presence of neutral objects (Carlson, Marcus-Newhall, & Miller, 1990); this so-called **weapons effect** occurs when people are already aroused. Thus, if family members have become angry, the presence of a handgun in the home creates the possibility of a weapons effect.

Aggressive cues may reduce the inhibitions that normally prevent aggressive behavior. As noted above, viewing aggressive behavior that is not punished increases the likelihood an individual will behave aggressively.

Reducing Aggressive Behavior

Aggressive behavior is often costly to individuals as well as to groups and the society to which they belong. For this reason, there is great interest in identifying ways to reduce the frequency of this behavior. Four methods that hold some promise are reducing frustrations, punishing aggressive behavior, providing nonaggressive models, and providing opportunities for catharsis.

Reducing Frustration

We noted that one cause of aggression is frustration, and the stronger the frustration the more likely aggression is to occur. Thus one way to reduce aggression is to reduce the frequency or strength of frustration. A major source of frustration in American society is inadequate resources. Many cases of robbery, assault, and murder are motivated simply by desire for money or property. Two studies compared the homicide rates of American cities and states (Land, McCall, & Cohen, 1990) and nations (Gartner, 1990) in an attempt to identify the determinants of homicide. Both studies found that economic deprivation was the best predictor. Frustration and, therefore, aggression could be reduced if everyone had access to necessities (for example, through guaranteed income and universal health insurance programs).

Many of the frustrations we experience arise from conflicts with other people. The resulting anger often leads to aggressive attacks, directed either at the target or at others we perceive as similar to the target.

Thus another way to reduce aggressive behavior is to provide people with alternative means of resolving interpersonal conflicts. Some people take their conflicts to lawyers; others call the police. Recent innovations in dispute resolution involve the increasing use of professionally trained mediators and the training of selected community members in conflict-resolution techniques.

Punishment to Suppress Aggression

Because punishment is widely used to control aggression, we might assume that punishment or the threat of punishment are effective deterrents. In fact, however, threats are effective in eliminating aggression only under certain narrowly defined conditions (Baron, 1977). For threats to inhibit aggression, the anticipated punishment must be great and the probability that it will occur very high. Even so, threatened punishment is largely ineffective when potential aggressors are extremely angry and when they believe they will gain by being aggressive.

For actual punishment (in contrast to mere threats) to control aggression, certain other conditions must be met (Baron, 1977): (1) The punishment must follow the aggressive act promptly; (2) it must be seen as the logical outcome of that act; and (3) it must not violate legitimate social norms. Unless these conditions are met, people perceive punishment as unjustified, and they respond with anger.

The criminal justice system often fails to meet many of these conditions. The probability that any single criminal act will be punished is low simply because most criminals are not caught. Even when criminals are caught, punishment rarely follows the crime promptly. Moreover, few criminals see the punishment as a logical or legitimate outcome of their act. Finally, they often have much to gain through their aggression. A longitudinal study of a sample of 1,497 adult offenders found that perception of risk of sanctions by the system was not related to criminal activity. The most significant predictor of crime was perception of opportunities to gain economically by breaking the law (Piliavin et al., 1986). As a result, the justice system is not very effective in deterring criminal aggression.

Nonaggressive Models

Just as aggressive models may increase aggression, nonaggressive models may reduce it. Mahatma Gandhi, who led the movement to free India of British colonialism, used pacifist tactics that have since been imitated by protestors around the world. Laboratory research has also demonstrated the restraining influence of nonaggressive models. In one study (Baron & Kepner, 1970), participants observed an aggressive model deliver many more shocks to a confederate than required by the task. Other participants observed a nonaggressive model who gave the minimum number of shocks required. A control group observed no model. Results showed that subjects who observed the nonaggressive model displayed less subsequent aggression than subjects in the control group. Subjects who observed the aggressive model displayed more subsequent aggression than the control group. Other research shows that nonaggressive models not only reduce aggression, but also can offset the influence of aggressive models (Baron, 1971).

Catharsis

Infuriated by a day of catering to the whims of her boss, Ruth turned on her teenage son as he drove her home. "Why must you drive like a maniac?" she snapped. Ralph was stunned. He was driving a sedate 25 miles per hour and had done nothing to provoke his mother's aggression.

Did Ruth feel better after venting her anger on Ralph? Many people believe that letting off steam is better than bottling up hostility. A very old psychological concept, catharsis, captures this idea (Aristotle, *Poetics*, Book 6). **Catharsis** is the reduction of aggressive arousal brought about by performing aggressive acts. The catharsis hypothesis states that we can purge ourselves of hostile emotions by intensely experiencing these emotions while performing aggression. A broader view of catharsis suggests that by observing aggression as an involved spectator to drama, television, or sports, we also release aggressive emotions.

Numerous studies support the basic catharsis hypothesis: Aggressive acts directed against the source of anger reduce physiological arousal (Geen & Quanty,

One way to reduce interpersonal aggression is to provide alternative methods of conflict resolution. In many communities, police officers are trained to use special techniques to resolve disputes, such as this one involving neighbors.

1977). We usually feel relieved after letting off steam against a tormentor. However, often we cannot direct aggression at the source. Even when we can, aggression may not bring catharsis because the fear of retaliation keeps us aroused. Aggression against a tormentor is also unlikely to reduce arousal under two other conditions: if we feel our aggression is inappropriate to the situation and will make us look foolish, or if our internalized values oppose the aggression and make us feel guilty about it. Thus catharsis results from aggression directed at the source only under limited conditions.

A second hypothesis that extends the original catharsis idea asserts that catharsis reduces subsequent aggression (Dollard et al., 1939; Freud, 1950). That is, once people act aggressively and release their anger, they are less inclined to future aggression. This hypothesis underlies advice such as "Put on the gloves and get the fight over with once and for all." Yet, with

few exceptions, research has shown that performing aggressive acts will increase future aggression, not reduce it. This is true whether the initial aggression is a verbal attack, a physical attack, or even aggressive play. One particularly telling study, detailed in Figure 11.2, shows a clear catharsis effect after retaliating against a tormentor, followed by increased aggression against that same tormentor (Geen, Stonner, & Shope, 1975).

Initial aggression promotes further aggression in several ways. First, initial aggressive acts produce **disinhibition,** the reduction of ordinary internal controls against socially disapproved behavior. Disinhibition is reflected in the reports of murderers as well as by soldiers who comment that killing was difficult the first time but became easier thereafter. Second, initial aggressive acts serve to arouse our anger even further. Third, they give us experience in harming others. Finally, the catharsis that follows aggression

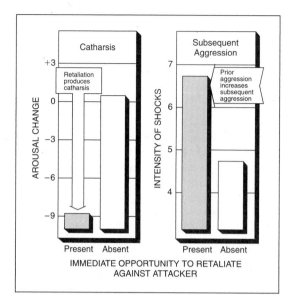

Figure 11.2 Aggression, Catharsis, and Subsequent Aggression

Participants in a learning experiment who were antagonized by a confederate became physically aroused. Some participants had an opportunity to retaliate immediately, whereas others did not. Those who retaliated experienced catharsis—a sharp drop in arousal indicated by a drop in blood pressure. Those who did not retaliate remained aroused (left). Later, both groups of participants had a chance to attack the confederate (right). Those who had retaliated earlier delivered more intense shocks to the confederate than those who had not retaliated earlier. Thus the group that experienced catharsis earlier subsequently behaved more aggressively. These findings contradict the idea that releasing pent-up anger through retaliation reduces future aggression against a target. Instead, as this study showed, catharsis increases subsequent aggression.

SOURCE: Adapted from Geen, Stonner, and Shope, 1975.

and the pleasure of thwarting our tormentor reward aggression, and rewarded behaviors are repeated more frequently.

Aggression in Society

In recent years, there has been increasing recognition that aggressive behavior is at the heart of several major social problems. This awareness is due, in part, to the widespread publicity given to certain incidents, such as the murderous assault by Colin Ferguson on homeward-bound commuters. But, fortunately, mass murders are rare. Much more common are other types of *interpersonal violence,* in which one person directs physical aggression toward another with the intent to injure or kill the target. This section discusses three aspects of interpersonal violence. First, it looks at the causes and consequences of sexual assault. Next, it examines the question of whether pornography contributes to sexual assault. Finally, it discusses whether television programming contributes to violence.

Sexual Assault

In this section, we consider three aspects of sexual assault. These are characteristics of society that encourage assault, characteristics of those who commit assault, and characteristics of the victims.

Sexual assault refers to sexual contact or intercourse without consent, accomplished by threat or use of force. The greater the force used or the resulting injury, the more severe the assault. Most cases involve a male offender and a female victim from the same racial or ethnic group. In some cases, the offender is motivated by sexual desire. In other cases, however, the offender's intent is to dominate, humiliate, or injure the victim.

Rape-Prone Society What causes sexual aggression? There are several answers to this question. One is that a specific set of cultural beliefs and practices creates conditions that encourage rape. In a *rape-prone society,* the sexual assault of women by men is allowed or overlooked (Sanday, 1981). Such societies share several characteristics. First, there are high levels of interpersonal violence. Second, women are viewed as property and generally are subservient to men. Third, men and women are regularly separated, such as during religious rituals. The United States is a rape-prone society. Rates of violent crime are high. Until recently, men dominated women politically, economically, and sexually. Also, there is continuing separation of men and women, for example,

11.1 Rape Myths

Among the causes of sexual aggression are cultural beliefs that encourage rape. These beliefs are **rape myths,** prejudicial, stereo-typed, and false beliefs about rape, rape victims, and persons who commit rape (Burt, 1980). Examples of these myths are "Only bad girls get raped," "Women ask for it," and "Rapists are sex starved, insane, or both." These beliefs create a climate that both encourages sexual assault and is suspicious of and hostile toward victims.

An attitude scale widely used to assess these beliefs is the Rape Myth Acceptance Scale, developed by Burt (1980). Some of the items that make up the scale are reproduced here.

Read each statement and circle the appropriate response: Strongly Agree (StA), Agree (A), Slightly Agree (SlA), Don't Know (?), Slightly Disagree (SlD), Disagree (D), or Strongly Disagree (StD).

1. A woman who goes to the home or apartment of a man on their first date implies she is ready to have sex.
   StA   A   SlA   ?   SlD   D   StD
2. Any healthy woman can successfully resist a rape if she wants to.
   StA   A   SlA   ?   SlD   D   StD
3. When women go around braless or wearing short skirts and tight tops, they are just asking for trouble.
   StA   A   SlA   ?   SlD   D   StD
4. If a girl engages in necking or petting and she lets things get out of hand, it is her own fault if her partner forces sex on her.
   StA   A   SlA   ?   SlD   D   StD
5. If a woman gets drunk at a party and has intercourse with a man she's just met there, she should be considered "fair game" to other males at the party.
   StA   A   SlA   ?   SlD   D   StD

Answer the following questions by selecting the best answer.

6. What percentage of women who report a rape would you say are lying because they are angry and want to get back at the man they accuse?
   Almost all   About 75%   About 50%   About 25%   Almost none

A person comes to you and claims they were raped. How likely would you be to believe their statement if the person were:

7. Your best friend?
   Always   Frequently   Sometimes   Rarely   Never
8. A neighborhood woman?
   Always   Frequently   Sometimes   Rarely   Never
9. A black woman?
   Always   Frequently   Sometimes   Rarely   Never
10. A white woman?
    Always   Frequently   Sometimes   Rarely   Never

As you can see, the scale is made up of two types of items. One type, items 1 through 5, asks the respondent whether rape is justified in particular situations. Agreement with an item indicates acceptance of that rape myth. The other type, items 6 through 10, asks the respondent whether he or she believes the claims of women who have been raped. For item 6, the belief that women who report rape are lying is considered a rape myth. For items 7 through 10, the belief that particular types of women (for example, blacks) lie about rape is a rape myth.

A high score on this scale indicates that the respondent believes most sexual activity is a direct result of the woman's behavior and not due to threat or force. A male respondent with a high score believes that if he has sexual intercourse with a woman who comes home with him on the first date, it is not rape even if she offers some resistance. This is one of the dangers of these myths: They provide scripts that legitimize sexual activity to which the woman may not have overtly consented. The other type of rape myth—that claims of rape are not true—creates an environment in which such claims are not believed, and therefore, sexual assault is not punished.

There have been many studies of the correlates of endorsing rape myths. One review summarizes the findings of 72 studies (Anderson, Cooper, & Okamura, 1997). Men, older persons, and persons from lower socioeconomic class backgrounds are more likely to hold such attitudes. Acceptance of rape was associated with traditional beliefs about gender roles, an adversarial view of male-female relationships, and conservative political beliefs. These results are consistent with the theory that rape myth acceptance is the result of socialization to sex-typed, conservative beliefs.

in athletic programs and in many workplaces (see Chapter 18).

Perpetrators of Sexual Assault Of course, individuals rather than societies commit rape. A second approach to determining what causes sexual assault is to identify characteristics of men that may be related to the behavior. Sexual assault is one type of aggressive behavior; others include physical and verbal aggression. Research suggests that some men are *sexually aggressive,* that is, they rely on aggressive behaviors in their relationships with women (Malamuth, Heavey, & Linz, 1993). These men attain high scores on measures of the desire to dominate and of hostility toward women. They have a variety of attitudes that facilitate aggression toward women (Malamuth, 1984). These attitudes include various *rape myths,* such as women secretly desire to be raped and enjoy it, victims cause rape, and other men are prone to rape (see Box 11.1, "Rape Myths"). These men are sexually aroused by portrayals of rape. In laboratory studies, such men are more likely to aggress against a woman who has mildly insulted or rejected them (Malamuth, 1983).

The tendency of men to be sexually aggressive is stable over time. Researchers collected data on 423 young men, including measures of hostility toward women and attitudes supportive of violence. Ten years later, they reinterviewed 176 of the men and 91 of their female partners (Malamuth et al., 1995). The characteristics measured 10 years earlier predicted which men were sexually aggressive toward their partners, as reported both by the men and by their partners.

Other research focuses specifically on the tendency of a man to use aggression to obtain sexual gratification. This tendency is measured by asking men whether they have ever used verbal threats or physical force to obtain sexual activity with women (Koss & Oros, 1982). Men who report the use of threat or force to obtain intercourse are termed *sexually assaultative.* Men who report the use of threat or force to obtain other types of intimacy (for example, breast fondling) are intermediate in the use of sexual aggression (see Table 11.3). Koss and Leonard (1984) report that a high level of sexually aggressive behavior is associated with the acceptance of rape myths and with the belief that sexual aggression is normal. A study of motivational factors found that men who had been sexually assaultative were angry toward women and had feelings of low power or inadequacy (Lisak & Roth, 1988).

Table 11.3 A Typology of Sexually Assaultative Men

Type	Behavior(s)	Percent
Sexually nonaggressive	Consensual sexual activity	59.0%
Sexually coercive	Accomplished intercourse by: a) threatening to end the relationship; or b) after "continual discussion and argument"; or c) saying things you didn't mean.	22.4%
Sexually abusive	Accomplished kissing or petting by use of physical force; or Threatened or used force in an unsuccessful attempt to obtain intercourse.	4.9%
Sexually assaultative	Accomplished vaginal, oral, or anal intercourse by threat or use of force.	4.3%

Sample: 1,846 male college students

SOURCE: Adapted from Koss and Leonard, 1984.

This research suggests that men who commit sexual assault have learned a script for heterosexual interactions that includes the use of force (Huesmann, 1986). This script suggests the use of verbal abuse or physical force to exercise influence over or obtain sexual gratification from a woman. Once learned, it is used to regulate behavior in various situations. Research suggests that it is learned in childhood. It is most likely to be learned when the child observes aggression frequently, is reinforced for aggressive behavior, and is the object of aggression.

Victims of Sexual Assault   The victims of sexual assault are often women between the ages of 15 and 24. Some women, probably a minority of all victims, are assaulted by men they do not know. Assaults by strangers often occur outdoors, in parks or deserted parking lots, or in the victim's residence. The offenders in these cases are often opportunistic, attacking any woman who is available or appears to be vulnerable.

Many women are assaulted by someone they know. It may be a man they are dating—date rape—or a neighbor or co-worker—acquaintance rape. The victims in most cases of date rape are young single women, often high school or college students. The couple goes out on a date. After dinner or a movie or dancing, they drive to a secluded spot or to his or her residence. The man initiates physical intimacy, perhaps with the woman's cooperation. At some point, the woman indicates she does not wish to continue, but the man persists and uses force. One contributing factor is alcohol, which is present in up to one half of all sexual assaults (Abbey et al., 1996); it lowers inhibitions that might otherwise prevent aggression. Another factor may be misinterpreted verbal or nonverbal messages. The woman may engage in some behavior that the male incorrectly interprets as a sexual invitation. Cultural beliefs are a third factor. According to one survey of 14- to 17-year-olds, both males and females believe a man is justified in forcing a woman to have intercourse if she gets him sexually excited, leads him on, or has dated him for a long time (Goodchilds & Zellman, 1984).

Many women who experience forced, nonconsensual sexual activity do not perceive the experience as rape (Kahn, Mathie, & Torgler, 1994). This

A rape victim is questioned by a female police officer. Many women do not report sexual assaults because they are afraid that rape myths will lead others to stereotype them (for example, as "bad girls").

may be because their experience—being assaulted by someone they know during a date after some sexual foreplay—does not match their *script* for rape—a violent attack by a stranger. Researchers asked women to write a description "of events before, during, and after a rape." Women who reported they had been forced to have sex but replied "No" to the question "Have you been raped?" were more likely to describe rape as an attack by a stranger than women who answered "Yes." This is a good example of the power of scripts to shape experience.

Pornography and Violence

Earlier, we suggested that some men learn scripts that encourage sexual aggression. One source of such scripts is growing up in an abusive family. Another is viewing or reading pornography.

On January 24, 1989, Florida prison officials executed convicted serial murderer Ted Bundy. Before his execution, Bundy admitted killing more than 24 young women. Authorities believe he killed many others. Bundy claimed he was addicted to violent pornography and that it had contributed to the murders he committed. Such claims create great interest in the connection between pornography and violence. Considerable research has been conducted on this link.

Nonaggressive Pornography

Various studies have shown that the effect of pornography on behavior depends on what the pornography portrays. Pornography that explicitly depicts adults engaging in consenting sexual activity is termed **nonaggressive pornography,** or *erotica.* Reading or viewing nonaggressive pornography creates sexual arousal (Byrne & Kelley, 1984), usually through the mechanism of cognitive and imaginative processing. Nonaggressive pornography by itself does not produce aggression toward women (Donnerstein, 1984). However, when the viewer's inhibitions are lowered, it may do so. Research has also demonstrated that when anger or frustration is induced in male subjects and they then view nonaggressive pornographic images, aggressive behavior toward a female may result (Donnerstein & Barrett, 1978). The mechanism is thought to be *transfer of arousal.* The sexual arousal that results from viewing pornography is added to the arousal induced by the anger and results in aggression.

Aggressive Pornography

The term **aggressive pornography** refers to explicit depictions of sexual activity in which force is threatened or used to coerce a woman to engage in sex (Malamuth, 1984). Unlike erotica, aggressive pornography has lasting effects on both attitudes and behavior.

In a study of the effects on attitudes (Donnerstein, 1984), male subjects viewed one of three films featuring aggression, nonaggressive sexual activity, or aggressive sexual activity. Following the film, the subjects completed several attitude scales, including one that measures acceptance of rape myths (see Box 11.1). Men who saw the films depicting aggression or aggressive sexual activity scored higher on the rape myth acceptance scale than men who saw the film depicting nonaggressive sexual activity. These men also indicated greater willingness to use force to obtain sex. The fact that both films depicting aggression affected attitudes more than the nonaggressive film suggests it is aggression rather than explicit portrayals of sex that influences attitudes. In another study, Malamuth and Check (1981) found that viewers of films which portray sexual aggression as having positive consequences showed greater acceptance of rape myths and of violence toward women. Finally, repeated exposure to sexually violent films is associated with more calloused attitudes toward female victims of domestic abuse (Mullin & Linz, 1995).

The mechanism by which viewing these films influences attitudes may be priming (Malamuth, 1984). *Priming* refers to the process whereby exposure to a stimulus makes psychologically related cognitions more accessible. Exposure to aggressive pornography makes cognitions related to sex and aggression more accessible, and these cognitions influence later judgments. The effect of aggressive pornography on attitudes is indirect (Malamuth & Briere, 1986). For most people, viewing such portrayals does not produce sexual arousal. (In fact, it may produce the aversive emotion of disgust.) It does, however, affect thought patterns in ways that make many viewers more tolerant of sexual aggression. If other influences occur, such as peer encouragement or hostility toward a woman, men with these attitudes may commit sexual assault (Malamuth, 1989).

Exposure to aggressive pornography also influences behavior, especially aggression toward females. This effect is evident in research by Donnerstein and Berkowitz (1981). Male subjects were either angered or treated neutrally by a male or female confederate. They next viewed one of four films: a neutral film, a nonaggressive pornographic film, or one of two aggressive pornographic films. In the latter films, a young woman is shoved around, tied up, stripped, and raped. In one version, she finds the experience disgusting; in the other she is smiling at the end. Following the film, the men were given an opportunity to aggress against the male or female confederate by delivering electric shocks. The films did not affect aggression toward the male. Subjects who saw the aggressive films delivered more intense electric shocks to the female confederate.

The fact that aggressive pornography produces aggressive behavior reflects three influences: sexual arousal, aggressive cues, and reduced inhibitions. Some men experience high levels of arousal in response to such portrayals. In addition, such pornography portrays women as targets of aggression. In the experiment conducted by Donnerstein and Berkowitz, the film created an association in the viewer's mind between the victim in the film and the female who angered him, suggesting aggression toward the latter. Note that aggressive films led to increased violence toward the female confederate and not the male, a finding consistent with this interpretation. Finally, these films also may reduce inhibitions to aggression, by suggesting that aggression directed toward females has positive outcomes.

One important question is whether we can generalize from the results of laboratory research to natural settings. Does the viewing of aggressive pornography in nonlaboratory settings contribute to violence against women? Some anecdotal evidence comes from sex offenders and therapists who work with them. These persons describe many actual sexual offenses that closely mimic activities depicted in pornographic materials available to the offender before the offense (Court, 1984). Another form of evidence comes from researchers who have studied the correlation between the availability of pornography and rates of violent crime. For instance, Baron and Straus (1984) obtained circulation figures for eight "sex magazines" (including *Playboy, Chic,* and *Hustler*) and the rates of assault, rape, and murder reported to police, for each state. Of several variables included in their study, the circulation index (number of magazines sold per 100,000 residents) was the strongest predictor of rates of rape.

Violent Television and Aggression

Evelyn Wagler was carrying a 2-gallon can of gasoline back to her stalled car. She was cornered by six young men who forced her to douse herself with the fuel. Then one of the men tossed a lighted match. She burned to death. Two nights earlier, a similar murder had been depicted on national television. The incident described at the beginning of this chapter, in which two men set a subway token booth on fire, mimicked a scene from the film *Money Train.*

Violence pervades television: stabbings, shootings, poisonings, and beatings. Not just humans, but even cartoon characters torment each other in astonishingly creative ways. Both heroes and villains perform aggression on television. There are five or six violent incidents per hour during prime time, and 20 to 25 incidents per hour during Saturday morning programs for children (American Psychological Association, 1991).

Do these depictions of violence encourage viewers to behave aggressively? Is our society more violent because of violence shown by the media? These are vexing questions, especially when the average American child spends more time watching television than engaging in any other waking activity (Huston & Wright, 1982). By age 18, the average American child is likely to have seen about 200,000 violent acts on television, including 40,000 homicides ("Violence in Our Culture," 1991). The two incidents cited above suggest that in some cases, media violence induces aggression. Yet establishing a causal link between watching television and subsequent aggression has been difficult and highly controversial (Milavsky et al., 1983).

Five processes explain why exposure to media violence might increase aggressive behavior (Huesmann & Moise, 1996): (1) *Imitation*—viewers learn specific techniques of aggression from media models. Social learning through imitation evidently played a part in both of the violent attacks cited above. Such incidents suggest that television portrayals of violence provide scripts that viewers may subsequently enact (Huesmann & Eron, 1989). (2) *Cognitive priming*—portrayals of violence activate aggressive thoughts, and proaggression attitudes. As noted in Chapter 6, activation of an attitude increases the likelihood it will be expressed in behavior. (3) *Legitimation/justification*—exposure to violence that successfully attains goals and has positive outcomes (e.g., punishes wrongdoers) legitimizes aggression and makes it more acceptable. (4) *Desensitization*—after observing violence repeatedly, viewers become less sensitive to aggression. This makes them less reluctant to hurt others and less inclined to ease others' suffering. (5) *Arousal*—viewing violence on television produces excitement and physiological arousal, which may amplify aggressive responses in situations that would otherwise elicit milder anger. Research has shown that all of these processes operate in linking media violence to aggression (Murray & Kippax, 1979).

By age 18, American children will see 200,000 violent acts on television. Such exposure directly contributes to aggressive behavior by the viewer.

To investigate these processes, researchers have conducted hundreds of laboratory and field experiments. Children, adolescents, and adults have been shown portrayals of aggression, both live and on film. Some of these scenes have been specially prepared, whereas others have been taken directly from popular television shows. With rare exceptions, these experiments show that observing violence increases subsequent aggression by viewers (Comstock, 1984; Friedrich-Cofer & Huston, 1986). Moreover, these results have been found in experiments using boys and girls of all ages, races, social classes, and levels of intelligence, in many countries (Huesmann & Moise, 1996).

These experiments demonstrate a causal link between viewing violence and aggression. However, our ability to generalize from these experiments to the effects of media violence on society has been challenged. For one thing, few of us watch several violent scenes within a brief period of time. For another, few of us watch movies or television programs in research settings. For these reasons, researchers have turned to studies of the correlation between exposure to violence on television and aggressive behavior in everyday settings, to complement the experimental approach.

Many studies have been carried out with thousands of subjects of widely varying age, social class, and ethnicity. They consistently report moderate positive correlations between television viewing and aggressive behavior (Friedrich-Cofer & Huston, 1986). In one correlational study (Singer & Singer, 1981), researchers measured the television viewing experiences and spontaneous aggression of preschoolers several times over a period of 1 year. Consistently, aggressive behavior during this period was linked to heavy viewing of aggressive action-adventure programs or cartoons. The level of aggressiveness these children exhibited 4 to 5 years later also correlated positively with the amount of violent television they had viewed as preschoolers. Another study (Eron, 1982; Huesmann et al., 1984) examined children's preferences for violent television and their levels of aggression in third grade and again 10 and 21 years later. This study revealed that viewing violent television in third grade correlated positively with later aggression.

The correlations in these studies may reflect a causal impact of viewing violence on subsequent aggression, but correlations do not prove causality. The correlation also might suggest that aggressive children prefer violent programs. The growing body of evidence suggests that the relationship between aggression and television viewing is circular (Friedrich-Cofer & Huston, 1986). Because aggressive children are relatively unpopular with their peers, they spend more time watching television. This exposes them to more violence, teaches them aggressive scripts and behaviors, and reassures them that their behavior is appropriate. When they then try to enact these scripts in interaction with others, they become even more unpopular and are driven back to television—and the vicious cycle continues (Huesmann, 1986; Singer & Singer, 1983). In a related study, researchers measured the aggressiveness of 210 men and 210 women, and then gave them descriptions of several films (Bushman, 1995). Men and women with high aggressiveness scores were more likely to chose a violent film than persons who scored low on aggression.

Based on the research, most observers agree that viewing media violence causes later aggression. Although no single study proves this connection beyond a doubt, the combined evidence from hundreds of studies is overwhelming (Huesmann & Moise, 1996). The inevitable question is this: What should be done

about violence on television? Comprehensive rating systems and V-chips that enable parents to exercise some control over what children watch are important steps. The television industry could make several changes as well (Kunkel, 1997). First, the number of violent incidents portrayed on both prime-time and Saturday morning programs for children should be reduced. Second, more portrayals of violent acts should include their negative consequences, such as pain and suffering for victims, family members, and the community. Third, programs should feature alternatives to the use of violence in resolving interpersonal conflicts, such as problem-focused discussion and mediation. Finally, when violence is portrayed, antiviolence themes should be included; for example, perpetrators could be shown as remorseful about the need to resort to violence. Changes such as these will not eliminate violence in interpersonal relationships. But they will help a great deal.

Summary

Aggression is behavior intended to harm another person, which that person wants to avoid.

Aggression and the Motivation to Harm

There are four main theories regarding motivation for aggression. (1) Aggression is based on a biological instinct that generates hostile impulses demanding release. (2) Aggression is a drive elicited by frustration. (3) Aggressive behavior is a response to aversive emotional arousal, such as anger. (4) Aggression is a learned behavior motivated by rewards.

Characteristics of Targets That Affect Aggression

Once aggressive motivation has been aroused, target characteristics influence whether aggressive behavior occurs. (1) Aggression is more likely if the target is of the same race or ethnicity. The target's gender also influences the response. (2) When we are attacked, our response is influenced by the attributions we make about the attacker's intentions. (3) We are less likely to engage in aggression toward a target that we perceive as capable of retaliating.

Situational Impacts on Aggression

Situational conditions are important influences on aggressive behavior. (1) Rewards that encourage aggression include material benefits, social approval, and attention. (2) Aggressive models provide information about available options, normative appropriateness, and consequences of aggressive acts. (3) The negative reciprocity norm encourages aggressive behavior in certain situations. (4) Aggressive behavior is more likely when stressors, such as high temperatures or loud noise, are present. (5) Aggressive behavior is more likely in the presence of aggressive cues, especially weapons.

Reducing Aggressive Behavior

(1) Reducing frustration levels by guaranteeing everyone the basic necessities would reduce aggression motivated by desire for these rewards. (2) Punishment is effective in controlling aggression only when it follows the aggressive act promptly, is seen as the logical outcome of that act, and does not violate social norms. (3) Nonaggressive models reduce the likelihood of aggression and can offset the effect of aggressive models. (4) Catharsis—the reduction of angry arousal—may follow aggressive acts, but such acts also may promote later aggression.

Aggression in Society

Interpersonal violence is a serious problem in American society. (1) Sexual assault is facilitated by societal characteristics, such as male domination of females, and by scripts that encourage male aggression toward women. (2) Nonaggressive pornography has little effect on attitudes or behavior. Exposing men to aggressive pornography in laboratory settings is associated with greater acceptance of rape myths and aggression toward women. (3) Experimental research shows that observing violence on film increases subsequent aggression. Correlational studies find a positive relationship between viewing violent programs and aggressive behavior in everyday settings. There are several changes that the television industry could make that would reduce aggressive behavior.

Key Terms

aggravated assault (p. 267)
aggression (p. 263)
aggressive pornography (p. 279)
aversive affect (p. 266)
catharsis (p. 273)
death instinct (p. 264)
disinhibition (p. 274)
frustration (p. 264)
frustration-aggression hypothesis (p. 264)
nonaggressive pornography (p. 279)
rape myths (p. 276)
sexual assault (p. 275)
target (p. 267)
weapons effect (p. 272)

CHAPTER 12

Interpersonal Attraction and Relationships

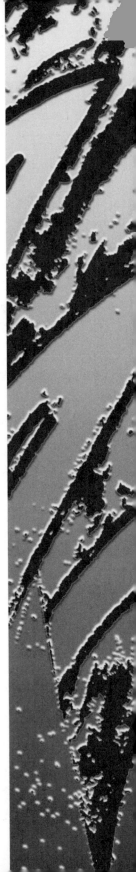

Introduction

Dan was looking forward to the new semester. Now that he was a junior, he would be taking more interesting classes. He walked into the lecture hall and found a seat halfway down the aisle. As he looked toward the front, he noticed a very pretty young woman removing her coat; as he watched, she sat down in the front row.

Dan noticed her at every class; she always sat in the same seat. One morning he passed up his usual spot and sat down next to her.

"Hi," he said. "You must like this class. You never miss it."

"I do, but it sure is a lot of work."

As they talked, they discovered they were from the same city and both were economics majors. When the professor announced the first exam, Dan asked Sally if she wanted to study for it with him. They worked together for several hours the night before the exam, along with Sally's roommate. Dan and Sally did very well on the test.

The next week he took her to a film at a campus theater. The week after she asked him to a party at her dormitory. That night, as they were walking back to her room, Sally told Dan that her roommate's parents had just separated and her roommate was severely depressed. Dan replied that he knew how she felt because his older brother had just left his wife. Because it was late, they agreed to meet the next morning for breakfast. They spent all day Sunday talking, about love, marriage, parents, and their hopes for the future. By the end of the semester, Sally and Dan were seeing each other two or three times a week.

At its outset, the relationship between Dan and Sally was based on **interpersonal attraction,** a positive attitude held by one person toward another person. Over time, however, the development of their relationship involved increasing interdependence and increasing intimacy, or pair relatedness (see Figure 12.1).

The development and outcome of personal relationships involves several stages. This chapter discusses each of these stages. Specifically, it considers the following questions:

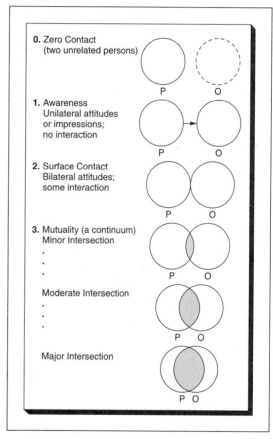

Figure 12.1 Levels of Pair Relatedness

A relationship between two persons develops through a series of stages, or levels. Before it begins, the two are unrelated. The next level occurs when one person becomes aware of the other. The third level involves interaction. Relationships that develop beyond the surface-contact stage involve progressively greater mutuality—disclosure of personal information, trust, and interdependence.

SOURCE: Adapted from Levinger, 1974.

1. Who is available? What determines whom we come into contact with?
2. Who is desirable? Of those available, what determines whom we attempt to establish relationships with?
3. What are the determinants of attraction or liking?
4. How do friendship and love develop between two people?
5. What is love?
6. What determines whether love thrives or dies?

Who Is Available?

Dozens or thousands of persons may go to school or live or work where you do. Most of them remain strangers, persons with whom you have no contact. Those persons with whom we come into contact, no matter how fleeting, constitute the field of **availables,** the pool of potential friends and lovers (Kerckhoff, 1974). What determines who is available? Is it mere chance that George rather than Bill is your roommate or that Dan met Sally rather than Heather? The answer, of course, is no.

Two basic influences determine who is available. First, institutional structures influence our personal encounters. The admissions office of your school, faculty committees that decide degree requirements, and the scheduling office all influence whether Dan and Sally enroll in the same class. Second, individuals' personal characteristics influence their choice of activities. Dan chose to take the economics class where he met Sally because of an interest in that field and the desire to go to graduate school in business. Thus institutional and personal characteristics together determine who is available.

Given a set of persons who are available, how do we make contact with one or two of these persons? Three influences progressively narrow our choices: routine activities, proximity, and familiarity.

Routine Activities

Much of our life consists of a routine of activities that we repeat daily or weekly. We attend the same classes and sit in the same seats, eat in the same places at the same tables, shop in the same stores, ride the same bus, and work with the same people. These activities provide opportunities to interact with some availables but not with others. More importantly, the activity provides a focus for our initial interactions. We rarely establish a relationship by saying "Let's be friends" at a first meeting. To do so is risky, because the other person may decide to exploit us. Or that person may reject such an opening, which may damage our self-esteem. Instead, we begin by talking about something shared—a class, an ethnic background, a school, or the weather.

Most relationships begin in the context of routine activities. A study of college students found that relationships began with a meeting in a class, a dorm, or at work (36%); with an introduction by a third person (38%); or at parties (18%) or bars (14%). A study of 3,342 adults ages 18 to 59 asked how respondents met sexual partners (Laumann et al., 1994). One third reported that they were introduced by a friend, and another third said they were introduced by family members or co-workers. Thus social networks play an important role in the development of relationships. Studies of the friendship patterns of city dwellers have found that friends are selected from relatives, co-workers, and neighbors (Fischer, 1984). Thus routine activities and social networks are important influences on the development of relationships.

Proximity

Although routines bring us into the same classroom, dining hall, or workplace, we are not equally likely to meet every person who is present. Rather, we are more likely to develop a relationship with someone who is in close physical proximity to us.

In classroom settings, seating patterns are an important influence on the development of friendships. One study (Byrne, 1961a) varied the seating arrangements for three classes of about 25 students each. In one class, they remained in the same seats for the entire semester (14 weeks). In the second class, they were assigned new seats halfway through the semester. In the third class, they were assigned new seats every 3.5 weeks. The relationships between students were assessed at the beginning and end of the semester. Few

When we think about where people meet available partners, we often picture the singles bar. However, one study of heterosexual relationships found that relatively few people met their partners at a bar. Much more common were meetings in classes, dorms, or workplaces.

relationships developed among the students in the class where seats were changed every 3.5 weeks. In the other two classes, students in neighboring seats became acquainted in greater numbers than students in nonneighboring seats. Moreover, the relationships were closer in the class where seat assignments were not changed.

Similar positive associations between physical proximity and friendship have been found in a variety of natural settings, including dormitories (Priest & Sawyer, 1967), married student housing projects (Festinger, Schachter, & Back, 1950), and business offices (Schutte & Light, 1978).

We are more likely to develop friendships with persons in close proximity because such relationships provide interpersonal rewards at the least cost. First, interaction is easier with those who are close by. It costs less in time and energy to interact with the person sitting next to you than with someone on the other side of the room. A second factor is the influence of social norms. In situations where people are physically close or interact frequently—such as in dormitories, classes, and offices—we are expected to behave in a polite, friendly way. Polite, friendly behavior provides increased rewards in these interactions. Failure to adhere to these norms might result in disapproval from others, which increases costs.

Familiarity

As time passes, people who take the same classes, live in the same apartment building, or do their laundry in the same place become familiar with each other. Having seen a person several times, sooner or later we will smile or nod. Repeated exposure to the same stimulus is sufficient to produce a positive attitude toward it; this is called the **mere exposure effect** (Zajonc, 1968). In other words, familiarity breeds liking, not contempt. This effect is highly general and has been demonstrated for a wide variety of stimuli—such as music, visual art, and comic strips—under many different conditions (Harrison, 1977).

Does mere exposure produce attraction? The answer appears to be yes. In one experiment, female undergraduates were asked to participate in an experiment on their sense of taste. They entered a series of booths in pairs and rated the taste of various liquids. The schedule was set up so that two subjects shared the same booth either once, twice, five times, ten times, or not at all. At the end of the experiment, each woman rated how much she liked each of the other subjects. As predicted, the more frequently a woman had been in the same booth with another subject, the higher the rating (Saegert, Swap, & Zajonc, 1973).

Who Is Desirable?

We come into contact with many potential partners, but contact by itself does not ensure the development of a relationship. Whether a relationship of some type actually develops between two persons depends on whether each is attracted to the other. Initial attraction is influenced by social norms, physical attractiveness, and processes of interpersonal exchange. If the attraction is mutual, the interaction that occurs is governed by *scripts*.

Social Norms

Each culture specifies the types of relationships that people may have. For each type, norms specify what kinds of people are allowed to have such a relationship. These norms tell us which persons are appropriate as friends, lovers, and mentors. In U.S. society there is a **norm of homogamy,** a norm requiring that friends, lovers, and spouses be characterized by similarity in age, race, religion, and socioeconomic status (Kerckhoff, 1974). Research shows that homogamy is characteristic of adolescent friendship pairs (Kandel, 1978) and married couples (Murstein, 1980). The NORC survey of 3,342 adults assessed the extent to which partners in relationships were similar on these dimensions (Laumann et al., 1994). Seventy-five to

83% were similar by age, 82% to 87% by education, 88% to 93% by race/ethnicity, and 53% to 72% by religion. Differences on one or more of these dimensions make a person less appropriate as a friend and more appropriate for some other kind of relationship. Thus a person who is much older but of the same social class and ethnicity may be appropriate as a mentor, someone who can provide advice about how to manage your career. Potential dates are single persons of the opposite sex who are of similar age, class, ethnicity, and religion.

Norms that define appropriateness influence the development of relationships in several ways. First, each of us uses norms to monitor our own behavior. We hesitate to establish a relationship with someone who is defined by norms as an inappropriate partner. Thus a low-status person is unlikely to approach a high-status person as a potential friend. For example, the law clerk who just joined a firm would not discuss her hobbies with the senior partner (unless the senior partner asked). Second, if one person attempts to initiate a relationship with someone who is defined by norms as inappropriate, the other person will probably refuse to reciprocate. If the clerk did launch into an extended description of the joys of restoring antique model trains, the senior partner would probably end the interaction. Third, even if the other person is willing to interact, third parties often enforce the norms that prohibit the relationship (Kerckhoff, 1974). Another member of the firm might later chide the clerk for presuming that the senior partner cared about her personal interests.

Physical Attractiveness

In addition to social norms that define who is appropriate, individuals also have personal preferences regarding desirability. Someone may be normatively appropriate but still not appeal to you. Physical attractiveness can have a significant impact on desirability.

The Impact of Physical Attractiveness
A great deal of evidence shows that given more than one potential partner, individuals prefer the one who is more physically attractive (Hendrick & Hendrick, 1992). A study of 752 first-year college students, for example, demonstrates that most individuals prefer

According to the matching hypothesis, people seek partners whose level of social desirability is about equal to their own. We frequently encounter couples who are matched—that is, who are similar in age, race, ethnicity, social class, and physical attractiveness.

more attractive persons as dates (Walster [Hatfield] et al., 1966). As part of the study, students were invited to attend a dance. Before the dance, each student had to purchase a ticket. As students waited in line, four ticket sellers secretly rated the person's physical attractiveness as low, average, or high. Before the dance, each student filled out several questionnaires and was told that dance partners would be selected by the computer. In fact, males and females were paired randomly. The couples arrived at the dance about 8 P.M. and danced until intermission at 10:30 P.M. During the intermission, students filled out a questionnaire on which they gave their impression of their date.

This study tested the **matching hypothesis,** the idea that each of us looks for someone who is of approximately the same level of social desirability. An important dimension of desirability is attractiveness. The researchers predicted that those students whose dates matched their own attractiveness would like their date most. Those whose dates were very different in attractiveness were expected to rate their date as less desirable. Contrary to the hypothesis, in this situation, students preferred a more attractive date, regardless of their own attractiveness. The matching

hypothesis is supported by analyses of singles ads (see Box 12.1, "Let's Make a Deal").

How can we explain the significance of attractiveness? One factor is simply aesthetic; generally, we prefer what is beautiful. Although beauty is, to a degree, "in the eye of the beholder," cultural standards influence our aesthetic judgments. A study of female facial beauty found substantial agreement among male college students about which features are attractive (Cunningham, 1986). These men rated such features as large eyes, small nose, and small chin as more attractive than small eyes, large nose, and large chin. What male features do females find attractive? Female college students rate men with large eyes, prominent cheekbones, and a large chin as more attractive (Cunningham, Barbee, & Pike, 1990). Research has also found a high level of agreement among men that certain female body shapes are more appealing than others (Wiggins, Wiggins, & Conger, 1968). There is also agreement among women about which male body shapes are attractive (Beck, Ward-Hull, & McLear, 1976).

A second factor is that we anticipate more rewards when we associate with attractive persons. A man with an extremely attractive woman receives

12.1   Let's Make a Deal

Reasonably good-looking professional man, 34, wishes to meet attractive, professional, sensitive woman, 24–30, to share interest in music, hockey, people, and conversation.
Reply to Box 3010.

Attractive Gemini, 36, blond, blue eyes, 5'7", 125 lbs., tanned, loves sports, playing piano, guitar. Interested in meeting man 26–46, physically fit, financially secure, witty.
Reply to Box 2407.

Ads like these can be found in most daily newspapers in the United States. Some papers carry dozens or even hundreds of these ads in each issue. If we look at these ads in terms of what they offer and what they seek, we can see that people are looking for a *complementary* exchange.

One study analyzed 800 such ads in an attempt to identify patterns in the "revelations" and "stipulations" contained in them (Harrison & Saeed, 1977). The ads were taken from a national weekly over a 6-month period. An ad was excluded if the person was under 20 or over 60 or if age was not given. Results of the study showed that most women offered physical attractiveness and were seeking financial security and an older companion. Conversely, most men offered financial security and were seeking an attractive, younger woman. Thus each group offered complementary resources.

Results also indicated that people sought partners whose level of social desirability was approximately equal to their own. This supports the *matching hypothesis* (Berscheid et al., 1971). The researchers calculated an "overall desirability score" for both the offerer and the person being sought. Attractiveness, financial security, and sincerity were each worth 1 point on the offerer's index, whereas seeking each of these was worth 1 point on the other index. Points were added to obtain desirability

scores, which ranged from 0 to 3. There was a significant correlation between the level of desirability offered and the level of desirability sought. In addition, most persons who stated they were physically attractive were seeking good-looking partners, evidence of the importance of physical attractiveness.

A study of 400 ads from two weekly newspapers replicated the finding that women offered physical attractiveness and were seeking status (security), whereas men offered status and were seeking attractiveness (Koestner & Wheeler, 1988). In addition, women offered low body weight and were seeking tall men, whereas men offered height and were seeking light women. Finally, women offered instrumental traits—independent, competent, bright—valued by men, and men offered expressive traits—warm, gentle, emotional—valued by women. Advertisers seem to understand attraction, offering precisely the traits sought by the opposite gender.

Singles ads have been appearing in newspapers for over 25 years. Have they changed over time? Research comparing ads published in 1986 and 1991 suggest they have not (Willis & Carlson, 1993). In both years, differences between ads placed by men and by women replicated the results summarized above.

In addition to advertisements, people can use telephone messages, computer bulletin boards, and videodating services in their search for partners. What do the people who use these mechanisms want to accomplish? A study of clients of a videodating service provides some answers (Woll & Young, 1989). The men believed women are looking for financial security and tried to convey in their video that they were secure. The women believed men are looking for physically attractive and independent women, so they tailored their self-presentation accordingly. Both men and women said they tried to create a video that would appeal to one special person, Ms. or Mr. Right.

Did they succeed? Most said, "No."

---

more attention and prestige from other persons than if he is seen with an unattractive woman, and vice versa (Sigall & Landy, 1973).

The Attractiveness Stereotype   A third

factor is the **attractiveness stereotype,** the belief that

"what is beautiful is good" (Dion, Berscheid, & Walster [Hatfield], 1972). We assume that an attractive person possesses other desirable qualities. In one study, male and female college students were shown three photographs and asked to rate the personality of the person photographed on various dimensions. The people in the photos varied in physical attractiveness; half of the male and female subjects were shown pictures of men, whereas the other half were shown pictures of women. Physically attractive people were rated as more sensitive, kind, interesting, strong, modest, and sexually responsive than less attractive persons (Sigall & Landy, 1973). They were also considered more talented than unattractive people (Landy & Sigall, 1974).

There are limits to the influence of this stereotype. An analysis of more than 70 studies found that attractiveness has a moderate influence on judgments of social competence—how sensitive, kind, and interesting a person is (Eagly et al., 1991). It has less influence on judgments of adjustment and intelligence, and no influence on judgments of integrity or concern for others. Also, the influence of attractiveness on judgments of intellectual competence is reduced when other information about the person's competence is available (Jackson, Hunter, & Hodge, 1995).

When we believe another person possesses certain qualities, those beliefs influence our behavior toward that person. Our actions may lead him or her to behave in ways that are consistent with our beliefs. In one experiment, males were shown photographs of either an attractive or an unattractive woman. They were then asked to interact with that woman via intercom for 10 minutes. (The woman was actually a student volunteer.) Each conversation was tape-recorded and rated by judges. Women who were perceived as attractive by the men were rated as behaving in a more friendly, likable, and sociable way compared with women who were perceived as unattractive. This happened in part because the men gave the target person opportunities to act in ways that would confirm their expectations based on the attractiveness stereotype (Snyder, Tanke, & Berscheid, 1977).

Judgments of attractiveness seem to be based on several components. College students were asked to sort photographs of 95 female fashion models. Analyses suggested that both men and women distinguished three dimensions in their judgments—sexy, cute (youthful), and trendy (up-to-date in clothing and grooming) (Ashmore, Solomon, & Longo, 1996).

Each of us knows that physically attractive people may receive preferential treatment. As a result, we spend tremendous amounts of time and money trying to increase our attractiveness to others. Men and women purchase clothing, jewelry, perfumes, colognes, and hair rinses in an effort to enhance their physical attractiveness. Our choice of products reflects current standards of what looks good. Increasingly, people are using cosmetic surgery to enhance their appearance. Plastic surgeons can lift your eyelids; pin your ears; fill your wrinkles; reshape your nose, jaw, or chin; enlarge your breasts, pectorals, or penis; and suck the fat from your abdomen, thighs, or ankles ("Body Rebuilding," 1997).

Not everyone prefers attractive persons. Many of us were taught, and some of us believe, that "Beauty is only skin deep" and "You can't judge a book by its cover." What kinds of people are influenced by another's attractiveness, and what kinds of people "read the book" before making a judgment?

People who hold traditional attitudes toward men and women are those whose judgments are much more likely to be influenced by beauty. In one study (Touhey, 1979), students were given the Macho scale, a series of questions used to assess to what degree a person endorses sexist/traditional attitudes. Men and women who obtained high and low scores (a high score reflecting more traditional attitudes) were given photographs and biographical descriptions of several members of the opposite sex. Each subject indicated probable liking for and willingness to date the person in the photograph and rated the person on various traits. Overall, men and women with high Macho scores liked attractive persons more than unattractive ones, whereas low scorers liked attractive and unattractive persons equally well. High scorers rated the attractive person as more dutiful and respecting of authority, a better dancer, and a better potential parent than the unattractive person. The low scorers' overall ratings for attractive versus unattractive persons were not significantly different.

The Evolutionary Perspective on Attractiveness According to the evolutionary perspective, men and women want to mate with healthy individuals so they will produce healthy offspring, who will in turn mate and pass on their genetic code. Physically attractive people are more likely to be healthy than unattractive people (Singh, 1995). Carrying the argument one step further, health is important when one is selecting a long-term mate, but sexual availability is more important when one is seeking a short-term mate. A study of mating strategies found that physical attractiveness and possession of resources were judged important in selecting a long-term mate, whereas sexual availability and giving gifts were judged more important in selecting a partner for a "one-night stand" (Schmitt & Buss, 1996). It is not surprising, according to this perspective, that singles ads emphasize attractiveness and resources.

Attractiveness Isn't Everything   Physical attractiveness may have a major influence on our judgments of others because it is readily observable. When we meet someone for the first time, one characteristic we can assess quickly is his or her attractiveness. If other relevant information is available, it might reduce or eliminate the impact of attractiveness on our judgments. In fact, an analysis of 70 studies found that when perceivers have other personal information about the target person, the effect of the attractiveness stereotype is smaller (Eagly et al., 1991).

Exchange Processes

How do we move from the stage of awareness of another person to the stage of contact? Recall that in our introduction, Dan noticed Sally at every lecture. Because she was young and not wearing a wedding ring, Dan hoped she was available. She was certainly desirable—she was very pretty and seemed like a friendly person. What factors did Dan consider when deciding whether to initiate contact? One important factor in this decision is the availability and desirability of alternative relationships (Backman, 1990). Thus, before Dan chose to initiate contact with Sally, he probably considered whether there was anyone else who might be a better choice.

Choosing Friends   We can view each actual or potential relationship—whether involving a friend, co-worker, roommate, or date—as promising rewards but entailing costs. Rewards are the pleasures or gratifications we derive from a relationship. These might include a gain in knowledge, enhanced self-esteem, satisfaction of emotional needs, or sexual gratifica-

tion. Costs are the negative aspects of a relationship, such as physical or mental effort, embarrassment, anxiety, and money.

Exchange theory proposes that this is, in fact, the way people view their interactions (Blau, 1964; Homans, 1974). People evaluate interactions and relationships in terms of the rewards and costs that each is likely to entail. They calculate likely **outcomes** by subtracting the anticipated costs from the anticipated rewards. If the expected outcome is positive, people are inclined to initiate or maintain the relationship. If the expected outcome is negative, they are unlikely to initiate a new relationship or to stay in an ongoing relationship. Dan anticipated that dating Sally would be rewarding; she would be fun to do things with, and others would be impressed that he was dating such an attractive woman. At the same time, he anticipated that Sally would expect him to be committed to her, and he would have to spend money to take her to movies, plays, and restaurants.

What standards can we use to evaluate the outcomes of a relationship? Two standards have been proposed (Kelley & Thibaut, 1978; Thibaut & Kelley, 1959). One is the **comparison level** (CL), the level of outcomes expected based on the average of a person's experience in past relevant relationships. Each relationship is evaluated as to whether it is above or below that person's CL—that is, better or worse than the average of past relevant relationships. Relationships that fall above a person's CL are satisfying, whereas those that fall below it are unsatisfying.

If this was the only standard, we would always initiate relationships that appeared to promise outcomes better than those we already experienced and avoid relationships that appeared to promise poorer outcomes. Sometimes, however, we use a second standard. The **comparison level for alternatives** (CLalt) is the lowest level of outcomes a person will accept in light of available alternatives. A person's CLalt varies depending on the outcomes that he or she believes can be obtained from the best of the available alternative relationships. The use of CLalt explains why we may sometimes turn down opportunities that appear promising or why we may remain in a relationship even though we feel the other person is getting all the benefits.

Whether or not a person initiates a new relationship depends on both the CL and the CLalt. An individual usually avoids relationships whose anticipated outcomes fall below the comparison level. If a potential relationship appears likely to yield outcomes above a person's CL, then initiation will depend on whether the outcomes are expected to exceed the CLalt. Dan believed that a relationship with Sally would be very satisfying. He was casually dating another woman, and that relationship was not gratifying. Thus the potential relationship with Sally was above both CL and CLalt, leading Dan to initiate contact.

Whereas CL is an absolute, relatively unchanging standard, several factors influence a person's CLalt. These factors include the extent to which routine activities provide opportunities to meet people, the size of the pool of eligibles, and one's skills in initiating relationships.

Making Contact   Once we decide to initiate interaction, the next step is to make contact. Sometimes we use technology, such as the telephone or computer mail. Often we arrange to get physically close to the person. At parties and bars, people often circulate, which brings them into physical proximity with many of the other guests.

Once in proximity, strangers frequently use a direct gaze as a signal that they are interested in conversation. In one study, pairs of undergraduates were brought together in a waiting room and videotaped for 5 minutes. In the room were two chairs placed side by side, with a magazine table between them. Most subjects gave an initial look at the other subject. If the gaze was mutual, continuous conversation during the 5-minute period was likely; when the gaze was not mutual, conversation was unlikely (Carey, 1978). This pattern is probably repeated many times every day in airplanes, on trains and buses, and in classes and waiting rooms.

Scripts   The development of relationships is influenced by an *event schema,* or script. A **script** specifies (1) the definition of the situation (a date, job interview, or sexual encounter), (2) the identities of the social actors involved, and (3) the range and sequence of permissible behaviors (see Chapter 5).

The initiation of a relationship requires an opening line. Often it is about some feature of the situation. At the beginning of this chapter, Dan initiated conversation by commenting that Sally never missed the class. Two people waiting to participate in a psychology experiment may begin talking by speculating over the purpose of the experiment. The weather is a widely used topic for openings. The opening line often includes an *identification display,* a signal that we believe the other person is a potential partner in a specific kind of relationship (Schiffrin, 1977). When Dan commented about Sally's attitude toward the class, it conveyed an interest in friendship. A different message would have been sent if he had asked whether she knew the woman sitting next to her. The person who is approached, in turn, decides whether she is interested in that type of relationship. If she is, she engages in an **access display,** a signal that further interaction is permissible. Thus Sally responded warmly to Dan's opening line, encouraging continued conversation.

Once initiated, scripts specify the permissible next steps. American society, or at least the subculture of college students, is characterized by a specific script for "first dates" (Rose & Frieze, 1993). When asked to describe "actions which a woman (man) would typically do," both men and women identified a *core action sequence:* dress, nervous, pick up (date), leave (meeting place), confirm plans, get to know, evaluate, talk, laugh, joke, eat, attempt to make out/accept or reject, take (date) home, kiss, go home. In general, both men and women ascribed a proactive role to the male and a reactive role to the female.

Actual first dates, of course, are characterized by departures from the script. A study of college students focused on the extent to which the roles of male and female are changing (Lottes, 1993). Both men and women were asked about the extent to which they had experienced the woman initiating a date, initiating sexual intimacy, and paying for a date. Increasing proportions of women are engaging in these traditionally male activities.

How do we learn these scripts, and departures from them? One source is the mass media. Both men and women learn about relationships and how to handle them from popular magazines. A study of magazines oriented toward women (*Cosmopolitan,*

Glamour, and *Self*) and men (*Playboy, Penthouse,* and *GQ*) found that they portray relationships in similar terms (Duran & Prusank, 1997). The dominant focus in both types was sexual relationships. In women's magazines (January 1990 to December 1991), the themes were (1) women are less skilled and more anxious about sex, and (2) sex is enjoyed most in caring relationships. In men's magazines during the same period, themes were (1) men are under attack in sexual relationships, and (2) men have natural virility and strong sexual appetites. Also, articles in women's magazines portray men as incompetent about relationships.

The Determinants of Liking

Once two people make contact and begin to interact, several factors determine the extent to which each person will like the other. Three of these factors are considered in this section: similarity, shared activities, and reciprocal liking.

Similarity

How important is similarity? Do "Birds of a feather flock together"? Or is it more the case that "Opposites attract"? These two aphorisms about the determinants of liking are inconsistent and provide opposing predictions. A good deal of research has been devoted to finding out which one is more accurate. The evidence indicates that birds of a feather do flock together; that is, we are attracted to people who are similar to ourselves. Probably the most important kind of similarity is **attitudinal similarity,** the sharing of beliefs, opinions, likes, and dislikes.

Attitudinal Similarity   A widely employed technique for studying attitudinal similarity is the attraction-to-a-stranger paradigm, initially developed by Byrne (1961b). Potential subjects fill out an attitude questionnaire that measures their beliefs about various topics, such as life on a college campus. Later, subjects receive information about a stranger as part of a seemingly unrelated study. The information they

receive describes the stranger's personality or social background and may include a photograph. They also are given a copy of the same questionnaire they completed earlier, ostensibly filled out by the stranger. In fact, the stranger's questionnaire is completed by the experimenter, who systematically varies the degree to which the stranger's supposed responses match the subject's responses. After seeing the stranger's questionnaire, subjects are asked how much they like or dislike the stranger and how much they would enjoy working with that person.

In most cases, the subject's attraction to the stranger is positively associated with the percentage of attitude statements by the stranger that agree with the subject's own attitudes (Byrne & Nelson, 1965; Gonzales et al., 1983). We rarely agree with our friends about everything; what matters is that we agree on a high proportion of issues. This relationship between similarity of attitudes and liking is very general; it has been replicated in studies using both men and women as subjects and strangers under a variety of conditions (Berscheid & Walster [Hatfield], 1978).

In the attraction-to-a-stranger paradigm, the subject forms an impression of a stranger without any interaction. This allows researchers to determine the precise relationship between similarity and liking. But what do you think the relationship would be if two people were allowed to interact? Would similarity have as strong an effect?

A study attempting to answer this question arranged dates for 44 couples (Byrne, Ervin, & Lamberth, 1970). Researchers distributed a 50-item questionnaire measuring attitudes and personality to a large sample of undergraduates. From these questionnaires, they selected 24 male-female couples whose answers were very similar (66% to 74% identical) and 20 couples whose answers were not similar (24% to 40% identical). Each couple was introduced, told they had been matched by a computer, and asked to spend the next 30 minutes together at the student union; they were even offered free sodas. The experimenter rated each subject's attractiveness before he or she left on the date. When they returned, the couple rated each other's sexual attractiveness, desirability as a date and marriage partner, and indicated how much they liked each other. The experimenter also recorded the physical distance between the two as they stood in front of his desk.

Results of this experiment showed that both attitudinal similarity and physical attractiveness influenced liking. Partners who were attractive and who held highly similar attitudes were rated as more likable. In addition, similar partners were rated as more intelligent and more desirable as a date and marriage partner. The couples high in similarity stood closer together after their date than couples low in similarity, another indication that similarity creates liking.

At the end of the semester, 74 of the 88 subjects in this study were contacted and asked whether they (1) could remember their date's name, (2) had talked to their date since their first meeting, (3) had dated their partner, or (4) wanted to date their partner. Researchers compared reports of subjects in the high attractiveness/high similarity condition with those in the low attractiveness/low similarity condition. Those in the former condition were more likely to remember their partner's name, to report having talked to their partner, and to report wanting to date their partner.

The story of Dan and Sally at the beginning of this chapter illustrates the importance of similarity in the early stages of a relationship. After their initial meeting, they discovered they had several things in common. They were from the same city. They had chosen the same major and held similar beliefs about their field and about how useful a bachelor's degree would be in that field. Each also found the other attractive; like the subjects in the high attraction/high similarity condition, Dan and Sally continued to talk after their first meeting.

Why Is Similarity Important?   Why does attitudinal similarity produce liking? One reason is the desire for consistency between our attitudes and perceptions. The other reason focuses on our preference for rewarding experiences.

Most people desire *cognitive consistency,* consistency between attitudes and perceptions of whom and what we like and dislike. If you have positive attitudes toward certain objects and discover that another person has favorable attitudes toward those objects, your cognitions will be consistent if you like that person

(Newcomb, 1971). When Dan discovered that Sally had a positive attitude toward his major, his desire for consistency produced a positive attitude toward Sally. Our desire for consistency attracts us to persons who hold the same attitudes toward important objects (see Chapter 6).

Similarity in attitudes is an important influence on the development of friendships. This was demonstrated in a study of the friendships that developed among 17 college men who shared a rooming house for a semester (Newcomb, 1961). Each man filled out several questionnaires before moving into the house. During the semester, the men expressed their liking for each other several times. The liking ratings stabilized by the middle of the semester, and they were best predicted by the degree of similarity. Another study reported similar findings among women (Hill & Stull, 1981).

Another reason we like persons with attitudes similar to our own is because interaction with them provides three kinds of reinforcement. First, interacting with persons who share similar attitudes usually leads to positive outcomes (Lott & Lott, 1974). At the beginning of this chapter, Dan anticipated that he and Sally would get along well because they shared similar likes and dislikes.

Second, similarity validates our own view of the world. We all want to evaluate and verify our attitudes and beliefs against some standard. Sometimes physical reality provides objective criteria for our beliefs. But often there is no physical standard, and so we must compare our attitudes with those of others (Festinger, 1954). Persons who hold similar attitudes provide us with support for our own opinions, which allows us to deal with the world more confidently (Byrne, 1971). Such support is particularly important in areas like political attitudes, where we realize others hold attitudes dissimilar to our own (Rosenbaum, 1986).

Research indicates that similarity in mood is also an important influence on attraction. In the attraction-to-a-stranger setting, nondepressed subjects prefer nondepressed strangers (Rosenblatt & Greenberg, 1988). In an experiment, male and female students interacted with a depressed or nondepressed person of the same sex. People in homogeneous pairs (both depressed or both not depressed) were more satisfied with the interaction than people in mixed pairs (Locke & Horowitz, 1990). In another study, researchers measured the depression levels of people and of their best friends. Depressed people had best friends who were also depressed (Rosenblatt & Greenberg, 1991).

Third, we like others who share similar attitudes because we expect that they will like us. In one experiment, college students were given information about a stranger's attitudes and the stranger's evaluation of them (Condon & Crano, 1988). The subject's perceptions of the stranger's similarity to and evaluation of him or her were also assessed. The students were attracted to strangers who they perceived as evaluating them positively, and that accounted for the influence of similar attitudes.

Shared Activities

As people interact, they share activities. Recall that after Sally and Dan met, they began to sit together in class and to discuss course work. When the professor announced the first exam, Dan invited Sally to study for it with him. Sally's roommate was also in the class; the three of them reviewed the material together the night before the exam. Sally and Dan both got A's on the exam, and each felt that studying together helped. The next week they went to a movie together. Several days later, Sally invited Dan to a party.

Shared activities provide opportunities for each person to experience reinforcement. Some of these reinforcements come from the other person; Sally finds Dan's interest in her very reinforcing. Often the other person is associated with a positive experience, which leads us to like the person (Byrne & Clore, 1970). Getting an A on the examination was a very positive experience for both Dan and Sally. The association of the other person with that experience led to increased liking for the other.

Thus, as a relationship develops, the sharing of activities contributes to increased liking. This was shown in a study in which pairs of friends of the same sex both filled out attitude questionnaires and listed their preferences for various activities (Werner & Parmalee, 1979). The duration of the friendships varied from 3 months to 20 years, with an average of 5 years. Thirteen male and 11 female pairs were included in the study. Results of the study showed similarity between friends in both activity preferences and attitudes. In addition, people were able to predict their friend's activity preferences more accurately than their friend's attitudes. A study of romantic relationships found that sharing of tasks or activities was a strong predictor of liking (Stafford & Canary, 1991). Thus participation in mutually satisfying activities is a strong influence on the development and maintenance of relationships. As Dan and Sally got to know each other, their shared experiences—studying, seeing movies, going to parties—became the basis for their relationship, supplementing the effect of similar attitudes.

Reciprocal Liking

One of the most consistent research findings is the strong positive relationship between our liking someone and the perception that the other person will like us in return (Backman, 1990). In most relationships, we expect reciprocity of attraction; the greater the liking of one person for the other, the greater the other person's liking will be in return. But will the degree of reciprocity increase over time, as partners have greater opportunities to interact? To answer this question, one study obtained liking ratings from 48 persons (32 males and 16 females), who had been acquainted for 1, 2, 4, 6, or 8 weeks (Kenny & La Voie, 1982). Results showed a positive correlation between each person's liking rating for the other, and reciprocity of attraction increased somewhat with the length of the acquaintanceship. Some subjects in this study were roommates rather than friends; they would be expected to like each other because of the proximity effect. When roommate pairs were eliminated from the results, the correlation between liking ratings increased substantially.

The Growth of Relationships

We have traced the development of relationships from the stage of zero contact through awareness (who is available), surface contact (who is desirable), to mutuality (liking). At the beginning of this chapter, Dan and Sally met, discovered that they had similar attitudes and interests, and shared pleasant experiences—such as doing well on an examination, going to a movie, and later, to a party.

Many of our relationships remain at the "minor" level of mutuality (see Figure 12.1). We have numerous acquaintances, neighbors, and co-workers that we like and interact with regularly but to whom we do not feel especially close. A few of our relationships grow closer; they proceed through "moderate" to "major" mutuality. Three aspects of the growth of relationships are examined in this section: self-disclosure, trust, and interdependence. As the degree of mutuality increases between friends, roommates, and co-workers, self-disclosure, trust, and interdependence also increase.

Self-Disclosure

Recall that when Dan and Sally returned from the party, Sally told Dan that her roommate's parents had just separated and her roommate was very depressed. Sally said she didn't know how to help her roommate, that she felt unable to deal with the situation. At this point, Sally was engaging in *self-disclosure,* the act of revealing personal information about oneself to another person. Self-disclosure usually increases over time in a relationship. Initially, people reveal things about themselves that are not especially intimate and that they believe the other will readily accept. Over time, they disclose increasingly intimate details about their beliefs or behavior, as well as information they are less certain the other will accept (Backman, 1990).

Self-disclosure increases as a relationship grows. In one study, same-sex pairs of previously unacquainted college students were brought into a laboratory setting and asked to get acquainted (Davis, 1976). They were given a list of 72 topics. Each topic had been rated earlier by other students on a scale of intimacy from 1 to 11. Subjects were asked to select topics from this list and to take turns talking about each topic for at least 1 minute while their partner remained silent. The interaction continued until each partner had spoken on 12 of the 72 topics. Results showed that the intimacy of the topic selected increased steadily from the 1st to the 12th topic chosen. The average intimacy of topics discussed by each couple increased from 3.9 to 5.4 over the 12 disclosures.

When Sally told Dan about the situation with her roommate, Dan replied that he knew how she felt because his older brother had just separated from his wife. This exchange reflects reciprocity in self-disclosure; as one person reveals an intimate detail, the other person usually discloses information at about the same level of intimacy (Altman & Taylor, 1973). In the Davis study just discussed, each subject selected a topic at the same level of intimacy as the preceding one or at the next level of intimacy. However, reciprocity decreases as a relationship develops. In one study, a researcher recruited students to be subjects and asked each to bring an acquaintance, a friend, or a best friend to the laboratory (Won-Doornink, 1985). Each dyad was given a list of topics that varied in degree of intimacy. Each was instructed to take at least four turns choosing and discussing a topic. Each conversation was tape-recorded and later analyzed for evidence of reciprocity. The association between the stage of relationship and the reciprocity of intimate disclosures was curvilinear; there was greater reciprocity of intimate disclosures among friends than among acquaintances and less reciprocity among best friends than among friends (see Figure 12.2).

Not all people divulge increasingly personal information as you get to know them. You have probably known people who were very open—who readily disclosed information about themselves—and others who said little about themselves. In this regard, we often think of men as less likely to discuss their feelings than women. However, research has shown that self-disclosure depends not only on gender but also on the nature of the relationship. In casual relationships (with men or women), men are less likely to disclose personal information than women (Reis, Senchak, & Solomon, 1985). In dating couples, the amount of disclosure is related more to sex-role orientation than it is to gender. Men and women with traditional orientations disclose less to their partners than those with egalitarian orientations (Rubin et al., 1980). Traditional sex roles are more segregated, with each person responsible for certain tasks, whereas egalitarian

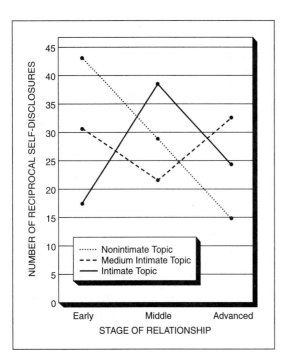

Figure 12.2   The Relationship Between Reciprocity and Intimacy

Reciprocity—picking a topic of conversation that is as intimate as the last topic introduced by your partner— is the process by which relationships become more intimate. The extent of reciprocity depends on the intimacy of the topic and the stage of the relationship. Students talked with an acquaintance (early stage), friend (middle stage), or best friend (advanced stage). With topics that were not intimate (such as the weather), reciprocity declined steadily as the stage increased. With intimate topics, in contrast, reciprocity was greatest at the middle stage, less at the advanced stage, and least at the early stage of a relationship.

SOURCE: Adapted from Won-Doornink, 1985, figure 4.

orientations emphasize sharing. An emphasis on joint activity leads to greater self-disclosure. In intimate heterosexual relationships, men and women do not differ in the degree of self-disclosure (Hatfield, 1982).

Trust

Why did Dan confide in Sally that his brother had just left his wife? Perhaps he was offering reciprocity in self-disclosure. Because Sally had confided in Dan,

she expected him to reciprocate. But had he been suspicious of Sally's motives, he might not have. This suggests the importance of trust in the development of a relationship.

When we **trust** someone, we believe that person is both honest and benevolent (Larzelere & Huston, 1980). We believe the person tells us the truth—or at least does not lie to us—and that his or her intentions toward us are positive. One measure of interpersonal trust is the interpersonal trust scale reproduced in Table 12.1. The questions focus on whether the other person is selfish, honest, sincere, fair, or considerate. We are more likely to disclose personal information to someone we trust. How much do you trust *your* partner? Answer the questions on the scale and determine your score. Higher scores indicate greater trust.

To study the relationship between trust and self-disclosure, researchers recruited men and women from university classes, from a list of people who had recently obtained marriage licenses, and by calling persons randomly selected from the telephone directory. Each person was asked to complete a questionnaire concerning his or her spouse or current or most recent date. The survey included the interpersonal trust scale in Table 12.1. Researchers averaged the trust scores for seven types of relationships, as shown in Figure 12.3. Note that as the relationship becomes more exclusive, trust scores increase significantly. Is there a relationship between trust and self-disclosure? Each person was also asked how much he or she had disclosed to the partner in each of six areas—religion, family, emotions, relationships with others, school or work, and marriage. Trust scores were positively correlated with disclosure—that is, the more the person trusted the partner, the greater the degree of disclosure.

Other research on interpersonal trust suggests that, in addition to honesty and benevolence, reliability is an important aspect of trust. We are more likely to trust someone we feel is reliable, who we can count on (Johnson-George & Swap, 1982), and who is predictable (Rempel, Holmes, & Zanna, 1985).

Interdependence

Earlier in this chapter, we noted that people evaluate potential and actual relationships in terms of the

Table 12.1 Interpersonal Trust Scale

	Strongly Agree	Agree	Slightly Agree	?	Slightly Disagree	Disagree	Strongly Disagree
1. My partner is primarily interested in his or her own welfare.	_____	_____	_____	–	_____	_____	_____
2. There are times when my partner cannot be trusted.	_____	_____	_____	–	_____	_____	_____
3. My partner is perfectly honest and truthful with me.	_____	_____	_____	–	_____	_____	_____
4. I feel I can trust my partner completely.	_____	_____	_____	–	_____	_____	_____
5. My partner is truly sincere in his or her promises.	_____	_____	_____	–	_____	_____	_____
6. I feel my partner does not show me enough consideration.	_____	_____	_____	–	_____	_____	_____
7. My partner treats me fairly and justly.	_____	_____	_____	–	_____	_____	_____
8. I feel my partner can be counted on to help me.	_____	_____	_____	–	_____	_____	_____

Note: For items 1, 2, and 6, Strongly Agree = 1, Agree = 2, Slightly Agree = 3, and so on. For items 3, 4, 5, 7, and 8 the scale is reversed.

SOURCE: Adapted from Larzelere and Huston, 1980.

outcomes (rewards minus costs) they expect to receive. Dan initiated contact with Sally because he anticipated positive outcomes. Sally encouraged the development of a relationship because she, also, expected the rewards to exceed the costs. As their relationship developed, each discovered the relationship was rewarding. Consequently, they increased the time and energy devoted to their relationship and decreased their involvement in alternative ones. As their relationship became increasingly mutual, Sally and Dan became increasingly dependent on each other for various rewards (Backman, 1990). The result is strong, frequent, and diverse interdependence (Kelley et al., 1983).

Increasing reliance on one person for gratifications and decreasing reliance on others is called **dyadic withdrawal** (Slater, 1963). One study of 750 men and women illustrates the extent to which such withdrawal occurs. Students identified the intensity

of their current heterosexual relationships, then listed the names of persons whose opinions they considered important. They also indicated how important each person's opinions were and how much they had disclosed to that person (Johnson & Leslie, 1982). As predicted, the more intimate his or her current heterosexual relationship, the smaller the number of friends listed by the subject; there was no difference in the number of relatives listed. In addition, as the degree of involvement increased, the proportion of mutual friends of the couple also increased (Milardo, 1982). Other studies have found that as heterosexual relationships become more intimate, each partner spends less time interacting with friends, and interaction with relatives may increase (Surra, 1990).

One study of the development of interdependence asked newlyweds to reconstruct their relationship from the time they met until the time they married (Surra, 1985). Each person reported the

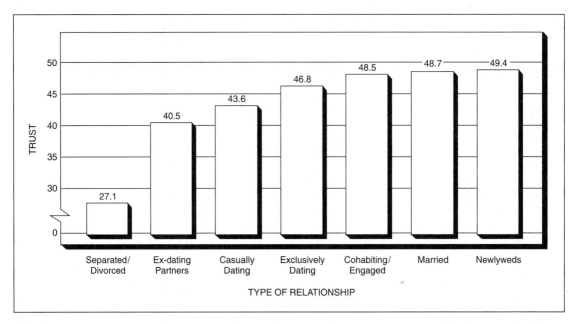

Figure 12.3   Average Interpersonal Trust Scores for Seven Types of Heterosexual Relationships

Trust involves two components: the belief that a person is honest and that his or her intentions are benevolent. More than 300 persons completed the interpersonal trust scale (see Table 12.1) for their current or most recent heterosexual partner. Results showed a strong relationship between the degree of intimacy in a relationship and the degree of trust.

SOURCE: Adapted from Larzelere and Huston, 1980, table 3.

extent to which he or she shared affectional, instrumental, and leisure activities with the future spouse as the relationship grew. The length of time between meeting and marrying (from a few months to 6.5 years) was closely related to how quickly the two people reduced their dependence on others for these activities and became interdependent.

Interdependence evolves out of the process of negotiation (Backman, 1990). Each person offers various potential rewards to the partner; the partner accepts some and rejects others. As the relationship develops, the exchanges stabilize. Shared activities are an important potential source of rewards. Each person has activity preferences. As the relationship develops, the couple must blend their separate preferences into joint activities. A study of dating couples found that men liked sex, games, and sports better than women; women preferred companionship, entertainment, and cultural activities (Surra & Longstreth,

1990). Some couples achieved a blend by taking turns, alternately engaging in activities preferred by each. Others cooperated, engaging in activities they both liked, such as preparing food and running errands. Some couples experienced continuing conflict over what to do.

A potential reward in many relationships is sexual gratification. As relationships develop and become more mutual, physical intimacy increases as well. The couple negotiates the extent of sexual intimacy, with the woman's preferences having a greater effect on the outcome (Peplau, Rubin, & Hill, 1977). How important is sexual gratification in dating relationships? A study of 149 couples assessed the importance of various rewards in relationships of increasing intimacy (preferred date, going steady, engaged, living together, and married). Among intimate couples, sexual gratification was much more likely to be cited as a major basis for the relationship (Centers, 1975). Other surveys indicate that the longer couples have been dating, the more likely they are to engage in sexual intimacy (De-Lamater & MacCorquodale, 1979).

Love and Loving

It is fair to say that what we feel for our friends, room-mates, co-workers, and some of the people we date is attraction. But is that all we feel? Occasionally, at least, we experience something more intense than a positive attitude toward others. Sometimes we feel and even say, "I love you."

How does loving differ from liking? We summa-rized much of the research in social psychology on at-traction or liking earlier in this chapter. By contrast, there has been much less research on love. Four views of love are considered next: the distinction between liking and loving, passionate love, romantic love, and love as story.

Liking Versus Loving

One of the first empirical studies of love distinguished between liking and loving (Rubin, 1970). *Love* is something more than intense liking: It is the attach-ment to and caring about another person (Rubin, 1974). Attachment involves a powerful desire to be with and be cared about by another person. Caring involves making the satisfaction of another person's needs as significant as the satisfaction of your own.

Based on this distinction, Rubin developed scales to measure both liking and love. The liking scale eval-uates one's dating partner, lover, or spouse on various dimensions, including adjustment, maturity, respon-sibility, and likability. The love scale measures attach-ment to and caring for one's partner, and intimacy (self-disclosure). These scales were completed by each member of 182 dating couples, both for her or his partner and best friend of the same sex (Rubin, 1970). Results showed a high degree of internal consistency within each scale and a low correlation between scales. Thus the two scales measure different things.

If the distinction between liking and loving is valid, how do you think you would rate a dating part-ner and your best friend on these scales? Rubin pre-dicted high scores on both liking and love for the dat-ing partner, lover, or spouse and a high liking score and lower love score for the (platonic) friend. The average scores of the 182 couples confirmed these predictions. Further work by Davis (1985) also dis-tinguishes between friendship and love. Friendship involves several qualities, including trust, understand-ing, and mutual assistance. Love involves all of these plus caring (giving the utmost to and being an advo-cate for the other) and passion (obsessive thought, sexual desire).

Passionate Love

Love certainly involves attachment and caring. But is that all? What about the agony of jealousy and the ec-stasy of being loved by another person? An alternative view of love emphasizes emotions such as these. It fo-cuses on **passionate love,** a state of intense longing for union with another and of intense physiological arousal (Hatfield & Walster, 1978).

Cognitive and emotional factors interact to pro-duce passionate love. Each of us learns about love from parents, friends, movies, and popular music. We learn who it is appropriate to fall in love with, how it feels, and how we should behave when we are in love. We experience an emotion only when we are physio-logically aroused. Thus we experience passionate love

The United States is one of a small number of societies in which love is widely accepted as a basis for getting married. In many other societies, marriages reflect political and economic influences, not romance.

when we experience intense arousal and the circumstances fit the cultural definitions we have learned.

Passionate love has three components: cognitive, emotional, and behavioral (Hatfield & Sprecher, 1986). The cognitive components include a preoccupation with the loved one, idealization of the person or the relationship, and a desire to know the other and be known by him or her. Emotional components include physiological arousal, sexual attraction, and desire for union. Behavioral elements include serving the other and maintaining physical closeness to him or her. A scale designed to measure passionate love is reproduced in Box 12.2, "Passionate Love." Notice that each item deals with one of these components.

Research using the passionate love scale finds that the items are closely related; all of them measure a single factor (Hendrick & Hendrick, 1989). A study of 60 men and 60 women found that scores on the scale are related to the stage of the relationship. Passionate love increases substantially from the early stage of dating to the stage of an exclusive relationship. It does not increase as the relationship moves from exclusively dating to living together or engaged (Hatfield & Sprecher, 1986).

Passionate love is associated with other intense emotions. When our love is reciprocated and we experience closeness or psychological union with the other person, we experience fulfillment, joy, and ecstasy. Conversely, positive emotional experiences—excitement, sexual excitement—can enhance passionate love. Unrequited love, in contrast, is often associated with jealousy, anxiety, or despair. Loss of a love can be emotionally devastating.

The Romantic Love Ideal

The studies and theories of love discussed so far assume that love consists of a particular set of feelings and behaviors. Furthermore, most of us assume that we will experience this emotion toward a member of the opposite sex at least once in our lives. But these are

---

12.2 Passionate Love

Think of the person you love most passionately right now. If you are not in love right now, think of the last person you loved passionately. If you have never been in love, think of the person you came closest to caring for in that way. Keep that person in mind as you complete this questionnaire. (The person you choose should be of the opposite gender if you are heterosexual and of the same gender if you are homosexual.) Try to record how you felt at the time when your feelings were the most intense.

Use the following scale to answer each item:

```
1    2    3    4    5    6    7    8    9
Not at all true  Moderately true  Definitely true
```

1. I would feel deep despair if _____ left me.
2. Sometimes I feel I can't control my thoughts; they are obsessively on _____.
3. I feel happy when I am doing something to make _____ happy.
4. I would rather be with _____ than anyone else.
5. I'd get jealous if I thought _____ were falling in love with someone else.
6. I yearn to know all about _____.
7. I want _____ physically, emotionally, mentally.
8. I have an endless appetite for affection from _____.
9. For me, _____ is the perfect romantic partner.
10. I sense my body responding when _____ touches me.
11. _____ always seems to be on my mind.
12. I want _____ to know me—my thoughts, my fears, and my hopes.
13. I eagerly look for signs indicating _____'s desire for me.
14. I possess a powerful attraction for _____.
15. I get extremely depressed when things don't go right in my relationship with _____.

SOURCE: Adapted from Hatfield and Sprecher, 1986.

very culture-bound assumptions. There are societies in which the state or experience we call love is unheard of. In fact, U.S. society is almost alone in accepting love as a major basis for marriage.

In U.S. society, we are socialized to accept a set of beliefs about love, beliefs that guide much of our behavior. The following five beliefs are known collectively as the **romantic love ideal:**

1. True love can strike without prior interaction ("love at first sight").
2. For each of us, there is only one other person who will inspire true love.
3. True love can overcome any obstacle ("love conquers all").
4. Our beloved is (nearly) perfect.
5. We should follow our feelings—that is, we should base our choice of partners on love rather than on other (more rational) considerations. (Lantz, Keyes, & Schultz, 1975)

Researchers have developed a scale to measure the extent to which individuals hold these beliefs (Sprecher & Metts, 1989). When the scale was given to a sample of 730 undergraduates, the results indicated that the first four beliefs are held by many young people. Interestingly, male students are more likely to hold these beliefs than female students.

Research suggests that the fourth belief, *idealization* of the partner, is an important influence on relationship satisfaction. Two studies of Dutch adults found that many of them believed that their relationship was better than the relationships of others, and this belief was associated with reported happiness (Buunk & van der Eijnden, 1997). In another study, researchers asked members of dating (98) and married (60) couples to rate themselves, their partner, and their ideal partner on 21 interpersonal characteristics (Murray, Holmes, & Griffin, 1996a). Analyses indicated that the ratings of the partner were more similar to the ratings of the self and the ideal partner than to the partner's self-ratings. Furthermore, people who idealized their partner and whose partner idealized them were happier. A longitudinal study found that over a 1-year period, partners came to share the individual's idealized image of him or her (Murray, Holmes, & Griffin, 1996b).

This ideal has not always been popular in the United States. A group of researchers conducted an analysis of best-selling magazines published during four historical periods (Lantz, Keyes, & Schultz, 1975; Lantz, Schultz, & O'Hara, 1977). They counted the number of times the magazines mentioned one or more of the five beliefs that make up the romantic ideal. The number of times the ideal was discussed increased steadily over time, as shown in Figure 12.4. Their findings suggest that American acceptance of

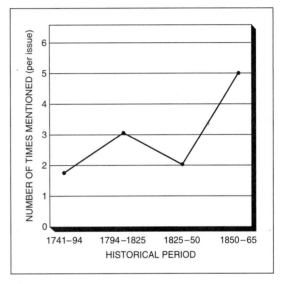

Figure 12.4 Occurrence of the Romantic Love Ideal in Major American Magazines, 1741–1865

One way to measure the influence of the romantic love ideal on American society is to determine the frequency with which it is mentioned in popular magazines. A team of researchers selected a sample of the best-selling magazines from four historical periods and counted the number of times each of the five romantic ideals was mentioned—including (1) idealization of the beloved, (2) love at first sight, (3) love conquers all, (4) there is one and only one for each of us, and (5) we should follow our hearts. They discovered that the number of times the ideals were mentioned increased more than 300% from 1741–94 to 1850–65.

SOURCE: Adapted from Lantz, Keyes, and Schultz, 1975; Lantz, Schultz, and O'Hara, 1977.

the romantic ideal occurred gradually from 1741 to 1865. The romantic ideal first really came into its own about the time of the Civil War.

Love as a Story

When we think of love, our thoughts often turn to the great love stories: Romeo and Juliet, Cinderella and the Prince (Julia Roberts and Richard Gere), King Edward VIII and Wallis Simpson, and *Pygmalion/My Fair Lady.* According to Sternberg, (1998), these stories are much more than entertainment. They shape our beliefs about love and relationships, and our beliefs in turn influence our behavior.

> Zach and Tammy have been married 28 years. Their friends have been predicting divorce since the day they were married. They fight almost constantly. Tammy threatens to leave Zach; he tells her that nothing would make him happier. They lived happily ever after.
>
> Valerie and Leonard had a perfect marriage. They told each other and all of their friends that they did. Their children say they never fought. Leonard met someone at his office and left Valerie. They are divorced. (Adapted from Sternberg, 1998)

Wait a minute! Aren't those endings reversed? Zach and Tammy should be divorced, and Valerie and Leonard should be living happily ever after. If love is merely the interaction between two people, how they communicate and behave, you're right; the stories have the wrong endings. But there is more to love than interaction; what matters is how each partner *interprets* the interaction. To make sense out of what happens in our relationships, we rely on our love stories.

A **love story** is a story (*script*) about what love should be like; it has *characters, plot,* and *theme.* There are two central characters in every love story, and they play roles that complement each other. The plot details the kinds of events that occur in the relationship. The theme is central; it provides the meaning of the events that make up the plot, and gives direction to the behavior of the principals. The story guiding Zach and Tammy's relationship is the "War" story. Each views love as war; a good relationship involves constant fighting. The two central characters are war-

riors, doing battle, fighting for what they believe. The plot consists of arguments, fights, threats to leave—in short, battles. The theme is that love is war; one may win or lose particular battles, but the war continues. Zach and Tammy's relationship endures because they share this view, and because it fits their temperaments. Can you imagine how long a wimp would last in a relationship with either of them?

According to this view, *falling in love* occurs when you meet someone with whom you can create a relationship that fits your love story. Further, we are satisfied with relationships in which we and our partner match the characters in our story (Beall & Sternberg, 1995). Valerie and Leonard's marriage looked great on the surface, but it didn't fit Leonard's love story. He left when he met his "true love," that is, a woman who could play the complementary role in his primary love story.

Where do our stories come from? Many of them have their origins in the culture, in folk tales, literature, theater, films, and television programs. The cultural context interacts with our own personal experience and characteristics to create the stories that each of us has (Sternberg, 1996). As we experience relationships, our stories evolve, taking account of unexpected events. Each person has more than one story; the stories often form a hierarchy. One of Leonard's stories was "House and Home"; home was the center of the relationship, and he (in his role of Caretaker) showered attention on the house and kids (not on Valerie). But when he met Sharon with her aloof air, ambiguous past, and dark glasses, he was hooked; she elicited the "Love Is a Mystery" story that was more salient to Leonard. He could not explain why he left Valerie and the kids; like most of us he was not consciously aware of his love stories. It should be obvious from the examples that love stories derive their power from the fact that they are self-fulfilling. We create in our relationships events according to the plot and then interpret those events according to the theme. Our love relationships are literally social constructions. Because our love stories are self-confirming, they can be very difficult to change.

Sternberg and his colleagues have identified five categories of love stories found in United States culture, and several specific stories within each category. They have also developed a series of statements that reflect the themes in each story. People who agree with the statements "I think fights actually make a relationship more vital" and "I actually like to fight with my partner" are likely to hold the "War" story. Sternberg and Hojjat studied samples of 43 and 55 couples (Sternberg, 1998). They found that couples generally held similar stories. The more discrepant the stories of the partners, the less happy the couple was. Some stories were associated with high satisfaction, for example, the "Garden" story, in which love is a garden that needs ongoing cultivation. Two stories associated with low satisfaction were the "Business" story (especially the version in which the roles are Employer and Employee?), and the "Horror" story in which the roles are Terrorizer and Victim.

Breaking Up

Few things last forever. Roommates who once did everything together lose touch after they finish school. Two women who were best friends gradually stop talking. Couples fall out of love, break up, and divorce. What causes the dissolution of relationships? Research suggests two answers: unequal outcomes and unequal commitment.

Unequal Outcomes and Instability

Earlier, this chapter discussed the importance of outcomes in establishing and maintaining relationships. Our decision to initiate a relationship is based on what we expect to get out of it. In ongoing relationships, we can assess our actual outcomes; we can evaluate the rewards we are obtaining relative to the costs of maintaining the relationship. A survey of college students examined the impact of several factors on satisfaction with a relationship; one factor was the value of overall outcomes compared with a person's comparison level (CL) (Michaels, Edwards, & Acock, 1984). Analyzing the reports of both men and women involved in exclusive relationships, the outcomes being experienced were most closely related to satisfaction with the relationship. Several other studies report the same results (Surra, 1990).

The comparison level for alternatives (CLalt) is also an important standard used in evaluating out-

comes. Are the outcomes from this relationship better than those obtainable from the best available alternative? One dimension on which people may evaluate relationships is physical appearance. A relationship with a physically attractive person may be rewarding. Two people who are equally attractive physically experience similar outcomes on this dimension. What about two people who differ in attractiveness? The less attractive person benefits from associating with the more attractive one, whereas the more attractive person experiences less positive outcomes. Because attractiveness is a valued and highly visible asset, the more physically attractive person is likely to find alternative relationships available and to expect some of them to yield more positive outcomes.

This reasoning was tested in a study of 123 dating couples. Photographs of each person in the study were rated by fiive men and five women for physical attractiveness, and a relative attractiveness score was calculated for each member of each couple. Both men and women who were more attractive than their partners reported having more friends of the opposite sex (that is, alternatives) than men and women who were not more attractive than their partners. Follow-up data collected 9 months later indicated that dating couples who were rated as similar in attractiveness were more likely to be still dating each other (White, 1980). These results are consistent with the hypothesis that persons experiencing outcomes below CLalt are more likely to terminate the relationship. Another study of 120 couples asked each to rate the relationship on 16 dimensions (Attridge, Berschied, & Simpson, 1995). Six months later the researchers determined whether the couple was still together. The best predictor of whether a couple broke up was the assessment of the "weak link," the person who gave the relationship lower ratings than the partner.

But not everyone compares their current outcomes with those available in alternative relationships. Individuals in the study by White who were committed—that is, cohabiting, engaged, or married—did not vary in the number of alternatives they reported. Also, their relative attractiveness was not related to whether they were still in the relationship 9 months later. Persons who are committed to each other may be more concerned with equity than alternatives.

Equity theory (Walster [Hatfield], Berscheid, & Walster, 1973) postulates that each of us compares the rewards we receive from a relationship to our costs or contributions. In general, we expect to get more out of the relationship if we put more into it. Thus we compare our outcomes (rewards minus costs) to the outcomes our partner is receiving. The theory predicts that **equitable relationships**—those in which the outcomes are equivalent—will be stable, whereas inequitable ones will be unstable.

This prediction was tested in a study involving 537 college students who were dating someone at the time (Walster [Hatfield], Walster, & Traupmann, 1978). Each student read a list of things that someone might contribute to a relationship, including good looks, intelligence, loving, understanding, and helping the other make decisions. Each student also read a list of potential consequences of a relationship, including various personal, emotional, and day-to-day rewards and frustrations. Each student was then asked to evaluate the contributions he or she made to the relationship, the contributions the partner made, the things he or she received, and the things the partner received. Each evaluation was made using an 8-point scale that ranged from extremely positive (+4) to extremely negative (−4). The researchers calculated the person's overall outcomes by dividing the rating of consequences by the rating of contributions. They calculated the perceived outcomes of the partner by dividing the rating of the consequences the partner received by the rating of the contributions the partner was making. By comparing the person's outcomes with the perceived partner's outcomes, the researchers determined whether the relationship was equitable.

Students were interviewed 14 weeks later to assess the stability of their relationships. Stability was determined by whether they were still dating their partner and how long they had been going together (or how long they had gone together). The results clearly demonstrated that inequitable relationships were unstable. The less equitable the relationship was at the start, the less likely the couple was to be still dating 14 weeks later. Furthermore, students who perceived their outcomes did not equal their partner's outcomes reported that their relationships were of shorter duration.

Differential Commitment and Dissolution

Are outcomes (rewards minus costs) the only factor we consider when deciding whether or not to continue a relationship? What about emotional attachment or involvement? We often continue a relationship because we have developed an emotional commitment to the person and feel a sense of loyalty to and responsibility for that person's welfare. The importance of commitment is illustrated by the results of a survey of 234 college students (Simpson, 1987). Each student was involved in a dating relationship and answered questions about 10 aspects of the relationship. Three months later, each person was recontacted to determine whether he or she was still dating the partner. The characteristics most closely related to stability included length and exclusivity of the relationship and having engaged in sexual intimacy; all three are aspects of commitment.

When both persons are equally committed, the relationship may be quite stable. But if one person is less involved than the other, the relationship may break up. The importance of equal degrees of involvement is illustrated in another study. Couples were recruited from four colleges and universities in the Boston area (Hill, Rubin, & Peplau, 1976). Each member of 231 couples filled out an initial questionnaire and completed three follow-up questionnaires, 6 months, 1 year, and 2 years later. At the time the initial data were collected, couples had been dating an average of 8 months; most were dating exclusively, and 10% were engaged. Two years later, researchers were able to determine the status of 221 of the couples. Some were still together, whereas others had broken up.

What distinguished couples who were together 2 years later from those who had broken up? Some major differences are summarized in Table 12.2. Couples who were more involved initially—those who were dating exclusively, who rated themselves as very close, who said they were "in love," and who estimated a high probability that they would get married—were more likely to be together 2 years later. Of those couples who reported equal involvement initially, only 23% broke up in the following 2 years. But of the

Table 12.2 Differential Involvement and Dissolution of a Relationship

Characteristics of Relationships in 1972	Status 2 Years Later			
	Women's Reports		Men's Reports	
	Together	Breakups	Together	Breakups
Mean Ratings				
Self-report of closeness (9-pt. scale)	7.9	7.3**	8.0	7.2**
Estimate of marriage probability (as percentage)	65.4	46.4**	63.1	42.7**
Love scale (max. = 100)	81.2	70.2**	77.8	71.5**
Liking scale (max. = 100)	78.5	74.0*	73.2	69.6
Number of months dated	13.1	9.9*	12.7	9.9*
Percentages				
Couple is "in love"	80.0	55.3*	81.2	58.0**
Dating exclusively	92.3	68.0**	92.2	77.5**
Seeing partner daily	67.5	52.0	60.7	53.4
Had sexual intercourse	79.6	78.6	80.6	78.6
Living together	24.8	20.4	23.1	20.4

Note: N = 117 together, 103 breakups for both men and women. Significance by t tests or chi-square for together-breakup differences; p = probability of obtaining such a result by chance.

*$p < .05$.

**$p < .01$.

SOURCE: Adapted from Hill, Rubin, and Peplau, 1976.

couples who reported unequal involvement initially, 54% were no longer seeing each other 2 years later.

Earlier in this chapter, we discussed the importance of similarity in establishing relationships. How important is similarity in determining whether a relationship persists over time? Among the 221 couples, both couples who stayed together and those who broke up were initially similar in their sex-role attitudes, approval of premarital sexuality, importance of religion, and in the number of children they wanted. Thus, although attitudinal similarity seems to determine the formation of relationships, it does not distinguish couples whose relationships persist from those whose relationships dissolve.

Not surprisingly, the breakup of a couple was usually initiated by the person who was less involved. Of those whose relationships ended, 85% reported that one person wanted to break up more than the other. There was also a distinct pattern in the timing of breakups; they were much more likely to occur in May–June, September, and December–January. This suggests that factors outside the relationship, such as graduation, moving, and arriving at school, led one person to initiate the breakup.

The dissolution of a relationship is often painful. But breaking up is not necessarily undesirable. It can be thought of as a part of a filter process through which people who are not suited to each other terminate their relationships.

Responses to Dissatisfaction

Not all relationships that involve unequal outcomes or differential commitment break up. What makes the difference? The answer is, in part, the person's reaction to these situations. The level of outcomes a person experiences and his or her commitment to the relationship are the main influences on *satisfaction* with that relationship (Bui, Peplau, & Hill, 1996; Rusbult, Johnson, & Morrow, 1986). As long as the person is satisfied, whatever the level of rewards or commitment, he or she will want to continue the relationship. People who are satisfied are

12.3   Are You Lonely Tonight?

Did you feel lonely when you first entered school here? If you did, you weren't alone. People entering a college or university are likely to feel lonely for the first several weeks or months (Cutrona, 1982). In fact, most people have experienced loneliness sometime during their lives.

Loneliness is an unpleasant, subjective experience that results from the lack of social relationships satisfying in either quantity or quality (Perlman, 1988). Loneliness is different from being alone or social isolation. Social isolation is an objective situation, whereas loneliness is a subjective, internal experience. You can feel lonely in the midst of a family reunion, and you can be alone in your room and yet feel connected to others.

Loneliness is different from shyness. Shyness is a personality trait that reflects characteristics of the person rather than the state of one's social ties. *Shyness* is defined as "discomfort and inhibition in the presence of others" (Jones, Briggs, & Smith, 1986). A study of several measures of shyness found that the common element in these measures is distress in and avoidance of interpersonal situations. When shy people interact with others, they are afraid they are being evaluated by the other person and are more likely to think they are making a negative impression on the other person (Asendorpf, 1987).

There are two types of loneliness (Weiss, 1973), which differ in their cause. One is social loneliness, which results from a lack of social relationships or ties to others. Several studies have found that people with few or no friends and few or no ties to family are more likely to feel lonely (Stokes, 1985). Thus events that disrupt

ties to social networks can cause loneliness (Marangoni & Ickes, 1989). The other type is emotional loneliness, which results from the lack of emotionally intimate relationships. One study of adolescents found a strong association between self-disclosure and loneliness; greater self-disclosure to others was associated with reduced loneliness (Davis & Franzoi, 1986). Thus shyness can cause loneliness by inhibiting self-disclosure. There is evidence that loneliness in men is the result of having few or no relationships with others, whereas in women it is the result of having no intimate relationships (Stokes & Levin, 1986). Clearly, loneliness is tied to the state of one's interpersonal relationships.

Because loneliness is related to the number and quality of interpersonal relationships, we can predict that people in some circumstances are more likely to experience it. In general, people undergoing a major social transition are at greater risk of loneliness. The transition from school to work may be accompanied by feelings of loneliness, especially when this transition involves a geographic move. Second, living arrangements are related to feeling lonely. A study of 554 adult men and women found that living alone was the most important determinant of these feelings (de Jong-Gierveld, 1987). Third, one's marital status is important. We described earlier in this chapter the increasing self-disclosure and interdependence that accompanies the development of romantic relationships; people who are engaged, cohabiting, or married should be less likely to experience emotional loneliness. Conversely, people who have recently gone through the termination of an intimate relationship—through breaking up, divorce, or death of a partner—may be especially vulnerable to loneliness.

---

more likely to engage in *accommodation,* to respond to potentially destructive acts by the partner in a constructive way (Rusbult et al., 1991). A study of 60 students and 36 married couples found that another important influence on satisfaction is the perception that your partner supports your attempts to achieve goals important to you (Brunstein, Dangelmayer, & Schultheiss, 1996).

An individual in an unsatisfactory relationship has four basic alternatives (Goodwin, 1991; Rusbult, Zembrodt, & Gunn, 1982): exit (termination), voice (discuss it with your partner), loyalty (grin and bear it), and neglect (stay in the relationship but do not contribute much). Which of these alternatives the person selects depends on the anticipated costs of breaking up, the availability of alternative relationships, and the level of rewards obtained from the relationship in the past.

To assess the costs of breaking up, the individual weighs the costs of an unsatisfactory relationship against the costs of ending that relationship. There are three types of barriers or costs to leaving a relationship: material, symbolic, and affectual (Levinger, 1976). Material costs are especially significant for persons who have pooled their financial resources. Breaking up requires agreeing on who gets what, and it may produce a lower standard of living for each person. Symbolic costs include the reactions of others. A survey of 254 persons, 123 of whom were in relationships, measured perception of friends' and family members' support for the relationship and commitment to it (Cox et al., 1997). Persons who perceived more support were more committed, in both dating and married couples. Will close friends and family members support or criticize the termination of the relationship? A longitudinal study of dating couples found that lower levels of support by friends for the relationship was associated with later termination of it (Felmlee, Sprecher, & Bassin, 1990). Affectual costs involve changes in one's relationships with others. Breaking up may cause the loss of friends and reduce or eliminate contact with relatives; that is, it may result in *loneliness* (see Box 12.3, "Are You Lonely Tonight?"). One may conclude that the costs of breaking up are too great and stay in the relationship.

A second factor in this assessment is the availability of alternatives. The absence of an attractive alternative may lead the individual to maintain an unrewarding relationship, whereas the appearance of an attractive alternative may trigger the dissatisfied person to dissolve the relationship.

Interestingly, people who are in relationships perceive opposite-sex persons of the same age as less physically and sexually attractive than do people who are not in relationships (Simpson, Gangestad, & Lerma, 1990). This devaluation of potential partners contributes to relationship maintenance. However, a longitudinal study found that the perceived quality of alternative partners increased among persons whose relationships subsequently ended (Johnson & Rusbult, 1989).

A third factor is the level of rewards experienced before the relationship became dissatisfying. If the relationship was particularly rewarding in the past, the individual is less likely to decide to terminate it.

How important are each of these three factors—that is, which factors are most important in determining whether a dissatisfied person responds by discussing the situation with his or her partner, waiting for things to improve, by neglecting the partner, or by terminating the relationship? In one study, subjects were given short stories describing relationships in which these three factors varied. They were asked what they would do in each situation (Rusbult, Zembrodt, & Gunn, 1982). Results showed that the lower the prior satisfaction—that is, the less satisfied and the less positive their feelings and caring for their partner—the more likely they were to neglect or terminate the relationship. The less the investment—that is, degree of disclosure and how much a person stands to lose—the more likely subjects were to engage in neglect or termination. Finally, the presence of attractive alternatives increased the probability of terminating the relationship. A later study of ongoing relationships yielded the same results (Rusbult, 1983).

A study of the stability of relationships of 167 couples over a 15-year period also found that satisfaction, level of investments, and quality of alternatives predicted commitment. Relationships in which commitment was high were more likely to endure (Bui, Peplau, & Hill, 1996).

Summary

Interpersonal attraction is a positive attitude held by one person toward another person. It is the basis for the development, maintenance, and dissolution of close personal relationships.

Who Is Available? Institutional structures and personal characteristics influence who is available to us as potential friends, roommates, co-workers, and lovers. Three factors influence who we select from this pool. (1) Our daily routines make some persons more accessible. (2) Proximity makes it more rewarding and less costly to interact with some people rather than others. (3) Familiarity produces a positive attitude toward those with whom we repeatedly come in contact.

Who Is Desirable?

Among the availables, we choose based on several criteria. (1) Social norms tell us what kinds of people are appropriate as friends, lovers, and mentors. (2) We prefer a more physically attractive person, both for aesthetic reasons and because we expect rewards from associating with that person. Attractiveness is more influential when we have no other information about a person. (3) We choose based on our expectations about the rewards and costs of potential relationships. We choose to develop those relationships whose outcomes we expect will exceed both comparison level (CL) and comparison level for alternatives (CLalt). We implement our choices by making contact, using an opening line that often indicates the kind of relationship in which we are interested.

The Determinants of Liking

Many relationships—between friends, roommates, co-workers, or lovers—involve liking. The extent to which we like someone is determined by three factors. (1) The major influence is the degree to which two people have similar attitudes. The greater the proportion of similar attitudes, the more they like each other. Similarity produces liking because we prefer cognitive consistency and because we expect interaction with similar others to be reinforcing. (2) Shared activities become an important influence on our liking for another person as we spend time with him or her. (3) We like those who like us; as we experience positive feedback from another, it increases our liking for that person.

The Growth of Relationships

As relationships grow, they change on three dimensions. (1) There may be a gradual increase in the disclosure of intimate information about the self. Self-disclosure is usually reciprocal, with each person revealing something about themselves in response to revelations by the other. (2) Trust in the other person—a belief in his or her honesty, benevolence, and reliability—also increases as relationships develop. (3) Interdependence for various gratifications also increases, often accompanied by a decline in reliance on and number of relationships with others.

Love and Loving

(1) Whereas liking refers to a positive attitude toward an object, love involves attachment to and caring for another person. Love also may involve passion, a state of intense absorption in the other and of intense physiological arousal. (2) The experience of passionate love involves cognitive, emotional, and behavioral elements. (3) The concept of love does not exist in all societies; the romantic love ideal emerged gradually in the United States and came into its own about the time of the Civil War. (4) Love stories shape our beliefs about love and relationships, and our beliefs influence how we behave in and interpret relationships.

Breaking Up

There are three major influences on whether a relationship dissolves. (1) Breaking up may result if one person feels that outcomes (rewards minus costs) are inadequate. A person may evaluate present outcomes against what could be obtained from an alternative relationship. Alternatively, a person may look at the outcomes the partner is experiencing and assess whether the relationship is equitable. (2) The degree of commitment to a relationship is an important influence on whether it continues. Someone who feels a low level of emotional attachment to, and concern for, the partner is more likely to break up with that person. (3) Responses to dissatisfaction with a relationship include exit, voice, loyalty, or neglect. Which response occurs depends on the anticipated economic and emotional costs, whether there are attractive alternatives, and the level of prior satisfaction in the relationship.

Key Terms

access display (p. 293)

attitudinal similarity (p. 294)

attractiveness stereotype (p. 290)

availables (p. 286)

comparison level (p. 290)

comparison level for alternatives (p. 290)

dyadic withdrawal (p. 299)

equitable relationship (p. 306)

interpersonal attraction (p. 285)

loneliness (p. 308)

love story (p. 304)

matching hypothesis (p. 289)

mere exposure effect (p. 287)

norm of homogamy (p. 288)

outcomes (p. 292)

passionate love (p. 301)

romantic love ideal (p. 303)

script (p. 293)

trust (p. 298)

CHAPTER 13
Group Cohesion and Conformity

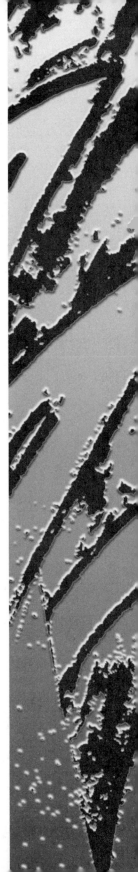

Introduction

Groups are everywhere. We all participate in them, and many of us spend a significant portion of each day engaging in group activities. Families, work groups, sports teams, street gangs, classes and seminars, therapy and rehabilitation groups, classical quartets and rock groups, small military units, neighborhood clubs, church groups—these are only some of the many groups we encounter.

Groups are important to us because they provide social support, a cultural framework to guide performance, and rewards and resources of all kinds. Without them, most individuals would be lost. The character Robinson Crusoe was depicted as living on an island, isolated and alone; he managed to survive, but his plight was less than ideal, and he was sustained primarily by a dream of rejoining his fellow humans. Without groups, most individuals would be isolated, unloved, disoriented, relatively unproductive, and very possibly hungry.

What Is a Group?

A **group** is a social unit that consists of two or more persons and has the following attributes:

1. *Membership.* The persons comprising the group are "members" of that group. To be a member, a person must think of himself or herself as belonging to the group and must also be recognized by other members as belonging to the group.
2. *Interaction among members.* Group members interact with one another. This means, specifically, that they communicate with one another and influence one another.
3. *Goals shared by members.* Group members are interdependent with respect to goal attainment, in the sense that progress by one member toward their objectives makes it more likely that other members will also reach his or her objectives.
4. *Norms held by members.* Group members hold a set of normative expectations (i.e., norms or rules)

that place limits on members' behavior and provide a blueprint for action.

As this definition suggests, groups are not mere collections of individuals; rather, they are organized systems in which relations among individuals are structured and patterned. Not all social units involving two or more persons are groups. For instance, persons in a theater crowd escaping in panic from a fire would not constitute a group. Although some communication may take place among the individuals in the crowd, there are no explicit normative expectations or shared membership with the others present. Likewise, a commercial transaction between yourself and a store clerk selling you a bag of groceries would not qualify as a group interaction. There is no common goal or explicit basis for group membership. To count as a full-fledged group under the definition, a social unit needs to have all four of the properties listed.

Of course, in everyday life we sometimes encounter social units that have some but not all of these properties. Units of this type, which may be considered "near-groups" or "semi-groups," are often interesting and worthy of study. We occasionally include them in the discussion in this chapter and the following.

Social psychologists use the term **small group** to refer to any group sufficiently limited in size that its members can interact face-to-face with one another. A group is not considered small if its size precludes interaction among all its members. Thus, small groups range from 2 to perhaps 20 members (Crosbie, 1975). A discussion group with as many as 30 or even 40 members could still qualify as a small group, however, provided its members were positioned so they could interact directly with one another. Most of the groups discussed in this chapter are small groups.

Framework for Analysis of Groups

To analyze interaction among group members, it helps to have a general framework of groups as systems. One such framework is provided by the **input-process-output model** (Gladstein, 1984; McGrath, 1964; O'Connor, 1980; Scott, 1981). Although abstract in nature, this model identifies important features of groups and calls attention to forms of social interaction that are crucial to the operation of groups.

The Input-Process-Output Model Depicted in Figure 13.1, the input-process-output treats the group as a social unit that transacts with entities (individuals, other groups, or even large bureaucratic organizations) in the environment. Any group has *boundaries* that separate it from the environment. These boundaries are real in the sense that they delineate membership (who is a member of the group and who is not) and ownership (what belongs to the group and what does not). Although restrictive, a group's boundaries may also be permeable in the sense that

they permit certain things to flow across. Thus a group usually receives some *inputs* from its environment and supplies various *outputs* to units in the environment (see Figure 13.1).

Important inputs to a group from its environment include personnel, materials, money, and information. These can be viewed as raw materials that the group combines or transforms in some manner to produce outputs. A group's outputs are the various products created by the group's members. Sometimes these outputs are consumed by the group's own members, but more often they are meant for individuals or other groups in the larger environment. Beyond any doubt, the many groups in our society produce a great variety of outputs. For example, a work group in an automobile factory assembles a new car for shipment to a dealer's showroom. In a courtroom setting, a jury deliberates a case and reaches a verdict regarding guilt or innocence. A team of surgeons in a hospital performs a bypass operation and thereby extends a patient's life. A rock group not only stages musical performances to entertain audiences but also releases new albums. As these examples indicate, outputs from groups include not only material products, but also services and performances benefiting others, and various cognitive products (decisions made, problems solved, and so on).

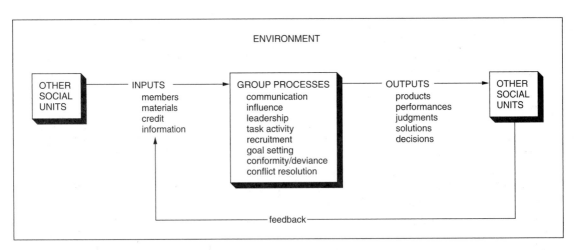

Figure 13.1 Input-Process-Output Model of a Group

The quantity and quality of the outputs from a group depend in part on the *processes* among its members. Important group processes include communication, interpersonal influence, group goal setting, leadership, conformity and consensus building, training of new members, and the like (see Figure 13.1). Not only do these processes affect the nature of group outputs, they also determine members' satisfaction with the group and relationships among the members. We consider them in more detail below.

Once a group generates various products, it may receive feedback from its environment (Figure 13.1). *Feedback* is information indicating whether the group's output has met the environment's requirements or achieved certain targets. By analyzing feedback, a group can decide whether or not to change its products, improve its procedures, refine its goals, or alter its basic direction.

In this chapter and the two that follow (Chapters 14 and 15), we discuss various aspects of group process, composition, structure, and performance. Chapter 14 focuses on communication, status, and equity in groups. Chapter 15 treats factors affecting group performance and decision making. The present chapter discusses group cohesion, goals, norms, and conformity.

Questions About Groups

This chapter examines the forces that give unity and integrity to a group. It addresses the following questions:

1. What factors hold a group together as a unit? That is, what produces cohesion—or the lack of it—in groups?
2. What are group goals? How do they differ from individual goals, and how do they relate to group functioning?
3. What are group norms, and in what ways do they regulate the behavior of members?
4. Under what conditions will group members conform to the standards of the majority? How do group majorities react to deviance by individual members?
5. Under what conditions can a minority coalition within a group successfully influence the majority?

Group Cohesion

A recreational baseball club—the Jaguars—has a long record of championships in its city league. The Jaguar players take pride in their performance and are very committed to their team. At practice and in games, this team is a model of enthusiasm and coordination. On the rare occasion when they have a losing streak, all the team members voluntarily hold extra practice sessions to sharpen their skills and teamwork. The players like each other, and they enjoy playing together and celebrating their victories. Although they do not always agree on strategy, the Jaguars resolve their differences quickly. Several of the players consider their teammates best friends, and they often spend some time together off the field. The Jaguars team rarely loses any of its players—not even its second stringers—and everyone turns out enthusiastically for practices and other meetings.

The players on another team in the league—the Penguins—provide a very different story. Less than a model of competence, the Penguins have finished in last place in the league standings for three seasons running. Occasionally, the Penguins have to forfeit a game because they cannot even field a team of nine players. The players don't even seem to like playing together very much—they are always busy with other activities and prefer not to spend much time practicing. The last time the team agreed to hold a practice, some players forget about it or did not bother to show up; this may be related to the fact that the players seldom run into one another outside of team activities. The Penguins' planning session last spring dissolved into chaos when the players could not agree on how to pay for some new equipment. Ever since then, a lot of friction has occurred among certain members, and there is some doubt whether the team will even participate in the league next year.

The Jaguars and the Penguins obviously differ in various respects. For one thing, the Jaguars win a lot more than the Penguins. The teams also differ notably with respect to their members' willingness to participate. The Jaguar players care about their membership on the team and want to participate, whereas the Penguin players seem to care much less. The Jaguars have a stronger grip on members' loyalty than the Penguins, and the team is bonded together more firmly. Stated in other terms, the Jaguars have a higher level of group cohesion than the Penguins.

Group cohesion refers to the extent to which members of a group desire to remain in that group and resist leaving it (Balkwell, 1994; Cartwright, 1968). A highly cohesive group, in general, maintains a firm hold over its members' time, energy, loyalty, and commitment. Cohesive groups are marked by strong ties among members and by a tendency for members to perceive events in similar terms (Braaten, 1991; Evans & Jarvis, 1980). Because members of a cohesive group desire to belong, the interaction among them typically has a positive, upbeat character and reflects a "we" feeling.

Interaction among happy sorority sisters on pledge day reflects a high level of group cohesiveness. Many groups safeguard cohesiveness by selecting new members carefully.

The Nature of Group Cohesion

There are many different motives for joining and staying in a group. For instance, members may belong to a group because they like the tasks they perform, they enjoy interacting with the other members, they take comfort in belonging to a group that has values and ideologies similar to their own, or they believe membership is a way of getting something else that they want (such as prestige, money, future opportunities, protection, or social contacts).

Different motives for belonging at the individual level lead to different types of cohesion at the group level. This fact has led some theorists to conclude that group cohesion is best understood as a complex, multidimensional phenomenon rather than as a single variable (Cota et al., 1995; Mullen & Copper, 1994; Tziner, 1982).

The most fundamental types of cohesion are social cohesion and task cohesion. A group is considered to have *social cohesion* if its members stay in the group primarily because they like one another as persons and desire to interact with one another. If a group consists of members who know and like one another,

it will have more social cohesion than one in which the members do not particularly care for one another or enjoy one another's company (Aiken, 1992; Lott & Lott, 1965). Other things being equal, social cohesion is greater when a group is homogeneous in composition. Because similarity often increases liking, social cohesion tends be greater when members are similar with respect to such attributes as education, ethnicity, and status, and when members hold similar attitudes rather than dissimilar ones.

The other major type of cohesion is *task cohesion*. When a group has high task cohesion, its members remain together primarily because they are heavily involved with the group's task(s). Task cohesion is greater if members find the group's tasks intrinsically valuable, interesting, and challenging, rather than not. It is also greater if the group's objectives (and the tasks contributing to attainment of those objectives) are sharply defined, rather than vaguely defined (Raven & Rietsema, 1957). Groups that succeed at achieving their goals (like the Jaguars) often have higher task cohesion than groups that fail repeatedly (like the Penguins).

Consequences of Group Cohesion

What are the consequences of group cohesion? That is, what difference does it make whether a group is highly cohesive or not? First, the level of cohesion affects the amount of interaction that occurs among group members. Given the opportunity, members of highly cohesive groups communicate more with one another than do members of less cohesive groups. This holds true for a wide variety of groups, ranging from student organizations to industrial training groups (Moran, 1966). Cohesion affects not only the amount of interaction in groups but also its form. Interaction among members in highly cohesive groups is usually friendlier, more cooperative, and entails more attempts to reach agreements and to improve coordination (Shaw & Shaw, 1962).

The exercise of influence tends to be greater in high-cohesion groups than in low-cohesion groups. That is, persons in high-cohesion groups try to influence other members more than do persons in low-cohesion groups (Lott & Lott, 1965). Moreover, members of high-cohesion groups tend to be more open to influence from, and are more influenced by, their fellow members than is the case in low-cohesion groups. Persons in highly cohesive groups conform more to expectations of their fellow members than do persons in less cohesive groups (Sakurai, 1975; Wyer, 1966). Under high cohesion, members care about belonging and want their group to perform well, so they naturally incline toward using influence to bring about coordination and consensus in the group.

In addition to its impact on influence and communication, group cohesion also affects the level of a group's productivity and performance. Several recent meta-analyses have concluded that group productivity is greater in high-cohesion groups than in low-cohesion groups (Evans & Dion, 1991; Gully, Devine, & Whitney, 1995; Mullen & Copper, 1994). Although not all studies have shown this effect, the preponderance of evidence supports it. The meta-analysis by Mullen and Copper (1994), based on many tests of the cohesion-performance effect, concluded that the relationship between cohesion and productivity is highly significant in a statistical sense, albeit small in magnitude.

One important qualification, however, is that group productivity depends on the form of cohesion holding the group together. Task cohesion (i.e., members' commitment to the group's task) has a significant, positive effect on productivity, but other forms of cohesion (such as social cohesion and group pride) appear to have little or no effect on productivity (Mullen & Copper, 1994). In groups held together primarily by social cohesion rather than by task cohesion, the members may prefer to spend their time socializing with one another rather than producing. (For more on ties among members, see Box 13.1, "Primary Groups.")

Group Goals

A **group goal** is an outcome or end state viewed by members as desirable and important for their group to attain. This desired end state is usually different from and preferred to the group's current state. As an example, a team of surgeons may pursue the goal of separating a pair of conjoined twins into two individuals capable of pursing normal lives. A high school basketball team with strong players might pursue the goal of winning the state basketball championship in its division. A new management team may set out to save a business from bankruptcy and restore its financial viability. Although some groups have just one major goal, many others have multiple goals that they pursue simultaneously.

Group goals can be defined at different levels of specificity. One approach distinguishes among missions, values, and objectives (Hare, 1992; Nadler, 1981; Tolle, 1988). A group's *mission* statement indicates, on the most general level, what the group does and why it exists. Its *values* statement indicates, broadly, the ideal states to be achieved by the group at some unspecified time in the future. But it remains for the *objectives* statement to spell out the dimensions or indicators in terms of which group performance will actually be measured and to establish specific targets to be achieved on these dimensions. Once the objectives have been developed, the group can then break these down further and assign specific task responsibilities to members. Task assignments indicate who is to do what, when, and where to help achieve the group's objectives.

13.1   Primary Groups

A **primary group** is a small group in which the emotional ties among members are strong. Members of primary groups usually know one another very well and like each other. Primary groups frequently have a distinctive subculture with informal rules and expectations that control the actions of members (Dunphy, 1972; Kimberly, 1984).

Primary groups exist in a variety of forms. Among the most common in our society are (1) nuclear families, (2) peer groups, such as children's play groups, teen delinquent gangs, and Wednesday night poker clubs, (3) informal work groups existing in organizational settings, such as factory work groups, classroom groups, and small military units, and (4) groups whose goal is to resocialize their members, such as rehabilitation groups, therapy groups, and self-analytic groups.

Members of a primary group usually possess similar values and attitudes, and they care about their mutual friendships. The interaction among members in primary groups is typically spontaneous, informal, and personal. This is not to say that it always takes a smooth course. The members of a family can have heated arguments, and children often fight with others in their peer group. Still, the crucial characteristic of primary groups is that members care about one another as individuals.

Primary groups are usually contrasted with **secondary groups,** which are groups whose members have relatively few emotional ties with one another. In secondary groups, members may not really care very much about one another as individuals, and they instead relate to one another as occupants of roles. Interaction in secondary groups tends to be formal, impersonal, and nonspontaneous. One common type of secondary group is the bureaucratic, structured work group that affords its members few opportunities for personal contact.

In our urban, bureaucratic society, primary groups perform an important function for their members. They provide an arena for friendship and deep personal commitments. They also serve to integrate the individual with larger organizations in society.

Group Goals and Individual Goals

It is useful to distinguish between group goals and individual goals; although related, they are not one and the same (Cartwright & Zander, 1968; Mackie & Goethals, 1987). Although group goals are outcomes that members view as desirable for the group to attain, individual goals are outcomes desired by members for themselves. Consider the case of three boys who team up to build and run a lemonade stand, each for different motives. The first boy's individual goal is to make enough money to buy a baseball glove; the second boy's goal is to make use of the carpenter tools that he received as a birthday gift; the third boy's goal is simply to spend time with his friends and participate in their activities. These individual goals differ not only from one another but also from the group goal of building and running the lemonade stand. A group goal is not necessarily the simple sum of individual goals, nor can it always be inferred directly from individual goals.

Most groups function best when there is substantial similarity, or isomorphism, between group goals and the individual goals of its members. The term **goal isomorphism** refers to a state in which group goals and individual goals are similar in the sense that actions leading to attainment of group goals also lead simultaneously to attainment of individual goals. Some research findings indicate that when there is little discrepancy or incompatibility between group goals and individual goals, members have more motivation to pursue the group goals and to contribute resources toward the group goals than when the discrepancy or incompatibility is large (Sniezek & May, 1990).

For this reason, a group may wish to heighten the level of isomorphism between individual goals and group goals. There are several ways to do this. First, a group can recruit selectively—that is, admit as members only persons who strongly support the group's main goal(s) from the outset. Second, a group can attempt to gain increased acceptance by members of group goals through such activities as socialization and training; of course, this approach may cost the group both time and effort. Third, a group can heighten goal isomorphism by increasing members' awareness that they belong to the group and making their identity as members more salient (Mackie & Goethals, 1987). This may be done by increasing the proximity of members to one another, increasing the experiences they share in common, increasing the amount of social contact and communication among them, or using a common designation to label them (Dion, 1979; Turner, 1981).

Goal Commitment and Group Performance

The goals established by a group have an important impact on the level of performance achieved by the group's members. In general, we can say that (1) groups working toward an explicit, challenging objective perform better than groups working without an explicit objective, and (2) the target level of the objective is an important determinant of the performance achieved, with greater performance occurring in groups that have more demanding targets or levels of aspiration. These effects have been demonstrated by many studies, using a variety of different tasks, with goals pertaining to quantity and quality of performance, in both naturally occurring groups and ad hoc laboratory groups (O'Leary-Kelly, Martocchio, & Frink, 1994; Pritchard et al., 1988; Weldon & Weingart, 1993).

These effects are understandable because a group generally is not going to achieve excellent performance unless it intentionally sets out to do so. For instance, a mountain-climbing team may include skilled climbers, but it is not going to achieve excellence unless it intentionally sets out to climb the highest and most demanding peak. Note, however, that performance depends on more than merely selecting a high target level. Weldon and Weingart (1993) maintain that group performance also depends heavily on members' goal commitment. **Goal commitment** refers to the strength of members' determination to reach specific group objectives. If members have a high level of goal commitment, they are more likely to work

hard and to persist in the face of setbacks than if they have a low level of goal commitment. Thus, to achieve strong performance, a group needs both a high target level and high goal commitment by members.

Because members are not always highly committed to a group's goals, it raises this question: What determines the level of members' commitment to group goals? Many studies over the years have investigated factors that affect members' level of goal commitment. Summarizing these findings in general terms, Weldon and Weingart (1993) conclude that members' goal commitment will be high only if (1) the members value the goal itself and find the prospect of attaining the goal highly attractive, and (2) the members hold a strong expectation that the goal can actually be attained by the group. In other words, goal commitment will be high only if members value the goal highly and believe it is not too difficult for them to achieve.

Many specific factors enter into this. Members tend to value group goals more highly—and therefore have greater commitment to them—when there is a high level of isomorphism between their individual goals and the group's goal. Goal commitment also tends to be higher when members identify strongly with their group, rather than weakly (Breckler & Greenwald, 1986). And it is greater if they themselves have participated in establishing the group's goal or perceive the goal has been established by an authority who behaved in a legitimate and unbiased way (Locke & Latham, 1990; Tyler & Lind, 1992).

Goal commitment is also higher when members believe they have the ability to meet the demands of the task. Thus, goal commitment of members is greater when they perceive the group has a high a level of skill and efficacy, rather than a low level (Bandura, 1988; Mesch, Farh, & Podsakoff, 1994). Goal commitment is greater when members construe the goal as realistic and attainable, rather than extraordinarily difficult and formidable. It also tends to be higher when the path to achieve the goal is clear and well defined, rather than unclear or vaguely defined (Raven & Rietsema, 1957).

Another important factor in goal commitment is the feedback the group receives regarding its performance while it is working to reach its target. Perhaps surprisingly, negative feedback is sometimes more effective than positive feedback in maintaining members' goal commitment and goal pursuit (Mesch, Farh, & Podsakoff, 1994). One illustration of the energizing effect of moderately negative feedback occurred in a study of a U.S. expedition climbing Mount Everest (Emerson, 1966). This expedition took 92 days to climb Everest, which, at 29,028 feet above sea level, is the world's tallest peak. At this altitude, the air is very thin, and even the simplest actions require enormous effort and will. A climber can honestly believe he wants to climb Everest but still fail to get out of his sleeping bag soon enough to get important work under way.

Serving as a member of this expedition, the investigator was able to observe the group's efforts to reach its goal. Feedback from group leaders to other members was crucial in sustaining members' goal commitment and motivation. Leaders were selective—sometimes even intentionally misleading—in passing along information to the climbers. To dampen overoptimism, they occasionally gave negative feedback to members on the group's progress. This kept the team challenged so members exerted their best efforts under the harsh and uncertain conditions of the mountain. Moderate levels of negative feedback fortified goal commitment and striving by the group.

Group Norms

A **norm** is a rule or standard that specifies how group members are expected to behave under given circumstances. Most groups develop a variety of norms that regulate the activities of their members. To illustrate, a factory work group may have norms specifying what time of the day workers are expected to start on the assembly line and how much they are expected to produce; norms of this type obviously have an impact on the level of productivity achieved by the work group. A group of college admissions officers may have norms that regulate how judgments are made by the officers; the nature of these norms indirectly affects which applicants will be admitted and which not. A family may have norms regulating who washes the dishes or

Group norms channel the behavior of members and may cause them to engage in conduct they would not otherwise do. Here, Shriners parade their miniature Tin Lizzies down a city boulevard in California.

mows the lawn, as well as norms specifying who can and cannot have sexual relations with whom (for example, the incest taboo). A teenage street gang may have norms regulating various actions of its members. For instance, one study of gangs of adolescent boys in the Southwest found they had norms governing what kinds of information members could reveal to outsiders (for example, parents and police) and how members should behave during street fights with rival gangs (Sherif & Sherif, 1964).

In this section, we discuss group norms in some detail. We look first at the functions that norms serve for groups and their members. Then we discuss the return potential model of group norms.

Functions of Norms

What functions do norms serve for groups? First, norms foster coordination among members in pursuit of group goals. Because norms usually reflect a group's fundamental value system, they prescribe behaviors that foster attainment of important goals. When members conform to group norms, they know what to expect of one another, and this strengthens coordination among them. Thus, norms direct members to carry out actions that facilitate goal attainment and to refrain from other actions that inhibit or block goal attainment.

Second, norms provide a cognitive frame of reference through which group members interpret and judge their environment. That is, norms provide a basis for distinguishing good from bad, important from unimportant, tenable from untenable. They are especially useful in novel or ambiguous situations, where they serve as pointers on how to behave. Because they are anchored in the group's values and culture, norms bring predictability and coherence to group activities; they permit members to predict how others in the group will behave.

Third, norms define and enhance the common identity of group members. This is especially true when group norms require members to behave differently from persons outside the group. Norms that prescribe distinctive dress (for example, clothing or hairstyles) or distinctive speech patterns (for example, dialects or vocabulary) differentiate group members from nonmembers. These norms demarcate group boundaries and reinforce the group's distinctive identity.

Return Potential Model of Norms

Group norms can be viewed as having structure and shape. One useful approach is the **return potential model** (Jackson, 1965), which treats norms as having two dimensions—the behavior dimension and the evaluation dimension. The *behavior dimension* specifies the frequency or amount of behavior regulated by the group norm, whereas the *evaluation dimension* refers to the response to that behavior by other members. Others' evaluation of behavior can range from positive to negative.

To illustrate, Figure 13.2 displays three norms (labeled norms A, B, and C). These are drawn as

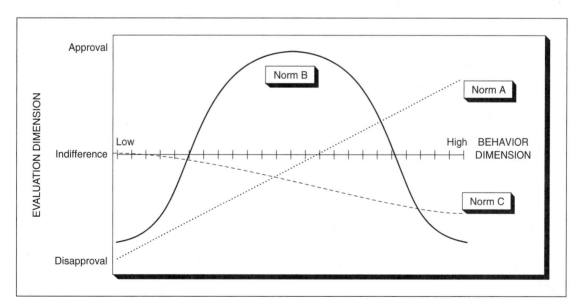

Figure 13.2 Return Potential Curves

The curves in this figure represent three group norms. In each case, varying amounts of behavior are met with some specified amount of approval or disapproval from group members. Under norm A, for example, low amounts of behavior are disapproved, whereas high amounts are approved. Under norm B, intermediate amounts of behavior are approved, whereas very low and very high amounts are disapproved. Under norm C, high amounts of behavior are disapproved, whereas low amounts are neither approved nor disapproved.

SOURCE: Adapted from Jackson, 1965.

curves in a space defined by the behavior dimension and the evaluation dimension. Because the curves represent different norms, they have different shapes; yet in each case they show that evaluation is a function of behavior.

Suppose that norm A (in Figure 13.2) pertains to productive activity by a group of editors working on a magazine. Under norm A, members encourage and reward higher levels of a given behavior, such as editing pages of manuscript per day. Under this norm, the more pages a member edits, the more she is rewarded. Any member who produces at a very low level on this behavior dimension receives a negative evaluation from other group members. She may be criticized or castigated or perhaps face a monetary fine. Only by producing at a higher level does she receive a positive evaluation.

Norm B offers an interesting contrast with norm A. Whereas norm A basically said "the more the better," norm B says that there is such a thing as producing too much. Under norm B, a member will be rewarded for producing in the middle range on the behavior dimension. Productivity at very low levels or at very high levels will be negatively evaluated and discouraged. When members notice that one among them is working much harder than the others, they may disapprove of this "overproduction."

A norm of this type was evident in a classic early study of an industrial work group (Roethlisberger & Dickson, 1939). A group of 14 men, known as the Bank Wiring Group, assembled banks of telephone switching equipment for the Western Electric Company. The management of the company had recently instituted a wage incentive plan rewarding higher levels of productivity by individual workers. Managers believed this plan would bring about an increase in the productivity of the Bank Wiring Group. Surprisingly, however, the workers continued to produce at the same rate as before, despite the opportunity to increase their earnings. This happened because as subsequent investigation revealed, the group had a firmly established production norm that resembled norm B in Figure 13.2. Workers feared that if they began to produce at higher levels, the company might eventually switch the pay schedule and require still higher levels of productivity for a given wage. Therefore, the group enforced its own production norm on individual members. If a man produced too much, he was ridiculed by other members as a "rate buster." If he produced too little, he was disparaged as a "chiseler." If an offender did not respond to these verbal reprimands, he was soon subjected to another form of sanction termed "binging," whereby other members would punch him in the arm. Although this may seem unusual, groups often develop their own special form of sanctions. What matters more is not the form, but whether the sanction is effective in regulating the behavior of its members.

The norms we have discussed to this point—A and B—involve both positive and negative evaluation. Other norms, however, may pertain to behaviors that group members consider simply undesirable. For instance, under norm C in Figure 13.2, group members may tolerate low levels of the undesirable behavior, but their reaction becomes more negative as the behavior becomes more frequent or intense. Although a norm like C is unlikely to pertain to productive activities, it might apply to such behaviors as talking too much, importuning others, making a nuisance of oneself, and so on.

Every norm entails a *range of tolerable behavior*—that portion of the behavior dimension that members of a group approve and evaluate positively. For example, in Figure 13.2, norm A delimits a range of tolerable behavior at the higher end of the behavior dimension, whereas norm B delineates a range of tolerable behavior in the middle of the behavior dimension and norm C delineates a range of tolerable behavior only at the lowest end of the behavior dimension. Behaviors outside this range are not tolerated or approved, whereas behaviors inside this range are.

Another characteristic of a norm is its *intensity,* which refers to the range of the return potential curve from its highest to its lowest point, both above and below the point of indifference. The intensity of a norm reflects the strength of group feelings (whether approval or disapproval) regarding that behavior. In Figure 13.2, group members feel more strongly about the behavior regulated by norm B than about that regulated by norm C. Norms that affect attainment of important group goals often have high intensity, whereas norms about matters of personal taste (such as style of dress or manner of speaking) have lower intensity.

Majority Influence and Conformity

The term **conformity** means adherence by an individual to group norms or standards. We say that a group member conforms to a given norm when his or her behavior falls within the range of tolerable behavior for that norm. No doubt a great proportion of behavior that we witness in daily life involves conformity to one group norm or another. Group members often change their behavior expressly so it will conform to group norms.

Nevertheless, conformity cannot be taken for granted; group members conform often but not always. In this section, we first consider some classic studies of conformity based on a line-judgment paradigm. Then we discuss two forms of influence—normative influence and informational influence—that produce conforming behavior in groups. Next, we consider some factors that affect the amount of conformity occurring in groups. Finally, we look at reactions by majorities to nonconformity in groups.

The Asch Conformity Paradigm

In groups, influence flows in many directions—members influence other members and are influenced in turn. Of special importance, however, is the influence exercised by the group's majority over the behavior of individual members. Social psychologists use the term **majority influence** to refer to the process(es) by which a group's majority pressures an individual member to conform or to adopt a specific position on some issue. Majority influence is important, for it gives a group integrity and continuity over time. Of course, the amount of influence exerted by the majority on individual members varies from group to group. In some cases, it is slight to moderate, whereas in others it is great.

The impact of majority influence on individual members was illustrated in a series of classic experiments by Asch (1951, 1955, 1957). Using a laboratory setting, Asch created a situation in which an individual was confronted by a majority that agreed unanimously on a factual matter (spatial judgments) but was obviously in error. These studies showed that, within limits, groups can pressure their members to change their judgments and conform with the majority's position, even when that position is self-evidently wrong.

In the basic Asch experiment, a group of eight persons participated in an investigation of "visual discrimination." In fact, all but one of these persons were confederates working for the experimenter. The remaining individual was a naive subject. In front of the experimental room, large cards displayed a standard line and three comparison lines, as shown in Figure 13.3. The subject's objective was to decide which of the three comparison lines was closest in length to the standard line.

The task was simple and straightforward, because one of the comparison lines was the same length as the standard and the other two were very different. The group repeated this task 18 times, using a different set of lines each time; on each trial, the standard line matched one of the three comparison lines. During each trial, the confederates announced their judgments publicly, one after another. The subject also announced his opinion publicly. The group was seated so the confederates responded prior to the real subject.

Although this task seems easy, it turned out to be difficult for the naive subject. On 6 of the 18 trials, the confederates gave a correct response, but on the other 12 trials, the confederates responded incorrectly. The erroneous responses by the confederates put the subject in a difficult position. On the one hand, he knew (or thought he knew) the correct response based on his own perception of the lines. On the other hand, he heard all the other persons (whom he believed to be sincere) unanimously announcing a different and incorrect judgment on 12 trials.

The purpose of the study was to observe how the naive subject would behave during the 12 critical trials. Results indicate that the incorrect opinion expressed by the majority strongly influenced the judgments announced by the naive subjects. In

Group norms can extend to any aspect of behavior, including dress. Bikers have a different dress code from corporate executives, but conformity is high within each group.

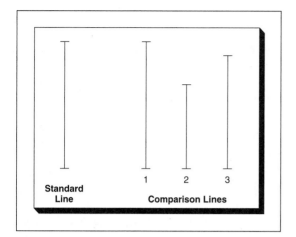

Figure 13.3 Judgmental Task Employed in Asch Conformity Studies

In the Asch paradigm, naive subjects are shown one standard line and three comparison lines. The task is to judge which of the three comparison lines is closest in length to the standard line. By itself, this task appears easy. However, subjects are surrounded by other persons (supposedly also naive subjects, but actually experimental confederates) who publicly announce erroneous judgments regarding the match between lines. Such a situation imposes pressure on the subject to conform to their erroneous judgments.

SOURCE: Adapted from Asch, 1952.

the 12 critical trials, nearly one third of the responses by subjects were incorrect (Asch, 1957). This compares with an error rate of less than 1% in a control condition in which no confederates were present and subjects recorded their judgments privately on paper. Although about one quarter of the subjects showed no conformity and remained independent throughout, the remainder conformed, at least to some degree. One third of the subjects conformed on 50% or more of the critical trials.

Interviews conducted after the experiment revealed that most of the subjects were quite aware of the discrepancy between the group's judgments and their own. They felt puzzled or under pressure and tried to figure out what might be happening. Some wondered whether they had misunderstood the experimental instructions; others began to look for other

explanations or to question their eyesight. Even those subjects who did not conform to the majority felt some apprehension, but they eventually decided the problem rested more with the majority than with themselves. The interviews indicated that conformity by subjects in this study was of a particular type. To a fair extent, it involved public compliance without private acceptance: Although many subjects conformed publicly, they did not believe or accept the majority's judgment privately. In effect, they viewed public compliance as the best choice in a difficult situation.

Dependence and Influence in Conformity

The occurrence of majority influence and conformity in groups can be explained generally by the fact that individual members are dependent on the majority both cognitively and socially. For one thing, members seek information about social reality, and they depend on the majority to validate their understanding of and opinions about the group and the world. For another, individual members want to obtain various rewards and benefits—not the least of which is acceptance of their continuing membership in the group—and they depend heavily on the majority for these outcomes.

Dependence of members on the majority leads naturally to the exercise of influence by the majority in groups. Influence by the majority can take various forms. One analysis distinguishes between normative influence and informational influence (Deutsch & Gerard, 1955; Kaplan, 1987; Turner, 1991, chapter 2). **Normative influence** occurs when a member conforms to expectations held by others (that is, to norms) in order to receive the rewards and/or avoid the punishments that are contingent on meeting these expectations. Being liked and accepted by other members is one important reward in normative influence. To exercise influence of this type, a group needs to maintain at least some degree of surveillance over the behavior of members. The impact of normative influence is heightened, for instance, when members respond publicly rather than anonymously (Insko et al., 1983, 1985). Normative influence is perhaps the most fundamental source of uniformity within groups, although informational influence is also very important.

Informational influence occurs when a group member accepts information from others as valid evidence about reality. This type of influence occurs especially when members need to reduce uncertainty, as in situations that involve ambiguity or entail an absence of objective standards to guide judgment. More concretely, informational influence often occurs in situations in which members are trying to solve a complex problem unfamiliar to them (Kaplan & Miller, 1987); members considered more expert or knowledgeable are especially likely to exercise informational influence during such tasks. It also occurs frequently in crisis situations when members must act immediately but lack knowledge about appropriate action. The common element in all these situations is that the group exerts influence on individual members by providing information that defines reality and serves as a basis for making judgments or decisions.

With respect to the Asch line-judgment task, it seems that normative influence was operating prominently in the situation. Of those subjects who conformed in the Asch experiment, many did so to avoid being embarrassed, ridiculed, or laughed at by the majority. They were seeking acceptance by the majority (or at least, they were seeking to avoid outright public rejection). It is hard to argue that Asch's majorities exercised informational influence to any great extent; after all, most persons have some skill in comparing line lengths and are not very dependent on others for this. Moreover, in one variation, Asch retested his subjects on the same stimuli with the group no longer present, and they gave correct answers; their experience of judging lines in the presence of the majority did not permanently alter their understanding of the lines' lengths.

Autokinetic Effect Studies Although informational influence was not very central in the Asch situation, we should not underestimate its importance in other situations. A famous study by Sherif (1935, 1936)—conducted years before Asch did his line-judgment research—dramatically illustrates the impact of informational influence under conditions of uncertainty. Sherif's study utilized a physical phenomenon known as the *autokinetic effect* (meaning "moves by itself"). The autokinetic effect occurs when a person stares at a stationary pinpoint of light located at a distance in a completely dark room. For most people, this light appears to move in an erratic fashion, even though the light is not actually moving at all. Sherif used the autokinetic effect as a basis for studying informational influence in groups. First, he placed subjects in a laboratory setting by themselves and asked them to estimate how far the light moved. In making these judgments, subjects were literally in the dark—they had no external frame of reference. From their individual estimates, the researcher was able to determine a stable range for each subject. Subjects differed quite a bit in this respect. Whereas some thought the light was moving only an inch or two, others believed it was moving as much as 8 or 10 inches.

Shortly thereafter, Sherif put the same subjects together in groups of three and placed them back in the autokinetic situation. Although the estimates they had made when alone were different, the estimates they made in groups converged on a common standard. This change in judgments by members provides evidence for the operation of informational influence. Lacking an external frame of reference and being uncertain about their own judgment, group members began to use one another's estimates as a basis for defining reality. Each group established its own (arbitrary) standard, and members used this as a frame of reference. This process of norm formation can be quite subtle; in fact, other research (Hood & Sherif, 1962) has shown that subjects involved in an autokinetic experiment are often unaware that their judgments are being influenced by other members.

One interesting finding from Sherif's original study emerged when, a week or two after their initial exposure, the subjects were again placed alone in the autokinetic situation. Results showed that subjects used the group norm as the framework for their new, individual judgments. Although not all studies have found evidence of enduring norm internalization, at least one study retested individual subjects in the autokinetic task a year after their initial exposure to the group norm and found evidence that the group norm still influenced the subjects' judgments, despite the passage of time (Rohrer et al., 1954).

Factors Affecting Conformity

When discussing the Asch line-judgment study, we saw that pressure from the majority can substantially influence the behavior of individual members. But an individual's tendency to conform will be greater under some conditions than under others. Investigators have tried to identity factors that have important effects on the amount of conformity occurring in groups, and we review some of these factors here.

Size of the Majority   Consider again the Asch situation in which a single subject is confronted by a majority. If the majority is unanimous—that is, if all the members of the majority are united in their position—then the size of the majority will have an impact on the behavior of the subject. As the size of the unanimous majority increases, the amount of conformity by subjects increases (Asch, 1955; Rosenberg, 1961). For example, a subject confronted by one other person in an Asch-type situation will conform very little; he or she will answer independently and correctly on nearly all trials. However, when confronted by two persons, the subject will experience more pressure and will agree with the majority's erroneous answer more of the time. Confronted by three persons, the subject will conform at a still higher rate. In his early studies, Asch (1951) found that conformity to unanimous false judgments increased with majority size up to three members and then remained essentially constant beyond that point. Although some research (Gerard, Wilhelmy, & Conolley, 1968) has questioned the exact point at which the effect of majority size begins to level off, increases in the size of the majority beyond three persons generally have little added impact.

Breach in the Majority   What happens when the group's majority is not unanimous? Basically, lack of unanimity among majority members has a liberat-

ing effect on behavior by subjects. A subject will be less likely to conform if a member breaks away from the majority (Gorfein, 1964; Morris & Miller, 1975). One explanation for this is that the member who abandons the majority provides validation and social support for the subject. In an Asch experiment, for example, if one or several members abandon the majority and announce correct judgments, their behavior will reaffirm the subject's own perception of reality and reduce his or her tendency to conform to the majority.

Beyond this, however, any breach in the majority—whether it provides social support or not—will reduce the pressure on the subject to conform (Allen & Levine, 1971; Levine, Saxe, & Ranelli, 1975). In one study (Allen & Levine, 1969), subjects participated in groups of five persons, four of whom were confederates. The subjects made judgments on a variety of items. These included visual tasks similar to those used by Asch, as well as informational items (for example, "In thousands of miles, how far is it from San Francisco to New York?") and opinion items for which there were no correct answers ("Agree or disagree: 'Most young people get too much education'"). Depending on experimental conditions, subjects were confronted with either a unanimous majority of four persons (the control condition), a majority of three persons and a fourth person who broke from the majority and gave the correct answer (the social support condition), or a majority of three persons and a fourth person who broke from the majority but gave an answer even more erroneous than that of the majority (the extreme erroneous dissent condition).

The results of this study are shown in Figure 13.4. The control condition, which involved a unanimous majority, produced a high level of conformity. The social support condition, in which the dissenter joined the subject, produced significantly less conformity than the control condition. Even the extreme erroneous dissent condition, in which the dissenter gave an answer that was more extreme and incorrect than the majority's, produced significantly less conformity. Thus any breach in the majority reduced conformity. Subjects in the various conditions, however, had very different impressions of dissenters. Under the social support condition, subjects held a positive impression

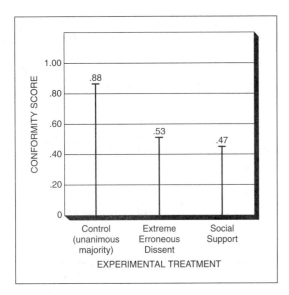

Figure 13.4   Conformity Scores and Dissent From the Majority

In an Asch-type experiment, subjects conformed less when one member broke from the majority and gave the same answer as the subject (the social support condition) than when faced with a unanimous majority (the control condition). Conformity was also lower when one member broke from the majority but gave an answer even more erroneous than the majority's (the extreme erroneous dissent condition). Thus, a breach in the majority of any kind had a liberating effect on the subjects and reduced conformity.

SOURCE: Adapted from Allen and Levine, 1969.

of the dissenter, whereas under the extreme erroneous dissent condition, subjects had a negative impression and viewed the dissenter as unlikable, stupid, insincere, and badly adjusted. Either way, the breach in the majority had the same effect: It called into question the correctness of the majority's position and reduced the subject's tendency to conform.

Attraction to a Group   Members who are highly attracted to a group conform more to group norms than members who are less attracted to it (Kiesler & Kiesler, 1969; Mehrabian & Ksionzky, 1970). One explanation for this maintains that when individuals are attracted to a group, they also wish to be accepted personally by its members. Because acceptance and friendship are strengthened when mem-

bers hold similar attitudes and standards, individuals highly attracted to a group conform more to the views held by the others (Feather & Armstrong, 1967; McLeod, Price, & Harburg, 1966). However, attraction to a group increases conformity only if that conformity leads to acceptance by others in the group (Walker & Heyns, 1962).

Commitment to Future Interaction
Members are more likely to conform to group norms when they anticipate that their relationship with the group will be permanent or enduring, as opposed to short-term. This was demonstrated by a study in which groups of five persons were asked to discuss such problems as urban affairs, international relations, pollution, and population (Lewis, Langan, & Hollander, 1972). Some of the participants were led to believe they would review these problems with group members again in the future, whereas others were not given this expectation. Those who anticipated future interaction conformed more to the majority's opinion than those who did not anticipate future interaction.

Commitment to future interaction affects conformity whether members are attracted to a group or not. For instance, one investigation showed that even if members do not especially like the others in a group, they conform at a high level provided they are already committed to continuing in the group and cannot readily leave. Such persons are likely to experience distress or dissonance at having to interact with others they do not like, and they may resolve this problem by bringing their own attitudes and beliefs into line with group standards (Kiesler & Corbin, 1965; Kiesler, Zanna, & DeSalvo, 1966).

Competence   Another factor affecting conformity is an individual member's level of expertise relative to that of other members. If a member skilled at the group's task holds a view different from the majority's, he or she will resist pressure to the degree that he or she believes himself or herself to be more competent than the other group members (Ettinger et al., 1971). Interestingly, the extent to which a person believes he or she is competent may be more important than actual level of competence (Stang, 1972). Persons who, in fact, are not competent will still resist conformity pressure if they believe they have more

13.2   Barriers to Independent Behavior

Conformity to norms is essential to group functioning. Without it, any group risks a serious lack of coordination among members and the prospect of failing to achieve important goals. There are, of course, some cases in which an individual might desire to act independently rather than to conform to group norms. A person might disagree with the majority's position and prefer an innovative departure. Activity of this type, however, is often disruptive from the standpoint of group operations, and groups use a variety of methods to discourage independent behavior. One theoretical approach (Carpenter & Hollander, 1982; Hollander, 1975) points to six major barriers that inhibit independent behavior by group members:

1. *Risk of disapproval from other group members.* Most groups establish norms governing central activities. These norms specify a range of tolerable behavior. To the extent a member's independent behavior falls outside that range, the member risks disapproval from others in the group. If he or she deviates too far, a member may face outright rejection.
2. *Lack of perceived alternatives.* Individuals frequently are unaware of alternatives other than those specified by group norms. Unless someone else speaks out and proposes alternatives, a member may not realize there is any choice other than conforming to the majority.
3. *Fear of disrupting the group's operations.* Because group norms usually reflect underlying group goals, an individual knows that departing from established norms may "rock the boat." Thus, he or she avoids

acting independently due to fear that it may block the attainment of group goals. This effect is especially strong if a member is afraid of being held personally responsible for group failure, should it occur.
4. *Absence of communication among group members.* Even if a number of members privately dissent from group standards, each may hesitate to express his or her reservations publicly. But without such communication, they do not discover that others are thinking along the same lines, and collective ignorance prevails. Lacking information that others would join in the nonconforming action, each avoids going out on a limb alone.
5. *No feeling of responsibility for group outcomes.* In some cases, where a group's established norms and procedures are deficient and produce poor results, members may actually cause a group to fail by conforming to the norms. Although some may realize that conformist behavior is producing poor outcomes, they may hesitate to take the initiative, speak out, and try to turn the situation around. This can occur especially when individuals feel they are not personally responsible for the group's success or failure.
6. *A sense of powerlessness.* If a member feels unable to change a situation, he or she is unlikely to try anything new. The apathy becomes self-fulfilling—no one tries anything different, and consequently nothing improves.

All these barriers to independent behavior serve to increase conformity in groups. If several are operating simultaneously, the probability of conformity is only increased further.

---

skill than other members. This is because group members who believe themselves competent rely less on the judgments of others; when confronted by a majority, they usually try to persuade other members to change their positions.

Gender   The gender composition of a group also affects the amount of conformity of its members. The balance of evidence gathered over the past four decades indicates that, within groups having members all of the same sex, females yield slightly more to con-

formity pressures than males (Becker, 1986; Cooper, 1979; Eagly & Carli, 1981). In other words, females yield somewhat more to pressures from other females than males yield to pressures from other males. We note, however, that the magnitude of this gender effect on conformity is quite small, and some studies have shown no difference at all between the sexes in conformity or influenceability.

Several explanations have been advanced for this difference. One explanation holds that it stems from differences in sex-role expectations (Eagly, 1987). The traditional female role in the United States rewards submissiveness, nurturance, passivity, and person orientation—all qualities related to conformity. The traditional male role rewards aggressiveness, assertiveness, dominance, and task orientation (Wiley, 1973). Consequently, sex-role expectations may lead to greater conformity among females than males. Some support for this view comes from research by Eagly and Chrvala (1986) and by Goldberg (1974, 1975). In these studies, women who accepted traditional sex roles conformed more than women who rejected traditional roles or who actively supported the women's movement. This pattern was particularly marked when the test items were traditionally feminine in nature.

A second explanation for gender differences in conformity stresses the nature of the tasks employed in conformity experiments themselves. At least some studies of conformity have employed tasks and opinion items involving male-typed activities. Thus, differences in conformity that appeared to result from differences in sex may actually be confounds that resulted from differences in familiarity with or competence in the tasks used in some of the studies.

Several studies have investigated this possibility (Karabenick, 1983; Sistrunk & McDavid, 1971). For instance, Sistrunk and McDavid reasoned that if women are more conformist on activities regarded as traditionally male, then men might be more conformist on activities regarded as traditionally female. Whereas women may be more conformist about cars and politics, for example, men may be more conformist about child care and fashion. These researchers compiled a list of statements, including opinions items and everyday matters of fact. Some of these items were of greater interest and familiarity to females, and other items were of greater interest and familiarity to males. They presented these items (along with some neutral filler items) to several groups of male and female subjects and included information on typical responses from peer majorities. The results indicated that males conformed more than females on the feminine items, and females conformed more than males on the masculine items. Both sexes conformed about the same amount on the neutral items. These findings suggest that the amount of conformity depends on the nature of the task as well as the gender of the subject. Both females and males yield to pressure more when the task is one with which they are unfamiliar.

Thus, both sex-role orientation and task competence serve as factors mediating the impact of gender on conformity. Although studies show that women conform somewhat more than men in same-sex groups, this may hold true primarily when the women accept traditional sex roles or when they lack competence in and familiarity with the task.

Minority Influence

Although influence exercised by the majority over individual members is important, this should not blind us to the activities of minorities within groups. In some instances, a small subgroup may disagree with the majority on some issue and advocate a different position. A *dissenting minority* is a coalition of members that holds a viewpoint different from the majority on some important issue(s) and presses for adoption of its position by the group. If the members of a dissenting minority are able to persuade or induce majority members to accept their viewpoint, they are said to exercise **minority influence.** For instance, a minority coalition may attempt influence if it wishes to change a judgment or decision by the majority or to change group norms or procedures. A significant change of this type is sometimes referred to as an *innovation.*

Attempts at influence by minority individuals or coalitions prove effective in some cases, but in many others they do not (Kalven & Zeisel, 1966, chapter 38). Throughout history, we encounter many instances of minority groups or individuals in scientific or religious contexts advocating viewpoints that were met initially with resistance or derision. Some of these minorities eventually succeeded in persuading or converting others—Galileo, Darwin, Pasteur, Freud, and Jesus Christ are examples—but many other minorities have been less successful and their messages are now forgotten or lost in oblivion. Viewpoints advocated by minorities often encounter resistance (or, at least, nonacceptance) from others, and sometimes these ideas are even considered to be deviant or seditious, with the potential to impair the group's established way of life. This raises a basic question: Under what conditions are attempts to exercise influence by a minority coalition likely to be successful? We turn to this question now.

Effectiveness of Minority Influence

According to the *conversion theory of minority influence* (Moscovici, 1980, 1985b), when minority coalitions exercise influence the process involved is usually conversion, not compliance. In other words, minority coalitions try to win over the other members by changing their underlying attitudes and beliefs, rather than by compelling compliance. To achieve this end, minority coalitions typically use persuasion or exemplification; this contrasts with majority coalitions, which more often use coercive pressure or (veiled) threats to get their way.

More specifically, this theory holds that for a minority coalition to be successful in exercising influence, it must do certain things. First, the minority needs to create conflict and disrupt the established order, thereby producing doubt and uncertainty in the minds of other group members. Sometimes this can be accomplished merely by stubbornly resisting pressure to conform to the majority's view. In doing this, the minority makes itself visible and salient, focusing attention on itself. Second, the minority needs to offer a constructive alternative—a coherent point of view different from that of the majority—and indicate it has strong confidence in the correctness of this viewpoint. Third, it must signal its intention not to compromise or abandon this view, with the implication that if other members wish to reestablish consensus and stability within the group, they must shift their own position toward that of the minority.

Consistency of the Minority
A central hypothesis of conversion theory is that a minority coalition has a greater chance of influencing the judgments or opinions of others if it takes a distinctive position and holds it consistently in the face of pressure. By holding the position consistently, the minority coalition can demonstrate commitment to and confidence in the position. This hypothesis has received support in empirical studies.

For instance, one study (Moscovici, Lage, & Naffrechoux, 1969) used a color perception paradigm (the "blue-green paradigm") to show that even if a minority coalition lacks power or status, it can influence the judgments of other members by maintaining a consistent position over time. Groups in this study consisted of six members; four of these were naive subjects, but the other two were confederates who constituted the minority coalition. These groups performed a simple color perception task. The members sat in front of a screen that displayed a set of six blue slides differing in luminousness. They saw these slides in six different orders; each slide was shown for 15 seconds. The members judged the color of each slide and announced their judgments aloud. In fact, all the slides used were blue. The minority coalition, however, voiced some judgments that departed from this standard. In one experimental condition, the two confederates in the minority coalition responded consistently and always labeled the blue slides "green." In another experimental condition, the two confederates responded inconsistently—they labeled the slides "green" 24 times and "blue" 12 times.

Results show that the behavior of the minority coalition influenced the judgments by the naive subjects. In the consistent-minority condition, 8.42% of all answers from naive subjects were "green," and as many as 32% of the naive subjects reported seeing a "green" slide at least once during the session. In the inconsistent-minority condition, only 1.25% of the responses from the naive subjects were "green." (These results contrast with the error rate in control groups consisting of six naive subjects, which was virtually zero. The control subjects always saw the slides as blue.) Thus, overall, the consistent minority exerted more influence than the inconsistent minority, as predicted by conversion theory. Results similar to these have been found in subsequent studies (Moscovici & Lage, 1976; Nemeth, Swedlund, & Kanki, 1974; Nemeth, Wachtler, & Endicott, 1977).

Behavioral Style
Another factor that affects the success of a dissenting minority is the extent to which its negotiating style is flexible rather than rigid. Minorities are most influential when their position is consistent but their behavioral style is not rigid, that is, when their style of presentation is flexible and multifaceted. A dissenting minority using this style is more likely to be successful than one that adopts a

rigid, hard-line, single-note approach (Mugny, 1982, 1984; Papastamou & Mugny, 1985). A flexible negotiating style tends to connote competence and honesty, whereas a rigid negotiation style, with its attendant refusal to make any concessions, is more likely to cause others to perceive the minority's consistency as dogmatic (and, hence, as idiosyncratic).

Size of the Minority

Another factor that affects a minority coalition's capacity to influence others is its size. In general, minority coalitions having many members can exert more influence than those having just a few members. For instance, one study found that a minority coalition consisting of two members is more influential than a minority of one (Arbuthnot & Wayner, 1982). In another study, the size of the majority was fixed at six persons, and the size of the minority coalition was varied experimentally. Results indicated that a minority of three or four persons exerted more influence than a minority of one or two (Nemeth, Wachtler, & Endicott, 1977). Reviewing findings across studies, Wood and colleagues (1994) reported that a minority coalition's influence increases as a function of its size. This effect was especially apparent on measures of public change and direct private change.

Social Identity of the Minority

Before going further, we should recognize that the term *minority* has at least two distinct meanings. On one hand, a minority is simply a subgroup that lacks numeric predominance. If 51% of the members of a group favor position A on some issue and 49% favor position B, then those favoring B constitute a numerical minority. Although this meaning of the term *minority* is appropriate for certain purposes (such as committee decision making), it neglects an important connotation—namely, that in some contexts, to be a minority means to be oppressed or underprivileged. When groups are composed heterogeneously of persons with different social characteristics (race, religion, ethnicity, and the like), then those persons considered minority or out-group members may be evaluated negatively or discriminated against.

Work on minority influence has begun to address these different meanings of minority, and recognition is now given to the distinction between a double minority and a single minority. A **double minority** is a person who not only advocates a stance on some issue which is different from that held by the majority, but who also has an out-group social identity (defined in terms of ethnicity, race, nationality, or other ascriptive traits). Persons who are double minorities are less likely than single minorities (i.e., in-group members who advocate minority viewpoints) to succeed in exercising influence vis-à-vis the majority. In an extension of conversion theory, several theorists have proposed that a minority coalition (or person) advocating a position different from the majority will be more influential when its social identity is the same as, or similar to, that of the target than when it is different (Mugny, 1984; Mugny & Papastamou, 1982; Turner, 1982). If the majority members see the minority as having attributes and category memberships similar to themselves, they will be more receptive to influence from the minority than if they see the minority as having different attributes and category memberships.

Research findings largely support this hypothesis, and they indicate that targets are particularly reluctant to agree with out-group minorities. Several studies have found that minority persons belonging to a manifestly different social category—for example, different gender, different school affiliation, different sexual orientation, and the like—have less impact on others' opinions or subjective judgments than minority members having social identities like that of the majority (Clark & Maass, 1988; Martin, 1988; Perez & Mugny, 1987). A study by Maass and Clark (1984), for instance, reported that an ostensibly gay minority arguing for gay rights exerted less influence on heterosexuals than did a heterosexual minority arguing for gay rights. This resulted in part because the gay minority was perceived as being more self-interested in the issue, which led to discounting.

Differences Between Minority and Majority Influence

One current controversy is whether minority influence (innovation) and majority influence (conformity) entail processes that are fundamentally similar or different. Some theorists have hypothesized that influence exercised by a minority and influence exercised by a majority involve the same underlying process (Doms, 1984; Latané w& Wolf, 1981; Tanford & Penrod, 1984). In this view, minority and majority influence may differ in quantity (with majorities exercising more) but not in quality. Several studies have reported results consistent with this single-process view (Doms & Van Avermaet, 1980; Personnaz, 1981; Wolf, 1985).

Nevertheless, the single-process view is not universally accepted. Some other theorists (Maass, West, & Cialdini, 1987; Moscovici, 1985a, 1985b) have proposed a dual-process view, which maintains that majorities and minorities differ qualitatively in the ways they influence their targets. Under this theory, majority influence is based on a *social comparison process* in which a member compares his or her own response with that of the majority in an attempt to achieve consensus, whereas minority influence is based on a *validation process* in which a member tries to comprehend the minority's viewpoint in an attempt to figure out what is valid or real about the world. Majority influence is largely normative in nature and involves compliance based on the target's desire to gain social rewards and approval; minority influence is largely informational, hence involves conversion (i.e., private agreement and belief change).

This dual-process model is controversial, and some writers have concluded that evidence supporting it is weak (Kruglanski & Mackie, 1990). Other work, however, provides at least some support for it. A meta-analysis of studies addressing this issue (Wood et al., 1994) concluded that majorities and minorities have different impacts on targets' public and private responses to influence. Majorities tend to exert more influence than minorities on measures of public change and direct private change, whereas minorities exert equal or greater influence than majorities on measures of indirect private change.

Also relevant to the dual-process model, some research shows that persons confronted by a dissenting minority tend to think more carefully and more creatively about the issue than they otherwise would (Nemeth & Kwan, 1985, 1987). Although they may not be persuaded to adopt precisely the position advocated by the minority, they do think in more divergent terms, attend to more aspects of the situation, and seek novel solutions (Maass, West, & Cialdini, 1987). Thus, the effectiveness of a dissenting minority is perhaps best measured in terms of its capacity to indirectly change opinions and viewpoints, not its capacity to force behavioral compliance.

Summary

A group is a social unit that consists of two or more persons and has certain defining attributes, including recognized membership, interaction among members, shared goals and objectives, and norms that guide members' behavior. The input-process-output model provides a framework for the analysis of groups.

Group Cohesion A cohesive group is one that can strongly attract and hold its members. (1) Two important types of cohesion are social cohesion and task cohesion. A group will be more cohesive if its members like one another, are similar to one another, find the group's tasks interesting and involving, believe the group's goals are consonant with their own individual goals, and gain rewards (prestige, money) by belonging to the group. (2) The level of a group's cohesion affects the interaction among members. Members in highly cohesive groups communicate more than those in less cohesive groups; they also exert more influence over one another, and their interaction is friendlier and more cooperative. Groups with high task cohesion tend to be somewhat more productive than groups with low task cohesion.

Group Goals A group goal is a desirable outcome that members strive collectively to bring about. Members can define group goals at different levels of specificity—missions, goals, objectives, and tasks. (1) Group goals differ from individual goals, which are outcomes desired by members for themselves. Most groups function best when there is substantial similarity, or isomorphism, between group goals and members' individual goals. (2) Groups working toward an explicit, challenging objective usually perform better than groups working without an explicit objective. Group performance also depends heavily on members' goal commitment; in turn, goal commitment will be high only if the members value the goal itself and hold a strong expectation that the goal can actually be attained by the group.

Group Norms

A norm is a rule or standard that specifies how group members are expected to behave under given circumstances. (1) Group norms serve several functions. They bring about coordination among members, provide a frame of reference that enables members to interpret their environment, and define the common identity of group members. (2) The return potential model of norms describes the relationship between behavior and subsequent reward or punishment.

Majority Influence and Conformity

Conformity means adherence by an individual to group norms and expectations. (1) The Asch conformity paradigm uses a simple visual discrimination task to investigate conditions that produce conformity by individuals to the majority's judgment. (2) Group majorities can utilize both normative influence and informational influence to exert pressure on individual members. Sherif's autokinetic effect studies illustrate the impact of informational influence on members. (3) Many factors affect the amount of conformity in Asch-type situations. Conformity increases with group size up to three, and it is greater when the majority is unanimous than when it is not. Conformity is also greater when members are highly attracted to a group and when conformity leads to liking and acceptance by other members. Commitmenr to future interaction affects conformity; conformity is greater when members believe their relationship with the group will be relatively permanent. Task competence affects conformity; members who oppose the majority's view will resist conformity pressures to the extent that they believe themselves to be more competent than other members. Finally, gender affects conformity. Females conform more than males in same-sex groups, although the magnitude of this difference is small.

Minority Influence

Minority influence refers to efforts by a dissenting minority to persuade majority members to accept a new viewpoint and adopt a new position. (1) A minority will be more influential if it maintains its position consistently over time, adopts a flexible negotiating style, has many members, and consists of members with an in-group identity similar to that of the majority. (2) Although controversial, some evidence indicates that minorities and majorities exert different types of influence. Although a majority can often compel members to comply, a minority must rely primarily on persuasion to change others' viewpoints.

Key Terms

conformity (p. 323)
double minority (p. 332)
goal commitment (p. 319)
goal isomorphism (p. 318)
group (p. 313)
group cohesion (p. 316)
group goal (p. 317)
informational influence (p. 326)
input-process-output model (p. 314)
majority influence (p. 323)
minority influence (p. 330)
norm (p. 320)
normative influence (p. 325)
primary group (p. 318)
return potential model (p. 321)
secondary group (p. 318)
small group (p. 313)

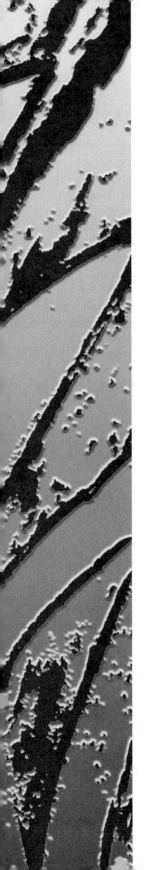

CHAPTER 14
Group Structure and Interaction

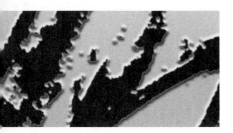

Introduction

Consider a sales group working for a small California company that manufactures laboratory equipment. This group consists of 11 people—one vice president, two managers, and eight salespersons. Their basic objective is to market equipment to industrial laboratories in the northern and southern regions of the state.

The members of this sales group have different backgrounds, personalities, and skills, and they contribute in different ways to the group's operations. The most influential person is the group is a woman who serves as the vice president. She has a lot of experience in the industry, and over the years she has amassed an outstanding record as a salesperson. In the role of vice president, she is responsible for maintaining the group's sales performance at a high level. She establishes sales goals for the group and closely monitors its performance each week, and in addition, she communicates with executives in other divisions about which products are selling and why. Any major changes made within the sales group must receive her approval before they are implemented.

Next in importance within the group are the two managers. Both have excellent skills and extensive experience in sales. One manager is responsible for the group's sales performance in the northern region of the state; the other, for sales in the southern region. The managers are well paid, although they receive lower salaries than the vice president.

The eight salespersons are less central than the managers, but they do perform important activities for the group. They are responsible for making direct contact with customers in their area and for providing products that meet their customers' needs. Younger and less experienced than the managers, the salespersons are paid partly on salary and partly on commission, and they earn less than the managers.

The positions in this group are diagrammed in Figure 14.1. As this figure indicates, the group's structure is hierarchical, with the vice president having more status and exercising more influence than the managers, and the managers more than the salespersons.

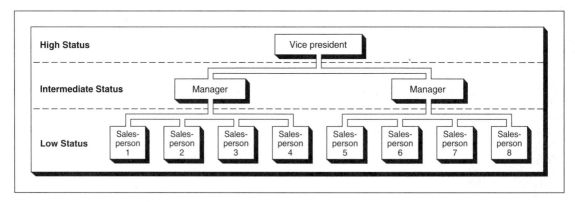

Figure 14.1   Status Structure in a Work Group

This work group has a three-level status hierarchy. The high-status member (vice president) exercises the most control and receives the greatest rewards. Reporting to the vice president are two intermediate-status members (managers). The eight low-status members (salespersons) exercise the least control and receive the least rewards.

Elements of Group Structure

By watching the equipment sales group over time, we can discover some patterns in the interaction among members. For one thing, there is a division of labor—the members perform different tasks—and they perform these tasks recurrently, week after week. In addition, there are some regularities in communication and influence within the group. Certain members talk more than others, and some consistently exercise more influence than others. There also exists a pattern of sentiments among members—certain members are friends, but others are not.

What matters about these patterns of task performance, communication, influence, and sentiment is not merely that they exist but that they recur and endure over time. They are so intrinsic to the equipment sales group that if someone left the group—if one of the managers, say, took a job with another firm—the remaining members would take steps to keep the patterns intact. They would find someone to "fill the shoes" of the departed manager, perhaps by promoting one of the salespersons.

The patterned interaction among members in the sales group is certainly not accidental. It is produced in part by the group's rules and roles, which give structure to the interaction (Bates & Harvey, 1975; Fararo & Skvoretz, 1984). When considering **group structure**—especially the structure of formal groups like the sales unit—it is useful to think in terms of *positions*. The 11 persons in the sales group occupy different positions (vice president, manager, salesperson), and they differ substantially from one another in the roles they enact.

From one perspective, a **role** is set of functions that a member performs for the group. From another perspective, a role is a cluster of rules or expectations indicating the set of duties to be performed by a member occupying a given position within a group. Group members hold *role expectations* regarding one another's performance, and they feel justified in making demands on each other. For instance, members in the equipment sales group expect the salespersons to contact potential customers, to identify customers' needs, and to offer customers products that meet these needs. If, for some reason, one of the salespersons suddenly stopped contacting customers, other members of the sales group would view that as a violation of role expectations and would doubtless take action to correct the situation.

In addition to differing in roles, positions may also differ in **status**—that is, they differ in prestige or social rank. Members can evaluate each position in terms of its importance or contribution to the group. Some positions are more highly evaluated than others, and these, of course, have higher status. Moreover, because the evaluation of a position often transfers to the occupant of that position, it is natural to talk (elliptically) about the status of a group member, rather than of the position per se. Within the equipment sales group, the woman who is vice president holds the highest status, the two managers have intermediate status, and the eight salespersons have the lowest status.

Status differences in groups are manifested in various ways. Among the most typical are (1) the level of contribution a member occupying a given position makes toward the group's overall performance, (2) the level of rewards received by that member, and (3) the level of authority or control exercised by that member over group decisions and activities. Although there surely are exceptions, a member's standing on one of these dimensions usually correlates with his or her standing on the others. Thus the vice president contributes the most important skills to the sales group, receives the greatest rewards, and exercises the highest level of control over important group decisions.

Questions About Group Structure   This chapter examines group structure and interaction. Specifically, it focuses on the causes and consequences of role and status differentiation among group members. The questions to be addressed include the following:

1. In a newly forming group that has no established structure, how do differences in roles and status arise among members? That is, what processes lead to role differentiation and status emergence in a new group?

2. In what ways do status characteristics (such as gender, race, education, and occupation) affect a member's standing in a group?

3. What criteria do groups use when distributing rewards to members? In what ways do differences among members in skill and performance affect the rewards received by members? How do group members react if there a mismatch between contributions and rewards?

4. What factors affect the willingness of members to support and endorse a group's formal leader? What circumstances lead to the formation of revolutionary coalitions that seek to overturn authority?

Role Differentiation in Newly Formed Groups

The group of equipment salespersons, discussed above, is obviously a formal, structured group that has been operating continuously for some time. We consider groups of this kind in a moment, but we begin our discussion here with a more elementary case—groups that are just beginning to organize. We first focus on communication in newly forming task groups and then consider the processes involved in the emergence of roles in such groups.

Communication in Task Groups

Consider a set of five undergraduates meeting for the first time. They are members of a task group discussing a human relations case. The case involves a delinquent juvenile who has committed a serious crime but who comes from an underprivileged background. The group members have been instructed to read a summary of the boy's history, discuss it, and reach a collective decision about his case. At one extreme, they might decide he should be punished severely for the crime; at the other extreme, they might decide he should be treated leniently. Because the case is complex and ambiguous, there is no obvious best decision.

All the members of this group are very similar in social attributes—that is, they are the same age, same race, same gender, and same basic background. The group has not been assigned a formal leader, so the five members enter the group on an equal footing.

Interaction in groups such as this one—newly formed, homogeneous, problem-solving groups working on human relations problems—have been extensively studied by investigators (Bales, 1965; Slater, 1955). When members in groups like this start to discuss their problem, certain things typically happen. For one, the initial equality among members disappears, and distinctions quickly arise among them. Some members participate more than others and exercise more influence regarding the group's decision. Members develop different expectations about their own and others' activities in the group. Most often, one or more persons start to exercise leadership within the group. Role differences begin to emerge among the members.

IPA Coding System   Although there are many ways to study interaction in groups, one useful technique is to monitor the flow of communicative acts among group members. By analyzing the frequency, direction, and types of messages that occur, it is possible to get a clear picture regarding the pattern of interaction among group members. Of course, one drawback of this approach is that groups naturally have different concerns and face different tasks, so their members send very different types of messages. A group of civil engineers, for example, might want to discuss the best way to repair an abandoned bridge; a group of high school teachers might be concerned with techniques for teaching algebra; a group of basketball players might ponder a way to increase rebounds and reduce turnovers. Because the groups' problems are so different, the content of communication among members differs greatly from group to group. To deal with this, investigators studying communicative acts need a coding system that is both sufficiently general to apply to many different groups, yet sufficiently specific to break the flow of communication among members into discrete and interpretable acts.

One coding system that tries to do this is the **interaction process analysis (IPA)**, developed by Bales (1950, 1970). The IPA system uses a small number of categories that apply to all problem-solving groups. Although the content of messages may differ from group to group, any communicative act can be classified into one of the IPA categories.

Underlying the IPA system is a particular theoretical viewpoint. This viewpoint, a functional theory, holds that any working group must solve two major sets of problems if it is to continue in existence. The first set, called *task problems,* refers to those that a group faces when attempting to reach its goals. For instance, the group must figure out how to combine inputs from members and produce a satisfactory product or output. The second set, called *social-emotional problems,* refers to interpersonal problems that arise among group members. The group must figure out how to restrict interpersonal friction and assure that members are sufficiently satisfied to remain in the group. In this view, if a group is unable to solve both its task problems and its social-emotional problems, it will eventually fail or cease to exist.

The particular categories included in the IPA coding system derive from this functional theory. Specifically, the IPA system consists of the 12 categories shown in Figure 14.2. These 12 categories may be grouped into clusters. The broadest and most important distinction is that between task acts and social-emotional acts. **Task acts** (categories 4–9) are instrumental behaviors that push a group toward the realization of its goals. **Social-emotional acts** (categories 1–3 and 10–12) are emotional reactions, both positive and negative, directed toward other group members.

Another breakdown of IPA categories is indicated by the letters *a* through *f* in Figure 14.2. These letters represent issues that must be addressed by any group. Within the area of task acts are issues of (a) communication, (b) evaluation, and (c) control. Issues of communication involve analyzing a situation; issues of evaluation refer to determining members' attitudes toward that situation; and issues of control refer to suggestions for action within the situation. Within the area of social-emotional acts are issues of (d) decision, (e) tension reduction, and (f) reintegration. Issues of decision refer to acceptance or rejection of proposed courses of action, whereas issues of tension reduction and reintegration refer to establishing emotional relations within the group.

In practice, the IPA system might be used as follows. Let's say a task group of five persons is meeting in a room to discuss a problem. Members are sitting around a table; a number has been placed in front of

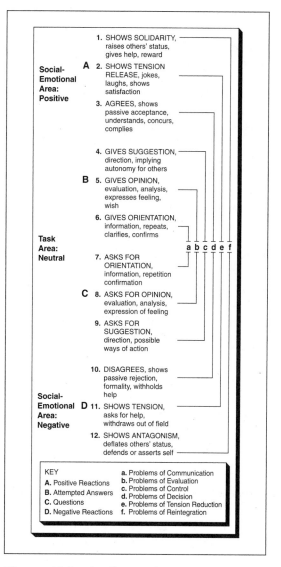

Figure 14.2 Coding Categories in Interaction Process Analysis

Interaction process analysis is a method of analyzing patterns of communication among group members. Researchers observe a group in action and code members' acts into 12 distinct categories. Because these categories are highly general, they apply regardless of the specific topic or issue under discussion by members. Half of the categories pertain to task, or instrumental, acts, whereas the other half pertain to social-emotional, or affective, acts.

SOURCE: Adapted from Bales, 1950.

each person. The room itself has one unusual feature—it is equipped with a one-way mirror along one wall so observers seated behind the wall can watch interaction among group members. The observers record the interaction in terms of who speaks to whom and what types of acts are communicated. For example, if member 2 turns to member 5 and asks 5's opinion on an issue, the observers score this by writing "2–5" in IPA category 8 (asks for opinion). By following this procedure through an entire group session (which might run an hour or two), the observers can obtain a complete record of communication within the group according to the IPA system.

Who Talks in Groups?

Investigators have used the IPA system to study communication patterns in a wide variety of problem-solving groups. Included among these are groups consisting of college students, military enlisted men, patients in therapy, prison inmates, jury members, third-grade boys, and so on (Bales & Hare, 1965).

One finding that regularly emerges in these studies is that group members do not participate equally in a discussion; some members talk far more than others. Although it can vary, the most talkative person in a problem-solving group typically initiates 40% to 45% of all communicative acts. The second most active person initiates approximately 20% to 30% of the acts. This pattern is apparent in Table 14.1, which summarizes initiated acts for groups ranging from three to eight members. As the size of the group increases, the most talkative person still initiates a consistently large percentage of communicative acts, but the less talkative persons are crowded out almost completely (Bales, 1970).

Another finding from IPA studies concerns the stability of participation: The group member who initiates the most communication during the beginning minutes of interaction is very likely to continue doing so throughout the life of the group. If a problem-solving group meets for several sessions, for example, the member who ranked highest in participation during the first session is likely to rank highest during subsequent sessions. In general, the ranking of members in terms of participation is stable over time (Fisek, 1974).

Table 14.1 Percentage of Total Acts Initiated by Each Group Member as a Function of Group Size

Member Number	Group Size					
	3	4	5	6	7	8
1	44	32	47	43	43	40
2	33	29	22	19	15	17
3	23	23	15	14	12	13
4		16	10	11	10	10
5			6	8	9	9
6				5	6	6
7					5	4
8						3

Note: Data are based on a total of 134,421 acts observed via the IPA system in 167 groups consisting of 3 to 8 members.

SOURCE: Adapted from Bales, 1970, pp. 467–474.

Types of Acts Occurring

The IPA system provides a way to assess the frequency with which various speech acts occur in groups. As we have noted, the IPA system consists of 12 categories. Of these, 6 pertain to task-oriented acts (categories 4–9), whereas the other 6 pertain to social-emotional acts (categories 1–3 and 10–12). Although groups differ, Table 14.2 summarizes the mean percentage of IPA acts (calculated across many problem-solving groups) falling into each of the 12 categories. As this table shows, about two thirds of the acts in a group session are task oriented and about one third are social-emotional (Bales & Hare, 1965). Within the task-oriented categories, most acts fall into categories 4 through 6 (gives suggestion, gives opinion, gives orientation), and fewer acts fall into categories 7 through 9 (asks for orientation, asks for opinion, asks for suggestion). Within the social-emotional categories, most acts fall into the positive categories 1 through 3 (shows solidarity, shows tension release, agrees); fewer acts fall into the negative categories 10 through 12 (disagrees, shows tension, shows antagonism). This pattern is reasonable, in the sense that a problem-solving group would surely defeat

Table 14.2 Interaction Profile: Mean Percentage of IPA Category Acts

Type of Act (IPA)	Mean Percentage
1. Shows solidarity	2.97
2. Shows tension release	8.17
3. Agrees	10.70
4. Gives suggestion	6.56
5. Gives opinion	22.24
6. Gives orientation	28.72
7. Asks for orientation	5.89
8. Asks for opinion	3.27
9. Asks for suggestion	0.60
10. Disagrees	4.73
11. Shows tension	3.43
12. Shows antagonism	2.41

SOURCE: Adapted from Bales and Hare, 1965.

itself if there were more questions than solution attempts and more negative emotional reactions than positive ones.

Task Specialists and Social-Emotional Specialists

IPA has proved useful in documenting the emergence of role specialization and leadership within task groups. One study (Bales, 1953) investigated some groups, each consisting of five men, that met to discuss a problem in a laboratory setting. Group interaction was scored by observers using the IPA system. At the end of the discussion period, members filled out questionnaires and rated each other. Items included such questions as "Who had the best ideas in the group?", "Who did the most to guide the group discussion?", and "Which group member was the most likable?" Typically, there was high agreement among group members in their answers regarding ideas and guidance but less agreement in their answers regarding liking.

This study's results showed a high correlation between participation rank as measured by the IPA system and members' perceptions of one another as measured by the questionnaire. The person who initiated the most acts was perceived by others as providing the most guidance and offering the best ideas. In short, the person initiating the most acts was perceived as the group's task leader. Results of this type have been observed in other studies as well (Reynolds, 1984; Sorrentino & Field, 1986).

But this is not the entire picture. Often the person initiating the most acts was not the best liked member and, indeed, he was sometimes the least liked member. More typically, the second highest initiator was the best liked person. Why does this occur?

In general, the highest initiator is someone who drives the group toward the attainment of its goals. A high proportion of acts initiated by this person are task oriented (IPA categories 4–6). For this reason, we can call the high initiator the group's **task specialist.** This person contributes many ideas and suggestions and pushes the group toward its objectives. In the effort to get things done, however, the task specialist tends to be pushy and openly antagonistic in some instances. He or she makes the most impact on the group's opinion, but this aggressive behavior often creates tension. It remains for some other member, the **social-emotional specialist,** to ease the tension and soothe hurt feelings in the group. The acts initiated by this person are likely to be acts showing tension, tension release, and solidarity (IPA categories 11, 2, and 1). The social-emotional specialist is the one who exercises tact or tells a joke at just the right moment. This person helps release tension and maintain good spirits within the group. Not surprisingly, the social-emotional specialist is often the best liked member of the group.

Thus problem-solving groups have two basic functions: getting things done and keeping relations pleasant. IPA results suggest that these functions are typically performed by different members. When members divide up functions in this manner, we say that **role differentiation** has occurred in the group. Frequently, the task specialist and the social-emotional specialist work closely together and complement one another. They tend to interact more and to agree more with each other than with other members of the group (Burke, 1972).

Role differentiation into task specialist and social-emotional specialist occurs frequently, but not invariably. In some groups, a single member performs both the task and the social-emotional functions. Observations suggest that for problem-solving groups in laboratory settings, a single member successfully performs both functions in about 20% to 30% of the cases (Lewis, 1972). For groups in natural, nonlaboratory settings, the incidence may be higher (Rees & Segal, 1984).

Status Characteristics and Social Interaction

To this point, we have considered interaction in newly formed problem-solving groups that consist of members who are initially very similar. Results based on the IPA technique show that despite initial equality among members, role and status differences quickly emerge within a group setting. Members differ in their rate of participation, their influence over group

14.1 Sociometry

Interaction process analysis can be used to measure not only status differences among members but also the flow of conversation in a group. These measurements reveal the naturally occurring communication channels in a group and, therefore, provide some insight into group structure. Communication is only one aspect of group structure, however. Other aspects include patterns of influence (who gives orders to whom) and patterns of attraction (who likes whom). Consider, for instance, the measurement of attraction in groups. One of the first empirical approaches designed to measure patterns of attraction was **sociometry** (Hallinan, 1981; Moreno, 1960). This technique—which can be used in groups such as classes, committees, and residential units—usually involves a short questionnaire. Each person in a group is asked to name others in the group he or she likes according to some criterion. For example, students in a classroom may be asked who they want to sit next to, work with, or play with during recess. In residential settings, persons may be asked who they want to eat with or have as a roommate. Sometimes, they are also asked the opposite question: Who do you dislike; who do you not want as a roommate? The resulting choices can then be diagrammed in a distinctive figure called a *sociogram.*

Conventionally, each person in a sociogram is represented by a small circle. Choices made by persons are represented by arrows; solid arrows indicate a positive choice or liking, whereas broken arrows indicate dislik-

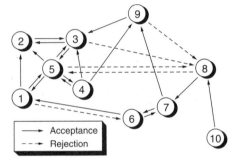

A Typical Sociogram

SOURCE: Adapted from Jahoda, Deutsch, and Cook, 1951.

ing. The distance between circles represents the closeness of a relationship. Two persons who choose one another are positioned close together (see persons 1 and 5 in the sociogram in this box), whereas two persons who reject one another are far apart (see persons 5 and 8).

Various aspects of group structure are reflected in a sociogram. For instance, person 5, who is liked by many other members, is a "star." Person 10, who was chosen by no one, is an "isolate." Persons 1, 2, 3, 4, and 5—who chose one another and form a tight subgroup—constitute a "clique." Patterns such as these affect the emotional climate in a group. In general, groups characterized by many reciprocal positive choices usually have a happy, positive climate. In contrast, groups with several isolates or several distinct cliques usually have a chilly climate.

decisions, and the types of acts they contribute. But what about interaction in newly formed groups whose members are not identical in social attributes? In particular, what about groups composed of members who differ in gender, race, age, education, and occupation? We encounter such groups every day—PTAs, student committees, neighborhood associations, juries, church groups, and so on. In fact, groups of this type are probably more common than groups consisting of members who are similar or homogeneous.

In this section, we discuss how status differences emerge in groups having heterogeneous membership. We first consider status characteristics and status generalization, and then we look at expectation states theory. Finally, we consider ways of overcoming status generalization in task groups.

Status Characteristics

A **status characteristic** is any social attribute of a person around which evaluations and beliefs about that person come to be organized. We often take status characteristics into consideration when establishing expectations regarding others in groups. To illustrate, suppose members of a new group are meeting for the first time. They have not had prior contact, so they do not know one another. In the course of meeting, they start to develop beliefs and expectations about each other; this may include expectations about who will be able to contribute most toward the group's task and who will be able to provide leadership.

In forming these expectations, members initially may not have much information to go on. Usually, however, they will know or observe such attributes as the age, gender, and race of the other members, so they may try to draw some inferences from these attributes. The term *diffuse status characteristics* refers to attributes that provide an indirect indication of a member's level of ability on the group's task. Attributes such as age, gender, race, ethnicity, education, and physical attractiveness serve as diffuse status characteristics in many contexts (Berger, Cohen, & Zelditch, 1972; Jackson, Hunter, & Hodge, 1995). In addition to these, members may take note of others' *specific status characteristics,* which are attributes that

The status hierarchy in this church is apparent, as a priest, bishop, altar boy, and members of the choir participate in a service. Each person's function in the recessional, as well as his or her physical location, depends on formal status.

more directly and precisely indicate someone's level of ability on the task to be performed by the group. For example, a player's height could be a specific status characteristic on a basketball team; prior experience in writing advertising copy would be a specific status characteristic in a work group developing the advertising campaign for a new product.

In many cases, specific status characteristics provide a more secure basis for inference than diffuse status characteristics. Expectations and evaluations based on diffuse status characteristics tend to reflect prevailing cultural stereotypes. In the United States, for example, despite a general ideology of equality, many persons consider it preferable to be male rather than female, white rather than black, adult rather than juvenile, and white-collar rather than blue-collar. Although we may not think of ourselves as holding such views, we are nevertheless very sensitive to attributes such as age, gender, and race. In group settings, we develop evaluations and expectations of others not only from their specific status characteristics but also from their diffuse status characteristics (Berger, Rosenholtz, & Zelditch, 1980; Ellard & Bates, 1990; Meeker, 1981).

Status Generalization

Status characteristics can significantly affect interaction among members in newly forming groups. Studies show that persons with high standing on status characteristics are accorded more respect and esteem than other members, and they are chosen more frequently as leaders. Their contributions to group problem solving are evaluated more positively, they are given more chance to participate in discussions, and they exert more influence over group decisions (Balkwell, 1991; Berger, Cohen, & Zelditch, 1972; Webster & Foschi, 1988). The tendency for members' status characteristics to affect group structure and interaction is called **status generalization.**

When status generalization occurs, a member's status outside a group affects his or her status inside that group. That is, the members who hold higher status in the group are those who hold higher status in society at large (Cohen & Zhou, 1991). Thus status generalization establishes a correspondence between a member's standing within small groups and the system of stratification in the wider society.

As an illustration, consider the impact of status generalization in a jury. The members of a jury, who usually differ in status characteristics, must discuss the facts of a case and reach a verdict. A study by Strodtbeck, Simon, and Hawkins (1965) demonstrated that status characteristics such as occupation and gender affect jury deliberations. Because the investigators were precluded by law from observing actual jury deliberations, they studied mock juries composed as authentically as possible. These juries, selected from the voter registration list in an urban area, consisted of men and women who had occupations of varying status, including proprietors, clerical workers, and skilled and unskilled laborers. During the study, they listened to a tape recording of a trial, after which they were instructed to do everything a real jury does—select a foreperson, deliberate on the case, and reach a verdict. Researchers recorded the content of the jurors' deliberations. At the end of the session, jurors completed a questionnaire to indicate their impressions of one another.

From the recording of the deliberations, it was possible to determine which members talked the most. Table 14.3 shows the rates of participation by jury members as a function of their occupation

Table 14.3 Rates of Participation During Deliberation by Occupation and Sex of Juror

Sex	Occupation				Combined Average
	Proprietor	Clerical	Skilled	Laborer	
Male	12.9	10.8	7.9	7.5	9.6
Female	9.1	7.8	4.8	4.6	6.6
Combined average	11.8	9.2	7.1	6.4	8.5

Note: Entries in this table are percentage rates of participation. Because there were 12 persons on a jury, the "average" juror would theoretically have a rate of participation of 8.3%. This can be used as a frame of reference against which to compare tabled values. (The "combined-combined" value in the table is 8.5 rather than 8.3 because 26 of 588 jurors in the study were not satisfactorily classified by occupation and were omitted from the analysis.)

SOURCE: Adapted from Strodtbeck, Simon, and Hawkins, 1965.

and gender. The results indicate that men initiated more interaction than women in the mock juries. The data also show the impact of occupational status: The higher the occupational status, the greater was the rate of participation. This held for both males and females.

The questionnaire completed by jurors at the end of the session provided information on their perceptions of one another. Table 14.4 shows the votes received by the jury members for being "most helpful in reaching the verdict." This measure reflected the amount of influence each member had over the group decision, as perceived by other members. The findings are very similar to those on the rates of participation. On average, male jurors were perceived as more helpful than female jurors, and jurors of high occupational status were perceived as more helpful than those of lower occupational status. These findings illustrate the status generalization effect.

Overall, this jury study revealed the impact of status generalization: Persons with higher standing in terms of gender and occupation became the group members with the higher status inside the group. They participated at a higher rate, were perceived as contributing more to the group's problem-solving efforts, and were more frequently chosen as the formal leader.

Although the findings in this study seem clearcut, the interpretation in terms of status generalization is admittedly open to criticism. A critic might argue, for example, that a person's status inside a group is not a function of his or her status outside but is instead caused by the same qualities or personal traits that determine the person's status outside the group. One might hypothesize, for instance, that if people of high intelligence are better able to achieve high occupational status than those of lower intelligence, they also (by virtue of intelligence) will be better able to contribute within a group. If this were the case, a person's standing inside a group would not be caused by his or her external (occupational) status; rather, both of these would be caused by a third factor, intelligence.

To check this possible confound, several studies have manipulated status characteristics experimentally. One of these studies (Moore, 1968) investigated small, two-person groups of female subjects. Both women were shown a series of large figures made up of smaller black and white rectangles. Their task was to judge which of the two colors, black or white, covered the greater area in each of the large figures. This task was difficult because the areas were in fact approximately equal, making the figures ambiguous. The subjects, who were seated so they could not see or talk with each other, signaled their preliminary judgments by pressing buttons on consoles in front of them. Each subject's answer was revealed to the other through a system of lights. The subjects knew that after seeing each other's initial judgments, they would have an opportunity to make a final judgment. Because their goal was to make accurate final judgments, subjects had been told they should weigh their own answers against the answers of their partners. The subjects did not know, however, that the connections on the consoles were, in fact, controlled by the experimenter. He could manipulate how often one subject perceived that her partner disagreed with her.

Table 14.4 Average Votes Received as "Helpful Juror" by Occupation and Sex of Juror

Sex	Occupation				Combined Average
	Proprietor	Clerical	Skilled	Laborer	
Male	6.8	4.2	3.9	2.7	4.3
Female	3.2	2.7	2.0	1.5	2.3
Combined average	6.0	3.4	3.5	2.3	3.6

SOURCE: Adapted from Strodtbeck, Simon, and Hawkins, 1965.

A jury listens skeptically as a lawyer presents her case. Although formally of equal status, some jury members exercise more influence than others in determining the verdict. Characteristics such as occupation, gender, and race can affect the patterns of influence among jurors during deliberations.

Between their first judgment and their final judgment, subjects were free to reverse their decisions when their partners appeared to disagree with them.

All the subjects in this experiment were junior college students. As a manipulation of status, one half of the subjects were told their partner was a high school student (low-status partner), whereas the other half were told their partner was from Stanford University (high-status partner). The results show that the women who believed their partner to be of higher status changed their answers on the judgmental task more often than those who thought themselves to have higher status than their partner. That is, group members were receptive to influence from a higher status person but relatively unreceptive to influence from a lower status person. The random assignment of subjects to experimental treatments eliminated the possibility that subjects differed systematically in intelligence or ability on the judgmental task. These findings, like those from the mock jury research, support the case for status generalization.

Expectation States Theory

Why does status generalization occur? That is, why do the status characteristics of group members affect the interaction occurring within the group? One answer to this question comes from **expectation states theory** (Berger, Conner, & Fisek, 1974; Berger et al., 1986; Knottnerus, 1988). This theory proposes that at the outset of interaction in a task group, members form expectations regarding one another's potential performance and contribution toward the group's task. These expectations, which are formed through an attribution process, affect subsequent interaction among members. In particular, members are more likely to defer to and accept influence from those they expect to perform well than from those they expect to perform poorly.

Because members form these expectations in the early stages of group interaction, they often base them on limited, sketchy information. Status characteristics (such as age, race, gender, and occupation) are readily apparent or easily learned, so they influence the expectations that members form regarding one another. These status characteristics are most likely to influence performance expectations when group members (1) have no prior history of interaction, (2) have no information about one another except for their standing on status characteristics, and (3) have no special experience or information about the group's task.

Burden of Proof It might seem reasonable that when forming performance expectations from status characteristics, members would insist on proof that the characteristics are relevant to the task at hand. In fact, however, it works the other way around. When developing performance expectations from status characteristics, group members behave as if the *burden of proof* is placed on demonstrating that these characteristics are not relevant to the task at hand rather than demonstrating that they are relevant. In the absence of such a demonstration, group members often treat status characteristics as relevant even when they are not.

To illustrate, suppose a group is developing plans to take a weekend trip. Although diffuse status characteristics (for example, race and gender) may not, in fact, be relevant to the performance on this planning task, the group's members will likely behave as if they are relevant. The members usually ignore these characteristics only after there has been some explicit demonstration or proof that they are not relevant to performance on the group's task. This phenomenon has been confirmed in several studies (Freese, 1976; Webster & Driskell, 1978; Wilke, Van Knippenberg, & Bruins, 1986).

Multiple Status Characteristics As already noted, each member in a newly forming group possesses a variety of status characteristics. How, then, do members combine information on two or more status characteristics to form performance expectations regarding a given person? What performance expectations result if that person's standing on one status characteristic is inconsistent with his or her standing on another?

Consider, for example, the case of a female physician. Such a person has inconsistent status characteristics. On the one hand, she is a physician, which is a high-status characteristic. On the other hand, she is a female, and some studies show this is perceived as lower status than being male (Berger, Rosenholtz, & Zelditch, 1980; Ridgeway, 1982). How will group members respond to her? Clearly, there are several possibilities. They might ignore one of the status characteristics and focus on the other; that is, they might respond to her only as a physician or only as a woman. Alternatively, they might combine both status characteristics to form some kind of intermediate expectation. Results of several studies show that, in general, members form an intermediate expectation in these cases rather than focusing on only one or another of these status characteristics (Markovsky, Smith & Berger, 1984; Pugh & Wahrman, 1983; Zelditch, Lauderdale, & Stublarec, 1980). Therefore, confronted with a female physician and lacking other specific information, group members will most likely consider her more competent than female nonphysicians but less competent than male physicians.

Overcoming Status Generalization

As indicated above, group members often treat diffuse status characteristics as relevant to performance expectations even when in actuality they are not, and they place the burden of proof on others to demonstrate irrelevance. Thus, despite recent societal changes in sexist and racist attitudes, status generalization can work to an individual's disadvantage. In a mixed setting with both males and females, for example, the females may find they are not permitted to influence the group's decision significantly even though they are as qualified (or maybe more qualified) as males with respect to the problem under discussion. Without a clear demonstration that gender is irrelevant to performance, verbal protests regarding gender equality may be to no avail (Pugh & Wahrman, 1983). Likewise, in an interracial interaction between blacks and whites, the blacks may feel they are treated as low-status, minority members. Because irrelevant diffuse status characteristics can so easily place someone at a disadvantage, we ask whether status generalization can be overcome or eliminated in face-to-face interaction.

Some have suggested that the best way to overcome status generalization is by direct methods—that is, to raise the expectations of lower status persons regarding their own performance on group tasks so that they can, in turn, force a change in other people's expectations regarding their performance. Unfortunately, this approach does not work very well. In one study, for example, blacks at a northern university were trained in assertiveness techniques and then participated with whites in biracial groups. The training raised blacks' expectations for themselves, and they behaved in an assertive and confident manner. But because the whites' expectations regarding the blacks' performance were not affected by the training, what ensued was not smooth interaction but a status struggle. The whites thought the blacks were arrogant and overreaching in light of their "limited ability," whereas the blacks viewed the whites as engaging in racist bigotry (Katz, 1970).

To overcome status generalization, it is necessary to change not only the expectations held by low-status members but also those held by high-status members. Demonstrating the right to high status through successful performance is usually more effective than merely claiming or asserting it (Freese & Cohen, 1973; Martin & Sell, 1985). A key to overcoming status generalization is to satisfy the burden-of-proof requirement by supplying group members with information that contradicts performance expectations inferred from a diffuse status characteristic (Berger, Rosenholtz, & Zelditch, 1980).

For instance, in one study (Cohen & Roper, 1972), investigators taught black junior high school students how to build a radio, then showed them how to teach another pupil to build a radio. This created two specific status characteristics inconsistent with students' conception of race. Investigators then had the blacks train white pupils to build a radio, thereby establishing the relative superiority of the blacks on this task. Finally, they informed some of the students that the skills involved in building the radio and teaching others to build it were relevant to another, entirely different task. This new task was a decision-making game, which the boys subsequently played. Results of the study indicated that this pattern of training modified the performance expectations held by both black and white boys. The change in expectations produced a significant increase in equality between blacks and whites, as indicated by who exercised influence over decisions when the boys played the game. The overall conclusion is that status generalization can be overcome, provided the expectations of both low-status and high-status persons are modified simultaneously. Similar findings appear in related studies (Cohen, 1982; Riordan & Ruggiero, 1980).

Equity and Reward Distribution

So far, we have considered the sources of status differences among group members in newly formed groups. Status differences arise from variations in those skills and motivations that affect group goal attainment. They also arise from status characteristics (age, gender, race, or occupation) that create expectations regarding performance in group contexts.

In this section, we change our focus and look at the consequences of status differences. In particular, we discuss the distribution of rewards among group members. When members contribute to the attainment of group goals, they typically receive rewards such as money and approval. Although there is often a direct relation between the amount of the contribution and the amount of the benefits, this is not always the case. The distribution of rewards within a group concerns all members because it raises questions of justice and fairness. Most individuals care not only about their own rewards but also about the rewards received by their fellow members.

In the following, we first look at various justice principles that govern reward distribution. Then we examine one of these principles—the equity principle—in closer detail. Finally, we consider the nature of responses to an inequitable distribution of rewards in a group.

Justice Principles in Reward Distribution

There are many criteria—**distributive justice principles,** as they are called—that members can use to judge the fairness and appropriateness of the distribution of rewards (Deutsch, 1985; Elliott & Meeker, 1986; Saito, 1988). Three of the most important are the equality principle, the equity principle, and the relative needs principle. When group members use the *equality principle,* they distribute rewards equally among members, regardless of members' contributions. When group members follow the *equity principle,* they distribute rewards in proportion to members' contributions. When they follow the *relative needs principle,* they distribute rewards according to members' personal needs, regardless of contributions. Although there are other possible criteria, the justice principles of equality, equity, and need are among the most important and widely used criteria in reward distribution within groups (Lamm & Schwinger, 1980).

When allocating rewards among members, a group may rely exclusively on one of these justice principles or may apply several of them simultaneously. These justice principles are often contradictory in the sense that they lead to different distributions of rewards, but what really matters is the relative importance (or weighting) accorded each principle. Not surprisingly, their relative importance varies from group to group and from situation to situation. For instance, the equality principle often prevails when members are concerned with solidarity and wish to avoid conflict (Leventhal, Michaels, & Sanford, 1972). It also prevails in cultural settings that are relationship oriented, rather than economically oriented (Mannix, Neale, & Northcraft, 1995). Some evidence indicates that females favor the equality principle (over the equity principle) more than males (Leventhal & Lane, 1970; Watts, Messe, & Vallacher, 1982). Friends are more likely to follow the equality norm than strangers (Austin, 1980), and members of small (3-person) groups are more likely to use it than members of large (12-person) groups (Allison, McQueen, & Schaerfl, 1992). In contrast, the equity principle is often used in work situations, where many persons want their share of rewards to reflect the importance of their contribution. The needs principle is frequently salient in close or intimate relationships involving friends, lovers, and relatives. However, this principle has also been invoked in other contexts. Karl Marx, for example, advocated the adoption of the needs principle in communist societies. He proposed that individuals would contribute according to their abilities and receive according to their needs.

Equity Theory

The equity criterion for reward allocation is used by many groups. A state of **equity** exists when members receive rewards in proportion to the contributions they make to the group. For example, in an industrial work group in the United States, a worker normally would expect to receive better outcomes (salary, benefits) than another if his or her job required more inputs (higher skill, more hours per week, and so on). Likewise, a wife would probably feel some inequity if she contributes more to the family than her husband but receives little help or love in return. As these examples suggest, equity judgments are made when one group member compares his or her own outcomes and inputs against those of another member.

In an effort to formalize the nature of these comparisons, theorists have developed *equity theory* (Greenberg & Cohen, 1982; Homans, 1974; Walster, Walster, & Berscheid, 1978). Although there is some disagreement regarding how equity should be operationalized (Alessio, 1980; Harris, Messick, & Sentis, 1981), the fundamental idea underlying equity can be expressed in terms of the following equation

$$\frac{\text{Person A's outcomes}}{\text{Person A's inputs}} = \frac{\text{Person B's outcomes}}{\text{Person B's inputs}}$$

This equation states that equity exists when the ratio of person A's outcomes to inputs is equal to the ratio of person B's outcomes to inputs. What matters is not merely the level of outcomes or inputs but the equality between the ratios of outcomes to inputs.

To make this more concrete, consider the case of two women employed by the same industrial work group. One of the women (person A) receives a high outcome—a salary of $60,000 a year, 4 weeks of paid vacation, reserved parking in the company's lot, and a fancy corner office with thick rugs and a great view. The other woman (person B) is about the same age but receives a lesser outcome—a salary of $30,000 per year, no paid vacation, no reserved parking, and a cramped, noisy office with no windows.

Do persons A and B feel this distribution of rewards is equitable? They may or they may not; it depends on their relative inputs. If their inputs to the company are identical, then the arrangement will almost certainly be experienced as distressing, especially by person B. For example, if both A and B work a 40-hour week, have only high school educations, and approximately equal experience, there is little basis for

Equity is relevant in marital relationships just as in work relationships. For this couple, equity means sharing the housework. What would happen to this relationship if either the husband or the wife insisted on doing less around the house?

paying person A more than person B. Person B will probably feel angry because the reward distribution is inequitable, and person A may feel uncomfortable or guilty.

Suppose instead that person A's inputs are much greater than B's. Say that person A works a 60-hour week, holds an advanced degree such as an MBA, and has 12 more years of relevant experience than person B. Suppose, also, that person A's job involves a high level of stress because it entails the risk of serious failure and financial loss for the company. In this event, A not only has greater "investments" (that is, education and experience) but is also bearing greater immediate "costs" (60 hours of work a week, plus high stress). Person A may receive better outcomes, but she also contributes much more to the company. Under these conditions, both A and B may feel that their outcomes, although not equal, are nevertheless equitable.

This example highlights the basic difference between an equitable and an inequitable relationship, but precise calculations regarding equity can be difficult to make in everyday life. Two persons may view the same situation in different ways. They may disagree, for instance, over how to evaluate particular inputs and outcomes. If one worker holds an advanced university degree and another worker has 7 years' more seniority, who is contributing the more important input? Moreover, persons may disagree over exactly which inputs and outcomes are to be included in the equity calculation. Should a worker's gender be considered an input when calculating equitable pay allocations? Many say it should not; yet, in our society, women often perform the same activities as men but receive less pay for their efforts.

Responses to Inequity

Inequity produces not only emotional reactions of distress (anger, guilt) but also direct attempts to change the conditions that produce it. By eliminating inequity, group members can rid themselves of emotional distress. There are two distinct types of inequity: underreward and overreward. **Underreward** occurs when a person's outcomes are too small relative to his or her inputs; **overreward** occurs when a person's outcomes are too large relative to inputs.

Responses to Underreward

Persons who are underrewarded typically become dissatisfied or angry (Austin & Walster, 1974; Sweeney, 1990; Wall & Nolan, 1987). The greater the degree of underreward, the greater is their dissatisfaction and desire to reestablish equity. To illustrate, suppose person A is underrewarded in a relationship. In this case, inequity may be expressed as follows:

$$\frac{\text{Person A's outcomes}}{\text{Person A's inputs}} < \frac{\text{Person B's outcomes}}{\text{Person B's inputs}}$$

This states that person A's ratio of outcomes to inputs is less than B's ratio. Given that A is underrewarded, any of various actions could restore equity between A and B. These include (1) increasing the outcomes received by A, (2) reducing the inputs from A, (3) reducing the outcomes received by B, or (4) increasing the inputs from B.

Studies show that underrewarded persons often take direct steps to reduce inequity. If the situation permits, such a person would attempt to reduce inequity by increasing his or her own outcomes. For example, this person might aggressively seek a pay raise. Alternatively, if this person has the ability to reallocate rewards in the group directly, he or she would do so (Schmitt & Marwell, 1972). If an industrial employee paid on a piecework basis feels underrewarded, he or she might reduce the quality of effort on each piece in order to increase the total number of pieces produced per hour. This would increase the worker's outcomes without increasing his or her input (Andrews, 1967; Lawler & O'Gara, 1967).

Responses to Overreward

What happens when a person receives more than his or her fair share in a relationship? Will he or she be content just to enjoy the benefits? Although overreward is apparently less troubling to individuals than underreward, it can still create feelings of inequity, often in the form of guilt rather than anger (Perry, 1993; Sweeney, 1990). A person who feels guilty about overreward may attempt to rectify the inequity (Austin & Walster, 1974). Suppose person A is overrewarded in a relationship. In this case, inequity is expressed as follows:

$$\frac{\text{Person A's outcomes}}{\text{Person A's inputs}} > \frac{\text{Person B's outcomes}}{\text{Person B's inputs}}$$

This states that person A's ratio of outcomes to inputs is greater than person B's ratio. Given that A is overrewarded, any of the following changes could restore equity: (1) reducing the outcomes received by A, (2) increasing the inputs from A, (3) increasing the outcomes received by B, and (4) reducing the inputs from B.

Research findings show that, in some situations, overrewarded persons sacrifice some of their rewards to increase those of others. However, the extent of the redistribution often will not be complete, and equity may be only partially restored (Leventhal, Weiss, & Long, 1969). Some evidence indicates that overrewarded members prefer to restore equity by increasing their inputs. For example, in a work situation, overrewarded members can strive to produce more or better products as a means of reducing inequity (Goodman & Friedman, 1971; Patrick & Jackson, 1991); this enables them to restore equity without sacrificing any of the outcomes they receive.

This process was investigated in a classic study in which students were hired to work as proofreaders (Adams & Jacobsen, 1964). In one condition, subjects were told that they were not really qualified for the job (due to inadequate experience and poor test scores) but they would nevertheless be paid the same rate as professional proofreaders (30 cents per page). In a second condition, subjects were told that, due to their lack of qualifications, they would be paid a reduced rate (20 cents per page). In a third condition, subjects were told that they had adequate experience and ability for the job and they would be paid the full rate (30 cents per page). Thus subjects in the first condition viewed themselves as overrewarded, whereas those in the second and third conditions saw their pay as equitable. Measures of the quality of the students' work showed that the overrewarded students caught significantly more errors than the equitably paid students. In fact, the overrewarded students were so vigilant, they often challenged the accuracy of material that was correct. These results indicate that the overrewarded students increased their inputs, thereby restoring equity. Similar findings appear in related studies (Adams & Rosenbaum, 1962; Goodman & Friedman, 1969).

Other Responses to Inequity   We have noted that inequity—both underreward and overreward—can be resolved by changing outcomes and/or inputs. There are other ways of coping with inequity, however. Although these methods do not rectify the conditions producing inequity, they make the inequity itself more tolerable. For instance, a person who is underrewarded might choose a different person for comparison. Person A, instead of comparing her inputs and outcomes with those of person B, might compare them with person C. By adjusting her frame of reference, person A would see the situation in a different light and experience less inequity.

Another way to cope with inequity is to withdraw from the situation entirely. An individual might choose to resign from her job, for example, rather than continue to perform under severe inequity. One study found evidence that among blue-collar workers, those experiencing inequity (underreward) are somewhat more likely to quit the job, be absent, or report sick than those not experiencing inequity (Van Yperen, Hagedoorn, & Geurts, 1996). If it is extreme, inequity can be so distressing it produces behavior that seems economically irrational. For instance, one study found that workers were often willing to accept an alternative position which paid less in order to remove themselves from severe inequity (Schmitt & Marwell, 1972).

Still another way to reduce inequity is through perceptual distortion or bias. By distorting inputs and outcomes, equity can be reestablished in a psychological sense. For example, an employer might convince himself that his overworked and underpaid secretary

is, in fact, being treated equitably by minimizing the secretary's inputs ("You wouldn't believe how incompetent she is!") or exaggerating her outcomes ("Work gives her a chance to socialize with all her friends"). Among other things, this spares the executive any need to reduce the secretary's workload or increase her salary (Austin & Hatfield, 1980). Bias of this type is especially convenient for those who are overrewarded. If a person is receiving more than he deserves, he might simply conclude that his inputs are greater than he originally believed; this eliminates any need to reallocate rewards (Gergen, Morse, & Bode, 1974).

Stability and Change in Authority

To this point, many of the groups we have considered in this chapter were *informal groups*. In groups of this type, interaction among members is not very structured or regulated, role definitions are not highly explicit, and collective tasks are often loosely defined or even nonexistent. Groups of this type often form spontaneously, and they may usually endure for only a short period. We turn now to consider *formal groups*, in which interaction is more structured and regulated, role definitions and authority are explicit, and tasks are well specified. Formal groups frequently exist within larger organizations or bureaucracies, and they often endure for extended periods (even many years). The equipment sales group mentioned at the beginning of this chapter provides an illustration of a formal group.

One of the more important features of formal groups is that members' activities are regulated by means of legitimate authority. Recall that **authority** refers to the capacity of one member to issue orders to others—that is, to direct or regulate the behavior of other members by invoking rights that are vested in his or her role. In formal groups such as the equipment sales group, members differ with respect to the status and authority they hold, and these differences usually persist over time. The persons who exercised authority yesterday (the vice president and two managers) are the same persons who exercise it today. Yet, the pattern of status and authority in formal groups cannot be taken for granted, for it is potentially subject to dispute and change. Persons of high status wishing to remain in control must take steps to assure they have the endorsement of other group members, and they must build coalitional alliances that will support their authority in the event a challenge arises.

In this section, we first consider factors that affect the level of endorsement accorded formal leaders by group members. Then we consider how various alliances—conservative coalitions and revolutionary coalitions—can strengthen or weaken the position of formal leaders.

Endorsement of Formal Leaders

Role systems in formal groups are structured various ways. Some involve a chain of command structured in a hierarchy or pyramid; others more nearly resemble networks or matrices. For instance, the sales group mentioned above was structured hierarchically into three levels (eight salespersons who reported to two managers, who in turn reported to a vice president). The person at the "top" of such a pyramid holds high status and typically performs various leadership functions for the group. These leadership functions—planning, organizing, and controlling the activities of group members—are essential if a group is to achieve its goals (Bass, 1990).

Groups usually maintain a pyramid structure like this through a tacit exchange between high-status members, who fulfill leadership functions, and the other group members, who accept direction from the high-status members (Hollander, 1985; Hollander & Julian, 1970). The high-status members, through their leadership efforts, help the other group members attain their collective goals. Because achievement of these goals is valuable and rewarding, the lower status members reciprocate by giving rewards, benefits, and privileges to the high-status group members. Thus, in return for providing effective leadership, high-status members receive rewards and benefits, and, more than this, they also receive support for their authority and control over the group's activities. Specifically, they receive support for their actions (plans, decisions, orders) and reaffirmation of their right to occupy high-status positions within the group.

This support is essential if the status and authority structure within a group is to remain stable. Without it, formal leaders are usually ineffective and vulnerable to removal from office. Moreover, without support, the very legitimacy of the group's status structure may be called into question (Walker, Thomas, & Zelditch, 1986).

Endorsement as Support Support for a formal leader can come from either or both of two sources within a group (Zelditch & Walker, 1984). When it comes from members having still higher status in the group, it is usually referred to as authorization. When it comes from members having equal or lower status, it is termed endorsement. We can view **endorsement** as an attitude held by a lower status member indicating the extent to which he or she supports the group's formal leader (Hollander & Julian, 1970; Michener & Tausig, 1971).

Endorsement can be measured in various ways. Most endorsement scales, however, include items that tap satisfaction with various aspects of the leader's behavior. For instance, a member's endorsement of a formal leader would be high if that member is satisfied with the leader's use of power in arriving at group decisions, satisfied with the performance of the leader in directing the group, satisfied the leader is occupying that position legitimately, and happy with the prospect that the leader may continue to head the group in the future (Michener & Lawler, 1975).

Lower status group members are not always unanimous in their attitudes regarding persons in position of authority; some members may endorse a formal leader more strongly than others. Moreover, the endorsement accorded a high-status member often fluctuates over time. With a high level of endorsement from members, a person may feel confident when exercising vigorous leadership; with a low level of endorsement, however, exercising leadership may be difficult or impossible.

Factors Affecting Endorsement Why do some formal leaders receive high levels of endorsement from members whereas others receive only low levels? Formal leaders generally receive endorsement in proportion to the benefits they deliver to group members. Thus the extent to which a group achieves its goals is one factor affecting endorsement. If members perceive a group as moving toward the attainment of its goals, endorsement of high-status members usually remains high; if members see the group as failing, endorsement declines. A distinction must be drawn, however, between initial failure and repeated failure on group objectives. When a group fails initially, members may actually increase their endorsement and rally around the leaders in hopes of improving the situation (Hollander, Fallon, & Edwards, 1977). Repeated failure, however, usually indicates that something is fundamentally wrong. If the responsibility for failure is attributed to faulty leadership, endorsement surely declines. In this case, group members may even attempt to remove high-status members from their positions or redistribute authority within the group (Michener & Lawler, 1975; Wit, Wilke, & Van Dijk, 1989). Circumstances like this sometimes arise in voluntary associations and industrial work groups; they also occur in military organizations (especially during wartime) and in professional sports teams (especially at the end of a losing season).

Failure to attain collective goals leads to low endorsement primarily because members infer that the high-status member is not competent as a leader. Repeated failure is evidence that the high-status member is not sufficiently skilled to move the group toward attainment of its goals. In addition, group members may construe other unrelated behaviors, such as difficulty on high-skill tasks or failure on written tests, as evidence of incompetence. Thus behavior of this type by a high-status member also lowers endorsement (Suchner & Jackson, 1976).

Another determinant of endorsement is the level of consideration a formal leader shows toward other members. For instance, if a high-status member controls the allocation of rewards and decides to share these rewards equitably, this demonstrates that person's concern for the welfare of other group members, and it will result in strong endorsement. In contrast, if a high-status member—even one who is very competent—behaves selfishly and inequitably when allocating rewards, that behavior tends to diminish endorsement (Michener & Lawler, 1975; Wit

A foreman describes plans to construction workers at a building site. A person in command will find it easier to exercise influence if he or she is strongly endorsed by the workers who must perform the tasks.

& Wilke, 1988). Group members are sensitive not only to the level of rewards they receive but also to the procedures used by a high-status member in allocating rewards. Formal leaders receive higher levels of endorsement if they allocate rewards by procedures viewed as fair rather than by procedures viewed as unfair or arbitrary (Tyler & Caine, 1981; Tyler & Folger, 1980).

Revolutionary and Conservative Coalitions

Lack of endorsement poses serious problems for an established high-status member, and it can render precarious his or her position in the group. Under certain conditions, low levels of endorsement lead to the formation of a **revolutionary coalition,** which is an alliance of medium- and low-status members who oppose the existing leadership (Crosbie, 1975). By combining forces, members of a revolutionary coalition hope to displace the high-status members and their adherents. Revolutionary coalitions are often instrumental in mutinies, revolts, and coups d'état.

Mobilization of Revolutionary Coalitions

Revolutionary coalitions are usually difficult to mobilize. Low-status members must first have reason to end their allegiance to the established status order and then agree to take joint action to overthrow that order. Aboard a ship at sea, for example, sailors must have a strong basis for discontent with the ship's captain before they join a mutiny. They must feel that existing channels of appeal provide no remedy for their discontent, and they must be willing to face the risk of punishment if the mutiny fails. For the mutiny to succeed, those involved must find a way to mobilize without tipping their hand to the authorities too soon; this may entail secret meetings, covert communications, and surreptitious planning. In addition, the mutineers must settle on who will be the new commander if the mutiny succeeds. All these factors make mobilization difficult.

Certain conditions increase the likelihood that a revolutionary coalition will form. These are essentially the same as those that produce low levels of endorsement. For instance, repeated failure to achieve group goals increases the level of discontent and may cause a revolutionary coalition to form. Members may then attempt to reverse the group's fortunes by removing the high-status member from power and establishing a new order. Once in power, the revolutionary coalition will reallocate important responsibilities among group members and change some of the group's operational procedures (Michener & Lawler, 1971).

Inequitable treatment of group members by high-status persons may also encourage formation of a revolutionary coalition. If an authority shows favoritism or selfishness in allocating rewards, group members are more likely to chafe under the inequity and form a revolutionary coalition than if the authority is more even-handed in allocating rewards (Michener & Lyons, 1972; Ross, Thibaut, & Evenbeck, 1971; Webster & Smith, 1978). Likewise,

perceived indiscretion or unethical personal or professional behavior by an authority may provoke an overthrow (Deluga, 1987).

Similarity of interest and attitudes among lower status members is yet another factor in the emergence of a revolutionary coalition. Sharing common interest heightens members' expectations of support from other members. This, in turn, increases the probability of individuals joining an emergent revolutionary coalition (Lawler, 1975a, 1975b).

Reactions to Revolutionary Coalitions

High-status members have a lot to lose by the emergence of a revolutionary coalition; once they discover an upheaval in the making, they usually take action to suppress the coalition and bolster their own position. For instance, the attempted revolt by some of Hitler's generals during World War II resulted ultimately in their own deaths, not Hitler's. Once established leaders learn that a revolutionary coalition is a prospect, they usually engage in various counterstrategies to thwart the coalition. One of these counterstrategies is simply to threaten to punish insurgent members. If the threats are sufficiently large and credible, this may effectively quash the revolt.

Another useful counterstrategy is to mobilize a **conservative coalition,** an alliance of members who support the existing status order against revolutionaries. Conservative coalitions usually consist of members who view the status differences in the group as legitimate and who (consistent with expectation states theory) hold performance expectations based on these status differences (Wilke, 1985). Thus, conservative coalitions form more readily in groups in which members differ substantially on central status attributes (Wilke, Van Knippenberg, & Bruins, 1986).

A third counterstrategy to thwart revolt is to co-opt the revolutionaries. By definition, **cooptation** is a strategic attempt to weaken the bond among potential revolutionaries by singling out one or several lower status members for favored treatment. For instance, a high-status member might offer several low-status members the prospect of future promotion. This would increase their investment in the existing status order, thereby making it more difficult for other revolutionaries to influence them or to recruit them as coalition members (Lawler, 1983; Lawler, Youngs, & Lesh, 1978). High-status members find cooptation especially useful when they fear an upheaval in the future and they wish to forestall any such event.

Summary

Members' positions in most any group differ with respect to roles (required performances) and status (rank or prestige in the group). In this chapter, we discussed both the origins and the consequences of role and status differences among members.

Role Differentiation in Newly Formed Groups

In task groups, role differences among members usually develop during the course of interaction. (1) Interaction process analysis (IPA) is a technique for investigating communication and interaction in groups. Studies using this technique have shown not only that some group members consistently send and receive more messages than others but also that patterns of communication stabilize over time. (2) Studies using IPA have also analyzed the emergence of status differences in groups. Differentiation of leadership functions—task specialists and social-emotional specialists—appears regularly in problem-solving groups.

Status Characteristics and Social Interaction

(1) A status characteristic is any social attribute of a person around which beliefs about that person come to be organized. Age, race, gender, and occupation are important diffuse status characteristics. (2) A member's status characteristics outside the group can substantially determine his or her status inside the group. Studies of juror interaction, for instance, have shown that characteristics such as occupation and gender affect which members participate the most, exert the most influence, and hold positions of leadership within the jury. The tendency for members' status characteristics to affect group structure

and interaction is called status generalization. (3) Expectation states theory holds that group members develop expectations regarding one another's performance. Status generalization in this form is most likely to occur when members have no prior history of interaction, no experience with the group's task, and no explicit proof to the contrary. (4) To overcome or eliminate status generalization, one must change not only the expectations of low-status members but also those of high-status members. Status generalization can be eliminated only when both low-status and high-status members believe a status characteristic is not relevant to performance of the group's task.

Equity and Reward Distribution

(1) Groups use various distributive justice principles when determining the level of rewards and benefits that members receive. Prominent justice principles include equality, equity, and relative need. (2) A state of equity exists when members receive rewards in proportion to the contributions they make to the group. If inequity occurs within a group, members typically react with distress (anger for underreward or guilt for overreward) and initiate efforts to restore equity. (3) Efforts to restore equity usually entail increasing the outcomes or reducing the inputs of underrewarded members or, alternatively, increasing the inputs or reducing the outcomes of overrewarded members.

Stability and Change in Authority

A formal leader can exercise authority only with the support of other group members. (1) In return for helping the group achieve its objectives, a leader receives endorsement as well as special rewards and privileges. Endorsement declines if the group fails repeatedly to achieve its goal(s), if the leader is judged incompetent, or if the level of consideration the leader shows toward members is low. (2) Without endorsement, members may attempt to oust an established leader by forming a revolutionary coalition. Coalitions of this type are difficult to mobilize, but they are most likely to emerge when a group fails to achieve its goals or when the leader distributes rewards inequitably among members. Revolutionary coalitions are not always successful because they may be thwarted by conservative coalitions or by a leader's use of cooptation.

Key Terms

authority (p. 352)
conservative coalition (p. 355)
cooptation (p. 355)
distributive justice principle (p. 348)
endorsement (p. 353)
equity (p. 349)
expectation states theory (p. 346)

group structure (p. 337)
interaction process analysis (IPA) (p. 338)
overreward (p. 351)
revolutionary coalition (p. 354)
role (p. 337)
role differentiation (p. 341)
social-emotional acts (p. 339)
social-emotional specialist (p. 341)
sociometry (p. 342)
status (p. 337)
status characteristic (p. 343)
status generalization (p. 344)
task acts (p. 339)
task specialist (p. 341)
underreward (p. 351)

CHAPTER 15

Group Productivity and Task Performance

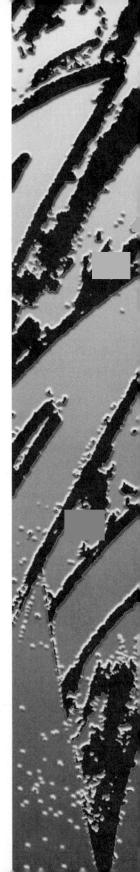

Introduction

Many groups in society exist primarily to attain specific goals. A work group in a clothing factory strives to meet its production quotas for swimsuit apparel; an airline crew seeks to transport passengers to their destinations on time; a police detective squad works around the clock to crack an urgent murder case; a professional ball club tries to win games and take the league championship; a jury deliberates a case carefully with the intent of reaching a correct verdict. Not every group, of course, pursues explicit goals or produces something tangible. Some groups—a Wednesday night poker club, for example—exist largely for its own sake, for simple pleasure and sociability. Even so, the incidence of task-oriented groups in our society is quite high, and a significant percentage of groups exist primarily to produce something valuable or to bring about some desired end state.

Group Productivity

If we were studying the productivity of a particular group, we surely would focus on observable outputs or products that result from the group's activity. Both the quantity and the quality of these outputs are of concern. Of course, the specific measures we use to assess the group's outputs would depend on the goals of that group. For instance, for an industrial work group that manufactures clothing, we might measure the number of garments produced per unit of time, the number of "seconds" or rejects, the amount of material wasted, and so on. In contrast, for a police detective squad investigating crimes, we might measure the number or percentage of cases cleared by arrest, time elapsed between offenses and arrests, and percentage of charges that result in a conviction (a measure reflecting the quality of the evidence gathered by police).

Measures such as these would give us a good start toward assessing the quantity and quality of a group's output. Note, however, that group output is not the same as group productivity. Rather, **group produc-**tivity refers to a group's output (per unit of time) gauged relative to something else, such as the level of resources utilized by the group or the group's targeted objective (Gilbert, 1978; Pritchard & Watson, 1992). This definition leads us naturally to a viewpoint regarding group efficiency and group effectiveness. A task group is considered highly *efficient* if it generates a lot of output from a small amount of input. For example, if a clothiers' work group produces a large number of garments per day and requires only a small number of workers to do it, that group is more efficient than another group which produces the same number of garments per day but requires many workers to do it. In contrast, a group is considered highly *effective* if its output meets or exceeds its goal or targeted objective. For example, if a basketball team establishes the objective of winning the league championship but finishes the season losing more games than it won, it obviously has a low level of effectiveness. To fairly compare groups in terms of their effectiveness, we must assume they have similar or identical goals.

From everyday observation, we know groups differ greatly in their levels of productivity. Even when groups are identical in size and similar in talent, they may perform at significantly different levels. Some groups produce more than others, win competitions more frequently, achieve better solutions to problems, work at a higher speed, commit fewer errors, and create less waste.

Questions About Group Productivity

This chapter focuses on productivity in a wide variety of groups. It is concerned not only with industrial work groups (such as in a clothing factory), but also with problem-solving, decision-making, and brain-storming groups that produce intellectual or cognitive products. The observation that seemingly similar groups often perform at different levels gives rise to several questions about group productivity. Specifically, this chapter addresses the following questions:

1. What types of tasks do groups typically face? In what ways are group processes and group productivity affected by the characteristics of the task(s) at hand?
2. To what extent is group productivity affected by such characteristics as group size, group goals, reward allocation, and communication channels?
3. How does the style of leadership in a group affect productivity? Are some leadership styles better than others in fostering a high level of group productivity?
4. In what ways do group processes either enhance or inhibit the quantity and quality of original ideas generated in a brainstorming group?
5. How do group processes affect the quality of group decisions? In what ways do conformity pressures and group discussion affect the character of group decisions and the risk-taking propensity of members?

Group Task Demands

Groups in our everyday environment perform a diversity of tasks. There are great differences in the tasks faced by such groups as a basketball team, a jazz band, a work group on an assembly line, a flight crew in a space shuttle, a study commission, a jury in deliberation, and an infantry squad in combat. In our desire to understand group productivity, we cannot afford to ignore differences in group tasks. Although a group's productivity depends on various factors—including the group's size, reward structure, goal specifications, and communication channels—it also depends crucially on the task itself.

Group tasks differ greatly in the demands they impose on groups and group members. For instance, tasks differ in their level of difficulty, in the mix of skills required for successful performance, and in the procedures and methods for carrying them out. They differ in the ways they restrict and channel interaction among members. And they also differ in the extent to which they permit a division of labor to develop among members. Given the importance of task demands in group productivity, we must first discuss some task properties before considering other factors in performance.

Task Divisibility

Group tasks differ with respect to their degree of divisibility (Shiflett, 1979; Steiner, 1972, 1974). A **unitary task** is one that is not divisible; that is, it requires all members to perform identical activities. These tasks obviously do not permit the group to establish a division of labor. Examples of unitary tasks include members pushing a car from a ditch, loading a heavy piano onto a truck, rowing in a boat race, and participating in a rope-pulling contest. Certain mathematical and logical problems, where all group members work simultaneously to figure out the solution, also qualify as unitary tasks.

In contrast, a **divisible task** is one that can be broken down into distinct subtasks, such that group members perform different, although complementary, activities. One example would be flying a commercial aircraft, where the pilot, navigator, and flight attendants all perform different activities between takeoff and landing. Another example is building a house. In this divisible task, some workers pour the foundation, others install the plumbing and the wiring, and still others erect the walls and construct the roof.

Both unitary and divisible tasks can pose problems for groups, in that the members must have the appropriate ability, motivation, tools, and information to do the job. However, divisible tasks raise some special hurdles not imposed by unitary tasks. To handle a divisible task successfully, a group must match its members' skills with subtask requirements—in other words, put the right person in the right job. The group must also establish a relatively high level of coordination among its members; otherwise, it will not

In a relay race, teams of runners face a unitary additive task. Each team member must complete a lap, and the team with the least total time wins.

be able to blend various subtask performances into a group product.

Combinatorial Procedures

Another important characteristic of group tasks is the procedure used to combine members' contributions. Whether it faces a divisible task or a unitary task, a group must employ some procedure to merge the separate contributions of its members, yielding an overall group product. Various combinatorial procedures are widely used, such as summing the products of the individual workers or using the output of the most proficient worker as the group's product.

Based on these procedures, we can distinguish among additive tasks, conjunctive tasks, and disjunctive tasks. An **additive task** is one in which all the members perform the same activity and final group product is simply the sum of their individual performances. One example of an additive task is group

members stuffing envelopes for a publicity campaign. All members perform the same activity (stuffing and sealing envelopes), and the group's productivity is the total number of envelopes ready for mailing at the end of the day. Another example is a cheerleading squad, the members of which are trying to make as much noise as possible in hopes of getting the crowd to respond. The squad's product is, in effect, the sum of the noise made by each of its members at a given moment. Yet another example is a four-member track team participating in a relay race where each member must complete a lap and then pass the baton to a teammate. The team's productivity score is the total time required for the four members to complete their respective laps (i.e., the sum of their individual lap times), where the team is striving to minimize the time.

Another type of task is a **conjunctive task,** where a group's productivity depends entirely on the capability of its weakest or least proficient member. To illustrate, consider a mountain-climbing expedition striving to ascend a steep precipice. Even if all

15.1   Types of Group Tasks

Several investigators have attempted to develop classificatory schemes for group tasks (Hackman & Morris, 1975; Laughlin, 1980; Steiner, 1972). McGrath (1984) proposed a taxonomy of group tasks that is quite general and consistent with other approaches. McGrath suggests group tasks can be divided usefully into the following eight types:

1. *Planning tasks.* In tasks of this type, a group's major goal is to generate an action-oriented plan to achieve some objectives on which the members have already agreed. The group must consider alternative paths to the goal, contemplate constraints imposed by resource availability, and develop a viable program. Examples include planning a military maneuver or mapping out an advertising campaign for a new product.
2. *Creativity tasks.* A group is required to generate new, original ideas. Usually, group interaction emphasizes creativity or entails brainstorming to generate new ideas, alternatives, or images. Examples include brainstorming to develop new product concepts or new comedy routines.

3. *Intellective tasks.* A group is required to solve a problem for which there is (or is believed to be) a correct answer. To solve problems of this type, group members typically must consider many alternative solutions and discard unsuitable ones. The group must strive to discover, select, or compute the right answer. Once uncovered, the answer usually seems intuitively right and compelling. Examples include arithmetic or logical reasoning problems, puzzles, cryptograms, bus routing problems, and computer programs.
4. *Decision-making tasks.* A group is required to solve a problem for which there is no inherently right answer. Through interaction, the group must consider a variety of alternatives and reach a consensus regarding which alternative is preferred. Examples include jury deliberations, investment decisions, and decision tasks used in risky shift and group polarization studies.
5. *Cognitive conflict tasks.* Group members hold varying viewpoints and strive to resolve their differences. Conflicts of this type occur not because members have different interests and goals but because they

continued on next page

members of the team have prior experience, they will differ in strength and endurance. Because all the members of the team are linked together by a rope, the team can move no faster than its slowest member. The team's productivity (as measured by speed and height of ascent) is limited by its least proficient member. The same characteristic is also apparent in other conjunctive tasks, such as performing music in concert. A quartet playing classical music cannot generate a sound better than its least skilled player.

Another type is the **disjunctive task,** where a group's productivity depends entirely on the capability of its strongest or most proficient member. Many problem-solving tasks are conjunctive. To illustrate, suppose we gave the following problem to a five-person group and measured group productivity in terms of the speed with which the solution is found:

Three missionaries and three cannibals are standing on one side of a wide river. The missionaries want to cross the river to the other side. One rowboat is available, but it can hold no more than two persons at a time. All of the missionaries and one of the cannibals can row the boat. The cannibals pose a danger to the missionaries, because if more cannibals than missionaries remain together on either bank of the river (when the boat is not immediately present), those cannibals will eat the missionaries. Problem: How can all six persons cross the river alive?*

---

*Here is the answer to the missionaries (M) and cannibals (C) problem: (1) M1 and C1 cross in the boat; M1 returns. (2) M1 and C2 cross; C2 (who can row) returns. (3) M2 and C2 cross; C2 returns. (4) C2 and C3 cross; C2 returns. (5) M3 and C2 cross. Everyone is now on the other side of the river.

continued from previous page

disagree about the use of information. That is, they disagree regarding what information is relevant to the group's goals and how the information should be weighted and combined. A resolution of these differences usually entails discussion and negotiation among members. Examples include some jury tasks and cognitive conflict tasks used in social judgment studies.

6. *Mixed-motive tasks.* Group members face an underlying conflict of interest with respect to conditions for reward. To resolve this conflict, group members must negotiate and bargain with one another. Examples include labor-management wage negotiations, prisoner's dilemma studies, and reward allocation tasks.

7. *Contests/battles.* Group members compete as a unit against an external opponent or enemy. The outcome is interpreted in terms of a winner and a loser, with corresponding payoffs. Examples include competition between sports teams, battles between military combat groups, and other winner-take-all conflicts.

8. *Performances/psychomotor tasks.* Group members are required to exercise manual or psychomotor skills to bring about desired results. Much of the heavy work of the world—lifting, connecting, extruding, digging, pushing—falls into this category. These tasks can be subclassified in a myriad of ways, including by the type of material being worked on, the type of activity involved, and the intended product from the activity. In general, when performing these psychomotor tasks, groups strive to meet objective or absolute standards of excellence. Examples include laying a pipeline, loading a ship to meet a deadline, or achieving high output on an assembly line.

Underlying the eight types of tasks proposed by McGrath are four general processes: generating ideas and alternatives (task types 1 and 2), choosing among options (types 3 and 4), negotiating resolutions to conflict (types 5 and 6), and executing manual and psychomotor tasks (types 7 and 8). Note also that four task types are basically cognitive or conceptual in nature (types 2, 3, 4, and 5), and four are behavioral or action oriented (types 1, 6, 7, and 8). Similarly, four task types are essentially cooperative in nature (types 1, 2, 3, and 8), and four involve at least some degree of conflict (types 4, 5, 6, and 7).

For the group to solve this problem, only one member need see the solution. As soon as any one member finds the solution, he or she can explain it to the others. Assuming they accept it, the group's productivity score is just the time required for solution by its fastest member. Because the task is disjunctive, solutions by its less proficient members do not affect the group's productivity.

An alternative, different approach to task demands is described in Box 15.1, "Types of Group Tasks."

Group Properties Affecting Productivity

Group productivity is a matter of considerable importance in everyday life. Groups that perform well are praised and prized; those that perform poorly are often treated as problematic and as needing improvement. Why do some groups perform better than other, seemingly similar groups? What group properties (in addition to the task itself) determine the level of productivity?

Addressing this issue, Hackman and Morris (1975) proposed that three general factors have an impact on task productivity by groups: (1) the motivation and effort brought to bear on the task by group members, (2) the knowledge and skill of members applied to the task, and (3) the task performance strategies used by group members in carrying out the task. Hackman and Morris suggest if we could control these three factors, then we could substantially determine the level of productivity of a group working on almost any task.

Structural properties of groups such as size, goal specification, reward allocation, and communication

A rope-pulling contest among high school students. Tasks such as this often involve the Ringelmann effect (i.e., per capita performance declines as a function of group size) because people get in one another's way (faulty coordination).

networks can have a substantial impact on the three factors highlighted by Hackman and Morris. That is, a group's properties can affect the level of effort by its members, the availability of knowledge and skills, and the adequacy of its task performance strategies. The implication is that a group's structural properties have an indirect impact on the quality and quantity of group productivity. In this section, we look more closely at group properties that affect productivity.

Group Size

How does the size of a group affect its level of productivity? Are large groups more productive than smaller ones? Large groups often have the advantage of greater resources (information, skills), which may lead to more productivity. But large groups can have more difficulty than small ones in establishing adequate coordination among members, and this may lead to diminished productivity. To understand the effects of a group's size on productivity, we must take into

account the type of task facing the group. We begin by looking at group size in the context of additive tasks, and then we consider disjunctive and conjunctive tasks.

Additive Tasks   In a classic study published before World War I, a French agricultural engineer named Ringelmann investigated the effects of group size on the total pulling power exerted by the members of a group (Kravitz & Martin, 1986; Moede, 1927; Ringelmann, 1913). Ringelmann asked individuals to pull as hard as possible on a rope that was attached to heavy weights, and he used a gauge to measure the amount of pull exerted. The subjects in this study worked either by themselves or as members of a group. The groups were of various sizes, up to a maximum of eight persons. Results showed that although a group of eight men could pull harder than a smaller group or a single individual, the average contribution of members declined as the group's size increased (see Table 15.1). Working alone, each individual pulled at

Table 15.1 Actual and Potential Pulling Power of Groups of Various Sizes

No. of Workers in Group	Actual Pull (kilograms)	Potential Pull (kilograms)	Ratio of Actual Pull to Potential Pull (in percent)
1	63	63	100
2	118	126	93
3	160	189	85
4	194	252	77
5	221	315	70
6	238	378	63
7	248	441	56
8	248	504	49

SOURCE: Reconstructed from Ringelmann, 1913; Moede, 1927; and Kravitz and Martin, 1986.

100% of his capacity (larger groups produced in toto more than smaller ones, the output per member in the larger groups was lower than that in the smaller groups).

Several explanations have been advanced for the **Ringelmann effect.** One is based on *coordination loss,* that is, losses in productivity due to faulty coordination among group members. In Ringelmann's rope-pull task, the possibility of poor coordination (unintentionally pulling in slightly different directions or failing to pull at precisely the same time) increases with group size. Ringelmann himself thought the decline in per capita productivity was due to coordination loss, specifically, to the lack of simultaneity of members' efforts (Ringelmann, 1913, p. 9). Although Ringelmann did not investigate this issue in detail, several studies have shown coordination loss does occur on tasks of this type (Ingham et al., 1974; Latané, Williams, & Harkins, 1979).

Nevertheless, coordination loss does not account for all of the decline in per capita productivity that accompanies increases in group size. Some research indicates the Ringelmann effect will occur even on tasks structured so that there is no possibility of loss from faulty coordination (Harkins & Petty, 1982). Thus, theorists have pointed to another factor—motivation loss—to explain the Ringelmann effect. *Motivation loss* refers to a circumstance where members reduce their work effort and slack off, thereby producing less than they otherwise would.

In recent times, social psychologists have looked extensively at **social loafing,** a form of motivation loss that occurs when there is no clear way to evaluate the contributions of individual members toward the group's product. Social loafing is surprisingly widespread, and many studies have documented a falloff in effort by group members on additive tasks (Harkins, 1987; Harkins & Szymanski, 1987; Latané, Williams, & Harkins, 1979).

Because social loafing can negatively impact group productivity, a major concern is how to reduce it. A number of factors affect the level of social loafing on shared tasks (Karau & Williams, 1993, 1995). First, social loafing is lower under conditions of *identifiability.* That is, members tend to work harder and avoid slacking off if each member can be distinctly identified with his or her own contribution to the group's performance than if he or she cannot be so identified (Hardy & Latané, 1986; Kerr & Bruun, 1981; Williams, Harkins, & Latané, 1981). However, the pivotal factor in suppressing social loafing is not just identifiability per se, but rather the *evaluation* that often accompanies identifiability. Persons loaf less under conditions of identifiability when reduced effort will result in negative evaluation than when it will not result in negative evaluation (Harkins & Jackson, 1985; Harkins & Szymanski, 1989; Szymanski & Harkins, 1987; Williams et al., 1989). The negative evaluation can be from others, or it can be negative self-evaluation based on social standards.

Individuals are less likely to slack off when they perceive that their contributions to the group are unique and indispensable, rather than merely redundant with the inputs from other members (Kerr & Bruun, 1981, 1983). For instance, when a group is working on tasks that are disjunctive in nature (i.e., tasks such that if any one group member succeeds in reaching a performance criterion, the group as a whole succeeds), members with unique abilities may feel their contributions are essential to a high-quality group product; under these conditions, members are more likely to exert a substantial effort and avoid loafing.

Another factor that affects social loafing is the valence of the group's task. When a group's members are working on an involving task—one that is interesting, meaningful, and challenging—they are less likely to loaf than when working on an uninvolving task (Brickner, Harkins, & Ostrom, 1986; Harkins & Petty, 1982; Zaccaro, 1984). Closely related to this is the impact of members' attraction to the group itself on their tendency to loaf. Individuals are less likely to engage in social loafing when they strongly value the group and their membership in it than when they do not (Karau & Williams, 1993).

Disjunctive and Conjunctive Tasks

Group size impacts group productivity not only on additive tasks, but also on disjunctive and conjunctive tasks. According to a theory proposed by Steiner (1972), the impact of group size on disjunctive tasks is different from that on conjunctive tasks. Specifically, the theory predicts productivity will (1) increase as a function of group size when the task is disjunctive, and (2) decrease as a function of group size when the task is conjunctive.

Consider the reasoning beneath these predictions. When a task is disjunctive, a group's productivity is determined by its most proficient member. On the assumption that groups are composed at random, a large group has more chance than a small group of containing members of very high ability. The larger the group, the more likely it is to contain the necessary skills. In consequence, as size increases, productivity should also improve. Of course, there may be an

upper limit to this effect—some point beyond which additional members will have little or no added impact. If a group already has 20 members, for example, the 21st member may not add much.

When a group's task is conjunctive, the predicted effects of size are different. Group productivity on conjunctive tasks can be no better than that of its least proficient member. Again assuming that groups are composed at random, a large group is more likely to have very weak members than a small group. These weak members will be slower or less competent at their task than other group members. Consequently, large groups should perform less well than small ones on conjunctive tasks.

To test these predictions, one study investigated the effects of group size on conjunctive and disjunctive tasks by comparing the productivity of groups with two, three, five, and eight members (Frank & Anderson, 1971). Each group worked on a series of tasks for 15 minutes; productivity was measured by the number of tasks completed successfully. The tasks were intellective in nature, requiring group members to generate ideas or images. (Example: "Write three points pro and con on the issue of legalizing gambling.") By manipulating the instructions, the investigators made these tasks either disjunctive or conjunctive. Some groups received disjunctive instructions, which stated the group was to work on various tasks in sequence. As soon as any member had completed the task, the group could move on to the next task. Other groups received conjunctive instructions, which stated the group was not permitted to move to the next task until every member had completed the task. Thus, a group in the disjunctive condition could proceed at the speed of its most competent member, whereas a group in the conjunctive condition could move no faster than its slowest member.

The findings from this study are shown in Figure 15.1. The observed results generally support the predictions indicated above. On disjunctive tasks, large groups performed better than small ones. On conjunctive tasks, large groups performed worse than small ones, although this difference was rather small. Findings of other studies have also provided some support for these predictions, especially with respect

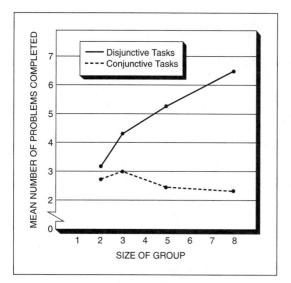

Figure 15.1 Effects of Disjunctive and Conjunctive Tasks on Group Productivity

On disjunctive tasks—tasks in which a group's performance depends on the performance of its best member—large groups were shown to perform better than small ones. However, on conjunctive tasks—tasks in which a group's performance is restricted by the performance of its worst member—large groups performed slightly less well than small ones.

SOURCE: Adapted from Frank and Anderson, 1971.

to disjunctive tasks (Bray, Kerr, & Atkin, 1978; Littlepage, 1991).

Group Goals

The goals established by a group have an important impact on the level of productivity by the group's members (Zander, 1985). In general, we can say that if a group establishes explicit, demanding objectives with respect to the group's performance, and if the group's members are highly committed to those objectives, then the group will perform at a higher level than if it does not do these things. These effects have been demonstrated by many studies, using a variety of different tasks, with goals pertaining to quantity and quality of productivity, in both naturally occur-

ring groups and ad hoc laboratory groups (O'Leary-Kelly, Martocchio, & Frink, 1994; Pritchard, Jones, Roth, Stuebing, & Ekeberg, 1988). This phenomenon is often referred to as the **group goal effect** on productivity.

The magnitude of the goal effect on productivity is often rather substantial. For instance, one study (Gowen, 1986) investigated the effect on group productivity of setting individual goals and group goals in a laboratory work setting. Results showed that setting explicit individual goals led to a 19% increase in group performance relative to no goals; and setting group goals led to a 12% increase in group performance relative to no goals. The combination of compatible individual goals and group goals led to a 31% increase in group productivity, clearly a large effect.

Why does the fact of establishing explicit, challenging goals generally lead to higher levels of group productivity? There seem to be several reasons (Weldon & Weingart, 1993). First, setting a goal tends to strengthen members' work efforts. With an explicit target, they have a focus for their efforts; this channels their performance, resulting in more productivity. Second, setting an explicit goal often makes clear the need for planning by the group. This is particularly true when the task is complex rather than simple and when work-flow interdependence among members is high rather than low. Planning by the group often leads to greater efficiency, hence to more total productivity. Third, the presence of an explicit and demanding goal often leads to higher levels of cooperation among members. This is sometimes reflected in a high level of morale-building communication among members, the effect of which is to heighten members' enthusiasm and sense of group efficacy. Finally, establishing an explicit goal often reduces members' concern for aspects of performance unrelated to the goal. This leads to less wasted motion, hence higher productivity.

Reward Allocation and Task Interdependence

Task groups differ markedly with respect to the allocation of rewards to members. This matters because a

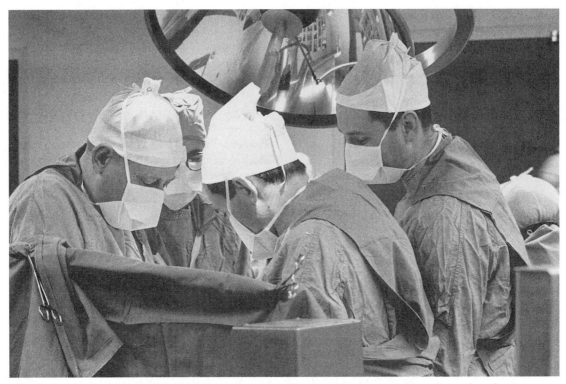

A team of surgeons performs a high-stakes operation. For success, this task requires an extraordinary degree of coordination among team members. Coordination is facilitated by members' high level of training, as well as by the unchallenged authority of the head surgeon.

group's level of productivity can be influenced by the distribution of rewards among its members. The rewards or benefits allocated to members affect their motivation and effort on group tasks. It goes without saying that not all groups allocate their earnings equally to their members. In many groups, certain members are rewarded more than others, depending on such factors as differences in contribution, seniority of membership, degree of task interdependence, and the like (Chen & Church, 1993). When members are not all rewarded equally, we say a condition of *differential rewarding* exists in the group.

Members often make judgments regarding what level of rewards they should receive for participating in a group. Reflecting a concern with fairness or equity, these judgments take the into account

such factors as the level of their own contributions relative to those of other members and the degree of task interdependence among members (Miller & Komorita, 1995). We say that a high degree of *task interdependence* exists when the task demands a high level of communication, coordination, and mutual performance monitoring among members for successful completion. (For instance, conjunctive tasks usually create a high degree of interdependence, whereas disjunctive tasks create a lower degree of interdependence.)

In cases of high task interdependence, group productivity may suffer if rewards are allocated very unequally among members. A high level of differential rewarding may cause some members to feel the (small) rewards they receive are not fair in light of others' dependence on them. If this happens, their work effort declines and the group's productivity suffers accordingly.

The effects of task interdependence and differential rewarding on group productivity were demonstrated vividly in a study by Miller and Hamblin (1963). This study investigated three-person groups in a laboratory setting where members worked on a series of problems. The members were placed in separate booths connected by a system of electrical lights and switches. The group's task was to determine which one of 13 numbers had been selected by the experimenter. Each member was privately informed of 4 numbers that had *not* been selected by the experimenter. These sets of 4 numbers were different for each member; thus, if the three members pooled their information (by the electrical signaling system), they would immediately see the correct answer. The group's productivity was measured in terms of time: The faster it discovered the correct answer, the higher the score it received.

Two independent variables were manipulated in the study. One was the degree of task interdependence among group members. In the high-interdependence condition, guessing was discouraged by a substantial penalty. Group members, therefore, were forced to coordinate their efforts and share information to solve the problem. In the low-interdependence condition, there was no restriction on guessing, making coordination among members unnecessary. To solve the problem, individuals could merely continue to guess until they hit the correct solution.

The other independent variable in the study was the degree of differential rewarding among group members. Rewards for group members were based both on their own efforts and on those of the group as a whole. Each group started with 90 points, and 1 point was subtracted for every second that elapsed before all 3 members had solved the problem. For example, if the group took 60 seconds to reach a solution, it scored 30 points. The group's points were then allocated among its members. In one experimental condition (no differential rewarding), each member received one third of the group reward. In a second condition (medium differential rewarding), the member who first solved the problem received one half of the group's points, the member who finished second received one third, and the one who finished third received one sixth. In a third condition (high differential rewarding), the member who first solved the problem received two thirds of the group's points, the member who finished second received one third, and the one who finished third received none.

The results of this study are displayed in Figure 15.2. Under conditions of low interdependence (where all three individuals could freely guess at the answer), differential rewarding did not significantly affect group productivity. But under conditions of high interdependence (where guessing was prohibited and members had to share information), higher levels

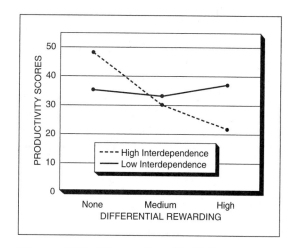

Figure 15.2 Group Productivity as a Function of Interdependence and Degree of Differential Rewarding

Under low task interdependence, differential rewarding (that is, unequal reward distribution) had little effect on group performance. However, under high task interdependence, a high level of differential rewarding induced low group productivity as well as competition among members, and a low level of differential rewarding induced high productivity and cooperation.

SOURCE: Adapted from Miller and Hamblin, 1963.

of differential rewarding sharply reduced the group's level of productivity.

To interpret these results, note that under conditions of high task interdependence and no differential rewarding, members realized their best strategy for achieving higher rewards was to cooperate with each other. They shared information quickly, and, as a result, the group got a high productivity score. Under conditions of high task interdependence and high differential rewarding, however, members realized they could achieve high rewards for themselves only by outperforming others in their group. This created a competitive atmosphere in which members tried to hinder one another. For example, one member might try to obtain information from others without yielding information in return, thereby blocking others' performance. This would diminish coordination, slow the group's progress, and inevitably result in poor group productivity.

Under conditions of low task interdependence, in contrast, members could not use a blocking strategy on one another because guessing was permitted. To increase their rewards, they focused on raising their performance independently of the other members (via guessing). This led to moderate group productivity regardless of the degree of differential rewarding. Related research shows similar results (Rosenbaum et al., 1980).

Communication Networks

We often hear about the frustrations of working within a large organization like a military or industrial bureaucracy, where communication must go through prescribed channels. Channels help to ensure messages are sent to the "right" person, but at the same time they restrict who can talk with whom. To some degree, this holds true for small groups as well. The term **communication network** refers to regularized channels of information exchange or communication opportunities within a group or organization. In some large groups, for instance, members may adopt a hierarchical communication network that prescribes how information will flow between superiors and subordinates. In other groups, members may have no formal rules governing the flow of information but still establish an information communication network over time. The structure of a group's communication network will influence actual communication among group members (that is, who talks with whom, how frequently, and about what). It can also affect the quantity and quality of group productivity on various tasks.

In recent years investigators have studied the impact of networks in various settings, including families, nonprofit organizations, and military units (Craddock, 1985; Hartman & Johnson, 1990; Monge, Edwards, & Kirste, 1983). The earliest systematic studies of communication networks, however, were based on small groups in laboratory settings.

The focus of those studies was the impact of different communication networks on a group's problem-solving productivity (Bavelas, 1950; Leavitt, 1951). In studies such as these, group members are seated in separate cubicles and given problems to solve. They are not permitted to talk to one another and can communicate only by passing notes through slots in the cubicle walls. When all of these slots are open simultaneously, each member can communicate directly with every other member; when some of the slots are closed, certain channels are eliminated and the flow of information restricted. This method enables investigators to study the effects of various communication networks on group problem-solving efficiency.

Figure 15.3 displays several different communication networks for a five-person group. Although the *comcon* (or "*com*pletely *con*nected") network allows each member to talk freely with all the other members, the other networks impose restrictions on communication. For example, in the *wheel* network, one person is at the hub, and all messages must pass through him or her. In a *chain* network, all information transmitted from a person at one end must pass through a series of others to reach the other end; people located near the middle of the chain are therefore more central to the flow of communication than those located at the extremes.

In what ways does a group's communication network affect its productivity on problem-solving tasks? This question arises in part from practical

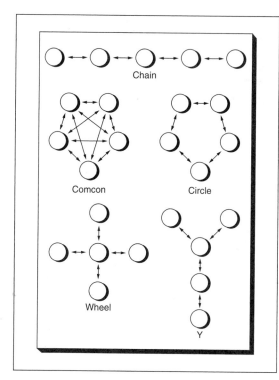

Figure 15.3 Communication Networks in Five-Person Groups

This figure shows different types of communication networks that can exist within five-person groups. Three of the networks shown—the wheel, the chain, and the Y—are centralized. That is, messages flowing from one end of the network to the other must pass through a central person (or hub). The other two networks—the circle and the comcon—are decentralized; messages need not pass through a single hub.

SOURCE: Adapted from Leavitt, 1951.

considerations, for people want to know how to organize work groups, committees, and teams in the most efficient way. In typical research studies on this issue, each group is assigned a problem-solving task that requires communication among members for a successful resolution. In some cases, these are simple intellective tasks, such as identifying which of various symbols (stars, triangles, circles, and so on) appear on cards. In other cases, these are more complex tasks that require the members to transform information

in some manner, such as arranging words, constructing sentences, and performing arithmetic. Whether the task is simple or complex, a group's efficiency in problem solving is measured by the time it takes to achieve a solution, by the number of messages sent by members in order to achieve a solution, or by the number of errors made while attempting to solve a problem.

Because a group's network restricts how group members can communicate while attempting to solve problems, it also affects how they organize themselves to work on problems. An early study (Leavitt, 1951) used a symbol-identification task to compare the organization and effectiveness of five-person groups under the wheel, chain, Y, and circle networks. Results indicate that groups having wheel, chain, and Y networks adopted a centralized organization in which the person occupying the central position (for example, the "hub" in the wheel) usually emerged as highly influential; those members occupying the peripheral positions did not often emerge as influential. In contrast groups having a circle network adopted a decentralized organization in which all members communicated with and influenced one another. Other studies have investigated the comcon network and have found that groups under comcon adopt a decentralized organization (Guetzkow & Dill, 1957; Shaw & Rothschild, 1956). In general, a group's communication network determines whether its organization is centralized or decentralized.

Centrality of organization impacts such outcomes as member satisfaction and participation (Mullen, Johnson, & Salas, 1991b). It also has important consequences for a group's productivity on problem solving tasks. Shaw (1964, 1978) compiled the findings from numerous experiments on the communication networks of problem-solving groups. He classified each study as involving either centralized networks (wheel, chain, Y) or decentralized networks (circle, comcon) and as entailing either simple tasks (collating information) or complex tasks (carrying out operations on the information). The summary of results appears in Table 15.2. This summary shows that for simple problems, centralized communication networks led to superior productivity. Groups with centralized networks solved simple problems

Table 15.2 Comparisons of Centralized (Wheel, Chain, Y) and Decentralized (Circle, Comcon) Networks' Efficiency as a Function of Task Complexity

	Simple Problems*	Complex Problems**	Total
Time			
Centralized faster	14	0	14
Decentralized faster	4	18	22
Messages			
Centralized sent more	0	1	1
Decentralized sent more	18	17	35
Errors			
Centralized made more	0	6	6
Decentralized made more	9	1	10
No difference	1	3	4
Satisfaction			
Centralized higher	1	1	2
Decentralized higher	7	10	17

*Simple problems: symbol-, letter-, number-, and color-identification tasks.
**Complex problems: arithmetic, word arrangement, sentence construction, and discussion tasks.
SOURCE: Adapted from Shaw, 1964.

faster, sent fewer messages, and made fewer errors than groups with decentralized networks. For complex problems, however, the findings were reversed: Groups with decentralized networks were more efficient. Even though they sent more messages, they solved problems faster and tended to make fewer errors than groups with centralized networks.

Numerous studies support the generalization that centralized networks are more efficient for simple tasks, whereas decentralized networks are more efficient for complex ones (Lawson, 1964; Morrissette, 1966). One explanation for this pattern lies in the concept of **information saturation,** which is the degree of communication overload experienced by group members occupying positions within communication networks (Gilchrist, Shaw, & Walker, 1954; Shaw, 1978). If a large number of complex communications are routed through a position in a network, the person in that position will not be able efficiently to monitor, process, collate, and route the incoming and outgoing messages. Information saturation can occur in a group with a decentralized network, but it

is more likely to occur in a group with a centralized network working on a complex problem. Persons in centralized networks—especially those persons in the most central positions—experience more information saturation than persons in decentralized networks. The greater the level of saturation, the less efficient a group is at solving problems.

To illustrate, consider the wheel network (see Figure 15.3). In this structure, the hub position is central. If the group is working on a simple task, communication requirements are not very demanding. The hub position does not become saturated, and the person occupying this position can work quickly and efficiently. If the group faces a more complex problem, however, the communication requirements placed on the hub are substantial, and the person occupying that position may become saturated and overburdened. When this happens, the group's problem-solving efficiency is low. Decentralized networks like the circle or comcon do not have this problem, because no position in the network is subject to extreme saturation. This topic is discussed

15.2   Communication Networks in an Organization

Communication networks have a significant impact not only in laboratory groups but also in organizational settings. In one case study (Mears, 1974), a group of division representatives within an aerospace firm tried out several different communication networks. This group consisted of an administrative officer and representatives from several divisions, such as manufacturing, quality control, procurement, contracts, and engineering. At first this group was organized in a *comcon network*. This permitted virtually unrestricted communication among all members. Satisfaction was very high, but only a modest amount of work was accomplished because members wasted a great deal of time in useless discussion and debate.

Top management grew dissatisfied with the group's productivity and restructured it in a *wheel network,* with the administrative officer in a position of author-ity at the hub. This restriction in communication reduced worker motivation and satisfaction. It also lowered productivity because the hub position became saturated, which caused many errors in relaying complex information.

Finally, the group was reorganized in a *modified comcon network,* which permitted each member to communicate only with persons who were directly involved in the task at hand. This reduced the communication overload on the administrative officer and also protected the time of members who did not need to be consulted. As a result, the group experienced high levels of satisfaction and productivity.

As this case study shows, communication networks are not theoretical abstractions found only in research laboratories. They come into existence—either through natural evolution or by intentional design—in real-world work groups and organizations. And just as with laboratory groups, they can significantly affect the productivity and satisfaction of workers.

further in Box 15.2, "Communication Networks in an Organization."

Leader Activities and Leadership Effectiveness

As we have seen, a group's productivity depends on several factors, including the nature of the task, communication structure, and reward structure. Another factor affecting productivity is the type of leadership exercised within the group.

By definition, **leadership** is the process whereby one group member (the leader) influences and coordinates the behavior of other members in pursuit of group goals (Yukl, 1981). In return for support from the others, a leader provides guidance, specialized skills, and environmental contacts that help the group attain its goals. Some of the factors that determine whether a person will become a leader are discussed in Chapter 14. Here, we first consider the activities of leaders—what leaders actually do—and then look at factors that determine whether or not a leader will be effective in helping a group achieve its goals.

Activities of Leaders

A person serving as a leader fulfills certain functions necessary for successful group performance. These functions include planning, organizing, and controlling the activity of group members (Bass, 1990). In formal groups, where roles are organized in a status hierarchy, leadership functions are typically fulfilled by high-status members. These members have both the right and the responsibility to provide leadership for the group. In informal groups, however, these functions may be fulfilled by one or several persons who emerge during interaction as task leaders and social-emotional leaders.

Exactly what do leaders do? That is, what behaviors do they engage in when providing leadership

A drill instructor uses a strict authoritarian style to direct his troops. Would the drill instructor's leadership style be effective in another setting, such as conducting a symphony orchestra?

for a group? Investigators have studied the behavior of leaders in many settings, including military units, industrial work groups, training groups, and laboratory groups (Schriesheim & Kerr, 1974; Wofford, 1970). They find, in broad terms, that leaders plan, organize, and control the activity of group members. More specifically, leaders usually do some or all of the following: (1) formulate a clear conception of the group's goals and objectives, and communicate this to group members; (2) develop specific strategies for the attainment of group goals; (3) specify role assignments and standards of productivity for members; (4) establish and maintain channels of communication among members; (5) recruit new members (if needed) and train members in crucial skills; (6) interact with members personally to maintain good relations; (7) influence the task activities of group members by means of persuasion, rewards, and punishments; (8) monitor the group's progress toward its goals and take corrective steps if the group is off track; (9) resolve conflict among members to reduce tension and maintain harmony; and (10) serve as a representative of the group to outside agencies and organizations. In any given group, some of these leadership behaviors are more important than others. Which are most crucial at any one time depends in part on whether the group has been successful or unsuccessful in attaining its goals and whether there are serious interpersonal problems and conflicts among members.

As this list suggests, leadership often involves a tacit exchange between the leader and the other group members. By fulfilling the planning, organizing, and controlling functions in a group, the leader helps move the group toward the attainment of its goals. In return, the leader receives support for continued control, as well as special rewards and privileges. Leadership of this type, based on an exchange between the leader and group members, is termed **transactional leadership** (Hollander, 1985; Hollander & Julian, 1970; Homans, 1974).

In some circumstances, however, leaders do more than merely mediate rewards and goal attainment for group members in exchange for power and privilege. They strengthen group productivity by changing the way members view their group, its opportunities, and its mission. Some leaders foster high levels of group productivity by conveying an extraordinary sense of mission to group members, arousing new ways of thinking within the group, and stimulating new learning by members. Leadership in this form, termed **transformational leadership,** often creates structural change and institutionalizes new practices within the group and thereby strengthens group productivity (Hater & Bass, 1988; Seltzer & Bass, 1990).

Contingency Model of Leadership Effectiveness

Depending on the situation and on their personalities, leaders use different techniques to plan, organize, and control group activities. For example, some adopt an authoritarian leadership style, whereas others use a democratic leadership style. A leader's effectiveness in directing a group depends both on his or her style and on the characteristics of the situation. Someone with a given leadership style may be effective in one situation but not in another.

This basic notion underlies the **contingency model of leadership effectiveness** (Fiedler, 1978a, 1978b, 1981). The contingency model pertains primarily to groups working on divisible tasks that require coordination among members for successful performance. This model is highly general and has been applied to a variety of groups in military, educational, and industrial settings.

Leadership Style: The LPC Score   There are four independent variables in the contingency model of leadership effectiveness. One of these is leadership style; the other three are properties of the situation in which a leader performs. The contingency model characterizes leadership style as either relationship oriented or task oriented. To determine leadership style, a leader is first asked to recall all the people with whom he or she has ever worked in a group setting and to select the one who was most difficult to get along with. This person is designated the *least preferred co-worker* (LPC). Next, the leader is asked to rate this person on dimensions such as pleasant-unpleasant, helpful-frustrating, cooperative-uncooperative, and efficient-inefficient. Some leaders give very low, negative ratings to their least preferred co-worker; these are called *low-LPC leaders.* Other leaders find positive qualities even in their least preferred co-worker and give high ratings; these are *high-LPC leaders.* Thus, LPC is a measure of leadership style. Some studies show low-LPC leaders care primarily about successful task productivity, whereas high-LPC leaders care primarily about establishing congenial interpersonal relations with others (Fiedler, 1978a; Rice et al., 1982).

Situational Factors   A leader's effectiveness depends not only on leadership style but on various characteristics of the situation. According to Fiedler's contingency model, three characteristics of the situation are crucial. In order of importance, these are (1) the leader's personal relations with other group members (good or poor), (2) the degree of structure in the group's task (structured or unstructured), and (3) the leader's position of power in the group (strong or weak).

In the contingency model, a leader's *personal relations* with other members is assumed to be the most important factor determining the leader's influence in the group. A leader whom the other members trust, like, and respect is in a more favorable situation than one with poor rapport, and he or she usually has the support of group members. The second factor, *task structure,* refers to how clearly defined the task requirements of a group are. Is the group's goal clearly spelled out? Is there a single path by which this goal can be achieved? Is there only one correct solution or decision? Is there some method of verifying the accuracy of the solution or decision? The more these things are characteristic of the task, the greater the task's structure and the more favorable the situation for the leader. The third factor, *position power,* has to do with a leader's formal authority over group members and the degree to which he or she can buttress directives with rewards and punishments. A leader who wields more power is in a more favorable situation.

For analytic convenience, each of these factors is treated as dichotomous: Leader–member relations are either good or poor, the group's task is either structured or unstructured, and the leader's power is either strong or weak. Taken together, these three dichotomies produce eight possible situations. These situations are shown in Figure 15.4, listed according to their overall favorableness for the leader.

From a leader's standpoint, a situation is favorable if it enables him or her to exercise a great deal of influence. Thus, a situation involving good leader-member relations, a highly structured task, and strong leader power is favorable for the leader. At the other extreme, a situation involving poor leader-member relations, an unstructured task, and weak leader power is very unfavorable for the leader.

Predictions   According to the contingency model, task-oriented (low-LPC) leaders are effective in situations that are either highly favorable or highly unfavorable, but they are ineffective in other situations. In contrast, relationship-oriented (high-LPC) leaders are effective in situations that are moderately favorable, but they are ineffective in the other situations. These predictions are shown in Figure 15.4.

When conditions are extremely unfavorable, a group needs strong task leadership, and it is the task-oriented (low-LPC) leader who best provides this. A low-LPC leader is willing to overlook interpersonal conflicts in order to concentrate on the group's task. A high-LPC leader attempts to smooth over interpersonal problems rather than work on the group's fundamental task.

However, when conditions for leadership are intermediate in favorableness, the high-LPC leader is the one who best provides what the group needs. Under these conditions, some situational factors are positive and others negative. Interpersonal problems may come to the surface, and the relationship-oriented (high-LPC) leader will be more effective than the task-oriented (low-LPC) leader.

Finally, when conditions are highly favorable, the low-LPC leader again emerges as most effective. Under these circumstances, low-LPC leaders feel they can relax because goal attainment is not problematic. They turn their attention to interpersonal relations with their co-workers because they know the task will get done in due course. In other words, low-LPC leaders undergo a change in behavior that enables them to be highly effective when conditions are favorable. Under these same conditions, however, relationship-oriented (high-LPC) leaders are relatively ineffective.

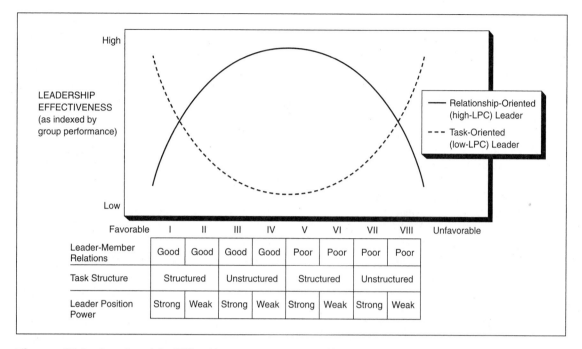

Figure 15.4 Leadership Effectiveness as a Function of Leadership Style and Situational Factors

The contingency model predicts that group performance depends on both leadership style and situational factors. According to this model, task-oriented (low-LPC) leaders are most effective in situations that are either highly favorable (that is, involving good leader-member relations, a structured task, and strong leader power) or highly unfavorable (that is, involving poor leader-member relations, an unstructured task, and weak leader power). In contrast, relationship-oriented (high-LPC) leaders are most effective in situations that are of medium favorableness.

SOURCE: Adapted from Fiedler, 1978a.

They start looking for things to do and become bossy and less concerned with the feelings and opinions of their co-workers, which diminishes their effectiveness (Fiedler, 1978a; Larson & Rowland, 1973).

Tests of the Contingency Model Although there are some exceptions, research generally supports the contingency model's predictions (Chemers, 1983; Peters, Hartke, & Pohlmann, 1985; Strube & Garcia, 1981). One study (Chemers & Skryzpek, 1972) used West Point cadets as subjects and manipulated all three dimensions of situational favorableness. Groups of four men were assembled on the basis of information about preexisting relations among members. In half of the groups the relations between leader and

members were good, whereas in the other half they were poor. Each group performed one structured task (converting blueprints from metric units to inches) and one unstructured task (discussing an issue and making policy recommendations). In half of the groups the leader had strong power—that is, members believed the leader would assess their performance and this would affect their standing at the academy. In the other half the leader had only weak power—members believed he would have little effect on their standing. As predicted by the contingency model, the task-oriented (low-LPC) leaders proved to be more effective when the situation was either very unfavorable or very favorable. In contrast, the relationship-oriented (high-LPC) leaders were more effective when the situation was of intermediate favorableness. Similar findings have emerged in studies of college and elementary school students (Hardy, 1975; Hardy, Sack, & Harpine, 1973) and naval personnel (Fiedler, 1966).

Despite this support, several studies have reported findings that contradict the contingency model (Graen, Orris, & Alvares, 1971; Vecchio, 1977). One discrepancy is that in the case where relations between leader and members are good, the group's task is structured, and the leader's power is weak, group productivity under high-LPC leaders is sometimes greater than predicted by the contingency model (Fiedler, 1978a). The contingency model has merit, but its predictions are not fully accurate in all eight situations (Peters, Hartke, & Pohlmann, 1985; Vecchio, 1983).

Idea Generation and Brainstorming

To this point in discussing productivity, we have considered various groups working on a wide array of tasks. For instance, we have seen groups working on such tasks as pulling a rope attached to weights (Ringelmann), creating noise by clapping and shouting, manufacturing clothing in an industrial setting, and striving to climb a mountain. We have also seen groups performing activities that are more fundamentally cognitive, such as solving problems or puzzles and discussing issues to make decisions or policy recommendations.

In everyday life, many cognitive products—solutions, decisions, verdicts, and judgments—are generated by groups. So prevalent is this activity that some persons have even suggested most of the truly important decisions and judgments in society are made by groups, not by individuals. Be that as it may, groups surely are a major source of important cognitive products. In the remainder of this chapter, we discuss various cognitive products from groups. Here, we focus on groups generating one special type of cognitive output, namely, novel or innovative ideas.

Suppose an advertising executive has the responsibility to develop a strong media campaign to promote a new product, such as a toothpaste that not only cleans but also strengthens teeth. Should she ask staff members to work individually on this task, or should she bring them together in a group and ask them to work jointly on the task? Some years ago, Alex Osborn (1963) proposed groups are better at generating creative new ideas than individuals working alone. In particular, Osborn maintained groups employing a specific procedure called **brainstorming** will be able to generate a large number of high-quality novel ideas in a brief period.

Osborn's approach requires that a group follow a specific behavioral protocol when generating new ideas. At the outset of a brainstorming session, members are given a focal problem to discuss. For instance, if members are to develop advertising slogans for the new toothpaste, their objective would be to think up as many different advertising themes or slogans as possible. Whenever a member has a new idea, it is written down or recorded in some manner. Osborn maintained that members should adhere to certain norms during the brainstorming session, including the following: (1) Members should express any idea that comes to mind, no matter how fanciful or far-out it may seem. The wilder the idea, the better. (2) Members should withhold criticism and defer judgment on all ideas until later. Members should not even criticize their own ideas; instead, they should let ideas flow freely without precensoring them. (3) Quantity is desirable—the more ideas, the better. Members should try to generate as many novel ideas as possible because this improves the odds of getting some good ones. (4) Members should try to build on ideas suggested by others; that is, members may generate new ideas by combining, modifying, or extending the ideas of others.

Although Osborn claimed considerable success with the brainstorming technique, systematic empirical studies are not positive in their results. For instance, one early study (Taylor, Berry, & Block, 1958) asked four-person brainstorming groups to work on various problems (e.g., how to increase tourist travel in the United States). Results showed that the number of novel suggestions generated by four-person groups was greater than the number of suggestions by single individuals working alone. This result might seem to support Osborn's thesis, but it ignores the fact that the four-person groups involved four times the person power. To take this into account, the investigators compared the productivity of the four-person brainstorming groups with the combined productivity of four persons who had worked alone (these pooled groups are called "nominal groups"). This comparison showed that the four-person nominal groups produced almost twice as many original ideas per time period as the four-person face-to-face groups operating under brainstorming instructions. Another study (Bouchard & Hare, 1970) used groups of five, seven, or nine subjects and asked them to discuss the following problem: What would happen if people had an extra thumb? The participants wrote down as many implications of this anatomical change as they could imagine. Answers generated by these face-to-face groups were compared with the pooled answers from five-, seven-, or nine-person nominal groups composed of individuals who had worked alone. Once again, the results showed that more solutions were generated by the nominal groups than by the interacting brainstorming groups. Moreover, this difference increased as the size of the groups increased.

These results illustrate the general finding that emerges from studies of brainstorming: Brainstorming groups generate fewer novel ideas than the same number of individuals working alone (Lamm & Trommsdorff, 1973; Mullen, Johnson, & Salas, 1991a). This effect is unmistakable, with nominal groups outperforming interacting groups by as much as twice the number of nonredundant ideas (Diehl & Stroebe, 1987; Street, 1974). In addition, the quality and the originality of the ideas produced by brainstorming groups is, on average, no higher than that of ideas produced by individuals working alone (Bouchard, Barsaloux, & Drauden, 1974).

Production Blocking

Why aren't brainstorming groups more effective? One prominent answer is that a phenomenon called production blocking limits the ideas generated by brainstorming groups (Diehl & Stroebe, 1987, 1991;

Lamm & Trommsdorff, 1973). **Production blocking** refers to a circumstance in which participants of a brainstorming group are unable to express their ideas due to turn taking among group members. Because most individuals follow the norm that persons should take turns when talking, members in brainstorming sessions often wait for others to stop speaking before they voice their own ideas. This naturally puts a limit on the rate at which new ideas can be expressed. In addition, because suggestions from others can be distracting, members may have to concentrate on remembering or "rehearsing" their own ideas while others are speaking; this further limits productivity because members who are rehearsing cannot turn their attention to generating more new ideas.

Production blocking is illustrated in a study by Diehl and Stroebe (1987). This study (experiment 4) utilized four-person groups on a brainstorming task and manipulated whether or not subjects could speak when others were speaking. In one condition, the four persons participated in an actual interacting brainstorming group where suggestions were tape-recorded; members were placed in separate rooms, but they could hear one another over an intercom; a signaling system of lights indicated when others were speaking, and members could speak into their own microphones only when others were not speaking (i.e., when others' lights were not on). In a second condition, the set-up was similar, except that participants could not actually hear each other's voices over the intercom. In a third condition, the participants could not hear one another over the intercom; they could see the lights, but were told to ignore these; and they were free to express their ideas into the microphone whenever they wanted. Results indicated that group production of ideas in the first two conditions (where subjects could speak only when the light indicated that others were not speaking) was much lower than production of ideas in the third condition (where subjects could express their ideas whenever they wanted, irrespective of whether others were speaking). The interference caused by the requirement that members speak in turn (when lights permitted) blocked production of ideas and reduced total group output.

Group Decision Making

Beyond the matter of idea generation or brainstorming, another important group process is decision making. A *decision* is a choice among options made on the basis of preferences or values. Typically, when a group makes a decision, its members consider the merits of several mutually exclusive options and then select one in preference to the others. Consider some everyday examples: Managers of a business firm decide which new products to develop and which to abandon; experts on a review committee decide which research proposals to support and which to turn down; family members decide what place to visit during their annual vacation together; participants on a high-ranking governmental board decide which of various foreign policy options to pursue. In every instance, the end product resulting from interaction among group members is a decision—a choice among alternative courses of action.

Reduced to its essentials, decision making is fairly easy and involves several basic steps (Janis & Mann, 1977). To make a decision effectively, group members need to define a set of possible options, gather all the relevant information about these options, share this information among themselves, carefully assess all the potential consequences of each option under consideration, and then calculate the overall value of each option. Once this is done, it remains only to select the most attractive option as the group's choice.

In practice, however, group decision making is not always so easy or straightforward, for the decision-making process can go awry in various ways. Information regarding certain options may prove hard to obtain, leading to incomplete or inadequate consideration of these options. Even if the individual members do have all the relevant information, they may fail to share it fully with one another (Stasser, 1992; Stasser & Titus, 1987). If members hold different values, they may disagree regarding which options are most attractive; this can spawn arguments and block consensus within the group. Then, too, conformity pressures within the group may impel members to abbreviate or short-circuit the deliberation processes; if this happens, group discussion may lead to ill-considered or unrealistic decisions. To explore

these issues further, in this section we look at various processes involved in group decision making.

Groupthink and Dysfunctional Decision Making

Aberrations in decision making can plague any group, even those at the highest levels of business and government. The history of American foreign policy provides many examples. The decisions by the United States to invade the Cuban Bay of Pigs, to cross the 38th parallel in the Korean conflict, and to escalate the Vietnam War were all made by committees. The infamous Bay of Pigs invasion, for example, was planned by a small group of top government officials immediately after President John Kennedy took office in 1961. The group included what some considered the nation's "best and brightest": McGeorge Bundy, Dean Rusk, Robert McNamara, Douglas Dillon, Robert Kennedy, Arthur Schlesinger, Jr., and President Kennedy himself, with representatives of the Pentagon and the Central Intelligence Agency. This group decided to invade Cuba in April 1961, using a small band of 1,400 Cuban exiles as troops. The invasion was to be staged at the Bay of Pigs and assisted covertly by the U.S. Navy and Air Force and the CIA. As it turned out, the invasion was poorly conceived. The matériel and reserve ammunition on which the exiles were depending never arrived because Castro's air force sank the supply ships. The exiles were promptly surrounded by 20,000 well-equipped Cuban soldiers, and within 3 days virtually all had been captured or killed. The United States suffered a humiliating defeat in the eyes of the world, and Castro's communist government became more strongly entrenched on the Caribbean island.

How could it happen? How could a group of such capable and experienced men make a decision that turned out so poorly? In a post hoc analysis, Janis (1982) suggests that a process termed *groupthink* may have produced the defective decision. **Groupthink** refers to a faulty mode of thinking by group members whereby their desire to evaluate realistically alternative courses of action is overwhelmed by pressures for unanimity within the group. That is, concerned that they not disrupt apparent group consensus, the members neglect to appraise alternatives critically and to weigh the pros and cons carefully. Once groupthink sets in, the typical result is an ill-considered decision.

Symptoms of Groupthink

How can one detect whether groupthink is present in a group discussion? Janis (1982) suggests that certain symptoms indicate groupthink is operating. These include the following:

1. *Illusions of invulnerability.* Group members may think they are invulnerable and cannot fail, and therefore display excessive optimism and take excessive risks.
2. *Illusions of morality.* Members may display an unquestioned belief in the group's inherent superior morality, which may incline them to ignore (undesirable) ethical consequences of their decisions.
3. *Collective rationalization.* Members may discount warnings that, if heeded, would cause them to reconsider their (incorrect) assumptions.
4. *Stereotyping of the adversary.* Especially in the political sphere, the group may develop a stereotyped view of enemy leaders as too evil to warrant genuine attempts to negotiate or as too weak to mount effective counteractions.
5. *Self-censorship.* Members may engage in self-censorship of deviation from the apparent group consensus, with each member inclining to minimize the importance of his or her own doubts.
6. *Pressure on dissenters.* The majority may exert direct pressure on any member who dissents or argues against any of the group's stereotypes, illusions, or commitments.
7. *Mindguarding.* There may emerge in the group some self-appointed "mindguards"—members who protect against information that might shatter the complacency about the effectiveness and morality of the group's decisions.
8. *Apparent unanimity.* Despite personal doubts, members may share an illusion that unanimity regarding the decision exists within the group.

Janis suggests some of these symptoms were present during the decision-making process for the Bay of Pigs invasion. For example, there was an assumed air of consensus that caused members of the decision-making group to ignore some glaring defects in their plan. Although several of Kennedy's senior advisers had strong doubts about the planning, the group's atmosphere inhibited them from voicing criticism. Several members emerged as "mindguards" within the group; they suppressed opposing views by arguing that the decision to invade had already been made, and everyone should help the president instead of distracting him with dissension. Open inquiry and clearheaded exploration were discouraged. Even the contingency planning was unrealistic. For instance, if the exiles failed in their primary military objective at the Bay of Pigs, they were supposed to join the anti-Castro guerrillas known to be operating in the Escambray Mountains. Apparently no one was troubled by the fact that 80 miles of impassable swamp and jungle stood between the guerrillas in the mountains and the exiles.

Groupthink might not be such a concern, except for the recognition that it can occur and recur in many groups. Janis notes that the Bay of Pigs invasion is not the only fiasco in which groupthink was implicated. He suggests groupthink was also involved in other high-level governmental decisions, including the decision to cross the 38th parallel and invade North Korea during the Korean War, the failure to defend Pearl Harbor adequately on the eve of World War II, the decision to escalate the Vietnam War, and the decision to engage in the Watergate cover-up. There is also some evidence that groupthink may have been involved in NASA's decision to hurry the launch of the *Challenger* shuttle, which led to the midair explosion (Moorhead, Ference, and Neck, 1991).

Causes of Groupthink

According to Janis's theory, groupthink is more likely to occur in high-cohesion groups than in low-cohesion groups. It is caused by various factors, including homogeneity of members, insulation of the group from its environment, lack of clear-cut rules to guide decision-making behavior within the group, and high levels of group stress. Another contributing factor is promotional leadership (that is, a leader who actively promotes his or her own favored solution to the problem facing the group, to the neglect of other possible solutions).

According to Janis, each of these factors contributes to groupthink, and their simultaneous occurrence makes groupthink very probable. The groupthink framework is displayed in Figure 15.5.

Laboratory and case study research on groupthink supports some of Janis's hypotheses (Park, 1990). Several studies have shown promotional leadership contributes to the occurrence of groupthink symptoms in business decisions (Leana, 1985; Moorhead & Montanari, 1986), personnel decisions (Flowers, 1977), and foreign policy decisions (McCauley, 1989). Other studies have shown groupthink symptoms are greater when a group is insulated from its environment (and therefore gets little information or criticism) than when it is not insulated (Hensley & Griffin, 1986; Manz & Sims, 1982; Moorhead & Montanari, 1986).

Although cohesion is important in Janis's theory, there is no more than limited support for the hypothesis that group cohesion contributes to groupthink symptoms (Aldag & Fuller, 1993; Michener & Wasserman, 1995; Park, 1990). Although a few studies support this hypothesis (Callaway & Esser, 1984;

Courtright, 1978), many others simply do not (Flowers, 1977; Fodor & Smith, 1982; Leana, 1985; Moorhead & Montanari, 1986). This inconsistency may arise, in part, from differences in the manipulations and/or measurements of cohesion.

More recently, on the basis of available data, Mullen and colleagues (1994) have proposed that cohesion affects groupthink by means of interaction with other independent variables. Specifically, these analysts propose that when other antecedent conditions (such as promotive leadership or insulation) are set up to promote groupthink, high cohesiveness diminishes the quality of decision making, whereas when other antecedent conditions are set up to prevent groupthink, high cohesiveness enhances the quality of decision making.

Avoiding Groupthink   If groupthink produces poor decisions and outcomes, how can one guard against it? There are several ways to prevent groupthink from occurring (Janis, 1982). Basically, these methods increase the probability that a group will obtain all the information relevant to a decision

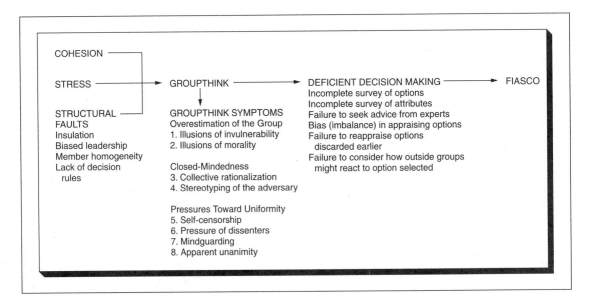

Figure 15.5   Janis's Model of Groupthink

and then evaluate that information with care. First, a group's leader should encourage dissent and call on each member to express any objections and doubts. Second, a leader should be impartial and not announce a preference for any particular option or plan. By describing a problem, rather than recommending a solution, a leader can foster an atmosphere of open inquiry and impartial exploration. Third, a group should divide itself into several independent subgroups, each working on the same problem but carrying out its deliberation independently. This prevents the premature development of consensus within the main group. Fourth, after a tentative consensus has been reached, a group should hold a "second chance" meeting at which each member can express any remaining doubts before a firm decision is taken. The net result of these steps is a better, more realistic decision.

Polarization in Decisions Involving Risk

Even when group decision making is not dysfunctional and follows a fairly rational course, it can still produce surprising consequences. For instance, some evidence indicates that discussion in groups causes individuals to favor courses of action riskier than what they would choose if they made the decision alone. This was demonstrated by Stoner (1961), in a study where subjects responded individually to a series of 12 problems called *choice dilemmas.* In each problem, the subjects were asked to advise a fictional character how much risk he or she should assume. The following item illustrates this task:

> Mr. A, an electrical engineer who is married and has one child, has been working for a large electronics corporation since graduating from college five years ago. He is assured of a lifetime job with a modest, although adequate, salary and liberal pension benefits upon retirement. On the other hand, it is very unlikely that his salary will increase much before he retires. While attending a convention, Mr. A is offered a job with a small, newly founded company which has a highly uncertain future. The new job would pay more to start and would offer the possibility of a share in the ownership if the company survived the competition of the larger firms. Imagine that you are advising Mr. A. Listed below are several probabilities or odds of the new company proving financially sound. Please check the lowest probability that you would consider acceptable to make it worthwhile for Mr. A to take the new job.

> _____ The chances are 1 in 10 that the company will prove financially sound.

> _____ The chances are 3 in 10 that the company will prove financially sound.

> _____ The chances are 5 in 10 that the company will prove financially sound.

> _____ The chances are 7 in 10 that the company will prove financially sound.

> _____ The chances are 9 in 10 that the company will prove financially sound.

> _____ Place a check here if you think Mr. A should not take the new job no matter what the probabilities. (Kogan & Wallach, 1964)

Other choice dilemmas were similar to this one, in that they posed a choice between a certain option and a risky option. For instance, in one of the dilemmas the captain of a college football team, in the final seconds of a game with the college's traditional rival, can choose to run a conservative play or a risky play. The conservative play would produce a tied score, whereas the risky play would lead either to victory (if successful) or defeat (if unsuccessful). In another dilemma, the president of an American corporation planning to expand could build a new plant in the United States or in a foreign country that has a history of political instability. Return on investment for the U.S. plant would be moderate, whereas return for the foreign plant would be either very high (if politics remain stable) or very low (if politics become unstable).

After working as individuals and responding to the 12 choice dilemma items, the participants assembled in groups of six and discussed each item until they reached a unanimous decision. The subjects were then separated and asked again to review each item and indicate an individual decision. The basic finding was that the group decisions following discussion were, on the average, riskier than the decisions made by individual members prior to the discussion. Moreover, the responses made individually after participating in the group were also riskier on average than the responses prior to discussion. This tendency to advocate more risk following a group discussion is termed a **risky shift.** This phenomenon has been observed in many studies (Cartwright, 1971; Dion, Baron, & Miller, 1970).

Other studies using similar tasks, however, have revealed something directly opposite to the risky shift. On certain issues when members are cautious or risk avoidant, group discussion actually causes members to become even more cautious than they were initially (Fraser, Gouge, & Billig, 1971; Stoner, 1968; Turner, Wetherell, & Hogg, 1989). This move away from risk following a group discussion is termed a **cautious shift.** So, although group discussion leads to more extreme decisions, these are not necessarily riskier decisions.

Both risky shift and cautious shift are forms of an underlying phenomenon called **group polarization.** Polarization occurs when group members shift their opinions toward a position that is similar to, but more extreme than, their opinions before group discussion. Thus, if members initially favored a moderately risky position prior to a group discussion, polarization would occur if they shifted toward greater risk following the discussion. Likewise, if they initially favored a moderately cautious position, polarization would occur if they shifted in the direction of even greater caution after the group discussion (Myers & Lamm, 1976).

The tendency for group discussion to create polarization is quite general. That is, discussion produces polarization not only on decisions involving risk, but also on judgments and attitudes in general. Polarization has been observed with respect to political attitudes (Paicheler & Bouchet, 1973), jury verdicts (Isozaki, 1984; Myers & Kaplan, 1976), satisfaction with new consumer products (Johnson & Andrews, 1971), judgments of physical dimensions (Vidmar, 1974), ethical decisions (Horne & Long, 1972), perceptions of other persons (Myers, 1975), and interpersonal bargaining and negotiating (Lamm & Sauer, 1974).

Why does group discussion lead to polarization? That is, what causes group members to shift their risk-taking responses toward an extreme position? Several basic explanations have been proposed (Isenberg, 1986; Myers & Lamm, 1976). According to one theory, group polarization results from a process of *social comparison* (Goethals & Zanna, 1979; Jellison & Riskind, 1970). This theory suggests people often value opinions more extreme than those they personally advocate. They fail to adopt these ideal (extreme) positions as their own because they fear being labeled extremist or deviant. However, during a group discussion in which members compare their positions, these persons may discover that other members hold opinions closer to their ideal position than they had realized. This motivates the moderate members to adopt more extreme positions. The overall result is a polarization of opinions.

Although controversial, the social comparison theory has been supported by various studies. The major source of support stems from demonstrations that exposure to simple information about other group members' positions can, by itself without any persuasive argumentation, produce polarization effects (Baron & Roper, 1976; Blascovich, Ginsburg, & Veach, 1975; Myers et al., 1980).

A second theory explains group polarization as the result of *persuasive argumentation* (Burnstein, 1982; Burnstein & Vinokur, 1973). According to this view, group polarization occurs whenever the preponderance of compelling arguments advanced during a group discussion favors a position more extreme than that held initially by the average member. Discussion within a group serves to persuade members who, because they had been unaware of the arguments, initially chose relatively moderate positions. After discussion, the moderate members shift their opinions in the direction of the most compelling, and relatively extreme, arguments. This produces group polarization.

Some research supports the persuasive argumentation theory. First, studies have shown polarization does result from the exchange of persuasive arguments. In particular, the greater the number of novel and valid arguments exchanged during discussion, the greater the influence of those arguments on members and the greater the polarization (Kaplan, 1977; Vinokur & Burnstein, 1978). Second, the greater the proportion of arguments favoring a particular point of view, the greater is the shift of opinion in its direction (Ebbesen & Bowers, 1974; Lamm, 1988; Madsen, 1978). Thus, subjects who are exposed to mostly risky arguments become more risk taking, whereas those who hear mostly cautious arguments become more risk avoidant.

Overall, then, there is support for both the social comparison and the persuasive argumentation theories. Both of these processes—individually or in combination—produce polarization, although the effects of persuasive argumentation tend to be stronger than those of social comparison (Isenberg, 1986).

Summary

This chapter has discussed group productivity and decision making. *Group productivity* refers to a group's output (per unit of time) gauged relative to something else, such as the level of resources utilized by the group or the group's targeted objective.

Group Task Demands

(1) Group tasks can be categorized as unitary or divisible. On unitary tasks, members perform identical activities, so there is no division of labor. This contrasts with divisible tasks, on which members perform complementary activities. (2) Tasks can also be conjunctive, disjunctive, or additive. Productivity on conjunctive tasks depends on the group's slowest or weakest member; productivity on disjunctive tasks depends on the fastest or strongest member; and productivity on additive tasks depends on the sum (or average) of members' contributions.

Group Properties Affecting Productivity

Several structural factors determine how well a group will perform. (1) Group size interacts with task type to affect productivity. On disjunctive tasks, group productivity increases directly with group size; on conjunctive tasks, it decreases with group size. On additive tasks, the Ringelmann effect occurs; that is, total productivity increases directly with group size, but

productivity per member decreases. (2) If a group establishes explicit, demanding objectives with respect to the group's performance, and if the group's members are highly committed to those objectives, then the group will perform at a higher level than if it does not do these things. (3) The group's reward structure affects productivity. Under low task interdependence, differential rewarding has little effect on group productivity. Under high task interdependence, a high level of differential rewarding induces low group productivity, whereas a low level of differential rewarding induces high group productivity. (4) Centralized communication networks in groups lead to faster, more accurate performance than decentralized ones on simple problems. Decentralized networks lead to superior performance on complex problems because they prevent excessive saturation.

Leader Activities and Leadership Effectiveness

Leadership is an important factor affecting group productivity and decision making. (1) In most groups, the activities of leaders include planning, organizing, and controlling. Transactional leadership entails an exchange between a leader and group members, whereas transformational leadership fosters high levels of group productivity by conveying an extraordinary sense of mission to group members and arousing new ways of thinking. (2) The contingency model of leadership effectiveness maintains group productivity is a function not only of leadership style but also of the situation in which a leader performs. According to this model, task-oriented (low-LPC) leaders are most effective both in situations that are highly favorable (that is, involving good leader-member relations, a structured task, and strong leader power) and in situations that are highly unfavorable (that is, involving poor leader-member relations, an unstructured task, and weak leader power). In contrast, relationship-oriented (high-LPC) leaders are most effective in situations that are moderately favorable. With some exceptions, studies generally support the contingency model.

Idea Generation and Brainstorming

(1) Despite some expectations to the contrary, brainstorming groups generate fewer novel ideas than the

same number of individuals working alone (a nominal group). In addition, the quality and the originality of the ideas produced by brainstorming groups is, on average, no higher than that of ideas produced by individuals working alone. (2) The main reason for unfavorable levels of productivity in brainstorming groups is the phenomenon of production blocking, in which participants are unable to express their ideas due to turn taking among group members.

Group Decision Making Although many decisions made by groups are good ones, the process of group decision making entails potential hazards that can lead to poor or inferior choices. (1) One factor affecting decisions is groupthink—a mode of thinking that occurs when pressures for unanimity overwhelm members' motivation to appraise alternative actions realistically. Groupthink is especially likely to occur in groups that have promotional leadership and are insulated from outside influences. Groupthink can be prevented or reduced if group leaders not only strive to obtain all information relevant to a decision but also encourage an atmosphere of impartial exploration of alternatives. (2) Group members in decision-making groups often shift toward a more extreme position following a group discussion; this phenomenon is termed *group polarization*. In decisions involving risk, polarization can cause members to shift toward higher or lower levels of risk.

Underlying group polarization are several distinct processes, including social comparison and persuasive argumentation.

Key Terms

additive task (p. 361)
brainstorming (p. 378)
cautious shift (p. 384)
communication network (p. 370)
conjunctive task (p. 361)
contingency model of leadership
    effectiveness (p. 375)
disjunctive task (p. 362)
divisible task (p. 360)
group goal effect (p. 367)
group polarization (p. 384)
group productivity (p. 359)
groupthink (p. 380)
information saturation (p. 372)
leadership (p. 373)
production blocking (p. 379)
Ringelmann effect (p. 365)
risky shift (p. 384)
social loafing (p. 365)
transactional leadership (p. 375)
transformational leadership (p. 375)
unitary task (p. 360)

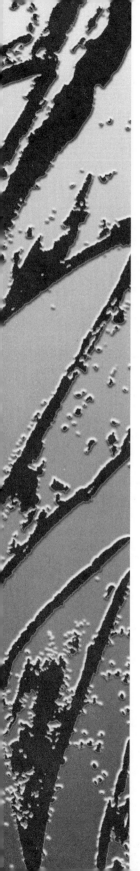

CHAPTER 16
Intergroup Conflict

Introduction

Kanawha County in West Virginia includes the state capital of Charleston as well as several smaller communities. Some years ago, the Kanawha County School Board—which had five members, elected at large—became embroiled in a controversy regarding the selection of textbooks. The issue at stake was whether the school board should adopt 325 new liberal, language-arts textbooks (English, composition, journalism, and speech) or instead select more traditional textbooks of the type already in use. The conflict was instigated by a member of the school board, the articulate wife of a fundamentalist minister. During the spring of that year, she opposed adopting the new books and argued they were excessively liberal in viewpoint. At the April 11 meeting of the school board, she objected to the method of textbook selection. At the June 2 meeting, she again objected to the new books and observed that little in the texts supported a traditional, fundamentalist conception of God, the Bible, and religion. On June 23 she spoke in opposition to the books to the congregation of a local Baptist church. These events were covered by the local news media.

Other persons, concerned about the larger issue of what curriculum should be taught in the schools, joined the board member's crusade. On June 27 more than 1,000 textbook protestors appeared at the regularly scheduled school board meeting. Nevertheless, after hours of testimony, the board voted formally to adopt the disputed books.

During July and August, the fundamentalist protestors organized their ranks and developed a strategy. There were several distinct groups within the protest movement. One of these, the Concerned Citizens of Kanawha County, was a large coalition of church congregations. A second group was the Businessmen and Professional People's Alliance for Better Textbooks, a middle-class group composed mainly of businesspeople, teachers, and other professionals. A third protest group was the Christian American Parents. These groups held marches and rallies, circulated petitions, appealed to elected officials, and planned a boycott of the school system for September.

Opposing these groups were the Kanawha County School Board and the Citizens for Quality Education. Further support for the new textbooks came from such liberal organizations as the American Civil Liberties Union and the National Association for the Advancement of Colored People (NAACP) and from teachers and school administrators.

On September 3, the new school term started for some 45,000 students. Protesting parents withheld their children (10,000 by some estimates) from the schools and prevented school and city buses from operating. On September 12, the school board closed the schools for a 3-day cooling-off period. The board also removed the controversial textbooks from classrooms pending review by a special citizens' committee.

During this period, some violence broke out. Police received reports of random gunfire and sniping near the schools. Vandalism of school property was commonplace. The fundamentalist protestors demanded that members of the school board and the superintendent resign and that the liberal textbooks be banned permanently. On October 28, the citizens' review committee recommended that the school board restore all but 35 of the 325 controversial books to the classroom. The protestors, extremely displeased by this recommendation, reacted with force and dynamited the county board building on the night of October 30. The explosion partially destroyed the building.

On November 9, the school board voted 4 to 1 to reintroduce most of the textbooks into the classrooms. In response, the protestors filed a formal complaint with civil authorities, and the police arrested four board members and the school superintendent on November 15 for "contributing to the delinquency of minors." During the remainder of the school term, many parents continued to withhold their children from the public schools. Some enrolled their children in newly created, private Christian schools that stressed a fundamentalist curriculum.

By the end of the year, the protestors appeared to have won several concessions from the school board. The board issued new guidelines for the selection of textbooks, and it planned to open some "alternative" elementary schools with a more traditional curriculum the following semester. Although the vehemence of the protest subsided, the anger of the protestors lingered on (Page & Clelland, 1978).

Intergroup Conflict

The Kanawha County textbook controversy is an instance of **intergroup conflict,** a circumstance in which groups take antagonistic actions toward one another to control some outcome important to each. The major participants in the textbook controversy were organized groups. These included the Concerned Citizens of Kanawha County, the Businessmen and Professional People's Alliance for Better Textbooks, and the Christian American Parents on one side and the Kanawha County School Board and the Citizens for Quality Education on the other.

The textbook controversy illustrates how easily intergroup conflict can expand and escalate. The issues at stake become larger and more apparent as the conflict progresses. Members of opposing groups develop antagonistic attitudes toward each other; distrust and hostility grow. Overt actions by participants become more damaging and destructive. Groups commit themselves to various positions, and the conflict becomes increasingly difficult to resolve.

Conflict between organized groups typically involves competing values, beliefs, and norms. For instance, a protesting mother in Kanawha County who intentionally blocks the movement of a school bus may, in the eyes of the school board, be performing an unlawful and deviant act. But she does not view the act as deviant because she is conforming to a different set of norms—the fundamentalist norms of the anti-textbook coalition. In intergroup conflict, behavior considered appropriate by members of one group is

The Kanawha County textbook controversy began with disagreement over the selection of textbooks. It escalated into a classroom boycott by students and parents, destructive attacks on the school board's building, and even the creation of alternative schools.

often unacceptable to members of another. Thus the conflict is rooted not merely in individual behavior but in the different value and belief systems of opposing groups.

The term *intergroup conflict* is used in several distinct ways. First, we use it when referring to conflict between organized groups that consist of members who interact with one another, who have well-defined role relationships, and who have interdependent goals. Second, we also use *intergroup conflict* to refer to what might be more correctly described as conflict between persons belonging to different social categories, not organized groups. Although not necessarily members of organized groups, these persons perceive themselves as members of the same social category and are involved emotionally in this common definition of themselves. For instance, conflict between members of ethnic or racial categories (such as neighborhood conflict between blacks and Hispanics in Miami) is usually considered intergroup conflict even though the conflictants may not in all instances belong to organized groups. Throughout this chapter, we use the term *intergroup conflict* to refer both to conflict between organized groups and to conflict between members of different social categories.

Intergroup conflict is widespread, and many instances are reported every day on TV and in the newspapers. Street fights between teen gangs, strife between different religious groups such as the Catholics and Protestants in Northern Ireland, confrontations between the Ku Klux Klan and the black community, strikes by organized labor against management and shareholders, economic competition and rivalry among ethnic groups, and long-standing family feuds—all these are instances of intergroup conflict.

In this chapter, we address the following questions:

1. What causes relations between groups to shift from peaceful to hostile? That is, what factors cause the development and escalation of intergroup conflict?
2. What elements sustain intergroup conflict? When conflict persists over a long time—as it often does—what mechanisms support its persistence?
3. What is the effect of intergroup conflict on relationships among members within the participating groups? That is, when a group is involved in conflict, what impact does this have on the group's structure and on the way its members relate to one another?
4. How can intergroup conflict be de-escalated or resolved?

Development of Intergroup Conflict

Intergroup conflict has several origins. Overt conflict can develop because groups have an underlying opposition of interests. When this opposition prevents

them from achieving their goals simultaneously, it can lead to antagonism and friction, and eventually to open conflict between them. Likewise, conflict can develop because members of one group view themselves as different in important ways from members of another group, and act in a prejudicial and discriminatory way toward members of that other group. Or conflict may occur because one group suddenly threatens or deprives another group of valuable outcomes, and thereby provokes an aggressive reaction. These factors are not mutually exclusive; in fact, they often work together to cause open conflict between groups (Taylor & Moghaddam, 1987, chapter 10). In this section, we consider each of these factors in intergroup conflict.

Realistic Group Conflict

Years ago, Muzafer Sherif and his colleagues conducted an important study of intergroup conflict at Robbers Cave State Park in Oklahoma (Sherif, 1966; Sherif et al., 1961; Sherif & Sherif, 1982). The participants in this experiment were well-adjusted, academically successful, white, middle-class American boys, ages 11 and 12. These boys attended a 2-week experimental summer camp and participated in camp activities, unaware their behavior was under systematic observation. Throughout the 2-week period, the boys were divided into two groups, called the Eagles and the Rattlers. The overall objective of this research was to investigate how an underlying opposition of interest can lead to overt intergroup conflict.

The experiment progressed in several stages. The first stage, which lasted about a week, produced a high level of cohesion within each of the groups. The boys arrived at the camp on two separate buses and settled into cabins located a considerable distance apart. By design, contact among boys within each group was high, but contact between the two groups was minimal.

The boys within each group engaged in various activities, many of which required cooperative effort for achievement. They camped out, cooked, worked on improving swimming holes, transported canoes over rough terrain to the water, and played various games. As they worked together, the boys in each group pooled their efforts, organized duties, and divided tasks of work and play. Eventually, the boys identified more and more with their own groups, and each unit developed a high degree of group cohesion and solidarity. This completed the first stage of the experiment.

Next, the experimenters began the second stage, in which conflict was induced between the groups. The researchers set up several competitive situations in which one group could attain its goal only at the other's expense. Specifically, the camp staff arranged a tournament of games, including baseball, touch football, tug-of-war, and a treasure hunt. In this tournament, prizes were awarded only to the victorious group.

The tournament started in the spirit of good sportsmanship, but as it progressed, the positive feelings faded. The good sportsmanship cheer that one customarily gives after a game, "2-4-6-8, who do we

appreciate," turned into "2-4-6-8, who do we appreci-HATE." Intergroup hostility intensified, and members of each group began to refer to their rivals as "sneaks" and "cheats." After suffering a stinging defeat in one game, the Eagles burned a banner left behind by the Rattlers. When the Rattlers discovered this "desecration," they confronted the Eagles and a fistfight nearly broke out. The next morning, the Rattlers seized the Eagles' flag. Name calling, threats, physical scuffling, and cabin raids by the opposing groups became increasingly frequent. When asked by the experimenters to rate each other's characters, a large proportion of the boys in each group gave negative ratings to all the boys in the other group. When the tournament was finally over, members of the two groups refused to have anything to do with each other.

In later stages of the study, when the level of intergroup antagonism was high, the experimenters tried various strategies for reducing strife. Several of these

A police barricade defines the point of confrontation between gay-pride marchers and Christian fundamentalist counter-protestors. Strongly held social identities elicit group stereotypes, intensifying the underlying conflict.

techniques failed, but the experimenters did succeed in reducing conflict by introducing important goals that required cooperation between groups for attainment (Sherif et al., 1961).

This study is a classic illustration of **realistic group conflict theory,** a well-established theory that provides one explanation for the development of intergroup conflict. The basic propositions of realistic group conflict theory are as follows: First, when groups are pursuing objectives where a gain by one group necessarily results in a loss by the other, they (by definition) have an *opposition of interest.* Second, this opposition of interest causes members of each group to experience frustration and to develop antagonistic attitudes toward the other group. Third, as members of one group develop negative attitudes and unfavorable perceptions regarding members of the other group, they become more strongly identified with and attached to their own group. Fourth, as solidarity and cohesion within each group increases, the likelihood of overt conflict between groups increases, and even a very slight provocation can trigger direct action by one group against another.

The pattern of conflict in the relationship between the Eagles and the Rattlers is consistent with this theory. So is the conflict between fundamentalists and liberals in Kanawha County regarding the content of textbooks and the social values taught in public schools. Intergroup conflict stemming from an underlying opposition of interest is also apparent in the everyday struggle for economic survival, such as competition between ethnic groups for access to jobs, housing, and schooling (Bobo, 1983; Olzak, 1992).

Social Identity

Beyond opposition of interest, another factor in intergroup conflict is the extent to which members identify with the group to which they belong. When opposition of interest is present, strong group identification by members can intensify conflict between groups. But even when an underlying opposition of interest is not present, strong group identification can, by itself, produce biased behavior toward outside groups.

In this section, we discuss the link between group identification and intergroup conflict. First, we consider the phenomenon of ethnocentrism. Next, we discuss minimal conditions for the activation of group identification and discrimination. Finally, we review the social identity theory of intergroup behavior.

In-Group Identification and Ethnocentrism   Years ago, Sumner (1906) noted that people tend to like their own group (the **in-group**) and to dislike competing or opposing groups (the **outgroups**). He hypothesized that members who strongly identify with the in-group are especially prone to hold positive attitudes toward the in-group and to hold negative attitudes toward out-groups. Sumner's term for this phenomenon was **ethnocentrism**—the tendency to regard one's own group as the center of everything and as superior to out-groups. Ethnocentrism involves a pervasive and rigid distinction between the in-group and one or more out-groups. It entails stereotyped positive imagery and favorable attitudes regarding the in-group, as well as stereotyped negative imagery and hostile attitudes regarding the out-groups. Group members can express ethnocentrism in many ways. Table 16.1 presents a summary of the in-group and out-group orientations in ethnocentrism. These include seeing the in-group as superior and the out-group as inferior, viewing the in-group as strong and the out-group as weak, and construing the in-group as honest and peaceful and the out-group as treacherous and hostile (LeVine & Campbell, 1972; Wilder, 1981).

Although ethnocentrism often plays a part in intergroup conflict, not all facets of ethnocentrism appear in every conflict. In some instances, only several of the orientations listed in Table 16.1 occur (Brewer, 1986; Brewer & Campbell, 1976).

Ethnocentric attitudes not only cause in-group members to devalue and demean out-group members, they also lead to discrimination. The term **discrimination** refers to overt acts, occurring without apparent justification, that treat members of certain out-groups in an unfair or disadvantageous manner. In circumstances entailing competition and/or a direct opposition of interest between groups, ethnocentric attitudes often produce discriminatory responses

Table 16.1   Ethnocentric Orientations Toward the In-Group and the Out-Group

Member Orientations Toward the In-Group	Member Orientations Toward the Out-Group
See themselves as virtuous and superior	See the out-group as contemptible, immoral, and inferior
See their own standards of value as universal and intrinsically true	Reject out-group values
See themselves as strong	See the out-group as weak
Maintain cooperative relations with other in-group members	Refuse to cooperate with the out-group
Obey authorities within the in-group	Disobey authorities in the out-group
Demonstrate a willingness to retain membership in the in-group	Reject membership in the out-group
Trust in-group members	Distrust and fear out-group members
Show positive affect toward other in-group members	Show negative affect and hate toward out-group members
Take credit for in-group successes	Blame the out-group for in-group troubles

SOURCE: Adapted with modifications from LeVine and Campbell, 1972.

toward the out-group. More striking, even when an underlying opposition of interest is not present, the mere categorization of persons as belonging to an out-group can lead to discriminatory responses by in-group members.

Discrimination in the Minimal Intergroup Situation

The simple process of social categorization—placing people into arbitrarily defined groups that have no important meaning—is sufficient to produce intergroup discrimination. This effect was demonstrated by studies using an experimental paradigm called the *minimal intergroup situation* (Tajfel, 1982b; Tajfel & Billig, 1974). A typical experiment (Tajfel et al., 1971) involved two distinct stages and was portrayed as a study of visual perception. Subjects were English schoolboys, ages 14 to 16, who knew one another prior to the study. In the first stage, groups of eight boys gathered in a lecture hall and were shown slides of clusters of dots. The exposure time for each slide was brief, and the boys wrote down their estimates of the number of dots in each cluster. While these estimates were being scored for accuracy, the experimenter told the subjects that on this task some

people consistently overestimate the number of dots ("overestimators") whereas others consistently underestimate the number ("underestimators"). After completing the estimation task, the boys were placed in separate rooms. Each boy was assigned a code number to keep his personal identity unknown to others, and then he was informed either that he was an overestimator (i.e., among those who had most overestimated the number of dots) or an underestimator. In fact, unknown to the boys, this assignment to a category was done at random and had nothing to do with their actual judgments. The experimenter wanted simply to divide the participants arbitrarily into two categories.

In the second stage of the study, each boy performed a task that involved allocating points to others. In this task, each boy had to allocate points between pairs of other boys participating in the study. They received no specific directions on how to award these points, but they were told that at the end of the experiment, each of them would receive money based on the number of points allotted to them by the other subjects. In doing this task, each boy was given information about the code number and the category membership (overestimator or underestimator) of the others to whom he could allocate points. Of course, this information was not especially revealing or helpful because no boy knew who had which code numbers (and therefore did not know the names or identities of the persons to whom he was allocating points).

We might think that in this circumstance, the fairest allocation would be to give an equal or nearly equal number of points (money) to each of the others. Yet the results show that allocations made by the boys were not equal. When the two persons to whom allocations could be made belonged to different categories (i.e., one being an overestimator and the other an underestimator), the boys heavily favored their own in-group. That is, boys who thought of themselves as overestimators gave relatively more money to others in the overestimator category, whereas boys who thought of themselves as underestimators gave relatively more money to others in the underestimator category. This effect was widespread, and approximately 70% of the subjects showed a bias favoring their own group.

This response is all the more remarkable when we note the conditions under which it occurred. First, such a response had no utilitarian value for the subjects themselves because they were distributing money to other people, not themselves. Second, there was no personalism involved, for subjects did not know the personal identities of other boys in their own group. Third, the category distinction (i.e., overestimator vs. underestimator) had no great importance or special meaning. Fourth, there was no communication or social interaction among the boys, either within a group or between groups. Fifth, there was neither an opposition of interest nor any previous hostility between the groups.

Other studies in this basic paradigm have extended these results. Some studies have used complete strangers as subjects and formed groups using the most trivial criteria imaginable. In one instance, subjects were openly assigned to categories at random based on a coin toss. The results still show the same pattern: Subjects discriminate in favor of their own in-group and against the out-group. This bias is reflected both in attitudinal and evaluation measures and in the allocation of money and other rewards (Brewer, 1979; Oakes & Turner, 1980; Tajfel, 1981).

Social Identity Theory   Because the subjects' discriminatory responses in these studies had no direct utilitarian value, it is very hard to explain the results in terms of an opposition of interest between groups. Realistic group conflict theory does not offer a convincing explanation for discrimination here. A more satisfactory explanation comes from the **social identity theory of intergroup behavior,** developed by Tajfel and others (Tajfel, 1981, 1982a; Tajfel & Turner, 1986). This theory begins with the assumption that individuals want to achieve and maintain a positive self-concept. According to this view, the self-concept has two components, a personal identity and a social identity. Any person can enhance his or her self-concept by improving the positive evaluation of either (or both) of these components. The social identity component depends primarily on the groups or social categories to which one belongs. The evaluation of one's own group is determined in part by a comparison with other groups. Thus, positive social identity depends on whether the comparisons made between one's membership group (in-group) and some relevant out-groups are favorable or not. Consistent with this, some surveys show that the importance respondents ascribe to belonging to a group correlates with the group's perceived rank relative to other groups (Gaskell & Smith, 1986).

The desire to maintain a positive self-concept creates pressures to evaluate one's own group positively. In the minimal intergroup situation, when an individual is assigned to a group, he or she begins to think of that group (the in-group) as better than the other (the out-group). He or she will also engage in actions to support this idea, such as allocating money to members of the in-group.

Natural settings often provide a stronger basis for intergroup discrimination than the minimal intergroup laboratory setting. If group boundaries are based on such attributes as skin color or language, a person's group membership will be salient and easily discernible. For groups in natural settings, Tajfel and others suggest that individuals will respond to a negative or unsatisfactory social identity in any of several ways. First, individuals might react by leaving an existing group and joining some group that is more positively evaluated. Second, individuals might pass themselves off as something else so others will not recognize them as members of a particular group (such as homosexuals passing as straight, blacks passing as white in South Africa, and so on). Third, individuals might engage in social protest and collective action to elevate the status or welfare of the group to which they belong. This alternative, of course, may provoke overt conflict with other groups.

Aversive Events and Escalation

Sometimes members of one group hold antagonistic attitudes toward members of another but refrain from engaging in open conflict. By staying away from one another as much as possible, they can minimize confrontation. But if underlying antagonism and latent conflict are present, a single provocative incident or aggressive act may be sufficient to trigger open conflict. For example, several years ago on Long Island, New York, an unexpected defeat in a crucial high school basketball game led to an argument between some fans of the competing teams. This argument quickly escalated into a serious brawl involving most of the persons (high school students) attending the game. The fistfight started on the gym floor and then spread into the streets. These fans had different racial and ethnic identities—those supporting the losing home team were largely black, whereas those supporting the visitors were mostly Irish and Italian. During the fight, some persons were beaten and several school buses were overturned. Squads of police arrived, and eventually they were able to bring the brawl to an end. But the conflict among groups of teenagers from the two communities continued for weeks, with recurrent outbreaks of street violence.

As this example shows, a single aversive event can provoke open hostilities between groups (Berkowitz, 1972; Konecni, 1979). An **aversive event** is a behavioral episode caused by (or attributed to) an out-group that entails negative or undesirable outcomes for members of an in-group. The unexpected loss of the basketball game was an aversive event for fans of the home team, and it triggered conflict extending far beyond the hardwood court. Although aversive events can assume many forms, they always involve painful outcomes that people would prefer to avoid, and they include being physically or verbally attacked, being slighted or humiliated, and being subjected to a loss of income or property.

As another illustration of how aversive events can trigger overt hostility, consider again the Kanawha County textbook controversy. In Kanawha County, a long-standing latent difference existed between fundamentalist church groups and the more liberal school board. But it took an aversive event—the decision to adopt new, liberal textbooks—to galvanize the fundamentalists into action. Once the protests began, groups supporting the different ideologies and interests mobilized quickly, and the conflict escalated.

The idea that aversive events trigger overt intergroup conflict is based on the general *frustration-aggression hypothesis.* This hypothesis holds that frustration leads to annoyance or anger, which can quickly turn into aggression if situational conditions are conducive (Berkowitz, 1989; Gustafson, 1989). The hypothesis is pertinent not only at the individual level, but also at the group level: If provoked by an aversive event seen to be caused by an out-group, an in-group will mobilize and attack the out-group. This response is most likely to happen when an underlying opposition of interest exists between groups, when obvious characteristics (such as language, religion, or skin color of members) serve as a basis for differentiation between groups, and when members of one group already hold antagonistic attitudes and negative stereotypes regarding the other.

Persistence of Intergroup Conflict

Probably the most famous family feud in the history of the United States was the long-enduring conflict between the Hatfields and the McCoys. In the early days, these mountain people lived peacefully on opposite sides of a narrow river, the Hatfields in West

Virginia and the McCoys in Kentucky. The feud began one day in 1873 when Floyd Hatfield drove a razorback sow and her piglets into his pigsty on the McCoy side of the river. The pigs settled in comfortably, but trouble broke out a few days later when Randolph McCoy, Floyd's brother-in-law, came up to the pigsty and accused Floyd of stealing the pigs. A furious argument broke out, and the dispute eventually wound up in a backwoods court. During the trial, witness upon witness went to the stand to testify regarding the ownership of the pigs. All those witnesses named Hatfield swore that Floyd owned the pigs, whereas all those named McCoy pointed to Randolph as the rightful owner. When the trial ended, Floyd Hatfield retained possession of the animals.

This was only the beginning, however, for the McCoys were very angry regarding the trial's outcome. Their familial affection toward the Hatfields quickly turned to hatred. The feud grew and deepened in intensity. Time passed and the pigs were forgotten, but the entire McCoy family joined the fight against the Hatfield clan. Several months later,

a McCoy murdered Ellison Hatfield; in retaliation, the Hatfields shot three McCoy boys. Hatred intensified and the vicious conflict escalated. Eventually the civil authorities of West Virginia and Kentucky became involved as they attempted to maintain order and protect human rights. According to one estimate, more than 100 persons lost their lives during the long feud. It was not until 1928, when Tennis Hatfield and Uncle Jim McCoy shook hands in public, that the conflict ended (Jones, 1948).

The Hatfield-McCoy feud is an interesting and puzzling event. It began as a simple disagreement over the ownership of some pigs and escalated into a conflict that lasted 55 years and cost many lives. The feud illustrates a fundamental point about intergroup conflict, namely, that processes internal to a conflict can cause it to persist over time. Even without outside intervention or provocation, intergroup conflicts are often self-sustaining and feed on themselves.

What are some processes that cause intergroup conflict to persist? It is possible to identify several, including biased perception of the character

This clinic has been the scene of demonstrations both for and against abortions. Demonstrators on both sides are likely to hold strong negative stereotypes and attitudes regarding members of the other group.

of out-group members and changes in the structure of the relationship between adversaries. We consider each of these below.

Biased Perception of the Out-Group

In intergroup conflict, it is not uncommon for members of an in-group to harbor unrealistic impressions regarding out-group members. When in-group members hold mistaken perceptions of the out-group, disputes become increasingly difficult to resolve. Mistaken impressions arise from certain biases inherent in group perception, including the illusion that the out-group is homogeneous, an excessive reliance on stereotypes, errors in causal attribution, and incorrect evaluation of in-group performance relative to that of the out-group. In the following, we examine these perceptual biases in more detail.

Out-Group Homogeneity In-group members tend to overestimate the degree of homogeneity among out-group members (Linville, Fischer, & Salovey, 1989; Quattrone, 1986). Research findings indicate that individuals usually perceive less variability among members of the out-group than among members of their own group (Mullen & Hu, 1989; Rothbart, Dawes, & Park, 1984). In other words, although in-group members perceive and appreciate the diversity within the in-group, they tend to perceive the out-group members as "all alike." This effect is referred to as the **illusion of out-group homogeneity.**

This perceptual bias is quite general and widespread. It has been observed in the mutual perceptions of men and women (Park & Rothbart, 1982), of students attending rival universities (Quattrone & Jones, 1980), and of young and elderly persons (Brewer & Lui, 1984).

Quattrone (1986) has suggested that out-group members are perceived as more homogeneous than in-group members because perceivers have limited contact with out-group members, whereas they have fuller contact with in-group members. They may interact with only a few members of the out-group, and/or they may see out-group members only in special situations or circumstances. This makes it less likely that the perceivers will have a chance to see or

appreciate the extent to which out-group members differ from one another in important ways.

Group Stereotypes and Images In-group members often make use of and rely on stereotypes of the out-group. By the term *stereotype,* we mean a fixed set of characteristics attributed to all members of a social category or group. Stereotypes are cognitive simplifications that guide how we perceive others and that affect how we process social information. In our society, stereotypes are commonly associated with such social categories as blacks, Latinos, elderly, gays, lawyers, liberals, right-wingers, jocks, and feminists. These stereotypes are widely known, and respondents can usually identify the characteristics stereotypically associated with various social and nationality groups (Devine & Elliot, 1995; Karlins, Coffman, & Walters, 1969).

Although stereotypes do have certain virtues (e.g., they make it possible to process information quickly), reliance on them can foster mistaken impressions of the out-group and its members. For one thing, stereotypes often exaggerate or accentuate the differences between an in-group and an out-group; they make the groups seem to differ from one another to a greater extent than they really do (Eiser, 1984). Moreover, many stereotypes are depreciatory, and they often ascribe negatively valued traits or characteristics to out-group members.

Images of the out-group can turn very negative when groups are perceived to have a large conflict of interests. One study of Israeli adults (primarily persons who were secular or traditional Jews) measured their impressions of an out-group (ultraorthodox Jews), whose incursion into residential neighborhoods was viewed as a threat at the time. Results indicated that the greater the extent to which subjects considered the out-group threatening, the greater their tendency to view the out-group as deficient in moral virtue—unfriendly, dishonest, and untrustworthy—and as hostile toward their welfare. Sentiments such as these mediated support for overt aggressive action directed at the out-group (Struch and Schwartz, 1989). Other investigators have reported a tendency during conflict for each side to perceive itself as relatively peaceful and cooperative and to perceive the adversary as aggressive and competitive (Bronfen-brenner, 1961).

Another important characteristic of stereotypes is that they tend to have low schematic complexity—that is, they are oversimple and unrealistic. The in-group's stereotype of the out-group is usually less complex—sometimes far less complex—than the in-group's image of itself. This occurs in part because in-group members have less information about the out-group than they do about their own group (Linville & Jones, 1980). Because of this lower schematic complexity, in-group members are at risk of neglect-ing or misinterpreting new information about the ac-tivities of the out-group, especially if it is inconsistent with the stereotype. For instance, if a peaceful initia-tive by the out-group is difficult to understand in light of the in-group's stereotype of the out-group, the in-group members may (incorrectly) interpret the ini-tiative as a new threat and react with hostility. This behavior only prolongs the conflict.

Ultimate Attribution Error Several studies have revealed a perceptual bias that Pettigrew (1979) has called the **ultimate attribution error.** When a member of our own in-group behaves in a positive or desirable manner, we are likely to attribute that behavior to the member's internal stable characteris-tics (such as positive personality dispositions). If that same person behaves in a negative or undesirable man-ner, we tend to discount it and attribute it to external unstable factors (she was operating under unusual stress or having a bad day). However, when perceiving a member of an out-group, we display the opposite bias. Positive behaviors by out-group members are at-tributed to unstable external factors such as situational pressures or luck. Negative behaviors are attributed to stable internal factors such as undesirable personal traits or dispositions. In other words, an in-group ob-server blames the out-group for negative outcomes but does not give it credit for positive outcomes (Cooper & Fazio, 1986; Hewstone, 1990; Taylor & Jaggi, 1974). This attribution bias serves to maintain favorable stereotypes of the in-group and unfavorable stereotypes of the out-group.

One study illustrating this attributional bias was based on the recurring intergroup violence in North-ern Ireland (Hunter, Stringer, & Watson, 1991). This study examined how real instances of violence were in-terpreted by persons identifying with different groups in the conflict. Catholic and Protestant university stu-dents were shown newsreel footage depicting scenes of in-group and out-group violence. One film showed a Protestant attack on mourners at a (Catholic) fu-neral, and the other showed a Catholic attack on a car containing two (Protestant) soldiers. Subjects were asked to provide explanations for the attackers' actions in these films, and the investigators then classified their explanations into internal and external attri-butions. Results show that subjects ascribed negative behavior by their own in-group to external causes, whereas they ascribed negative behavior by the out-group to internal causes. In other words, both Catho-lics and Protestants were more likely to attribute out-group violence than in-group violence to internal causes (such as strongly held hostile attitudes). Such attributions tend to maintain each side's negative view regarding the character and motives of the other side.

Biased Evaluation of Group Performance Another common bias is that in-group members rate the performance of their own group more favorably than that of the out-group, even when there is no ob-jective basis for this difference (Hinkle & Schopler, 1986). And this bias increases as the distinction be-tween in-group and out-group becomes more salient (Brewer, 1979). One illustration of this bias appeared in the Robbers Cave study, discussed earlier (Sherif et al., 1961). When antagonism between the Eagles and the Rattlers was at its peak, investigators arranged for the boys to participate in a bean-collecting con-test. They scattered beans on the ground, and the boys collected as many as possible in 1 minute. Each boy stored his beans in a sack with a narrow opening, so he could not check the number of beans in it. Later, the experimenters projected a picture of the beans gath-ered by each boy on a screen in a large room. Boys from both groups tried to estimate the number of beans in each boy's collection. The projection time was very short and precluded counting. In reality, the experimenters projected the same number of beans (35) each time, although in different arrangements. The boys' estimates revealed a strong in-group bias;

they overestimated the number of beans collected by members of their own group and underestimated the number collected by the out-group. This bias was more pronounced among members of the group that had won the preceding tournament of competitive sports events.

Bias in the evaluation of group performance can produce a variety of consequences. This bias can serve as a positive motivational device that strengthens the in-group's effort, boosts group morale, and helps members avoid complacency (Worchel, Lind, & Kaufman, 1975). However, overvaluation of an in-group's relative performance can lead to faulty decision making or groupthink (Janis, 1982). Overestimation of its own capacity relative to that of an adversary may cause the in-group to become overconfident, hence too willing to continue a fight that, realistically, should be abandoned or settled.

Changes in Relations Between Conflicting Groups

Once a conflict is under way, changes occur in the relationship between the conflicting groups. Often the issues under dispute expand in number. The relationship between groups becomes more polarized, and intergroup communication may break down. These changes, in turn, make it more difficult to resolve the conflict.

Expansion of the Issues When groups are disputing a particular matter, they often compound the conflict by introducing additional issues. The process of expansion usually moves from very specific issues to more general ones. The Kanawha County textbook controversy, discussed earlier in this chapter, provides a good illustration. Initially, participants in this conflict were concerned with the issue of what textbooks should be adopted by the school system. This quickly expanded to the larger issue of what curriculum should be taught. Soon the conflict broadened further to include the issue of who should serve on the school board and make decisions for the school system. Opposing sides also raised the larger, more embracing issue of what broad viewpoint—liberal or traditional—should be taught in the public schools. The conflict then expanded to include another issue: Should schools teaching the "wrong" viewpoint be permitted to continue operating without interference, or should they be shut down by force? Some extremists went still further and asked whether legitimacy be withdrawn entirely from the public school system and invested instead in a new system of alternate schools teaching the fundamentalist viewpoint.

Expansion of issues occurs for several reasons. First, as relations between conflicting groups deteriorate, latent issues that members may have previously suppressed or ignored come to the fore. Conflicting parties feel less inhibition about raising these issues, and they may even view the overt conflict as a good occasion to get even or "settle accounts" with the out-group (Ikle, 1971). Second, if one group takes aggressive steps against another, those actions become issues themselves. In the Kanawha County conflict, certain actions by the participants—blocking the passage of school buses, bringing charges leading to the arrest of school board members, and dynamiting the county board's building—were highly provocative. These actions became issues in themselves and pulled new participants into the fray.

Polarization and Communication Breakdown As a conflict expands, the relationship between adversaries typically becomes more polarized. Members' identification with their own groups intensifies. Their attitudes harden, growing more negative toward the out-group. Any positive bonds that may have existed among the members of different groups before the conflict begin to weaken and disintegrate. Many relationships between in-group members and out-group members are severed. Instead of having both positive and negative ties with one another, the groups are left with only negative ones. As a result, they become increasingly polarized and have fewer reservations about using coercive tactics.

Moreover, as the conflict expands, patterns of communication change. Often there is a *communication breakdown*—that is, a sharp decrease in communication between the conflicting groups (Dube-Simard, 1983; Giles & Coupland, 1991). In some cases, a communication breakdown takes the form of an actual decrease in the quantity or frequency of messages sent and received between conflicting groups. In other cases, it takes a different form wherein groups continue to send and receive messages, but the impact of these messages is nil. Once a conflict is under way, stereotypes exert considerable influence over how an in-group interprets messages received from the out-group (Hewstone & Giles, 1986).

A communication breakdown causes the in-group and the out-group to become even more isolated from one another, of course, and this in turn makes it more difficult for participants to resolve or de-escalate the conflict. Members of the opposing groups have fewer opportunities to communicate openly about issues. People on one side lack accurate information about the plans and desires of the other side. Fresh proposals and new ideas go uncommunicated or unheeded. Even if one group sincerely wants to resolve or de-escalate the conflict, it may have a hard time signaling that intention. Efforts by one side to reduce the conflict may be interpreted by the other side as a trick or trap, as stereotypes come into play. Overall, polarization and the accompanying communication breakdown make the conflict more difficult to resolve.

Impact of Conflict on Intragroup Processes

Intergroup conflict produces changes in the internal structure of the groups participating in the conflict. Once a struggle has begun, each group in the conflict undergoes changes that tend to promote escalation or make conflict resolution more difficult. In this section, we consider the effects of conflict on group cohesion, the militancy of group leaders, and the normative structure of the in-group.

Group Cohesion

Theorists have long recognized that overt conflict and external threats can produce changes in the internal structure of groups (Coser, 1967). Various studies have supported the hypothesis that when a group engages in conflict against another group, or is threatened by another group, its level of cohesion rises (Dion, 1979; Ryen & Kahn, 1975; Worchel & Norvell, 1980). During conflict, a group's boundaries become more firmly etched, and its members generally show higher levels of loyalty and commitment to it. For instance, consider again the Robbers Cave study, where groups of preadolescent boys at a summer camp engaged in competitive activities (Sherif, 1966; Sherif et al., 1961). As conflict between the Eagles and the Rattlers escalated, various measures reflecting group cohesion—such as cooperativeness and friendship choice among group members—rose to high levels. The Eagles and the Rattlers each became more cohesive internally and more antagonistic toward the other group as they participated in the win-or-lose competition.

Of course, there are some limits to this effect. If a group is embroiled in a conflict where it cannot possibly prevail, members may give up all hope. If this occurs, cohesion declines and some members may leave the group. But under conditions where success is still possible, in-group cohesion usually increases under conflict.

Why does intergroup conflict usually lead to high levels of in-group cohesion? First, as conflict escalates, members often endow their cause with additional significance, and they increase their commitment to it; if this happens, group cohesion rises. Second, intergroup conflict frequently entails threats; if an out-group issues a threat, that action quickly identifies the out-group as an enemy. Having a common enemy heightens perceived similarity among in-group members and increases cohesion (Holmes & Grant, 1979). In one study (Samuels, 1970), researchers varied the degree of cooperation and competition within groups separately from the presence or absence of competition between groups. The results indicated that groups facing intergroup competition were more cohesive than those not facing intergroup competition, regardless of whether members within a given group related cooperatively or competitively with each other. Thus the common antagonism of group members

A demonstrator blocking the entrance to a building is physically removed by the police. During intergroup conflict, conformity to the standards of one's own group may lead to actions that violate the law.

toward an opposing group overshadowed any friction among themselves.

What are the consequences of heightened in-group cohesion during intergroup conflict? As noted in Chapter 13, if a group is cohesive, its members desire to remain in it and resist leaving it. A highly cohesive group, in general, maintains a firm hold over its members' time, energy, loyalty, and commitment. Conformity and cooperation tend to be greater in high-cohesion groups than in low-cohesion groups (Sakurai, 1975). For this reason, cohesive groups are capable of taking well-coordinated action in pursuit of their goals. In the context of intergroup conflict,

high-cohesion groups are often more vigorous and contentious than low-cohesion groups.

Leadership Militancy

The activities of group leaders under conditions of intergroup conflict are somewhat different from those of leaders under conditions of peace. Under conflict, leaders have to direct the charge against the adversary. They plan the group's strategic moves, obtain resources that gear the group for the conflict, coordinate members' actions, and serve as spokespersons for the group in negotiations with the adversary. How well these activities are performed has an important impact on a group's success or failure in intergroup conflict.

It is not uncommon for groups embroiled in heavy conflict to change leaders. If the campaign against an opposing group is not progressing well, rivals for leadership may emerge within the in-group. Frequently, these rivals are angrier, more radical, and more militant than the existing leader(s). This challenge from rivals places the existing leaders under pressure. To defend against this threat, they may react by adopting a harder line and taking stronger action against the out-group. Although existing leaders do not always react to rivals in this manner, they are especially prone to do so when their own position within the group is insecure or precarious (Rabbie & Bekkers, 1978). In this manner, competition for leadership within a group increases group militancy and intensifies the level of conflict with outside groups (Kriesberg, 1973).

The impact of competition for leadership is illustrated by a study of civil rights leaders in 15 U.S. cities in the mid-1960s (McWorter & Crain, 1967). At that time, civil rights organizations were trying to bring about societal changes favorable to blacks and other minorities. Researchers interviewed civil rights leaders to determine the extent to which there was rivalry for leadership within civil rights groups. Results showed that organizations with higher levels of rivalry also had greater militancy. Militancy was apparent both in the extremity of attitudinal responses and in the frequency of civil rights demonstrations in the cities. Thus a process internal to these civil rights groups—rivalry for leadership leading to greater militancy—led to intensified conflict between groups.

Normative Structure and Conformity

Intergroup conflict not only increases cohesion and leadership militancy, but it also changes group norms and goals. Once serious intergroup hostilities have begun, group members grow concerned with winning (or surviving) the conflict. Some behaviors and activities that the group considered valuable prior to the conflict may now seem useless or even detrimental to success in the conflict. If this happens, the group will reorder goal priorities and favor those behaviors by members that can help it win the conflict.

As part of this, the group may reassess the importance of various tasks and make corresponding changes in members' role definitions and task assignments. This can result in a redistribution of status and rewards among members that, judged by preconflict standards, may not appear fair. The reallocation of tasks may impose an unequal sharing of costs and hardships, and it may not reflect members' seniority or past contributions. Changes such as these can increase tension among members within the group (Leventhal, 1979). But if the conflict is intense, concerns about group effectiveness and survival will overshadow concerns about equity and fairness.

Under severe conflict, members increase their demands on one another for conformity to group norms and standards. Pressure for enhanced coordination and task performance increases, especially as these bear on the group's success in the conflict. There is also pressure to adopt the group's negative attitudes and stereotypes regarding the adversary—a form of "right thinking." The importance of loyalty to the ingroup increases, and members increasingly expect one another to display a distrusting, competitive orientation toward the out-group.

These conformity pressures may well impinge on the rights and liberties of individual members. Yet, the group will care less about these rights than it did before the conflict, and there will be less tolerance of dissent (Korten, 1962). If internal dissent does occur, the majority will likely react by suppressing it or forcing the dissidents out of the group, especially if they suspect them of sympathizing with the adversary or engaging in behavior that jeopardizes the group's chance of victory.

Resolution of Intergroup Conflict

If a conflict expands beyond rational bounds, as occurred in the Kanawha County textbook controversy and the Hatfield-McCoy family feud, it can consume vast amounts of human energy, time, and other resources. Conflicts often begin as small disagreements and then grow in scope and intensity, pulling in new participants. Because intergroup conflicts are potentially dangerous and costly, many theorists have wondered how to stop them in the early stages, before they escalate beyond control.

This problem is surprisingly complex. One cannot resolve intergroup conflict merely by "reversing" the processes that initially caused it. It is often impossible to eliminate underlying opposition of interest, to diminish ethnocentric identification with the ingroup, or to forestall aversive events.

Nevertheless, investigators and practitioners have developed various techniques to reduce or resolve intergroup conflict. In this section, we discuss four of them. These are, first, establishing a superordinate goal to induce collaborative action between the in-group and the out-group; second, increasing contact and communication between the in-group and the out-group; third, using outside parties to mediate or arbitrate the dispute between groups; and fourth, initiating unilateral conciliatory moves in the hope the out-group will respond in a manner that lessens hostility. Each of these techniques has practical value and merits consideration.

Superordinate Goals

One of the most effective techniques for resolving intergroup conflict is to interpose one or more superordinate goals. A **superordinate goal** is an objective held in common by all groups in a conflict that cannot be achieved by any one group without the supportive efforts of the others. Research findings indicate that once introduced, superordinate goals usually reduce in-group bias and intergroup conflict (Bettencourt et al., 1992; Sherif et al., 1961).

The Robbers Cave study, discussed earlier, provides a clear illustration that superordinate goals can reduce conflict. After a high level of conflict had developed between the Eagles and the Rattlers, the researchers introduced a series of superordinate goals. These goals involved important shared needs. First, the researchers arranged for the system that supplied water to both groups to break down. To find the source of the problem and restore water to the camp, the two groups of boys had to work together. Next, the food delivery truck became stuck along the roadway. If the boys were to eat, they had to work together to free the heavy vehicle and push it up a steep grade. By inducing some cooperation between the groups, the superordinate goal structure also reduced hostility (Sherif et al., 1961).

The impact of superordinate goals on conflict reduction is not immediate, but gradual and cumulative. Results are stronger when several goals are introduced one after another, rather than a single goal. Because superordinate goals are cumulative in effect, they have greater impact when they are massed (Blake, Shepard, & Mouton, 1964; Sherif et al., 1961).

Superordinate goals reduce intergroup conflict for several reasons. First, they serve as a basis for restructuring the relationship between groups. Superordinate goals create cooperative interdependence between the in-group and the out-group. By changing a hostile win-lose situation into one of collaborative problem solving with the possibility of a win-win outcome, a superordinate goal reduces friction between groups. The activities of out-group members become a source of positive value for in-group members, and vice versa, because members of one group are contributing to outcomes valued by the other.

Second, introduction of a superordinate goal often leads to an increase in the level of interaction between in-group and out-group members. Increased contact by itself is generally not sufficient to assure a reduction in intergroup bias or hostility. But if some of the interaction with the out-group members is personalized (rather than just task oriented), or if it provides the type of information that eventually reduces stereotyping, the superordinate goal will reduce bias and hostility (Bettencourt et al, 1992; Brewer & Miller, 1984; Worchel, 1986).

Third, introduction of a superordinate goal can generate a new, superordinate identity shared by all members. That is, the superordinate goal induces a recategorization of all members so the sharp distinction between the in-group ("us") and the out-group ("them") is erased, and a new common identity applying to all persons is substituted instead. One theory of recategorization, termed the *common in-group identity model,* proposes that when persons belonging to separate social groups come to view themselves as members of a single social unit or category, their attitudes toward one another become more positive (Gaertner et al., 1989; Gaertner et al., 1993). Former out-group members increase in attractiveness, and the favoritism that in-group members originally afforded their own group is now extended to the whole collective. Recategorizing all members and weakening the original us-them distinction eventually reduces intergroup bias and hostility.

Intergroup Contact

Some theorists have hypothesized that an increase in contact and communication between members of opposing groups will reduce intergroup conflict. Increased contact will eradicate stereotypes and reduce bias and, consequently, lessens antagonism between groups. Called the **intergroup contact hypothesis,** this has been proposed primarily with respect to relations between different racial and ethnic groups (Amir, 1976; Cook, 1985; Stephan, 1987).

Support for this hypothesis is mixed. Research findings show that although intergroup contact reduces prejudice and conflict between groups in some cases, it does not do so in others (Brewer & Kramer, 1985; Hewstone & Brown, 1986; Riordan, 1978). School desegregation, for instance, has increased contact between black and white children, but it has not always produced positive changes in intergroup relations (Cook, 1984; Gerard, 1983). In some instances, increasing the level of intergroup contact can actually increase conflict (Brewer, 1986). Given this state of affairs, investigators have focused their attention on identifying the conditions under which intergroup contact leads to reduced bias and conflict, and the conditions under which it does not. We consider this issue here.

Sustained Close Contact　Contact between members of different groups is more likely to bring about a reduction in prejudice and conflict if the contact is sustained rather than brief and personal rather than superficial (Amir, 1976; Brown & Turner, 1981). Brief or superficial contact may not be sufficient to produce a change in intergroup attitudes. Some early experiments on interracial contact found that it required interaction over as much as a 20-day period between prejudiced whites and a black coworker to bring about change in the whites' attitudes toward the co-worker (Cook, 1985). Laboratory studies with groups of differing college affiliations have also shown that contact involving repeated sessions with an out-group under favorable conditions is more effective in decreasing intergroup bias than contact involving just a single session (Wilder & Thompson, 1980).

The extent of closeness or personalization of intergroup contact also affects the extent to which attitudes are changed and stereotyping reduced. Close, personal contact between members of different groups can increase liking for one another and reduce stereotyping, especially if the contact is sustained over time rather than fleeting. In contrast, if contact is superficial and involves only low levels of intimacy, it has little effect on intergroup prejudice and stereotyping (Segal, 1965).

There are several reasons why sustained close contact tends to reduce prejudice and stereotyping. First, persons engaging in close intergroup contact may experience cognitive dissonance, which may produce attitude change. If individuals with negative attitudes find themselves subject to situational pressures, and if they consequently engage in positive actions toward members of an out-group, then their behavior will be inconsistent with their attitudes,

which may create a state of cognitive dissonance. The theory of cognitive dissonance predicts these persons will end up changing their attitudes (becoming more positive toward the out-group) as a means of justifying their new behavior to themselves.

Second, during close contact, members of the different groups may engage in self-disclosure. Higher levels of self-disclosure generally promote interpersonal liking, provided the attributes revealed by one person are viewed positively (or at least not negatively) by the other (Collins & Miller, 1994). If members of different groups discover through self-disclosure that they hold certain fundamental values in common, liking may be enhanced.

Third, sustained close contact between members of different groups can serve to break down or disconfirm stereotypes. Of course, contact with a single representative or "token" member of an out-group is usually not sufficient to change group stereotypes, because that person can too easily be viewed by in-group members as an exceptional individual who is unrepresentative of the out-group (Weber & Crocker, 1983). But close contact with multiple members of an out-group sustained over time may provide enough contrary information to compel a change in old stereotypes. Out-group members will be perceived as less homogeneous than before, which makes it harder to sustain the stereotype.

Equal-Status Contact　Intergroup contact is more likely to reduce prejudice when in-group and out-group members occupy positions of equal status than when they occupy positions of unequal status (Riordan, 1978; Robinson & Preston, 1976). One early demonstration of equal-status contact comes from a classic study conducted in the military during World War II (Mannheimer & Williams, 1949). At that time, the U.S. Army was still largely segregated by race; only a few companies were integrated. This study showed that white soldiers changed their attitudes toward black soldiers after the two racial groups fought in combat as equals, side by side. When asked how they felt about their company including black as well as white platoons, only 7% of the white soldiers from integrated units reacted negatively. In contrast, 62% of the soldiers in completely segregated white companies reacted negatively to the prospect of having black platoons in their unit.

Equal-status contact has been effective in reducing prejudice in other situations as well. For instance, it has been a factor in reducing prejudice among black and white children at interracial summer camps (Clore et al., 1978) and in interracial housing situations (Hamilton & Bishop, 1976). This point is discussed further in Box 16.1, "Equal-Status Contact and Interracial Attitudes."

To see why equal-status contact is important, consider what happens when contact is not based on equal status (Cohen, 1984). When status is unequal, members of a higher status group may refuse to accept influence or to learn from a lower status group. They can justify this to themselves on the grounds that the lower status group has lesser skill or experience. With one side unwilling to accept influence, expectations of lesser competence appear to be supported and stereotypes are all the more difficult to overcome. To have any impact, the lower status group needs repeatedly to demonstrate to the higher status group that it is as good as the other in relevant respects. For all these reasons, intergroup contact is more effective in reducing prejudice and conflict if members of the different groups enter a situation on an equal footing.

16.1   Equal-Status Contact and Interracial Attitudes

According to various theorists (Allport, 1954; Amir, 1969), contact between ethnic and racial groups is likely to reduce prejudice when members of different groups have equal status during the encounter. In U.S. society, however, equal-status contacts may be more the exception than the rule. Because blacks historically have not enjoyed the same economic position as whites, interracial contacts often involve status inequality instead of status equality. Nevertheless, any contacts that are structured in terms of equal status have the potential to change interracial attitudes. One demonstration of this occurred in an experimental interracial summer camp for children (Clore et al., 1978).

This camp was structured to foster equal-status contact among children ages 8 to 12. Half of the children were white, and half were black. In addition, half of the counselors and administrative staff were white, and half were black. Living assignments in the camp ensured that each unit was half white and half black. The living situation provided an opportunity for intimate acquaintanceships rather than the casual associations typical of many integrated social settings. All children had equal privileges and duties around the camp. Moreover, counselors provided tasks—such as fire building and cooking—that required cooperative efforts among the children.

Approximately 200 children attended the camp during the summer. They attended in groups of 40, with each group staying for 1 week. Interracial attitudes were measured in several ways. First, researchers asked the children how they felt toward persons of the opposite race; the children responded in terms of evaluative scales such as good-bad, clean-dirty, pleasant-unpleasant, valuable-worthless, and so on. Second, researchers assessed the extent of interracial liking and friendship by means of games that required the children to indicate their interpersonal choices. For instance, in the "name game," children designated the others they knew well by circling names on a card with a pencil. At the end of the week, children indicated the names of three other campers whose telephone numbers and addresses they wanted to have.

Results of this study showed that a change in attitude occurred for girls but not for boys. Boys began camp sessions with neutral attitudes toward persons of the opposite race and did not have much room for change. Girls began with negative attitudes and shifted in a positive direction as a function of the interracial contact. Similarly, there was a significant increase in cross-race interpersonal choices from the beginning of the camp to the end. This increase was more pronounced for girls than for boys. Overall, the results of this study suggest that prolonged, intimate, equal-status contact across races can lead to change in interracial attitudes.

Institutionally Supported Contact   Intergroup contact is more likely to reduce stereotyping and create favorable attitudes if it is backed by social norms that promote equality among groups (Adlerfer, 1982; Cohen, 1980; Williams, 1977). If the norms support openness, friendliness, and mutual respect, the contact has a greater chance of changing attitudes and reducing prejudice than if they do not.

Institutionally supported intergroup contacts—that is, contacts sanctioned by an outside authority or by established customs—are more likely to produce positive changes than unsupported contacts. Without institutional support, members of an in-group may be reluctant to interact with outsiders because they feel it is deviant or simply inappropriate. In the presence of institutional support, however, they may view contact between groups as appropriate, expected, and worthwhile. For instance, with respect to desegregation in elementary schools, there is evidence that students are more highly motivated and learn more in classes conducted by teachers (that is, authority figures) who support rather than oppose desegregation (Epstein, 1985).

In sum, intergroup contact tends to reduce conflict when it is anchored by institutional or authoritative support, when it is based on equal rather than unequal status, and when it is personal rather than superficial in character.

Mediation and Third-Party Intervention

Disputes of certain types—such as strife between organized labor and management or conflict between community groups—are sometimes most easily resolved through the intervention of third parties, such as mediators or arbitrators. A **mediator** is any third party who serves as a go-between and who helps

A mediator tries to settle a running dispute between teen gangs in California. Efforts by mediators are more likely to be successful when the mediator is trusted by the conflicting groups.

groups in conflict to identify issues and agree on some resolution. Typically, mediators are independent third parties, although they may have some personal stake in seeing a settlement to the conflict or dispute. Mediators generally serve as advisers, rather than as decision makers in the dispute. In contrast to a mediator, an **arbitrator** is a neutral third party who has the power to decide how a conflict will be resolved. An arbitrator listens to arguments from conflicting parties and then makes a decision that is binding on the conflicting groups. This differs from mediation, where a resolution may or may not be attained and, even if attained, may not be binding. Mediators or arbitrators are found in divorce and small claims courts, labor-management negotiations, community and neighborhood disputes, and other situations including international disputes.

Under what conditions is mediation most effective in resolving conflicts? In research on mediation, one feature stands out: The greater the magnitude of the conflict and the worse the parties' relationship, the dimmer the prospects that mediation will prove successful (Carnevale & Pegnetter, 1985; Kressel & Pruitt, 1985). Conflicts that have endured a long time, include a large number of disputed issues, or entail destructive or violent tactics are difficult to mediate.

Mediation has a better chance of success when each of the conflicting groups trusts the mediator. Without trust, a mediator will find it difficult to obtain access to the groups and to conduct candid discussions with their members. Of course, in some conflicts, persons trusted by all sides may be few and far between.

Mediators can engage in various actions that improve the chances of resolving the conflict. One of the most crucial acts is to facilitate communication between the conflicting groups (Donohue, Allen, & Burrell, 1988; Thoennes & Pearson, 1985). Some mediators prefer to meet individually with each side to explore the issues in the dispute; later, they bring the conflicting groups together to discuss possible resolutions. Other mediators engage in "shuttle diplomacy" and carry messages between groups. In cases where the parties in a conflict might otherwise refuse to reply to messages from the other side, mediators can encourage a response.

Another action by mediators that aids conflict reduction is to diagnose the conflict carefully. Diagnosis enables the mediator to offer insight and clarification regarding the roots of the conflict; it also enables the mediator to separate the issues and suggest trade-offs that facilitate resolution (Carnevale & Pegnetter, 1985; Thoennes & Pearson, 1985). In some cases, the mediator can help the parties design their own resolution to the conflict; this may better suit the participants' needs than an imposed settlement. In other cases, where the dispute involves a lot of tension and hostility, the mediator may need to exert some pressure on one party or another.

Although aggressive or assertive tactics may seem alien to good mediation, mediators frequently do pressure the parties to accept proposals on specific issues (Hiltrop, 1985; Kressel & Pruitt, 1985; Zartman & Touval, 1985). One tactic used by mediators is threatening to quit unless the conflicting parties begin to make concessions. Research on collective bargaining in Great Britain indicates that in the context of rancorous disputes, use of the threaten-to-quit tactic by a mediator increased the chances of a settlement being reached (Hiltrop, 1985). Mediators are most likely to use assertive tactics when they think there is little likelihood of agreement without it and when they do not have a strong concern for the parties' own aspirations (Carnevale & Henry, 1989).

Even at its best, mediation is successful in resolving conflicts only some of the time. For instance, mediation of labor-management disputes produces successful settlements about 50% of the time (Hiltrop, 1985; Kressel & Pruitt, 1985; Lewin, Feuille, & Kochan, 1977, chapter 5). Even though it is not completely successful in all cases, mediation is still an important element in the tool kit of conflict resolution techniques.

Unilateral Conciliatory Initiatives

The approaches discussed to this point—superordinate goals, intergroup contact, and mediation by third parties—often suffice to reduce bias or resolve conflict. But if a conflict has already escalated or reached a stalemate, the opposing sides may not be open to further intergroup contact or mediation. In such a case, any group wishing to reduce tensions can do little except take steps unilaterally. Resolving intergroup conflict by making unilateral conciliatory initiatives is a difficult undertaking; even if one group does make sincere conciliatory moves, the other may react with hostility or demand total capitulation. Still, use of conciliatory initiatives can often be successful if carried out a certain way.

One approach to conflict reduction through unilateral initiatives is a strategy called **GRIT,** which stands for "Graduated and Reciprocated Initiatives in Tension-reduction" (Osgood, 1962, 1979, 1980). Originally formulated during the cold war between the United States and the Soviet Union, GRIT arose from a desire to reduce tension between the superpowers (Granberg, 1978). Despite its origins, the GRIT strategy is general in character; it can be applied not only to conflicts between nations but also to those between smaller organizations and groups.

GRIT assumes that each side in a conflict has an interest in reducing tension and would like to reallocate resources from the conflict to better purposes. At base, GRIT is a kind of "tit-for-tat" strategy involving de-escalatory moves. One side initiates de-escalatory steps in the hope that the other side will reciprocate; if reciprocation occurs, the initiator then quickly responds with still further de-escalation, and so on. More specifically, the principles of the GRIT strategy are as follows:

1. To begin, the group using the strategy (i.e., the in-group) issues a public statement that describes its plan to reduce intergroup tension through subsequent actions.
2. Prior to making any unilateral move, the in-group should publicly announce the move and label it as part of the overall strategy.
3. With each announcement of a unilateral move, the in-group should explicitly invite reciprocation in some form by the out-group.
4. To bolster credibility, each unilateral move should be carried out by the in-group in accordance with the announced timetable.
5. Initiatives by the in-group should be continued for some time, even without reciprocation by the out-group. This entails risk, but it also intensifies pressure on the out-group to reciprocate.

6. The initiatives by the in-group should be clear-cut, unambiguous, and open to verification by the out-group.
7. The in-group should not unilaterally disarm and should retain the capacity to retaliate if the out-group reacts with hostility to the in-group's initiatives. Ideally, initiatives should be sufficiently risky that they are vulnerable to exploitation, but not so risky that they seriously impair the capacity of the in-group to retaliate against an attack from the out-group.
8. Further initiatives by the in-group should be graduated to match the responses by the out-group. If the out-group responds in a friendly manner, the in-group should react by taking further risk. If the out-group responds in a hostile manner, the in-group should react with retaliation (but avoid escalation).
9. Unilateral moves by the in-group should be diversified in form, so all they have in common is their conciliatory nature. This demonstrates the in-group's underlying peaceful intentions, and it also illustrates to the out-group that a variety of moves could be made in reciprocation. It also avoids developing a large gap in the in-group's defenses.
10. The out-group should be rewarded for cooperating, with the level of reward depending on the level of cooperation (Lindskold, 1986; Osgood, 1980).

The GRIT strategy tries not only to change the out-group's perceptions of the initiating group, but also to change its behavior through the principles of reinforcement. If the out-group reciprocates positively, it is rewarded with further conciliatory initiatives; but if the out-group responds with aggressive action, it is punished with retaliation.

How effective is the GRIT strategy in practice? At the level of international relations, GRIT has not been fully tried or tested. During the cold war, however, several events provided partial tests of the GRIT principles. Notable among these was the unilateral initiative in June 1963 by U.S. President John Kennedy to halt all atmospheric nuclear testing. Kennedy began by unilaterally halting U.S. testing and followed this up with some further concessions on other issues such as the status of the Hungarian delegation at the United Nations. These concessions were reciprocated by Premier Khrushchev of the Soviet Union. This led immediately to renewed activity regarding a test ban treaty, which had been stalled for a long period. Kennedy's GRIT-like strategy proved successful and, within just a few months, the players agreed on a treaty (Etzioni, 1967).

Further evidence regarding the effectiveness of GRIT comes from simulation studies of conflict and from experimental games. These studies show that many of the principles in the GRIT strategy do contribute to tension reduction. For instance, the act of announcing intentions in advance (principles 1 and 2) is more likely to elicit reciprocal cooperation from the out-group than not announcing them (Lindskold, Han, & Betz, 1986). Conciliatory moves without such an opening announcement may be incorrectly construed by the out-group as indicating weakness or passivity and therefore lead to exploitation (Oskamp, 1971). Repeated truthful announcements that one intends to cooperate on the next move produces greater reciprocation from the out-group than does the same rate of cooperation unaccompanied by announcements (Lindskold & Finch, 1981; Voissem & Sistrunk, 1971).

Some studies show that carrying out initiatives as announced (principle 4) increases the credibility of the initiator (Ayers, Nacci, & Tedeschi, 1973; Schlenker et al., 1973), whereas failure to carry out initiatives reduces credibility and diminishes reciprocation (Gahagan & Tedeschi, 1968). In addition, maintaining the capacity to retaliate (principle 7) is important because if one side makes conciliatory moves but fails to maintain the capacity to retaliate, it will quickly weaken itself and create an imbalance of power. Cooperation drops off when a power imbalance exists (Aronoff & Tedeschi, 1968; Michener & Cohen, 1973). Studies regarding reciprocity (principle 8) show that GRIT with communication is superior to simple tit-for-tat responses (Han & Lindskold, 1983).

A major limitation of GRIT: It is not very effective between groups that differ greatly in power. The GRIT strategy works best under conditions of approximate power equality. When groups differ in power, initiatives from the strong group are likely to be met with concessions by the weaker one, but initiatives from the weaker group are not likely to be met with concessions by the strong one (Lindskold & Aronoff, 1980; Michener et al., 1975). The GRIT strategy works best when the conciliatory initiatives come from a group having equal or superior power in the conflict.

Summary

Intergroup conflict is a circumstance in which groups engage in antagonistic actions toward one another to control some outcome important to them.

Development of Intergroup Conflict Intergroup conflict has several origins. (1) Often a fundamental opposition of interests underlies conflict between groups. This opposition prevents them from achieving their goals simultaneously and leads to friction, hostility, and overt conflict. (2) A high level of in-group identification, accompanied by ethnocentric attitudes, may create discrimination and ill will between groups and foster actions that escalate conflict. (3) One group, by threatening or depriving another, may create an aversive event that turns latent antagonism into overt conflict. In many intergroup conflicts, two or more of these factors operate simultaneously.

Persistence of Intergroup Conflict Although some conflicts between groups dissipate quickly, others extend for a long time. Several mechanisms support the persistence of intergroup conflict. (1) Perception of the out-group by in-group members is often biased. This bias, caused by insufficient information regarding the out-group and excessive reliance on stereotypes, produces an incorrect understanding of characteristics and intentions of out-group members and an overestimation of in-group capabilities. (2) Once a conflict escalates, changes occur in the relationship between the opposing sides. The number of issues under dispute often expands and the number of parties in the conflict may increase. Relations between adversaries polarize, communication across groups declines, and trust becomes harder to establish.

Impact of Conflict on Intragroup Processes

Intergroup conflict changes the internal structure of the in-group. (1) Conflict increases the level of cohesion of the in-group as members endow their cause with increased commitment and unite to face a common adversary. (2) Conflict may produce rivalry for leadership among in-group members, and this rivalry can exert pressures that make the group's leadership more militant. (3) Conflict often changes the normative structure of the in-group. Standards of fairness may shift as the in-group reorders its priorities to achieve a strategic advantage in the conflict. Conflict also increases the pressure on in-group members to conform and lessens the majority's tolerance of dissenters.

Resolution of Intergroup Conflict

Intergroup conflict has the potential to escalate to severe levels, dissipating resources and imposing large losses. Several approaches can be used to reduce intergroup conflict. (1) One approach is to introduce superordinate goals into the conflict. Because goals of this type can be achieved only through the joint efforts of opposing sides, they promote cooperative behavior and serve as a basis for restructuring the relationship between groups. (2) Another approach is to increase intergroup contact. This approach is more effective in reducing bias and conflict when contact is sustained, close, based on equal status, and supported institutionally. (3) Another approach is to use third parties as mediators in the dispute. Mediators improve communication between conflicting parties; they help analyze the dispute and develop possible resolutions. Use of mediators is most effective when the conflict is not extreme and when the mediator is trusted by all parties involved. (4) A further approach is to use unilateral conciliatory initiatives (i.e., the GRIT strategy) to reduce conflict. Under GRIT, one side takes a firm but conciliatory stance and initiates de-escalatory steps in the hope the other side will reciprocate. The GRIT approach works best when the initiating group is at least as strong as its opponent.

Key Terms

arbitrator (p. 407)

aversive event (p. 395)

discrimination (p. 392)

ethnocentrism (p. 392)

GRIT (p. 408)

illusion of out-group homogeneity (p. 397)

in-group (p. 392)

intergroup conflict (p. 389)

intergroup contact hypothesis (p. 403)

mediator (p. 406)

out-group (p. 392)

realistic group conflict theory (p. 392)

social identity theory of intergroup behavior (p. 394)

superordinate goal (p. 402)

ultimate attribution error (p. 398)

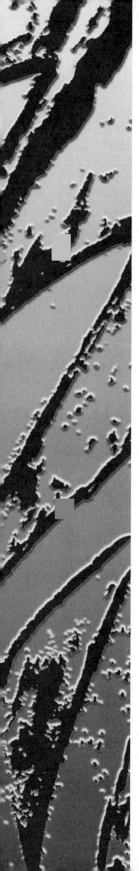

CHAPTER 17

Life Course and Gender Roles

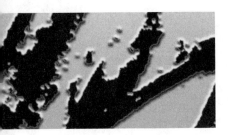

Introduction

"I still can't get over Liz," said Sally. "I sat next to her in almost every class for 3 years, and still, I hardly recognized her. Put on some weight since high school, of course, and dyed her hair. But mostly it was the defeated look on her face. When she and Hank announced they were getting married, they were the happiest couple ever. But that lasted long enough for a baby. Then there were years of underpaid jobs. She works part-time in sporting goods at Sears now. Had to take that job when her real estate work collapsed in the recession. Pity, just when she was beginning to get on her feet!"

Jim had stopped listening. How could he get excited about Sally's Lincoln High School reunion and people he'd never met? But Sally's mind kept racing. A lot had happened in 20 years:

John—Still larger than life. Football coach at the old school and assistant principal too. Must be a fantastic model for the tough kids he works with. That scholarship to Indiana was the break he needed.

Frank—Hard to believe he's in a mental hospital! He started okay as an engineer. Severely burned in a helicopter crash and then hooked on painkillers. Just fell apart. And we voted him Most Likely to Succeed.

Andrea—Thinking about a career in politics. She didn't start college until her last kid entered school. Now she's an urban planner in the mayor's office. Couldn't stop saying how she feels like a totally new person.

Tom—Head nurse at Westside Hospital's emergency room. Quite a surprise. Last I heard, he was a car salesman. Started his nursing career at 28. Got the idea while lying in the hospital for a year after a car accident.

Julie—Right on that one, voting her Most Ambitious. Finished Yale Law, clerked for the New York Supreme Court, and just promoted to senior partner with Wine and Zysblat. Raised two kids at the same time. Having a husband who writes novels at home made life easier. Says she was lucky things were opening up for women just when she came along.

Linda—Too bad she quit journalism school to put her husband through med school. She was a great yearbook editor. Still, says she enjoys writing stories as a stringer for the *News.* Leaves time for family and travel.

Sally's reminiscences show how different lives can be and how unpredictable. When we think about people like Liz or Tom or Frank, change seems to be the rule. There is change throughout life for all of us. But there is continuity too. Julie's string of accomplishments is based on her continuing ambition, hard work, and competence. John is back at Lincoln High—once a football hero, now the football coach. Although not a journalist as planned, Linda writes occasionally for a paper, and she may yet develop a serious career in journalism over the next 20 years. Even Frank had started on the predicted path to success before his tragic helicopter crash.

As we look into the future, we cannot project with any certainty what will happen to us. But people's lives have patterns that allow us to make sense of them. Each of us will experience a life characterized both by continuity and change. This chapter examines the **life course,** the individual's progression through a series of age-linked social roles embedded in social institutions (Elder & O'Rand, 1995), and the important influences that shape the life course we each experience.

Our examination of the life course is organized around four broad questions:

1. What are the major components of the life course?
2. What are the major influences on progression through the life course? That is, what causes people's careers to follow the paths they do?
3. What are the typical courses of life for men and women in American society? What happens when people depart from the typical patterns?
4. In what ways do historical events and trends modify the typical life-course pattern?

Components of the Life Course

Lives are too complex to study in all their aspects. Consequently, we will focus on the three main components of the life course: (1) careers, (2) identity and self-esteem, and (3) stress and satisfaction. By examining these components, we trace the continuities and changes that occur in what we do throughout the life course.

Careers

A **career** is a sequence of roles, each with its own set of activities, that a person enacts during his or her lifetime. Our most important careers are in three major social domains: family and friends, education, and work. The idea of careers comes from the work world, where it refers to the sequence of jobs held. Liz's work career, for example, consisted of a sequence of jobs as waitress, checkout clerk, clothing salesperson, real estate agent, and sporting goods salesperson.

The careers of one person differ from those of another in several ways—in the roles that make up the careers, in the order in which the roles are performed, and in the timing and duration of role-related activities. For example, one person's family career may consist of roles as infant, child, adolescent, spouse, parent, grandparent, and widow. Another woman's family career may include roles as stepsister and divorcée but exclude the parent role. A man's career might include the roles of infant, child, adolescent, partner, and uncle. The order of roles also may vary. "Parent before spouse" has very different consequences from "spouse before parent." In addition, the timing of career events is important. Having a first child at age 36 has different life consequences from having a first child at 18. Finally, the duration of enacting a role may vary. For example, some couples end their marriage before the wedding champagne has gone flat, whereas others go on to celebrate their golden wedding anniversary.

Societies provide structured career paths that shape the options available to individuals. The cultural norms, social expectations, and laws that organize life in a society make various career options more or less attractive, accessible, and necessary. In the United States, for example, educational careers are socially structured so that virtually everyone attends kindergarten, elementary school, and at least a few years of high school. Thereafter, educational options are more diverse—night school, technical and vocational school, apprenticeship, community college, university, and so on. But individual choice among these options is also socially constrained. The norms and expectations of our families and peer groups strongly influence our educational careers.

A person's total life course consists of intertwined careers in the worlds of work, family, and education (Elder, 1975). The shape of the life course derives from the contents of these careers, from the way they mesh with each other, and from their interweaving with those of family members. Sally's classmates, Julie and Andrea, enacted similar career roles: Both finished college, held full-time jobs, married, and raised children. Yet the courses of their lives were very different. Julie juggled these roles simultaneously, helped by a husband who was able to work at home. Andrea waited until her children were attending school before continuing her education and then adding an occupational role. The different content, order, timing, and duration of intertwining careers make each person's life course unique.

Identities and Self-Esteem

As we engage in career roles, we observe our own performances and other people's reactions to us. Using these observations, we construct *role identities*—conceptions of the self in specific roles. The role identities available to us depend on our location in society and on the career paths we are following. When Liz's work in real estate collapsed, she got a job in sales at Sears; she was qualified to sell sporting goods because of her prior work experience.

As we enact major roles, especially familial and occupational ones, we evaluate our performances and thereby gain or lose *self-esteem*—one's sense of how good and worthy one is. Self-esteem is influenced by our achievements; Julie has high self-esteem as a consequence of being a senior partner in a prestigious law firm. Self-esteem is also influenced by the feedback we receive from others.

Identities and self-esteem are crucial guides to behavior, as discussed in Chapter 4. Therefore we consider identities and self-esteem as the second component of the life course.

Stress and Satisfaction

Performing career activities often produces positive feelings, such as satisfaction, and negative feelings, including stress. These feelings reflect how we experience the quality of our lives. Thus we see stress and satisfaction as the third component of the life course.

Changes in career roles, such as having a baby/adopting a child, or changing jobs, place emotional and physical demands on the person. Life events, such as moving, or serious conflict with a parent or lover, may have similar effects. At times, the demands made on a person exceed the individual's ability to cope with them; such a discrepancy is called **stress** (Dohrenwend, 1961). People who are under stress often experience psychological (anxiety, tension, depression) and physical (fatigue, headaches, illness) consequences (Wickrama et al., 1997).

These feelings vary in their intensity in response to life-course events. Levels of stress, for example, change as career roles become more or less demanding (parenting roles become increasingly demanding as children enter adolescence), as different careers compete with each other (family versus occupational demands), and as unanticipated setbacks occur (one's employer goes bankrupt). Levels of satisfaction vary as career rewards change (salary increases or cuts) and as we cope more or less successfully with career demands (meeting sales quotas, passing exams) or with life events (a heart attack or skiing accident).

The extent to which particular events or transitions are stressful depends on several factors. First, the more extensive the changes associated with the event, the greater the stress. For example, a change in employment that requires a move to an unfamiliar city is more stressful than a new job located across town. Second, the availability of social support, in the form of advice and emotional and material aid, increases our ability to cope successfully with change. To help their members, families reallocate their resources and reorganize their activities. Thus parents lend money to young couples, and older adults provide care for their grandchildren so their children can work.

Personal resources and competence influence how one copes with stress. Coping successfully with earlier transitions prepares individuals for later transitions. Men who develop strong ego identities in young adulthood perceive events later in their lives as less negative (Sammon, Reznikoff, & Geisinger, 1985). Conversely, early experiences of failure reduce an individual's sense of competence (Duncan & Morgan, 1980) and may leave people with a sense of helplessness when they face later events and transitions.

Influences on Life-Course Progression

At the beginning of this chapter, we noted many events that had an important impact on the lives of Sally's classmates: loss of a job because of an economic recession, a helicopter crash, a car accident, having a baby, and graduating from a prestigious law school. These are **life events,** episodes that mark transition points in our lives and involve changes in roles. They provoke coping and readjustment (Hultsch & Plemons, 1979). For many young people, for example, the move from home to college is a life event marking a transition from adolescence to young adulthood. This move initiates a period during which students work out new behavior patterns and revise their self-expectations and priorities.

There are three major influences on the life course: (1) biological aging, (2) social age grading, and (3) historical trends and events. These influences act on us through specific life events (Brim & Ryff, 1980). Some life events are carefully planned—a trip to Europe, for example. Other events, no less important, occur by chance—like meeting one's future spouse in an Amsterdam hostel (Bandura, 1982a).

Biological Aging

Throughout the life cycle, we undergo biological changes in body size and structure, in the brain and central nervous system, in the endocrine system, in our susceptibility to various diseases, and in the acuity of our sight, hearing, taste, and so on. Changes are rapid and dramatic in childhood. Their pace slows considerably after adolescence, picking up again in old age. Even in the middle years, however, biological changes may have substantial impacts. The shifting hormone levels associated with menstrual periods in women and with aging in men and women, for example, are thought by many to affect mood and behavior (Doering, 1980; Hoyenga & Hoyenga, 1979).

Biological aging is inevitable and irreversible, but it is only loosely related to chronological age. Puberty may come at any time between 8 and 17, for example, and serious decline in the functioning of body organs may begin before age 40 or after age 85. The neurons of the brain die off steadily throughout life and do not regenerate. Yet intellectual functioning, long assumed to be determined early in life and to decline with aging, is now known to be capable of increasing over the life course. Even in old age, mental abilities can improve with opportunities for learning and practice (Baltes & Willis, 1982).

Biologically based capacities and characteristics limit what we can do. Their impacts on the life course depend, however, on the social significance we give them. How does the first appearance of gray hair affect careers, identities, and stress, for instance? For some this biological event is a painful source of stress. It elicits dismay, sets off thoughts about mortality, and instigates desperate attempts to straighten out family relations and to make a mark in the world before it is too late. Others take gray hair as a sign to stop worrying about trying to look young, to start basing their priorities on their own values, and to demand respect for their experience. Similarly, the impacts of other biological changes on the life course—such as the growth spurt during adolescence or menopause in middle age—also depend on the social significance given them.

Social Age Grading

Which members of a society should raise children and which should be cared for by others? Who should attend school and who should work full-time? Who should be single and who should marry? Age is the primary criterion that every known society uses to assign people to such activities and roles (Riley, 1987). Throughout life, individuals move through a sequence of age-graded social roles. Each role consists of a set of expected behaviors, opportunities, and constraints. Movement through these roles shapes the course of life.

Each society prescribes a customary sequence of age-graded activities and roles. In American society, many people expect a young person to finish school before he or she enters a long-term relationship. Many people expect a person to marry before she or he has or adopts a child. There are also expectations about the ages at which these role transitions should occur. People are expected to complete their education at ages 18 to 22 and to marry at ages 19 to 25 (Neugarten, Moore, & Lowe, 1965). These age norms serve as a basis for planning, as prods to action, and as brakes against moving too fast (Neugarten & Datan, 1973).

Pressure to make the expected transitions between roles at the appropriate times means that the life course consists of a series of normative life stages. A **normative life stage** is a discrete period in the life course during which individuals are expected to perform the set of activities associated with a distinct age-related role. The order of the stages is prescribed, and people try to shape their own lives to fit socially approved career paths. In addition, people perceive deviations from expected career paths as undesirable.

Not everyone experiences major transitions in the socially approved progression. Consider the transition to adulthood: The normative order of events is leaving school, possibly performing military service, getting a job, and getting married. Analyzing data about the high school class of 1972 collected between 1972 and 1980, researchers found that half of the men and women experienced a sequence that violated the "normal" path (Rindfuss, Swicegood, & Rosenfeld, 1987). Common violations included entering military service before one finished school and returning to school after a period of full-time employment.

In some cases, violating the age norms associated with a transition has lasting consequences. The transition to marriage is expected to occur between the ages of 19 and 25. Research consistently finds that making this transition earlier than usual has long-term effects on marital as well as occupational careers. A survey of 63,000 adults allowed researchers to compare men who married as adolescents with men of

Violating the age norms associated with a major transition, such as the transition to parenthood, may have lasting consequences. Having a baby at the age of 16 may force a young woman to leave school and limit her to a succession of poorly paid jobs.

made by all or most members of a defined population (Cowan, 1991). Although most members undergo this institutional passage, each individual's experience of it may be different, reflecting his or her past experience. Normative transitions are often marked by a ceremony, such as graduation or a wedding. But the transition is a process that may occur over a period of weeks or months. This process involves both a restructuring of the person's cognitive and emotional makeup and of his or her social relationships.

Transitions from one life stage to another influence a person in three ways. First, they change the roles available for building identities. The transition to adulthood brings major changes in roles. Those who marry or have their first child begin to view themselves as spouses and parents, responsible for others. Second, transitions modify the privileges and responsibilities of persons. Age largely determines whether we can legally drive a car, be employed full-time, or serve in the military. Third, role transitions change the nature of socialization experiences. The content of socialization shifts from regulating biological drives in childhood, to instilling values in adolescence, to transmitting role-related norms for behavior in adulthood (Brim, 1966). The power differences between socializee and socializing agents also diminish as we age and move into higher education and occupational organizations. As a result, adults are more able to resist socialization than children (Mortimer & Simmons, 1978).

similar age who married as adults (Teti, Lamb, & Elster, 1987). Because the sample included people of all ages, the researchers could study the careers of men who married 20, 30, and 40 years earlier. Men who married as adolescents completed fewer years of education, held lower status jobs, and earned less income. In addition, the marriages of those who married early were less stable. These effects were evident 40 years after marriage. Early marriage has similar effects on women. Women who marry before age 20 experience reduced educational and occupational attainment, and are more likely to get divorced (Teti & Lamb, 1989).

Movement from one life stage to another involves a **normative transition,** socially expected changes

Historical Trends and Events

Recall that Sally's classmate Julie attributed her rapid rise to senior law partner to lucky historical timing. Julie applied to Yale Law School shortly after the barriers to women had been broken, and she sought a job just when affirmative action came into vogue at the major law firms. Sally's friend Liz attributed her setback as a real estate broker to an economic recession coupled with high interest rates that crippled the housing market. As the experiences of Julie and Liz illustrate, historical trends and events are another major influence on the life course. The lives of individuals are shaped by *trends* that extend across historical periods (such as the increasing equality of men and women and improved nutrition) and by *events* that occur at particular points in history (such as recessions, wars, and earthquakes).

Birth Cohorts   To aid in the understanding of how historical events and trends influence the life courses of individuals, social scientists have developed the concept of cohorts (Ryder, 1965). A **birth cohort** is a group of people who were born during the same period. The period could be 1 year or several years, depending on the issue under study. What is most important about a birth cohort is that its members are all approximately the same age when they encounter particular historical events. The birth cohort of 1950, for example, was entering college at the height of the protests against the Vietnam War in 1967 and 1968. Many of these young people were profoundly influenced by those events. Members of the birth cohort of 1970 don't remember Vietnam protests but were entering college when the protests against racism erupted in 1987 and 1988. Some of them were profoundly influenced by those events.

A person's membership in a specific birth cohort locates that person historically in two ways. First, it points to the trends and events the person is likely to have encountered. Second, it indicates approximately where an individual is located in the sequence of normative life stages when historical events occur. Life-stage location is crucial because historical events or trends have different impacts on individuals who are in different life stages.

To illustrate, consider the effects of the widespread corporate downsizing that occurred between 1990 and 1995. Large numbers of workers and managers were laid off. Some people in their 50s found it impossible to get new jobs, perhaps because of age discrimination, and they experienced prolonged unemployment. Some persons in their 30s and 40s returned to school, and subsequently entered new fields. Workers who survived were left with insecurity and increased workloads. Persons just finishing college, the birth cohort of 1970, found fewer employment opportunities than those who graduated in 1985 or 1995. Of course, not all members of a cohort experience historical events in the same way. Members of the class of 1992 who majored in liberal arts faced

more limited opportunities than those earning professional degrees.

Placement in a birth cohort also affects access to opportunities. Members of large birth cohorts, for example, are likely to be disadvantaged throughout life. They begin their education in overpopulated classrooms. They then must compete for scarce openings in professional schools and crowded job markets. As they age, they face reduced retirement benefits because their numbers threaten to overwhelm the Social Security system. Table 17.1 presents examples of how the same historical events affect members of cohorts in distinct ways. These historically different experiences mold the unique values, ideologies, personalities, and behavior patterns that characterize each cohort through the life course. Within each cohort there are differences too. For example, war led to a father's absence for some children but not for others.

Cohorts and Social Change   Because of differences in their experiences, each birth cohort ages in a unique way. Each cohort has its own set of collective experiences and opportunities. As a result, cohorts differ in their career patterns, attitudes, values, and self-concepts. As cohorts age, they succeed one another in filling the social positions in the family, as well as in political, economic, and cultural institutions. Power is transferred from members of older cohorts with their historically based outlooks to members of younger cohorts with different outlooks. In this way, the succession of cohorts produces social change. It also causes intergenerational conflict around issues on which successive cohorts disagree (Elder, 1975).

In this section, we have provided an overview of changes during the life course. Based on this discussion, it is useful to think of ourselves as living simultaneously in three types of time, each deriving from a different source of change. As we age biologically, we move through *developmental time* in our own biological life cycle. As we pass through the intertwined sequence of roles in our society, we move through *social time*. And as we respond to the historical events that impinge on our lives, we move together with our cohort through *historical time*.

We have emphasized the changes that occur as individuals progress through the life course. However,

Table 17.1 History and Life Stage

	Cohort of 1950–1955		Cohort of 1975–1980	
	Life Stage When Event Occurred	Some Life-Course Implications of the Event	Life Stage When Event Occurred	Some Life-Course Implications of the Event
Vietnam War (1964–1973)	Adolescence	Increased political awareness, activity. Disrupted educational career.	—	—
Women's Movement (1972–1978)	Young adulthood	For women, increased work opportunities, delayed marriage. For men, increased competition for some jobs.	Infancy	Substantial time spent in day care, preschool. More socially skilled, independent.
Recession (1980–1982)	Adulthood	Increase in dual-earner couples to maintain living standard. Blue-collar unemployment.	Childhood	"Latch key" children. Non-traditional gender-role socialization. Increased time spent with peers.
Economic Expansion (1992–1997)	Middle adulthood	For women, increased occupational achievement, self-confidence. For men, reduced economic responsibilities.	Adolescence	Raised in dual-career family. Greater opportunities for girls.

there is also stability. Normative transitions usually involve choices, and individuals usually make choices that are compatible with preexisting values, selves, and dispositions (Elder & O'Rand, 1995). More than 90% of all Americans experience the normative transition of marriage. Most persons choose who they marry. Longitudinal research indicates that we choose a spouse compatible with our own personality, thus promoting stability over time (Caspi & Herbener, 1990).

Stages in the Life Course: Age and Gender Roles

People in every society experience a standard sequence of normative life stages. In this section, we examine some typical life-course patterns in American society and some important variations on these patterns. We consider experiences that distinguish males and females as well as experiences common to both genders.

Table 17.2 Postchildhood Life Stages

Stage	Major Challenge	Conventional Labels	Age Range
I	Achieving independence	Youth, Late adolescence	16–23
II	Balancing family and work commitments	Young adulthood	18–40
III	Performing adult roles	Adulthood, Maturity, Middle age	35–70
IV	Coping with loss	Late maturity, Old age	60–90

Note: The overlap in ages between the stages shows that age is only a rough indicator of life stage.

Every society expects certain role behaviors, values, and attributes of males and others of females (Doyle & Paludi, 1991). Differences in expectations for males and females are least pronounced at very young ages, increase through adolescence, and become even sharper in young adulthood. Before children are 10, for example, similar amounts of self-reliance, obedience, and leadership are usually expected of both girls and boys. By age 18, however, males and females typically face gender-differentiated expectations for these and other attributes as well as for choices of college major, occupation, and lifestyle (Hyde, 1996). Expectations remain strongly differentiated by gender throughout adulthood. With retirement and old age, expectations for males and females become more similar again.

The four postchildhood stages are shown in Table 17.2. Each stage is labeled according to the major social task or challenge characterizing it. The labels point to the fact that these stages are socially derived rather than the direct products of biological or cognitive development. In discussing each stage, we will focus on three major components of the life course—careers, identity and self-esteem, stress and satisfaction. We also note the normative transitions inherent in moving through these stages and consider how people cope with the problems posed by these transitions.

Stage I: Achieving Independence

Most college students are in the stage of achieving independence. This is a period of transition from lives centered psychologically and economically around parents to lives in which we stand on our own. This stage challenges us to disengage from parents and take responsibility for ourselves. For both men and women, achieving independence means acting in ways that are competent, persevering, and task oriented. In addition, women—but not men—often view the expression of warmth, kindness, and empathy as signs of independence (Johnson et al., 1975). In contrast, evidence indicates that as college men become more competent and task oriented, they see themselves as increasingly less warm, open, and interested in others (Mortimer, Finch, & Kumka, 1982).

Several major social transitions are typically associated with this life stage: leaving the family home, leaving school, entering the workforce, getting married, and establishing an independent household. A century ago, these transitions were usually spread over many years, occurring one at a time in a fixed order. More recently, these transitions have been compressed into the age range of 17 to 25 for most Americans. Individuals are freer now to choose when to make each transition, but they are constrained in these choices by the requirements of getting a formal education and preparing for an occupation. Individuals who make transitions too early or too late pay a price. Being out of step can lead to lost income and lost occupational and marital opportunities (Hogan, 1981), as noted earlier.

Individuals not only choose when to make these transitions, but they make choices about the amount and type of education they complete, the occupation they pursue, and who they marry. We might expect that women and men who are socially competent as adolescents will make wiser choices. Wise

Socially important role transitions are often marked by rites of passage—public ceremonies that affirm the individual's new status. These ceremonies signify both to the individual and to others that he or she now has a new identity and that new behaviors, rights, and duties are now appropriate.

choices should, in turn, result in more stable occupational careers and marriages. Researchers in the San Francisco area collected data on more than 500 children and adolescents between 1928 and 1931. Follow-up data were obtained from 281 of them at ages 53 to 62. Those who got higher scores on measures of competence in adolescence indeed had more stable lives throughout young and middle adulthood (Clausen, 1991). It is also the case that wise choices in adolescence—for example, entering the military after completing high school—can counter the effects of deprivation in childhood or early adolescence (Caspi, 1992).

Careers   During this stage, individuals explore the fit between their personal abilities and interests and available work options. Following high school, many youths join the workforce in entry-level jobs (as stock clerks, workers in fast-food restaurants, waitpersons, and so on). Others, especially minority group members in large cities, are frustrated in their search for jobs. They find only sporadic employment during this life stage. Still others enter military service. About one third of all youths attend college, and about one half of these youth work as well (Sweet & Bumpass, 1987; U.S. Bureau of the Census, 1987). Most young people try out a variety of jobs during this life stage, acquiring skills and preferences that eventually lead to more permanent employment.

College students benefit from institutional support in their movement toward independence. College provides a partially protected environment in which one can develop self-reliance with the support of peers. College also instills in many students the motivation to pursue long-range socially approved goals—such as studying 8 years to earn an advanced degree—even if this is not personally attractive. Such motivation is very helpful in the struggle for occupational success (Becker, 1964). The academic interests of students often reflect the gender typing of subject areas (physics and engineering for males, humanities and social work for females). Gender-differentiated preferences for subjects arise in middle school when boys or girls see particular subjects as useful in their future occupational choices (Eccles, 1987).

One ultimate goal for most men and women is to establish a long-term intimate relationship with a member of the opposite sex. In American society, dating is the mechanism by which potential partners get to know each other. As discussed in Chapter 12, the conventions of dating encourage men to take the initiative and women to be more passive and dependent. If followed, this pattern fosters a sense of autonomy among men, but it inhibits the achievement of independence among women.

Identities and Self-Esteem   A central challenge of this life stage is to solidify a personal identity—to develop a firm sense of continuity and direction in one's life (Erikson, 1968). Because men and women face different adult role expectations, they tend to build their identities on different bases. National surveys reveal that males tend to construct their identities around their anticipated or actual occupational roles. By anchoring their identities in their

17.1   Gender Differences: Myth or Reality?

In Chapter 5, we discussed the research on gender stereotypes. Broverman and colleagues (1972) created a scale that included 20 personality traits. They asked various groups of adults to indicate how much they thought each trait characterized adult men and adult women. In numerous studies, people cited consistent differences between males and females on the 20 traits. Men were more likely to be seen as independent, aggressive, and ambitious, whereas women were seen as excitable, emotional, and easily influenced by others.

Do men and women actually differ in the ways suggested by these stereotypes? The picture is not clear. There are some differences between males and females in social behavior. But the stereotypes exaggerate both the number and size of the differences. Here is a summary of some major conclusions reached in critical surveys of the research.

Aggression

Males are, indeed, more physically aggressive than females. Starting with more hitting and shoving at age 2 or 3, and progressing through more physical fights, violent crime, and spouse beating, males exhibit more physical aggression through the life course (Eagly, 1987). This difference appears in many different cultures, suggesting that biological factors may be involved. Note, however, that although males are more likely to initiate physical aggression, females are no less aggressive than males when they are directly provoked and when aggression is socially acceptable (Frieze et al., 1978; Maccoby & Jacklin, 1974). Also, the difference in aggressiveness is largest in studies of children under age 6 and smallest in studies of adults (Hyde, 1984).

Emotionality

The stereotype holds that women are more fearful, anxious, and easily upset than men. The majority of more than 30 studies based on people's descriptions of their own and others' traits supports this stereotype. Women and girls tend to describe themselves as more fearful and anxious, and other raters also tend to describe women as more anxious than men. However, studies in which researchers observe how people actually behave in fear-arousing situations and that measure physiological signs of emotionality (pulse, heartbeat, respiration), have revealed no consistent differences between males and females. Taken together, the evidence on emotionality is inconclusive: Women may score higher than men on self-reports of emotionality because women are given more freedom and encouragement to express their feelings in our society (Frieze et al., 1978; Tavris & Offir, 1984).

Dependency

Implicit in the stereotype that women are more dependent is the idea that women rely more on others for protection and help and are more easily influenced and persuaded. Studies offer only weak support for this stereotype (Becker, 1986). Young girls and boys do not differ in clinging to their parents, resisting separation, or wandering about freely. Studies of adult behavior indicate that women are more likely than men to accept suggestions, comply with requests, and be persuaded in face-to-face interaction. These gender differences may occur because the presence of others makes gender role more salient, so that in face-to-face interaction, women yield more readily than men (Eagly, 1987). Also, women often hold low-status positions, or are assumed to have lower status if no other status indications are available; persons of lower status are expected to yield to influence attempts by higher status persons.

Sociability

Are women more interested in people, friendlier, and more capable of establishing interpersonal relationships than men? Recent evidence suggests that neither men nor women consistently seek more social contact. The number of friends people have depends mainly on opportunities to meet and spend time with others, not on gender. Thus college women and men—having equal social opportunities—have about the same number of friends. The same holds for unmarried women and men. Young married women whose social contacts are typically more confined have fewer friends. The studies reveal differences in the quality of male and female

continued on next page

---

continued from previous page

friendships, however. Same-sex friendships tend to be more intimate and spontaneous for women than for men. Women are more likely to talk about their feelings and concerns, men to engage in activities such as sports (Caldwell & Peplau, 1982; Fischer, 1980; Rubin, 1983).

Research on other social behaviors and on intellectual abilities have revealed few differences between males and females (Hyde & Linn, 1986) except those that emerge after age 10 (such as females' greater verbal ability, males' greater visual-spatial ability). Future research may reveal additional differences, however. The causes of most of the differences we observe in everyday behavior of men and women are clear; they are the product of socialization, of responses to social expectations, and of the channeling of opportunities in the family, school, and workplace.

---

own vocational choices, young men can gain a concrete sense of self-direction (Lowenthal, Thurnher, & Chiriboga, 1975).

Most men do not perceive their occupational career as dependent on future family roles. In contrast, many females consider their anticipated roles as wives and mothers in constructing their identities (Hyde, 1996). Young women's identity commitments may be more tentative and ambiguous than men's because the exact nature of a woman's future roles depends in part on when and who she will marry. In recent years, however, an increasing number of young women hold nontraditional views of gender roles, particularly women whose mothers held a nontraditional gender-role ideology when they were young (Moen, Erickson, & Dempster-McClain, 1997). These women anchor their identities in their own vocational aspirations.

Studies of self-evaluation during this life stage show that self-esteem may drop after high school as youths struggle to achieve independence. Illustrating this drop, a sample of male college undergraduates from Michigan rated themselves as less competent, successful, active, and strong in their senior year than they did in their freshman year (Mortimer, Finch, & Kumka, 1982). Self-esteem rises later, when people successfully adapt to family and work roles. For the Michigan sample, self-ratings rose over the 10 years following college graduation.

Stress and Satisfaction Both men and women experience this period of achieving independence as stressful. Women are more likely than men to feel frightened, to feel that life is hard, and to feel financially insecure (Campbell, Converse, & Rodgers, 1976). These sentiments may well reflect the fact that some young women perceive less control than men over the important directions their lives are taking. Once they set their life directions by marrying, however, the stress women feel is usually reduced. Other women seriously pursue occupational goals at this time; their experience of stress may be similar to that of career-oriented men. If they enter traditionally male occupations and professions, they may experience social disapproval.

Both overall life satisfaction and satisfaction with specific aspects of life tend to be lower for women and men during this life stage than during most later stages (Campbell et al., 1976; Gould, 1978; Lowenthal et al., 1975). Individuals in this stage are unsure of their objectives, of their abilities, and of their futures. Aspirations for occupational and marital success may be high, but individuals worry about translating their dreams into realities. As people make firm career commitments and solidify their identities—and thereby move into the next stage—life satisfaction increases.

Stage II: Balancing Family and Work Commitments

During their 20s and early 30s, most men and women hold worker, spouse, and parent roles. The central challenge of this stage is to establish oneself firmly in these roles—to forgo other options and commit one's energy, time, and self-definition to a particular job and to a particular mate. Priorities also must be set for work and family roles. Women usually give first

Table 17.3 Percentage of Men and Women Single by Age Group in 1970, 1980, 1991, and 1995

	Women					Men			
Age	1970	1980	1991	1995	Age	1970	1980	1991	1995
18–19	75.6	82.8	90.4	91.6	18–19	92.8	94.3	96.6	97.7
20–24	35.8	50.2	64.1	66.8	20–24	54.7	68.8	79.7	80.7
25–29	10.5	20.9	32.3	35.3	25–29	19.1	33.3	46.7	51.0
30–34	6.2	9.5	18.7	19.0	30–34	9.4	15.9	27.3	28.2

SOURCE: U.S. Bureau of the Census, 1992, 1996.

priority to their roles as wife and mother, men to their work roles (Bielby & Bielby, 1989).

Careers Both familial and occupational roles are developed during this stage. During the past century, the family careers of over 90% of Americans have included marriage. Young adults are marrying later now than in recent decades (see Table 17.3). In 1980 half of the women had married by age 22.5 and half of the men had married by age 24.5. In 1995 half of the women were married by age 27.0, and half of the men were married by age 29.5. Blacks postpone marriage longer than adults of other races. Contributing factors include high unemployment rates and more time devoted to education (Glick, 1997).

Survey data clarify the nature of this change. Three fourths of the decline in the proportion of people marrying in their 20s is due to the increasing number of couples who are living together. This trend includes both white and black couples. Young people today set up such living arrangements at the same ages at which members of earlier cohorts married (Bumpass, Sweet, & Cherlin, 1991). At least one third of young adults live with a member of the opposite sex for 6 months or more before marriage (Cherlin, 1981). Forty percent of these couples break up; 50% of them marry within 5 years. Finally, remaining single for an extended period has become a viable lifestyle (Bernard, 1981).

Divorce rates reveal that many young adults have difficulties establishing firm family commitments. If recent trends continue, about half of those who marry this year will eventually divorce, most within the first 10 years of marriage. Analyses indicate a strong cohort effect on divorce rates; the rate among persons who married between 1960 and 1964 is much lower than the rate among persons who married between 1975 and 1979 (Morgan & Rindfuss, 1985). A longitudinal study of marriages found that experiencing certain problems substantially increases the likelihood of divorce. These problems include infidelity, spending money in ways the spouse considers foolish, and drinking and drug use; each increases the risk of divorce by 20% (Amato & Rogers, 1997). Also, adult children of divorce and adults who never lived with a father perceive greater likelihood of divorce than adults from two-parent families (Webster, Orbuch, & House, 1995). These adults were more likely to report patterns of marital interaction (arguing, shouting, and hitting) that strain a relationship.

People who get divorced typically remarry and they tend to do so within 3 years. As a result, most divorces and remarriages occur before people reach age 35 (Cherlin, 1981).

The most striking transition of this stage is the transition to parenthood. For about 13% of couples, the first child arrives within 7 months of marriage. For 26% of married couples, the first child arrives between 8 and 23 months into the marriage. The probability of having a child slowly declines as the length of marriage increases beyond 2 years (Sweet & Bumpass, 1987). Newly married couples need to develop priorities, a shared lifestyle, and commitments to each other. When the first child arrives, it brings the challenge of developing a parental role and a division of labor regarding child care. If the first child arrives shortly after marriage, couples are forced to make both transitions—to marriage and to parenthood—at virtually the same time. A longitudinal study of African American and white couples who had a child within the first 2 years of marriage compared them with couples who remained childless (Crohan, 1996). New parents reported more frequent conflicts after the transition than before, and they reported lower marital happiness than childless couples.

For men, parental and occupational roles are relatively independent. Most men work full-time throughout stage II, barring problems of unemployment or disability. A study comparing parents and nonparents of similar age and marital duration found no differences in career orientation or job characteristics between men who had children and men who did not (Waite, Haggstrom, & Kanouse, 1986). In 1993, however, 1.9 million fathers were their preschool child(ren)'s primary caregivers (U.S. Bureau of the Census, 1993). Some fathers do so because they want to be involved in their children's lives. Others do so because of their wives' greater earning capacity.

For many women, parental and occupational roles are intertwined. Most women work full-time until they have their first child, stay home to care for their young child(ren), and then return to work as the youngest child grows out of infancy. Statistics for women's employment in 1995 reflected this pattern. Over 80% of childless women ages 25 to 34 (single or married) but only 63% of women with children under age 6 were in the labor force. Regular jobs outside the home were held by 76% of the women with children ages 6 to 17. Many women with young children choose to stay home. But women are also constrained to stay home because their housework load typically doubles with the birth of a child and because child care is both difficult to arrange and expensive.

With the arrival of the first child comes the need to develop a division of home-related work between husband and wife. Studies of household work consistently find that women spend more time performing it than do men. According to one survey, women spend almost 3 hours per day on housework, compared to about 1 hour by men (Bielby & Bielby, 1988). Furthermore, there are differences in the kind of tasks performed. Typically, women are responsible for and

spend more time on cooking and cleaning; men do laundry and repairs (Biernat & Wortmann, 1991). Also, working mothers spend 3.5 hours per day on child care, compared with 1.75 hours by working fathers. Interestingly, working husbands and wives view the woman's often substantial contribution to family income as minimal, and they view the man's often minimal contribution to child care as substantial (Thompson & Walker, 1989).

What influences how much time men spend on household work? One influence is gender-role attitudes; nontraditional attitudes, particularly of the husband, are positively related to men's participation in housework (Starrels, 1994). The relative contributions to the family's economic situation are also important; wives who have greater income than their husbands spend fewer hours per week in housework (Shelton & John, 1996). The time each has available is a third influence; the person who does fewer hours of paid work spends more hours doing housework (South & Spitze, 1994). Another influence is the family life cycle; although wives consistently perform more housework than husbands, the gap is larger when there are young children at home, as illustrated in Figure 17.1 (Rexroat & Shehan, 1987).

Several authors (South & Spitze, 1994; Zvonkovic et al., 1996) suggest that the household division of labor reflects the culture's gender norms and is constructed via a complex process of negotiation heavily influenced by the couple's relationship. Thus relative contributions to household income and time available are resources that partners use in this negotiation.

Findings with regard to the relationship between race and the household division of labor are inconsistent (Shelton & John, 1996). One study of dual-earner African American couples found that mothers spent more time caring for their infant and doing housework, and that the father participated in both (Hossain & Roopnarine, 1993). Father's involvement in both did not vary by the number of hours the mother worked.

Identities and Self-Esteem   Men committed to a particular line of work become increasingly caught up in their occupational careers during this stage. Men build their identities around their

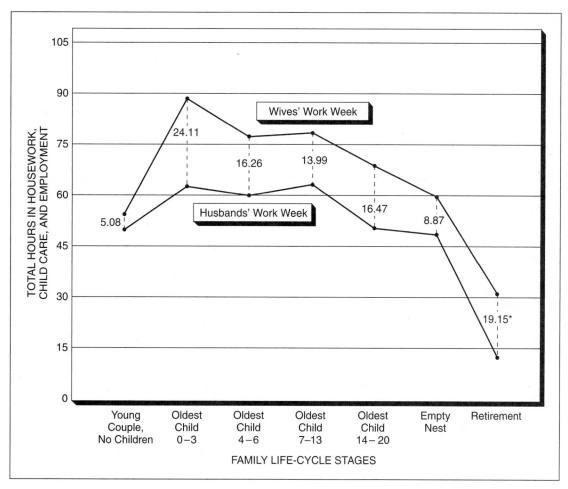

Figure 17.1   Total Hours (per week) Spent by Working Spouses in Housework, Child Care, and Employment

*The retirement stage includes spouses who are not employed.

SOURCE: Rexroat and Shehan, 1987, figure 1.

performance at work and around the future advancement they anticipate (Maines & Hardesty, 1987). They are likely to think of themselves as "rising executives," "budding craftsmen," "maturing scholars," or "salesmen on the move." Success in their occupational careers brings a rise in self-esteem during this stage. In contrast, men who fail to establish themselves in a work career by the end of this stage experience confusion about their identities and have relatively low self-esteem (Van Maanen, 1976).

Most women become increasingly committed to their families during this stage. Even if they are employed, married women tend to build their identities primarily around their relationships with their husbands and children (Bielby & Bielby, 1989). They are likely to think of themselves as wives and mothers, as persons who try to provide a pleasant, supportive home for their children and husbands. Women who enter business and the professions often do construct firm identities around their occupational careers. They are sometimes hindered in this, however, by the treatment they receive as a minority in the upper reaches of large organizations. Women are more often excluded from informal male peer networks and find it harder to be taken seriously than men (Kanter, 1976).

It is evidently more difficult to build a sense of self-esteem on work as a homemaker than on work outside the home (Mackie, 1983). In fact, when we talk about "work" we are usually referring to activities for which people get paid. We devalue housework because it is unpaid labor (Daniels, 1987). There are no clear criteria for judging the quality of homemaking and no raises or promotions to signify a job well done. In contrast, jobs outside the home usually provide supportive social contacts, and the regular paycheck indicates that other people value one's work. Thus even low-prestige jobs can bolster self-esteem. The effects of outside work on self-esteem can be seen clearly in a survey of working-class women. Three out of four housewives felt incompetent at running their homes, whereas over half the employed wives said they were "extremely good at their jobs" and not one said she felt incompetent (Ferree, 1976). A study of educated middle-class women also showed the effects of work on self-esteem: Housewives said they felt "worthless" almost twice as often as employed wives (Shaver & Freedman, 1976).

Stress and Satisfaction   Marriage brings a drop in feelings of stress for most women but an increase in stress for men. Newly married husbands are more likely than their wives to feel rushed, to feel life is hard, and to worry about paying their bills (Campbell et al., 1976). Data suggest that these differences are getting smaller (Lee, Seccombe, & Shehan, 1991).

The birth of a first child increases stress to the highest level in the life course for both husbands and wives. Strain between spouses increases. One study followed couples who were having their first child from late pregnancy until 9 months after the baby was born (Belsky, Lang, & Rovine, 1985). Marital quality declined significantly over the period, particularly as reported by wives. The level of stress associated with the transition to motherhood is influenced by the de-gree to which the birth disrupts the wife's relationship with the husband (Stemp, Turner, & Noh, 1986). Also, wives with a nontraditional gender-role ideology are more distressed by an unequal distribution of child-care responsibilities (Ross & Van Willigan, 1996). The transition is less stressful, however, if the new mother's social network includes other parents (McCannell, 1988) and if she perceives network members as concerned and caring. We can see that intimate relationships are an important buffer during stressful transitions.

Levels of satisfaction are only partly determined by stress; hence patterns of satisfaction differ somewhat across the life course. Figure 17.2 portrays satisfaction levels for men and women in different life-cycle stages. Young married women report higher levels of satisfaction with their lives than any other group. Men are also very happy during the childless married years. The arrival of children reduces satisfaction, especially among women. When young children are present, there is less satisfaction with standards of living, with savings, with housing, and with the marriage relationship. It is the divorced and separated men and women who are the least satisfied, however.

In dual-earner families, satisfaction is influenced by time spent together; couples who spend more time talking, eating meals, and having fun together report higher levels of satisfaction with the marriage (Kingston & Nock, 1987). In contrast, couples in which the wife works more than 40 hours per week outside the home are characterized by higher levels of marital instability (Booth et al., 1984), especially if the wife has a nontraditional gender ideology (Greenstein, 1995). Work and family demands are obviously intertwined and must be balanced if satisfaction is to be kept high and stress kept low.

Several recent studies have analyzed data from blacks. With regard to global satisfaction, married black men and women are more satisfied than nonmarried ones (Taylor et al., 1990; Zollar & Williams, 1987). The distinction between male as provider and female as homemaker is less rigid in black families than in white ones probably because black women have historically been more likely to work outside the home due to economic necessity. In fact, one study found that black men whose wives are homemakers

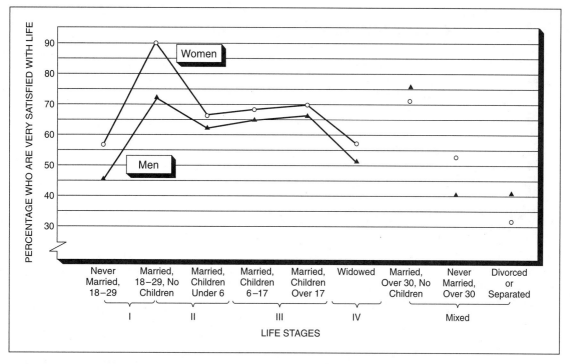

Figure 17.2 Satisfaction Over the Life Course

A national sample of Americans responded to the question "How satisfied are you with your life as a whole these days?" Responses were scaled from 1 (completely dissatisfied) to 7 (completely satisfied). This chart shows the percentages of men and women who indicated that they were very satisfied (by choosing 6 or 7 on the scale). The sample has been divided into life-cycle categories based on age, marital status, and age of youngest child. The first six categories, arranged in sequence from the left, represent the standard life-course pattern, beginning with the young unmarried status, through marriage and parenthood, to widowhood. The curves connecting the responses of these groups show the trends of satisfaction across the life course. The final three categories represent people who diverge from the common pattern. The chart also shows the match between the life-cycle categories and the life stages.

SOURCE: Adapted from Campbell et al., 1976, p. 398.

have poorer psychological well-being than those married to working and career-oriented women (Orbuch & Custer, 1995). The former have the burden of providing sufficient financial resources for their families with no help from their wives. Also, among blacks, men who report doing most of the housework are less satisfied with their family life (Broman, 1991).

Stage III: Performing Adult Roles

What is the major challenge around which people of your parents' age organize their lives? For most adults,

the major challenge from their late 30s until into their 60s is to put their lives to a useful purpose—to make a meaningful social contribution (Erikson, 1968). Of course, few people talk about making "meaningful social contributions." Instead, people try to be good workers, parents, and spouses; that is, they try to meet high standards for performance in the adult roles to which they are committed.

Careers Most men and childless career women spend the first part of this stage working their way

17.2   Gender Differences in Communication

Two of the best selling books of the 1990s, *You Just Don't Understand* (Tannen, 1990) and *Men Are From Mars, Women Are From Venus* (Gray, 1992), proclaim there are important differences in the way men and women communicate. Indeed, numerous studies document asymmetries in cross-gender communication (Thorne, Kramerae, & Henley, 1983). Do any of the following patterns appear in your own interactions?

Regardless of status, women are more likely than men to be addressed by first name rather than by title and last name. This use of lower status address forms has been observed in such work settings as hospitals, universities, and newsrooms. Women are less assertive than men when introducing topics into a conversation. Women are much more likely to ask a question ("Did you see the report on . . . ?"), whereas men declare their interest in a topic ("I saw a report on . . . "). Women also reveal their lower status by checking more frequently whether their conversational partner is still listening. They do this by adding unnecessary questions onto their assertions ("Isn't it?") and by inserting "You know" into their remarks (Fishman, 1978, 1980).

A woman who disagrees with a man is often more tentative in her speech. She may use disclaimers ("I may be wrong, but . . . "), hedge ("sort of true"), or use intensifiers ("It's too dangerous") (Carli, 1990). Interestingly, such speech is likely to make her more influential with men than if she speaks assertively. The reverse is true if her conversational partner is another woman.

Another difference is in the use of *back-channel feedback,* small vocal comments a listener makes while a speaker is talking. Women use less intrusive responses than men to indicate attention or agreement during conversation. Women prefer head nods and "M-hmn" rather than the more assertive "Yeah" or "Right." In conversations between men and women, men interrupt more and are more likely to yield the floor when interrupted by another man than by a woman. Also, men interrupt other men with supportive comments, whereas they interrupt women with neutral or disagreeing remarks (Smith-Lovin & Brody, 1989).

There are also gender differences in nonverbal behavior. Men tend to signal dominance through freer staring, pointing, and walking slightly ahead of the women they are with. Women are more likely to avert or lower their eyes and move out of a man's way when they are passing him (LaFrance & Mayo, 1978; Leffler, Gillespie, & Conaty, 1982). An observational study of 799 instances of intentional touch found that in public situations—at shopping malls, outdoors on a college campus—men are more likely to touch women. In greeting or leave-taking situations—at bus stations and airports—there was not an asymmetry by gender (Major, Schmidlin, & Williams, 1990).

Why do men use higher status, more dominant styles of communication than women? One explanation points to gender roles. In Western cultures, men are socialized from early childhood to be dominant and women to be submissive. As a result, men and women use speech styles that reflect the personalities and self-

continued on next page

up the occupational ladder. Devoting themselves primarily to their jobs, they seek the increased responsibility, respect, and financial rewards available in their occupations.

The occupational experiences of women and men are likely to be different. More than half of all employed women are in clerical (secretarial, typing, data entry), retail sales, and service occupations. Men are much more likely to be managers, craftsmen (carpenters, plumbers, electricians), or machinery operators (Reskin & Hartmann, 1986; U.S. Bureau of the Census, 1992). Although almost equal percentages of working women and working men are in the professions, women are frequently in the lower status professions such as teaching and nursing. These occupational differences produce differences in the prestige and income that men and women derive from employment.

At some point in their 40s or 50s, many workers recognize that their occupational life has reached a plateau which may extend to retirement. Taking stock of where they are, these workers often become less

continued from previous page

images they have acquired through socialization (Lakoff, 1979). A study of 91 college students compared the influence of sex and of gender identity on speech style. The gender identity of each student was assessed; later, 30 men and 26 women participated in two-person conversations. The results showed that the more masculine the student's gender identity, regardless of sex, the greater the likelihood that the person interrupted and engaged in overlapping conversation (Drass, 1986). Furthermore, the more masculine one's gender identity, the more likely she or he is to challenge statements made by the other conversationalist (Spencer & Drass, 1989).

A second explanation points to differences in power rather than differences in identity. It proposes that the different communication styles are situational adaptations to the gender-linked distribution of power in society. Because women are granted lower status and power than men in most situations, women lack the power needed to control communication through dominant styles. Instead, women adopt verbal and nonverbal styles that enable low-power individuals to gain some control over communication (Fishman, 1980; O'Barr & Atkins, 1980).

One study of the influence of power on communication style focused on couples who were living together. The power of each person relative to the partner was measured by an eight-item scale. Couples were selected so that in half of them power was shared equally, whereas in the other half one person had greater power. There were five male-female couples, five male-male couples, and five female-female couples. The results showed that interruptions and back-channel feedback

("M-hmn," "Yeah") were linked to power rather than sex; the more powerful person interrupted more and gave less feedback. The amount of time the person talked was linked to both sex and power; males and women with greater relative power talked more (Kollock, Blumstein, & Schwartz, 1985).

In another study, researchers manipulated power by giving one person greater relevant information or control over rewards. Both males and females who had more power displayed visual dominance, looking at the other person more while speaking to him or her than while listening. Low-power persons looked at the other person more while listening than while speaking (Dovidio et al., 1988).

A study of communication in same-race, mixed-gender groups of adolescents compared African American and white eighth graders (Filardo, 1996). On several measures, there was greater equality in the black than in the white groups. White females' speech was more tentative, conciliatory, and polite than the speech of the other three race x gender groups. Analyses suggested that the equality in the black groups was due to assertive behavior by black females.

A comprehensive review of the literature on gender differences in communication (Aries, 1996) indicates that speech patterns vary not only by gender and status but by characteristics of the context, such as the goals of the interaction and the roles of the participants. Anyone is capable of displaying "masculine" or "feminine" styles of communication when it is appropriate.

---

concerned with their own achievements and more interested in promoting others (Gould, 1978; Lowenthal et al., 1975). Thus lawyers, machinists, managers, or researchers who earlier enjoyed the guidance of mentors may now take pleasure in guiding younger associates themselves.

Workers who want greater self-determination may respond to the occupational plateau by pursuing new experiences and challenges, often with the objective of increasing their income. Some seek jobs similar to their current ones in firms or institutions where

the path to further advancement is still open (such as leaving one automobile company for another, or a local government job for one in Washington). Others return to school or study on their own in order to launch new careers (such as switching from teaching to stock brokerage, or from banking to public law). Starting one's own business, moonlighting at a second job, and turning a hobby into a money-making venture are other self-determining responses at this life stage.

Some employed mothers devote themselves primarily to their families when their children are young, which causes them to fall behind men in their work careers (Reskin & Hartmann, 1986). Thus women who seek jobs in stage III, after several years at home, are often at a disadvantage; they have no established record of reliable employment, nor have they developed or maintained marketable skills. Consequently, they must often settle for jobs that are below their educational level and less than fulfilling (Sewell, Hauser, & Wolf, 1980). As a result, these women may pursue occupational advancement and personal achievement most intensively during their 40s and 50s, after their children become teenagers. Paradoxically, this may be just when their husbands are beginning to feel less driven or to perceive restricted opportunities for further advancement.

During stage III, another type of family role may become important—the role of sibling. Brothers and sisters may provide emotional support and direct services throughout one's life (Goetting, 1986). Geographic proximity and emotional closeness are the main influences on how often siblings see each other (Lee, Mancini, & Maxwell, 1990). In midlife, the need to care for elderly parents may lead to increased sibling interaction and cooperation. A study of 50 pairs of sisters noted distinct styles of participation in parental care, ranging from routinely providing care to providing no assistance at all (Matthews & Rossner, 1988). The results of interviews with the women suggest that birth order and geographic proximity are important influences on style; oldest siblings and those who live closest to the parent(s) usually provide the routine care.

Identities and Self-Esteem   It is widely assumed that men continue to anchor their identities in their occupational roles throughout this midlife stage. People commonly gauge their success by whether they receive promotions and raises early, late, or on time compared with others in similar occupational roles (Levinson, 1978; Lowenthal et al., 1975). However, the results of two surveys of national samples of men and women challenge these assumptions (Pleck, 1985). On measures of role involvement, including items such as "My main satisfaction in life comes from my work," employed men and women report

higher levels of involvement in family roles than in work roles. In addition, satisfaction with family was more closely associated with overall well-being than satisfaction with work.

Some women and men anchor their identities primarily in their family roles. For these parents, loss of the parental role when children leave home may therefore weaken sense of identity temporarily and undermine self-esteem. But the ensuing period of freedom is usually accompanied by an expanding sense of competence, maturity, and self-assurance (Harris, Ellicott, & Holmes, 1986).

Menopause was also once thought to weaken women's self-assurance and esteem. Perhaps this was true when menopause signified the end of childbearing—the loss of what was considered women's most important capacity. But menopause no longer has this social meaning in the United States because women end childbearing earlier today and cultivate alternative roles around which to build identities. There is even cross-cultural evidence that following menopause, women become more assertive, dominant, and autonomous, whereas men at this age become more nurturant and affiliative (Freedman, 1979; Guttman, 1977).

Being employed increases the self-esteem of married women (Baruch & Barnett, 1986; Kessler & McCrae, 1982). In particular, women who complete college and work as professionals have a firmer sense of identity and higher self-esteem. This was illustrated in a study that compared the sense of identity and self-esteem of a group of married female professionals, single female professionals, and housewives (Birnbaum, 1975). Single female professionals resembled married female professionals on both dimensions, although the group of singles felt as lonely and unattractive as the housewives. The responses of each group suggest that employment for women contributes to a firmer identity and sense of competence and that the combination of employment and marriage combats feelings of loneliness and being unattractive to men.

Stress and Satisfaction   For both married men and married women, stage III is a period of declining psychological stress. For parents the following types of stress are greatest when children are under 6, and decline steadily as children grow older: feeling tied down, feeling life is hard, worrying about having a nervous breakdown, and—especially—worrying about finances (Campbell et al., 1976). There is a marked increase in marital happiness when the youngest child leaves home (White & Edwards, 1990). Reduced work responsibilities are also associated with an increase in marital satisfaction in later life (Orbuch et al., 1996).

In addition to characteristics of one's roles, personality is an important source of well-being at midlife. A longitudinal study of women found that quality of work and family roles were positively associated with life satisfaction at age 48. Further, generativity—concern with teaching others—was also related to satisfaction (Vandewater, Ostrove, & Stewart, 1997).

As the midlife stage progresses, two other sources of stress gradually increase: physical illness and the death of parents or close friends. People ages 45 to 64 reported three times more anxiety about physical illness than people ages 21 to 34 in one national survey (Gurin, Veroff, & Feld, 1960). In a more recent study, women's reports of deteriorating health (eyesight, hearing, teeth, hair, and so on) increased greatly after age 45 (Rossi, 1980). The death of a parent or close friend can be stressful because it increases one's own sense of mortality, deprives one of important roles as child or friend, may increase fears about health, and may impose added financial burdens.

A major source of stress is caring for someone who is physically or mentally ill for a prolonged period. A family may be providing such care for an adult child, an ailing parent, or a sibling of one of the adults. The primary care provider usually experiences increased stress, impaired relationships with other family members, and a decrease in social activities (Dura & Kiecolt-Glaser, 1991). Providing long-term care also can result in severe financial strain on the family.

Stage IV: Coping With Loss

Most of us will enter the final life stage during our 60s. Retirement or the onset of a major physical disability

Many elderly people participate in organized activities, such as this exercise group. As long as they stay healthy and economically independent, most elderly people maintain their social involvements, activities, and self-esteem.

are the key markers of the transition into this stage. The central challenge of stage IV is to cope with a series of practically unavoidable losses: loss of one's occupational role through retirement, of significant relationships through death, and of health, energy, income, and independence. Despite the severity of these losses, most older people say they are satisfied with life and are coping well (Harris et al., 1982).

Most people who reach age 65 in the United States today will spend as many years in this final stage of life as they spent in childhood and adolescence. Men who reach 65 can expect to live another 14 years, women another 18.

Through many of these years, older people can actively enjoy a wide range of activities because most remain reasonably healthy at least until age 75. Yet the aged, as a group, are often mistakenly believed to be narrow-minded, unteachable, not very bright, uninterested in sex, and not good at getting things done (Harris et al., 1975). Thus, in addition to coping with losses, older people must often cope with **ageism**—prejudice and discrimination against the elderly based on negative beliefs about aging.

Careers   Most older people maintain strong primary relationships. These include ties to their adult children. As children grow into adulthood, parents gradually give up responsibility for their offsprings' personal care, work, and financial status; the relationship evolves toward status equality (Blieszner & Mancini, 1987). Parents and children usually continue to have strong emotional ties, communicate regularly, and may provide various kinds of help to each other. Parents are selective in providing aid, giving it to the offspring who needs it—for example, temporary financial support in the event of unemployment (Aldous, 1987).

More than half of the elderly live with their spouses, and marriage continues to be their most important social tie. But the marital relationship often becomes more egalitarian in this stage. Women typically continue the shift toward greater assertiveness and independence begun in the preceding life stage, and men continue their shift toward greater nurturance and expressiveness (Guttman, 1977). Retirement adds to equality by reducing differences between spouses' activities. The happiest marital relationships among the elderly are those characterized by relative equality, mutual emotional support, and flexible sharing of household tasks (Sinnot, 1977).

Death of one's spouse is the most severe trauma the elderly confront. Women are typically widowed because they have longer life expectancies than their husbands. Half of married women lose their husbands by age 70. When husbands outlive their wives, they usually become widowers only after age 85. Death of a spouse causes many types of loss. It severs the deepest of emotional bonds, takes away the main companion in day-to-day activities, frustrates the fulfillment of sexual needs, removes the key significant other for affirming one's identity, and—especially for women—produces economic loss.

How well a widow copes with her husband's death depends on several factors. First, the death of a spouse is more stressful if it is sudden and unexpected, if other family members or friends die within weeks of the spouse's death, or if the surviving spouse is in poor health prior to the death (Sanders, 1988). Further, a woman whose identity and lifestyle were built on her marital role will experience greater disorganization when he dies (Lopata, 1988). Finally, women with strong social support networks cope with bereavement more successfully. Continuing interaction with married sisters seems to be a particularly important contributor to a widow's well-being (O'Bryant, 1988).

Although there are a few tragic exceptions, older Americans—married or not—are rarely abandoned by their families. In fact, more than half see at least one of their adult children almost every day, and the vast majority have contact with a child or sibling every week (Harris et al., 1982; Shanas, 1979). The image of the elderly as isolated in institutions such as nursing homes is inaccurate. Only about 1 in 4 is likely to spend any part of this life stage in an institution (Tubin, 1980). Three quarters of the elderly live in independent households (Hoyert, 1991). A few older people live with their adult children; those who do are typically single and are unable to maintain an independent residence because of poor health or small incomes.

Occupations, the second main career line, end for most older people with retirement between ages 60 and 65. Ideally, retirement would be a gradual process loosely linked to aging because occupational abilities and inclinations diminish only gradually. People who have control over giving up their occupational careers (for example, top management and the self-employed) do indeed withdraw more gradually (Hochschild, 1975). Retirement brings many losses; income, prestige, a sense of competence and usefulness, and social contacts may all decline. Retired persons may need to fill free time, develop new everyday routines, and—if married—adjust to spending more time with their spouses. Men experience substantial role discontinuity upon retiring. Women experience less discontinuity because they retain at least their homemaking responsibilities.

The main fear of retirees is that they will be cut off from social participation. Does this happen?

Usually people continue the level of social involvement they developed in their preretirement years. Those who were constantly busy and involved with people find new outlets in voluntary activities, hobbies, and social visits. Individuals who were uninvolved remain so. Few withdraw further (Palmore, 1981).

Reduced financial resources do, however, disrupt social participation, especially among retirees from the working class (Robson, 1982). Retirees who must worry about finances hesitate to spend money traveling to visit friends and entertaining them. Nor can they buy "proper" clothes or tickets for social events.

Identities and Self-Esteem

Retirement, widowhood, and declining health deprive people of many of the central roles and relationships around which their identities have been built. Considering these losses, identity change in this stage is less than one might expect (Atchley, 1980). The relative stability of identities occurs because older people continue to think of themselves in terms of their former roles. Although retired, individuals still think of themselves as nurses, accountants, or musicians, for example. Although widowed, they remain "John's wife" or "Sarah's husband" in their own eyes. Interactions with siblings often reinforce continuing identification with roles occupied earlier in life, for example, through reminiscences (Goetting, 1986). Identities are also preserved because many personal qualities remain stable. Individuals are likely to see themselves as unchanged in their honesty, outspokenness, religiousness, and so on.

Neither widowhood nor retirement alone do serious damage to self-esteem (Atchley, 1980). Several aspects of aging are, however, associated with self-esteem loss. Whatever deprives older people of their independence and of control over their own lives—such as ill health or falling into poverty—weakens self-esteem. Moving into an institution or family residence where one becomes highly dependent also undermines self-esteem if caretakers make all the decisions for an older person. These well-meaning actions communicate the assumption that the older person is mentally and physically incompetent.

Stress and Satisfaction

An important determinant of satisfaction levels in older persons is continued participation in activities. A survey of 618 persons, average age 65, assessed participation in eight clusters of activities. Those who reported participation in community service and social activities had higher levels of life satisfaction (Harlow & Cantor, 1996). Participation in social activities was especially important for retirees, and more so for men than women.

Loss of one's spouse is a major source of stress for older persons (Zantra, Reich, & Guarnaccia, 1990). Becoming widowed often results in declines in mental and physical health and in income (Brubaker, 1990). Financial worries are a source of stress for retirees living on fixed incomes and those who have experienced a recent divorce. Elderly parents whose adult children experience serious problems—for example, impaired physical or mental health, abuse of alcohol—experience increased stress (Pillemer & Suitor, 1991).

Physical health is a primary predictor of well-being among older persons (Brubaker, 1990). Stress from failing health rises throughout this life stage, especially after age 75. Older people become anxious and depressed when they experience reduced energy levels, lack of motivation, memory loss, a slowdown in their ability to process information, and chronic or acute diseases. Depression induced by ill health is often misdiagnosed as mental deterioration and confusion. As a result, depressed older people often fail to receive the psychotherapeutic treatment that could restore them to alertness and satisfactory functioning.

Historical Variations

Throughout this chapter, we have based our description of stages in the life course on recent findings and on projected future trends. But unique historical events—wars, depressions, medical innovations—change life courses. And historical trends—fluctuating birth and divorce rates, rising education, varying patterns of women's work—also influence the life courses of individuals born in particular historical periods.

No one can predict with confidence the future changes that will result from historical trends and events. What can be done is to examine how major events and trends have influenced life courses in the past. We present two examples: the historical trend toward greater involvement of women in the occupational world, and the effects of military service during the Vietnam War on the lives of men. The goals of this section are (1) to emphasize the influence of historical trends on the typical life course, and (2) to illustrate how to analyze the links between historical events and the life course.

Women's Work: Gender-Role Attitudes and Behavior

The percentage of women who work outside the home in the United States has increased substantially. We consider the role of attitudes and of economic changes in this trend.

Gender-Role Attitudes

In the past three decades, attitudes toward women's roles in the world outside the family have changed dramatically. The historical trend in attitudes has been away from the traditional division of labor (paid occupations for men and homemaking for women) to a more egalitarian view.

Consider the following statements. Do you agree with them?

1. It is better if the man is the achiever outside the home and the woman takes care of home and family.
2. Most of the important decisions in life should be made by the man of the house.
3. If her husband can support the family, a woman should not work for pay.
4. It is wrong for a woman to be very active in clubs, politics, and other outside activities before her children are grown.
5. Preschool children are likely to suffer if their mother works.

These are typical of attitude statements included in one or more large-scale surveys of adult women dur-ing the 1960s and 1970s. In the 1960s, about two thirds or more of women surveyed agreed with these statements. However, by the late 1970s, three of these attitudes had become minority views (2, 3, 4), and statements (1) and (5) were endorsed only by about half the women. This shift from traditional to egalitarian gender-role attitudes has been quite strong among women (Spitze & Huber, 1980; Thornton & Freedman, 1979).

Workforce Participation

This historical trend is not limited to attitudes. Women's actual participation in the workforce has been on the increase for almost a century. Figure 17.3 shows the percentage of women employed outside the home since 1960. The proportion of married women who are employed has been growing steadily, with a slight acceleration of growth between 1975 and 1985. Among young single women, the employment level, already very high in 1960, has remained high. The proportion of women who work during pregnancy and who return to work while their child is still an infant has also grown steadily over this time period (Sweet & Bumpass, 1987). These trends cut across racial lines; in 1995 almost equal percentages of white, black, and Hispanic women, controlling for age and family status, were employed outside the home (U.S. Bureau of the Census, 1996).

Why have women joined the workforce in ever greater numbers throughout the 20th century? Has the spread of egalitarian attitudes been an important source of influence? Probably not. The idea that wives and mothers should not work except in cases of extreme need was widely held until the 1940s. Yet women's employment increased steadily between 1900 and 1940. The change in gender-role attitudes occurred largely in the 1970s, yet women's employment rose rapidly during the two decades preceding these attitude changes. It therefore seems likely that gender-role attitude changes have not been a cause of the increased employment of women but a response to it—an acceptance of what more and more women are, in fact, doing.

What, then, are the causes? Perhaps most convincing is the argument that the types of industries

Some parents are able to blend work roles and family roles by working at home. As further advances occur in telecommunication, more women and men may choose this option.

and occupations demanding female labor are the ones that have expanded most rapidly in this century. Light industries like electronics, pharmaceuticals, and food processing have grown rapidly, for example, and service jobs in education, health services, and secretarial and clerical work have multiplied. Many of these occupations were so strongly segregated by gender that men were reluctant to enter them (Oppenheimer, 1970). In addition, male labor has been scarce during much of the century because of rapidly expanding industry and commerce. The majority of the slack was taken up by a large pool of unemployed married women. These women could be pulled into the workforce at a lower wage because they were often supplementing their family income.

The changes noted in the preceding paragraph led to increased job opportunities for women. Other factors influenced women's desire to work outside the home. One of these was continuing inflation and rising interest rates; in many families, two incomes became necessary to make ends meet. Other factors that may have promoted the increased employment of women include rising divorce rates, falling birthrates, rising education levels, and the invention of labor-saving devices for the home. None of these factors alone can explain the continuing rise in the employment of women over the whole century. However, at one time or another, each of these factors probably strengthened the historic trend, along with changes in gender-role attitudes.

The specific changes in women's work behavior demonstrate that the timing of a person's birth in history greatly influences the course of his or her life. Whether you join the workforce depends in part on

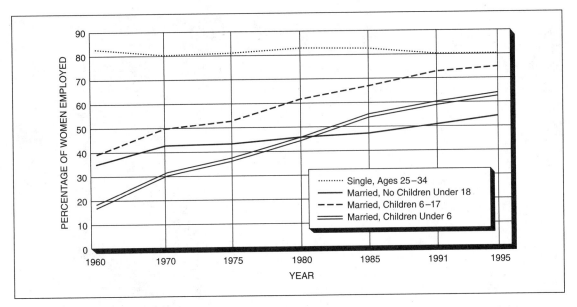

Figure 17.3 Women's Employment: 1960–1995

The percentage of married women who are employed has risen steadily since 1960. Young single women have maintained virtually the same high level of employment throughout this period. Among married women, the level of employment has risen slowly for those with no children and more rapidly among those with children under age 17.

SOURCES: U.S. Bureau of the Census, 1996; U.S. Bureau of Labor Statistics, 1989.

historical trends during your lifetime. So does the likelihood that you will get a college education, marry, have children, divorce, die young or old, and so on.

Effects of Historical Events: The War in Vietnam

The Vietnam War affected the lives of millions of Americans, some directly and others indirectly. One group whose lives were directly touched by the war were the hundreds of thousands of men and women who served in the military during the period from 1964 to 1973, when American combat troops fought in Vietnam.

There is a growing literature on the effect of military service in Vietnam, much of it focused on the association between exposure to traumatic experiences during the war and later psychological and social difficulties. To assess the impact of an historical event such as this, we must compare those who were directly

involved in the war with similar persons who were not. In this case, we want to compare men who served in Vietnam with men who are similar in age, race, and education but did not serve. It is possible, of course, that serving in the military has certain effects regardless of whether the person is involved in combat. We can take this into account by studying a group of veterans who were in the military service at the same time but did not serve in Vietnam. Comparing these three groups—nonveterans, non-Vietnam veterans, and Vietnam veterans—is a comparison *within* a birth cohort.

Laufer and Gallops (1985) carried out a within-cohort analysis of this type. Their data are based on a sample of 1,257 men who were eligible for the draft (ages 18 to 26) during the war. The data were collected in 1977 and 1979, providing information on the life experiences of the veterans for up to 12 years following their discharge from the military. What was the effect of military service and combat experience

Table 17.4 Effects of Military Service and Combat Experience on Marital Patterns

| Marital Patterns | Military Experience | | |
	Nonveterans (N = 590)	Non-Vietnam Veterans (N = 341)	Vietnam Veterans (N = 326)
Rates of marriage	70%	81%	84%
Percentage married by 25	56%	58%	67%
Rates of divorce	24%	27%	20%

SOURCE: Adapted from Laufer and Gallops, 1985.

on familial careers? Table 17.4 summarizes the results. Looking first at likelihood of marriage, veterans, especially those who served in Vietnam, were more likely to have married by the time of the survey. Also, veterans married for the first time at somewhat younger ages, with 67% of those who served in Vietnam married by age 25, compared with 56% of the nonveterans. Additional analyses of the data indicate that the veterans married quickly after they left the service. Thus, at the same time they were experiencing the transition back to civilian life and trying to settle into an occupational role, they also made the transition to marriage. Such a pileup of life events and transitions can be especially stressful.

What happened to these marriages? We might expect that, because of continuing combat-related stress, the veterans' marriages were more likely to end quickly in divorce. The non-Vietnam vets did experience the highest divorce rate; by contrast, those who served in Vietnam had the lowest divorce rate. However, when those who served in Vietnam were grouped according to amount of combat experience, the results indicate an increasing likelihood of divorce as exposure to combat increased. Additional analyses indicated that Vietnam veterans with combat experience reported the largest number of symptoms of stress in the year prior to the interview and were most likely to report drinking heavily during the 2 years before the survey.

From this study, we draw two general conclusions. First, the greater the exposure of individuals to an historic event, the greater its impact on their lives. For example, the more directly involved a man was in the war, the greater its effect on the likelihood of divorce. Second, historical events influence individuals by changing everyday interactions, socialization experiences, and/or life chances. Men who served in the military in Vietnam seemed to have rushed into marriage, creating a pileup of stressful life events.

Summary

This chapter has discussed the life course and gender roles in American society.

Components of the Life Course This chapter focuses on three components of the life course. (1) The life course consists of careers— sequences of roles and associated activities. The principal careers involve work, family, and friends. (2) As we engage in career roles, we develop role identities, and evaluations of our performance contribute to self-esteem. (3) The emotional reactions we have to career and life events include feelings of stress and of satisfaction.

Influences on Life-Course Progression There are three major influences on progression through the life course. (1) The biological growth and decline of body and brain set limits on what we can do. The impacts of biological developments on the life course, however, depend on the social meanings we give them. (2) Each society has a customary, normative sequence of age-graded roles and activities. This normative sequence largely determines the bases for building identities, the responsibilities and privileges, and the socialization experiences available to individuals of different ages. (3) Historical trends and events modify an individual's life course. The impact of an historical event depends on the person's life stage when the event occurs.

Stages in the Life Course: Age and Gender Roles

Four broad life stages characterize the life course beyond childhood. (1) Achieving independence (age 16 to 23) entails crucial transitions in education, work, and family life. Stress is high and satisfaction with life relatively low during this stage. Young men emphasize jobs in building their identities; young women emphasize family ties. (2) Balancing commitments to family and work (age 18 to 40) is a major life task. The patterns and timing of family and work careers have undergone major changes in recent decades. For most men and some women, identity and self-esteem are tied to occupational success; for many women and some men, family is central. Satisfaction rises with marriage, but birth of a first child increases stress and reduces satisfaction. An important source of stress is the unequal division between men and women of the household labor. (3) Performing adult roles competently (age 35 to 70) leads to an occupational plateau for many employed adults, who then seek other challenges. Adults who have been family oriented turn more to occupational advancement or to community involvements as family demands recede. Self-esteem is supported by occupational achievement for both women and men. Stress eases during this stage because resources grow faster than needs. (4) Coping with loss (age 60 to 90) is the key challenge for the elderly. Most cope well with retirement and even with death of a spouse, unless or until they experience financial difficulties, failing health, or reduced social participation. As long as the elderly retain independence, their self-esteem remains stable and their identities change little.

Historical Variations

The historical timing of one's birth influences the life course through all stages. (1) Over the past 30 years, women's participation in the workforce has increased dramatically and attitudes toward women's employment have become much more favorable. The likelihood that women will experience pressures and opportunities to work outside the home is now greater at every life stage. (2) Military service in Vietnam, particularly combat experience, had long-term effects on the familial careers of men. Those who served in Vietnam married at younger ages, and those with combat experience were more likely to experience divorce and other difficulties a decade after the war.

Key Terms

ageism (p. 432)

birth cohort (p. 417)

career (p. 413)

life course (p. 412)

life event (p. 414)

normative life stage (p. 415)

normative transition (p. 416)

stress (p. 414)

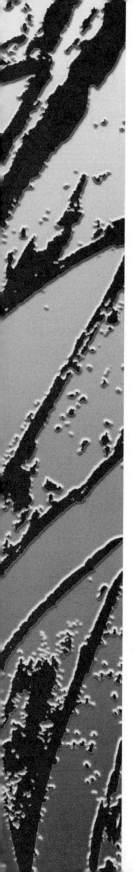

CHAPTER 18

Social Structure and Personality

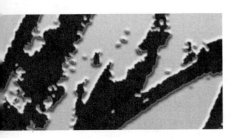

Introduction

Fred is 38, married, the father of two children, and sells pacemakers and artificial joints to hospitals. He travels 2 or 3 days a week and works at home the rest of the time in his $150,000 house in the suburbs. He earns $65,000 a year. Because his income is based entirely on commission, Fred worries about his sales falling off; but, on the whole, he is satisfied with his life. His values are conservative, and he votes for Republican candidates.

Larry is also 38 and has a wife and two children. He runs a service station, works 6 days a week from early morning until 6 or 7 P.M. Larry and his family live in a small three-bedroom house. Last year he made about $28,000. He worries a lot about money and has been very tense the past year. He has liberal values and usually votes for Democratic candidates.

Marie is 39. She is head nurse in a hospital pediatric ward. Last year her salary was $31,400. Although she enjoys her patients, she hates all the paperwork and the personnel problems. Some of her values are conservative, whereas others are liberal; she considers herself an Independent.

Fred, Larry, and Marie are three very different people. Each has a different occupation, which produces differences in income and lifestyle. They differ in their values—in what they believe is important—and in the amount of stress they feel.

Where do these differences come from? Often, they are the result of one's location in society. Every person occupies a **position,** a designated location in a social system (Biddle & Thomas, 1966). The ordered and persisting relationships among these positions in a social system make up the **social structure** (House, 1990). This chapter considers the impact of social structure on the individual. There are three ways in which social structure influences a person's life. First, every person occupies one or more positions in the social structure. Each position carries a set of expectations about the behavior of the occupant of that position, called a **role** (Rommetveit, 1955). Role expectations are anticipations of how a person will behave based on knowledge of his or her position. Through socialization and personal experience, each of us knows the role expectations associated with our positions (Heiss, 1990). For example, Fred enacts several roles, including salesman, husband, and father. The expectations associated with these roles are a major influence on his behavior.

A second influence on the individual is **social networks,** the sets of relationships associated with the various positions a person occupies. Each of us is woven into several networks, including those involving co-workers, family, and friends. Two examples of social networks are depicted in Figure 18.1. Both networks center on a single individual, Mary. The network on the left depicts the pattern of relationships among Mary and her friends that have evolved out of their shared experience. For example, the ties between Mary and Margie and John developed because they took classes together, whereas Mary got to know Kathy at work. Mary described this network when asked to name the people she is closest to, excluding relatives. When asked to include relatives, Mary described the network on the right of Figure 18.1. The most striking difference is the greater number of direct ties between the persons in the network on the right. This network is *dense*—most of the persons in it know each other independently of their ties to Mary (Milardo, 1988). In both networks, a tie between Mary and another person reflects a **primary relationship,** one that is personal, emotionally involving, and of long duration. Such relationships have a substantial effect on one's behavior and self-image (Cooley, 1902).

A third way that social structure influences the individual is through **status,** the social ranking of a person's position. In every society, some positions are accorded greater prestige than others. Differences in ranking indicate a person's relative standing—his or her status—in the social structure. Each of us occupies several positions of differing status. In the United States, occupational status is especially influential. It

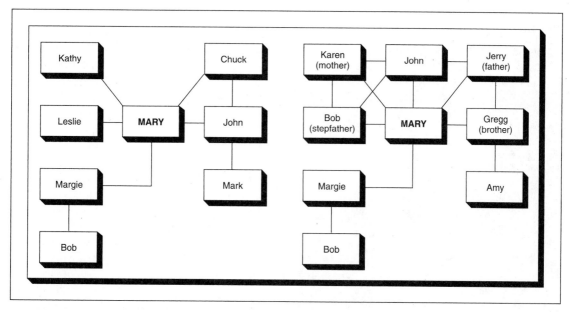

Figure 18.1 Social Networks

These two networks are focused on Mary. When asked to name her best friends, excluding relatives, Mary described the network on the left. John is her boyfriend, whereas Mark is John's roommate. Mary met Chuck through John, and since then she and Chuck have become friends. Mary's three best women friends are Kathy (a co-worker), Leslie (whose parents are friends of Mary's parents), and Margie (a classmate). Margie has recently married Bob. This is a loosely knit network, because few of Mary's friends are close to each other. The network on the right represents the people to whom Mary feels closest including her relatives. Notice that only Margie and John remain from the friendship network. Mary's intimates now include Gregg (her brother) and Amy (Gregg's girlfriend). They also include her mother (Karen), stepfather (Bob), and her father (Jerry). This network is dense, because most of its members are close to Mary and to each other.

is the major determinant of income, which has a substantial effect on one's lifestyle. One of the obvious differences between Fred and Larry, for instance, is their annual income.

This chapter focuses on how social structure influences the individual via roles, social networks, and status. It considers the impact of social structure on four areas: achievement, values, physical and mental health, and a person's sense of belonging. Specifically, it considers these four questions:

1. How does location in society affect educational and occupational achievement?
2. How does social position influence people's values?
3. How does social location influence a person's physical and mental health?
4. How does position influence a person's sense of belonging in society, or the lack thereof?

Status Attainment

The individual's relative standing in the social structure—status—is perhaps the single most important influence on his or her life. Status determines access to resources—to money and influence over others. In the United States, occupation is the main determinant of status. This section considers the nature of occupational status, the determinants of the status particular individuals attain or achieve, and the impact of social networks on the attainment of status.

Occupational Status

Occupational status is a key component of social standing and a major determinant of income and lifestyle. Fred is a sales representative for a company that

Uniforms are a very efficient way of communicating one's personal roles. We can tell at a glance what role this woman is enacting, and as a result, we know how to behave toward her.

makes artificial hip and elbow joints, pacemakers, and other medical equipment. These items are in great demand, and few companies make them. Fred sells a single pacemaker for $3,000 and keeps half of the money as his commission. He only needs to be on the road 2 or 3 days per week to earn $65,000 a year. He has a beautiful suburban home and two cars. Larry, by contrast, owns a service station. He works from morning until night pumping gas and repairing cars. His station is in a good location, but his overhead is high; he only earned $28,000 last year, and he worries that this year that figure will be lower. Larry and his family live in a smaller, older house and have a 6-year-old car.

The benefits Fred and Larry receive from their occupational statuses are clearly different. First, Fred earns more than twice as much money as Larry. This determines the quality of housing, clothing, and medical care his family receives. Fred also has much greater control over his own time. Within limits, he can choose which days he works and how much he works; this, in turn, affects the time he can spend with family and friends. Larry doesn't have much free time. Finally, Fred receives a great deal of respect from the people he works with. He controls a scarce resource, so doctors and hospital personnel generally treat him well. Larry, however, deals with people who are usually preoccupied or angry because their cars are not running properly.

In addition to these tangible benefits, occupational status is associated with prestige. Several surveys in the United States have found widespread agreement about the prestige ranking of specific occupations. In these studies, respondents typically are given a list of occupations and asked to rate each occupation in terms of its "general standing" or "social standing." The average rating is often used as a measure of relative prestige. The prestige scores for the United States shown in Table 18.1 were taken from an occupational prestige scale of 0 to 100 (Nakao & Treas, 1994). Surprisingly, there is considerable agreement across diverse societies in the average ranking of occupations. Even adults in China give rankings similar to those displayed in Table 18.1 (Lin & Xie, 1988).

The social structure of the United States can be viewed as consisting of several groups or social classes. A *social class* consists of persons who share a common status in the society. There are various views regarding the nature of social classes in the United States. One view of social class emphasizes occupational prestige, in conjunction with income and education, in defining class boundaries. This approach ordinarily classifies people into upper-upper, upper-lower, upper-middle, lower-middle, working, and lower classes (Coleman & Neugarten, 1971). A very different approach emphasizes the control, or lack thereof, an individual has over his or her work and co-workers as the main determinant of class standing (Wright et al., 1982).

Intergenerational Mobility

When a person moves from an occupation lower in prestige and income to one higher in prestige and income, he or she is experiencing **upward mobility.** To

Table 18.1   Occupational Prestige in the United States

Occupation	Nakao-Treas Prestige Score
Physician	86
Lawyer	75
College or university professor	74
Registered nurse	66
Electrical engineer	64
Elementary school teacher	64
Police officer	60
Social worker	52
Dental hygienist	52
Office manager	51
Electrician	51
Housewife	(51)
Office secretary	46
Data entry keyer	41
Farmer	40
Auto mechanic	40
Beautician	36
Assembly-line worker	35
Housekeeper (private home)	34
Precision assembler	31
Truck driver	30
Cashier	29
Waitress/waiter	28
Garbage collector	28
Hotel chambermaid	20
Househusband	(14)

SOURCE: Hauser and Warren, 1996.

what extent is upward mobility possible in the United States? On the one hand, we have the Horatio Alger rags-to-riches imagery in our culture: Anyone who is determined and works hard can achieve economic success. This imagery is fueled by stories about the astonishing success of the woman who founded Mary Kay Cosmetics, the man who borrowed from his friends to establish Motown Records, the man who founded the first electronic data processing company, and so on. On the other hand, some argue that America is not an open society, that our eventual occupational and economic achievements are fixed at birth by our parents' social class, our ethnicity, and gender. To be sure, every city has families who have been

wealthy for generations and families who have been poor for as long. This suggests that the United States is characterized by *castes,* groups whose members are prevented from changing their social status.

These two views of upward mobility in American society are concerned with *intergenerational mobility,* the extent of change in social status from one generation to the next. To measure intergenerational mobility, we compare the social status of persons with that of their parents. If the rags-to-riches image is accurate, we should find that a large number of adults attain a social status significantly higher than their parents'. If the caste-society image is correct, we should find little or no upward mobility.

What are the influences on upward (intergenerational) mobility in American society? In this section, we consider the impact of three factors: socioeconomic background, gender, and occupational segregation.

Socioeconomic Background   Occupational attainment in American society rests heavily on educational achievement. To be a doctor, dental assistant, computer programmer, lawyer, or business executive, one needs the required education. To become a registered nurse, Marie (who we met at the beginning of the chapter) had to complete nursing school. Fred, our medical equipment salesman, earned a bachelor's degree in business.

Beyond education, what other factors influence occupational attainment? To answer this question effectively, we need to trace the occupational careers of individuals over their life course. Such longitudinal data are available from a research project begun in the 1950s (Sewell & Hauser, 1980). In 1957 all high school seniors in Wisconsin were surveyed about their post–high school plans. From this population, a random sample of 10,317 was selected for continuing study. In 1964 researchers obtained information from students' parents about post–high school education, military service, marital status, and current occupation. Later, they obtained information about the students' earnings and about the colleges or universities they attended. In 1975, 97% of the original sample were located and most were interviewed by telephone. The interview focused on post–high school education, work history, and family characteristics. Data from this study enabled researchers to trace the impact of characteristics of high school seniors on subsequent education, occupation, earnings, and work experience.

Figure 18.2 presents a diagram of the relationships found among the variables studied. The arrows indicate causal impacts. Variables are arranged from left to right, reflecting the order in which the variables affect the person through time. These results indicate that children from more affluent homes have greater ability, higher aspirations, and receive more educa-

tion. Children with higher ability get better grades, which reward them for their academic work and reinforce their aspirations. Children who do well are also encouraged by significant others, such as teachers and relatives, which also contributes to high aspirations. These children are likely to choose courses that will prepare them for college. They are likely to spend more time on academic pursuits and less time on dating and social activities (Jessor et al., 1983). As a result, they are likely to continue their education beyond high school and perhaps beyond college. Finally, high ability, encouragement of significant others, and

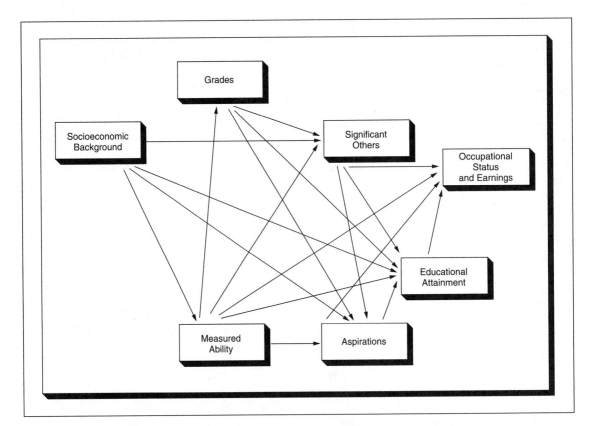

Figure 18.2 The Determinants of Occupational Status Attainment

This figure summarizes the influences that determine educational and occupational status over the life course. Socioeconomic background (parents' education, occupation, and income) influences ability, aspirations, and educational attainment. Ability influences grades, which, in turn, affect encouragement from significant others and aspirations for educational attainment. Occupational status is affected by education and also by ability, aspirations, and significant others.

SOURCE: Adapted from Sewell and Hauser, 1980.

high educational attainment lead to greater occupational status and earnings.

Note that socioeconomic background and grades have an indirect effect on occupational status and a direct effect on educational attainment. This does not mean that parental socioeconomic status and an individual's grades are unrelated to occupational status. Rather, it indicates that status and grades influence occupational attainment through other variables—like aspirations—which have a direct impact on occupational attainment (Sewell & Hauser, 1975).

In the research summarized in Figure 18.2, the family characteristics studied are mothers' and fathers' socioeconomic standing—education, occupation, and income. How is it that variables such as your father's education and your mother's income influence your educational attainment? Parents often use their resources to create a home environment that facilitates doing well in school (Teachman, 1987). Thus they provide such aids as a quiet place to study and encyclopedias. In addition, they may provide cultural enrichment activities, such as attending concerts and arts events (DiMaggio & Mohr, 1985). A study of the daily activities of children aged 3 to 11 found that children of highly educated parents spent more time reading and studying and less time watching TV (Bianchi & Robinson, 1997).

One review of the research on the intergenerational transmission of poverty concludes that children raised in poor families complete fewer years of school, are less likely to attend college, and more likely to be poor as adults (Corcoran, 1995). Note that many black families, although not wealthy, do give their children the motivation and the skills to succeed in school (McAdoo, 1997).

Family structure also plays a role in the attainment process. A study of a national sample of 30- to 59-year-old men and women compared those raised in an original two-parent family with those raised in other family structures. Those raised in original families earned more as adults (Powell & Parcell, 1997). Among blacks, the presence of two parents, both employed outside the home, is essential to mobility (McAdoo, 1997).

The experiences of Fred and Larry clearly reflect the importance of these processes. Fred's parents were upper middle class; they sent him to preschool at age 4 and encouraged him to learn to read. Larry's parents were working class; they encouraged him to get out and play and not to waste time reading. Fred did well in school; his grades were always high. Larry struggled with his schoolwork, especially math. By eighth grade, Fred had an excellent record and his teachers gave him lots of encouragement; Larry's teachers, in contrast, didn't pay much attention to him. Fred worked hard in high school, got good grades, and, with the support of his teachers and family, he went to the university. After finishing high school, Larry went into the army, where he learned vehicle mechanics. When Fred finished college, he got a job in a medical equipment firm. Ten years after he graduated from high school, Fred was selling $200,000 worth of equipment per year and earning 20% commissions. After he finished his military service, Larry went to work in a gas station. Ten years after Larry graduated from high school, he was earning $20,000 per year working in a gas station.

Studies like the one described earlier indicate there is upward mobility in American society and that socioeconomic background does not fix one's occupational attainment and earnings. Through greater education, many persons achieve an occupational status and income larger than would be expected solely on their background. Thus America is not a caste society. At the same time, socioeconomic background is not irrelevant to one's educational and occupational attainment. Not everyone can be a doctor, lawyer, or engineer. Opportunities for upward mobility are not unlimited.

Gender   Is the process of status attainment different for men and women? According to the data obtained on Wisconsin high school students, the determinants of occupational status as depicted in Figure 18.2 are the same for both men and women, although the size of some relationships varies.

Most striking were the findings about occupational status. Using a prestige scale of 0 to 100, the first jobs held by women were, on the average, 6 points higher than the first jobs held by men. That is, women in general started in higher prestige jobs than men. Women's first jobs were concentrated within a narrow range of prestige, whereas there was much greater variation in the prestige scores of first jobs held by males (Sewell, Hauser, & Wolf, 1980). Table 18.1 reveals how this occurred. The first jobs women held included registered nurse, schoolteacher, social worker, dental assistant, and secretary. The prestige scores of these jobs range from 66 to 46. In contrast, men's first jobs ranged from physician (86) to garbage collector (28).

When the researchers looked at 1975 occupations, they found men had gained an average of 9 points in status (in the 18-year period since graduation from high school). Women, in contrast, had actually lost status; the average prestige of current occupations for women was 2 points lower than the average prestige of their first jobs. Men experience upward mobility because they work continuously. In addition, they are in occupations with possibilities of promotion and advancement. Women's work careers are often interrupted by marriage and by raising children; when they return to work, they often take up the same job. Thus women are often unable to build up enough continuous experience to gain promotions. Moving also interferes with women's advancement, especially when the purpose of the move is to further the man's career (Shihadeh, 1991). Advancement is also more limited in occupations held largely by women. The top positions in schools, social work, airlines, and sales are more often held by men than women. So the occupational status achieved by men and women differs over the course of their careers.

These differences are evident in the lives of Fred, Larry, and Marie, who were introduced at the beginning of this chapter. After college, Fred began in sales (prestige score 49) and his income increased substantially every year. If he wanted, he could move up in the company to regional sales manager, national sales manager, and perhaps vice president of sales. Larry has moved from gas station attendant (prestige score 21) to owner of a service station (prestige score 44). Like Fred, Marie went to college and earned a bachelor's degree. Her first job involved working on a surgical unit in a large hospital (prestige score 66). As head nurse in pediatrics, she works days now, gets weekends off, and earns more, but her occupation is

unchanged. She could move up to director of nursing, but she is unlikely to strive for the position because the added responsibility isn't balanced by added pay.

Occupational Segregation   In the preceding section, we saw that the influence of factors such as socioeconomic background and ability on occupational attainment is similar for men and women. At the same time, working men and women are not proportionately distributed across occupational categories. Look back at Table 18.1. As you look at each occupation in the list, what gender comes to mind? Chances are that when you think of engineers, carpenters, or auto mechanics, you picture males performing those jobs. In 1991, of those employed in these occupations, 92%, 99%, and 99%, respectively, were men (U.S. Bureau of the Census, 1992). Similarly, when you think of registered nurse or dental hygienist, you picture females in these roles; in 1991, of those employed in these occupations, 95% and 99%, respectively, were women. Many occupations consist overwhelmingly of either men or women; there is substantial *occupational-level segregation* by gender (Reskin & Padovic, 1994). Because occupation is the basis of prestige, and a major determinant of income, this segregation has serious consequences.

Experience with occupational segregation begins in adolescence. Data from a sample of 3,101 tenth- and eleventh-grade students in suburban high schools provide concrete evidence (Greenberger & Steinberg, 1983). Adolescents' first jobs are segregated by sex, with girls earning a lower hourly wage. These differences reflect differential opportunity; employers hire primarily girls or primarily boys for a particular job (for example, newspaper carrier, fast-food sales), and they pay boys more. The other contributor to occupational segregation is differences in the skills taught boys and girls.

Adults often experience gender segregation in the workplace. In a survey of 290 organizations, with a total of more than 50,000 employees, the results indicated that men and women rarely perform similar work in a single organization; when they do, they usually have different job titles (Bielby & Baron, 1986). There is little evidence that employers' practices in

Segregation of occupations by gender is widespread in the United States; for example, most water department repairmen are men, whereas most data entry operators are women. This segregation has serious consequences for a woman's earnings and occupational prestige.

this regard are a rational response to differences between men and women. Jobs held by men in one organization are held by women in other organizations. Moreover, differences in work performed or in job titles often result in large differences in pay. In 1995 the median weekly earnings of women employed full-time was $406, whereas the median for men was $538 (U.S. Bureau of the Census, 1996). Thus the median annual earnings of a woman were $21,112, and the earnings of a man were $27,976, a difference of $6,864.

A study of the sources of this gender gap in wages analyzed data from 16 industries and 10 professional and administrative occupations (Petersen & Morgan, 1995). The data indicate that within-job discrimination (different pay for the same work) is not a major contributor to the gap. Much more significant is seg-

regation by organization, wage differences between organizations, and occupational-level segregation.

There is also a black-white gap in earnings. One study estimates that, among workers 25 to 33, it was $1.41 per hour for females, and $2.66 per hour for males in 1985 (Cancio, Evans, & Maume, 1996). A substantial proportion of this gap is due to discrimination, not to differences in measures of workers' skills.

Social Networks

We have seen that socioeconomic background, ability, educational attainment, and earlier jobs influence occupational attainment over the life course. In part, this is because differences in experiences create differences in an individual's aspirations and abilities to cope with the occupational world. Varied experiences also move people into different social networks. This exposes them to varied social contacts, which have an important effect on their upward mobility. This section considers the ways in which position in social networks promotes entry into specific jobs.

Networks provide channels for the flow of information, including information about job opportunities. What types of networks are likely to provide information on finding new jobs? You might think it is networks characterized by strong ties, such as families or peer groups. Surprisingly, employment opportunities are often found through networks characterized by *weak ties*—relationships involving infrequent interaction and little closeness or emotional depth (Marsden & Campbell, 1984). Those to whom our ties are weak are involved in different groups and activities than we are. Consequently, they are exposed to information that is different from the information we already have. For this reason, new information is more likely to come via a weak tie than a strong one.

In one study (Granovetter, 1973), a random sample of persons who had recently changed jobs was asked how they found out about their new job. Those who heard about their job through another person were asked whether they had seen that contact at least twice a week—a strong tie—or less often— a weak tie. Of those who found jobs through contacts, only 17% were obtained through strong ties. The remainder found jobs through people they saw less than twice a week. The contacts were often friends from school, former co-workers, or former employers, people with whom they had little recent interaction. It was often a chance meeting or a reintroduction by a mutual friend that led to the individual learning about the job. Thus people were more likely to hear about jobs from those to whom they were weakly tied.

We noted earlier that women are less likely than men to experience upward occupational mobility during their careers. Might this occur because men and women differ in their access to networks that carry job information? Our ties to networks grow out of the activities we share with others. The organizations we belong to are a major setting for such activities (Feld, 1981). The larger the organization, the larger the potential number of weak rather than strong ties. If men belong to larger organizations than women, they would have more weak ties and, hence, better access to information useful in finding jobs.

To examine this possibility, 1,799 adults were asked the name and size of each organization to which they belonged (Miller-McPherson & Smith-Lovin, 1982). On the average, men belonged to organizations such as business and professional groups and labor unions, whereas women were more likely to belong to smaller charitable, church, neighborhood, and community groups. Moreover, job-related contacts are more likely to develop in business, professional, and union groups. Findings showed that males had an average of 170 job-related potential contacts, whereas females had an average of fewer than 35. Apparently, men are in networks that allow greater access to information about, and opportunities for, advancement.

Although finding a job is influenced by social networks, the status of the job one finds is influenced by the contact's status. You are more likely to get a management job at the phone company, for example, if your friend's father is a vice president than if he is a lineman. In a study of males ages 21 to 64 in the Albany, New York, area, 57% of the men reported using contacts to get their first job. The higher the contact's occupational status, the greater the prestige of the position the job seeker obtained (Lin, Ensel, & Vaughn, 1981).

Social networks also contribute to mobility within one's workplace. A longitudinal study of employees in one high-technology firm found that having a large network of informal ties was associated with promotions and salary increases (Podolny & Baron, 1997).

Individual Values

Last year, Fred, Larry, and Marie were each approached by a labor union organizer. Fred, the sales representative, was approached by a member of Retail Clerks International. The organizer explained that under a union contract, Fred would spend fewer days on the road and would receive a travel allowance from his employer. Larry was approached by a representative of the International Brotherhood of Teamsters. The organizer sympathized with the problems of independent service station owners and urged Larry to let the Teamsters represent his interests in dealing with his supplier. Marie was approached by the president of United Health Care Workers; she was promised higher wages and greater respect from physicians if she would join.

Fred flatly rejected the invitation, believing a union contract would limit his freedom and perhaps reduce his income. Larry's reaction was mixed. On the one hand, he feels he is at the mercy of "big oil." On the other hand, he is also a self-employed businessman; like Fred, he doesn't want to join a labor organization that might limit his ability to determine his prices and the pace at which he works. Marie reacted very favorably to her invitation and began to attend union meetings "to see what they are like." She thought a union might lead to higher pay and might force the hospital to give her more freedom in determining the pace at which she worked.

In making their decisions, Fred, Larry, and Marie used their personal **values,** which are enduring beliefs that certain patterns of behavior or end states are preferable to others (Rokeach, 1973). All three were concerned with protecting or enhancing their freedom and their wealth or income. These values provided criteria for making decisions. Thus each person weighed the potential effect of joining a union on freedom and income. Fred felt the effect on both would be negative. Larry was sure union membership would limit his freedom but uncertain about its effect on his income. Marie perceived a potential gain in both freedom and income, so she decided to explore union membership.

Each of us has his or her own values. Each of us believes that particular goals and modes of behavior are more desirable than others. Although our values are general, they influence many specific attitudes, choices, and behaviors. For example, values are related to our attitudes toward public policy. Thus the importance one places on personal property and on social equality is related to one's attitudes toward paying higher taxes to help the poor (Tetlock, 1986). Those who place greater value on property oppose higher taxes, whereas those who place greater importance on equality favor increasing taxes to help the poor. Those who feel these values are equally important should find it hard to decide.

Values are related to choices. A study of university students assessed their values and asked them to respond to 10 hypothetical scenarios. Each scenario required a choice between two options, each representing a different value. Respondents' choices were consistent with their values (Feather, 1995).

Values are related to behavior. A survey of shoppers at supermarkets and natural food stores assessed several individual values. Shoppers who valued self-fulfillment and self-respect reported shopping more frequently and spending more money at natural food stores than those who valued sense of belonging and security (Homer & Kahle, 1988).

How do value systems arise? They are influenced by our location in the social structure. This section examines two aspects of social position that affect individual values—occupational role and education.

Occupational Role

We spend up to half of our waking hours at work, so it is not surprising that our work influences our values. But occupational experiences vary tremendously. To determine their effect on values, we must identify the basic differences between occupations. Three important characteristics have been suggested (Kohn, 1969). The first is closeness of supervision—the extent to which the worker is under the direct surveillance and control of a supervisor. As a traveling salesman, Fred is rarely under close supervision, whereas Marie's work is supervised by the director of nursing and various physicians. The second occupational characteristic is routinization of work—the extent to which tasks are repetitive and predictable. Much of Larry's work is routine—pumping gas, tuning engines, relining brakes. But Larry's work is not highly predictable. From one day to the next, he never knows what kind of auto breakdown he will encounter or what unusual request some customer may make. The third characteristic is substantive complexity of the work—how complicated the work tasks are. Work with people is usually more complex than work with

Workers on an assembly line often experience alienation. Assembly-line jobs are monotonous, do not allow workers to exercise initiative, and give them no influence over working conditions.

data and work with objects. Marie's occupation as a nurse is especially complex because she must constantly cope with the problems posed by doctors, patients, and families.

All three of these characteristics were measured in several studies of employed men to determine the impacts of occupational role on values and personality (Kohn & Schooler, 1983). Results of these studies show a relationship between particular occupational characteristics and particular values: Men whose jobs were less closely supervised, less routine, and more complex placed especially high value on responsibility, good sense, and curiosity. Men whose work was closely supervised, routine, and not complex were more likely to value conformity. Thus the occupational conditions that encourage self-direction—less supervised, nonroutine, complex tasks—are associated with valuing individual qualities that facilitate adjustment and success in a self-directed environment—responsibility, curiosity, and good sense. Occupational conditions that encourage adherence to a prescribed routine—close supervision, and routine and simple tasks such as bolting bumpers on new cars—are associated with qualities that facilitate success in that environment, such as neatness and obedience. This pattern has emerged in studies of employed men and women (Miller et al., 1979), and in studies conducted in several countries including the

United States, Japan, and Poland (Slomczynski, Miller, & Kohn, 1981).

Early studies of the relationships between workers' values and their occupational conditions reveal that workers exposed to particular conditions hold particular values. However, these studies were unable to determine with certainty whether adjustment to occupational conditions actually *caused* people to value particular qualities. Perhaps men who value curiosity and desire responsibility select occupations that allow them to exercise these traits (Kohn & Schooler, 1973). In attempting to identify the causal order, researchers compared the men's values and occupational conditions in 1974 with their values and occupational conditions 10 years earlier (Kohn & Schooler, 1982). What they found indicated causal impacts in *both* directions between values and occupational conditions. Men who had valued self-direction highly in 1964 were more likely to be in work roles that were more complex, less routine, and less closely supervised 10 years later. Thus values influenced job selection. At the same time, men who were in occupations that allowed or required self-direction in 1964 placed greater value on responsibility, curiosity, and good sense in 1974. Thus their earlier job conditions influenced their later values.

A recent review of research on cross-national psychological differences concludes that opportunities for self-direction in one's work are consistently associated with differences in values (Schooler, 1996).

Education

Are differences in education also related to differences in an individual's values? The research by Kohn and his colleagues described in the preceding section demonstrated that men in jobs which are not closely supervised and are nonroutine and substantively complex value self-direction, whereas men in jobs with the opposite characteristics value conformity. Education is associated with the value one places on these characteristics; the greater one's education, the greater the value placed on self-direction.

Substantively complex occupations involve working independently with people, objects, or data. Such work requires intellectual flexibility, the ability to evaluate information or situations and to solve problems.

These abilities should be related to educational attainment, so education should be related to intellectual flexibility. Analyses of data from a sample of 3,101 men indicate that as education increases, so does intellectual flexibility (Kohn & Schooler, 1973). Thus education influences both the value placed on self-direction and the abilities needed for success in substantively complex occupations.

In fact, it is possible to identify variation among students in self-direction. In one study, researchers assessed the complexity of a student's course work, of his or her most recent term paper or project, and of extracurricular activities (Miller, Kohn, & Schooler, 1986). In a sample of students in seventh grade through fourth year of college, the greater the substantive complexity of the student's work, the greater the value he or she placed on self-direction. Thus the exercise of self-direction in school or at work increases the value one places on self-direction.

Social Influences on Health

Most of us attribute diseases to biological rather than social factors. But the transmission of disease obviously depends on people's interactions, and our physical susceptibility to disease is influenced by our lifestyles. This is true, for instance, with a disease such as AIDS. Similarly, our mental health is influenced by our relationships with relatives, friends, lovers, professors, supervisors, and so on. Thus social position affects both physical and mental health. This section examines the impact of occupation, gender, marital role, and social class on physical health. It also considers the relationship between these factors and mental health.

Physical Health

Occupational Roles What do the physician addicted to Demerol, the executive with an ulcer, the coal miner with black lung disease (chronic obstructive pulmonary disease), and the factory worker with a heart condition all have in common? The answer is a health problem that may be due largely to occupational role.

Occupational roles affect physical health in two ways. First, some occupations directly expose workers to health hazards. Miners who are exposed to coal dust, workers exposed to chemical fumes, and workers who process grain often suffer damage to lung tissue. Waitpersons, bartenders, and kitchen workers exposed to cigarette smoke may develop lung cancer. Workers exposed to various toxic chemicals may die of bladder cancer. Occupational conditions caused 60,300 deaths and 862,200 illnesses in 1992, about 165 deaths and 2,300 plus illnesses per day (Leigh et al., 1997).

Second, many occupational roles expose individuals to stresses that affect physical health indirectly. Each of the roles we play carries a set of obligations or duties. Meeting these demands requires time, energy, and resources. When these demands exceed the person's ability to meet them, the result is *stress* (Dohrenwend, 1961).

Stress is related to the number of physical ailments we experience. In one study, researchers asked adults to report the amount of stress in their lives and their health problems 20 times over a 6-month period. Those who reported higher levels of stress also reported more health problems, including sore throats, headaches, and flu. Increased stress was associated with more ailments on the same day as well as on subsequent days (DeLongis, Folkman, & Lazarus, 1988). The connection between stress and physical health may be via the body's immune system. A study of college students found that as their reports of stress increased, the concentration of antibodies in their saliva decreased (Jemmott & Magliore, 1988). The lower the level of antibodies, the more susceptible one is to illness.

Stress is often temporary. A move from one apartment to another, for example, will be stressful during the weeks all the arrangements are being made and during the move itself. As one becomes settled, however, the demands will decline. Also, the ability to respond to the demands of moving may increase as we learn how to cope with packing, disconnecting the old phone and getting a new one, and so on.

But stress may be continuous. Chronic stress may lead to physical illness. Excellent evidence of this link comes from a longitudinal study of two samples of adults (140 and 190 persons, respectively) employed at a large company (Maddi, Bartone, & Puccetti, 1987). Each person's level of stress was assessed by a carefully designed measure of stressful life events. One or 2 years later, each person completed a questionnaire regarding illness that included both mild (for example, influenza) and serious (for example, heart attack) conditions. There were strong associations between level of stress experienced initially and reported illness 1 or 2 years later.

Many people spend energy, time, and money jogging, playing tennis, or working out. Does it do any good? Evidence indicates that people who are physically fit are less likely to experience stress-related illness. One study of students obtained self-reports of time spent per week in each of 14 fitness activities. Researchers also assessed fitness directly, measuring blood pressure, aerobic capacity, and endurance. Higher levels of self-reported fitness were associated with higher levels of health. Greater fitness as measured directly was associated with fewer visits to the student health center (Brown, 1991b). However, although stress is related to health, it is not related to self-reported fitness (Roth et al., 1989). Fitness does not reduce the amount of stress one experiences but does reduce illness.

The most widely studied relationship between job characteristics and physical health is the impact of occupational stress on coronary heart disease. As workload increases—including perceived demands on one's time, number of hours worked, and feelings of responsibility—so does the incidence of coronary heart disease (House, 1974). Heart attacks are associated with a high level of serum cholesterol in the blood. Several studies report that the level of serum cholesterol rises among persons under high work-related stress (Sales, 1969). This suggests another tangible link between role demands and physical health.

An important aspect of jobs associated with an increased risk of heart attack is lack of control over work pace and task demands (Karasek et al., 1988). Occupations associated with the highest risk include cooks, waitpersons, assembly-line operators, and gas station attendants. These jobs are characterized by high demand—heavy workload and rapid pace—over which the worker has little or no control. Cashiers and waiters are four to five times more likely to

have a heart attack than foresters or civil engineers. A recent study assessed the contribution of lack of control on the job to coronary heart disease, controlling for other factors, including individual risk factors and the availability of social support (Marmot et al., 1997). Longitudinal data were obtained from 7,372 employed men and women at three points in time. Reported lack of control at time 1 was associated with increased incidence of self-reported chest pain and angina and doctor-diagnosed angina or heart attack at times 2 and 3.

Gender Roles   Who is more likely to experience coronary heart disease, cirrhosis of the liver, or lung cancer—men or women? You probably picked men, and if you did, you are right. Men are two to six times more likely than women to die from these conditions. Although there is evidence that genetic and hormonal factors play a role, traditional role expectations for males and females and occupational role segregation in our society are also significant factors.

Some gender differences in health are associated with reproductive roles (Macintyre, Hunt, & Sweeting, 1996). Health problems related to reproduction, such as premenstrual syndrome and pregnancy-related conditions, are most likely among women of childbearing age. Similarly, hormonal changes at menopause affect the physical health of some women, causing osteoporosis and an increased likelihood of broken bones. Older males are likely to experience enlarged prostate glands, and some die of prostate cancer. There are no significant gender differences in reports of a variety of other illnesses, including diabetes, epilepsy, cancer, and high blood pressure.

Role overload, in which the demands of one's role(s) exceed the amount of time, energy, and other resources the person has, is associated with coronary heart disease. Professionals such as physicians, lawyers, accountants, and so on, are especially vulnerable to overload; until recently, the persons holding these positions have been primarily male. Other studies have shown that heart attacks are correlated with certain personality traits known as Coronary Prone Behavior Patterns (Jenkins, Rosenman, & Zyzanski, 1974). People who exhibit these behavior patterns are work oriented, aggressive, competitive, and impatient. They often initiate two or more tasks simultaneously (Kurmeyer & Biggers, 1988). Males are much more likely to be characterized by this behavior pattern than females.

Men are more prone than women to have cirrhosis of the liver because they are four times more likely than women to be heavy drinkers (Calahan, 1970). Until recently, men were much more likely to smoke cigarettes, and therefore more likely than women to contract lung cancer and emphysema. They are also more likely to die in auto accidents, both because of higher rates of driving under the influence of alcohol and because of poor driving habits (Waldron, 1976).

Marital Roles Marriage is associated with physical health. Married men and women experience fewer acute and chronic health conditions (Ross, Mirowsky, & Goldsteen, 1990). Divorced and separated persons have the highest rates of illness and disability, and widowed and single persons, intermediate rates (Verbrugge, 1979). Rates of hospital residence are highest for singles and lowest for married persons. A study of 20,000 deaths revealed that singles had the highest death rate in any age group. Following singles were single "heads of households" (unmarried people with others dependent on them), followed by married persons without children, and married persons with children (Waldron, 1976).

Why is it that being married and having children protects people against illness and accidents? The most likely explanation is that married persons are less likely to engage in behaviors that expose them to illness and accidents. They probably eat and sleep better than unmarried persons. They are perhaps less likely to smoke and drink. They may take fewer risks, reducing the likelihood they will be involved in accidents. Finally, they may be more likely to seek medical care when ill (Verbrugge, 1979).

This explanation suggests that the health advantage of married people is the result of living with another person. To test this interpretation, data were analyzed from a national sample of women. Measures of illness included the number of days spent in bed and the number of doctor visits in the past year. Women who lived with another adult reported

no more illness than married women, regardless of whether they were single, separated, divorced, or widowed (Anson, 1989). When another adult is present, she or he can provide emotional support, help identify illness early, and provide care that encourages rapid recovery.

Is merely being married sufficient to reduce risk? Perhaps. But being happily married is even more beneficial. According to one study, married men and women who were satisfied with their marital roles reported better physical health than those who were dissatisfied (Wickrama et al., 1995). In addition, traditional gender-role definitions influenced perceptions of health. Men who reported greater involvement in their work role and women who reported greater involvement in their maternal role reported better physical health.

Social Class We noted earlier that status has a major impact on lifestyle. One aspect of this impact is the effect of class on physical health.

A model of the influences on health is presented in Figure 18.3 (Williams, 1990). In this model, socioeconomic status is one of three influences on health. Whether education, occupation, or income is used as the indicator of status, lower socioeconomic-status groups in the United States experience higher death rates. A study of 18,733 deaths in 1986 found that social class was strongly correlated with mortality (Rogers, 1995). The highest mortality was found among those who were single and poor. Controlling for class, there were few differences by race. Rates of chronic illness are higher among lower status persons, as are rates of various disabilities. Rates of infant mortality are also negatively related to social class; the rate of mortality among black infants is twice as high as the rate for white infants.

Several factors have been identified as causes of the negative relationship between class and health. First, those with higher status are more likely to be employed full-time, to have subjectively rewarding jobs, and to have higher income. Second, higher status persons are more likely to have a sense of control on the job, and of control over their lives and health. Finally, higher status persons are less likely to engage in health risk behaviors—smoking and heavy

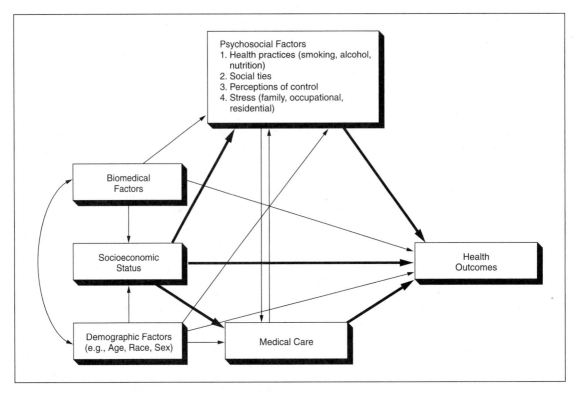

Figure 18.3 A Model of the Influences on Health

There are several influences on a person's physical and mental health. Biomedical, socioeconomic (such as occupation), and demographic (including age, race, gender, and marital status) factors all influence health, both directly and indirectly. All influence such immediate psychosocial factors as health practices and stress. Socioeconomic status is the major influence on the amount and quality of medical care available to the person: The availability of medical care, in conjunction with the other factors, influences how physically and mentally healthy one is.

SOURCE: Adapted from Williams, 1990.

drinking—and more likely to eat properly, exercise, and utilize health care services. In other words, social class is associated with several of the factors discussed earlier. Analyses of data from two national probability samples found that full-time employment, sense of control on the job, and lifestyle were all related to self-reported health (Ross & Wu, 1995). A study of mortality differences between men and women found that differences in death rates by class (income, occupation) were as large for women as for men (Koskinen & Martelin, 1994). The exception was among married women, whose death rates varied little by class;

that is, being married compensates for the health risks experienced by working-class single women.

Class differences in physical health vary across the life course (House et al., 1994). Differences in health associated with class are small in early adulthood (25 to 44), increase with age (45 to 54) until late in life when they become small again (75 and older). The large differences in later adulthood reflect greater exposure of lower-class persons to risk factors such as occupational health hazards, lack of control on the job, excessive alcohol consumption, and number of stressful life events.

Family members help us to cope with stressful events, such as the death of a relative or close friend. They are an important source of emotional support and may help by temporarily taking over some of our role responsibilities.

In sum, the relationship between social status and physical health is complex. Occupational roles may expose men and women directly to health risks, causing illness and death, or to the stressful effects of lack of control on the job. Gender differences in reproductive roles lead to differences in health risks. Role overload is associated with some occupations, and differentially affects men and women due to occupational segregation. Being married or living with another adult reduces one's health risk. Social class is also associated with differences in mortality because of its association with occupation and lifestyle.

Mental Health

At the beginning of the chapter, we introduced Larry, who owns a service station, works long hours, and earns about $28,000 per year. He has two children, a large mortgage on his home, and has trouble making ends meet. He comes home from work every day exhausted. He worries about the economy and whether or not there will be another energy crisis leading to inadequate supplies of gasoline or an oil glut leading to gasoline price wars. Either one would ruin his business, because more than one half of his income is from gasoline sales.

Like many Americans, Larry finds his life situation very demanding. His customers expect him to do high-quality repair work at low prices, his wife expects him to support the family and spend time with her, and his children want more toys than he can afford. At times, the demands made on him by others exceed his ability to cope with them, causing psychological *stress*. People who are under stress often become tense, anxious, are troubled by poor appetite, or experience insomnia. A widely used questionnaire designed to measure stress-related symptoms is reproduced in

18.1 How Do You Respond to Stress?

Stress is a discrepancy between the demands on a person and his or her ability to successfully respond to those demands. Individuals under stress experience a variety of physical and psychological symptoms. A widely used measure of these symptoms is reproduced here. Read the instructions and complete the scale.

Here is a list of ways you might have felt or behaved in *the past week*. Read each of the following statements and then circle the appropriate number to the right of the statement to indicate how often you have felt this way *during the past week*.

During the past week:	Rarely or none of the time (less than 1 day)	Some or a little of the time (1–2 days)	Occasionally or moderate amount of the time (3–4 days)	Most or all of the time (5–7 days)
1. I was bothered by things that usually don't bother me	0	1	2	3
2. I did not feel like eating; my appetite was poor	0	1	2	3
3. I felt I could not shake off the blues even with help from my family and friends	0	1	2	3
4. I felt that I was just as good as other people	0	1	2	3
5. I had trouble keeping my mind on what i was doing	0	1	2	3
6. I felt depressed	0	1	2	3

continued on next page

Box 18.1, "How Do You Respond to Stress?" Many of the items on this scale measure behavior, thoughts, and feelings associated with depression.

Stress is a major influence on mental health. Short-term stressors, such as final exam week or an approaching deadline, may produce a temporary increase in stress-related symptoms or depression. As soon as the exams are over or the deadline passes, mood may return to normal. Long-term stressors, such as the continuing economic worries that Larry experiences, may produce impaired psychological functioning. Neuroses, schizophrenia, and affective disorders such as depression are among the mental illnesses associated with severe stress. The experience of stress and impaired psychological functioning varies by occupation, by gender, by marital and work roles, by membership in social networks, and by social class.

Occupational Roles Work-related stress not only affects physical health but also can affect mental health. We noted earlier that occupations involving heavy workloads and those in which workers have little or no control over work pace are stressful.

Beyond the stresses associated with specific occupations, one's economic circumstances are an important source of stress (Voydanoff, 1990). Economic hardship—insufficient income to meet basic needs—is stressful. Interviews with more than 2,000 adults

continued from previous page

During the past week:	Rarely or none of the time (less than 1 day)	Some or a little of the time (1–2 days)	Occasion-ally or moderate amount of the time (3–4 days)	Most or all of the time (5–7 days)
7. I felt that everything I did was an effort	0	1	2	3
8. I felt hopeful about the future	0	1	2	3
9. I thought my life had been a failure	0	1	2	3
10. I felt fearful	0	1	2	3
11. My sleep was restless	0	1	2	3
12. I was happy	0	1	2	3
13. I talked less than usual	0	1	2	3
14. I felt lonely	0	1	2	3
15. People were unfriendly	0	1	2	3
16. I enjoyed life	0	1	2	3
17. I had crying spells	0	1	2	3
18. I felt sad	0	1	2	3
19. I felt that people disliked me	0	1	2	3
20. I could not get "going"	0	1	2	3

These questions measure depression, a common response to stress. This is the CES-D (Center for Epidemiological Studies-Depression) scale.

You can determine your score by adding up the numbers you circled (0, 1, 2, or 3) for all of the items *except* items 4, 8, 12, and 16. Notice that these four items refer to positive feelings, whereas the other items refer to negative ones. To score items 4, 8, 12, and 16, give yourself 0 if you circled 3, 1 if you circled 2, 2 if you circled 1, and 3 if you circled 0. Total scores on the scale may range from 0 to 60. If your score is more than 16 points, you could be diagnosed as depressed.

showed that not having enough money to provide food, clothing, and medical care for self or family was the major variable associated with depression scores (Pearlin & Johnson, 1977).

Economic uncertainty—concern over one's prospects of finding or keeping a job—is also stressful. A study of 7,095 workers found that the level of unemployment in an industry as a whole is associated with the level of distress experienced by employed workers in that industry (Reynolds, 1997). The relationship was stronger for workers in complex, rewarding jobs. Unemployment is especially debilitating (Vinokur, Price, & Caplan, 1996). It is associated with anxiety, depression, and admission to mental hospitals (Voydanoff, 1990). In addition to economic rewards, one's work is a highly salient role identity for many people. The meaning the individual attaches to work influences its importance to psychological well-being (Simon, 1997). The impact of unemployment is greater on men than on women, probably because the work-role identity is more salient for men.

In some instances, the stress associated with unemployment leads to violence. Each year there are incidents in which a fired employee returns to the workplace with guns or rifles, often seeking the person believed responsible for his or her being fired. These incidents can lead to injuries and deaths, and to suicide by the former employee.

Gender Roles   Adult women in the United States have somewhat poorer mental health than men. On measures of distress, women attain significantly higher scores than men. For example, Macintyre, Hunt, and Sweeting (1996) found that women report greater malaise (sleep problems, difficulty concentrating, worry, and fatigue) than men at all ages. Are women under greater stress? Or are they more likely to report symptoms of distress than men? One study assessed frequency of experience of emotions such as anger, sadness, and happiness, and levels of distress in men and women (Mirowsky & Ross, 1995). The results indicated that women expressed emotions more freely than men, but this did not account for the differences in stress scores. Overall, women experienced stress about 30% more often than men.

There also may be a gender difference in how people respond to stress. Men experiencing high stress report higher rates of illness, whereas women experiencing high stress report higher rates of impaired psychological functioning. In a survey of 3,131 adults, stress was measured by the number of life events experienced by the person, or someone close to him or her, in the prior 6 months. Respondents were also asked questions about participation in various behaviors and psychological functioning. Men experiencing stress were more likely to report alcohol or drug use or dependence, and women experiencing stress reported increased anxiety and emotional disorders (Aneshensel, Rutter, & Lachenbruch, 1991).

Marital Roles   Just as married men and women are physically healthier than people who are not married, they are characterized by greater psychological well-being and less depression than single, separated, divorced, or widowed persons (Ross, Mirowsky, & Goldsteen, 1990). Again, it appears it is the presence of another adult in the residence, rather than being married per se, that is associated with being healthy. A study of a representative national sample of 6,573 adults compared the mental health of persons who were married, cohabiting, or living alone. Married persons reported greater happiness and less depression than cohabitors, who reported better emotional health than persons living alone (Kurdek, 1991).

Of course, it might be that people who have higher levels of well-being are more likely to marry or cohabit, and persons with lower levels remain single. A longitudinal study of 18- to 24-year-old men and women, some of whom married while others remained single, found that marriage did improve well-being (Horowitz, White, & Howell-White, 1996).

A study of a representative sample of more than 13,000 adults assessed the relationship between roles and mental health in four ethnic groups (Jackson, 1997). Occupying the spousal role was associated with greater well-being among blacks, Mexican Americans, Puerto Ricans, and non-Hispanic whites. Occupying other family roles, especially sibling, was related to better mental health in all groups except Puerto Ricans.

The greater well-being of married persons reflects the beneficial effects of social support. A spouse can provide social support—care, advice, and aid in times of stress—and emotional support. Do husbands and wives share equally in receiving these benefits? Apparently not. Married men are characterized by better mental health than married women (Kurdek, 1991). Does this gender difference reflect differences in the stresses experienced by single and married women and men? Using data from a large representative sample, Wu and DeMaris (1996) constructed measures of six different types of strain: work, parental, housework, economic, caregiving, and strain due to physical illness. Married men had the lowest depression scores, followed by married women and then single men; single women had the highest scores. These differences in depression were associated with differences in the nature and kinds of strains reported by the participants. Family-based strains (hours of housework, housework strain, and strain due to illness) and economic strain were associated with the greater depression of women. Men are less likely to experience these strains. Married women are likely to be or feel responsible for housework and taking care of someone who is ill. Married women may experience economic strain because they depend on a man's income, or because they work but earn less (see the earlier discussion of the gender gap in wages).

Some researchers focus on the interrelations of work and family roles. Of particular interest is *work-family conflict,* the extent to which the demands associated with one role are incompatible with the other. One common circumstance is **spillover,** in which the stress experienced at work or in the family is carried into the other domain (Bolger et al., 1989). A study of air traffic controllers and their wives documented the impact of work stress on marital interaction (Repetti, 1989). As the controllers' daily workload increased (larger number of planes handled, poorer visibility), wives reported that the men were more withdrawn at home. A longitudinal study of 166 married couples obtained completed diaries from both partners for either 28 or 42 consecutive days. Each person reported on stressors at work (too much to do, arguments with co-workers) and at home (too much to do, arguments with spouse, arguments with child). For both husbands and wives, increased stress at work was associated with increased stress at home (Bolger et al., 1989).

Stress associated with marital roles can influence work-role performance. The research by Bolger and colleagues (1989) also found that, for husbands, increased stress at home was associated with increased stress at work. Forthofer and colleagues (1996) analyzed data from a study of 8,098 persons ages 15 to 54; they focused on 1,431 employed married men and 1,138 employed married women. The results indicated that problems within the marriage were related to work loss among both men and women.

If marital strains can lead to reduced performance at work, can positive experiences at home enhance work experience? Two studies suggest the answer is "Yes." Barnett (1994) studied 300 full-time employed women in dual-earner couples. Positive experiences in the roles of partner or parent buffered the effects of negative job experiences on distress. A survey of 221 managers, both male and female, found that positive involvement in family or community had a positive effect on work experience (Kirchmeyer, 1993).

Do married employed women and married employed men respond differently to work-family conflict? Apparently they do not. A study of 200 dual-earner couples found that high levels of parental stress and of occupational stress had the same effect on reports of depression symptoms by husbands and wives (Windle & Dumenci, 1997).

Work-family conflict can affect the quality of the marital relationship by influencing the couple's interaction. The study of air traffic controllers discussed earlier found that as stress experienced at work increased, husbands became more withdrawn at home. Other research demonstrates the impact of conflict and the resulting distress on two dimensions of marital interaction, hostility and warmth. As distress increases, both the person and the spouse report greater hostility and less warmth (Matthews, Conger, & Wickrama, 1996).

How do people cope with work-family conflict? They use one or more of several strategies. In one study, wives reported the use of planning and cognitive restructuring—changing their definition of the situation (e.g., deciding the house does not need to be cleaned every week). Husbands reported restructuring and withdrawing from interaction (Padden & Buehler, 1995). The managers (both men and women) in the study discussed earlier reported the use of prioritizing, reducing their personal standards (restructuring), asking others for help, and ending involvement in one or more roles (e.g., in community organizations) (Kirchmeyer, 1993).

Thus the relationships among occupational, gender, and marital roles and psychological well-being are complex (Ross, Mirowsky, & Goldsteen, 1990). The demands of work roles may lead to distress, especially when work demands are high and not under the person's control. Economic hardship and unemployment cause distress. Men typically have somewhat better mental health than women, in part because men react to stress by drug and alcohol use, whereas women respond psychologically. Married men report greater well-being than married women, apparently because wives are more likely to experience family-related strains. Stresses experienced at work can spill over and affect marital relationships; conversely, strain at home can produce losses at work. In any case, the social roles we occupy are major influences on mental health.

Social Networks To this point, we have reviewed some evidence showing that our relationships with others, that is, our membership in social networks, can be major sources of stress. At the same time, social networks can serve as an important resource in coping with stress (Wellman & Worley, 1990). Networks provide us with continuing social support and with help during stressful events. Also, network members may teach us strategies for coping with stressors.

First, a network of close friends and kin eases the impact of stressful events by providing various types of support (Cooke et al., 1988; House, 1981). One type is emotional support, letting us know that they care for and are concerned about us. Emotional support is an important buffer for negative psychological states like depression (Harlow & Cantor, 1995). A second type is esteem support, providing us with positive feedback about our abilities and worth as a person. A poor grade, for example, is less stressful if our friends let us know they think we are a good student. Informational support from others prepares us to avoid problems or to handle them when they arise. Advice from friends on how to handle job interviews, for example, improves our ability to cope with this situation. Finally, network members provide each other with instrumental support—money, labor, and time. Research shows that people who report poor well-being tend to seek out others who can provide the type of support they need (Harlow & Cantor, 1995).

The presence or absence of support is a major determinant of the impact of a stressful life event. A study of 882 women seeking an abortion obtained longitudinal data from 615 of them. Before the abortion, each woman rated the degree to which she received positive (expressed concern, offered help), and negative (argued, criticized) support from her partner, mother, and friends. Perceptions of positive support from each source were associated with greater well-being following the abortion (Major et al., 1997).

Research has documented the impact of supportive relationships on the individual's ability to cope with stress. A longitudinal study of a representative sample of 900 adults focused on the relationship between social network membership and physical health (Seeman, Seeman, & Sayles, 1985). Persons who reported in the initial interview that they had instru-

mental support available were in better physical health 1 year later. (These were persons who, when ill, had others who would call, express concern, and offer help.) Another longitudinal study assessed the impact of family support on mental health (Aldwin & Revenson, 1987). The sample consisted of 245 men and 248 women from randomly selected families in an urban area. The availability of support from one's family at the time of the initial survey was associated with better psychological adjustment 1 year later. Other research indicates that individuals with family support are more likely to cope with stressful events by using active strategies rather than avoidance or withdrawal strategies (Holohan & Moos, 1990). Finally, a longitudinal study of the relation between coping strategies and mental health found that people who used active strategies at the time of the initial survey reported fewer psychological symptoms on the second survey (Aldwin & Revenson, 1987).

A second way social networks reduce stress is by teaching us strategies for coping with stressful events or crises when they occur. When members of a group are all subjected to similar stressors, the group may develop coping strategies. A study of interns and residents in a hospital found they were subjected to long hours of demanding work in often poor facilities (Mizrahi, 1984). These physicians coped with stress by minimizing time spent with each patient, limiting interaction with patients to "relevant topics," and by treating patients as nonpersons, for example, by focusing exclusively on the illness. These strategies were passed on from experienced group members to new ones.

Several types of relationships can provide support, including primary kin (parents, siblings, adult children), secondary kin, and friends and neighbors. The kind of support provided depends on the type of relationship. Persons to whom we have strong ties provide emotional support and companionship. Primary kin provide us with financial aid and services, and friends and neighbors give us services and emotional support (Wellman & Worley, 1990). Research suggests that blacks receive less overall social support than whites and rely more on secondary kin (uncles, aunts, cousins) (Magdol, 1989).

Social Class   The lower a person's socioeconomic status, the greater the amount of stress reported (Mirowsky & Ross, 1986). Education, occupation, and income are the principal measures of socioeconomic status. Does low standing on each of these components contribute to stress independently? Or is stress related to only one or two of these components?

An analysis of data from surveys of eight quite diverse samples (Kessler, 1982) shows a consistent pattern: Low education, low occupational attainment, and low income contribute separately to stress. The relative importance of these three components as sources of stress is different for men and women. For men, income appears most important; for women (employed or not), education appears to be the most important component. Occupational attainment is the least important determinant of stress for both sexes.

How does education affect stress? Earlier in this chapter we summarized research showing that people with more financial resources, more control over their work, and partners who provide support are in better physical and mental health. Research shows that people who are well educated have lower levels of distress, primarily because of paid work and financial resources (Ross & Van Willigen, 1997).

In the United States, a large percentage of the lower class are black. As a result, we might expect blacks to have poorer mental health than whites. However, the results of research comparing the psychological functioning of blacks and whites are inconsistent: Some studies find higher average symptom scores among blacks, and others do not (Vega & Rumbaut, 1991).

Further analyses have sought to identify the causes of the negative relationship between social status and stress. Are lower-class persons exposed to greater stress, or are they simply less able to cope effectively with stressful events? The answer is both (Kessler & Cleary, 1980). On the one hand, lower-class persons are more likely to experience economic hardship—not having enough money to provide adequate food, clothing, and medical care (Pearlin & Radabaugh, 1976). They also experience higher rates of a variety of physical illnesses (Syme & Berkman,

1976). Both economic hardship and illness increase the stress an individual experiences. Furthermore, persons who are low in income, education, and occupational attainment lack the resources that would enable them to cope with these stresses effectively. Low income reduces their ability to cope with illness. In addition, low-status persons are less likely to have a sense of control over their environment and less access to political power or influence. For this reason, they are less likely to attempt to change stressful conditions or events.

If stress increases as socioeconomic status decreases, we would expect persons lower in status to have poorer mental health. Research over the past 30 years has consistently confirmed this expectation; there is a strong correlation between social class and serious mental disorders (Eaton, 1980). This correlation has been found in studies conducted in numerous countries (Dohrenwend & Dohrenwend, 1974). In general, persons in the lowest socioeconomic class have the highest rates of mental illness.

The differences by social class in rates of mental disorders are due in part to differences in stress. Persons in low-status occupations are more likely to experience lack of control over work. They may also experience economic uncertainty because of risk of layoff or seasonal variations in employment opportunities. This stress is likely to spill over into family interaction patterns, causing familial relations to become an added source of stress rather than a buffer. Unemployment is especially stressful, and rates of unemployment are highest in the lowest class (U.S. Bureau of the Census, 1996, table 649). Finally, the members of one's social networks have fewer economic and emotional resources.

Alienation

Jim dragged himself out of bed and headed for the shower. As the water poured over him he thought, "Thursday . . . another 10-hour shift . . . if the line doesn't shut down. I'll bolt 500 bumpers . . . sick of car frames . . . I'd rather do almost anything else. If only I'd finished high school . . . Damn the money!

Let 'em take the job and shove it, but what else pays a guy who quit school $17.36 an hour?"

Jim is experiencing **alienation,** the sense that one is uninvolved in the social world or lacks control over it. Several types of alienation have been identified (Seeman, 1975). We discuss two here: self-estrangement and powerlessness.

Self-Estrangement

Jim's hatred for his job reflects **self-estrangement,** the awareness he is engaging in activities that are not rewarding in themselves. Work is an important part of our waking hours. When work is meaningless, the individual perceives the self as devoting time and energy to something unrewarding—that is, something "alien." Although social background and individual characteristics have some influence, alienation from work is primarily determined by the occupational and organizational conditions of work (Mortimer & Lorence, 1995).

What makes a job intrinsically rewarding? Perhaps the most important feature is autonomy. Work that requires the individual to use judgment, exercise initiative, and surmount obstacles contributes to self-respect and a sense of mastery. A second feature is variety in the tasks that the person performs. Jim has no autonomy; his job does not allow him to exercise judgment or initiative. It also has no variety; it is monotonous and boring.

Four features of industrial technology produce self-estrangement. First, self-estrangement is higher if the worker has no connection with the finished product itself. Second, it is higher if the worker has no control over company policies. Third, it is higher if the worker has little influence over the conditions of employment—over which days, which hours, or how long he or she works. Finally, it is higher if the worker has no control over the work process—for example, the speed with which he or she must perform tasks (Blauner, 1964). Notice that alienation, like stress, is caused by lack of control over the conditions of work.

These features are especially characteristic of assembly-line work, in which each person performs the same highly specialized task many times per day. Thus

workers on assembly lines should be more likely to experience self-estrangement than other workers. A study testing this hypothesis (Blauner, 1964) compared assembly-line workers in textile and automobile plants with skilled printers and chemical industry technicians. As expected, assembly-line workers were more alienated than skilled workers, who had jobs that were more varied and involved the exercise of judgment and initiative.

Work in bureaucratic organizations—like large insurance companies or government agencies—may also produce self-estrangement. In many bureaucratic organizations, workers have little or no control over the work process and do not participate in organizational decision making. Thus workers at the lowest levels of such organizations should experience self-estrangement or dissatisfaction with their work. Conversely, workers who are involved in decision making should be less alienated. A survey of 8,000 employees in 100 companies located in the United States or Japan found that workers involved in participatory decision-making structures had higher commitment to their work (Lincoln & Kalleberg, 1985). Such workers were willing to work harder and were proud to be employed by and wanted to remain with the company.

More generally, the extent to which workers are alienated depends on the system of production within which they work. Hodson (1996) identifies five systems: craft, where each worker produces a product; direct supervision; assembly line; bureaucratic; and participatory. A review of studies of all five types of workplaces reveals a U-shaped relationship between workers' attitudes and the system of production (see Figure 18.4). Both craft and participatory systems are associated with high job satisfaction and pride in one's work; direct supervision is the most alienating system. Results such as these have led many large firms to introduce participatory systems, including General Motors.

As noted earlier, individual characteristics do influence reactions to work (Mortimer & Lorence, 1995). Surveys indicate that job satisfaction and involvement are most stable among workers ages 30 to 45. Women are as committed to work as men, although they place greater emphasis on the quality of

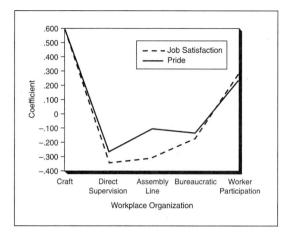

Figure 18.4   Systems of Production and Alienation From Work

The way in which work is organized, or the system of production, is a major influence on people's attitudes toward their work. Historically, five different systems of production have been used: craft, where each worker has considerable autonomy; direct supervision, where another person monitors one's work; assembly line, where the work activity is determined by the organization and speed of the line; bureaucratic, where many aspects of work are governed by impersonal rules; and participatory organization, where teams of managers and workers make decisions.

A review of the research literature suggests that the pride workers have in their work and their job satisfaction vary depending on the system in which they work. The graph displays these variations, indicating that although participative systems are less alienating than the assembly lines and bureaucratic forms they replaced, they are not as satisfying as the craft system of production.

SOURCE: Hodson, 1996.

interpersonal relations in the workplace. Among those holding comparable jobs, blacks are as committed as whites to their jobs and employers.

According to the theory developed years ago by Karl Marx (Bottomore, 1964), whether a person experiences self-estrangement is determined by his or her relation to the means of production. The most alienated employees are hypothesized to be those who have no autonomy, who do not have the freedom to solve nonroutine problems, and who have no subordinates. Marx referred to such workers as the *proletar-*

iat. In contemporary society, assembly-line workers, salesclerks, file clerks, and laborers are all in occupations that have these characteristics. A recent survey of 1,499 working adults found that 46% were in jobs of this type (Wright et al., 1982). Several studies have found that men whose jobs are characterized by lack of autonomy and complexity typically have high scores on measures of self-estrangement (Kohn, 1976) and low scores on measures of job involvement (Lorence & Mortimer, 1985).

There is some evidence that characteristics of the work environment influence psychological well-being. One researcher assessed the common environment of office workers by averaging the ratings of all of those employed in each of 37 branch offices; each worker's own ratings were used as a measure of his or her immediate environment (Repetti, 1987). Workers who rated his or her branch more positively (on interpersonal climate, and support and respect from co-workers) reported lower levels of anxiety and depression. Aggregate ratings by the workers of the environment in the branch were also related to anxiety and depression scores.

Powerlessness

Consider the fact that vandalism is widespread in certain sections of large cities, that many middle- and upper-class adults do not vote in presidential elections, and that some people on welfare make no effort to find a job. These facts all have something in common. They reflect, at least in part, people's sense of **powerlessness,** the sense of having little or no control over events.

Powerlessness is a generalized orientation toward the social world. People who feel powerless believe they have no influence on political affairs and world events; this is different from feeling a lack of control over events in day-to-day life. A typical measure of powerlessness is reproduced in Table 18.2. Most people's scores on measures of powerlessness are quite stable over a period of many years (Neal & Groat, 1974). There is some evidence that a sense of powerlessness develops during childhood (Seeman, 1975). Interestingly, a sense of powerlessness is not associated

Table 18.2   A Measure of Powerlessness

	Strongly Agree	Agree	Disagree	Strongly Disagree
1. People like me can change the course of world events if we make ourselves heard.	SA	A	D	SD
2. I think each of us can do a great deal to improve world opinion of the United States.	SA	A	D	SD
3. There's very little that persons like myself can do to improve world opinion of the United States.	SA	A	D	SD
4. The average citizen can have an influence on government decisions.	SA	A	D	SD
5. This world is run by the few people in power, and there is not much the little guy can do about it.	SA	A	D	SD
6. It is only wishful thinking to believe that one can really influence what happens in society at large.	SA	A	D	SD
7. A lasting world peace can be achieved by those of us who work toward it.	SA	A	D	SD
8. More and more, I feel helpless in the face of what's happening in the world today.	SA	A	D	SD

Note: Agreement with statements 3, 5, 6, and 8 indicates a sense of powerlessness, as does disagreement with statements 1, 2, 4, and 7. How powerless do you feel?

SOURCE: Adapted from Zeller, Neal, and Groat, 1980.

with social class—that is, income, occupation, or education.

Questions that measure powerlessness, such as "People like me have no say" and "Politicians don't care what I think," were included in several surveys between 1952 and 1980. Analysis of patterns of agreement with these items shows that powerlessness or political alienation declined from 1952 to 1960, rose steadily from 1960 to 1976, and then declined (Rahn & Mason, 1987). The increase in the 1960s and 1970s was associated with increased concern about such political and social issues as civil rights for blacks, the war in Vietnam, and the Watergate political scandal. Thus fluctuations in powerlessness reflect, at least in part, events in the larger society.

Although the sense of powerlessness is found in all classes, upper and lower classes may have different means of expressing it. Middle- and upper-class persons may be more likely to stay home on election

day or to feel apathetic about political affairs or organizations that influence public policy. Lower-class persons may be more likely to have a hostile attitude toward city officials and to vandalize city buses, subway trains, and businesses in their neighborhoods. Thus how an individual expresses frustration over lack of influence on the world may depend on his or her social position.

Summary

This chapter considers the impact of social structure on four areas of a person's life: achievement, values, physical and mental health, and sense of belonging in society. Social structure influences the individual through the expectations associated with one's roles, the social networks to which one belongs, and the status associated with one's positions.

The graffiti gracing buildings in some American cities are responses by youth to alienation. Spray-painted messages reflect the lack of control over their lives that many youth feel.

Status Attainment An individual's status determines access to resources—to money, lifestyle, and influence over others. Three generalizations can be made about status in the United States: (1) An individual's status is closely tied to his or her occupation. (2) Occupational attainment is influenced directly by the individual's educational level and ability and indirectly by socioeconomic background. Among women, occupational status and income is limited by gender segregation. (3) Information about job opportunities is often obtained via social networks, especially those characterized by weak ties.

Individual Values Two aspects of the individual's position in society influence his or her values. (1) Particular values are reliably associated with certain occupational role characteristics. Men and women whose jobs are closely supervised, routine, and not complex value conformity, whereas those whose jobs are less closely supervised, less routine, and more complex value self-direction. (2) A formal education influences values. Higher education is associated with placing greater value on self-direction and with greater intellectual flexibility.

Social Influences on Health Physical health is influenced by occupation, gender, marital roles, and social class. (1) Occupational roles determine the health hazards individuals are exposed to and whether they experience role overload. (2) The traditional role expectations for males and females make males more vulnerable than females to illnesses such as coronary heart disease. (3) Marriage protects both men and women from illness and premature death. (4) Lower status groups experience higher rates of illness, disability and death.

Mental health is also influenced by social factors. (1) Economic hardship, uncertainty, and unemployment are associated with poor mental health. (2) Women have somewhat poorer mental health than men. (3) Although marriage is associated with reduced stress for both men and women, married men have better mental health, perhaps because they experience less family-related strain. Spillover of stress from work can influence family relationships, and vice versa. (4) Social networks are an important resource in coping with stress; they provide the person with emotional, esteem, and informational support, as well as instrumental aid. (5) Lower-class persons report greater stress and experience a higher incidence of mental illness.

Alienation Two types of alienation are self-estrangement and powerlessness. (1) Self-estrangement is associated with work roles that do not allow workers a sense of autonomy, such as assembly-line jobs. (2) Powerlessness, which seems to develop in childhood, is a generalized sense that one has little or no control over the world.

Key Terms

alienation (p. 462)
position (p. 440)
powerlessness (p. 463)
primary relationship (p. 440)
role (p. 440)
role overload (p. 452)
self-estrangement (p. 462)
social network (p. 440)
social structure (p. 440)
spillover (p. 459)
status (p. 440)
upward mobility (p. 442)
values (p. 449)

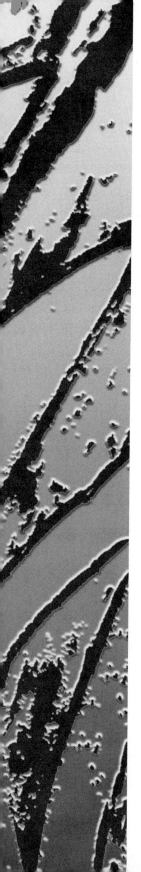

CHAPTER 19

Deviant Behavior and Social Reaction

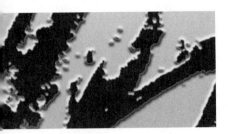

Introduction

Virginia and Susan wandered through the department store, stopping briefly to look at blouses and then going to the jewelry counter. Each looked at several bracelets and necklaces. Susan kept returning to a 18-karat gold bracelet with several jade stones, priced at $299.50. Finally, she picked it up, glanced quickly around her, and dropped the bracelet into her shopping bag.

The only other shopper in the vicinity, a well-dressed man in his 40s, saw Susan take the bracelet. He looked around the store, spotted a security guard, and walked toward him. Virginia stammered to Susan, "I, uh, I don't think we should do this." "Oh, it's okay. Nothing will happen," Susan replied, before walking quickly out of the store. Moments later, Virginia followed her. As Susan entered the mall, the security guard stepped up to her, took her by the elbow, and said, "Come with me, please."

Shoplifting episodes like this one occur dozens of times every day in the United States. Shoplifting is one of many types of **deviant behavior**—behavior which violates the norms that apply in a given situation. In addition to crime, deviance includes cheating, substance use or abuse, fraud, corruption, delinquent behavior, harassment, and behavior considered symptomatic of mental illness.

There are two major reasons social psychologists study deviant behavior, one theoretical and one practical. First, social norms and conformity are the basic means by which the orderly social interaction necessary to maintain society is achieved. By studying nonconformity, we learn about the processes that produce social order. For example, we might conclude that

Susan took the bracelet because there were no store employees nearby, suggesting the importance of surveillance in maintaining order. Second, social psychologists study deviant behavior to better understand its causes. Deviant behaviors such as alcoholism, drug addiction, and crime are perceived as serious threats to society. Once we understand their causes, we may be able to develop better programs that reduce or eliminate deviance or help people change their deviant behavior.

This chapter addresses four fundamental questions:

1. What are the causes of deviant behavior?
2. How important for deviant behavior is the reaction of observers? That is, does someone have to react to behavior in particular ways for it to be considered deviant?
3. Why do some people engage in deviance regularly? Why do they adopt a lifestyle that involves participation in deviant activities?
4. What determines how authorities and agents of social control deal with incidents of deviance? Is their reaction influenced by the deviant's gender, social status, or other characteristics of the situation?

The Violation of Norms

When we read or hear that someone is accused of murder or embezzling money from a bank, we often ask "Why?" In Susan's case, we would ask, "Why did she take that bracelet?" In this section, we consider first the meaning of norms and then look at several theories about the causes of deviant behavior. These include anomie, control, differential association, and routine activities theories.

Norms

Most people would regard Susan's behavior in the department store as deviant because it violated social norms. Specifically, she violated laws that define taking merchandise from stores without paying for it as a criminal act. Thus deviance is a social construction; whether a behavior is deviant or not depends on the norms or expectations for behavior in the situation in which it occurs.

In any situation, our behavior is governed by norms derived from several sources (Suttles, 1968). First are purely "local" and group norms. Thus roommates and families develop norms about what personal topics can and cannot be discussed. Second are subcultural norms that apply to large numbers of persons who share some characteristic. For example, racial or ethnic group norms govern the behavior of blacks or persons of Polish descent that do not apply to other Americans. A subculture that is particularly relevant to the discussion of deviance is the *subculture of violence,* discussed later in the chapter. Third are societal norms, such as those requiring certain types of dress or those limiting sexual activity to certain relationships and situations. Thus the norms that govern our daily behavior have a variety of origins, including family, friends, socioeconomic or religious or ethnic subcultures, and the society in general.

The repercussions of deviant behavior depend on which type of norm an individual violates. Violations of local norms may be of concern only to a certain group. Failing to do the dishes when it is your turn may result in a scolding from your roommate, although your friends may not care about that deviance. Subcultural norms are often held in common by most of those with whom we interact, whether friends, family members, or co-workers. Violations of these norms may affect most of our day-to-day interactions. Violations of societal norms may subject a person to action by formal agencies of control such as the police or the courts. Earlier in this book, we discussed the violation of local norms (Chapter 9) and group norms (Chapter 13). In this chapter we focus on the violation of societal norms and reactions to norm violations.

Anomie Theory

The **anomie theory** of deviance (Merton, 1957) suggests that deviance arises when people striving to achieve culturally valued goals such as wealth find

they do not have any legitimate way to attain these goals. These people then break the rules, often in an attempt to attain these goals illegitimately.

Anomie Every society provides its members with goals to aspire to. If the members of a society value religion, they are likely to socialize their youth and adults to aspire to salvation. If the members value power, they teach people to seek positions in which they can dominate others. U.S. culture extols wealth as the appropriate goal for most members of society. In every society are also norms that define acceptable ways of striving for goals, called **legitimate means.** In the United States, legitimate means for attaining wealth include education, working hard at a job to earn money, starting a business, and making wise investments.

A person socialized into U.S. society most likely desires material wealth and strives to succeed in a desirable occupation—to become a teacher, nurse, business executive, doctor, or the like. The legitimate means of attaining these goals are to obtain a formal education and to climb the ladder of occupational prestige. The person who has access to these means—who can afford to go to college and has the accepted skin color, ethnic background, and gender—can attain these socially desirable goals.

What about those who do not have access to the legitimate means? As Americans, these people desire material wealth like everyone else, but they are blocked in their strivings. Because of the way society is structured, certain members are denied access to legitimate means. Government decisions regarding budgeting and building or closing schools determine the availability of education to individuals. Similarly, certain members of society are denied access to jobs. Not only individual characteristics, such as lack of education, but also social factors, such as the profitability of making steel in Ohio, determine who is unemployed.

A person who strives to attain a legitimate goal but is denied access to legitimate means will experience *anomie*—a state that reduces commitment to norms and the pursuit of goals. There are four ways a person may respond to anomie; each is a distinct type

Most Americans are socialized to strive for economic success. But some people do not have access to legitimate employment, so they seek wealth by alternative, sometimes illegal means, such as prostitution.

of deviance. First, an individual may reject the goals, give up trying to achieve success, but continue to conform to social norms. This adaptation is termed *ritualism.* The poorly paid stock clerk who never misses a day of work in 45 years is a ritualist. He is deviant because he has given up the struggle for success. Second, the individual might reject both the goals and the means, withdrawing from active participation in society by *retreatism.* This may take the form of drinking, drug use, withdrawal into mental illness, or other kinds of escape. Third, one might remain committed to the goals but turn to disapproved or illegal ways of achieving success. This adaptation is termed *innovation.* Earning a living as a burglar, prostitute, or loan shark is an innovative means of attaining wealth. Finally, one might attempt to overthrow the existing system and create different goals and means through *rebellion.*

Shoplifting is a form of innovation. Like other types of economic crime, it represents a rejection of the normatively prescribed means (buying what you want) while continuing to strive for the goal (possessing merchandise). According to anomie theory, Susan, the shoplifter, has been socialized to desire wealth but does not have access to a well-paying job because of her poor education. As a result, she steals what she wants because she does not have the money to pay for it.

Another influence on an individual's adaptation is access to deviant roles. Utilizing a means of goal achievement—whether legitimate or illegitimate—requires access to two structures (Cloward, 1959). The first is a **learning structure**—an environment in which an individual can learn the information and skills required. A shoplifter needs to learn how to conceal objects quickly, to spot plainclothes detectives, and so forth. The second is an **opportunity structure**—an environment in which an individual has opportunities to play a role, which usually requires the assistance of those in complementary roles. Anomie theory assumes anyone can be an innovator—through shoplifting, prostitution, or professional theft. But not everyone has access to the special knowledge and skills needed to succeed as a prostitute (Heyl, 1977) or a black market banker (Wiegand, 1994). Just as access to legitimate means to achieve goals is limited, so is access to illegitimate means. Only those who have both the learning and opportunity structures necessary to become a shoplifter, prostitute, or embezzler can utilize these alternative routes to success (Coleman, 1987).

The opportunities for deviance available to a person depend on age, sex, kinship, ethnicity, and social class (Cloward, 1959). These characteristics, with the possible exception of class, are beyond the individual's control. Thus prostitution in our society primarily involves young, physically attractive persons. People who do not have access to the learning or opportunity structures necessary for deviance are double failures. They can succeed neither through legitimate nor illegitimate means. Double failure often produces retreatism. Drug addicts, alcoholics, and mentally ill persons may be losers in both the conventional and criminal worlds.

Anomie and Social Class Anomie theory emphasizes access to education and employment. Those who have access to both should not engage in deviant behavior. Those who do not have access to one or both should experience anomie and are likely to engage in deviance. A survey of 1,614 youths ages 15 to 18 measured commitment to success goals ("making a lot of money") and perceived access to college education (Farnworth & Lieber, 1989). Those who said they wanted to make a lot of money but did not expect to complete college were much more likely to report delinquent behavior.

One measure of access to legitimate means is the unemployment rate. According to the theory, as unemployment increases, rates of deviance also should increase. One study analyzed the relationship between unemployment rates and crime rates in the United States for each year of the 1948 to 1985 period (Devine, Sheley, & Smith, 1988). There was a strong relationship; as unemployment increases, so does crime. The relationship is stronger for economic crime (burglary) than violent crime (murder). Another study looked at the relationship between job availability (unemployment) and job quality (pay per hour) and arrest rates by state, for 1977 through 1980 (Allan & Steffensmeier, 1989). The results indicate that unemployment was strongly associated with juvenile arrest rates. Job quality was highly correlated with adult arrest rates.

Two studies suggest it is relative socioeconomic standing that determines whether one experiences anomie. A study of arrest rates for burglary and robbery from 1957 to 1990 found that as income inequality among blacks increased, so did black arrest rates (LaFree & Drass, 1996). Similarly an analysis of the number of Latinos murdered in 1980 found that the amount of income inequality among Latinos was an important factor (Martinez, 1996). Thus it is one's economic standing relative to similar others that matters, not standing in the society as a whole.

Anomie theory directs our attention to the importance of social class. Because lower-class persons are more frequently excluded from quality education and jobs, the theory predicts they will commit more crimes. Data collected by police departments and the FBI generally confirm this prediction, showing a dis-

proportionate number of those arrested for crimes are poor minority males. This has led some to conclude that crime and social class are inversely related—that the highest crime rates are found in the lower social strata (Cloward, 1959).

However, a class bias is built into official statistics on crime. Not all illegitimate economic activities are included in these statistics. Data on burglary, robbery, and larceny are compiled by police departments, but data on income tax evasion, price-fixing, and stock swindles are not. Police and FBI statistics are much more likely to include "street" crimes than the kinds of economic crimes committed by wealthy corporate executives and stockbrokers. The latter are called *white-collar crime,* activities that violate norms of trust, usually for personal gain (Shapiro, 1990). To embezzle or misappropriate funds or engage in stock fraud, one needs access to a position of trust. Such positions usually are filled by middle- and upper-class persons. These crimes are facilitated by the social organization of trust; the acts of trustees are invisible, hidden in a network of connections between organizations. Thus although specific crimes may vary by class, illegitimate economic activity may be common in all classes.

Box 19.1, "How Deviant Are You?" includes a typical self-report measure of deviant behavior. Research using such measures finds that these behaviors are common across social classes.

Control Theory

If you were asked why you don't shoplift clothing from stores, you might reply, "Because my parents (or lover or friends) would kill me if they found out." According to **control theory,** social ties influence our tendency to engage in deviant behavior. We often conform to social norms because we are sensitive to the wishes and expectations of others. This sensitivity creates a bond between the individual and other persons. The stronger the bond, the less likely the individual is to engage in deviant behavior.

There are four components of the social bond (Hirschi, 1969). The first is *attachment*—ties of affection and respect for others. Attachment to parents is

19.1 How Deviant Are You?

Researchers frequently use self-report to study the incidence of deviant behavior. By asking people direct questions, researchers hope to avoid the biases found in official statistics which suggest that crime is concentrated in the lower class. The following is a typical questionnaire on deviance. Take a few minutes and fill it out.

Most people have done at least a few things that others would consider wrong. For each item in the table, circle the number of times you have engaged in the activity in the past 2 years.

Behavior	Number of Times					
1. Taken an item from a store without paying for it	0	1	2	3	4	5+
2. Taken things without permission from someone else's room or home that did not belong to you	0	1	2	3	4	5+
3. Bought, kept, or used something that you knew had been stolen from someone else	0	1	2	3	4	5+
4. Damaged, destroyed, or mistreated property on purpose	0	1	2	3	4	5+
5. Beaten up or hurt someone on purpose	0	1	2	3	4	5+
6. Threatened to beat up or hurt someone unless they did what you wanted	0	1	2	3	4	5+
7. Smoked or used marijuana	0	1	2	3	4	5+
8. Used cocaine	0	1	2	3	4	5+
9. Used drugs such as LSD, pep pills ("uppers," "speed"), tranquilizers ("downers"), or sleeping pills without a doctor's prescription	0	1	2	3	4	5+
10. Cheated on an exam or quiz in class	0	1	2	3	4	5+
11. Worked with other students on homework or a take-home project when you weren't supposed to	0	1	2	3	4	5+

SOURCE: Adapted from Tittle and Villemez, 1977.

continued on next page

especially important because they are the primary socializing agents of a child. A strong attachment to them leads the child to internalize social norms. The second component is *commitment* to long-term educational and occupational goals. Someone who aspires to go to law school is unlikely to commit a crime because a criminal record would be an obstacle to a career in law. The third component is *involvement*. People who are involved in sports, scouts, church groups, and other conventional activities simply have less time to engage in deviance. The fourth component is *belief*—a respect for the law and persons in positions of authority.

We can apply control theory to the shoplifting incident described at the beginning of this chapter. Susan does not feel attached to law-abiding adults; therefore, she was not concerned about their reactions to her behavior. Nor did she seem deterred by commitment when she said, "Nothing will happen." Susan's deviant act reflects the absence of a bond with conventional society.

The relationship between delinquency and the four components of the social bond has been the focus of numerous studies. Several studies have found a relationship between a lack of attachment and delinquency; young people from homes characterized by

continued from previous page

Answers to these questions from a sample of college sophomores, juniors, and seniors (97 males and 196 females) are reproduced here. The sample was obtained from a survey of two large undergraduate classes at the University of Wisconsin (Zimmerman & DeLamater, 1983). These responses indicate that students commit some crimes frequently, such as taking things without permission, vandalism, and using marijuana. Compare your responses to the questionnaire with theirs.

Behavior		Number of Times					
		0	1	2	3	4	5+
1. Taken something from a store	males	81%	8	4	3	0	3
	females	85	7	4	2	0	3
2. Taken something from someone's room or home	males	56	12	12	10	2	7
	females	54	17	14	5	2	8
3. Bought or used stolen property	males	61	20	15	2	1	1
	females	85	6	4	1	1	4
4. Vandalized property	males	58	16	13	6	1	5
	females	77	14	6	1	0	2
5. Beaten up someone	males	86	8	2	2	1	1
	females	89	8	3	0	0	0
6. Threatened someone	males	79	7	8	2	1	2
	females	91	1	4	0	1	3
7. Used marijuana	males	36	10	9	6	8	29
	females	34	16	8	12	9	20
8. Used cocaine	males	69	5	2	2	2	19
	females	75	5	3	4	3	11
9. Used other drugs	males	72	5	4	0	0	19
	females	66	8	6	6	1	15
10. Cheated in class	males	55	25	8	3	4	5
	females	58	18	12	7	5	1
11. Worked with other students when you weren't supposed to	males	65	12	10	4	3	5
	females	59	19	13	4	2	3

SOURCE: Zimmerman and DeLamater, 1983.

a lack of parental supervision, communication, and support report more delinquent behavior (Hirschi, 1969; Hundleby & Mercer, 1987; Messner & Krohn, 1990). Attachment to school, measured by grades, is also associated with delinquency. Boys and girls who do well in school are less likely to be delinquent. Regarding commitment to long-term goals, research indicates that youths who are committed to educational and career goals are less likely to engage in property crimes such as robbery and theft (Johnson, 1979; Shover et al., 1979). Findings relevant to the third component, involvement, are mixed. Whereas involvement in studying and homework is negatively associated with reported delinquency, participation in athletics, hobbies, and work is unrelated to reported delinquency. Involvement in religion, as reflected in frequent church attendance and rating religion as important in one's life, is associated with reduced delinquency (Sloane & Potrin, 1986). Finally, evidence suggests that conventional beliefs reduce the frequency of delinquent behavior (Gardner & Shoemaker, 1989).

Control theory asserts that attachment to parents leads to reduced delinquency. Implicitly, the theory assumes that parents do not encourage delinquent behavior. Although this assumption may be correct in

These urban Boy Scouts are nailing planks together as part of a work project. Participation in such group projects increases attachment to and involvement in conventional society, reducing the likelihood of delinquency.

most instances, there are exceptions. Studies suggest that some parents encourage some delinquent behaviors. Adolescent drinking is associated with parental alcohol consumption; parents who are heavy drinkers are more likely than nondrinking parents to have adolescents who are heavy drinkers (Barnes, Farrell, & Cairns, 1986). In this instance, parental attachment leads to increased delinquency.

Strength of the social bond is also related to adult criminal behavior. One study assessed month-to-month variations in circumstances that could strengthen or weaken the bond, and related this variation to the occurrence of criminal behavior (Horney, Osgood, & Marshall, 1995). The circumstances were starting/stopping school, starting/stopping work, and starting/stopping living with a girlfriend or wife. Interviews were conducted with 658 men in prison who had committed felonies. Increases in criminal behavior were closely related to changes that reduced the men's bond to others—stopping school or work, and stopping living with a girlfriend or wife. A study of Swedish men found that alcohol use and unemployment often preceded suicide (Norstrom, 1995). Both alcohol use and job loss reduce one's social integration, that is, social ties to other persons.

One longitudinal study indicates that the strength of the social bond influences whether adults engage in deviant behavior (Sampson & Laub, 1990). The researchers studied 500 boys ages 10 to 17 who

were in a correctional school and 500 boys the same age from public school. Each boy was followed until he was 32. Generally, strong ties to social institutions were associated with reduced rates of crime, alcohol abuse, gambling, and divorce. In adolescence, the important attachments were to family and school. In young adults, the influential ties were to school, work, or marriage. In later adulthood, the important ties were to work, marriage, and parenthood.

A study of the relationship between age and crime found large variation by age in the type of crime committed (Steffensmeier et al., 1989) (see Figure 19.1). Typical adolescent offenses include vandalism, auto theft, and burglary. Persons 18 to 28 are more likely to be involved in drug violations and homicide. Middle-age persons are more frequently involved in gambling offenses.

Differential Association Theory

Are all types of deviance explained by the absence of a social bond? Perhaps not. Sometimes people deviate from one set of norms because they are being influenced by a contradictory set of norms. U.S. society is composed of many groups with different values, norms, and behavior patterns. With respect to many behaviors, there is no single, society-wide norm. An adolescent's use of marijuana may deviate from her parents' norms, for example, but may conform to her friends' norms. Here the deviance involved in marijuana use reflects a conflict between the norms of two groups rather than an insensitivity to the expectations of others. In fact, the use of marijuana may reflect a high degree of sensitivity to the expectations of one's peers.

This view of deviance is the basis of **differential association theory,** developed by Sutherland. He argued that though the law provides a uniform standard for deviance, one group may define a behavior as deviant, whereas another defines it as desirable. Shoplifting, for example, is legally defined as a crime. Some groups believe it is wrong because (1) it leads to increased prices, which hurts everyone; (2) it violates the moral principle against stealing; and (3) it constitutes lawbreaking. Other groups, in contrast, believe shoplifting is acceptable because (1) businesses deserve

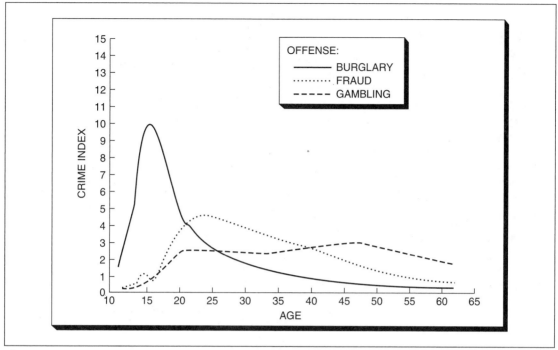

Figure 19.1　The Relationship Between Age and Crime

Involvement in criminal behavior is not constant across the life course. People of a certain age are much more likely to commit some crimes and not others. Youth ages 10 to 20 are especially likely to commit burglary; half of all those arrested for burglary are under 18. Fraud, in contrast, is much more likely to be committed by persons ages 20 to 40. Those arrested for gambling are as likely to be 50 as 20.

SOURCE: Steffensmeier et al., 1989, figure 1.

to have items taken because they overcharge; (2) the loss is covered by insurance; and (3) the shoplifter won't be caught. Susan's comment, "It's okay. Nothing will happen," reflects the latter belief.

Attitudes about behaviors are learned through associations with others, usually in primary group settings. People learn motives, drives, and techniques of engaging in specific behaviors. What they learn depends on whom they interact with—that is, on their differential associations. Whether someone engages in a specific behavior depends on how frequently he or she is exposed to attitudes and beliefs that are favorable toward that behavior.

The principle of differential association states that a "person becomes delinquent because of an excess of definitions favorable to violation of the law over definitions unfavorable to violation of the law" (Sutherland, Cressey, & Luckenbill, 1992). Studies designed to test this principle typically ask individuals questions about their attitudes toward a specific behavior and about their participation in that behavior. One study revealed that the number of definitions favorable to delinquency accurately predicted which young males reported delinquent behavior (Matsueda, 1982). The larger the number of definitions a youth endorsed, the larger the number of delinquent acts he reported having committed in the preceding year. A subsequent study found that associating with delinquent peers was also related to delinquent behavior (Heimer & Matsueda, 1994).

Certain groups within the United States hold a set of beliefs that justify the use of physical aggression

19.2   The Power of Suggestion

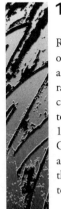

Rape, robbery, murder, and other types of deviant behavior receive a substantial amount of coverage in newspapers and on radio and television. One function of publicizing deviance is to remind us of norms—to tell us what we should not do (Erikson, 1964). But is this the only consequence? Could the publicity given particular deviant activities increase the frequency with which they occur? In some cases, the answer appears to be yes.

A study of the relationship between the publicity given suicides and suicide rates suggests the two are positively correlated (Phillips, 1974). This study identified every time a suicide was publicized in three major U.S. daily newspapers from 1947 to 1968. Next, the researchers calculated the number of expected suicides for the following month by averaging the suicide rates for that same month from the year before and the year after. For example, researchers noted that the suicide of a Ku Klux Klan leader on November 1, 1965, was widely publicized. They then obtained the expected number of suicides (1,652) by averaging the total number of suicides for November 1964 (1,639) and November 1966 (1,665). In fact, there were 1,710 suicides in November 1965; the difference between the observed

and expected rates (58) could be due to suggestion via the mass media.

Results of this study showed that suicides increased in the month following reports of a suicide in major daily papers. Moreover, the more publicity a story was given—as measured by the number of days the story was on the front page—the larger the rise in suicides. If a story was published locally—in Chicago but not in New York, for example—the rise in suicides occurred only in the area where it was publicized.

Why should such publicity lead other persons to kill themselves? There must be some factor that predisposes a small number of persons to take their own lives following a publicized suicide. That predisposing factor may be *anomie.* According to this theory, suicide is a form of retreatism, of withdrawal from the struggle for success. Persons who don't have access to legitimate means are looking for some way to adapt to their situation. Publicity given to a suicide may suggest a solution to their problem.

When we think of suicide, we think of shooting oneself, taking an overdose of a drug, or jumping off a building. We distinguish suicide from accidents when we presume the person did not intend to harm himself or herself. But the critical difference is the person's

continued on next page

in certain situations. This set of beliefs is referred to as the *subculture of violence.* Within this subculture, violence is considered appropriate when used as a means of self-defense and protection of one's home. A review of state laws governing spouse abuse, corporal punishment, and capital punishment found that southern states have laws more accepting of violence (Cohen, 1996). Several studies report a relationship between these beliefs and behavior. Felson and colleagues (1994) studied young males in 87 high schools. The young men were asked whether aggressive responses were appropriate in three situations involving insults or threats. Those young men who endorsed the use of violence were much more likely to report involvement in eight types of interpersonal

violence, including striking a parent or teacher, fighting, and using weapons in disputes. Endorsing the use of violence was also associated with delinquency within the school, including cheating, tardiness, and truancy.

The theory of differential association does not specify the process by which people learn criminal or deviant behavior. For this reason, Burgess and Akers (1966) developed a modified theory of differential association. This modified version emphasizes the influence of positive and negative reinforcement on the acquisition of behavior. Much of this reinforcement comes from friends and associates. Thus, if beliefs are learned through interaction with others, then people whose attitudes are favorable toward a behavior should

continued from previous page

intent, not the event itself. Some apparent accidents may be suicides. For example, when a car hits a bridge abutment well away from the pavement on a clear day with no evidence of mechanical malfunction, this may be suicide.

If some auto accidents are, in fact, suicides, we should observe the increase in motor vehicle accidents following newspaper stories about a suicide. In fact, data from newspapers and motor vehicle deaths in San Francisco and Los Angeles verify this hypothesis (Phillips, 1979). Statistics show a marked increase in the number of deaths due to automobile accidents 2 and 3 days after a suicide is publicized—especially accidents involving one vehicle. In the Detroit metropolitan area, an analysis of motor vehicle fatalities for the years 1973 to 1976 revealed an average increase in fatalities of 35% to 40% the third day after a suicide story appeared in the daily papers (Bollen & Phillips, 1981). Again, the more publicity, the greater the increase. Finally, if the person whose suicide is publicized was young, deaths of young drivers increase; whereas if the person killing himself was older, the increase in fatalities involves more older drivers.

Does an increase in suicide follow any publicized suicide, or are some suicides more likely to be imitated than others? Stack (1987) studied instances in which celebrities killed themselves. Each was classified according to whether the person was an entertainer, political figure, artist, member of the economic elite, or villain. The results showed that when entertainers and politicians took their own lives, there was an increase in suicide rates. Suicides by artists, members of the elite, and villains were not followed by an increase. Moreover, the findings suggest that the effect of publicized suicide is gender and race specific. Suicide by a male celebrity was followed by an increase in the number of males who killed themselves but not in the number of females who took their lives, and vice versa. Similarly, an increase in suicides by whites followed a publicized case involving a white celebrity, whereas rates among blacks were unaffected. The fact that the effects of publicized suicide are age, gender, and race specific is consistent with the concept of imitation.

A very different form of deviance, the hijacking of aircraft, also appears to be influenced by publicity. An analysis of all instances of air piracy in the United States for 1968 to 1972 (Holden, 1986) found that successful hijackings were followed by an increase in hijacking attempts. There was no increase following unsuccessful hijacking attempts.

Thus media reports of some types of deviance may suggest behavioral responses to some problem, suggestions that are acted on by some persons.

---

have friends who also have favorable attitudes toward that behavior. Alternatively, people whose attitudes are opposed to an activity should have friends who share those negative views.

A survey of 3,056 high school students was conducted to test these hypotheses (Akers et al., 1979). In particular, it assessed the relationship between differential association, reinforcement, and adolescents' drinking behavior and marijuana use. Differential association was measured by three questions: How many of your (1) best friends, (2) friends you spend the most time with, and (3) friends you have known longest smoke marijuana and/or drink? The survey also assessed students' definitions of drug and alcohol laws. Both social reinforcement (whether the adolescent expected praise or punishment for use from parents and peers) and nonsocial reinforcement (whether the effects of substance use were positive or negative) were measured. Findings of this survey showed that differential association was closely related to the use of alcohol and/or drugs. The larger the number of friends who drank and/or smoked marijuana, the more likely the student was to drink alcohol and/or smoke marijuana. Reinforcement was also related to behavior; those who used a substance reported it had positive effects. The students' definitions were also related to those with whom they associated; if their friends drank and/or used marijuana, they were more likely to have positive attitudes toward the behavior and negative attitudes toward laws defining that behavior as criminal. Finally, students' attitudes were consistent with their behavior. Those who opposed marijuana use and supported the marijuana laws were much less likely to use that substance.

A similar study (Akers et al., 1989) focused on drinking among older persons. Interviews were conducted with 1,410 people ages 60 and over. The measures used were the same or similar to those used with adolescents. The results were essentially the same. The drinking behavior of persons 60 and over was related to drinking behavior of spouse, family, and/or friends, reinforcements, and an individual's attitudes toward drinking.

Survey data collected at one point in time often cannot be used to test hypotheses about cause-and-effect relationships. However, survey data collected from the same people at two or more times can be. Stein and colleagues (1987) analyzed data from 654 young people who were surveyed three times at 4-year intervals that began when they were in junior high school. The measures included peer drug use, adult drug use, and community approval of drug use. The results showed that adolescents who believed that both peers and adults were using drugs were more likely to become drug users. Thus association with persons who use alcohol and drugs, especially in primary relationships, is one cause of substance use by adolescents.

Because each person usually associates with several groups, the consistency or inconsistency in definitions across groups is an important influence on behavior (Krohn, 1986). *Network multiplexity* refers to the degree to which individuals who interact in one context also interact in other contexts. When you interact with the same people at church, at school, on the athletic field, and at parties, multiplexity is high. When you interact with different people in each of these settings, multiplexity is low. When multiplexity is high, definitions of an activity are consistent across groups; when it is low, definitions may be inconsistent across groups. Thus associations should have the greatest impact on attitudes and beliefs when multiplexity is high. A survey of 1,435 high school students measured the extent to which individuals interacted with parents and with the same peers in each of several activities (Krohn, Massey, & Zielinski, 1988). Students who participated jointly with parents and

peers in various activities were less likely to smoke cigarettes.

Routine Activities Perspective

So far we have considered characteristics of the person (motivation, beliefs) and of his or her associations with others (parents, friends). These have been shown to be related to delinquency, assault, murder, burglary, economic crimes, suicide, and alcohol and drug use. The **routine activities perspective** focuses on instances of these behaviors and how they emerge from the routines of everyday life (Felson, 1994).

Each instance of deviant behavior requires the convergence of the elements necessary for the behavior to occur. Crimes such as burglary, larceny, or robbery require the convergence of an offender, a likely target (residence, store, or person), and the absence of some guardian who could intervene. In the illustration at the beginning of the chapter, the shoplifting incident involves such a convergence: Susan, the bracelet, and the absence of a clerk or security guard. Illegal consumption requires two offenders (seller, user), a substance, and a setting with no guardian; "crack houses" provide the latter in many large cities. Without such convergence, deviance will not occur. We can understand another aspect of deviance if we analyze everyday activity from the perspective of how it facilitates or prevents such convergences. This perspective calls our attention to the contributions of situations to behavior.

One class of situations that facilitates deviance is unstructured socializing with peers in the absence of an authority figure (Osgood et al., 1996). The presence of peers makes it likely that definitions will be shared, including definitions favorable to particular forms of deviance. The absence of an authority figure/guardian reduces the likelihood of punishment for deviance. Lack of structure makes time available for deviance. What situations have these characteristics? They include joyriding in a car with friends, going to parties, and "hanging out" with friends. Data from a longitudinal study of a national sample of 1,200 persons ages 18 to 26 allowed researchers to relate involvement in these situations to deviance. Frequency of participation in them was related to alcohol and marijuana use, dangerous driving, and criminal behavior. Changes across five waves of data collection in an individual's participation in these activities were related to changes in his or her involvement in deviance.

Researchers have consistently noted that men are much more likely to commit criminal acts than women. This is not only true of "street crime," but also of economic crimes involving violation of trust, such as stock fraud. The routine activities perspective explains this as due to gender-role socialization, which teaches women different norms/definitions; to lack of access to tutelage in various forms of deviance; and to restrictions on activities that keep women out of certain settings (Steffensmeier & Allan, 1996). Thus few women commit either burglary or stock fraud because of lack of access in everyday life to the apprenticeships where one learns these behaviors.

The anomie, control, differential association, and routine activities perspectives are not incompatible. Anomie theory suggests that culturally valued goals and the opportunities available to achieve these goals are major influences on behavior. Opportunities to learn and occupy particular roles are influenced by age, social class, gender, race, and ethnic background, that is, by the structuring of everyday life based on these variables. According to control theory, we are also influenced by our attachments to others and our commitment to attaining success. Our position in the social structure and our attachments to parents and peers determine our differential associations—the kinds of groups to which we belong. Within these groups, we learn definitions favorable to particular behaviors, and we learn that we face sanctions when we choose behaviors which group members define as deviant.

Reactions to Norm Violations

When we think of murder, robbery, or sexual assault, we think of cases we have read about or heard of on radio or television. We frequently refer to police and FBI statistics as measures of the number of crimes that have occurred in our city or county. Our knowledge of alcohol or drug abuse depends on knowing or hearing about persons who engage in these behaviors. All of these instances of deviance share another important characteristic as well. In every case, the behavior was discovered by someone who called it to the attention of others.

Does it matter that these instances involve both an action (by a person) and a reaction (by a victim or an observer)? Isn't an act equally as deviant whether others find out about it or not? Let's go back to Susan's theft of the bracelet. Suppose Susan had left the store without being stopped by the security guard. In this case, she and Virginia would have known she had taken the bracelet, but she would not have faced sanctions from others. She would not have experienced the embarrassment of being confronted by a store guard and accused of a crime. Moreover, she would have had a beautiful bracelet. But the fact is that she was stopped by the guard. She will be questioned, the police will be called, and she may be arrested. Thus the consequences for committing a deviant act are quite different when certain reactions follow. This reasoning is the basis of **labeling theory,** the view that reactions to a norm violation are a critical element in deviance. Only after an act is discovered and labeled "deviant" is the act recognized as such. If the same act is not discovered and labeled, it is not deviant (Becker, 1963).

If deviance depends on the reactions of others to an act rather than on the act itself, the key social psychological question becomes why particular audiences choose to label an act deviant (whereas other audiences may not). Labeling theory is an attempt to understand how and why acts are labeled deviant. In the case of the stolen bracelet, labeling analysts would not be concerned with Susan's behavior. Rather, they would be interested in the responses to Susan's act by Virginia, the male customer, and the security guard. Only if an observer challenges Susan's behavior or alerts a store employee does the act of taking the bracelet become deviant.

Reactions to Rule Breaking

Labeling theorists refer to behavior that violates norms as **rule breaking,** to emphasize that the act by itself is not deviant. Most rule violations are "secret" in the

The reactions of others to rule-breaking behavior depend on the characteristics of the actor. The dress and grooming of this shoplifter make it less likely that the man observing her behavior will report it.

behavior, for example, they often replied that they had not considered their husbands ill or in need of help (Yarrow et al., 1955). People react to isolated episodes of unusual behavior in one of four ways. A common response is *denial,* in which the person simply does not recognize that a rule violation occurred. In one study, denial was typically the first response of women to their husband's excessive drinking (Jackson, 1954). A second response is *normalization,* in which the observer recognizes that the act occurred but defines it as normal or common. Thus wives often reacted to excessive drinking as normal, assuming many men drink a lot. Third, the person may recognize the act as a rule violation but excuse it, attributing its occurrence to situational or transient factors; this reaction is *attenuation.* Thus some wives of men who were later hospitalized believed the episodes of bizarre behavior were caused by unusually high levels of stress or by physical illness. Finally, people may respond to the rule violation by *balancing* it, recognizing it as a violation but deemphasizing its significance due to the actor's good qualities.

The man who witnessed Susan's behavior looked around, spotted a security guard, and reported the act. In doing so, he labeled the actor. Labeling involves a redefinition of the actor's social status; the man placed Susan into the category of "shoplifter" or "thief." The security guard, in turn, probably defined Susan as a "typical shoplifter." Although labeling is triggered by a behavior, it results in a redefinition or typing of the actor. As we shall see, this has a major impact on people's perceptions of and behavior toward the actor.

Determinants of the Reaction

What determines how an observer reacts to rule breaking? Reactions depend on three aspects of the rule violation, including the nature of the actor, the audience, and the situation.

Actor Characteristics
Reaction to a rule violation often depends on who performs the act. First, people are more tolerant of rule breaking by family members than by strangers. The research cited earlier

sense that no one other than the actor (and on occasion, the actor's accomplices) is aware of them. Many cases of theft and tax evasion, many violations of drug laws, and some burglaries are never detected. These activities can be carried out by a single person. Other acts, such as robberies, assaults, and various sexual activities, involve other people who will know about them but may not label the act deviant.

How do members of an audience respond to a rule violation? It depends on the circumstances, but studies suggest that very often people *ignore* it. When wives of men hospitalized for psychiatric treatment were asked how they reacted to their husband's bizarre

revealed extraordinary tolerance of spouses for bizarre, disruptive, and even physically abusive behavior. Many of us probably know of a family who is attempting to care for a member whose behavior creates problems for them. Second, people are more tolerant of rule violations by persons who make positive contributions in other ways. In small groups, tolerance is greater for persons who contribute to the achievement of group goals (Hollander & Julian, 1970). We seem to tolerate deviance when we are dependent on the person committing the act, perhaps because if we punish the actor it will be costly for us. Third, we are less tolerant if the person has a history of rule breaking (Whitt & Meile, 1985).

Does gender affect reactions to behavior? An ingenious field experiment suggests that gender does not affect an audience's response to shoplifting. With the cooperation of store employees, shoplifting events were staged in the presence of customers who could see the event. The experiment was conducted in a small grocery store, a large supermarket in a shopping mall, and a large discount department store. Three aspects of the situation were varied: the gender of the shoplifter, the appearance of the shoplifter, and the gender of the observer. Neither the shoplifter's nor the customer's gender had an effect on the frequency with which the customer reported the apparent theft. The appearance of the shoplifter, however, had a substantial effect. If the man or woman who took an item was wearing soiled, patched clothing and had unkempt hair, the customer was much more likely to report the theft than if the shoplifter was neatly dressed and well groomed. Perhaps the customers balanced the theft against what they presumed were the good qualities of the well-dressed shoplifter (Steffensmeier & Terry, 1973).

More generally, analysts have suggested that we are less likely to label women than men for violations of criminal law (Haskell & Yablonsky, 1983). Research indicates that women are less likely to be kept in jail between arraignment and trial, and they receive more lenient sentences than men. One explanation for this differential treatment is that women are subject to greater informal control by family members and friends, and so are treated more leniently in the courts. A recent study of the influences on pretrial release and sentence severity found that both men and women with families received more lenient treatment; the effect was stronger for women (Daly, 1987).

However, research suggests that psychiatrists are more likely to label women as having a personality disorder than men (Dixon, Gordon, & Khomusi, 1995). Case histories were prepared that included symptoms of clinical disorders (as defined in the *Diagnostic and Statistical Manual III* of the American Psychiatric Association) and personality disorders (*DSM III*, Axis II). The latter disorders are generally less serious and more ambiguously defined. The histories were identical except for gender: male, female, or unspecified. Psychiatrists' diagnoses of clinical disorders was not influenced by gender, but they were more likely to diagnose women as having personality disorders than men with the same symptoms.

Audience Characteristics The reaction to a violation of rules also depends on who witnesses it. Because groups vary in their norms, audiences vary in their expectations. People enjoying a city park on a warm day react quite differently to a nude man walking through the park than a group of nudists in a nudist park do. Recognizing this variation in reaction, people who contemplate breaking the rules—by smoking marijuana, drinking in public, or jaywalking, for example—often make sure no one is around who will punish them.

An important influence on whether a witness will label a rule violation is the level of concern in the community about the behavior. Citizens who are concerned about drug use as a social problem are probably more alert for signs of drug sales and use and more likely to label someone as a user. A major determinant of the level of concern is the amount of activity by politicians, service providers, and the mass media calling attention to the problem (Beckett, 1994).

Officials who routinely deal with suspects react very differently to suspected offenders than do citizens. One study focused on officials working in a court-affiliated unit who evaluate suspected murderers following arrest. These officials had a stereotyped image of the type of person who commits murder (Swigert & Farrell, 1977). When lower-class male members of ethnic minorities committed murder, these officials believed it was in response to a threat on their masculinity. For example, if an Italian American truck driver was arrested for murder, they were likely to assume he had killed the other man in response to verbal insults. This labeling based on a stereotype had important consequences. Suspects who fit this image were less likely to be defended by a private attorney, more likely to be denied bail, more likely to plead guilty, and more likely to be convicted on more severe charges.

Consider the example of a student with a drinking problem seeking help at a university counseling center. The treatment will depend on how counselors view student "troubles." One study found that the staff of a university clinic believed students' problems could be classified into one of the following categories: problems in studying, choosing a career, achieving sexual intimacy, or handling personal finances; conflict with family or friends; and stress arising from sociopolitical activities. When a student came to the clinic because of excessive drinking, the therapist first decided which of these categories applied to this person's troubles, that is, which type of problem was causing this student to drink excessively. How the problem was defined in turn determined what the therapist did to try to help the student (Kahne & Schwartz, 1978).

Situational Characteristics

Whether a behavior is construed as normal or labeled deviant also depends on the definition of the situation in which the behavior occurs. Marijuana and alcohol use, for example, are much more acceptable at a party than at work (Orcutt, 1975). Various sexual activities expected between married persons in the privacy of their home would elicit condemnation if performed in a public park.

Consider so-called gang violence. In some major cities, incidents in which teenage gangs assault each other are common. News media, police, and other outsiders often refer to such incidents as "gang wars." These events often occur in the neighborhoods where the gang members live. How do their parents, relatives, and friends react to such incidents? According to a study of one Chicano community, it depends on the situation (Horowitz, 1987). Young men are ex-

pected to protect their families, women, and masculinity. When violence results from a challenge to honor, the community generally tolerates it. If the violence disrupts a community affair, however, such as a dance or wedding, it is not tolerated.

We often rely on the behavior of others to help us define situations. Our reaction to a rule violation may be influenced by the reactions of other members of the audience. The influence of the reactions of others is demonstrated in a field experiment of intrusions into waiting lines (Milgram et al., 1986). Members of the research team intruded into 129 waiting lines with an average length of six persons. One or two confederates approached the line and stepped between the third and fourth persons. In some cases, other confederates served as buffers; they occupied the fourth and fifth positions and did not react to the intrusion. When the buffers were present, others in the line were much less likely to react verbally or nonverbally to the intrusion.

A good deal of research suggests that interpersonal violence, especially assaults and murders, often involves two young men and is triggered by a verbal insult (Katz, 1988). But whether a remark is an insult is a matter of social definition. Not surprisingly, fights are more likely to erupt following a remark when there is a male audience and the men have been drinking (Felson, 1994). A remark is less likely to lead to a fight if the audience includes women.

Consequences of Labeling

Assume that an audience defines an act as deviant. What are the consequences for the actor and the audience? We consider four possible outcomes.

Institutionalization of Deviance

In some cases, individuals who label a behavior deviant may decide it is in their own interest for the person to continue the behavior. They may, in fact, reward that person for the deviant behavior. If you learn a good friend is selling drugs, you may decide to use this person as a source and purchase drugs from him. Over time, your expectations will change; you will come to expect him to sell drugs. If your drug-selling friend decides to stop dealing, you may then treat him as a rule breaker. Illegal activities by stockbrokers are likely to be ignored or encouraged by other employees and supervisors when all benefit economically from the activity (Zey, 1993). The process by which members of a group come to expect and support deviance by another member over time is called **institutionalization of deviance** (Dentler & Erikson, 1959).

Backtracking   Even when an audience reacts favorably to a rule violation, the actor may decide to discontinue the behavior. This second consequence of labeling is called *backtracking*. It may occur after the actor learns that others label his or her act deviant. Although some audiences react favorably, the actor may wish to avoid the reaction of those who would not react favorably and the resulting punishment. Many teenagers try substances like marijuana once or twice. Although their friends may encourage its continued use, some youths backtrack because they want to avoid their parents' negative reactions.

Effective Social Control   An audience that reacts negatively to rule breaking and attempts to punish the actor or threatens to do so may force the actor to give up further involvement in the activity. This third consequence of labeling is known as *effective social control*. This reaction is common among friends or family members who often threaten to end their association with an actor who continues to engage in deviance. Similarly, they may threaten to break off their relationship if the person does not seek professional help. In these instances, the satisfaction of the actor's needs is contingent on changing his or her behavior. Members of the audience also may insist the actor renounce aspects of his or her life that they see as contributing to future deviance (Sagarin, 1975). If excessive drinking is due to job-related stressors, for example, family members may demand that the person find a different type of employment. Displays of remorse may also lead to reduced punishment for an offense (Robinson, Smith-Lovin, & Tsobdis, 1994).

Unanticipated Deviance   Still another possibility is that the individual may engage in further or unanticipated deviance. Note the use of the word *unanticipated*. Negative reactions by members of an audience are intended to terminate rule-breaking activity. However, such reactions may, in fact, produce further deviance. This occurs when the audience's response sets in motion a process that leads the actor to greater involvement in deviance. This process and its outcomes are the focus of the next section.

Labeling and Secondary Deviance

Labeling a person deviant may set in motion a process that has important effects on the individual. The process of societal reaction produces changes in the behavior of others toward the individual and may lead to corresponding changes in his or her self-image. A frequent consequence of the process is involvement in secondary deviance and a deviant subculture. In this section, we consider this process in detail.

Societal Reaction

Earlier in this chapter, we mentioned that labeling is a process of redefining a person. By categorizing a person as a particular kind of deviant, we place that person in a stigmatized social status. The deviant (addict, pimp, thief) is defined as undesirable, not acceptable in conventional society, and frequently treated as inferior. This stigmatized status has two important consequences. First, it leads to changes in the behavior of others toward the person. Second, the person gradually comes to perceive himself or herself as deviant, and to behave in ways that are consistent with the label.

Changes in the Behavior of Others   When we learn someone is an alcoholic, homosexual, or mentally ill, our perceptions and behavior toward that person change. For example, if we learn someone has a drinking problem, we may respond to his or her request for a drink with "Do you think you should?" or "Why don't you wait?" to convey our objection. We may avoid jokes about drinking in the person's presence, and we may stop inviting him or her to parties or dinners where alcohol will be served.

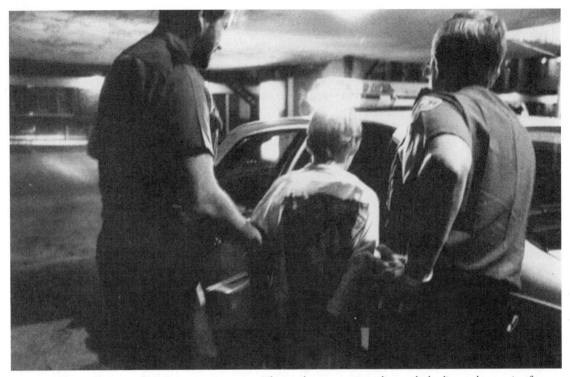

Being caught in a deviant act has important consequences. This youth may experience disrupted schooling and separation from family and friends as a result of being arrested.

A more severe behavioral reaction involves withdrawal from the stigmatized person (Kitsuse, 1964). For instance, the labeled shoplifter, alcoholic, or homosexual may be fired from his or her job. Behavioral withdrawal may occur because of hostility toward the deviant, or it may reflect a sincere desire to help the person. For example, the employer who fires an alcoholic may do so because he dislikes alcoholics or because he believes that relief from work obligations will reduce stress which may be causing the drinking problem.

Paradoxically, our reaction to deviance may produce additional rule breaking by the labeled person. We expect people who are psychologically disturbed to be irritable or unpredictable, so we avoid them to avoid an unpleasant interaction. The other person may sense he or she is being avoided and respond with anger or distrust. His anger may cause co-workers to talk about him behind his back; he may respond with suspicion and become paranoid. When members of an audience behave toward a person according to a label and cause the person to respond in ways that confirm the label, they have produced a **self-fulfilling prophecy** (Merton, 1957). Lemert (1962) documents a case in which such a sequence led to a man's hospitalization for paranoia.

Self-Perception of the Deviant   Another consequence of stigmatized social status is that it changes the deviant's self-image. A person labeled deviant often incorporates the label into his or her identity. This redefinition of oneself is due partly to feedback from others who treat the person as a deviant. In addition, the new self-image may be reinforced by the individual's own behavior. Repeated participation in shoplifting, for example, may lead Susan to define herself as a thief.

Redefinition is facilitated by the social programs and agencies that deal with specific types of deviant persons. Such agencies pressure persons to acknowledge they are deviant. Admitting that one is a thief often leads police and prosecutors to go easy on a shoplifter, especially if it is a first offense. Failure to acknowledge this may lead to a long prison sentence. Admitting one is mentally ill is often a prerequisite for psychiatric treatment (Goffman, 1959a). Mental health professionals often believe a patient cannot be helped until the individual recognizes his or her problem. Employees of an agency that provided jobs for unemployed persons viewed clients' employment problems as partly the result of individual failure (Miller, 1991). To receive agency services, clients had to agree with this view and change their behavior accordingly.

Thus the deviant experiences numerous pressures to accept a stigmatized identity. Acceptance of a stigmatized identity has important effects on self-perception. Everyone has beliefs about what people think of specific types of deviant persons. Accepting a label such as "thief," "drunk," or "crazy" leads a person to expect that others will stigmatize and reject him or her, which in turn produces self-rejection. Self-

rejection makes subsequent deviance more likely (Kaplan, Martin, & Johnson, 1986). In a study of junior high school students, data were collected three times at 1-year intervals. Self-rejection (feeling that one is no good, a failure, rejected by parents and teachers) was related to more favorable dispositions (definitions) toward deviance and an increased likelihood of associating with deviant persons 1 year later. A high disposition and associations with deviant peers were related to increased deviance—theft, gang violence, drug use, and truancy—1 year later (Kaplan, Johnson, & Bailey, 1987). Figure 19.2 summarizes these relationships. Delinquent behavior, in turn, is associated with reduced self-esteem (McCarthy & Hoge, 1984).

In short, labeling may set in motion a cycle in which changes in behavior produce changes in other people's behavior, which, in turn, changes the deviant's self-image and subsequent behavior. Self-fulfilling prophecies can also be positive. One study assessed the expectations of 98 sixth-grade math

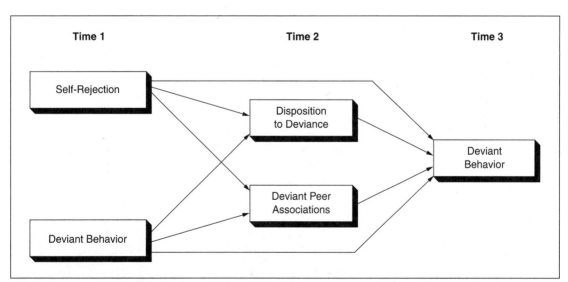

Figure 19.2 **The Relationship of Self-Rejection to Deviant Behavior**

A person who engages in deviant behavior anticipates that others will reject him or her, which, in turn, can lead to self-rejection. A longitudinal study collected data from junior high school students three times, each 1 year apart. At time 1, reported participation in deviance was positively related to self-rejection (feeling one is no good, a failure, rejected by parents and teachers). Self-rejection at time 1 was associated with more favorable dispositions (attitudes) toward deviance but a decreased likelihood of associating with other deviants 1 year later (at time 2). Favorable dispositions and deviant associations at time 2, as well as deviance at time 1, were related to increased deviance—theft, gang violence, drug use, and truancy—at time 3.

SOURCE: Kaplan, Johnson, and Bailey, 1987, figure 2.

teachers for their students (N = 1,539) (Madon, Jussim, & Eccles, 1997). Teachers' expectations (positive or negative) predicted performance much better for students who were low achievers. Also, teachers' overestimates, positive expectations, predicted actual achievement better than their underestimates. Perhaps positive expectations inspire underachievers.

Although more attention has been given to situations in which others label the person, some persons become committed to deviance without such labeling. For example, some persons voluntarily seek psychiatric treatment; some of these cases reflect *self-labeling* (Thoits, 1985). People know that others view certain behaviors as symptoms of mental illness. If they observe themselves engaging in those behaviors, they may label themselves mentally ill.

Secondary Deviance

A frequent outcome of the societal reaction process is **secondary deviance,** in which a person engages increasingly in deviant behavior as an adjustment to others' reactions (Lemert, 1951). Usually the individual becomes openly and actively involved in the deviant role, adopting the clothes, speech, and mannerisms associated with it. For example, initially a person with a drinking problem may drink only at night and on weekends to prevent his or her drinking from interfering with work. Once the person adopts the role of "heavy drinker" or "alcoholic," however, he or she may drink continually. For male homosexuals, "coming out" constitutes an act of self-labeling and a public commitment to homosexuality (Dank, 1971). Before coming out, the homosexual tries to limit his homosexual involvement. After coming out, he engages in sexual activity with males more openly and frequently.

As an individual becomes openly and regularly involved in deviance, he or she may increasingly associate with others who routinely engage in the same or related activity. The individual may join a **deviant subculture,** a group of people whose norms encourage participation in the deviance and who regard positively those who engage in it. Subcultures provide not only acceptance but also the opportunity to enact deviant roles. Through a deviant subculture the would-be drug dealer or prostitute can gain access to customers more readily.

Subcultural groups are an attractive alternative for deviants for two reasons. First, these people are often forced out of nondeviant relationships and groups through others' reactions. As family and friends progressively break off relationships with them, they are compelled to seek acceptance elsewhere. Second, membership in subcultural groups may result from deviants' desire to associate with people who are similar and who can provide them with feelings of social acceptance and self-worth (Cohen, 1966). Deviants are no different from others in their need for positive reflected appraisals.

Deviant subcultures help persons cope with the stigma associated with deviant status. We have already noted that deviants are often treated with disrespect and sanctioned by others for their activity. Such treatment threatens self-esteem and produces fear of additional sanctions. Subcultures help the deviant cope with these feelings. They provide a *vocabulary of*

Deviant subcultures create opportunities for people to enact roles not acceptable elsewhere in society. This camp provides a place where people can undress without attracting attention or being arrested.

Figure 19.3 Images of the Deviant

We often have very negative images of many types of deviant persons. For example, many people view marijuana users as dirty, unkempt dropouts, like the person on the left. Because these images are widely shared, persons who engage in some form of deviant behavior are usually aware that others look down on them. To counter this stigma, deviants attempt to create a positive self-image, which is reinforced by members of deviant subcultures. The user views himself as clean, cool, and in touch, like the person on the right. It is easier to view oneself as "normal" when others support that view.

motives—beliefs that explain and justify the individual's participation in the behavior.

The norms and belief systems of subcultures support a positive self-conception. In the early 1970s, a prostitute's rights group, COYOTE (Cast Off Your Old Tired Ethics), emerged in San Francisco. Although it did not obtain the legalization of prostitution, it did enhance the self-images of its members (Weitzer, 1991). Many people think nudists are exhibitionists who take off their clothes to get sexual kicks. Nudists, in contrast, consider themselves morally respectable and hold several beliefs designed to enhance that claim: (1) nudity and sexuality are unrelated, (2) there is nothing shameful about the human body, (3) nudity promotes a feeling of freedom and natural pleasure, and (4) nude exposure to the sun promotes

physical, mental, and spiritual well-being. There are also specific norms—"no staring," "no sex talk," and "no body contact"—designed to sustain these general beliefs (Weinberg, 1976). The contrast between the stigmatized image of the deviant and the deviant's self-image is illustrated in Figure 19.3. The belief systems of deviant subcultures provide the social support the person needs to maintain a positive self-image.

Joining a deviant subculture often stabilizes participation in one form of deviance. It also may lead to involvement in additional forms of deviant behavior. Black heroin users, for example, are encouraged by group norms and values to engage in a variety of other illegal, moneymaking activities (Finestone, 1964). Similarly, many prostitutes become drug users through participation in a subculture.

Formal Social Controls

So far, this chapter has been concerned with **informal social control**—the reactions of family, friends, and acquaintances to rule violations by individuals. Informal controls are probably the major influence on an individual's behavior. In modern societies, however, there are often elaborate systems set up specifically to process rule breakers. Collectively, these are called **formal social controls**—agencies given responsibility for dealing with violations of rules or laws. Typically, the rules enforced are written, and, in some cases, punishments also may be specified. The most prominent system of formal social control in our society is the criminal justice system, which includes police, courts, jails, and prisons. A second system of formal social control is the juvenile justice system, which includes juvenile officers, social workers, probation officers, courts, and treatment or detention facilities. A third system of formal social control deals with mental illness. It includes mental health professionals, commitment procedures, and institutions for the mentally ill and mentally impaired.

Formal Labeling and the Creation of Deviance

Most of us think of formal agencies as reactive, as simply processing individuals who have already committed crimes or who are mentally retarded or in need of psychiatric treatment. But these agencies do much more than take care of persons already known to be deviant. It can be argued that the function of formal social-control agencies is to select members of society and identify or certify them as deviant (Erikson, 1964).

In the 1990s, crime control has become big business in U.S. society. Federal and state governments have been providing funds to hire thousands of additional police officers, sheriff's deputies, and federal agents. Many states are building new prisons. Additional officers and new prisons require large investments in new equipment. It has been suggested that there is a crime control industry, with many people lobbying for its preservation and growth (Chamblis,

1994). More officers and prisons lead to more arrests and increases in prison populations. Is this expansion due to real increases in crime? No. Crime has not increased substantially in the past 25 years. In fact, in 1997, crime rates were declining. What has increased is political rhetoric and mass media attention to a stable level of crime, leading the public to perceive an increase and support the expansion of formal control systems.

Functions of Labeling   Of what value is labeling people "criminals," "delinquents," or "mentally ill"? There are three functions of labeling persons deviant: (1) to provide concrete examples of undesirable behavior, (2) to provide scapegoats for the release of tensions, and (3) to unify the group or society.

First, the public identification of deviance provides concrete examples of how we should not behave (Cohen, 1966). When someone is actually apprehended and sanctioned for deviance, the norms of society are made starkly clear. For instance, the arrest of someone for shoplifting dramatizes the possible consequences of taking things that do not belong to us. The controversy and publicity in 1997 surrounding claims of sexual harassment and assault in the U.S. armed forces heightened awareness of these behaviors.

According to the **deterrence hypothesis,** the arrest and punishment of some individuals for violations of the law deters other persons from committing the same violations. To what extent does general deterrence really affect people's behavior? Most analysts agree that the objective possibility of arrest and punishment does not deter people from breaking the law. Rather, conformity is based on people's perceptions of the likelihood and severity of punishment. Thus youths who perceive a higher probability that they will be caught and the punishment will be severe are less likely to engage in delinquent behavior (Jensen, Erickson, & Gibbs, 1978). Similarly, a study of theft of company property by employees found that those who perceived greater certainty and severity of organizational sanctions for theft were less likely to have taken property (Hollinger & Clark, 1983).

For punishment of some offenders to deter others, the punishment must be publicized. In recent

A man convicted of a crime talks to high school students about the nature and consequences of his deviance. By publicizing the penalties for crime, such programs attempt to deter others from breaking the law.

years, executions of murderers have been widely publicized. Does this publicity deter murder? Specifically, does coverage of executions on the evening news on network television lead to a reduction in homicide rates? A study of coverage and rates from 1976 through 1987 found no relationship (Bailey, 1990).

Perceived certainty of sanctions generally has a much greater effect on persons who have low levels of moral commitment (Silberman, 1976). People whose morals define a behavior as wrong are not as affected by the threat of punishment. For example, a person's moral beliefs are a more important influence on whether adults use marijuana than the fear of legal sanctions (Meier & Johnson, 1977). Adults who believe the use of marijuana is wrong do not use it, regardless of their perception of the likelihood they will be sanctioned for its use.

A second function of public identification of deviants is to provide a scapegoat for the release of tension. Many people face threats to the stability and security of their daily lives. Some fear the possibility that they will be victimized by aggressive behavior or the criminal activity of others. The existence of such threats arouses tension. Persons identified publicly as deviants provide a focus for these fears and insecurities. Thus the publicly identified deviant becomes the concrete threat we can deal with decisively.

This scapegoating process is illustrated among the Puritans, who came to New England during the 1600s to establish a community based on a specific Christian theology. As time passed, groups within the community periodically challenged the ministers' claims that they were the sole interpreters of the theology. In addition, the community faced the threat of Indian attacks and the problems of daily survival in a harsh environment. In 1692 a group of young women began to behave in such bizarre ways as screaming, convulsing, crawling on all fours, and barking like dogs. The community focused attention on these women. The physician defined them as "witches," representatives of Satan, and the entire community banded together in search of others who were under the "devil's influence." The community imprisoned many persons suspected of sorcery and sent 22 persons to their deaths. Thus the witch-hunt provided a scapegoat, an outlet for people's fears and anxieties (Erikson, 1966).

A third function of public identification of deviants is to increase the cohesion and solidarity of society. Nothing unites the members of a group like a common enemy (Cohen, 1966). Deviants, in this context, are "internal enemies," persons whose behavior threatens the morale and efficiency of a group. Should the solidarity of the group be threatened, it can be restored by identifying one member as deviant and imposing appropriate sanctions.

Suppose you are given the case study of a boy with a history of delinquency who is to be sentenced for a minor crime. You are asked to discuss the case with three other persons and decide what should be done. One member of the group argues for extreme discipline, but you and the other two favor leniency. Suddenly, an expert in criminal justice who has been sitting quietly in the corner announces that your group should not be allowed to reach a decision. How might you deal with this threat to the group's existence? The reasoning outlined above suggests that the person who took the extreme position will be identified as the cause of the group's poor performance and that other members will try to exclude him from future group meetings. A laboratory study used exactly this setup, contrasting the reaction of threatened groups to the person taking the extreme position with the reaction of nonthreatened groups. In the former condition, the person taking the extreme position was more likely to be stigmatized and rejected (Lauderdale, 1976).

Thus controlled amounts of deviant behavior serve important functions. If deviance is useful, we might expect control agencies to "create" deviance when the functions it serves are needed. In fact, the number of persons who are publicly identified as deviant seems to reflect the levels of stress and integration in society (Scott, 1976). When integration declines, there is an increased probability of deviance. Eventually, the level and severity of deviance may reach a point where citizens demand a "crackdown." Social-control agencies step up their activity, increasing the number of publicly identified deviants. This, in turn, increases solidarity and lowers stress, leading to an increase in the amount of informal control and a reduction in deviance.

The Process of Labeling   Labeling is not a simple one-step procedure for formal agencies. The processing of rule breakers usually involves a sequence of decisions. At each step, someone has to decide whether to terminate the process or to pass the rule breaker on to the next step. Figure 19.4 shows the sequence of steps involved in processing criminal defendants.

Each of the control agents—police officers, prosecutors, and judges—has to make many decisions every day. Like anyone else, they develop a cognitive *schema* and rules that simplify their decision making. A very common police-citizen encounter occurs when an officer stops a motorist who has been drinking. What determines whether a driver who has been drinking is labeled a "drunken driver"? Officers on the street have to rely on a variety of subjective data because the breathalyzer or blood or urine test may only be available at the police station. Research suggests that police officers develop a series of informal guidelines which they use in deciding whether to arrest the motorist. In one study of 195 police encounters with persons who had been drinking, arrests were more likely if the encounter occurred downtown and if the citizen was disrespectful (Lundman, 1974).

Prosecutors also develop informal rules that govern their decisions. For example, in one large midwestern city, taking an object worth less than $100 is a misdemeanor, and conviction normally results in a fine. Theft of a more valuable object is a felony and results in a prison sentence. Because felony theft cases require much more time and effort, the prosecutor has charged most persons arrested for shoplifting with misdemeanors, even if they have taken jewelry worth hundreds of dollars.

In many jurisdictions, probation officers are asked to prepare a presentencing report and to recommend a sentence for the convicted person. Research indicates that these officers have a set of typologies

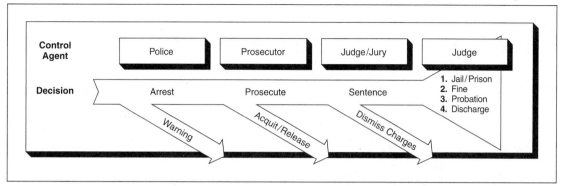

Figure 19.4 Formal Social Control: Processing Criminal Defendants

Formal social control often involves several control agents, each of whom makes one or more decisions. The first step in the criminal justice system is an encounter with a law enforcement officer. If you are arrested, the case is passed to a prosecutor, who decides whether to prosecute. If your case goes to court, the judge or jury decides whether you are guilty. Finally, the judge renders a sentence. These decision makers are influenced by their own personal attitudes, cognitive schemas, role expectations, and the attitudes of others regarding their decisions. Much research is devoted to the social psychological aspects of decision making in the criminal justice system.

or schemas into which they sort persons (Lurigis & Carroll, 1985). Semistructured interviews with probation officers in one community identified 10 schemas, including burglar, addict, gang member, welfare fraud, and con man. Each schema was associated with beliefs about the motive for the crime and the appropriate treatment and prognosis. When officers were asked to evaluate sample cases, those fitting a schema were evaluated more quickly and confidently. More experienced officers were more likely to use similar schemas (Drass & Spencer, 1987).

Each step in this process involves interaction between professionals and the alleged rule breaker, and often members of his or her family. The professional's goal is to have the rule breaker and other laypersons accept the label. Research on the labeling of children with developmental disabilities suggests that family members are more likely to accept a diagnosis if the professional elicits family members' schemas, and frames the diagnosis/label in those terms (Gill & Maynard, 1995).

Biases in Social Control Not all persons who violate the rules are labeled. Most social-control agencies process only some of those who engage in rule-breaking behavior. In the study of police encounters

with drunken persons, only 31% were arrested (Lundman, 1974). In some cases, control agents may be influenced by the demeanor of the rule breaker, by the agent's schema, or by where the violation occurs. This leads us to ask whether systematic biases exist in the social-control system.

It has been suggested that control agents are more likely to label those people who have the least power to resist certification as deviant (Quinney, 1970). This hypothesis predicts that people from the lower class and members of racial and ethnic minorities are more likely to be certified as deviant than upper-class, middle-class, and white persons. This hypothesis offers a radically different explanation for the correlation between crime and social class. Earlier in this chapter, we suggested that crime rates are higher for lower-class persons because they do not have access to nondeviant means of economic success. Here we are suggesting that crime rates are higher among lower-class persons because they are more likely to be arrested, prosecuted, and found guilty, even though the underlying rate of deviant activity may not vary as a function of social class.

Does social class or race influence how an individual is treated by control agents? One way to answer this question is by studying police-citizen encounters

Whether or not a police officer gives a citizen a traffic ticket depends partly on the demeanor of the citizen. Officers are more likely to ticket or arrest hostile, argumentative persons than polite and submissive ones.

through the "ride-along" method, in which trained observers ride in squad cars and systematically record data about police-citizen encounters. In the largest study of this kind, observers rode with some officers on all shifts every day for 7 weeks. Data were collected in Boston, Washington, and Chicago and included 5,713 encounters. There was no evidence that blacks were more likely to be arrested than whites. Rather, arrests were more likely when a third party demanded an arrest, when the evidence was strong, and when the crime was serious (Black, 1980). A study of how police officers managed violent encounters between citizens found that arrest was more likely if the incident involved white persons, two men (instead of one woman and one man, or two women), and if one person acted abusively toward the officer (Smith, 1987). Recent research that included ride-alongs in Washington, D.C., suggests, at least in that city, blacks are subjected to more intense police surveillance than other racial/ethnic groups (Chamblis, 1994).

What about decisions by prosecutors? Do they entail discrimination based on race or class? Prosecutors are generally motivated to maximize the ratio of convictions to trials. This may be one criterion citizens use in evaluating the performance of a district attorney. Prosecutors develop beliefs about which cases are "strong"—those likely to result in conviction. A study of a random sample of 980 defendants charged with felonies found that prosecutors are more likely to prosecute cases involving serious crimes where the evidence is strong and the defendant has a serious prior police record. Race was not generally influential (Myers & Hagan, 1979).

Does the social class of an arrested person influence how he or she is treated by the courts? Several studies of the handling of juvenile cases report little evidence of class or race bias. A study of cases in Denver and Memphis found that the seriousness of the offense and the youth's prior record were the major determinants of the sentence given (Cohen & Kleugel, 1978). Two longitudinal studies, of 9,945 boys in Philadelphia (Thornberry & Christenson, 1984) and of cases in Florida (Henretta, Frazer, & Bishop, 1986), found that the most important influence on the disposition of a charge was the disposition imposed for a prior offense or offenses.

A common practice in adult criminal cases is *plea bargaining,* in which a prosecutor and a defendant's lawyer negotiate a plea to avoid the time and expense of a trial. A single action frequently violates several laws. Thus, if a driver who has been drinking runs a red light and hits a pedestrian who later dies, that incident involves at last three crimes: drunken driving, failure to obey a signal, and vehicular manslaughter. These offenses vary in seriousness and thus in their associated sentences. The prosecutor may offer not to indict the driver for manslaughter if a plea of guilty is entered to a drunken driving charge. The attorney may accept the offer, provided the prosecutor also recommends a suspended sentence.

Are the members of certain groups more likely to be tried or to get bigger reductions in sentences? An analysis of charge reduction or plea bargaining in a sample of 1,435 criminal defendants found that women and whites received slightly more favorable reductions than men and blacks (Bernstein et al., 1977). Another study of 1,213 men charged with felonies found the characteristics of an offense—especially the seriousness of the crime and the strength of the evidence—were most important in determining the disposition. The outcomes of the cases were not related to age, ethnicity, or employment status (Bernstein, Kelly, & Doyle, 1977). A study of 296 women who killed another person found that, while they were all initially charged with murder, in two thirds of the cases the charge was reduced to manslaughter or a lesser offense (Mann, 1996). Women in southern cities and women who killed men were less likely to have the charges reduced, and received more severe sentences if convicted.

Among persons convicted, do we find a class or racial bias in the length of sentences given? One study focused on the sentences received by 10,488 persons in three southern states: North Carolina, South Carolina, and Florida (Chiricos & Waldo, 1975). Researchers examined sentences for 17 different offenses and found no relationship between socioeconomic status or race and sentence length. Again, the individual's prior record was the principal variable related to sentence length. A study of a random sample of 16,798 felons convicted during the years 1976 to 1982 in Georgia looked at racial differences in sentencing (Myers & Talarico, 1986). In general, the seriousness of a crime was the principal influence on the sentence length.

Earlier we discussed white-collar crime, which is often committed by middle- and upper-class persons. Are white-collar offenders more likely to receive lenient sentences? A study of persons charged with embezzlement and tax, lending, credit, postal, and wire fraud found that, within this group, high-status persons were no less likely to be imprisoned or to receive shorter sentences (Benson & Walker, 1988). The significant influences were total dollars involved and how widespread the offenses were. Blacks did receive longer sentences than whites. It is sometimes argued that judges are lenient on high-status offenders because they suffer serious informal sanctions, such as loss of job. A study of the likelihood of job loss and the influence of job loss on sentence severity found no relationship (Benson, 1989). However, class position did influence job loss; high-status offenders and those whose frauds were larger in scale were less likely to lose their jobs.

Long-Term Effects of Formal Labeling

How long does the official label of deviant stick to a person? Can it be shaken? In contrast with the trial or hearing in which a person is formally certified as deviant, there is no formal ceremony terminating one's deviant status (Erikson, 1964). People are simply released from prison or a mental hospital, or the final day of probation passes—with no fanfare. Does the individual regain his or her former status upon release, or does deviant status in our society tend to be for life?

Some argue that ex-convicts, ex-patients, and others who have been labeled as deviant face continuing pressures from family and friends that could prevent them from readjusting to normal life. Such pressures constitute a reminder of their former deviant status. However, two studies of former psychiatric patients (Greenley, 1979; Sampson et al., 1964) found no evidence of continuing stigmatization by members of their families.

Another area in which former prison inmates and mental patients might face discrimination is employment. This may occur because others continue to perceive these persons as deviant and expect them to behave in ways consistent with that label. In one study (Schwartz & Skolnick, 1964), researchers prepared four versions of a job application to be shown to prospective employers. All four applications were largely identical; all applicants had a succession of short-term jobs. The only variable was their legal record. In one condition, the applicant had been convicted of assault; in the second, he had been charged and acquitted. In the third condition, he had been acquitted and the application included a letter from the trial judge verifying the acquittal. In the fourth (control) condition, he had no legal record. Each of the four versions was shown to 25 employers. Compared with the control condition, employers were less likely to respond favorably when the person had been arrested. Employers were about equally likely to respond negatively whether the man had been acquitted or convicted. When the letter from the judge was included, the employer's interest in the applicant was higher.

A related concern is the impact of mental illness on occupational careers (Huffine & Clausen, 1979). A study of psychiatrically disturbed persons compared the income and employment status of those who had been treated (labeled) with the income and status of those who had not been treated. Treatment was negatively associated with both income and employment (Link, 1982). The impact seemed to depend partly on whether occupational competence was developed before or after the onset of the illness. Men who had no history of competent work performance had more difficulty obtaining employment following hospitalization. Men who had a history of occupational competence usually kept their jobs, even during periods when their work performance was seriously affected.

Some persons turn a career as a deviant into an occupational asset by becoming a "professional ex-" (Brown, 1991a). Individuals with histories of alcohol or drug abuse or other problem behaviors sometimes become counselors, working with others who are involved in these behaviors. Professionalizing rather than giving up the deviant identity is another way of going straight.

A study of the long-term impact of being labeled mentally ill suggests it is not the label by itself that has impact but the label combined with changes in self-perception (Link, 1987). The study compared samples of residents and clinic patients from the same area within New York City. Three samples involved people who had been labeled: first-treatment contact patients, repeat-treatment contact patients, and formerly treated community residents. The other two groups were untreated "cases" (people with symptoms) and a sample of residents. All participants completed a scale which measured the belief that mental patients are stigmatized and discriminated against. High scores on the measure were associated with reduced income and unemployment in the labeled groups but not in the unlabeled ones. Later research shows that when people enter treatment, those who expect discrimination use strategies such as keeping their condition secret or withdrawing from interaction (Link et al., 1989). This tends to cut them off from social support and interfere with their work performance.

The long-term effects of formal labeling on the reactions of others may be limited because persons who have been labeled in the past engage in various tactics to prevent others from learning about their stigma. These tactics include selective concealment of past labeling, preventive disclosure to close friends, and various deception strategies (Miall, 1986). Longitudinal research, however, suggests that persons who have been publicly labeled and treated continue to anticipate rejection from others even though they no longer engage in the symptomatic behavior (Link et al., 1997). Thus stigma may have lasting effects on the person's psychological well-being.

Summary

Deviant behavior is any act that violates the social norms which apply in a given situation.

The Violation of Norms   (1) Norms are local, subcultural, or societal in origin. The repercussions of deviant behavior depend on which type of norm an individual violates. (2) Anomie theory asserts that deviance occurs when persons do not have legitimate means available for attaining cultural success goals. Possible responses to anomie include ritualism, retreatism, innovation, and rebellion. (3) Control theory states that deviance occurs when an individual is not responsive to the expectations of others. This responsiveness, or social bond, includes attachment to others, commitment to long-term goals, involvement in conventional activities, and a respect for law and authorities. (4) Differential association theory emphasizes the importance of learning through interaction with others. Individuals often learn the motives and actions that constitute deviant behavior just as they learn socially approved behavior. (5) The routine activities perspective calls attention to situations that facilitate the convergence of offenders and targets, in the absence of a guardian.

Reactions to Norm Violations   Deviant behavior involves not only acts that violate social norms but also the societal reactions to these acts. (1) There are numerous possible responses to rule breaking. Very often we ignore it. At other times, we deny the act occurred, define the act as normal, excuse the perpetrator, or recognize the act but deemphasize its significance. Only after an act is discovered and labeled "deviant" is it recognized as such. (2) Our reaction to rule breaking depends on the characteristics of the actor, the audience, and the situation. People often have a stereotyped image of deviant persons; these stereotypes influence how audiences react to rule violations. (3) The consequences of rule breaking depend on the reactions of the audience and the response of the rule breaker. If members of the audience reward the person, the deviance may become institutionalized. Alternatively, the person may decide to avoid further deviance, in spite of others' encouragement. If the person is punished, he or she may give up the behavior or respond with additional rule violations.

Labeling and Secondary Deviance   The process of labeling has two important consequences. (1) It leads members of an audience to change their perceptions of and behavior toward the actor. If they withdraw from the stigmatized person, they may create a self-fulfilling prophecy and elicit the behavior they expected from the actor. (2) Labeling often causes the actor to change his or her self-image and come to define the self as deviant. This, in turn, may lead to secondary deviance—an open and active involvement in a lifestyle based on deviance. Such lifestyles are often embedded in deviant subcultures.

Formal Social Controls   Every society gives certain agents the authority to respond to deviant behavior. (1) In U.S. society, the major formal social control agents are the criminal justice, juvenile justice, and mental health systems. These agencies select persons and identify them as deviant through a se-

quence of decisions. Within the criminal justice system, the sequence includes the decisions to arrest, prosecute, and sentence the person. Various factors influence each step in decision making, including the strength of the evidence, the seriousness of the rule violation, and the individual's prior record. (2) Contrary to popular belief, people do not systematically stigmatize former deviants. Most families do not continue to stigmatize relatives following their release from mental hospitals, and most employers do not stigmatize ex-patients and ex-convicts who have established competent work records. But stigma may have long-term effects on the ex-deviant's psychological well-being.

Key Terms

anomie theory (p. 469)
control theory (p. 471)
deterrence hypothesis (p. 488)
deviant behavior (p. 468)
deviant subculture (p. 486)
differential association theory (p. 474)
formal social control (p. 488)
informal social control (p. 488)
institutionalization of deviance (p. 483)
labeling theory (p. 479)
learning structure (p. 470)
legitimate means (p. 469)
opportunity structure (p. 470)
routine activities perspective (p. 478)
rule breaking (p. 479)
secondary deviance (p. 486)
self-fulfilling prophecy (p. 484)

CHAPTER 20

Collective Behavior and Social Movements

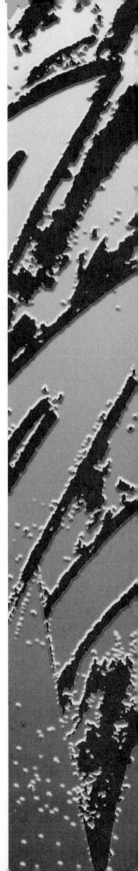

Introduction

Collective Behavior

Social Movements

Introduction

- Following the victory by the Denver Broncos in Super Bowl XXXII, thousands of excited fans poured into the 16th Street Mall in Denver. Within 30 minutes, the crowd turned violent, setting fires, breaking windows, and overturning cars.

- Rumors about the failure of a major South Korean bank set off waves of panicky selling on Wall Street and foreign stock markets, completely disrupting the flow of transactions on the New York Stock Exchange.

- In the wake of the shooting of a black teenager by police officers, thousands of blacks marched through the streets to city hall, and a series of speakers demanded changes in police practices.

Events such as these occur daily and sometimes receive national media coverage. In part because they occur frequently, and often have serious consequences, they have been of interest to social scientists since the turn of the 20th century.

Collective behavior refers to two or more persons engaged in behavior judged common or concerted on one or more dimensions (McPhail, 1991). This is an intentionally broad definition because a wide range of events have been studied by social scientists as examples of collective behavior, including the three described above.

Collective behavior has three dimensions: the spatial frame, the temporal frame, and the scale of social activity. With regard to *space,* collective behavior may occur at a single point (such as a street corner), or at a larger site (such as a football stadium), or across an entire state or nation. The *temporal duration* of collective behavior can vary from a few minutes (such as a violent attack by a gang), to several hours (such as a victory celebration), to several days (such as the racial disorder that occurred in Los Angeles in 1992).

Collective behavior also varies in terms of the third dimension, the *scale* of the activity. In fact, we usually only learn about large-scale events because newspapers and television news programs tend to report only the largest rallies, demonstrations, riots, victory celebrations, or political campaigns (McCarthy, McPhail, & Smith, 1996).

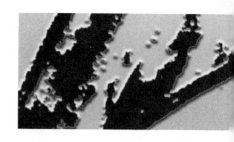

Most research on collective behavior focuses either on short-term, unorganized events—often referred to as crowds—or on long-term, relatively organized social movements.

The first part of this chapter is concerned with collective behavior. The second part discusses social movements. Specifically, this chapter addresses the following questions:

1. What social processes are involved in collective behavior?
2. What causes collective behavior? That is, what conditions facilitate it, and what conditions precipitate particular collective activities?
3. What factors influence the behavior of people when they gather in a crowd?
4. How do social movements develop? What are the processes by which social movements define issues and attract members?
5. How do movement organizations mobilize supporters? How is their operation affected by processes inside and outside the organization?
6. How do social movements affect the larger society?

Collective Behavior

Several years ago, on Saturday, October 30, the University of Wisconsin Badgers, playing a football game at home, beat a fabled opponent, the University of Michigan Wolverines, 13 to 10. Thousands of exuberant fans poured onto the field to congratulate the Wisconsin players and to celebrate. In Sections O, P, and Q of the stadium, an estimated 12,000 persons attempted to move forward onto the field, seemingly in unison. But their progress was blocked by a 3-foot-high iron railing in front of the stands and by a 6-foot-high chain-link fence just beyond it. As the people in the back pressed forward, those in front were crushed against the barrier, and many fell and were trampled. The force ripped the railing out of its concrete moorings and flattened the fence. At least 68 people were injured, and 16 were hospitalized for one night or more. Miraculously, no one was killed. The incident received nationwide publicity.

Several perspectives can be applied to help us understand incidents like these. The classic perspective has its roots in the earliest writings of social psychologists (LeBon, 1895). It uses such concepts as the *crowd,* anonymity or *deindividuation, contagion,* and *emergent norms* to analyze collective incidents. We begin our discussion using this perspective.

Crowds

A **crowd** is a temporary gathering of persons in close physical proximity, engaging in joint activity that is unconventional (Snow & Oliver, 1995). Participants may engage in one activity in common, such as listening to a speech or spontaneously singing a song, or in concerted action, such as vandalizing cars or rescuing victims from a collapsed building. Crowd incidents may be characterized by unanimity of feeling (Turner & Killian, 1972). For instance, in the surge at the Wisconsin football game, most persons in the stadium were Wisconsin fans and were elated at the unexpected victory over a powerful rival. Similarly, a crowd engaged in looting stores during a riot may share feelings of hostility toward the stores' owners.

According to the classic perspective, it is this unanimity of feeling that gives direction to the crowd's behavior. LeBon referred to this as "the mental unity of the crowd." Unity leads participants to think, feel, and act in ways that are different than they would if each were alone. Thus the elation shared by the Wisconsin fans led them to want to celebrate on the field. Many of them pushed forward at the same time, causing the surge. Obviously, if these persons all had different feelings, a surge would not have occurred.

Deindividuation   One influence on behavior in crowds is anonymity, or **deindividuation,** a temporary reduction in self-awareness and sense of personal responsibility (Festinger et al., 1952). When an individual participates in a crowd with others, many of whom may be strangers, there is a decline in the sense of responsibility for one's behavior. This makes it easier for the person to act on impulse and to engage in behavior that violates social norms—such as shoving hard against others or breaking windows and overturning cars.

Contagion   A second influence on behavior in crowds is **contagion,** the rapid spread through a crowd of visible and often unusual symptoms or behavior. The outbreak of violence in Denver during the celebration of the Broncos' victory apparently occurred when a group of people began kicking in the windows of an Athlete's Foot store (Griego, 1998). This behavior spread; others began to break windows in neighboring stores.

Another illustration of contagion occurred in a child-care center in Florida in 1989 (Centers for Disease Control, 1990). On July 26, 102 children ate lunch at the center; many of them ate a prepackaged lunch. During the meal, a 12-year-old girl complained her food tasted bad, said she felt nauseous, and vomited. Quite suddenly, within the next 40 minutes, 62 other children experienced one or more symptoms, including cramps (77%), nausea (75%), and headaches (51%). Forty-two of these children appeared to have vomited, although some may have only spit out their food. All children who experienced symptoms were examined in hospital emergency rooms within 2 hours; none showed any physical signs of illness. An investigation was subsequently conducted, including interviews with students and staff and tests of meal samples. No chemical or biological cause of illness could be identified.

This incident is an example of *hysterical contagion* (Kerckhoff, Back, & Miller, 1965). Although we are all frequently exposed to models of behavior, such as people stating they have a headache, we usually are not influenced by such models. However, contagion is more likely to occur when two other conditions are met: (1) the persons involved are experiencing tension, and (2) the behavior being modeled is relevant to their situation. Both conditions may have been present in the Florida day-care center. Perhaps the affected children had been stressed by incidents at home

Behavioral contagion often occurs in a crowd. One person acts, liberating others from the restraints that prevented them from performing the act. Behavioral contagion can produce something trivial like a "wave" at a football game—or something serious like this attack on a spectator at a soccer match.

or at the center. Certain patterns in the so-called epidemic reflect the influence of relevance. For example, the first person to become ill was a girl, and 75% of those affected were girls. Only children who ate the prepackaged lunch were affected. Moreover, many who became ill had been seated where they could see other children becoming ill.

Research on another epidemic, which involved fainting, nausea, and reports of pain with no identifiable cause, reveals that ties to social networks affected the spread of the epidemic. The epidemic occurred in a textile plant, at a time when there were rumors that the plant would close. As the so-called illness spread, it increasingly affected friends of those who had already been affected (Kerckhoff, Back, & Miller, 1965).

Emergent Norms   According to *emergent norm theory,* collective behavior occurs when people find themselves in an undefined or unanticipated situation (Turner & Killian, 1972). The situation may be novel, so there are no cultural norms to guide or direct action. For instance, in recent years there have been several incidents in which a person with a gun walked into a school and opened fire. On December 8, 1997, 14-year-old Michael Carneal walked into his high school in Paducah, Kentucky, carrying five guns. He stood in the hall waiting for a group of students to finish a prayer. Then he took a pistol from his backpack and fired 12 shots at the group, killing three girls and wounding five other students. At first, no one moved or attempted to stop Carneal. Such incidents are completely unexpected, and there are no behavioral guidelines; people don't know what to do. In other cases, the social structure may be temporarily disrupted by a natural disaster, such as a tornado, or by an event, such as a citywide strike by police officers. Another possibility is that there may be conflicting definitions

20.1   Reactions to Disasters

A **disaster** is an event that produces widespread physical damage or destruction of property accompanied by social disruption (Quarantelli & Dynes, 1977). Many disasters, such as explosions and fires, occur with no warning. Sometimes, however, there may be prior warning, such as increasingly strong tremors before an earthquake or warnings by authorities prior to a hurricane.

There are three types of reactions to the threat of disaster (Perry & Pugh, 1978). Denial is the most common when the likelihood of the disaster is perceived as small or the warning is ambiguous. Extreme emotion, such as terror and hysteria, is likely if the threat is accurately perceived but not imminent. Effective adaptation is most likely if there are repeated, accurate warnings that include information about what people can do to enhance their survival. Research shows that reactions to threat are more appropriate and effective when information is communicated through official channels; inappropriate and ineffective behavior is more likely when information is transmitted through unofficial, interpersonal channels such as rumor.

Little is known about how people behave at the time a disaster hits. What data there are indicate that most people do not panic (Perry & Pugh, 1978). In fact, critical thinking and problem solving may be enhanced (McPhail, 1991). Panic in response to a disaster occurs only under certain conditions. First, the situation must be defined as dangerous. Panic will not occur if most people deny there is any threat to themselves. Second, the danger must be perceived as escapable by some action. If there is a fire in a building, for example, at least some of those present must believe there are doors or windows through which they can escape. People often remain calm when it is obvious there is no possibility of escape, for example, in an emergency on an airplane at high altitudes. Thus, when a section of the roof of a commercial jetliner disintegrated at 26,000 feet, the passengers, although scared, remained seated until the plane landed about an hour later. Third, at least some people must believe escape routes are inadequate or will be cut off.

Under these conditions, people are influenced by the behavior of others. If there is an emergency during a concert, and the emcee calmly announces the procedures to be followed, everyone may leave in an orderly fashion. But if several people suddenly run toward the exits, others may attempt to do the same. This clearly illustrates the emergent norm model. An experiment (Kelley et al., 1965) in which all three conditions were simulated revealed that people were more likely to escape successfully when the threat was low rather than high and when the group was small (three or four) rather than large (five or six).

What happens after a disaster depends, in part, on the ability of existing emergency organizations to respond, including police and fire departments, hospitals, utility companies, and civil defense agencies (Quarantelli & Dynes, 1977). Such organizations often have drills to prepare their personnel to respond to emergencies rapidly and efficiently. At the scene of a disaster, an *emergency social system* emerges (Perry & Pugh, 1978). At first, residents of the impact area may work together in informal groups. They take responsibility for the recovery of victims, removal of debris, and the initial assessment of damages. A tremendous sense of community and high morale often develop. Gradually, the emergency organizations move in to supplement or supplant the emergency social system.

of how people should behave (see Box 20.1, "Reactions to Disasters"). To act in these situations, those present must develop a shared definition of the situation and the associated behavioral norms.

In all these circumstances, people want to find out what is going on or what they should do. Because they need information, conventional barriers to communication break down. Strangers talk to each other or to members of groups they usually avoid. In addition, the usual standards of judgment and morality may be suspended. **Rumor**—communication through informal and often novel channels that cannot be validated—can exert a major influence on the emerging definition of the situation. In some incidents, rumors are broadcast by radio and TV stations, making them appear to be true. In other incidents,

People do not always panic and run when a disaster occurs. When it is clear that escape is not possible, as it was when a section of this plane ripped off in flight, people will remain in their seats, stationary if not calm.

milling—the movement of persons within a setting and the consequent exchange of information between crowd members—is the primary method through which rumor is transmitted. Persons who are not physically present may learn about the emerging situation from radio or television reports, or through telephone calls (McPhail, 1991).

Diverse interpretations and action tendencies may be present in a crowd situation (Turner & Killian, 1972). Someone initiates an act, perhaps in the belief that others will support him or her. Once a person initiates an act, the support of those nearby determines whether that person will persist in attempts to influence others. If enough people reinforce that person's position or behavior, a consensus will emerge. The definition of the situation that results from interac-

tion in an initially ambiguous situation is termed an **emergent norm.** The emergent norm is usually not completely novel; it involves a modification or transformation of preexisting norms (Killian, 1984).

Once a definition of the situation develops, people are able to act purposively. In a crowd, behaviors consistent with the norms are encouraged, whereas behaviors inconsistent with the norms are discouraged. Thus there are normative limits on the behavior of crowd participants. Crowds celebrating a football championship do not engage in looting. Conversely, crowds of looters in the inner city do not congregate in bars for several hours, drinking alcoholic beverages and chanting for their victorious team.

A distinctive image of crowds emerges from this perspective. Crowds are viewed as emotional. Crowd activity reflects the rapid spread of a behavior, often one that violates social norms (such as looting stores), through the crowd. The spread is facilitated by anonymity; people can engage in deviance without fear of sanction because the others present don't know them. The result is large numbers engaging in the same behavior, often leading to collectively irrational outcomes such as the injuries suffered due to the surge at the Wisconsin-Michigan game.

Gatherings

Although the traditional view of crowds as emotional has a certain value, such a view is at best incomplete. Crowds are not uniquely emotional, nor are they unanimous in activity or always antisocial. In recent years, an alternative perspective has been developed that calls our attention to other aspects of crowd incidents and collective behavior (McPhail, 1991). This perspective uses such concepts as the *gathering*, the *phases* of a gathering, and *companion clusters* to analyze collective behavior.

According to this view, the social setting for many forms of collective behavior is a **gathering,** a temporary collection of two or more persons occupying a common space and time frame (McPhail, 1991). Gatherings are the basis of collective behavior. People may gather for a variety of reasons. Some gatherings are for purposes of *recreation* or "hanging out," as in parks, theaters, swimming pools, or at the scene of a fire, accident, or arrest. Other gatherings are *demonstrations* that involve two or more people meeting in public to protest or celebrate some person, principle, or condition; these may be political or religious in nature, or involve an athletic event. Still other gatherings are *ceremonies* intended to mark a change in status or a life-course transition; these may be semi-public or private events.

Behavior in Gatherings The behaviors of persons in gatherings reflect their purposes. Many people who attend share the stated purpose (e.g., to celebrate a victory). But others come with other purposes—to accompany a friend, to meet potential dates, or to pick someone's pocket. What occurs reflects two influences: (1) participants' purposes, and (2) features of the situation. Consider again the surge. At the end of the game, some of those in the student section wanted to go onto the field to celebrate. Others wanted to leave the stadium. Others wanted to get something to eat or drink. Each of these required movement toward the lower level of the stands. A situational feature, unknown to most of them, was the iron railing; because it was only 3 feet high, it was not visible to those standing more than a few rows back. It is the interaction of participants' purposes and the situational feature, the railing, that caused the undesirable outcome, injuries to 68 people.

Gatherings have three phases: assembling, activities, and dispersal (McPhail, 1991). We will briefly examine each in turn.

Assembling Any gathering is the result of people coming together in a common space and time frame (McPhail, n.d.). This process may involve convergence, or it may reflect the ecology of the location at which the gathering occurs.

Convergence refers to the situation in which those present at a gathering share certain qualities. The spectators at a football game are there to see the game. They may have other, more idiosyncratic purposes as well, but they are fans, and this fact influences their behavior. The surge at the Wisconsin-Michigan game occurred in part because many fans wanted to celebrate the unexpected victory. Convergence at a gathering is much more likely if the gathering has been publicized in advance. It is also more likely if the media broadcast news of it as it occurs.

More often, the composition of a gathering reflects the social ecology of the environment. Other factors being equal, the greater the density of an area, the larger the number of potential participants. Crowd events are much more likely to occur in central cities than in suburbs or rural areas. Organizers of demonstrations learned many years ago that they need to provide buses if sympathizers are not located near the site of the demonstration.

Research suggests that many persons present at gatherings come as part of a group of from two to five persons (McPhail, 1991). These small groups are usually made up of acquaintances, friends, or family members. This composition is important because the presence of others who know the person establishes some informal social control over his or her behavior. The classic perspective on crowds, which emphasized anonymity and the resulting lack of control over participants' behavior, was incorrect on this point. Although many participants in gatherings do not know each other, each participant is often part of a small group. These groups, not individuals, are really the elementary social units of collective behavior.

Activities The activities of participants in gatherings are not random. McPhail (1991) has identified "elementary forms of collective action." The most common is **companion clusters,** of family, friends, or acquaintances who remain together throughout the gathering. A second activity is the queue, for admission, access, or service. A third form is arcs or rings of participants around performers, speakers, or fights. A fourth is displays of evaluation—oohs, aahs, whistles, boos, applause.

Certain activities are common in particular types of gatherings. For instance, religious and sport gatherings involve celebration rituals—individual or collective chanting, singing, or praying, combined with symbolic gestures, such as the "wave." Most members of the culture are familiar with these rituals because of childhood socialization or exposure via mass media. Thus even a first-time participant in a religious service can act in unison with others. Ceremonial gatherings often involve singing, dancing, and musical performances; these are sometimes quite complex and require considerable advance planning and coordination. Retirement and farewell ceremonies and funerals often involve tributes to the person, and have a typical form.

Dispersal Gatherings end or disperse in one of three ways: routine, coerced, and emergency. By far the most common but least studied is the *routine* type in which those present leave the setting in an orderly fashion. People often queue as they leave an airplane, football stadium, or concert hall. They typically leave in the company of the same people they assembled with. When there are large numbers of persons or vehicles, officials may facilitate dispersal by directing traffic. In more open settings, such as a concert in a park, people may leave in clusters.

Coerced dispersal refers to the situation in which social-control agents, such as police officers or firefighters, direct people to leave before the intended purpose of the gathering is achieved. This occurs when authorities suspect that those gathered are in some danger. An example is police directing citizens to leave a stadium or concert because of a bomb threat. Another type of coerced dispersal occurred in downtown Denver, during the Super Bowl victory celebration, when the police used tear gas to force the revelers to leave the area.

The most frequently studied is the *emergency* dispersal. This refers to situations where people have to deal with a suddenly disrupted or dangerous environment. Behavior in disasters was discussed in Box 20.1, "Reactions to Disasters."

Underlying Causes of Collective Behavior

Having considered the internal dynamics of gatherings, we turn now to the causes of collective behavior. In some instances, collective behavior is simply a response to some event, such as a natural disaster, an athletic victory, or an assassination. Other types of collective behavior—demonstrations, boycotts, lynchings, lootings, and epidemics—frequently involve not only a specific event but also more basic underlying conditions in the larger society. Three such conditions are strain, relative deprivation, and grievances.

Strain Society may be viewed as normally in a state of equilibrium, maintaining a balance between the emphasis on achieving society's goals and the provision of the means to achieve them—education and jobs (Merton, 1957). At times, however, social change may disrupt this equilibrium, so one aspect of society is no longer in balance with other aspects. Advances in technology, for example, demand changes in occupational structure. Machines and robots have replaced many blue-collar workers in automobile plants. This has produced high unemployment in cities like Detroit that depend heavily on the auto industry. Such change produces strains in society that cause some individuals to experience stress (see Chapter 18). Although those who are affected may not recognize the source (for example, automation), they experience stress or frustration, which can contribute to the occurrence of collective behavior.

Historically, economic issues have frequently been at the heart of collective protest (Rude, 1964). Food riots to protest the lack of sufficient food, attacks on factories and businesses to prevent mechanization, and sabotage to disable machinery and other property are often economically motivated. These activities were common in preindustrial England and France. More recently, bank failures in Japan and Korea produced economic crises in several Southeast Asian countries, where currencies declined sharply in value. The reduced purchasing power that resulted led to widespread rioting in Indonesia in February 1998 (*New York Times,* 1998). Rioters frequently targeted businesses and homes of ethnic Chinese, whom they blamed for soaring prices. These protests reflect the strain caused by widespread unemployment and inadequate incomes.

Evidence that economic issues are related to strain comes from a study of support for radical political proposals (Plutzer, 1987). Telephone interviews were conducted with 912 adults living in the largest U.S. cities and with 458 unemployed men and women. Each was asked whether he or she supported several radical proposals. These included having the government limit personal income or corporate profits and doing away with capitalism, if necessary, to limit unemployment. Results showed that economic insecurity—concern about one's economic future—was associated with support for the proposals.

Relative Deprivation   In the 18th century, the revolt against the feudal socioeconomic structure occurred first in France. Yet France had already lost many feudal characteristics by the time the French Revolution began in 1789. The French peasant was free to travel, to buy and sell goods, and to contract services. In Germany, however, the feudal social structure was still intact. Thus, based on objective conditions, we would have expected a revolution to occur in Germany before it did in France. Why didn't it? One analyst (de Tocqueville, 1856/1955) argued that the decline of medieval institutions in France caused peasants to become obsessed with the ownership of land. The improvement in their objective situation created subjective expectations for further improvement. Peasant participation in the French Revolution was motivated by the desire to fulfill subjective expectations—to obtain land—rather than by a desire to eliminate oppressive conditions.

This basic insight into the causes of revolutions was expanded into a more systematic view. According to the **J-curve theory** (Davies, 1962, 1971), the "state of mind" of citizens determines whether there is political stability or revolution. Based on external conditions, individuals develop expectations regarding the satisfaction of their needs. Expectations may be derived from one's own experience or from a comparison with the experiences of other groups. Under certain conditions, persons expect continuing improvement in the satisfaction of their needs. If these expectations are met, people are content and political stability results. But if the gap between expectations and reality becomes too great, people become frustrated and engage in protest and rebellious activity.

Revolutions usually occur when the level of actual satisfaction declines following a period of rising expectations and their relative satisfaction (Davies,

1971). These relationships are summarized in Figure 20.1. Note the J shape of the actual need-satisfaction curve; as satisfaction declines, an intolerable gap between expected need satisfaction and actual need satisfaction is created.

The gap between expected and actual need satisfaction is called **relative deprivation.** As relative deprivation becomes greater, it is hypothesized that the likelihood of protests and social movements increases. As noted earlier, expectations may be based on one's past experience or on a comparison with the experiences of other groups. Although their freedom had increased dramatically, French peasants in prerevolutionary times experienced deprivation because they were not free to own land like the aristocracy.

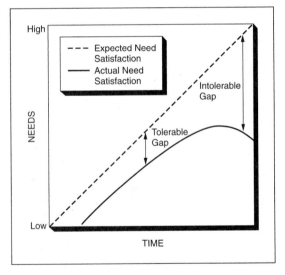

Figure 20.1   The J-Curve Model

One theory of the causes of revolutions is the J-curve theory. According to this model, revolutions occur when there is an intolerable gap between people's expectations of need satisfaction and the actual level of satisfaction they experience. In response to improved economic and social conditions, people expect continuing improvement in the satisfaction of their needs. As long as they experience satisfaction, there is political stability, even if there is a gap between expected and actual satisfaction (called relative deprivation*). If the level of actual satisfaction declines, the gap gets bigger; at some point it becomes intolerable, and revolution occurs.*

SOURCE: Adapted from Davies, 1962.

A number of studies have measured the individual's level of frustration or deprivation and analyzed the relationship between it and participation in collective action or protest. These studies found no differences in frustration or relative deprivation between participants and nonparticipants (McPhail, 1994).

Whether economic or other stresses lead to collective action may depend on the attributions individuals make about their causes. In a laboratory experiment, individuals were led to believe they had an easy or a hard task. Further, either their individual ability or their group membership was made salient. Participants' evaluations of outcomes reflected the salience manipulation; those whose group identity was made salient interpreted their disadvantage (the hard task) in terms of their group membership (Smith & Spears, 1996). Another experiment manipulated whether deprivation (loss of a promised $10 payment) was seen as due to individual failure or group membership. Participants who perceived it as due to group membership were more supportive of collective action (Foster & Matheson, 1995). Other research confirms it is the feeling that members of one's own group are deprived relative to members of other groups that is associated with collective protest (Begley & Alker, 1982; Guimond & Dubé-Simard, 1983).

A common form of collective protest in the contemporary United States is the strike, in which union members collectively withhold their labor in the hope of improving their economic position. Can strikes be explained by relative deprivation? The answer is yes, under certain conditions. Strikes are more likely when economic conditions are good and there is a large gap between what workers expect and what management offers—that is, high relative deprivation (Snyder, 1975). Furthermore, strikes are only effective when labor unions are institutionalized—that is, when there is ongoing labor-management accommodation, when union membership is large and stable, and when organized labor has influence in the national political system (Rubin, 1986).

Grievances

In any society, certain resources are highly valued but scarce. These resources include income or property, skills of certain types, and power and influence over others. Because of their scarcity,

such resources are unequally distributed. Some groups have more access to a given resource than others. When one group has a *grievance*—is discontent with the existing distribution of resources—collective behavior may occur to change that distribution (Oberschall, 1973). Attempts to change the existing arrangement frequently elicit responses by other groups that are designed to preserve the status quo. The result may be a series of actions by challengers and power holders.

There are three types of collective action (Tilly et al., 1975). *Competitive* action involves conflict between communal groups, usually on a local scale. One example is conflict and violence directed toward members of certain ethnic groups. Such incidents are more likely when members of two groups are competing for low-wage jobs or where there are sharp increases in immigration (Olzak, 1989). The high rates of lynching of blacks in the South in the 1890s is another example. From 1865 to 1880, blacks enjoyed large gains in political influence. By 1890, however, whites were attempting to regain political control. Between 1890 and 1900, several state legislatures discussed laws that would have taken the vote away from blacks. During these years, the number of blacks lynched in Alabama, Georgia, Louisiana, Mississippi, and South Carolina reached a peak (Wasserman, 1977). Lynching of blacks also increased during economic downturns, for example, when the price of cotton was declining (Beck & Tolnay, 1990).

A second type of collective action, called *reactive,* involves a conflict between a local group and agents of a national political system. Tax rebellions, draft resistance movements, and protests of governmental policy are reactive. Such behavior is a response to attempts by the state to enforce its rules (regarding military service, for example) or extend its control (such as imposing a new tax). Thus such events represent resistance to the centralization of authority.

A third type of collective action, called *proactive,* involves demands for material resources, rights, or power. Unlike reactive behavior, it is an attempt to influence rather than resist authority. Strikes by workers, demonstrations for equal rights or against abortion, and various nonviolent protest activities are all proactive. Most proactive situations involve broad coalitions rather than one or two locally based groups.

The three underlying conditions discussed in this section differ in their emphasis. The strain model emphasizes the individual's emotional state in explaining collective behavior. The relative deprivation view emphasizes the person's subjective assessment of need satisfaction. The grievance model suggests that collective behavior results from rational attempts to redistribute resources in society (Zurcher & Snow, 1990).

Origins of Collective Behavior

Conditions of strain, relative deprivation, and grievances may be present in a society over extended periods of time. By contrast, incidents of collective behavior are often sporadic. Frequently, there are warning signals that a group is frustrated or dissatisfied. Members of the dissatisfied group or third parties may attempt to convince those in power to make changes (Oberschall, 1973). If changes are not made, members may increasingly perceive legitimate chan-

nels as ineffective, leading to marches, protests, or other activities. Eventually, an incident may occur that adversely affects members of the group and symbolizes the problem, triggering collective behavior by group members; such an incident is referred to as a *precipitating event.* Such incidents appear to sharply increase the dissatisfaction of those who already have low levels of grievance prior to the event (Opp, 1988).

An incident is more likely to trigger collective behavior if it occurs in an area accessible to many members of the affected group; this facilitates the assembling process. It is also more likely to lead to collective action if it occurs in a location that has special significance to group members (Oberschall, 1973). An event that occurs in such a place may produce a stronger reaction than would the same incident in a less meaningful location.

In April 1992 a California jury acquitted four police officers charged with beating a black motorist, Rodney King. Word of the acquittal was broadcast

Civil disorders, which occur periodically in American cities, often involve members of disadvantaged groups. Looting and vandalism are common, with businesses whose owners are disliked as the likely targets.

throughout Southern California. Within minutes, a crowd of young men gathered at the intersection of Florence and Normandie in mostly black South Central Los Angeles. At first, some of the men shouted at and harassed passing motorists. As their numbers grew, others began stopping cars and beating occupants. Many of those present merely observed these activities. Violence and looting spread rapidly. The ensuing disorder lasted 3 days, resulting in 53 deaths and the destruction of 10,000 businesses.

The acquittal of the officers symbolized for many blacks their inferior position. Relations between black citizens and white police officers in Los Angeles (and other cities) have been characterized by hostility for many years. The videotaped beating of Rodney King was a clear example of the mistreatment many blacks had suffered. The verdict suggested that white police officers can abuse black citizens without fear of punishment. This increased the frustration felt by large numbers of blacks. Some of them acted, and others quickly joined in.

Empirical Studies of Riots

Because they are unpredictable, hostile crowd events such as the one in Los Angeles are difficult to study empirically. Nevertheless, extensive and sophisticated research has been conducted on past racial disturbances, such as those that occurred in many U.S. cities between 1965 and 1969. These studies support many of the theories presented earlier in this chapter.

In the first 9 months of 1967, there were 164 racial disturbances. In response, President Lyndon Johnson appointed the National Advisory Commission on Civil Disorders to study the causes of these incidents. In its report (1968), the Commission concluded the racial disturbances were caused by the underlying social and economic conditions affecting blacks in our society. The report pointed to the high rates of unemployment, poverty, and poor health and sanitation conditions in black ghettos; the exploitation of blacks by retail merchants; and the experience of racial discrimination, all of which produced a sense of deprivation and frustration among blacks.

The Commission studied 24 disorders in 23 cities in depth. It concluded that:

Disorder was generated out of an increasingly disturbed social atmosphere, in which typically a series of tension-heightening incidents over a period of weeks or months became linked in the minds of many in the Negro community with a reservoir of underlying grievances. At some point in the mounting tension, a further incident—in itself often routine or trivial—became the breaking point and the tension spilled over into violence. Violence usually occurred almost immediately following the occurrence of the final precipitating incident, and then escalated rapidly. Disorder generally began with rock and bottle throwing and window breaking. Once store windows were broken, looting usually followed. (*Report,* 1968, p. 6)

The precipitating event frequently involved contacts between police officers and blacks. In Tampa, Florida, a disturbance in 1967 began after a policeman shot a fleeing robbery suspect. A rumor quickly spread that the black suspect was surrendering when the officer shot him. In other cities, disorder was triggered by incidents involving police attempts to disperse a crowd in a shopping district or to arrest predominantly black patrons of a tavern selling alcoholic beverages after the legal closing time. To many blacks, police officers symbolize white society and are therefore a readily available target for grievances and frustration. When a police officer arrests or injures a black under ambiguous circumstances, it provides a concrete focus for discontent.

Severity of Disturbances   In some cities, racial disorders involved a few dozen people, and there was little property damage. In other cities, they involved thousands of persons, and millions of dollars worth of property was destroyed. What determined how severe a disorder was?

The Commission's report suggested that the deprivations experienced by blacks fueled the disorders. Numerous researchers have studied this hypothesis. One question is whether absolute or relative deprivation is more influential. Are grievances greater only when unemployment, poor housing, and poor health are widespread, or are grievances greater when conditions experienced by blacks are poorer than conditions experienced by whites?

Measures of both absolute and relative deprivation were included in a study of 322 incidents that occurred in 1967 and 1968 (Spilerman, 1976). The absolute level of deprivation was measured by the unemployment rate, the average income, and the average education of nonwhites in each city where a disturbance occurred. Relative deprivation was measured by the differences between white and nonwhite unemployment rates, average income, average education, and average occupational status. To measure the severity of disorders, the study used the composite riot severity scale reproduced in Table 20.1. This scale distinguishes four degrees of severity based on the amount of personal injury, property damage, crowd size, and number of arrests.

Both the severity and frequency of disturbances were associated with the size of the nonwhite population of a city. Neither absolute nor relative deprivation was associated with the severity of disorders.

The Commission sponsored a survey in early 1968. A probability sample of 200 blacks ages 15 to 65 was drawn in each of 15 major cities. This survey included several measures of relative deprivation, including perceived job discrimination, unresponsiveness of local government, and police abuse of blacks. It also included objective measures of these three aspects of deprivation (Carter, 1990). The measures of

Table 20.1   Riot Severity Scale

0   Low intensity—rock and bottle throwing, some fighting, little property damage. Crowd size < 125; arrests < 15; injuries < 8.

1   Rock and bottle throwing, fighting, looting, serious property damage, some arson. Crowd size 75–250; arrests 10–30; injuries 5–15.

2   Substantial violence, looting, arson, and property destruction. Crowd size 200–500; arrests 25–75; injuries 10–40.

3   High intensity—major violence, bloodshed, and destruction. Crowd size > 400; arrests > 65; injuries > 35.

SOURCE: Adapted from Spilerman, 1976.

relative deprivation (perceived conditions) did not correlate significantly with the objective measures of deprivation. Two measures of objective (economic) grievances were correlated with severity of disturbance. But the variable that was most strongly related to severity was the size of the black population of the city.

These results suggest that black protests were not due to local community conditions but to general features of the society, such as increased black consciousness, heightened racial awareness, and greater identification with other blacks because of the civil rights movement. Television may have contributed to the disorders of the late 1960s by providing role models: Blacks in one city witnessed and later copied the actions of those in other cities. Blacks who were engaged in vandalism, looting, and other collective behavior served as a model for blacks experiencing grievances and deprivation.

Ambient Temperature and Collective Violence   It is often suggested that high temperatures contribute to large-scale racial disturbances. The *Report of the National Advisory Commission on Civil Disorders* (1968) noted that 60% of the 164 racial disorders that occurred in U.S. cities in 1967 took place in July during hot weather. Of the 24 serious disturbances studied in detail, in most instances the temperature during the day on which violence first erupted was very high.

One study of the relationship between temperature and collective violence focused on 102 incidents that occurred between 1967 and 1971 (Baron & Ransberger, 1978). Results showed a strong relationship between temperature and the occurrence of violence. The incidents of collective violence were much more likely to have begun on days when the temperature was high (71–90 degrees). This relationship is depicted in Figure 20.2.

As the figure indicates, there were few riots on days when the temperature was greater than 90 degrees. Does this indicate it is too hot on such days or that there are very few days when the temperature is that high? To answer this question, researchers estimated the probability of a disturbance, controlling for the number of days in each temperature range.

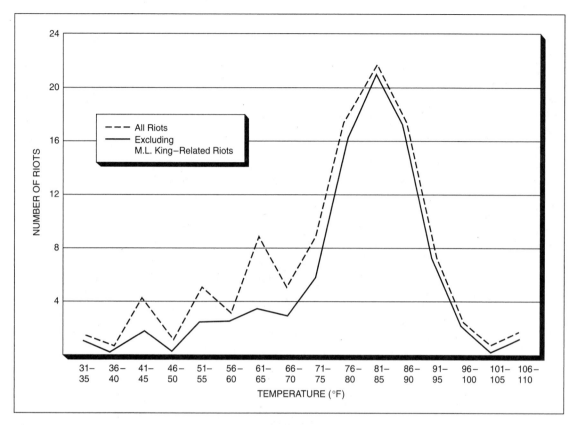

Figure 20.2 Ambient Temperature and Collective Violence

There is a strong relationship between mean temperature and the occurrence of collective disturbances. An analysis of 102 incidents between 1967 and 1971 generated this graph. As the temperature increased, the frequency of riots also increased. As the temperature increases, so does the number of people who are outside. Large numbers of people on the streets facilitate the development of a crowd.

SOURCE: Adapted from Baron and Ransberger, 1978.

The results show a direct relationship (Carlsmith & Anderson, 1979). In other words, the higher the temperature, the more likely a disturbance is to occur. One interpretation of this relationship is that in high-density neighborhoods with little air conditioning, the number of people on the streets increases with the temperature. Large street crowds facilitate the transmission of rumors and increase the likelihood of supportive responses to acts initiated by an individual or a small group.

If temperature is associated with the occurrence of collective violence, an obvious question is whether high temperature is related to other types of violent behavior. One readily available measure of behavior is the rate of violent crime—of murder, sexual assault, and assault. One study analyzed the relationship between rates of violent crime and average temperature in data from 260 cities for the year 1980 (Anderson, 1987). As expected, the higher the average temperature, the higher the rates of violent crime in a city were. Another study analyzed data from Dayton, Ohio, for a 2-year period (Rotton & Frey, 1985). The researchers looked at daily variations in the number of reports to police of assaults and family disturbances.

The number of each was positively associated with the temperature. These results suggest that people are more irritable in hot weather and, thus, more likely to engage in aggressive or violent behavior.

Selection of Targets

Looting during a civil disturbance does not occur randomly. During racial disorders, such as the one in Los Angeles in 1992, property damage is primarily to retail stores. Residences, public buildings such as schools, and medical facilities such as clinics and hospitals are usually unaffected. In addition, the looting and vandalism of businesses is often selective. Some stores are cleaned out, whereas others in the same block are untouched.

According to one survey (Berk & Aldrich, 1972), the reason for this discrepancy is that businesses with higher average prices of merchandise (that is, more attractive merchandise) were more likely to be attacked. A second factor was familiarity with the interior of the store. The larger the percentage of black customers, the more likely the store was to be looted. Retaliation was also a factor; stores whose owners refused to cash checks and give credit to blacks were more likely to be attacked. White ownership by itself was the least important factor.

Thus the selection of targets during a riot reflects the desire of participants to obtain expensive consumer goods and to retaliate against antiblack owners. This may reflect the operation of social control within the crowd. Emergent norms may define some buildings and types of stores as appropriate targets and others as inappropriate targets. These norms are probably enforced by members of the crowd itself (Oberschall, 1973).

Social Control and Collective Behavior

Social-control agents such as police officers strongly influence the course of a collective incident once it begins. In some cases, the mere appearance of authorities at the scene of an incident sets off collective action.

The importance of control agents is especially clear in protest situations. Protesters usually enter a situation with (1) beliefs about the efficacy of violence, and (2) norms regarding the use of violence (Kritzer,

1977). If participants' norms do not oppose violence, and if they believe violence may be effective, they are predisposed to choose violent tactics. Similarly, control agents enter a situation with (1) beliefs about what tactics the protesters are likely to use, and (2) informal norms regarding violence. If the police anticipate violence, they prepare by bringing specially trained personnel and special equipment. Based on these beliefs and expectations, either the control agents or the protesters may initiate violence. Violence by one group is likely to produce a violent response from the other. Data from 126 protest events support this view (Kritzer, 1977).

In some instances, the response of authorities determines the severity and duration of disorders (Spiegel, 1969). In any disturbance, there are two critical points at which undercontrol or overcontrol can cause a protest to escalate. First is the authorities' response to the initial phase. Undercontrol by police in reaction to the initial disorder may be interpreted by protesters as "an invitation to act." It suggests illegal behavior will not be punished. Overcontrol at this point, such as an unnecessary show of force or large numbers of arrests, may arouse moral indignation, which may attract new participants and increase violence. In Los Angeles in April 1992, the police initially maintained a low profile; officers assembled and remained in an area several blocks from Normandie and Florence avenues. The second critical point in a disturbance is the response to widespread disorder and looting during the second day. As it progresses, participants gradually become physically exhausted. Undercontrol by authorities may facilitate the collapse of the protest. Overcontrol may result in incidents that fuel hostility, draw in new participants, and increase the intensity of the disturbance.

The response of authorities to one incident also may affect the severity of subsequent disorders in the same city (Spilerman, 1976). One study investigating incidents of collective violence in France, Germany, and Italy between 1830 and 1930 (Tilly et al., 1975) found that episodes involving violence were often preceded by nonviolent, collective action. Moreover, a substantial amount of the violence consisted of the forcible *reaction* of authorities (often military or police forces employed by the government) to the nonviolent protests of citizens. Thus violence was not necessarily associated with attempts to influence authority. It was associated with reactions to such attempts by the agents of authority. A study of supporters of the Irish Republican Army found they supported the use of violence only when they viewed peaceful protest as ineffective and knew others who had experienced repressive acts by authorities (White, 1989). A survey of the death rate associated with political violence in 49 countries in the period from 1968 to 1977 found the death rate was higher in countries with moderate scores on an index of regime repressiveness (Muller, 1985).

Social Movements

The difference between collective behavior and social movements is one of degree. Both involve gatherings of people who engage in unconventional behavior—behavior that is inconsistent with some norms of society, or is unconventional for the social space and time within which it occurs (Snow & Oliver, 1995). Both are caused by social conditions that generate strain, frustration, or grievances. Their differences lie in degree of organization. Crowd incidents are often unorganized; they occur spontaneously, with no widely recognized leaders and no specific goals.

A **social movement** is collective activity that expresses a high level of concern about some issue (Zurcher & Snow, 1990). Participants are people who feel strongly enough about an issue to act. Persons involved in a movement take a variety of actions—sign petitions, donate time or money, talk to family or friends, participate in rallies and marches, engage in civil disobedience, or campaign for candidates. These activities are drawn from the repertoire of ordinary political activities in society, but are often outside the conventional two-party structure. Within the movement, an organization may emerge—a group of persons with defined roles who engage in sustained activity to promote or resist social change (Turner & Killian, 1972).

In this part of the chapter, we first discuss the development of a social movement. Then we consider the movement organization and some influences on how it operates. Finally, we discuss the consequences of social movements.

The Development of a Movement

Preconditions By itself, strain or grievances cannot create a social movement. For a movement to occur, people must perceive their discontent as the result of controllable forces external to themselves (Ferree & Miller, 1985). If they attribute their discontent to such internal forces as their own failings or bad luck, they are not predisposed to attempt to change their environment. In addition, people must believe they have a right to the satisfaction of their unmet expectations (Oberschall, 1973). These attributions are often the result of interaction with others in similar circumstances. The moral principles used to legitimate their demand may be taken from the culture or from a specific ideology or philosophy. Thus at the core of any social movement are beliefs rooted in the larger society.

A current social movement taking place in the United States involves abortion. In the late 1960s, an organized social movement developed that pressed for change in the laws restricting the availability of abortion. This movement culminated in the Supreme Court decision *Roe* v. *Wade* on January 22, 1973. This decision held that the state cannot interfere in an abortion decision by a woman and her physician during the first 3 months of pregnancy. The increasingly widespread availability of abortion created strain for others in our society. Many people view a fetus as a person and thus define abortion as murder. Drawing primarily on conservative Christian theology, these people gradually organized a movement in the mid-1970s to obtain legislation that would sharply restrict a woman's right to abortion.

In addition to perceiving unmet needs, people in a social movement must believe the satisfaction of their needs cannot be achieved through established channels. This perception may be based on lack of access to such channels (Graham & Hogan, 1990). Or it may result from the failure of attempts to bring about change through those channels. At first, the antiabortion movement emphasized lobbying and attempts to influence elections. As these activities did not produce the legislative or judicial action they sought, the movement increasingly adopted more aggressive tactics, including attempts to physically prevent women from entering abortion clinics. In the

1990s, some individuals have engaged in violent acts, including bombing clinics where abortions are performed and killing physicians who perform the procedure.

Another precondition may be a solution, action that people believe will ameliorate their discontent or redress their grievance (Willmoth & Ball, 1995). A case study of the movement to control world population suggests the development of birth control in the 1960s, especially oral contraceptives and surgical sterilization procedures, provided a feasible solution to the problem of overpopulation. As a result, population control efforts were more organized and successful in the 1970s than they had been in the 1950s.

Ideology As affected individuals interact, an ideology or generalized belief emerges. **Ideology** is a conception of reality that emphasizes certain values and justifies a movement (Turner & Killian, 1972; Zurcher & Snow, 1990). Ideologies are often developed by movement participants as the movement grows.

The antiabortion, or prolife, ideology rests on several assumptions. First, each conception is an act of God. Thus abortion violates God's will. Second, the fetus is an individual who has a constitutional right to life. Third, every human life is unique and should be valued by every other human being. Prolife forces view the status of abortion as temporary, a departure from the past when it was morally unacceptable. Persons and programs (such as sex education) are evaluated in terms of whether they support or undermine these beliefs. Any person or group who favors continued legal abortion is defined as immoral. In recent elections in many communities, a candidate's position on abortion has been a major political issue. Prolife activists believe that by opposing people and programs which encourage abortion, they will cause a sharp decline in its availability and redefine it as illegitimate.

Such an ideology fulfills a variety of functions (Turner & Killian, 1972). First, it provides a way of identifying people and events and a set of beliefs regarding appropriate behavior toward them. Ideology is usually oversimplified because it emphasizes one or a few values at the expense of others. A second func-

tion of ideology is that it gives a movement a temporal perspective. It provides a history (what caused the present undesirable situation), an assessment of the present (what is wrong), and a conception of the future (what goals can be attained by the movement) (Martin, Scully, & Levitt, 1990). Third, it defines group interests and gives preference to them. Finally, it creates villains. It identifies certain persons or aspects of society as responsible for the discontent. This latter function is essential because it provides the rationale for activity designed to produce change (Oberschall, 1973).

A common element of the ideology of contemporary groups seeking to produce change is the rhetoric of "the public good" (Williams, 1995). Like other elements of ideology, this rhetoric is drawn from the larger culture and provides a resource that movements utilize in making their claims.

Recruitment The development and continuing existence of any movement depends on *recruitment*—the process of attracting supporters. Some people are attracted to a movement because they share some distinctive attributes (Zurcher & Snow, 1990). In many instances, these are persons who experience the discontent or grievances at the base of the movement. A study comparing people who participated in the movement to prevent the reopening of Pennsylvania's Three Mile Island nuclear power plant with a group of nonparticipants found the activists had opposed commercial and military uses of nuclear energy before the accident and the accident had served to increase their discontent substantially (Walsh & Warland, 1983). There are limitations, however, to this grassroots view of recruitment (Turner & Killian, 1972). Many studies have found that supporters of a movement are not the most deprived or frustrated. Also, the goals of a movement may not be aimed at removing the sources of the discontent. The content of the ideology reflects several influences, not only a desire to eliminate a particular source of frustration. Once developed, people may be attracted by the ideology who do not share the discontent.

Recruitment depends on two catalysts: the ideology and existing social networks (Zurcher & Snow,

1990). The content of the ideology is what attracts supporters. The ideology spreads, at least in part, through existing social networks. Supporters communicate the ideology to their friends, families, and co-workers as they interact with them.

Research on recruitment into religious movements documents the importance of friendship and kinship ties (Stark & Bainbridge, 1980). Adherents to the Mormon religion, for example, are prone to proselytize. They establish friendship ties with nonmembers, then gradually introduce their beliefs to their new friends. Likewise, members of a doomsday cult who believed the earth would soon be destroyed often were relatives of other members. Those with kinship ties were less likely to leave the cult.

A study of conversion to Catholic Pentecostalism compared 150 converts with a control group of non-Pentecostal Catholics (Heirich, 1977). The major difference between the two groups was in their social networks. Converts had been introduced to the "born again" movement by someone they trusted—a parent, priest, or teacher. Converts also reported positive or neutral reactions to their participation from family and friends.

This demonstration is part of the social movement aimed at increasing the supply of affordable housing. Organizers of such events hope for coverage by the mass media, because reports of movement activities may attract additional supporters.

Sometimes entire groups are recruited all at once (Oberschall, 1973). The civil rights movement in the South in the 1950s is one example. Because of their religious views, black ministers were predisposed to support a movement whose ideology emphasized freedom and equality. These ministers recruited their entire congregations and communicated the ideology to other ministers. As a result, the movement spread rapidly. More recently, the prolife movement has grown by recruiting entire congregations of Catholics and Mormons. Such bloc recruitment is much more efficient than recruiting individuals (Jenkins, 1983).

An alternative mechanism for recruitment is the mass media. On October 3, 1970, an estimated 15,000 to 20,000 persons participated in Reverend Carl McIntire's March for Victory in Washington. McIntire, a fundamentalist pastor who supported the Vietnam War, communicated his views in a weekly radio program carried by 600 stations and a weekly newsletter. Interviews with 201 march demonstrators revealed that most had come from outside the Washington area, and the overwhelming majority learned of the march through McIntire's radio program or newsletter (Lin, 1974–1975).

Thus the media play an important role in social movements. In the McIntire case, prior political beliefs appear to have led persons to seek specific information about the organizational program and the march. Media reports convey a movement's ideology and attract members by providing role models or by providing information about the time and place of activities. It is no accident that movement groups devote considerable effort to getting reporters and camera crews to cover their activities. Media attention—time on television and radio, space in newspapers and magazines—is a scarce resource. To get it, movement groups may have to engage in novel or dramatic acts (Holgartner & Bosk, 1988). Such actions may lead to confrontation and violence.

Whether one is attracted by the ideology or activities of a movement, participation in a social movement is influenced by socialization. A study gathered data in 1965 from 1,669 high school seniors and 1,562 of their parents; 80.8% of the students were reinterviewed in 1973 (Sherkat & Blocker, 1994). Young people who participated in the civil rights, stu-

dent, or antiwar protests between 1965 and 1973 were more likely to have been taught political beliefs encouraging participation. Nonparticipants were more likely to have been taught religious beliefs that discouraged political action. A study of participation in protests against the Gulf War in 1991 found that protesters were more likely to have parents who participated in protests against the war in Vietnam (Duncan & Stewart, 1995).

Social Movement Organizations

If a movement is to have any impact, there must be some degree of organization to exert continuing pressure for (or resistance to) change. A group of persons with defined roles engaged in sustained activity that reflects a movement's ideology is called a **movement organization.** It develops through the mobilization of people and resources. Once developed, the organization is influenced by its environment and by its own internal processes.

An important influence on the development of a movement organization over time may be other movements (Meyer & Whittier, 1994). A case study of the impact of the women's movement on the peace movement found that the peace movement adopted parts of the ideology (e.g., the identification of masculine patriarchy as the villain), some of the tactics (e.g., nonviolent blockades), and some of the organizational structures (e.g., decentralized decison making) from the women's movement. This transmission from one movement to another occurred partly through overlapping communities of supporters and shared personnel.

Resource Mobilization   Having attracted supporters, a movement must induce some of them to become committed members (Zurcher & Snow, 1990). Commitment involves the creation of links between an individual's interests and those of the movement so the individual will be willing to contribute actively to the achievement of movement goals. Committed members are necessary if the movement is to become active and self-sustaining. **Mobilization** is the process through which individuals surrender personal resources and commit them to the pursuit of group or organizational goals (Oberschall, 1978). Resources can be many things: money or other material goods, time and energy, leadership or other skills, or moral or political authority. From the individual's viewpoint, mobilization may involve a rational decision. The person weighs the costs and benefits of various actions. If the potential rewards outweigh the potential risks, a particular course of action is taken. Mobilization also involves group processes, such as collective rituals and democratic decision making, which increase the individual's commitment to movement actions (Hirsch, 1990).

Leadership is obviously essential to an organization. Taking a leadership role in a movement organization may be risky but potentially very rewarding (Oberschall, 1973). If the movement is successful, leaders may attain prestige, visibility, a permanent, well-paid position with the organization, and opportunities to interact with powerful, high-status members of society. Leaders of movements are often persons with substantial education (such as lawyers, writers, professors, and students) and at least moderate status in society (Weed, 1990). They are frequently persons whose skills cannot be confiscated by authorities, who can expect social support, and who will be dealt with leniently if arrested. Thus their risk/reward ratio is favorable for involvement.

For a movement to succeed, others also must be induced to work actively in the organization (Zurcher & Snow, 1990). One basis of commitment is moral—anchoring an individual's worldview in the movement's ideology. Members who are attracted by the ideology tend to see their own interests as furthered by the achievement of movement goals. Many women become involved in proabortion organizations because preserving freedom of choice for all women will benefit them. At the same time, movement adherents clarify, extend, and even reinterpret the ideology as they attempt to persuade others to commit resources to the movement (Snow et al., 1986). A second basis of commitment is a sense of belonging, which is facilitated by collective rituals in which members participate. One advantage of recruiting preexisting groups, such as church congregations, is that this sense of belonging is already developed. A third basis of commitment is instrumental. If the organization has

enough resources at its disposal, it can provide utilitarian rewards for committed members. These rewards may be distributed equally among members or selectively to members who make a particular contribution (Oliver, 1980).

What determines the relative success of a movement organization in mobilizing resources? One study compared local MADD (Mothers Against Drunk Driving) organizations, assessing the effect of effort, strategy, organizational structure, and national affiliation (McCarthy & Wolfson, 1996). Effort (hours of work by officers, number of public appearances) predicted the mobilization of volunteer labor, revenue, and members. Organizational structure, as measured by the number of task committees, also predicted mobilization success.

Depending on its overall strategy, a movement organization can use moral, affective, or instrumental rewards as bases for building commitment. These are sufficient for most organizations to induce members to contribute time, materials, and other resources. Other organizations demand that members commit themselves to exclusive participation. They require that members renounce other roles and commitments and undergo **conversion**—the process through which a movement's ideology becomes the individual's fundamental perspective. This degree of commitment is required by some religious movements and by so-called utopian communities. Conversion involves persuasion and "consciousness raising"; it is an attempt to change the individual's worldview. Conversion is usually accomplished during a period of intensive interaction with other movement members. It is thus a very labor intensive mobilization strategy (Ferree & Miller, 1985).

Movement efforts to mobilize resources vary according to what potential supporters are willing to contribute. If there is considerable latent support for a movement's goals but supporters are unwilling to make large contributions, mobilization efforts may focus on monetary donations. Another low-cost means of translating weak support into a resource is through using initiative or referendum ballots, asking supporters to vote for a movement's goal, such as clean water or a nuclear free zone (Evans, 1977). Mobilization efforts also vary according to the resources being sought.

An organization seeking volunteers for high-risk activities, such as those involving the risk of arrest, must use powerful inducements. A study of volunteers for a project to register voters in rural Mississippi in 1964 found that several characteristics distinguished participants from volunteers who decided not to go (McAdam, 1986). Participants had intense ideological commitment, previous experience with activism, and ties to other activists (see Figure 20.3).

The other side of mobilization is loss of members and support, or *erosion*. A study of a mobilization campaign carried out by the Dutch peace movement involved telephone interviews with random samples before (May 1985) and after (November 1985) the campaign (Oegema & Klandermans, 1994). The factors associated with erosion of support were, in May, moderate (compared to strong) willingness to participate, declining willingness to engage in activities

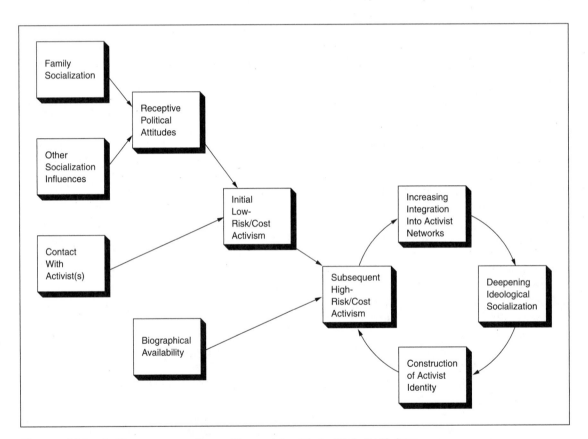

Figure 20.3 Influences on Recruitment to High-Risk Activism

Many social movement activities are low risk—involving little or no cost to participants. Such activities include donating money, distributing leaflets, and attending rallies. Sometimes, however, movement leaders decide to engage in activities that are high risk—involving the risk of injury or death. Political activism in the rural South in 1964 is an example of a high-risk activity. A study of the characteristics associated with the willingness to participate in this activity compared those who went with those who initially volunteered but decided not to go. Analyses suggested that a sequence of experiences, beginning with particular socialization experiences and including contact with other activists and prior activism, were associated with being available for and participating in the high-risk activity. Once involved, participation facilitated greater involvement.

SOURCE: McAdam, 1986, figure 1.

requiring greater effort, and a social environment perceived as less supportive.

Organization-Environment Relations The social environment is another influence on the development of a movement organization. First, changes in the environment can increase or decrease the size of the movement and the probability of its success. This is illustrated in a study comparing the evolution of the National Farm Labor Union (NFLU) movement from 1946 to 1955 and the United Farm Workers (UFW) movement from 1965 to 1972 (Jenkins & Perrow, 1977). Both movements had very similar ideologies and organizational structures. However, whereas the NFLU was generally unsuccessful, the UFW achieved major benefits for many farm workers. The difference was the amount of resource support from the environment due to a change in the political climate in the United States. An analysis of newspaper articles published during each period indicates that government, liberal political organizations, and organized labor gave much greater support to the UFW. Thus challenges to the established order are more likely to succeed when there is sustained support from the environment and when organized opposition is absent.

Second, the growth and effectiveness of a movement organization depends, in part, on whether there are other organizations with similar goals. The National Association for the Advancement of Colored People (NAACP) was the principal black civil rights organization before 1960. Its goal was racial integration, and it relied heavily on a combination of litigation and lobbying. From 1960 to 1973, other black organizations developed with more radical goals, methods, or both. The Congress of Racial Equality (CORE) and the Student Nonviolent Coordinating Committee (SNCC) favored direct action, and both later excluded whites from membership. Membership in the NAACP remained stable while the other organizations grew rapidly. The competition between these organizations for resources forced the NAACP to alter its short-term goals, to give greater priority to economic advances. The NAACP also shifted from relying on members for financial support to relying on donations from foundations and corporations (Marger, 1984). Paradoxically, the more radical programs of CORE and SNCC made it easier for more moderate organizations like the NAACP to attract financial support from whites, especially after the riots of 1967 and 1968 (Haines, 1984).

Under some circumstances, organizations with similar goals form coalitions. Coalitions are possible if organizations appear to have similar values and clearly defined common interests (Ferree & Miller, 1985). A study of cooperation and competition between movement organizations focused on 13 organizations and 6 coalitions involved in proabortion activity between 1966 and 1983 (Staggenborg, 1986). Coalitions were likely under conditions of exceptional opportunity or threat. However, abortion coalitions often dissolved because of differences over the use of civil disobedience or conflicts between the maintenance needs of the organizations.

Under other circumstances, organizations with similar goals may engage in sustained conflict. Organizations with conflicting ideologies may compete for influence over legislation (see Box 20.2, "Protest Against Pornography: The Vagaries of Collective Action"). For example, between 1890 and 1920, both Catholic and Protestant organizations endorsed the goal of providing shelters for neglected and homeless children. The organizations might have cooperated and built a single home in each city. However, each insisted that the children be raised in its own faith, which resulted in two or more homes in some cities (Sutton, 1990).

A third important aspect of the environment is countermobilization. The development of an organization that seeks change of some kind is likely to stimulate those who would be adversely affected by that change to mobilize as well. For instance, an analysis of the resistance by corporate managers to the union movement in the 1920s identified a wide range of tactics that were used to counter union efforts to mobilize workers (Griffin, Wallace, & Rubin, 1986). Managers tried to diminish the common interests of the workers by hiring immigrants and mechanizing factories. They attempted to reduce the power of the union movement by boycotting union goods and harassing other firms that recognized unions. Often these tactics were successful.

20.2   Protest Against Pornography: The Vagaries of Collective Action

The underlying causes of collective behavior are present more or less continuously, yet protests and other collective acts occur only occasionally. Consider the example of pornography. Many communities have adult, or X-rated, bookstores and theaters that show X-rated movies. Many of these businesses have been operating for years. There are undoubtedly persons in these communities who find the goods and services sold by these businesses offensive. Yet public attention and protests focus on these businesses only once or twice a year (or less often) and for only a few days at a time.

One measure of the level of public concern about such businesses is the number of newspaper articles published per month or per year. Generally, collective action aimed at the availability of pornography is reported in local newspapers. Madison, Wisconsin, is a community of 180,000 people. There are two daily newspapers, one more liberal in editorial policy and one more conservative. Thus it is likely that collective action concerning pornography would be reported in at least one of these papers.

A case study of the antipornography movement in Madison included a count and an analysis of newspaper articles reporting local collective action (Duesterhoeft, 1987). The results of the count for the years 1972 to 1986 are displayed in the accompanying figure. The study identified two distinct groups that engaged in collective action, one comprised of Christian women and one of feminists. In general, these two groups had been unable to form a coalition because of ideological differ-

ences. The Christian group defined as pornographic any depiction of sex outside of heterosexual, monogamous marriage. The feminists defined as pornographic any depiction that was degrading to women or children. The Christian group did not distinguish between acceptable portrayals ("erotica") and pornography, whereas the feminist group did. Because of these differences, the two groups engaged in separate activities. Most newspaper articles during the period of the study reported action by one group or the other but not of the two in coalition.

A look at the figure indicates that in 9 of the 16 years, there were fewer than 10 newspaper articles. These were years in which there was little protest activity. There were two periods in which numerous articles were published: 1980–1981 and 1984–1985. In the first of these periods, there were three small protests in response to the showing of notorious films *(Caligula* and *Windows)* and a single large protest march that focused on pornography and sexual assault.

The only instance of sustained protest activity during the 16-year period occurred in 1984–1985. In August 1984, local interest developed in passing a county ordinance that would define the sale or exhibition of pornography as a violation of civil rights, similar to legislation passed in Minneapolis. Protests, news conferences, forums, and other activities occurred regularly from August 1984 through 1985. The protests were aimed at the sale of magazines such as *Playboy* and *Penthouse* at both adult bookstores and other businesses such as convenience stores. Thus the introduction, debate, and, ultimately, action on the proposed ordinance

continued on next page

The woman's suffrage movement in the early 1900s and the Equal Rights Amendment (ERA) movement in the 1970s both provoked countermovements. The National Association Opposed to Women Suffrage (NAOWS) defended the homemaker's role and lifestyle and attempted to safeguard the material privileges enjoyed by well-to-do women with an "educational campaign" (Marshall, 1986). The anti-ERA movement in the 1970s used more overtly politi-

cal strategies than the NAOWS, including rallies, marches, and parades. The leader of the anti-ERA movement, Phyllis Schlafly, had considerable political experience working in the National Federation of Republican Women, and she used her experience and contacts (Marshall, 1985).

Internal Processes   Other influences on the development of a movement organization are internal

continued from previous page

increased public interest in the availability of pornography. Both the Christian and the feminist groups took advantage of this interest and engaged in a series of actions.

This case study illustrates several principles about collective behavior and social movements. First, although underlying causes are always present, collective action occurs sporadically. Second, collective action is often a response to a precipitating incident, such as the showing of a notorious film. Third, to sustain collective action over time, there must be support from the community—for example, people willing to attend speeches and demonstrations. Finally, when there is more than one organization, differences in ideology may prevent the formation of a coalition and, thus, limit the effectiveness of the protest.

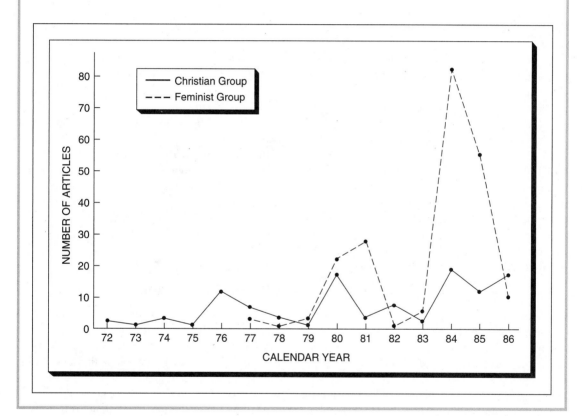

processes. These include factionalization, professionalization, and radicalization.

Factionalization is the emergence of subgroups within an organization. It is more likely to occur in a democratic movement organization. The Clamshell Alliance, an antinuclear protest group, was based on the ideology that every person is equal. This was an important element in mobilizing people to protest the activities of utilities and government agencies. As a result, every decision had to be unanimous; one dissenter could prevent the group from acting. Over a period of 5 years, sharp disagreements developed over strategy, leading to factionalism and reduced effectiveness (Downey, 1986).

Another internal process involves increasing reliance on paid professionals. Over time, as an organization develops, it is subject to increasing routinization. The organization develops an administrative structure. Roles are increasingly filled by secretaries, accountants, and lawyers, rather than by volunteers committed to the movement, which may inhibit, substitute for, or facilitate volunteer activism (Kleidman, 1994). This change also may produce an increasingly conservative stance among the movement's leadership, which may alienate members who are more radical in orientation.

A study analyzed the effects of these changes in the black protest movement from 1953 to 1980 (Jenkins & Eckert, 1986). The data indicate that as the flow of financial resources into movement organizations grew, there was increasing *professionalization* of organization staff. This process did not generate further growth and may have contributed to later decline. At the same time, professionalization does not appear to have produced changes in movement goals and tactics.

There are several other reasons social movements often become more conservative over time (Myers, 1971). First, the leader-follower relationship is characterized by increasing distance and formality. Second, there is increased bureaucracy, including a role hierarchy, rules and procedures, and membership criteria. Third, there is a tendency for goal displacement to occur; leaders become increasingly concerned with maintaining the organization and less concerned with the goals of the broader movement.

Increasing conservatism is not inevitable. An organization committed to democracy (not necessarily equality) may adopt practices that allow members to retain control. These include a system of open elections, an open decision-making process, and regular communication between leaders and members (Nyden, 1985). A study of the United Steelworkers of America documented one case in which a bureaucratized union organization was turned into a democratic one by a reform group that emerged from within.

Much more rare is a movement that undergoes increased *radicalization,* like the Committee of 100 in Great Britain (Myers, 1971). The Committee was formed to protest the increasing reliance of the British government on nuclear weapons for national defense. From the beginning, the group was committed to the use of civil disobedience to achieve its goals. This commitment led to greater radicalization rather than conservatism. In general, groups committed to illegal tactics attract members who are radical. This commitment also produces extreme demands on the time and energy of leaders, which leads to a predominance of younger leaders who have more energy and fewer competing commitments. Organizations of this type are less likely to achieve their goals because the use of illegal tactics tends to alienate the larger society and reduce the amount of support received. Such tactics may even elicit violence by control agents (Kritzer, 1977).

Other overlooked sources of decline in a social movement are the affective and sexual relationships that develop among members (Goodwin, 1997). A case study of the Huk rebellion in the Philippines suggests that the intensity and isolation of participants of this high-risk movement facilitated romantic and sexual relationships. These relationships weakened identification with the movement and competed with movement activity for participants' time and resources. Some members embezzled movement funds and stole other resources for these "families," and some of the women had babies and dropped out. The movement eventually disintegrated.

The Consequences of Social Movements

Once a movement has established an ideology, attracted supporters, and developed an organization that embodies its ideals, its ultimate success or failure often depends on the reaction of authorities outside the movement organization. Those in positions of power—whether political, economic, or religious—can use various strategies in dealing with a social movement (Oberschall, 1973).

One obvious strategy is to restrict a movement through the exercise of social control. For example, authorities can control access to social roles; they determine the educational, occupational, or religious roles available. A student protesting racism on campus may be expelled from school by administrators. A Catholic priest challenging his vow of celibacy may be removed from his parish by the bishop. Beyond this, authorities can manipulate material benefits and

Because of the efforts of dedicated participants in the woman's suffrage movement early in the 20th century, women today can routinely exercise their right to vote.

punishments, which may affect the risk-reward assessments of movement members. Authorities also can use physical power in the form of police or troops. Of course, this method often results in violence.

A second general strategy is conciliation. Authorities may open channels of communication and negotiate with movement leaders in an attempt to resolve grievances. Conciliation is more likely when a movement is relatively powerful and has the support of large numbers or influential members of society. It is also more likely if both sides are highly organized, united internally, and have strong leadership.

By responding to conciliation from authorities, a movement may achieve at least some of its objectives. As a result of negotiation, there may be changes in the distribution of resources or in power relationships in society. A movement that succeeds in this sense may itself become institutionalized. The changes it brings may ensure a continuing flow of resources that perpetuate the movement organization. This is what happened to the labor, civil rights, and women's movements in the United States.

These consequences do not occur overnight. Many movements span decades as they attempt to bring about change. The movement's founders are likely to maintain their collective identity over time.

But as succeeding cohorts join a movement over the years, they may have different identities, based on the environmental context at the time they join (Whittier, 1997). This process of replacement of members over time may lead to changes in the ideology and strategy, and influence whether the movement is successful.

Summary

This chapter discusses both collective behavior and social movements.

Collective Behavior (1) The classic perspective on collective behavior focuses on the crowd. Processes that contribute to crowd behavior include unanimity of feeling, deindividuation, and contagion. When people find themselves in undefined situations, such as disasters, norms and social organization may emerge from their collective efforts. (2) An alternative perspective considers the gathering. Gatherings have a purpose; participants' behaviors reflect that purpose, in interaction with features of the setting. (3) Three conditions have been studied as underlying causes of collective behavior: strain, relative deprivation, and grievances. (4) Collective behavior is often triggered by an event that adversely affects those already experiencing strain, deprivation, or grievances. (5) Empirical studies of riots suggest the severity of a disturbance is influenced by the number of potential participants. Once it begins, the course of a collective incident and the likelihood of future disorders are influenced by the behavior of police and other social-control agents.

Social Movements   A social movement is collective activity that expresses a high level of concern about some issue. (1) The development of a social movement rests on several factors. First, people must experience strain or deprivation, believe they have a right to satisfy their unmet needs, and believe satisfaction cannot be achieved through established channels. Second, as participants interact, an ideology must emerge that justifies collective activity. Third, to sustain the movement, additional people must be recruited by spreading the ideology, often through existing social networks. (2) A movement organization is a group of persons engaged in sustained activity that reflects the movement ideology. Development of a movement organization depends on resource mobilization—getting individuals to commit personal resources to the group. The organization's effectiveness depends in part on its external environment, on whether there are cooperating or competing organizations, and on whether countermobilization occurs. Over time, organizations may experience such internal processes as factionalization, professionalization, or radicalization. (3) The ultimate success or failure of a social movement depends on the reactions of authorities outside the movement—whether those in positions of power decide to exercise control or to negotiate and attempt to ameliorate the sources of strain or deprivation.

Key Terms

collective behavior (p. 497)
companion cluster (p. 502)
contagion (p. 498)
conversion (p. 515)
crowd (p. 498)
deindividuation (p. 498)
disaster (p. 500)
emergent norm (p. 501)
gathering (p. 501)
ideology (p. 512)
J-curve theory (p. 504)
mobilization (p. 514)
movement organization (p. 514)
relative deprivation (p. 504)
rumor (p. 500)
social movement (p. 511)

GLOSSARY

Access display Signal (verbal or nonverbal) from one person indicating to another that further social interaction is permissible.

Accounts Explanations people offer after they have performed acts that threaten their social identities. Accounts take two forms—excuses that minimize one's responsibility and justifications that redefine acts in a more socially acceptable manner.

Achievement motive Conscious or unconscious desire to reach high standards of excellence.

Activation (of an attitude) The bringing of an attitude from memory into conscious awareness (which may then influence behavior).

Actor-observer difference Bias in attribution whereby actors tend to see their own behavior as due to characteristics of the external situation, whereas observers tend to attribute actors' behavior to the actors' internal, personal characteristics.

Additive model Theoretical model for combining information about individual traits in order to predict how favorable the overall impression people form of another will be. This model adds together the favorability values of all the single traits people associate with the person. *See also* Averaging model; Weighted averaging model.

Additive task Type of unitary group task in which the group's performance is equal to the sum of the performances of its members.

Ageism Prejudice and discrimination against the elderly based on negative beliefs about aging.

Aggravated assault Attack by one person on another with the intent of causing bodily injury.

Aggression Behavior intended to harm another person that the other wants to avoid.

Aggressive pornography Explicit depiction—in film, video, photograph, or story—of sexual activity in which force is threatened or used to coerce a person to engage in sex. *See also* Nonaggressive pornography.

Alienation Sense that one is uninvolved in the social world or lacks control over it.

Aligning actions Actions people use to define their apparently questionable conduct as actually in line with cultural norms, thereby repairing social identities, restoring meaning to situations, and reestablishing smooth interaction.

Altercasting Tactics we use to impose roles and identities on others that produce outcomes to our advantage.

Altruism Actions performed voluntarily with the intention of helping someone else and that entail no expectation of receiving a reward or benefit in return (except possibly an internal feeling of having done a good deed for someone).

Anomie theory Theory that deviant behavior arises when people striving to achieve culturally valued goals find they do not have access to the legitimate means of attaining these goals.

Anticipatory socialization Process of learning the knowledge, skills, and values of a role that is not yet assumed. Unlike explicit training, anticipatory socialization is not intentionally designed as role preparation by socialization agents.

Arbitrator In situations of conflict, a neutral third party who has the power to decide how a conflict will be resolved. *See also* Mediator.

Archival research Research method that involves the acquisition and analysis (or reanalysis) of existing information collected by others.

Arousal/cost-reward model Theory of help-giving that proposes bystanders weigh the needs of the victim and their own needs and goals, and then decide whether helping is too costly in the circumstance. Bystanders often take into account not only the cost of giving direct help but also the cost to themselves of not giving help.

Attachment Warm, close relationship with an adult that provides an infant with a sense of security and stimulation.

Attitude Predisposition to respond to a particular object in a generally favorable or unfavorable way.

Attitude change Change in a person's attitudes about some issue, person, or situation.

Attitudinal similarity Sharing by two people of beliefs, opinions, likes, and dislikes.

Attractiveness stereotype Belief that "what is beautiful is good"; the assumption that an attractive person possesses other desirable qualities.

Attribution Process by which people make inferences about the causes of behavior or attitudes.

Authority Capacity of one group member to issue orders to others—that is, to direct or regulate the behavior of other members by invoking rights that are vested in his or her role.

Availables Those persons with whom we come into contact and who constitute the pool of potential friends and lovers.

Averaging model Theoretical model for combining information about individual traits in order to predict how favorable the overall impression people form of another will be. This model averages the favorability values of all the single traits people associate with the person. *See also* Additive model; Weighted averaging model.

Aversive affect Negative affect that the individual seeks to reduce or eliminate, for example, anger or pain.

Aversive event In intergroup relations, a situation or event caused by (or attributed to) an outside group that produces negative or undesirable outcomes for members of the target group.

Back-channel feedback The small vocal and visual comments a listener makes while a speaker is talking, without taking over the speaking turn. This includes responses such as "Yeah," "Huh?," head nods, brief smiles, and completions of the speaker's words. Back-channel feedback is crucial to coordinate conversation smoothly.

Back region Setting used to manage appearances. In back regions, people allow themselves to violate appearances while they prepare, rehearse, and rehash performances. Contrasts with front regions, where people carry out interaction performances and exert efforts to maintain appropriate appearances.

Balanced state State in which the sentiment relations among three cognitive elements are all positive, or in which one is positive and the other two are negative.

Balance theory Theory concerning the determinants of consistency in three-element cognitive systems.

Bilateral threat In bargaining, a situation where both bargainers can issue threats and inflict punishments on one another.

Birth cohort Group of people who were born during the same period of one or several years and who are therefore all exposed to particular historical events at approximately the same age.

Body language (kinesics) Communication through the silent motion of body parts — scowls, smiles, nods, gazes, gestures, leg movements, postural shifts, caresses, slaps, and so on. Because body language entails movement, it is known as kinesics.

Borderwork Interaction across gender boundaries that is based on and strengthens such boundaries.

Brainstorming In groups, a procedure intended to generate a large number of high-quality novel ideas in a brief period. Brainstorming is based on the principles that members should freely express any idea that comes to mind, withhold criticism and defer judgment until later, try to generate as many ideas as possible, and build on ideas suggested by others.

Bystander effect Tendency for bystanders in an emergency to help less often and less quickly as the number of bystanders present increases.

Bystander intervention In an emergency situation, a quick response by a person witnessing the emergency to help another who is endangered by events.

Career Sequence of roles, each role with its own set of activities, that a person enacts during his or her lifetime.

People's most important careers are in the domains of family and friends, education, and work.

Categorization Tendency to perceive stimuli as members of groups or classes rather than as isolated entities; the act of encoding stimuli as members of classes.

Catharsis Reduction of aggressive arousal by means of performing aggressive acts. The catharsis hypothesis states that we can purge ourselves of hostile emotions by intensely experiencing these emotions while performing aggression.

Cautious shift In group decision making, the tendency for decisions made in groups after discussion to be more cautious (less risky) than decisions made by individual members prior to discussion.

Cognition An element of cognitive structure. Cognitions include attitudes, beliefs, and perceptions of behavior.

Cognitive dissonance State of psychological tension induced by dissonant relationships between cognitive elements.

Cognitive processes Mental activities of an individual, including perception, memory, reasoning, problem solving, and decision making.

Cognitive structure Any form of organization among a person's concepts and beliefs.

Cognitive theory Theoretical perspective based on the premise that an individual's mental activities (perception, memory, and reasoning) are important determinants of behavior.

Collective behavior Emergent and extrainstitutional behavior that is often spontaneous and subject to norms created by the participants.

Communication Process through which people transmit information about their ideas and feelings to one another.

Communication accuracy Extent to which the message inferred by a listener from a communication matches the message intended by the speaker.

Communication network Pattern of communication channels or opportunities within a group; typical communication networks include the comcon, the wheel, the circle, and the chain.

Communication-persuasion paradigm Research paradigm that conceptualizes persuasion attempts in terms of source, message, target, channel, and impact — that is, who says what to whom by what medium with what effect.

Communicator credibility In persuasion, the extent to which the communicator is perceived by the target audience as a believable source of information.

Companion cluster Group of family, friends, or acquaintances who remain together throughout a gathering.

Comparison level Standard used to evaluate the outcomes of a relationship, based on the average of the person's experience in past relevant relationships.

Comparison level for alternatives Standard specifying the lowest level of outcomes a person will accept in light of available alternatives. The level of profit available to an individual in his or her best alternative relationship.

Compliance In social influence, adherence by the target to the source's requests or demands. Compliance may occur either with or without a concomitant change in attitudes.

Conditioning Process of learning in which, if a person performs a particular response and if this response is then reinforced, the response is strengthened.

Conflict spiral Form of escalation occurring in bilateral threat situations. In a conflict spiral, the participants exchange successively larger threats and counterthreats to bolster their demands and end up inflicting more and more damage on one another.

Conformity Adherence by an individual to group norms, so that behavior lies within the range of tolerable behavior.

Conjunctive task Type of group task in which the group's performance depends entirely on that of its weakest or slowest member.

Conservative coalition In groups, a union of medium- and low-status members who support the existing leadership and status order against revolutionaries.

Contagion Rapid spread through a group of visible and often unusual symptoms or behavior.

Content analysis Research method that involves a systematic scrutiny of documents or messages to identify specific characteristics and then making inferences based on their occurrence.

Contingency model of leadership effectiveness Middle-range theory of leadership effectiveness which maintains that group performance is a function of the interaction between a leader's style (task oriented or relationship oriented) and various situational factors such as the leader's personal relations with members, the degree of task structure, and the leader's position power.

Control theory Theory that an individual's tendency to engage in deviant behavior is influenced by his or her ties to other persons. There are four components of such ties: attachment, commitment, involvement, and belief.

Conversion Process through which the ideology of a social movement becomes the individual's fundamental perspective.

Cooling-out Response to repeated or glaring failures that gently persuades an offender to accept a less desirable, although still reasonable, alternative identity.

Cooperative principle Assumption that conversationalists ordinarily make that a speaker is behaving cooperatively by trying to be (1) informative, (2) truthful, (3) relevant to the aims of the ongoing conversation, and (4) clear.

Cooptation In groups, a strategy by which an existing leader singles out one or several lower status members for favored treatment; this is done to weaken the bonds among lower status persons who might otherwise form revolutionary coalitions.

Correspondence Degree to which the action, context, target, and time in a measure of attitude are the same as those in a measure of behavior.

Crowd Substantial number of persons who engage in behavior recognized as unusual by participants and observers.

Cultural routines Recurrent and predictable activities that are basic to day-to-day social life.

Death instinct In Freud's theory, the innate urge we carry within us to destroy. This instinct constantly generates hostile impulses that demand release and is the basis for aggression.

Definition of the situation In symbolic interaction theory, a person's interpretation or construal of a situation and the objects in it. An agreement among persons about who they are, what actions are appropriate in the setting, and what their behaviors mean.

Deindividuation Temporary reduction of self-awareness and sense of personal responsibility; it may be brought on by such situational conditions as anonymity, a crowd, darkness, and consciousness-altering drugs.

Dependent variable In an experiment, the variable that is measured to determine whether it is affected by the manipulation of one or more other variables (i.e., independent variables).

Deterrence State in which someone avoids performing an act due to fear or doubt. In the context of bilateral threat, mutual deterrence is a kind of armed stand-off, where all participants want to avoid suffering damage inflicted by the other and therefore refrain from initiating punishment.

Deterrence hypothesis The view that the arrest and punishment of some individuals for violation of laws deters other persons from committing the same violations.

Deviant behavior Behavior which violates the norms that apply in a given situation.

Deviant subculture Group of people whose norms encourage participation in a specific form of deviance and who regard positively those who engage in it.

Differential association theory Theory that deviant behavior occurs when people learn definitions favorable to the behavior through their associations with other persons.

Diffusion of responsibility Process of accepting less personal responsibility to act because responsibility is shared with others. Diffusion of responsibility among bystanders in an emergency is one reason why they sometimes fail to help.

Disaster Event that produces widespread physical damage or destruction of property, accompanied by social disruption.

Disclaimer Verbal assertion intended to ward off any negative implications of impending actions by defining these actions as irrelevant to one's established social identity. By using disclaimers, a person suggests that although the impending acts may ordinarily imply a negative identity, his or hers is an extraordinary case.

Discrepant message In persuasion, a message advocating a position that is different from what the target believes.

Discrimination Overt acts, occurring without apparent justification, that treat members of certain out-groups in an unfair or disadvantageous manner.

Disinhibition In aggression, the reduction of ordinary internal controls against socially disapproved behavior.

Disjunctive task Type of group task in which the group's performance depends entirely on that of its strongest or fastest member.

Displacement The release of pent-up anger in the form of aggression directed against a target other than the original cause of anger.

Display rules Culture-specific norms for modifying facial expressions of emotion to make them fit with the social situation.

Dispositional attribution Decision by an observer to attribute a behavior to the internal state(s) of the person who performed it, rather than to factors in that person's environment. *See also* Situational attribution.

Dissonance effect The greater the reward or incentive for engaging in counterattitudinal behavior, the less the resulting attitude change.

Distributive justice principle Criterion in terms of which group members can judge the fairness and appropriateness of the distribution of rewards. Three of the most important are the equality principle, the equity principle, and the relative needs principle.

Divisible task Group task in which members perform different, although complementary, activities.

Double minority In a group, a member who not only advocates a position on some issue which is different from that held by the majority, but who also has an out-group social identity (in terms of ethnicity, race, nationality, or other ascriptive traits).

Dyadic withdrawal Process of increasing reliance on one person for gratification and decreasing reliance on others.

Egoism Helping behavior motivated by a helper's own sense of self-gratification.

Elaboration likelihood model Model of persuasion which maintains that there are two basic routes through which a message may alter a target's existing attitudes—the central route and the peripheral route.

Embarrassment Feeling that people experience when interaction is disrupted because the identity they have claimed in an encounter is discredited.

Emergent norm Definition of the situation that results from interaction in an initially ambiguous situation.

Emotion work Individual's attempt to change the intensity or quality of emotions in order to bring them into line with the requirements of the occasion. Emotion work may be used to evoke feelings that are not present

but should be (psyching oneself up) or to suppress feelings that are present but should not be (calming oneself down).

Empathy Emotional response to others as if we ourselves were in that person's situation, feeling pleasure at another's pleasure or pain at another's pain.

Empathy-altruism model Theory proposing that adults can experience two distinct states of emotional arousal while witnessing another's suffering: distress and empathic concern. These states of emotional arousal give rise to different motivations for helping.

Encoder-decoder model Theory that views communication as a linear process in which the message is encoded by a transmitter, transmitted, and decoded by a receiver.

Endorsement Attitude held by a group member indicating the extent to which he or she supports the group's leader.

Equitable relationship Relationship in which the outcomes received by each person are equivalent.

Equity In exchange theory, a state of affairs that prevails in a dyad or group when people receive rewards in proportion to the contributions they make toward attainment of group goals.

Ethnocentrism In intergroup relations, the tendency to regard one's own group as the center of everything and to evaluate other groups in reference to it. The tendency to regard one's own group as superior to all out-groups.

Evaluation apprehension State of concern about what others expect of us and how others will evaluate our behavior. Evaluation apprehension inhibits helping when bystanders fear others will view their intervention as foolish; it promotes helping when bystanders believe others expect them to help.

Expectation states theory Theory proposing that status characteristics cause group members to form expectations regarding one another's potential performance on the group's task; these performance expectations affect subsequent interaction among members.

Experiment Research method used to investigate cause-and-effect relations between one variable (the independent variable) and another (the dependent variable). In an experiment, the investigator manipulates the independent variable, randomly assigns subjects to various levels of that variable, and measures the dependent variable.

External validity Extent to which it is possible to generalize the results of one study to other populations, settings, or times.

Extraneous variable Variable that is not explicitly included in a research hypothesis, but has a causal impact on the dependent variable.

Extrinsically motivated behavior Behavior that results from the motivation to obtain a reward (food, praise) or avoid a punishment (spanking, criticism) controlled by someone else.

Feeling rules   Social rules that dictate what an individual with a particular public identity ought to feel in a given situation.

Field study   Investigation that involves the collection of data about ongoing activity in everyday settings.

Flirting   Class of nonverbal facial expressions and behavior, exhibited by women, which serves to attract the attention and elicit the approach of a man. Flirting can also be exhibited by males and can involve two people of the same gender.

Focus of attention bias   Tendency is to overestimate the causal impact of whomever or whatever we focus our attention on.

Formal social control   Agencies that are given responsibility for dealing with violations of rules or laws.

Frame   Set of (widely understood) rules or conventions pertaining to a transient but repetitive social situation that indicate which roles should be enacted and which behaviors are proper.

Front region   Setting used to manage appearances. In front regions, people carry out interaction performances and exert efforts to maintain appropriate appearances. Contrasts with back regions, where they allow themselves to violate appearances while they prepare, rehearse, and rehash performances.

Frustration   Blocking of goal-directed activity. According to the frustration-aggression hypothesis, frustration leads to aggression.

Frustration-aggression hypothesis   Hypothesis that every frustration leads to some form of aggression and every aggressive act is due to some prior frustration.

Fundamental attribution error   Tendency to underestimate the importance of situational influences and to overestimate personal, dispositional factors as causes of behavior.

Gathering   Temporary collection of two or more people occupying a common space and time frame.

Gender role   Behavioral expectations associated with gender.

Generalized other   Conception of the attitudes and expectations held in common by the members of the organized groups with whom one interacts.

Goal commitment   In a group, the degree of members' resolve and determination to reach specific group objectives. If members have a high level of goal commitment, they are more likely than otherwise to work hard and to persist in the face of setbacks.

Goal isomorphism   In groups, a state in which group goals and individual goals held by a member are similar in the sense that actions leading to attainment of group goals also lead simultaneously to attainment of individual goals.

GRIT   Strategy for reducing intergroup conflict whereby one side initiates deescalatory steps in the hope that they will eventually be reciprocated by the other side; GRIT is an acronym for Graduated and Reciprocal Initiatives in Tension-reduction.

Group   Social unit that consists of two or more persons and has the following characteristics: shared goal(s), interaction (communication and influence) among members, normative expectations (norms and roles), and identification of members with the unit.

Group cohesion   Property of a group, specifically the degree to which members of a group desire to remain in that group and resist leaving it. A highly cohesive group will maintain a firm hold over its members' time, energy, loyalty, and commitment.

Group goal   Desirable outcome that group members strive collectively to accomplish or bring about.

Group goal effect   Empirical generalization regarding group productivity, namely, that if a group establishes explicit, demanding objectives with respect to the group's performance, and if the group's members are highly committed to those objectives, the group will perform at a higher level than if it does not do these things.

Group polarization   In group decision making, the tendency for group members to shift their opinions toward a position that is similar to, but more extreme than, the positions they held prior to group discussion; both the risky shift and the cautious shift are instances of group polarization.

Group productivity   Level of a group's output (per unit of time) gauged relative to something else, such as the level of resources utilized by the group or the group's targeted objectives.

Group self-esteem   Individual's evaluation of self as a member of a racial or ethnic group.

Group structure   Ordered arrangements among group members that define and regulate their behavior and provide a patterned constancy and stability to their behavior together.

Groupthink   Mode of thinking within a cohesive group whereby pressures for unanimity overwhelm the members' motivation to appraise realistically alternative courses of action.

Halo effect   Tendency of our general or overall liking for a person to influence our assessment of more specific traits of that person. The halo effect can produce inaccuracy in our ratings of others' traits and performances.

Helping   Any behavior that has the consequences of providing some benefit to or improving the well-being of another person.

Hypothesis   Conjectural statement of the relation between two or more variables. Some hypotheses are explicitly causal in nature, whereas others are noncausal.

Identity   The categories people use to specify who they are, to locate themselves relative to other people.

Identity degradation   Response to repeated or glaring failures that destroys the offender's current identity and transforms him or her into a "lower" social type.

Ideology   In the study of social movements, a conception of reality that emphasizes certain values and justifies the movement.

Illusion of out-group homogeneity   Tendency among in-group members to overestimate the extent to which out-group members are homogeneous or all alike.

Imbalanced state   State in which the sentiment relations among three cognitive elements are all negative, or in which two are positive and one is negative.

Imitation   Process of learning in which the learner watches another person's response and observes whether that person receives reinforcement.

Implicit personality theory   Set of assumptions people make about how personality traits are related—which ones go together and which do not. Theories of this type are implicit because people neither subject them to explicit examination nor ordinarily know their contents.

Incentive effect   The greater the reward or incentive for engaging in counterattitudinal behavior, the greater the resulting attitude change.

Independent variable   In an experiment, the variable that is manipulated by the investigator to study its effects on one or more other (dependent) variables.

Informal social control   Reactions of family, friends, and acquaintances to rule violations by individuals.

Informational influence   In groups, a form of influence that occurs when a group member accepts information from others as valid evidence about reality. Influence of this type is particularly likely to occur in situations of uncertainty or where there are no external or "objective" standards of reference.

Information saturation   In communication networks, the degree of overload experienced by members occupying central positions.

Informed consent   Voluntary consent by an individual to participate in a research project based on information received about what his or her participation will entail.

Ingratiation   Deliberate use of deception to increase a target person's liking for us in hopes of gaining tangible benefits that the target person controls. Techniques such as flattery, expressing agreement with the target person's attitudes, and exaggerating one's admirable qualities may be used.

In-group   In intergroup relations, one's own membership group; contrast with the out-group.

Input-process-output model   Abstract model of groups that focuses on the internal processes which mediate the conversion of resources (inputs) used by the group to products (outputs) generated by the group.

Institutionalization of deviance   Process by which members of a group come to expect and support deviance by another member over time.

Instrumental conditioning   Process through which an individual learns a behavior in response to a stimulus to obtain a reward or avoid a punishment.

Intentionalist model   Theory that views communication as the exchange of communicative intentions, and views messages transmitted as merely the means to this end.

Interaction process analysis (IPA)   Coding system of 12 categories used to measure and analyze communication patterns and processes in groups.

Intergroup conflict   State of affairs in which groups having opposing interests take antagonistic actions toward one another to control some outcome important to them.

Intergroup contact hypothesis   Hypothesis holding that in intergroup relations, increased interpersonal contact between groups will reduce stereotypes and prejudice and consequently reduce antagonism between groups.

Internalization   Process through which initially external behavioral standards become internal and subsequently guide an individual's behavior.

Internal validity   Extent to which the findings are free from contamination by extraneous variables.

Interpersonal attraction   Positive attitude held by one person toward another person.

Interpersonal spacing (proxemics)   Nonverbal communication involving the ways people position themselves at varying distances and angles from others. Because interpersonal spacing refers to the proximity of people, it is known as proxemics.

Interview survey   Method of research in which a person (i.e., an interviewer) asks a series of questions and systematically records the answers from the respondents. *See also* Questionnaire survey.

Intrinsically motivated behavior   Behavior which results from the motivation to achieve an internal state that an individual finds rewarding.

J-curve theory   Theory that revolutions occur when there is an intolerable gap between people's expectations of need satisfaction and the actual level of satisfaction they experience.

Labeling theory   View that reactions by others to a rule violation are an essential element in deviance.

Leadership   In groups, the process whereby one member influences and coordinates the behavior of other members in pursuit of group goals. The enactment of several functions necessary for successful group performance; these functions include planning, organizing, and controlling the activities of group members.

Learning structure   Environment in which an individual can learn the information and skills required to enact a role.

Legitimate means   Those ways of striving to achieve goals that are defined as acceptable by social norms.

Life course   Individual's progression through a series of socially defined, age-linked social roles.

Life event   Episode marking a transition point in the life course that provokes coping and readjustment.

Likert scale   Technique for measuring attitudes that asks a respondent to indicate the extent to which he or she agrees with each of a series of statements about an object.

Linguistic relativity hypothesis   Hypothesis proposed by Whorf that language shapes ideas and guides the individual's mental activity. The strong form of the hypothesis holds that people cannot perceive distinctions absent from their language. The weak form holds that languages facilitate thinking about events and objects which are easily codeable or symbolized in them.

Loneliness   Unpleasant, subjective experience that results from the lack of social relationships satisfying in either quantity or quality.

Majority influence   Process by which a group's majority pressures an individual to adopt a specific position on some issue.

Mass media   Those channels of communication (TV, radio, newspapers) that enable a source to reach and influence a large audience.

Matching hypothesis   Hypothesis that each person looks for someone to date who is of approximately the same level of social desirability.

Meaning   In symbolic interactionist theory, the implications that stem from an object or thing when it is judged in terms of a person's plan of action.

Media campaign   Systematic attempt by an influencing source to use the mass media to change attitudes and beliefs of a target audience.

Mediator   Third party who helps groups in conflict identify issues and agree on some resolution to the conflict. Mediators usually serve as advisers, rather than as decision makers.

Mere exposure effect   Repeated exposure to the same stimulus that produces a positive attitude toward it.

Methodology   Set of systematic procedures used to conduct empirical research. Usually these procedures pertain to how data will be collected and analyzed.

Middle-range theory   Narrow, focused theoretical framework that explains the conditions which produce some specific social behavior.

Minority influence   Attempt by an active minority within a group to persuade majority members to accept their viewpoint and adopt a new position.

Mobilization   Process through which individuals surrender personal resources and commit them to the pursuit of group or organizational goals.

Moral development   Process through which children become capable of making moral judgments.

Movement organization   Group of persons with defined roles who engage in sustained activity that reflects the ideology of a social movement.

Negotiation   Communication process between opposing sides in a conflict, whereby the parties make offers and counteroffers in hopes of eventually reaching an agreement.

Nonaggressive pornography   Explicit depictions—in film, video, photograph, or story—of adults engaging in consenting sexual activity. *See also* Aggressive pornography.

Nonstandard speech   Style of speech characterized by limited vocabulary, improper pronunciation, and incorrect grammar. The use of this style is associated with low status and low power. *See also* Standard speech.

Norm   In groups, a standard or rule that specifies how members are expected to behave under given circumstances. Expectations concerning which behaviors are acceptable and which are unacceptable for specific persons in specific situations.

Normative influence   In groups, a form of influence that occurs when a member conforms to group norms in order to receive the rewards and/or avoid the punishments that are contingent on adherence to these norms.

Normative life stage   Discrete period in the life course during which individuals are expected to perform the set of activities associated with a distinct age-related role.

Normative transition   Socially expected changes made by all or most members of a defined population.

Norm of homogamy   Social norm requiring that friends, lovers, and spouses be characterized by similarity in age, race, religion, and socioeconomic status.

Norm of reciprocity   Social norm stating that people should (1) help those who have previously helped them, and (2) not help those who have denied them help for no legitimate reason.

Observational learning   Acquisition of behavior based on observation of another person's behavior and of its consequences for that person. Also known as modeling.

Opportunity structure   Environment in which an individual has opportunities to enact a role, which usually requires the assistance of those in complementary roles.

Outcomes   Rewards expected from an interpersonal relationship minus the expected costs.

Out-group   In intergroup relations, a group of persons different from one's own group.

Overreward   In equity theory, a condition of inequity in which a group member receives more rewards than would be justified on the basis of his or her contribution to the group.

Panel study   Research method in which a given sample of respondents is surveyed at one point in time and then resurveyed at a later point (or several later points). Also known as a longitudinal survey.

Paralanguage   All the vocal aspects of speech other than words, including loudness, pitch, speed of speaking, pauses, sighs, laughter, and so on.

Passionate love   State of intense longing for union with another and intense physiological arousal.

Personal norm   Feeling of moral obligation to perform specific actions that stem from individuals' internalized system of values.

Perspective-taking model   Theory that views communication as the exchange of messages using symbols whose meaning is created by the interaction itself.

Persuasion   Effort by a source to change the beliefs or attitudes of a target person through the use of information or argument.

Population   Set of all people whose attitudes, behavior, or characteristics are of interest to the researcher.

Position   Designated location in a social system.

Powerlessness   Sense of having little or no control over events.

Prejudice   Strong like or dislike for members of a specific group.

Primacy effect   Tendency when forming an impression to be most influenced by the earliest information received. The primacy effect accounts for the fact that first impressions are especially powerful.

Primary group   Small group in which there exist strong emotional ties among members. *See also* Secondary group.

Primary relationship   Interpersonal relationship that is personal, emotionally involving, and of long duration.

Principle of consistency   In cognitive theory, a principle maintaining that if a person holds several ideas which are incongruous or inconsistent with each other, he or she will experience discomfort or conflict and subsequently change one or more of the ideas to render them consistent.

Principle of covariation   Principle that attributes behavior to the potential cause which is present when the behavior occurs and absent when the behavior fails to occur.

Principle of determinism   Principle that discoverable causes exist for all events in a science's domain of interest.

Production blocking   In brainstorming groups, a phenomenon that inhibits production of novel ideas. Production blocking occurs when participants in a brainstorming group are unable to express their ideas due to bottlenecks caused by turn-taking among members.

Promise   Influence technique taking this general form: "If you do X (which I want), then I will do Y (which you want)."

Prosocial behavior   Broad category of actions considered by society as being beneficial to others and as having positive social consequences. A wide variety of specific behaviors qualify as prosocial, including donation to charity, intervention in emergencies, cooperation, sharing, volunteering, sacrifice, and the like.

Prototype   In person perception, an abstraction that represents the "typical" or quintessential instance of a class or group.

Punishment   Painful or discomforting stimulus that reduces the frequency with which the behavior occurs.

Questionnaire survey   Method of research in which a series of questions appear on a printed questionnaire and the respondents read and answer them at their own pace. Usually no interviewer is present. *See also* Interview survey.

Random assignment   In an experiment, the assignment of subjects to experimental conditions on the basis of chance.

Rape myths   Prejudicial, stereotyped, and false beliefs about rape, rape victims, and persons who commit rape.

Realistic group conflict theory   Theory of intergroup conflict that explains the development and the resolution of conflict in terms of the goals of each group; its central hypothesis is that groups will engage in conflictive behavior when their goals involve opposition of interest.

Reinforcement   Any favorable outcome or consequence that results from a behavioral response by a person. Reinforcement strengthens the response—that is, it increases the probability it will be repeated.

Reinforcement theory   Theoretical perspective based on the premise that social behavior is governed by external events, especially rewards and punishments.

Relative deprivation   Gap between the expected level and the actual level of satisfaction of the individual's needs, in which the level expected by the individual exceeds the level of need satisfaction experienced.

Reliability   Degree to which a measuring instrument produces the same results each time it is employed under a set of specified conditions.

Response rate   In a survey, the percentage of people contacted who complete the survey.

Return potential model   Formal model of a social norm. This model treats norms as having two dimensions—the behavior dimension and the evaluation dimension. A norm is a function that specifies what level of evaluation will correspond to what behaviors.

Revolutionary coalition   In groups, a union of medium- and low-status members who oppose the existing leadership and wish to overturn it.

Ringelmann effect   Empirical generalization with respect to group performance. On additive tasks, large groups produce in toto more than smaller ones, but the output per member in the large groups is lower than that in the smaller groups.

Risk-benefit analysis   Technique that weighs the potential risks to research subjects against the anticipated benefits to subjects and the importance of the knowledge which may result from the research.

Risky shift   In group decision making, the tendency for decisions made in groups after discussion to be riskier than decisions made by individual members prior to discussion.

Role   Set of functions to be performed by a person on behalf of a group of which he or she is a member. A cluster of rules indicating the set of duties to be performed by a member occupying a given position within a group.

The set of expectations governing the behavior of an occupant of a specific position within a social structure.

Role differentiation   Emergence of distinct roles within a group. The division of labor within a group.

Role discontinuity   Discontinuity that results when the values and identities associated with a new role contradict those of earlier roles.

Role identity   An individual's concept of self in a specific social role.

Role overload   Condition in which the demands placed on a person by his or her roles exceed the amount of time, energy, and other resources available to meet those demands.

Role taking   In symbolic interactionist theory, the process of imaginatively occupying the position of another person and viewing the situation and the self from that person's perspective; the process of imagining the other's attitudes and anticipating that person's responses.

Role theory   Theoretical perspective based on the premise that a substantial portion of observable, day-to-day social behavior is simply persons carrying out role expectations.

Romantic love ideal   Five beliefs regarding love, including belief (1) in love at first sight, (2) that there is one and only one true love for each person, (3) that love conquers all, (4) that our beloved is (nearly) perfect, and (5) that one should follow his or her heart.

Routine activities perspective   Theory which considers how deviant behavior, such as crime and substance abuse, emerges from the routines of everyday life.

Rule breaking   Behavior that violates social norms.

Rumor   Communication via informal and often novel channels that cannot be validated.

Salience   Relative importance of a specific role identity to the individual's self-schema. The salience hierarchy refers to the ordering of an individual's role identities according to their importance.

Sample   Subset of a population selected for investigation.

Schema   Specific cognitive structure that organizes the processing of complex information about other persons, groups, and situations. Our schemas guide what we perceive in the environment, how we organize information in memory, and what inferences and judgments we make about people and things.

Script   Schema about a social event that specifies (1) the definition of the situation, (2) the identities of the social actors involved, and (3) a range and sequence of permissible behaviors.

Secondary deviance   Deviant behavior employed by a person as a means of defense or adjustment to the problems created by others' reactions to rule breaking by him or her.

Secondary group   Group whose members have few emotional ties with one another and relate in terms of limited roles. Interaction in secondary groups is formal, impersonal, and nonspontaneous. *See also* Primary group.

Self   The individual viewed as both the active source and the passive object of reflexive behavior.

Self-awareness   State in which we take the self as the object of our attention and focus on our own appearance, actions, and thoughts.

Self-disclosure   Process of revealing personal information (aspects of our feelings and behaviors) to another person. Self-disclosure is sometimes used as an impression management tactic.

Self-discrepancy   State in which a component of the individual's actual self is the opposite of a component of the ideal self or the ought self.

Self-esteem   Evaluative component of the self-concept. The positive and negative evaluations people have of themselves.

Self-estrangement   Awareness that one is engaging in activities which are not rewarding in themselves.

Self-fulfilling prophecy   When persons behave toward another person according to a label (impression) and cause the person to respond in ways that confirm the label.

Self-presentation   All conscious and unconscious attempts by people to control the images of self they project in social interaction.

Self-reinforcement   Individual's use of internalized standards to judge personal behavior and reward the self.

Self-schema   Organized structure of information that people have about themselves; the primary influence on the processing of information about the self.

Self-serving bias   In attribution, the tendency for people to take personal credit for acts that yield positive outcomes and to deflect blame for bad outcomes, attributing them to external causes.

Semantic differential scale   Technique for measuring attitudes that asks a respondent to rate an object on each of a series of bipolar adjective scales, that is, scales whose ends are two adjectives having opposite meanings.

Sentiment relations   In balance theory, a positive or negative relationship between two cognitive elements.

Sentiments   Socially significant feelings such as grief, love, jealousy, or indignation that arise out of enduring social relationships. Each sentiment is a pattern of sensations, emotions, actions, and cultural beliefs appropriate to a social relationship.

Sexual assault   Sexual contact or intercourse without consent, accomplished by threat or use of force.

Shaping   Learning process in which an agent initially reinforces any behavior that remotely resembles the desired response and subsequently requires increasing correspondence between the learner's behavior and the desired response before providing reinforcement.

Significant other   Person whose views and attitudes are very important and worthy of consideration. The reflected views of a significant other have great influence on the individual's self-concept and self-regulation.

Simple random sample   Sample of individuals selected from a population in such a way that everyone is equally likely to be selected.

Situated identity   Conception, held by a person in a situation, that indicates who he or she is in relation to the other people involved in that situation.

Situated self   Subset of self-concepts that constitutes the self people recognize in a particular situation. Selected from the person's various identities, qualities, and self-evaluations, the situated self depends on the demands of the situation.

Situational attribution   Decision by an observer to attribute a behavior to environmental forces facing the person who performed it, rather than to the internal state of the person who performed it. *See also* Dispositional attribution.

Situational constraint   Influence on behavior due to the likelihood that other persons will learn about that behavior and respond positively or negatively to it.

Small group   Any group of persons sufficiently limited in size that the members are able to interact directly or face-to-face. Groups of this type usually consist of no more than 20 persons.

Social-emotional act   In group interaction, emotional reactions, both positive and negative, directed by one member toward other members.

Social-emotional specialist   In groups, a person who strives to keep emotional relationships pleasant among members; a person who initiates acts that ease the tension and soothe hurt feelings.

Social exchange theory   Theoretical perspective, based on the principle of reinforcement, that assumes people will choose whatever actions maximize rewards and minimize costs.

Social identity   Definition of the self in terms of the defining characteristics of a social group.

Social identity theory of intergroup behavior   Theory of intergroup relations based on the premise that people spontaneously categorize the social world into various groups (especially in-groups and out-groups) and experience high self-esteem to the extent that the in-groups to which they belong have more status than the out-groups.

Social impact theory   Theoretical framework, applicable to both persuasion and obedience, which states that the impact of an influence attempt is a function of strength, immediacy, and number of sources present.

Social influence   Interaction process in which one person's behavior causes another person to change an opinion or to perform an action that he or she would not otherwise do.

Socialization   Process through which individuals learn skills, knowledge, values, motives, and roles appropriate to their positions in a group or society.

Social learning theory   Theoretical perspective maintaining that one person (the learner) can acquire new responses without enacting them simply by observing the behavior of another person (the model). This learning process, called imitation, is distinguished by the fact that the learner neither performs a response nor receives any reinforcement.

Social loafing   Tendency by group members to slack off and reduce their effort on additive tasks, which causes the group's output to fall short of its potential.

Social movement   Collective activity that expresses a high level of concern about some issue; the activity may include participation in discussions, petition drives, demonstrations, or election campaigns.

Social network   Sets of interpersonal relationships associated with the social positions a person occupies.

Social perception   Process through which we construct an understanding of the social world out of the data we obtain through our senses. More narrowly defined, the processes through which we use available information to form impressions of people.

Social psychology   Field that systematically studies the nature and causes of human social behavior.

Social responsibility norm   Widely accepted social norm stating that individuals should help people who are dependent on them.

Social structure   Ordered and persisting relationships among the positions in a social system.

Sociobiology   Theoretical perspective that focuses on the perpetuation of genes rather than on perpetuation of physical attributes; in this view, the fittest animal is one that passes its genes to subsequent generations.

Sociolinguistic competence   Knowledge of the implicit rules for generating socially appropriate sentences that make sense because they fit with the listeners' social knowledge.

Sociometry   Empirical research method, based on questionnaire nominations, used to measure patterns of interpersonal attraction in groups.

Source   In social influence, the person who intentionally engages in some behavior (persuasion, threat, promise) to cause another person to behave in a manner different from what he or she would otherwise do.

Spillover   Situation in which stress experienced at work or in the family is carried into the other domain.

Spoken language   Socially acquired system of sound patterns with meanings agreed on by the members of a group.

Standard speech   Speech style characterized by diverse vocabulary, proper pronunciation, correct grammar, and abstract content. The use of this style is associated with high status and power. *See also* Nonstandard speech.

Status   Social evaluation or ranking assigned to a position (in a group) indicating its prestige, importance, or value. Also, the evaluation or ranking of the person occupying that position; a member's relative standing vis-à-vis others.

Status characteristic Any property of a person around which evaluations and beliefs about that person come to be organized (properties such as race, occupation, age, sex, ethnicity, education, and so on).

Status generalization Process through which differences in members' status characteristics lead to different performance expectations and hence affect patterns of interaction in groups. The tendency for a member's status inside a group to reflect his or her status outside that group.

Stereotype Fixed set of characteristics that is attributed to all the members of a group. A simplistic and rigid perception of members of one group that is widely shared by others.

Stigma Personal characteristics that others view as insurmountable handicaps preventing competent or morally trustworthy behavior.

Stratified sample In survey research, a sampling design whereby researchers subdivide the population into groups according to characteristics known or thought to be important, select a random sample of groups, and then draw a sample of units within each selected group.

Stress Condition in which the demands made on the person exceed the individual's ability to cope with them.

Subjective expected value (SEV) With respect to threats, the product of a threat's credibility times its magnitude; with respect to promises, the product of a promise's credibility times its magnitude.

Subjective norm Individual's perception of others' beliefs about whether or not a behavior is appropriate in a specific situation.

Subtractive rule Rule which states that when making attributions about personal dispositions, the observer subtracts the impact of situational forces from the personal disposition implied by the behavior itself.

Summons-answer sequence Most common verbal method for initiating a conversation in which one person summons the other as with a question or greeting, and the other indicates availability for conversation by responding. This sequence establishes the mutual obligation to speak and to listen that produces conversational turn taking.

Superordinate goal In intergroup conflict, an objective held in common by all conflicting groups that cannot be achieved by any group without the supportive efforts of the others.

Survey Research method that involves gathering information by asking questions of individuals, either through face-to-face interviews or through questionnaires.

Symbol Form used to represent ideas, feelings, thoughts, intentions, or any other object. Symbols represent our experiences in a way that others can perceive with their sensory organs—through sounds, gestures, pictures, and so on.

Symbolic interaction theory Theoretical perspective based on the premise that human nature and social order are products of communication among people.

Tactical impression management Selective use of self-presentation tactics by a person who wishes to manipulate the impressions that others form of him or her.

Target In social influence, the person who is impacted by a social influence attempt from the source. In aggression, the person toward whom an aggressive act is directed.

Task acts In group interaction, instrumental behaviors that move the group toward realization of its goal.

Task specialist In groups, a member who pushes the group toward attainment of its goals; a person who contributes many ideas and suggestions to the group.

Theoretical perspective Theory that makes broad assumptions about human nature and offers general explanations of a wide range of diverse behaviors. *See also* Middle-range theory.

Theory Set of interrelated propositions that organizes and explains a set of observed facts; a network of hypotheses that may be used as a basis for prediction.

Theory of cognitive dissonance Theory concerning the sources and effects of inconsistency in cognitive systems with two or more elements.

Theory of reasoned action Theory that behavior is determined by behavioral intention, which in turn is determined by both attitude and subjective norm.

Theory of speech accommodation Theory that people express or reject intimacy with others by adjusting their speech behavior (accent, vocabulary, or language) during interaction. They make their own speech behavior more similar to their partner's to express liking, and more dissimilar to reject intimacy.

Threat Influence technique that is a communication taking this general form: "If you don't do X (which I want), then I will do Y (which you don't want)."

Trait centrality A personality trait has a high level of trait centrality when information about a person's standing on that trait has a large impact on the overall impression which others form of that person. The *warm-cold* trait, for instance, is highly central.

Transactional leadership Leadership in groups based on an exchange between the leader and other group members. The leader performs actions that move the group toward the attainment of its goals; in return, the leader receives support, endorsement, and rewards.

Transformational leadership Leadership that strengthens group performance by changing the way members view their group, its opportunities, and its mission. Leadership that conveys an extraordinary sense of mission to group members and arouses new ways of thinking within the group.

Trust Belief that a person is both honest and benevolent.

Ultimate attribution error Perceptual bias occurring in intergroup relations. Negative behaviors by out-group members are attributed to stable, internal factors such as undesirable personal traits or dispositions, but positive behaviors by out-group members are attributed to unstable, external factors such as situational pressures or luck. As a result, in-group observers blame the out-group for negative outcomes but do not give it credit for positive outcomes.

Underreward In equity theory, a condition of inequity in which a group member receives fewer rewards than would be justified on the basis of his or her contribution to the group.

Unitary task Group task in which all members perform identical activities.

Unit relations In balance theory, the extent of perceived association between two cognitive elements.

Upward mobility Movement from an occupation that is lower in prestige and income to one that is higher in prestige and income.

Validity Degree to which a measuring instrument actually measures the theoretical construct the investigator intends to measure.

Values Enduring beliefs that certain patterns of behavior or end states are preferable to others.

Weapons effect Increase in aggression caused by the presence of a gun in a situation where an individual is already frustrated.

Weighted averaging model Theoretical model for combining information about individual traits in order to predict how favorable the overall impression people form of another will be. This model averages the favorability values of all the single traits people associate with the person after first weighting each trait according to criteria such as its credibility, negativity, and consistency with prior impressions. *See also* Additive model; Averaging model.

REFERENCES

Abbey, A., Ross, L. T., McDuffie, D., & McAuslan, P. (1996). Alcohol and dating risk factors for sexual assault among college women. *Psychology of Women Quarterly, 20,* 147–169.

Abelson, R. P. (1981). The psychological status of the script concept. *American Psychologist, 36,* 715–729.

Acock, A., & Scott, W. (1980). A model for predicting behavior: The effect of attitude and social class on high and low visibility political participation. *Social Psychology Quarterly, 43,* 59–72.

Adams, J. S. (1963). Toward an understanding of inequity. *Journal of Abnormal and Social Psychology, 67,* 422–436.

Adams, J. S., & Jacobsen, P. R. (1964). Effects of wage inequities on work quality. *Journal of Abnormal and Social Psychology, 69,* 19–25.

Adams, J. S., & Rosenbaum, W. B. (1962). The relationship of worker productivity to cognitive dissonance about wage inequities. *Journal of Applied Psychology, 46,* 161–164.

Aderman, D., & Berkowitz, L. (1983). Self-concern and the unwillingness to be helpful. *Social Psychology Quarterly, 46*(4), 293–301.

Adler, P. A., & Adler, P. (1989). The gloried self: The aggrandizement and the construction of self. *Social Psychology Quarterly, 52,* 299–310.

Adler, P. A., & Adler, P. (1995). Dynamics of inclusion and exclusion in preadolescent cliques. *Social Psychology Quarterly, 58,* 145–162.

Adlerfer, C. P. (1982). Problems in changing white males' behavior and beliefs concerning race relations. In P. Goodman et al. (Eds.), *Change in organizations* (pp. 122–165). San Francisco: Jossey-Bass.

Aiken, L. R. (1992). Some measures of interpersonal attraction and group cohesiveness. *Educational and Psychological Measurement, 52*(1), 63–67.

Ainsworth, M. (1979). Infant-mother attachment. *American Psychologist, 34,* 932–937.

Ajzen, I. (1982). On behaving in accordance with one's attitudes. In M. Zanna, E. Higgins, & C. Herman (Eds.), *Consistency in social behavior: The Ontario Symposium* (Vol. 2). Hillsdale, NJ: Erlbaum.

Ajzen, I. (1985). From intentions to action: A theory of planned behavior. In J. Kulh & J. Beckman (Eds.), *Action control: From cognition to behavior.* Heidelberg: Springer.

Ajzen, I., & Fishbein, M. (1977). Attitude-behavior relations: A theoretical analysis and review of research. *Psychological Bulletin, 84,* 888–918.

Ajzen, I., & Fishbein, M. (1980). *Understanding attitudes and predicting social behavior.* Englewood Cliffs, NJ: Prentice-Hall.

Ajzen, I., & Holmes, W. H. (1976). Uniqueness of behavioral effects in causal attribution. *Journal of Personality, 44,* 98–108.

Akers, R. L., Krohn, M., Lanza-Kaduce, L., & Radosevich, M. (1979). Social learning and deviant behavior: A specific test of a general theory. *American Sociological Review, 44,* 636–655.

Akers, R. L., LaGreca, H. J., Cochran, J., & Sellers, C. (1989). Social learning theory and alcohol behavior among the elderly. *Sociological Quarterly, 30,* 625–638.

Aldag, R. J., & Fuller, S. F. (1993). Beyond fiasco: A reappraisal of the groupthink phenomenon and a new model of group decision processes. *Psychological Bulletin, 113,* 533–552.

Aldous, J. (1987). New views on the family life of the elderly and the near-elderly. *Journal of Marriage and the Family, 49,* 227–234.

Aldwin, C., & Revenson, T. (1987). Does coping help? A reexamination of the relationship between coping and mental health. *Journal of Personality and Social Psychology, 53,* 337–348.

Alessio, J. C. (1980). Another folly for equity theory. *Social Psychology Quarterly, 43,* 336–340.

Alexander, C. N., Jr., & Rudd, J. (1984). Predicting behaviors from situated identities. *Social Psychology Quarterly, 47,* 172–177.

Alexander, C. N., Jr., & Wiley, M. G. (1981). Situated activity and identity formation. In M. Rosenberg & R. Turner (Eds.), *Social psychology: Sociological perspectives.* New York: Basic Books.

Allan, E. A., & Steffensmeier, D. J. (1989). Youth underemployment and property crime: Differential effects of job availability and job quality on juvenile and young adult arrest rates. *American Sociological Review, 54,* 107–123.

Allen, H. (1972). Bystander intervention and helping on the subway. In L. Bickman & T. Henchy (Eds.), *Beyond the laboratory: Field research in social psychology.* New York: McGraw-Hill.

Allen, V. L., & Levine, J. M. (1969). Consensus and conformity. *Journal of Experimental Social Psychology, 5,* 389–399.

Allen, V. L., & Levine, J. M. (1971). Social support and conformity: The role of independent assessment of reality. *Journal of Experimental Social Psychology, 7,* 48–58.

Allen, V. L., & Van de Vliert, E. (1982). A role theoretical perspective on transitional processes. In V. L. Allen & E. Van de Vliert (Eds.), *Role transitions: Explorations and explanations* (pp. 3–18). New York: Plenum.

Allison, S. T., McQueen, L. R., & Schaerfl, L. M. (1992). Social decision making processes and the equal partitioning of shared resources. *Journal of Experimental Social Psychology, 28,* 23–42.

Allport, F. H. (1924). *Social psychology.* Cambridge, MA: Houghton Mifflin.

Allport, G. W. (1935). Attitudes. In C. Murchison (Ed.), *Handbook of social psychology* (pp. 798–844). Worcester, MA: Clark University Press.

Allport, G. W. (1954). *The nature of prejudice.* Reading, MA: Addison-Wesley.

Allport, G. W. (1961). *Pattern and growth in personality.* New York: Holt, Rinehart and Winston.

Allport, G. W. (1985). The historical background of social psychology. In G. Lindzey & E. Aronson (Eds.), *Handbook of social psychology* (3rd ed., Vol. 1, pp. 1–46). New York: Random House.

Altman, I., & Taylor, D. A. (1973). *Social penetration: The development of interpersonal relationships.* New York: Holt, Rinehart and Winston.

Alwin, D. F. (1986). Religion and parental child-rearing orientations: Evidence of Catholic-Protestant convergence. *American Journal of Sociology, 92,* 412–440.

Alwin, D. F., & Krosnick, J. A. (1991). Aging, cohorts, and the stability of sociopolitical orientations over the life span. *American Journal of Sociology, 97,* 169–195.

Al-Zahrani, S. S. A., & Kaplowitz, S. A. (1993). Attributional biases in individualistic and collectivistic cultures: A comparison of Americans with Saudis. *Social Psychology Quarterly, 56,* 223–233.

Amato, P. R. (1983). Helping behavior in urban and rural environments: Field studies based on a taxonomic organization of helping episodes. *Journal of Personality and Social Psychology, 45,* 571–586.

Amato, P. R. (1990). Personality and social network involvement as predictors of helping behavior in everyday life. *Social Psychology Quarterly, 53* (1), 31–43.

Amato, P. R., & Keith, B. (1991). Parental divorce and adult well-being: A meta-analysis. *Journal of Marriage and the Family, 53,* 43–58.

Amato, P. R., & Rogers, S. J. (1997). A longitudinal study of marital problems and subsequent divorce. *Journal of Marriage and the Family, 59,* 612–624.

Amir, Y. (1969). Contact hypothesis in ethnic relations. *Psychological Bulletin, 71,* 319–342.

Amir, Y. (1976). The role of intergroup contact in change of prejudice and ethnic relations. In P. A. Katz (Ed.), *Towards the elimination of racism* (pp. 245–308). New York: Pergamon.

Anderson, B. A., & Silver, B. D. (1987). The validity of survey responses: Insights from interviews of married couples in a survey of Soviet emigrants. *Social Forces, 66,* 537–554.

Anderson, C. A. (1987). Temperature and aggression: Effects on quarterly, yearly, and city rates of violent and nonviolent crime. *Journal of Personality and Social Psychology, 52,* 1161–1173.

Anderson, C. A., Anderson, K. B., & Denser, W. E. (1996). Examining an affective aggression framework: Weapon and temperature effects on aggressive thought, affect, and attitudes. *Personality and Social Psychology Bulletin, 22,* 366–376.

Anderson, C. A., & Sedikides, C. (1991). Thinking about people: Contribution of a typological alternative to associationistic and dimensional models of person perception. *Journal of Personality and Social Psychology, 60,* 203–217.

Anderson, K. B., Cooper, H., & Okamura, L. 1997. Individual differences and attitudes toward rape: A meta-analytic review. *Personality and Social Psychology Bulletin, 23,* 295–315.

Anderson, N. H. (1968). Likeableness ratings of 555 personality trait words. *Journal of Personality and Social Psychology, 9,* 272–279.

Anderson, N. H. (1981). *Foundations of information integration theory.* New York: Academic Press.

Andrews, B., & Brewin, C. R. (1990). Attributions of blame for marital violence: A study of antecedents and consequences. *Journal of Marriage and the Family, 52,* 757–767.

Andrews, I. R. (1967). Wage inequity and job performance: An experimental study. *Journal of Applied Psychology, 51,* 39–45.

Aneshensel, C. S., Rutter, C. M., Lachenbruch, P. A. (1991). Social structure, stress, and mental health: Competing conceptual and analytic models. *American Sociological Review, 56,* 166–178.

Anson, O. (1989). Marital status and women's health revisited: The importance of a proximate adult. *Journal of Marriage and the Family, 51,* 185–194.

Appleton, W. (1981). *Fathers and daughters.* New York: Doubleday.

Appleton (WI) *Post-Crescent* (1992, February 16). Rescuers respond on instinct, p. 1.

Arbuthnot, J., & Wayner, M. (1982). Minority influence: Effects of size, conversion, and sex. *Journal of Psychology, 111,* 285–295.

Archer, D., & Akert, R. (1977). Words and everything else: Verbal and nonverbal cues in social interpretation. *Journal of Personality and Social Psychology, 35,* 443–449.

Archer, R. L. (1980). Self-disclosure. In D. M. Wegner & R. R. Vallacher (Eds.), *The self in social psychology.* New York: Oxford University Press.

Arendt, H. (1965). *Eichmann in Jerusalem: A report on the banality of evil.* New York: Viking.

Aries, E. (1996). *Men and women in interaction.* New York: Oxford University Press.

Aristotle, *Poetics,* Book 6.

Aronfreed, J., & Reber, A. (1965). Internalized behavior suppression and the timing of social punishment. *Journal of Personality and Social Psychology, 1,* 3–16.

Aronoff, D., & Tedeschi, J. T. (1968). Original stakes and behavior in the prisoner's dilemma game. *Psychonomic Science, 12,* 79–80.

Aronson, E., Ellsworth, P. C., Carlsmith, J. M., & Gonzales, J. H. (1990). *Methods of research in social psychology* (2nd ed.). New York: McGraw-Hill.

Aronson, E., Turner, J. A., & Carlsmith, J. M. (1963). Communicator credibility and communication discrepancy as determinants of opinion change. *Journal of Abnormal and Social Psychology, 67,* 31–36.

Asch, S. E. (1946). Forming impressions of personality. *Journal of Abnormal and Social Psychology, 41,* 258–190.

Asch, S. E. (1951). Effects of group pressure upon the modification and distortion of judgments. In H. Guetzkow (Ed.), *Groups, leadership, and men.* Pittsburgh: Carnegie Press.

Asch, S. E. (1952). *Social psychology.* Englewood Cliffs, NJ: Prentice-Hall.

Asch, S. E. (1955). Opinions and social pressure. *Scientific American, 193,* 31–35.

Asch, S. E. (1957). An experimental investigation of group influence. *Symposium on preventive and social psychiatry.* Walter Reed Army Institute of Research, Washington, DC: U.S. Government Printing Office.

Asch, S. E., & Zukier, H. (1984). Thinking about persons. *Journal of Personality and Social Psychology, 46,* 1230–1240.

Asendorpf, J. B. (1987). Videotape reconstruction of emotions and cognitions related to shyness. *Journal of Personality and Social Psychology, 53,* 542–549.

Ashmore, R. D. (1981). Sex stereotypes and implicit personality theory. In D. L. Hamilton (Ed.), *Cognitive processes in stereotyping and intergroup behavior* (pp. 37–81). Hillsdale, NJ: Erlbaum.

Ashmore, R. D., Solomon, M. R., & Longo, L. C. (1996). Thinking about fashion models' looks: A multidimensional approach to the structure of perceived physical attractiveness. *Personality and Social Psychology Bulletin, 22,* 1083–1104.

Astone, N., & McLanahan, S. S. (1991). Family structure, parental practices, and high school completion. *American Sociological Review, 56,* 309–320.

Atchley, R. C. (1980). *The social forces in later life.* Belmont, CA: Wadsworth.

Atkin, C. K. (1981). Mass media information campaign effectiveness. In R. E. Rice & W. J. Paisley (Eds.), *Public communication campaigns.* Beverly Hills, CA: Sage.

Attridge, M., Berscheid, E., & Simpson, J. A. (1995). Predicting relationship stability from both partners vs. one. *Journal of Personality and Social Psychology, 69,* 254–268.

Austin, W. (1980). Friendship and fairness: Effects of type of relationship and task performance on choice distribution rules. *Personality and Social Psychology Bulletin, 6,* 402–408.

Austin, W. G., & Hatfield, E. (1980). Equity theory, power, and social justice. In G. M. Kula (Ed.), *Justice and social interaction.* Bern, Switzerland: Hans Huber.

Austin, W., & Walster, E. (1974). Participants' reactions to "equity with the world." *Journal of Experimental Social Psychology, 10,* 528–548.

Averill, J. R. (1982). *Anger and aggression: An essay on emotion.* New York: Springer-Verlag.

Ayers, L., Nacci, P., & Tedeschi, J. T. (1973). Attraction and reaction to noncontingent promises. *Bulletin of the Psychonomic Society, 1*(lB), 75–77.

Bacharach, S. B., & Lawler, E. J. (1981). *Bargaining: Power, tactics, and outcomes.* San Francisco: Jossey-Bass.

Backman, C. (1990). Attraction in interpersonal relationships. In M. Rosenberg & R. Turner (Eds.), *Social psychology: Sociological perspectives.* New Brunswick, NJ: Transaction.

Backman, C., & Secord, P. (1962). Liking, selective interaction, and misperception in congruent interpersonal relations. *Sociometry, 25,* 321–325.

Backman, C., & Secord, P. (1968). The self and role selection. In C. Gordon & K. J. Gergen (Eds.), *The self in social interaction.* New York: Wiley.

Bagozzi, R. P. (1981). Attitudes, intentions and behavior: A test of some key hypotheses. *Journal of Personality and Social Psychology, 41,* 607–627.

Bailey, W. C. (1990). Murder, capital punishment and television: Execution publicity and homicide rates. *American Sociological Review, 55,* 628–633.

Bales, R. F. (1950). *Interaction process analysis.* Reading, MA: Addison-Wesley.

Bales, R. F. (1953). The equilibrium problem in small groups. In T. Parsons, R. F. Bales, & E. A. Shils (Eds.), *Working papers in the theory of action.* New York: Free Press.

Bales, R. F. (1965). Task roles and social roles in problem solving groups. In I. D. Steiner & M. Fishbein (Eds.), *Current studies in social psychology* (pp. 321–333). New York: Holt, Rinehart and Winston.

Bales, R. F. (1970). *Personality and interpersonal behavior.* New York: Holt, Rinehart and Winston.

Bales, R. F., & Hare, A. P. (1965). Diagnostic use of the interaction profile. *Journal of Social Psychology, 67,* 239–258.

Balkwell, J. W. (1991). From expectations to behavior: An improved postulate for expectation states theory. *American Sociological Review, 56*(3), 355–369.

Balkwell, J. W. (1994). Status. In M. Foschi & E. J. Lawler (Eds.), *Group processes: Sociological analyses* (pp. 119–148). Chicago: Nelson-Hall.

Ball, D. W. (1976). Failure in sports. *American Sociological Review, 41,* 726–739.

Baltes, P., & Willis, S. (1982). Plasticity and enhancement of intellectual functioning in old age: Penn State's adult development and enrichment project. In F. Craik & S. Trehub (Eds.), *Aging and cognitive processes.* New York: Plenum.

Banaji, M., & Prentice, D. (1994). The self in social contexts. *Annual Review of Psychology, 45,* 297–332.

Bandura, A. (1965). Influence of models' reinforcement contingencies on the acquisition of imitative responses. *Journal of Personality and Social Psychology, 1,* 589–595.

Bandura, A. (1969). Social-learning theory of identificatory processes. In D. Goslin (Ed.), *Handbook of socialization theory and research* (pp. 213–262). Chicago: Rand McNally.

Bandura, A. (1973). *Aggression: A social learning analysis.* Englewood Cliffs, NJ: Prentice-Hall.

Bandura, A. (1977). *Social learning theory.* Englewood Cliffs, NJ: Prentice-Hall.

Bandura, A. (1978). The self system in reciprocal determinism. *American Psychologist, 33,* 344–358.

Bandura, A. (1982a). The psychology of chance encounters and life paths. *American Psychologist, 37,* 747–755.

Bandura, A. (1982b). Self-efficacy mechanism in human agency. *American Psychologist, 37,* 122–147.

Bandura, A. (1982c). The self and the mechanisms of agency. In J. Suls (Ed.), *Psychological perspectives on the self* (Vol. 1). Hillsdale, NJ: Erlbaum.

Bandura, A. (1988). Self-regulation of motivation and action through goal systems. In V. Hamilton, G. H. Bower, & N. H. Frijda (Eds.), *Cognitive perspectives on emotion and motivation* (pp. 37–61). Dordrecht: Kluwer.

Bandura, A., Ross, D., & Ross, S. (1961). Transmission of aggression through imitation of aggressive models. *Journal of Abnormal and Social Psychology, 63,* 575–582.

Barber, J. J., & Grichting, W. L. (1990). Australia's media campaign against drug abuse. *International Journal of the Addictions, 25,* 693–708.

Bargh, J. A., & Thein, R. D. (1985). Individual construct accessibility, person memory, and the recall-judgment link: The case of information overload. *Journal of Personality and Social Psychology, 49,* 1129–1146.

Barkan, S. E. (1997). *Criminology: A sociological understanding.* Upper Saddle River, NJ: Prentice-Hall.

Barker, R. G., Dembo, T., & Lewin, K. (1941). Frustration and regression: An experiment with young children. *University of Iowa Studies in Child Welfare, 18,* 1–34.

Barnartt, S. N., & Harris, R. J. (1982). Recent changes in predictors of abortion attitudes. *Sociology and Social Research, 66,* 320–334.

Barnes, G., Farrell, M., & Cairns, A. (1986). Parental socialization factors and adolescent drinking behavior. *Journal of Marriage and the Family, 48,* 27–36.

Barnett, M. A. (1987). Empathy and related responses in children. In N. Eisenberg & J. Strayer (Eds.), *Empathy and its development* (pp. 146–162). New York: Cambridge University Press.

Barnett, R. C. (1994). Home-to-work spillover revisited: A study of full-time employed women in dual-earner couples. *Journal of Marriage and the Family, 56,* 647–656.

Baron, L., & Straus, M.A. (1984). Sexual stratification, pornography, and rape in the United States. In N. M. Malamuth & E. Donnerstein (Eds.), *Pornography and sexual aggression.* Orlando, FL: Academic Press.

Baron, R., & Ransberger, V. (1978). Ambient temperature and the occurrence of collective violence: The "long, hot summer" revisited. *Journal of Personality and Social Psychology, 36,* 351–360.

Baron, R. A. (1971a). Behavioral effects of interpersonal attraction: Compliance with requests from liked and disliked others. *Psychonomic Science, 25,* 325–326.

Baron, R. A. (1971b). Reducing the influence of an aggressive model: The restraining effects of discrepant modeling cues. *Journal of Personality and Social Psychology, 20,* 240–245.

Baron, R. A. (1977). *Human aggression.* New York: Plenum.

Baron, R. A., & Kepner, C. R. (1970). Model's behavior and attraction toward the model as determinants of adult aggressive behavior. *Journal of Personality and Social Psychology, 14,* 335–344.

Baron, R. S., & Roper, G. (1976). A reaffirmation of a social comparison view of choice shifts, averaging, and extremity effects in autokinetic situations. *Journal of Personality and Social Psychology, 33,* 521–530.

Bar-Tal, D. (1976). *Prosocial behavior: Theory and research.* New York: Halsted.

Baruch, G., & Barnett, R. (1986). Role quality, multiple-role involvement, and psychological well-being in mid-life women. *Journal of Personality and Social Psychology, 51,* 578–585.

Bass, B. M. (1990). *Bass & Stogdill's handbook of leadership: Theory, research, and managerial applications* (3rd ed.). New York: Free Press.

Bates, E., O'Connell, B., & Shore, C. (1987). Language and communication in infancy. In J. Osofsky (Ed.), *Handbook of infant competence* (2nd ed.). New York: Wiley.

Bates, F. L., & Harvey, C. C. (1975). *The structure of social systems.* New York: Gardner Press (Halsted).

Batson, C. D. (1987). Prosocial motivation: Is it ever truly altruistic? In L. Berkowitz (Ed.), *Advances in experimental social psychology* (Vol. 20, pp. 65–122). New York: Academic Press.

Batson, C. D. (1991). *The altruism question: Toward a social psychological answer.* Hillsdale, NJ: Erlbaum.

Batson, C. D., & Coke, J. S. (1981). Empathy: A source of altruistic motivation for helping? In J. P. Rushton & R. M. Sorrentino (Eds.), *Altruism and helping behavior.* Hillsdale, NJ: Erlbaum.

Batson, C. D., Duncan, B., Ackerman, P., Buckley, T., & Birch, K. (1981). Is empathic emotion a source of altruistic motivation? *Journal of Personality and Social Psychology, 40,* 290–302.

Batson, C. D., Dyck, J. L., Brandt, J. R., Batson, J. G., Powell, A. L., McMaster, M. R., & Griffitt, C. (1988). Five studies testing two new egoistic alternatives to the empathy-altruism hypothesis. *Journal of Personality and Social Psychology, 55,* 52–77.

Batson, C. D., & Oleson, K. C. (1991). Current status of the empathy-altruism hypothesis. In M. S. Clark (Ed.), *Review of personality and social psychology: Vol. 12. Prosocial behavior* (pp. 62–85). Newbury Park, CA: Sage.

Batson, C. D., O'Quin, K., Fultz, J., Vanderplas, M., & Isen, A. M. (1983). Influence of self-reported distress and empathy on egoistic versus altruistic motivation to help. *Journal of Personality and Social Psychology, 45,* 706–718.

Bauer, R. (1964). The obstinate audience: The influence process from the point of view of social communication. *American Psychologist, 19,* 319–328.

Baum, A., Riess, M., & O'Hara, J. (1974). Architectural variants of reaction to spatial invasion. *Environment and Behavior, 6,* 91–100.

Baumeister, R. F. (1982). A self-presentational view of social phenomena. *Psychological Bulletin, 91,* 3–26.

Baumrind, D. (1980). New directions in socialization research. *American Psychologist, 35,* 639–652.

Bavelas, A. (1950). Communication patterns in task-oriented groups. *Journal of the Acoustical Society of America, 22,* 725–730.

Beck, E., & Tolnay, S. E. (1990). The killing fields of the deep South: The market for cotton and the lynching of blacks; 1882–1930. *American Sociological Review, 55,* 526–539.

Beck, J. G., & Davies, D. K. (1987). Teen contraception: A review of perspectives on compliance. *Archives of Sexual Behavior, 16,* 337–368.

Beck, S. B., Ward-Hull, C. I., & McLear, P. M. (1976). Variables related to women's somatic preferences of the male and female body. *Journal of Personality and Social Psychology, 34,* 1200–1210.

Becker, B. J. (1986). Influence again: Another look at studies of gender differences in social influence. In J. S. Hyde & M. C. Linn (Eds.), *The psychology of gender: Advances through meta-analysis.* Baltimore, MD: Johns Hopkins University Press.

Becker, H. S. (1963). *Outsiders: Studies in the sociology of deviance.* New York: Free Press.

Becker, H. S. (1964). What do they really learn at college? *Trans-Action, 1,* 14–17.

Beckett, K. (1994). Setting the public agenda: "Street crime" and drug use in American politics. *Social Problems, 41,* 425–447.

Beebe, L. M., & Giles, H. (1984). Speech accommodation theories: A discussion in terms of second language learning. *International Journal of the Sociology of Language, 46,* 5–32.

Begley, T., & Alker, H. (1982). Anti-busing protest: Attitudes and actions. *Social Psychology Quarterly, 45,* 187–197.

Bell, R. (1979). Parent, child and reciprocal influences. *American Psychologist, 34,* 821–826.

Bell, R. A., Zahn, C. J., & Hopper, R. (1984). Disclaiming: A test of two competing views. *Communication Quarterly, 32*(1), 28–36.

Belsky, J. (1990). Parental and nonparental child care and children's socioemotional development: A decade in review. *Journal of Marriage and the Family, 52,* 885–903.

Belsky, J., & Eggebeen, D. (1991). Early and extensive maternal employment and young children's socioemotional development: Children of the National Longitudinal Study of Youth. *Journal of Marriage and the Family, 53,* 1083–1110.

Belsky, J., Lang, M., & Rovine, M. (1985). Stability and change in marriage across the transition to parenthood: A second study. *Journal of Marriage and the Family, 47,* 855–865.

Bem, D. J. (1970). *Beliefs, attitudes and human affairs.* Belmont, CA: Brooks/Cole.

Benedict, R. (1938). Continuities and discontinuities in cultural conditioning. *Psychiatry, 1,* 161–167.

Bennett, M. (1990). Children's understanding of the mitigating function of disclaimers. *Journal of Social Psychology, 130,* 29–37.

Bennett, M., & Dewberry, C. (1989). Embarrassment at others' failures: A test of the Semin and Manstead model. *Journal of Social Psychology, 129,* 557–559.

Benson, M. L. (1989). The influence of class position on the formal and informal sanctioning of white-collar offenders. *Sociological Quarterly, 30,* 465–479.

Benson, M. L., & Walker, E. (1988). Sentencing the white-collar offender. *American Sociological Review, 53,* 294–302.

Benson, P. L., Karabenick, S. A., & Lerner, R. M. (1973). Pretty pleases: The effect of physical attraction, race, and sex on receiving help. *Journal of Experimental Social Psychology, 12,* 409–415.

Bergen, D., & Williams, J. (1991). Sex stereotypes in the United States revisited: 1972–1988. *Sex Roles, 24,* 413–423.

Berger, J., Cohen, B. P., & Zelditch, M., Jr. (1972). Status characteristics and social interaction. *American Sociological Review, 37,* 241–255.

Berger, J., Conner, T. L., & Fisek, M. H. (Eds.). (1974). *Expectation states theory.* Cambridge, MA: Winthrop.

Berger, J., Rosenholtz, S. J., & Zelditch, M., Jr. (1980). Status organizing processes. In A. Inkeles, N. J. Smelser, & R. H. Turner (Eds.), *Annual review of sociology* (Vol. 6, pp. 479–508). Palo Alto, CA: Annual Reviews.

Berger, J., Webster, M., Jr., Ridgeway, C., & Rosenholtz, S. J. (1986). Status cues, expectations, and behavior. In E. J. Lawler (Ed.), *Advances in group processes* (Vol. 3, pp. 1–22). Greenwich, CT: JAI Press.

Berk, R. A., & Aldrich, H. (1972). Patterns of vandalism during civil disorders as an indicator of selection of targets. *American Sociological Review, 37,* 533–547.

Berkowitz, L. (1972). Frustrations, comparisons, and other sources of emotion arousal as contributors to social unrest. *Journal of Social Issues, 28,* 77–91.

Berkowitz, L. (1978a). Decreased helpfulness with increased group size through lessening the effects of the needy individual's dependency. *Journal of Personality, 46,* 299–310.

Berkowitz, L. (1978b). Whatever happened to the frustration-aggression hypothesis? *American Behavioral Scientist, 21,* 691–708.

Berkowitz, L. (1989). Frustration-aggression hypothesis: Examination and reformulation. *Psychological Bulletin, 106,* 59–73.

Berkowitz, L., Klanderman, S. B., & Harris, R. (1964). Effects of experimenter awareness and sex of subject and experimenter on reactions to dependency relationships. *Sociometry, 27,* 327–337.

Berkowitz, L., & Troccoli, B. T. (1986). An examination of the assumptions in the demand characteristics thesis: With special reference to the Velten mood induction procedure. *Motivation and Emotion, 10,* 337–349.

Berkowitz, M. W., Mueller, C. W., Schnell, S. V., & Pudberg, M. T. (1986). Moral reasoning and judgments of aggression. *Journal of Personality and Social Psychology, 51,* 885–891.

Bernard, J. S. (1981). *The female world.* New York: Free Press.

Bernstein, I., Kelly, W., & Doyle, P. (1977). Societal reaction to deviants: The case of criminal defendants. *American Sociological Review, 42,* 743–755.

Bernstein, I., Kick, E., Leung, J., & Schulz, B. (1977). Charge reduction: An intermediary stage in the process of labelling criminal defendants. *Social Forces, 56,* 362–384.

Bernstein, W. M., Stephan, W. G., & Davis, M. H. (1979). Explaining attributions for achievement: A path analytic approach. *Journal of Personality and Social Psychology, 37,* 1810–1821.

Berry, D. S., & Landry, J. R. (1997). Facial maturity and daily social interaction. *Journal of Personality and Social Psychology, 72,* 570–580.

Berscheid, E. (1966). Opinion change and communicator-communicatee similarity and dissimilarity. *Journal of Personality and Social Psychology, 4,* 670–680.

Berscheid, E., Dion, K., Walster (Hatfield), E., & Walster, G. (1971). Physical attractiveness and dating choice: A test of the matching hypothesis. *Journal of Experimental Social Psychology, 7,* 173–189.

Berscheid, E., & Walster (Hatfield), E. (1978). *Interpersonal attraction* (2nd ed.). Reading, MA: Addison-Wesley.

Bertenthal, B. I., & Fischer, K. (1978). Development of self-recognition in the infant. *Developmental Psychology, 14,* 44–50.

Bettencourt, B. A., Brewer, M. B., Croak, M. R., & Miller, N. (1992). Cooperation and the reduction of intergroup bias: The role of reward structure and social orientation. *Journal of Experimental Social Psychology, 28*(4), 301–319.

Bianchi, S. M., & Robinson, J. (1997). What did you do today? Children's use of time, family composition, and the acquisition of social capital. *Journal of Marriage and the Family, 59,* 332–344.

Bickman, L. (1971). The effect of another bystander's ability to help on bystander intervention in an emergency. *Journal of Experimental Social Psychology, 7,* 367–379.

Biddle, B. J. (1979). *Role theory: Expectations, identities, and behaviors.* New York: Academic Press.

Biddle, B. J. (1986). Recent developments in role theory. In A. Inkeles, J. Coleman, & N. Smelser (Eds.), *Annual review of sociology* (Vol. 12, pp. 67–92). Palo Alto, CA: Annual Reviews.

Biddle, B. J., & Thomas, E. (Eds.). (1966). *Role theory: Concepts and research.* New York: Wiley.

Bielby, D., & Bielby, W. (1988). She works hard for the money: Household responsibilities and the allocation of work effort. *American Journal of Sociology, 93,* 1031–1059.

Bielby, W., & Baron, J. (1986). Men and women at work: Sex segregation and statistical discrimination. *American Journal of Sociology, 91,* 759–799.

Bielby, W., & Bielby, D. (1989). Balancing commitments to work and family in dual career households. *American Sociological Review, 54,* 776–789.

Biernat, M., & Wortmann, C. B. (1991). Sharing of home responsibilities between professionally employed women and their husbands. *Journal of Personality and Social Psychology, 60,* 844–860.

Birdwhistell, R. L. (1970). *Kinesics in context: Essays on body motion communications.* Philadelphia: University of Pennsylvania Press.

Birnbaum, J. A. (1975). Life patterns and self-esteem in gifted family-oriented and career committed women. In M. Mednick, S. Schwartz, & L. Hoffman (Eds.), *Women and achievement: Social and motivational analyses.* New York: Halsted.

Black, D. (1980). *The manners and customs of the police.* New York: Academic Press.

Blair, E., Sudman, S., Bradburn, N., & Stocking, C. (1977). How to ask questions about drinking and sex: Response effects in measuring consumer behavior. *Journal of Marketing Research, 14,* 316–321.

Blake, R. R., Shepard, H. A., & Mouton, J. S. (1964). *Managing intergroup conflict in industry.* Houston: Gulf.

Blanck, P. D., & Rosenthal, R. (1982). Developing strategies for decoding "leaky" messages. In R. S. Feldman (Ed.), *Development of nonverbal behavior in children* (pp. 203–229). New York: Springer-Verlag.

Blascovich, J., Ginsburg, G. P., & Veach, T. L. (1975). A pluralistic explanation of choice shifts on the risk dimension. *Journal of Personality and Social Psychology, 31,* 422–429.

Blass, T. (1991). Understanding behavior in the Milgram obedience experiment: The role of personality, situations, and their interactions. *Journal of Personality and Social Psychology, 60,* 398–413.

Blau, P. (1964). *Exchange and power in social life.* New York: Wiley.

Blauner, R. (1964). *Alienation and freedom.* Chicago: University of Chicago Press.

Blieszner, R., & Mancini, J. (1987). Enduring ties: Older adults' parental role and responsibilities. *Family Relations, 36,* 176–180.

Blom, J. P., & Gumperz, J. J. (1972). Social meaning and linguistic structure: Code-switching in Norway. In J. J. Gumperz & D. Hymes (Eds.), *Directions in sociolinguistics.* New York: Holt, Rinehart and Winston.

Blumenthal, M., Kahn, R. L., Andrews, F. M., & Head, K. B. (1972). *Justifying violence: Attitudes of American men.* Ann Arbor, MI: Institute for Social Research.

Blumer, H. (1962). Society and symbolic interactionism. In A. M. Rose (Ed.), *Human behavior and social processes.* Boston: Houghton Mifflin.

Blumer, H. (1969a). Elementary collective groupings. In A. M. Lee (Ed.), *Principles of sociology* (3rd ed.). New York: Barnes and Noble.

Blumer, H. (1969b). *Symbolic interactionism: Perspective and method.* Englewood Cliffs, NJ: Prentice-Hall.

Blumstein, P. W. (1974). The honoring of accounts. *American Sociological Review, 39,* 551–566.

Blumstein, P. W. (1975). Identity bargaining and self-conception. *Social Forces, 53,* 476–485.

Bobo, L. (1983). Whites' opposition to busing: Symbolic racism or realistic group conflict. *Journal of Personality and Social Psychology, 45,* 1196–1210.

Bochner, S., & Insko, C. A. (1966). Communicator discrepancy, source credibility, and opinion change. *Journal of Personality and Social Psychology, 4,* 614–621.

Bodenhausen, G., & Wyer, R., Jr. (1985). Effects of stereotypes on decision making and information-processing strategies. *Journal of Personality and Social Psychology, 48,* 267–282.

Bodenhausen, G. V., & Lichtenstein, M. (1987). Social stereotypes and information processing strategies: The impact of task complexity. *Journal of Personality and Social Psychology, 52,* 871–880.

Bohra, K. A, & Pandey, J. (1984). Ingratiation toward strangers, friends, and bosses. *Journal of Social Psychology, 122*(2), 217–222.

Bolger, N., DeLongis, A., Kessler, R. C., & Wethington, E. (1989). The contagion of stress across multiple roles. *Journal of Marriage and the Family, 51,* 175–183.

Bollen, K. A. (1989). *Structural equations with latent variables.* New York: Wiley Interscience.

Bollen, K. A., & Phillips, D. P. (1981). Suicidal motor vehicle fatalities in Detroit: A replication. *American Journal of Sociology, 87,* 404–412.

Booth, A., Johnson, D., White, L., & Edwards, J. (1984). Women, outside employment and marital instability. *American Journal of Sociology, 90,* 567–583.

Bord, R. J. (1976). The impact of imputed deviant identities in structuring evaluations and reactions. *Sociometry, 39,* 108–116.

Borkenau, P., & Ostendorf, F. (1987). Fact and fiction in implicit personality theory. *Journal of Personality, 55,* 415–443.

Bose, C. E., & Rossi, P. H. (1983). Gender and jobs: Prestige standings of jobs as affected by gender. *American Sociological Review, 48,* 316–330.

Bouchard, T. J., Barsaloux, J., & Drauden, G. (1974). Brainstorming procedure, group size, and sex as determinants of the problem-solving effectiveness of groups and individuals. *Journal of Applied Psychology, 59* (2), 135–138.

Bouchard, T. J., Jr., & Hare, M. (1970). Size, performance, and potential in brainstorming groups. *Journal of Applied Psychology, 54,* 51–55.

Boucher, J. D., & Ekman, P. (1975). Facial areas of emotional information. *Journal of Communication, 25,* 21–29.

Boulding, K. E. (1981). *Ecodynamics: A new theory of societal evolution.* Beverly Hills, CA: Sage.

Bourhis, R. Y., Giles, H., Leyens, J. P., & Tajfel, H. (1979). Psycholinguistic distinctiveness: Language diversity in Belgium. In H. Giles & R. N. St. Clair (Eds.), *Language and social psychology.* Oxford, England: Blackwell.

Bowlby, J. (1965). Maternal care and mental health (1953). In J. Bowlby (Ed.), *Child care and the growth of love.* London: Penguin.

Braaten, L. J. (1991). Group cohesion: A new multidimensional model. *Group, 15,* 39–55.

Bradley, G. W. (1978). Self-serving biases in the attribution process: A reexamination of the fact or fiction question. *Journal of Personality and Social Psychology, 36,* 56–71.

Bray, R. M., Kerr, N. L., & Atkin, R. S. (1978). Effects of group size, problem difficulty, and sex on group performance and member reactions. *Journal of Personality and Social Psychology, 36*(11), 1224–1240.

Breckler, S., & Greenwald, A. (1986). Motivational facets of the self. In R. Sorrentino & T. Higgins (Eds.), *Handbook of motivation and cognition* (pp. 145–164). New York: Guilford.

Brehm, J. W. (1956). Postdecision changes in the desirability of alternatives. *Journal of Abnormal and Social Psychology, 52,* 384–389.

Brehm, J. W., & Cohen, A. (1962). *Explorations in cognitive dissonance.* New York: Wiley.

Brewer, M. B. (1979). In-group bias in the minimal intergroup situation: A cognitive-motivational analysis. *Psychological Bulletin, 86,* 307–324.

Brewer, M. B. (1986). The role of ethnocentrism in intergroup conflict. In S. Worchel & W. G. Austin (Eds.), *Psychology of intergroup relations* (2nd ed., pp. 88–102). Chicago: Nelson-Hall.

Brewer, M. B., & Campbell, D. T. (1976). *Ethnocentrism and intergroup attitudes: East African evidence.* New York: Halsted.

Brewer, M. B., & Kramer, R. M. (1985). The psychology of intergroup attitudes and behavior. *Annual Review of Psychology, 36,* 219–243.

Brewer, M. B., & Lui, L. (1984). Categorization of the elderly by the elderly: Effects of perceiver's category membership. *Personality and Social Psychology Bulletin, 10,* 585–595.

Brewer, M. B., & Miller, N. (1984). Beyond the contact hypothesis: Theoretical perspectives on desegregation. In N. Miller & M. B. Brewer (Eds.), *Groups in contact: The psychology of desegregation* (pp. 281–302). Orlando, FL: Academic Press.

Brickman, P., Rabinowitz, V. C., Karuza, J., Coates, D., Cohn, E., & Kidder, L. (1982). Models of helping and coping. *American Psychologist, 37,* 368–384.

Brickner, M. A., Harkins, S. G., & Ostrom, T. M. (1986). Effects of personal involvement: Thought provoking implications for social loafing. *Journal of Personality and Social Psychology, 51,* 763–769.

Brim, O. G., Jr. (1966). Socialization through the life-cycle. In O. G. Brim, Jr., & S. Wheeler (Eds.), *Socialization after childhood.* New York: Wiley.

Brim, O. G., Jr., & Ryff, C. (1980). On the properties of life events. In P. Baltes & O. G. Brim, Jr. (Eds.), *Life-span development and behavior* (Vol. 3). New York: Academic Press.

Brody, E. M. (1990). *Women in the middle: Their parent-care years.* New York: Springer.

Broman, C. (1988a). Household work and family life satisfaction of blacks. *Journal of Marriage and the Family, 50,* 743–748.

Broman, C. (1988b). Satisfaction among blacks: The significance of marriage and parenthood. *Journal of Marriage and the Family, 50,* 45–51.

Broman, C. (1991). Gender, work-family roles, and psychological well-being of blacks. *Journal of Marriage and the Family, 53,* 509–520.

Bronfenbrenner, U. (1961). The mirror image in Soviet-American relations: A social psychologist's report. *Journal of Social Issues, 17*(3), 45–56.

Broverman, I., Vogel, S., Broverman, D., Clarkson, F., & Rosenkrantz, P. (1972). Sex-role stereotypes: A current appraisal. *Journal of Social Issues, 28*(2), 59–78.

Brown, J. D. (1991a). The professional ex-: An alternative for exiting the deviant career. *Sociological Quarterly, 32,* 219–230.

Brown, J. D. (1991b). Staying fit and staying well: Physical fitness as a moderator of life stress. *Journal of Personality and Social Psychology, 60,* 555–561.

Brown, J. D., Collins, R. L., & Schmidt, G. W. (1988). Self-esteem and direct vs. indirect forms of self-enhancement. *Journal of Personality and Social Psychology, 55,* 445–453.

Brown, P., & Elliott, R. (1965). Control of aggression in a nursery school class. *Journal of Experimental Child Psychology, 2,* 103–107.

Brown, R. (1964). The acquisition of language. In D. Rioch & E. Weinstein (Eds.), *Disorders of communication* (Vol. 42). Proceedings of the Association for Research in Nervous and Mental Disease. Baltimore, MD: Williams and Wilkins.

Brown, R. (1965). *Social psychology.* Glencoe, IL: Free Press.

Brown, R., & Fraser, C. (1963). The acquisition of syntax. In C. Cofer & B. Musgrave (Eds.), *Verbal behavior and learning.* New York: McGraw-Hill.

Brown, R. J., & Turner, J. C. (1981). Interpersonal and intergroup behavior. In J. Turner & H. Giles (Eds.), *Intergroup behavior* (pp. 33–65). Chicago: University of Chicago Press.

Brubaker, T. H. (1990). Families in later life: A burgeoning research area. *Journal of Marriage and the Family, 52,* 959–981.

Brunstein, J. C., Dangelmayer, G., & Schultheiss, O. C. (1996). Personal goals and social support in close relationships: Effects on relationship mood and marital satisfaction. *Journal of Personality and Social Psychology, 71,* 1006–1019.

Bryan, J. H., & Davenport, M. (1968). *Donations to the needy: Correlates of financial contributions to the destitute* (Research Bulletin No. 68-1). Princeton, NJ: Educational Testing Service.

Bryan, J. H., & Test, M. (1967). Models and helping: Naturalistic studies in aiding behavior. *Journal of Personality and Social Psychology, 6,* 400–407.

Buchanan, C. M., Maccoby, E. E., & Dornbusch, S. M. (1996). *Adolescents after divorce.* Cambridge, MA: Harvard University Press.

Bugenthal, D. E. (1974). Interpretations of naturally occurring discrepancies between words and intonation: Modes of inconsistency resolution. *Journal of Personality and Social Psychology, 30,* 125–133.

Bui, K-V. T., Peplau, L. A., & Hill, C. T. (1996). Testing the Rusbult model of relationship commitment and stability in a 15-year study of heterosexual couples. *Personality and Social Psychology Bulletin, 22,* 1244–1257.

Bumpass, L. L., Sweet, J. S., & Cherlin, A. (1991). The role of cohabitation in declining rates of marriage. *Journal of Marriage and the Family, 53,* 913–927.

Burgess, R. L., & Akers, K. L. (1966). A differential association-reinforcement theory of criminal behavior. *Social Problems, 14,* 128–147.

Burke, P. J. (1972). Leadership role differentiation. In C. G. McClintock (Ed.), *Experimental social psychology.* New York: Holt, Rinehart and Winston.

Burke, P. J., & Reitzes, D. (1981). The link between identity and role performance. *Social Psychology Quarterly, 44,* 83–92.

Burnstein, E. (1982). Persuasion as argument processing. In H. Brandstatter, J. H. Davis, & G. Stocher-Kreichgauer (Eds.), *Contemporary problems in group decision-making* (pp. 103–124). New York: Academic Press.

Burnstein, E., & Vinokur, A. (1973). Testing two classes of theories about group-induced shifts in individual choice. *Journal of Experimental Social Psychology, 9,* 123–137.

Burt, M. R. (1980). Cultural myths and supports for rape. *Journal of Personality and Social Psychology, 38,* 217–230.

Bush, D., & Simmons, R. (1990). Socialization processes over the life course. In M. Rosenberg & R. Turner (Eds.), *Social psychology: Sociological perspectives.* New Brunswick, NJ: Transaction.

Bushman, B. J. (1988). The effects of apparel on compliance: A field experiment with a female authority figure. *Personality and Social Psychology Bulletin, 14,* 459–467.

Bushman, B. J. (1995). Moderating role of trait aggressiveness in the effect of violent media on aggression. *Journal of Personality and Social Psychology, 69,* 950–960.

Buss, D. M. (1988). The evolution of human intrasexual competition: Tactics of mate attraction. *Journal of Personality and Social Psychology, 54,* 616–628.

Buunk, B. P., & van der Eijnden, R. (1997). Perceived prevalence, perceived superiority, and relationship satisfaction: Most relationships are good but ours is the best. *Personality and Social Psychology Bulletin, 23,* 219–228.

Byrne, D. (1961a). The influence of propinquity and opportunities for interaction on classroom relationships. *Human Relations, 14,* 63–69.

Byrne, D. (1961b). Interpersonal attraction and attitude similarity. *Journal of Abnormal and Social Psychology, 62,* 713–715.

Byrne, D. (1971). *The attraction paradigm.* New York: Academic Press.

Byrne, D., & Clore, G. L. (1970). A reinforcement model of evaluative responses. *Personality: An International Journal, 1,* 103–128.

Byrne, D., Ervin, C., & Lamberth, J. (1970). Continuity between the experimental study of attraction and real-life computer dating. *Journal of Personality and Social Psychology, 16,* 157–165.

Byrne, D., & Kelley, K. (1984). Introduction: Pornography and sex research. In N. M. Malamuth & E. Donnerstein (Eds.),

Pornography and sexual aggression. Orlando, FL: Academic Press.

Byrne, D., & Nelson, D. (1965). Attraction as a linear function of proportion of positive reinforcements. *Journal of Personality and Social Psychology, 1,* 659–663.

Cacioppo, J. T., Petty, R. E., Feinstein, J. A., & Jarvis, W. B. G. (1996). Dispositional differences in cognitive motivation: The life and times of individuals varying in need for cognition. *Psychological Bulletin, 119*(2), 197–253.

Cahill, S. (1987). Children and civility: Ceremonial deviance and the acquisition of ritual competence. *Social Psychology Quarterly, 50,* 312–321.

Calahan, D. (1970). *Problem drinkers.* San Francisco: Jossey-Bass.

Caldwell, M. A., & Peplau, L. (1982). Sex differences in same-sex relationships. *Sex Roles, 8,* 721–732.

Callaway, M. R., & Esser, J. K. (1984). Groupthink: Effects of cohesiveness and problem-solving procedures on group decision making. *Social Behavior and Personality, 12,* 157–164.

Callero, P. L. (1985). Role identity salience. *Social Psychology Quarterly, 48,* 203–214.

Callero, P. L., Howard, J. A., & Piliavin, J. A. (1987). Helping behavior as role behavior: Disclosing social structure and history in the analysis of prosocial action. *Social Psychology Quarterly, 50,* 247–256.

Campbell, A., Converse, P., & Rodgers, W. (1976). *The quality of American life.* New York: Russell Sage Foundation.

Campbell, D. T. (1967). Stereotypes in the perception of group differences. *American Psychologist, 22,* 817–829.

Campbell, D. T., & Stanley, J. C. (1963). *Experimental and quasi-experimental designs for research.* Chicago: Rand McNally.

Campbell, J. D. (1990). Self-esteem and clarity of self-concept. *Journal of Personality and Social Psychology, 59,* 538–549.

Cancio, A. S., Evans, T. D., & Maume, D. J., Jr. (1996). Reconsidering the declining significance of race: Racial differences in early career wages. *American Sociological Review, 61,* 541–556.

Cano, I., Hopkins, N., & Islam, M. R. (1991). Memory for stereotype-related material: A replication study with real-life social groups. *European Journal of Social Psychology, 21,* 349–357.

Cantor, J., Alfonso, H., & Zillmann, D. (1976). The persuasive effectiveness of the peer appeal and a communicator's first hand experience. *Communication Research, 3,* 293–310.

Caplan, F. (1973). *The first twelve months of life.* New York: Grosset & Dunlap.

Carey, M. (1978). The role of gaze in the initiation of conversation. *Social Psychology, 41,* 269–271.

Carli, L. L. (1989). Gender differences in interactional style and influence. *Journal of Personality and Social Psychology, 56,* 565–576.

Carli, L. L. (1990). Gender, language and influence. *Journal of Personality and Social Psychology, 59,* 941–951.

Carli, L. L., LaFleur, S. J., & Loeber, C. C. (1995). Nonverbal behavior, gender and influence. *Journal of Personality and Social Psychology, 68,* 1030–1041.

Carlsmith, J. M., & Anderson, C. (1979). Ambient temperature and the occurrence of collective violence. *Journal of Personality and Social Psychology, 37,* 337–344.

Carlson, M., Charlin, V., & Miller, N. (1988). Positive mood and helping behavior: A test of six hypotheses. *Journal of Personality and Social Psychology, 55*(2), 211–229.

Carlson, M., Marcus-Newhall, A., & Miller, N. (1990). Effects of situational aggression cues: A quantitative review. *Journal of Personality and Social Psychology, 58,* 622–633.

Carlson, M., & Miller, N. (1987). Explanation of the relation between negative mood and helping. *Psychological Bulletin, 102,* 91–108.

Carnevale, P. J., & Henry, R. A. (1989). Determinants of mediator behavior: A test of the strategic choice model. *Journal of Applied Social Psychology, 19,* 481–498.

Carnevale, P. J., & Pegnetter, R. (1985). The selection of mediation tactics in public sector disputes: A contingency analysis. *Journal of Social Issues, 41,* 65–81.

Carnevale, P. J., Pruitt, D. G., & Carrington, P. I. (1982). Effects of future dependence, liking, and repeated requests for help on helping behavior. *Social Psychology Quarterly, 45*(1), 9–14.

Carpenter, W. A., & Hollander, E. P. (1982). Overcoming hurdles to independence in groups. *Journal of Social Psychology, 117,* 237–241.

Carroll, J. M., & Russell, J. A. (1996). Do facial expressions signal specific emotions? Judging emotion from face in context. *Journal of Personality and Social Psychology, 70,* 205–218.

Carter, G. L. (1990). Black attitudes and the 1960s black riots: An aggregate-level analysis of the Kerner Commission's "15 cities" data. *Sociological Quarterly, 31,* 269–286.

Cartwright, D. (1968). The nature of group cohesiveness. In D. Cartwright & A. Zander (Eds.), *Group dynamics* (3rd ed., pp. 91–109). New York: Harper & Row.

Cartwright, D. (1971). Risk taking by individuals and groups: An assessment of research employing choice dilemmas. *Journal of Personality and Social Psychology, 20,* 361–378.

Cartwright, D., & Zander, A. (1968). Motivational processes in groups: Introduction. In D. Cartwright & A. Zander (Eds.), *Group dynamics* (3rd ed.). New York: Harper & Row.

Caspi, A. (1992). Surmounting childhood disadvantage: Turning points and pathways to change in the life course.

Caspi, A., & Herbener, E. S. (1990). Continuity and change: Assortative marriage and the consistency of personality in adulthood. *Journal of Personality and Social Psychology, 58,* 250–258.

Catalano, R., Novaco, R., & McConnell, W. (1997). A model of the net effect of job loss on violence. *Journal of Personality and Social Psychology, 72,* 1440–1447.

Catrambone, R., & Markus, H. (1987). The role of self-schemas in going beyond the information given. *Social Cognition, 5,* 349–368.

Centers, R. (1975). Attitude similarity-dissimilarity as a correlate of heterosexual attraction and love. *Journal of Marriage and the Family, 37,* 305–312.

Centers for Disease Control. (1990, May 11). Mass sociogenic illness in a day-care center—Florida. *Morbidity and Mortality Weekly Report, 39*(18), 301–304.

Chaffee, S. (1981). Mass media in political campaigns: An expanding role. In R. E. Rice & W. J. Paisley (Eds.), *Public communication campaigns.* Beverly Hills, CA: Sage.

Chaiken, S. (1980). Heuristic versus systematic information processing and the use of source versus message cues in persuasion. *Journal of Personality and Social Psychology, 39,* 752–766.

Chaiken, S. (1986). Physical appearance and social influence. In C. P. Herman, M. P. Zanna, & E. T. Higgins (Eds.), *Physical appearance, stigma, and social behavior: The Ontario Symposium* (Vol. 3, pp. 143–177). Hillsdale, NJ: Erlbaum

Chaiken, S., & Maheswaran, D. (1994). Heuristic processing can bias systematic processing: Effects of source credibility, argument ambiguity, and task importance on attitude judgment. *Journal of Personality and Social Psychology, 66,* 460–473.

Chaiken, S., & Yates, S. (1985). Affective-cognitive consistency and thought-induced polarization. *Journal of Personality and Social Psychology, 49,* 1470–1481.

Chambliss, W. J. (1994). Policing the ghetto underclass: The politics of law and law enforcement. *Social Problems, 41,* 177–193.

Chapman, R. S., Streim, N. W., Crais, E. R., Salmon, D., Strand, E. A., & Negri, N. A. (1992). Child talk: Assumptions of a developmental process model for early language learning. In R. S. Chapman (Ed.), *Processes in language acquisition and disorder.* Chicago: Mosby-Year Book.

Charon, J. M. (1995). *Symbolic Interactionism: An introduction, an interpretation, an integration* (5th ed.). Englewood Cliffs, NJ: Prentice-Hall.

Chebat, J-C., Filiatrault, P., & Perrien, J. (1990). Limits of credibility: The case of political persuasion. *Journal of Social Psychology, 130,* 157–167.

Chemers, M. M. (1983). Leadership theory and research: A systems-process integration. In P. B. Paulus (Ed.), *Basic group processes* (pp. 9–39). New York: Springer-Verlag.

Chemers, M. M., & Skrzypek, G. J. (1972). An experimental test of the contingency model of leadership effectiveness. *Journal of Personality and Social Psychology, 24,* 172–177.

Chen, Y. R., & Church, A. H. (1993). Reward allocation preferences in groups and organizations. *International Journal of Conflict Management, 4*(1), 25–59.

Cherlin, A. J. (1981). *Marriage, divorce, remarriage.* Cambridge, MA: Harvard University Press.

Cherulnik, P. D. (1983). *Behavioral research: Assessing the validity of research findings in psychology.* New York: Harper & Row.

Chicago Tribune. (1992, March 11). Wisconsin town thanks hero who saved boy, p. 9.

Chiricos, T., & Waldo, G. (1975). Socioeconomic status and criminal sentencing: An empirical assessment of a conflict proposition. *American Sociological Review, 40,* 753–772.

Cialdini, R., & Baumann, D. (1981). Littering: A new unobtrusive measure of attitudes. *Social Psychology Quarterly, 44,* 254–259.

Cialdini, R. B., Borden, R., Thorne, A., Walker, M., & Freeman, S. (1976). Basking in reflected glory: Three (football) field studies. *Journal of Personality and Social Psychology, 34,* 366–375.

Cialdini, R. B., Darby, B. L., & Vincent, J. E. (1973). Transgression and altruism: A case for hedonism. *Journal of Experimental Social Psychology, 9,* 502–516.

Cialdini, R. B., & Fultz, J. (1990). Interpreting the negative mood-helping literature via "mega"-analysis: A contrary view. *Psychological Bulletin, 107*(2), 210–214.

Cialdini, R. B., Kendrick, D. T., & Baumann, D. J. (1982). Effects of mood on prosocial behavior in children and adults. In N. Eisenberg (Ed.), *The development of prosocial behavior* (pp. 339–359). New York: Academic Press.

Cialdini, R. B., Schaller, M., Houlihan, D., Arps, K., Fultz, J., & Beaman, A. L. (1987). Empathy-based helping: Is it selflessly or selfishly motivated? *Journal of Personality and Social Psychology, 52,* 749–758.

Cialdini, R. B., Trost, M. R., & Newsom, J. T. (1995). Preference for consistency: The development of a valid measure and the discovery of surprising behavioral implications. *Journal of Personality and Social Psychology, 69,* 318–328.

Clark, E. V. (1976). From gesture to word: On the natural history of deixis in language acquisition. In J. S. Bruner & A. Gartner (Eds.), *Human growth and development.* Oxford, England: Clarendon.

Clark, M. S., Gotay, C. C., & Mills, J. (1974). Acceptance of help as a function of the potential helper and opportunity to repay. *Journal of Applied Social Psychology, 4,* 224–229.

Clark, R. D., III, & Maass, A. (1988). The role of social categorization and perceived source credibility in minority influence. *European Journal of Social Psychology, 18,* 381–394.

Clark, R. D., III, & Word, L. E. (1972). Why don't bystanders help? Because of ambiguity? *Journal of Personality and Social Psychology, 24,* 392–400.

Clausen, J. A. (1968). *Socialization and society.* Boston: Little, Brown.

Clausen, J. S. (1991). Adolescent competence and the shaping of the life course. *American Journal of Sociology, 96,* 805–842.

Clore, G. L., Bray, R. M., Itkin, S. M., & Murphy, P. (1978). Interracial attitudes and behavior at a summer camp. *Journal of Personality and Social Psychology, 36,* 107–116.

Cloward, R. (1959). Illegitimate means, anomie and deviant behavior. *American Sociological Review, 24,* 164–176.

Code of Federal Regulations. (1992). *Title 45, Part 46.101–46.124: Basic HHS policy for protection of human research subjects.* Washington, DC: U.S. Government Printing Office.

Cohen, A. (1966). *Deviance and control.* Englewood Cliffs, NJ: Prentice-Hall.

Cohen, B. P., & Zhou, X. (1991). Status processes in enduring work groups. *American Sociological Review, 56,* 179–188.

Cohen, C. E. (1981). Person categories and social perception: Testing some boundaries of the processing effects of prior knowledge. *Journal of Personality and Social Psychology, 40,* 441–452.

Cohen, D. (1996). Law, social policy and violence: The impact of regional cultures. *Journal of Personality and Social Psychology, 70,* 961–978.

Cohen, D., Nisbett, R. E., Bowdle, B. F., & Schwarz, N. (1996). Insult, aggression and the southern culture of honor: An experimental ethnography. *Journal of Personality and Social Psychology, 70,* 945–960.

Cohen, E. G. (1980). Design and redesign of the desegregated school: Problems of status, power, and conflict. In W. G. Stephan & J. Feagin (Eds.), *School desegregation.* New York: Academic Press.

Cohen, E. G. (1982). Expectation states and interracial interaction in school settings. In R. H. Turner (Ed.), *Annual review of sociology* (Vol. 8, pp. 209–235). Palo Alto, CA: Annual Reviews.

Cohen, E. G. (1984). The desegregated school: Problems in status, power and interethnic climate. In N. Miller & M. Brewer (Eds.), *Groups in contact: The psychology of desegregation* (pp. 77–96). Orlando, FL: Academic Press.

Cohen, E. G., & Roper, S. (1972). Modification of interracial interaction disability: An application of status characteristic theory. *American Sociological Review, 37,* 643–657.

Cohen, L., & Kluegel, J. (1978). Determinants of juvenile court dispositions: Ascriptive and achieved factors in two metropolitan courts. *American Sociological Review, 43,* 162–176.

Cohen, R. (1982). In the end, this son is the father. *Washington Post Syndicate.*

Cohn, E. G., & Rotton, J. (1997). Assault as a function of time and temperature: A moderator-variable time-series analysis. *Journal of Personality and Social Psychology, 72,* 1322–1334.

Cohn, R. M. (1978). The effect of employment status change on self-attitudes. *Social Psychology, 41,* 81–93.

Cole, E. R., & Stewart, A. J. (1996). Measures of political participation among black and white women: Political identity and social responsibility. *Journal of Personality and Social Psychology, 71,* 130–140.

Coleman, J. W. (1987). Toward an integrated theory of white-collar crime. *American Journal of Sociology, 93,* 406–439.

Coleman, R. P., & Neugarten, B. (1971). *Social status in the city.* San Francisco: Jossey-Bass.

Collett, P. (1971). On training Englishmen in the nonverbal behavior of Arabs: An experiment in inter-cultural communication. *International Journal of Psychology, 6,* 209–215.

Collins, N. L., & Miller, L. C. (1994). Self-disclosure and liking: A meta-analytic review. *Psychological Bulletin, 116*(3), 457–475.

Comstock, G. (1984). Media influences on aggression. In A. Goldstein (Ed.), *Prevention and control of aggression: Principles, practices and research.* New York: Pergamon.

Comstock, G. S., Chaffee, S., Katzman, N., McCombs, M., & Roberts, D. (1978). *Television and human behavior.* New York: Columbia University Press.

Condon, J. W., & Crano, W. D. (1988). Inferred evaluation and the relation between attitude similarity and interpersonal attraction. *Journal of Personality and Social Psychology, 54,* 789–797.

Condon, W. S., & Ogston, W. D. (1967). A segmentation of behavior. *Journal of Psychiatric Research, 5,* 221–235.

Conley, J. J. (1985). Longitudinal stability of personality traits: A multitrait-multimethod-multioccasion analysis. *Journal of Personality and Social Psychology, 49,* 1266–1282.

Cook, K. (Ed.). (1987). *Social exchange theory.* Newbury Park, CA: Sage.

Cook, S. W. (1984). The 1954 social science statement and school desegregation: A reply to Gerard. *American Psychologist, 39,* 819–832.

Cook, S. W. (1985). Experimenting on social issues: The case of school desegregation. *American Psychologist, 40*(4), 452–460.

Cooke, B., Rossmann, M., McCubbin, H., & Patterson, J. (1988). Examining the definition and measurement of social support: A resource for individuals and families. *Family Relations, 37,* 211–216.

Cooley, C. H. (1902). *Human nature and the social order.* New York: Scribner.

Cooley, C. H. (1908). A study of the early use of self-words by a child. *Psychological Review, 15,* 339–357.

Cooper, H. M. (1979). Statistically combining independent studies: A meta-analysis of sex differences in conformity research. *Journal of Personality and Social Psychology, 37,* 131–146.

Cooper, J., & Fazio, R. H. (1986). The formation and persistence of attitudes that support intergroup conflict. In S. Worchel & W. G. Austin (Eds.), *Psychology of intergroup relations* (2nd ed., pp. 183–195). Chicago: Nelson-Hall.

Cooper, W. H. (1981). Ubiquitous halo. *Psychological Bulletin, 90,* 218–244.

Coopersmith, S. (1967). *The antecedents of self-esteem.* San Francisco: Freeman.

Corcoran, M. (1995). From rags to riches: Poverty and mobility in the United States. *Annual Review of Sociology, 21,* 237–267.

Corsaro, W. A. (1992). Interpretive reproduction in children's peer cultures. *Social Psychology Quarterly, 55,* 160–177.

Corsaro, W. A., & Eder, D. (1995). Development and socialization of children and adolescents. In K. S. Cook, G. A. Fine, & J. S. House (Eds.), *Sociological perspectives on social psychology* (pp. 421–451). Boston: Allyn & Bacon.

Corsaro, W. A., & Rizzo, T. A. (1988). *Discussione* and friendship: Socialization processes in the peer culture of Italian nursery school children. *American Sociological Review, 53,* 879–894.

Coser, L. A. (1967). *Continuities in the study of social conflict.* New York: Free Press.

Cota, A. A., & Dion, K. L. (1986). Salience of gender and sex composition of ad hoc groups: An experimental test of distinctiveness theory. *Journal of Personality and Social Psychology, 50,* 770–776.

Cota, A. A., Evans, C. R., Dion, K. L., Kilik, L., & Longman, R. S. (1995). The structure of group cohesion. *Personality and Social Psychology Bulletin, 21*(6), 572–580.

Court, J. H. (1984). Sex and violence: A ripple effect. In N. M. Malamuth & E. Donnerstein (Eds.), *Pornography and sexual aggression.* Orlando, FL: Academic Press.

Courtright, J. A. (1978). A laboratory investigation of groupthink. *Communications Monographs, 45,* 229–246.

Cowan, P. A. (1991). Individual and family life transitions: A proposal for a new definition. In P. A. Cowan & M. Hetherington (Eds.), *Family transitions.* Hillsdale, NJ: Erlbaum.

Cox, C. L., Wexler, M. O., Rusbult, C. E., & Gaines, S. O., Jr. (1997). Prescriptive support and commitment processes in close relationships. *Social Psychology Quarterly, 60,* 79–90.

Cozby, P. C. (1972). Self-disclosure, reciprocity and liking. *Sociometry, 35,* 151–160.

Craddock, A. E. (1985). Centralized authority as a factor in small group and family problem solving: A reassessment of Tallman's propositions. *Small Group Behavior, 16*(1), 59–73.

Crano, W. (1997). Vested interests, symbolic politics, and attitude-behavior consistency. *Journal of Personality and Social Psychology, 72,* 485–491.

Crocker, J., Thompson, L. L., McGraw, K. M., & Ingerman, C. (1987). Downward comparison, prejudice, and evaluations of others: Effects of self-esteem and threat. *Journal of Personality and Social Psychology, 52,* 907–916.

Crocker, J., Voelkl, K., Test, M., & Major, B. (1991). Social stigma: The affective consequences of attributional ambiguity. *Journal of Personality and Social Psychology, 60,* 218–228.

Crohan, S. E. (1996). Marital quality and conflict across the transition to parenthood in African-American and white couples. *Journal of Marriage and the Family, 58,* 933–944.

Crosbie, P. V. (Ed.). (1975). *Interaction in small groups.* New York: Macmillan.

Cummings, K. M., Sciandra, R., Davis, S., & Rimer, B. (1989). Response to anti-smoking campaign aimed at mothers with young children. *Health Education Research, 4,* 429–437.

Cunningham, J. D. (1981). Self-disclosure intimacy: Sex, sex-of-target, cross-national and "generational" differences. *Personality and Social Psychology Bulletin, 7,* 314–319.

Cunningham, M. R. (1981). Sociobiology as a supplementary paradigm for social psychological research. In L. Wheeler (Ed.), *Review of personality and social psychology* (Vol. 2, pp. 69–101). Beverly Hills, CA: Sage.

Cunningham, M. R. (1986). Measuring the physical in physical attractiveness: Quasi-experiments on the sociobiology of female facial beauty. *Journal of Personality and Social Psychology, 50,* 925–935.

Cunningham, M. R., Barbee, A. P., & Pike, C. L. (1990). What do women want? Facialmetric assessment of multiple motives in the perception of male facial physical attractiveness. *Journal of Personality and Social Psychology, 59,* 61–72.

Cunningham, M. R., Steinberg, J., & Grev, R. (1980). Wanting to and having to help: Separate motivations for positive mood and guilt-induced helping. *Journal of Personality and Social Psychology, 38,* 181–192.

Cutrona, C. E. (1982). Transition to college: Loneliness and the process of social adjustment. In L. A. Peplau & D. Perlman (Eds.), *Loneliness: A resource book of current theory, research and therapy.* New York: Wiley.

Dabbs, J. M., Jr., & Leventhal, H. (1966). Effects of varying the recommendations in a fear-arousing communication. *Journal of Personality and Social Psychology, 4,* 525–531.

Daher, D. M., & Banikiotes, P. G. (1976). Interpersonal attraction and rewarding aspects of disclosure content and level. *Journal of Personality and Social Psychology, 33,* 492–496.

Daly, K. (1987). Discrimination in the criminal courts: Family, gender and the problem of equal treatment. *Social Forces, 66,* 152–175.

Daniels, A. (1987). Invisible work. *Social Problems, 34,* 403–415.

Dank, B. (1971). Coming out in the gay world. *Psychiatry, 34,* 180–197.

Darley, J. M., & Batson, C. D. (1973). From Jerusalem to Jericho: A study of situational and dispositional variables in helping behavior. *Journal of Personality and Social Psychology, 27,* 100–108.

Darley, J. M., & Fazio, R. H. (1980). Expectancy confirmation processes arising in the social interaction sequence. *American Psychologist, 35,* 867–881.

Darley, J. M., & Latané, B. (1968). Bystander intervention in emergencies: Diffusion of responsibility. *Journal of Personality and Social Psychology, 8,* 377–383.

Darley, J. M., Teger, A. I., & Lewis, L. D. (1973). Do groups always inhibit individuals' response to potential emergencies? *Journal of Personality and Social Psychology, 26,* 395–399.

Davidson, A. R., Yantis, S., Norwood, M., & Montano, D. (1985). Amount of information about the attitude object and attitude-behavior consistency. *Journal of Personality and Social Psychology, 49,* 1184–1198.

Davidson, L. R., & Duberman, L. (1982). Friendship: Communication and interaction patterns in same sex dyads. *Sex Roles, 8,* 809–822.

Davies, J. C. (1962). Toward a theory of revolution. *American Sociological Review, 27,* 5–19.

Davies, J. C. (Ed.). (1971). *When men revolt—and why.* New York: Free Press.

Davis, D., & Perkowitz, W. T. (1979). Consequences of responsiveness in dyadic interaction: Effects of probability of response and proportion of content-related responses on interpersonal attraction. *Journal of Personality and Social Psychology, 37,* 534–550.

Davis, F. (1961). Deviance disavowal: The management of strained interaction by the visibly handicapped. *Social Problems, 9,* 120–132.

Davis, J. D. (1976). Self-disclosure in an acquaintance exercise: Responsibility for level of intimacy. *Journal of Personality and Social Psychology, 33,* 787–792.

Davis, K. (1947). Final note on a case of extreme isolation. *American Journal of Sociology, 52,* 432–437.

Davis, K. E. (1985, February). Near and dear: Friendship and love. *Psychology Today,* pp. 22–30.

Davis, M. H., & Franzoi, S. L. (1986). Adolescent loneliness, self-disclosure, and private self-consciousness: A longitudinal investigation. *Journal of Personality and Social Psychology, 51,* 595–608.

Dawkins, R. (1976). *The selfish gene.* Oxford, England: Oxford University Press.

Deaux, K., & Lewis, L. (1983). Components of gender stereotypes. *Psychological Documents, 13*(2583), 25.

DeBono, K. G., & Telesca, C. (1990). The influence of source physical attractiveness on advertising effectiveness: A functional perspective. *Journal of Applied Social Psychology, 20,* 1383–1395.

Deci, E. (1975). *Intrinsic motivation.* New York: Plenum.

de Jong-Gierveld, J. (1987). Developing and testing a model of loneliness. *Journal of Personality and Social Psychology, 53,* 119–128.

DeLamater, J., & MacCorquodale, P. (1979). *Premarital sexuality: Attitudes, relationships, behavior.* Madison: University of Wisconsin Press.

DeLamater, J., & McKinney, K. (1982). Response-effects of question content. In W. Dijkstra & J. Van der Zouwen (Eds.), *Response behavior in the survey-interview.* London: Academic Press.

DeLongis, A., Folkman, J., & Lazarus, R. S. (1988). The impact of daily stress on health and mood: Psychological and social resources as mediators. *Journal of Personality and Social Psychology, 54,* 486–495.

Deluga, R. J. (1987). Corporate coups d'état: What happens and why. *Leadership and Organizational Development Journal, 8,* 9–15.

DeMartini, J. R. (1983). Social movement participation: Political socialization, generational consciousness, and lasting effects. *Youth and Society, 15,* 195–223.

Dembroski, T. M., Lasater, T. M., & Ramires, A. (1978). Communicator similarity, fear-arousing communications, and compliance with health care recommendations. *Journal of Applied Social Psychology, 8,* 254–269.

Demerath, N. J., III, Marwell, G., & Aiken, M. T. (1971). *Dynamics of idealism: White activists in a black movement.* San Francisco: Jossey-Bass.

Demo, D. H. (1992). The self-concept over time: Research issues and directions. *Annual Review of Sociology, 18,* 303–326.

Demo, D. H., & Acock, A. C. (1988). The impact of divorce on children. *Journal of Marriage and the Family, 50,* 619–648.

Demo, D. H., & Hughes, M. (1990). Socialization and racial identity among black Americans. *Social Psychology Quarterly, 53,* 364–374.

Dentler, R., & Erikson, K. (1959). The functions of deviance in groups. *Social Problems, 7,* 98–107.

Denton, K., & Krebs, D. (1990). From the scene to the crime: The effect of alcohol and social context on moral judgment. *Journal of Personality and Social Psychology, 59,* 242–248.

Denzin, N. (1977). *Childhood socialization: Studies in the development of language, social behavior, and identity.* San Francisco: Jossey-Bass.

Denzin, N. (1983). *On understanding emotion.* San Francisco: Jossey-Bass.

DePaulo, B. M. (1992). Nonverbal behavior and self-presentation. *Psychological Bulletin, 111,* 203–243.

DePaulo, B. M., & Fisher, J. D. (1980). The costs of asking for help. *Basic and Applied Social Psychology, 1,* 23–35.

DePaulo, B. M., Lassiter, G. D., & Stone, J. T. (1982). Attentional determinants of success at determining deception and truth. *Personality and Social Psychology Bulletin, 8,* 273–279.

DePaulo, B. M., Nadler, A., & Fisher, J. D. (1983). *New directions in helping: Help seeking* (Vol. 2). New York: Academic Press.

DePaulo, B. M., & Rosenthal, R. (1979). Ambivalence, discrepancy, and deception in nonverbal communication. In R. Rosenthal (Ed.), *Skill in nonverbal communication.* Cambridge, MA: Oelgeschlager, Gunn and Hain.

DePaulo, B. M., Rosenthal, R., Eisenstat, R. A., Rogers, P. L., &

Finkelstein, S. (1978). Decoding discrepant nonverbal cues. *Journal of Personality and Social Psychology, 36,* 313–323.

DePaulo, B. M., Stone, J. I., & Lassiter, G. D. (1985). Deceiving and detecting deceit. In B. R. Schlenker (Ed.), *The self and social life* (pp. 323–370). New York: McGraw-Hill.

DePaulo, B. M., Zuckerman, M., & Rosenthal, R. (1980). Detecting deception: Modality effects. In L. Wheeler (Ed.), *Review of personality and social psychology* (Vol. 1, pp. 125–162). Beverly Hills, CA: Sage.

Der-Karabetian, A., & Smith, A. (1977). Sex-role stereotyping in the United States: Is it changing? *Sex Roles, 3,* 193–198.

Derlega, V. J., Durham, B., Gockel, B., & Sholis, D. (1981). Sex differences in self disclosure: Effects of topic content, friendship, and partner's sex. *Sex Roles, 7,* 433–447.

Derlega, V. J., & Grzelak, J. (1979). Appropriate self-disclosure. In G. J. Chelune et al. (Eds.), *Self-disclosure: Origins, patterns, and implications of openness in interpersonal relationships* (pp. 151–176). San Francisco: Jossey-Bass.

Derlega, V. J., Metts, S., Petronio, S., & Margulis, S. T. (1993). *Self-disclosure.* Newbury Park, CA: Sage.

Dervin, B. (1981). Mass communicating: Changing conceptions of the audience. In R. E. Rice & W. J. Paisley (Eds.), *Public communication campaigns.* Beverly Hills, CA: Sage.

de Tocqueville, A. (1955). *The old regime and the French Revolution* (Stuart Gilbert, Trans.). Garden City, NY: Doubleday. (Original work published 1856)

Deutsch, M. (1985). *Distributive justice: A social psychological perspective.* New Haven, CT: Yale University Press.

Deutsch, M., & Gerard, H. B. (1955). A study of normative and informational social influences upon individual judgment. *Journal of Abnormal and Social Psychology, 51,* 629–636.

Deutsch, M., & Krauss, R. M. (1962). Studies of interpersonal bargaining. *Journal of Abnormal and Social Psychology, 61,* 181–189.

Devine, P. G. (1989). Stereotypes and prejudice: Their automatic and controlled components. *Journal of Personality and Social Psychology, 56,* 5–18.

Devine, P. G., & Eliot, A. J. (1995). Are racial stereotypes really fading? The Princeton trilogy revisited. *Personality and Social Psychology Bulletin, 21*(11), 1139–1150.

Devine, J. A., Sheley, J. F., & Smith, M. D. (1988). Macroeconomic and social-control policy influences on crime-rate changes, 1948–1985. *American Sociological Review, 53,* 407–420.

Dickoff, H. (1961). *Reactions to evaluations by others as a function of self-evaluation and the interaction context.* Unpublished doctoral dissertation, Duke University, Raleigh, NC.

Diehl, M., & Stroebe, W. (1987). Productivity loss in brainstorming groups: Toward the solution of a riddle. *Journal of Personality and Social Psychology, 53*(3), 497–509.

Diehl, M., & Stroebe, W. (1991). Productivity loss in idea-generating groups: Tracking down the blocking effect. *Journal of Personality and Social Psychology, 61,* 392–403.

DiMaggio, P., & Mohr, J. (1985). Cultural capital, educational attainment, and marital selection. *American Journal of Sociology, 90,* 1231–1257.

Dindia, K., & Allen, M. (1992). Sex differences in self-disclosure: A meta-analysis. *Psychological Bulletin, 112*(1), 106–124.

Dion, K. L. (1979). Intergroup conflict and intragroup cohesiveness. In W. G. Austin & S. Worchel (Eds.), *The social psychology of intergroup relations* (pp. 211–224). Monterey, CA: Brooks/Cole.

Dion, K. L., Baron, R., & Miller, N. (1970). Why do groups make riskier decisions than individuals? In L. Berkowitz (Ed.), *Advances in experimental social psychology* (Vol. 5, pp. 306–377). New York: Academic Press.

Dion, K. L., Berscheid, E., & Walster (Hatfield), E. (1972). What is beautiful is good. *Journal of Personality and Social Psychology, 24,* 285–290.

Dion, K. L., & Schuller, R. A. (1991). The Ms. stereotype: Its generality and its relation to managerial and marital status stereotypes. *Canadian Journal of Behavioural Science, 23,* 25–40.

Dixon, J., Gordon, C., & Khomusi, T. (1995). Sexual symmetry in psychiatric diagnosis. *Social Problems, 42,* 429–448.

Doering, C. H. (1980). The endocrine system. In O. Brim, Jr., & J. Kagan (Eds.), *Constancy and change in human development.* Cambridge, MA: Harvard University Press.

Dohrenwend, B. P. (1961). The social psychological nature of stress: A framework for causal inquiry. *Journal of Abnormal and Social Psychology, 62,* 294–302.

Dohrenwend, B. P., & Dohrenwend, B. S. (1974). Social and cultural influences on psychopathology. *Annual Review of Psychology, 25,* 417–452.

Dohrenwend, B. P., & Dohrenwend, B. S. (1976). Sex differences and psychiatric disorders. *American Journal of Sociology, 81,* 1447–1454.

Dollard, J., Doob, J., Miller, N., Mowrer, O., & Sears, R. (1939). *Frustration and aggression.* New Haven, CT: Yale University Press.

Doms, M. (1984). The minority influence effect: An alternative approach. In S. Moscovici & W. Doise (Eds.), *Current issues in European social psychology* (pp. 1–31). New York: Cambridge University Press.

Doms, M., & Van Avermaet, E. (1980). Majority influence, minority influence, and conversion behavior: A replication. *Journal of Experimental Social Psychology, 16,* 283–293.

Donnerstein, E. (1984). Pornography: Its effect on violence against women. In N. M. Malamuth & E. Donnerstein (Eds.), *Pornography and sexual aggression.* Orlando, FL: Academic Press.

Donnerstein, E., & Barrett, G. (1978). The effects of erotic stimuli on male aggression toward females. *Journal of Personality and Social Psychology, 36,* 180–188.

Donnerstein, E., & Berkowitz, L. (1981). Victim reactions in aggressive erotic films as a factor in violence toward women. *Journal of Personality and Social Psychology, 41,* 710–724.

Donohue, W. A., Allen, M., & Burrell, N. (1988). Mediator communication competence. *Communication Monographs, 55,* 104–119.

Dovidio, J. F. (1984). Helping behavior and altruism: An empirical and conceptual overview. In L. Berkowitz (Ed.), *Advances in experimental social psychology* (Vol. 17, pp. 362–427). New York: Academic Press.

Dovidio, J. F., Allen, J. L., & Schroeder, D. A. (1990). Specificity of empathy-induced helping: Evidence for altruistic motivation. *Journal of Personality and Social Psychology, 59*(2), 249–260.

Dovidio, J. F., & Ellyson, S. L. (1982). Decoding visual dominance: Attributions of power based on relative percentages of looking while speaking and looking while listening. *Social Psychology Quarterly, 45,* 106–113.

Dovidio, J. F., Ellyson, S. L., Keating, C. F., Heltman, K., & Brown, C. E. (1988). The relationship of social power to visual displays of dominance between men and women. *Journal of Personality and Social Psychology, 54,* 233–242.

Dovidio, J. F., & Gaertner, S. L. (1981). The effects of race, status, and ability on helping behavior. *Social Psychology Quarterly, 44,* 192–203.

Dovidio, J. F., & Gaertner, S. L. (1996). Affirmative action, unintentional racial biases, and intergroup relations. *Journal of Social Issues, 52*(4), 51–75.

Dovidio, J. F., Piliavin, J. A., Gaertner, S. L., Schroeder, D. A., & Clark, R. D., III (1991). The arousal:cost-reward model and the process of intervention: A review of the evidence. In M. S. Clark (Ed.), *Review of personality and social psychology: Vol. 12. Prosocial behavior* (pp. 86–118). Newbury Park, CA: Sage.

Downey, G. L. (1986). Ideology and the Clamshell identity: Organizational dilemmas in the anti-nuclear power movement. *Social Problems, 33,* 357–373.

Doyle, J. A., & Paludi, M. A. (1991). *Sex and gender: The human experience* (2nd ed.). Dubuque, IA: William C. Brown.

Drass, K. A. (1986). The effect of gender identity on conversation. *Social Psychology Quarterly, 49,* 294–301.

Drass, K. A., & Spencer, J. W. (1987). Accounting for pre-sentencing recommendations: Typologies and probation officers' theory of office. *Social Problems, 34,* 277–293.

Dreben, E. K., Fiske, S. T., & Hastie, R. (1979). The independence of evaluative and item information: Impression and recall order effects in behavior-based impression formation. *Journal of Personality and Social Psychology, 37,* 1758–1768.

Dubé-Simard, L. (1983). Genesis of social categorization, threat to identity, and perceptions of social injustice: Their role in intergroup communication breakdown. *Journal of Language and Social Psychology, 2,* 183–206.

Duesterhoeft, D. (1987). *An unholy alliance? A case study comparison of religious and feminist anti-pornography activists.* Unpublished master's thesis, University of Wisconsin.

Duncan, L. E., & Stewart, A. J. (1995). Still bringing the Vietnam War home: Sources of contemporary student activism. *Personality and Social Psychology Bulletin, 21,* 914–924.

Duncan, S., Jr., & Fiske, D. W. (1977). *Face-to-face interaction: Research methods and theory.* Hillsdale, NJ: Erlbaum.

Dunphy, D. (1972). *The primary group.* New York: Appleton-Century-Crofts.

Dura, J. A., & Kiecolt-Glaser, J. K. (1991). Family transitions, stress, and health. In D. A. Cowan & M. Hetherington (Eds.), *Family transitions.* Hillsdale, NJ: Erlbaum.

Duran, R. L., & Prusank, D. T. (1997). Relational themes in men's and women's popular nonfiction magazine articles. *Journal of Social and Personal Relationships, 14,* 165–189.

Dutton, D., & Aron, A. (1974). Some evidence for heightened sexual attraction under conditions of high anxiety. *Journal of Personality and Social Psychology, 30,* 510–517.

Dutton, D., & Lake, R. (1973). Threat of own prejudice and reverse discrimination in interracial situations. *Journal of Personality and Social Psychology, 28,* 94–100.

Dyck, R. J., & Rule, B. G. (1978). Effect on retaliation of causal attribution concerning attack. *Journal of Personality and Social Psychology, 36,* 521–529.

Dynes, R. R., & Quarantelli, E. L. (1980). Helping behavior in large-scale disasters. In D. H. Smith & J. Macaulay (Eds.), *Participation in social and political activities* (pp. 339–354). San Francisco: Jossey-Bass.

Eagly, A. (1987). *Sex differences in social behavior: A social-role interpretation.* Hillsdale, NJ: Erlbaum.

Eagly, A. H., Ashmore, R. D., Makhijani, M. G., & Longo, L. C. (1991). What is beautiful is good, but . . . : A meta-analytic review of research on the physical attractiveness stereotype. *Psychological Bulletin, 110,* 109–128.

Eagly, A. H., & Carli, L. L. (1981). Sex of researchers and sex-typed communications as determinants of sex differences in influenceability. *Psychological Bulletin, 90,* 1–20.

Eagly, A. H., & Chaiken, S. (1975). An attribution analysis of the effects of communicator characteristics on opinion change: The case of communicator attractiveness. *Journal of Personality and Social Psychology, 32,* 136–144.

Eagly, A. H., & Chrvala, C. (1986). Sex differences in conformity: Status and gender role interpretations. *Psychology of Women Quarterly, 10,* 203–220.

Eagly, A. H., & Crowley, M. (1986). Gender and helping behavior: A meta-analytic review of the social psychological literature. *Psychological Bulletin, 100*(3), 283–308.

Eagly, A., & Steffen, V. J. (1984). Gender stereotypes stem from distribution of women and men into social roles. *Journal of Personality and Social Psychology, 46,* 735–754.

Eagly, A. H., Wood, W., & Chaiken, S. (1978). Causal inferences about communicators and their effect on attitude change. *Journal of Personality and Social Psychology, 36,* 424–435.

Ebbesen, E. B., & Bowers, R. J. (1974). Proportion of risky to conservative arguments in a group discussion and choice shift. *Journal of Personality and Social Psychology, 29,* 316–327.

Eccles, J. S. (1987.) Gender roles and women's achievement related decisions. *Psychology of Women Quarterly, 11,* 135–172.

Edelmann, R. J. (1985). Social embarrassment: An analysis of the process. *Journal of Social and Personal Relationships, 2,* 195–213.

Edelmann, R. J. (1987). *The psychology of embarrassment.* Chichester, England: Wiley.

Edelmann, R. J., Evans, G., Pegg, I., & Tremain, M. (1983). Responses to physical stigma. *Perceptual and Motor Skills, 57*(1), 294.

Edelmann, R. J., & Iwawaki, S. (1987). Self-reported expression and consequences of embarrassment in the United Kingdom and Japan. *Psychologia, 30,* 205–216.

Eder, D., with Evans, C. C., & Parker, S. (1995). *School talk: Gender and adolescent culture.* New Brunswick, NJ: Rutgers University Press.

Edwards, K. (1990). The interplay of affect and cognition in attitude formation and change. *Journal of Personality and Social Psychology, 59,* 202–216.

Efran, M. G., & Cheyne, J. A. (1974). Affective concomitants of the invasion of shared space: Behavioral, physiological and verbal indicators. *Journal of Personality and Social Psychology, 29,* 219–226.

Eibl-Eibesfeldt, I. (1979). Universals in human expressive behavior. In A. Wolfgang (Ed.), *Nonverbal behavior: Applications and cultural implications.* New York: Academic Press.

Eisenberg, N., & Fabes, R. A. (1990). Empathy: Conceptualization, measurement, and relation to prosocial behavior. *Motivation and Emotion, 14*(2), 131–149.

Eisenberg, N., & Fabes, R. A. (1991). Prosocial behavior and empathy: A multimethod developmental perspective. In M. S. Clark (Ed.), *Review of personality and social psychology: Vol. 12. Prosocial behavior* (pp. 34–61). Newbury Park, CA: Sage.

Eisenberg, N., & Miller, P. A. (1987). The relation of empathy to prosocial and related behaviors. *Psychological Bulletin, 101,* 91–119.

Eiser, J. R. (Ed.). (1984). *Attitudinal judgment.* New York: Springer.

Ekman, P. (1972). Universals and cultural differences in facial expression of emotion. In J. K. Cole (Ed.), *Nebraska symposium on motivation, 1971.* Lincoln: Nebraska University Press.

Ekman, P., & Friesen, W. V. (1969). Nonverbal leakage and clues to deception. *Psychiatry, 32,* 88–106.

Ekman, P., & Friesen, W. V. (1974). Detecting deception from the body or face. *Journal of Personality and Social Psychology, 29,* 288–298.

Ekman, P., & Friesen, W. V. (1975). *Unmasking the face.* Englewood Cliffs, NJ: Prentice-Hall.

Ekman, P., Friesen, W. V., et al. (1987). Universals and cultural differences in the judgments of facial expressions of emotion. *Journal of Personality and Social Psychology, 53,* 712–717.

Ekman, P., Friesen, W. V., & O'Sullivan, M. (1988). Smiles when lying. *Journal of Personality and Social Psychology, 54,* 414–420.

Ekman, P., Friesen, W. V., & Scherer, K. R. (1976). Body movements and voice pitch in deceptive interaction. *Semiotica, 16,* 23–27.

Ekman, P., Friesen, W. V., & Tomkins, S. S. (1971). Facial affect scoring technique (FAST): A first validity study. *Semiotica, 3,* 37–58.

Ekman, P., & O'Sullivan, M. (1991). Who can catch a liar? *American Psychologist, 46*(9), 913–920.

Elder, G. H., Jr. (1975). Age differentiation and the life course. In A. Inkeles, J. Coleman, & N. Smelser (Eds.), *Annual review of sociology* (Vol. 1, pp. 165–190). Palo Alto, CA: Annual Reviews.

Elder, G. H., Jr., & O'Rand, A. M. (1995). Adult lives in a changing society. In K. S. Cook, G. A. Fine, & J. S. House (Eds.), *Sociological perspectives on social psychology* (pp. 452–475). Needham Heights, MA: Allyn & Bacon.

Elkin, R. A., & Leippe, M. (1986). Physiological arousal, dissonance, and attitude change: Evidence for a dissonance-arousal link and a "Don't remind me" effect. *Journal of Personality and Social Psychology, 51,* 55–65.

Ellard, J. H., & Bates, D. D. (1990). Evidence for the role of the justice motive in status generalization processes. *Social Justice Research, 4*(2), 115–134.

Ellemers, N., Van Rijswijk, W., Roefs, M., & Simons, C. (1997). Bias in intergroup perceptions: Balancing group identity with social reality. *Personality and Social Psychology Bulletin, 23,* 186–198.

Elliott, D. S., & Ageton, S. S. (1980). Reconciling race and class differences in self-reported and official estimates of delinquency. *American Sociological Review, 45,* 95–110.

Elliott, G. C., & Meeker, B. F. (1986). Achieving fairness in the face of competing concerns: The different effects of individual and group characteristics. *Journal of Personality and Social Psychology, 50,* 754–760.

Ellsworth, P. C., Carlsmith, J. M., & Henson, A. (1972). The stare as a stimulus to flight in human subjects. *Journal of Personality and Social Psychology, 21,* 302–311.

Emerson, R. M. (1966). Mount Everest: A case study of communication feedback and sustained group goal-striving. *Sociometry, 29,* 213–227.

Emmons, R. A., & Diener, E. (1986). Situation selection as a moderator of response consistency and stability. *Journal of Personality and Social Psychology, 51,* 1013–1019.

Emmons, R. A., Diener, E., & Larsen, R. J. (1986). Choice and avoidance of everyday situations and affect congruence: Two models of reciprocal interactionism. *Journal of Personality and Social Psychology, 51,* 815–826.

Emswiller, T., Deaux, K., & Willis, J. E. (1971). Similarity, sex, and requests for small favors. *Journal of Applied Social Psychology, 1,* 284–291.

Ennis, J. G., & Schreuer, R. (1987). Mobilizing weak support for social movements: The role of grievance, efficiency and cost. *Social Forces, 66,* 390–409.

Epstein, J. L. (1985). After the bus arrives: Resegregation in desegregated schools. *Journal of Social Issues, 41,* 23–44.

Erikson, E. H. (1968). *Identity: Youth and crisis.* New York: Norton.

Erikson, K. (1964). Notes on the sociology of deviance. In H. Becker (Ed.), *The other side* (pp. 9–21). New York: Free Press.

Erikson, K. (1966). *The wayward Puritans.* New York: Wiley.

Eron, L. D. (1982). Parent-child interaction, television violence, and aggression of children. *American Psychologist, 37,* 197–211.

Estes, R. L., & Wilensky, H. (1978). Life cycle squeeze and the morale curve. *Social Forces, 56,* 277–292.

Ettinger, R. F., Marino, C. J., Endler, N. S., Geller, S. H., & Natziuk, T. (1971). Effects of agreement and correctness on relative competence and conformity. *Journal of Personality and Social Psychology, 19,* 204–212.

Etzioni, A. (1967). The Kennedy experiment. *Western Political Quarterly, 20,* 361–380.

Evans, C. R., & Dion, K. L. (1991). Group cohesion and performance: A meta-analysis. *Small Group Research, 22,* 175–186.

Evans, N. J., & Jarvis, P. A. (1980). Group cohesion: A review and reevaluation. *Small Group Behavior, 11,* 359–370.

Faley, T., & Tedeschi, J. T. (1971). Status and reactions to threats. *Journal of Personality and Social Psychology, 17,* 192–199.

Fararo, T. J., & Skvoretz, J. (1984). Institutions as production systems. *Journal of Mathematical Sociology, 10,* 117–182.

Farhar-Pilgrim, B., & Shoemaker, F. F. (1981). Campaigns to affect energy behavior. In R. E. Rice & W. J. Paisley (Eds.), *Public communication campaigns* (pp. 161–180). Beverly Hills, CA: Sage.

Farina, A., Gliha, D., Boudreau, L. A., Allen, J. G., & Sherman, M. (1971). Mental illness and the impact of believing others know it. *Journal of Abnormal Psychology, 77,* 1–5.

Farnworth, M., & Leiber, M. J. (1989). Strain theory revisited: Economic goals, educational means, and delinquency. *American Sociological Review, 54,* 263–274.

Fazio, R. H. (1990). Multiple processes by which attitudes guide behavior: The MODE model as an integrative framework. In L. Berkowitz (Ed.), *Advances in experimental social psychology* (Vol. 23, pp. 75–109). New York: Academic Press.

Fazio, R. H., Powell, M., & Herr, P. (1983). Toward a process model of the attitude-behavior relation: Accessing one's attitude upon mere observation of the attitude object. *Journal of Personality and Social Psychology, 44,* 723–735.

Fazio, R. H., Sanbonmatsu, D. M., Powell, M.C., & Kardes, F. R. (1986). On the automatic activation of attitudes. *Journal of Personality and Social Psychology, 50,* 229–238.

Fazio, R. H., & Williams, C. J. (1986). Attitude accessibility as a moderator of the attitude-perception and attitude-behavior relations: An investigation of the 1984 presidential election. *Journal of Personality and Social Psychology, 51,* 505–514.

Fazio, R. H., & Zanna, M. (1981). Direct experience and attitude-behavior consistency. In L. Berkowitz (Ed.), *Advances in experimental social psychology* (Vol. 14). New York: Academic Press.

Feather, N. T. (1967). A structural balance approach to the analysis of communication effects. In L. Berkowitz (Ed.), *Advances in experimental social psychology* (Vol. 3). New York: Academic Press.

Feather, N.T. (1995). Values, valences, and choice: The influence of values on the perceived attractiveness and choice of alternatives. *Journal of Personality and Social Psychology, 68,* 1135–1151.

Feather, N. T., & Armstrong, D. J. (1967). Effects of variations in source attitude, receiver attitude and communication stand on reactions to source and contents of communications. *Journal of Personality, 35,* 435–455.

Feld, S. (1981). The focused organization of social ties. *American Journal of Sociology, 86,* 1015–1035.

Felipe, N. J., & Sommer, R. (1966). Invasions of personal space. *Social Problems, 14,* 206–214.

Felmlee, D., Sprecher, S., & Bassin, E. (1990). The dissolution of intimate relationships: A hazard model. *Social Psychology Quarterly, 53,* 13–30.

Felson, M. (1994). *Crime and everyday life: Insights and implications for society.* Thousand Oaks, CA: Pine Forge Press.

Felson, R. B. (1981). Ambiguity and bias in the self-concept. *Social Psychology Quarterly, 44,* 64–69.

Felson, R. B. (1985). Reflected appraisal and the development of self. *Social Psychology Quarterly, 48,* 71–78.

Felson, R. B. (1989). Parents and the reflected appraisal process: A longitudinal analysis. *Journal of Personality and Social Psychology, 56,* 965–971.

Felson, R. B., Liska, A. E., South, S. J., & McNulty, T. L. (1994). The subculture of violence and delinquency: Individual vs. school context effects. *Social Forces, 73*(1). 155–173.

Felson, R. B., & Reed, M. (1986). The effects of parents on the self-appraisals of children. *Social Psychology Quarterly, 49,* 302–308.

Felson, R. B., & Zielinski, M. A. (1989). Children's self-esteem and parental support. *Journal of Marriage and the Family, 51,* 727–735.

Fernandez-Dols, J. M., & Ruiz-Belda, M-A. (1995). Are smiles a sign of happiness? Gold medal winners at the Olympic Games. *Journal of Personality and Social Psychology, 69,* 1113–1119.

Ferree, M. M. (1976). Working-class jobs: Housework and paid work as sources of satisfaction. *Social Problems, 23,* 431–441.

Ferree, M. M., & Miller, F. D. (1985). Mobilization and meaning: Toward an integration of social psychological and resource perspectives on social movements. *Sociological Inquiry, 55,* 38–55.

Festinger, L. (1954). A theory of social comparison processes. *Human Relations, 7,* 117–140.

Festinger, L. (1957). *A theory of cognitive dissonance.* Stanford, CA: Stanford University Press.

Festinger, L., & Carlsmith, J. (1959). Cognitive consequences of forced compliance. *Journal of Abnormal and Social Psychology, 58,* 203–210.

Festinger, L., Pepitone, A., & Newcomb, T. (1952). Some consequences of de-individuation in a group. *Journal of Abnormal and Social Psychology, 47,* 382–389.

Festinger, L., Schachter, S., & Back, K. W. (1950). *Social pressures in informal groups.* New York: Harper & Row.

Fiedler, F. E. (1966). The effect of leadership and cultural heterogeneity on group performance: A test of the contingency model. *Journal of Experimental Social Psychology, 2,* 237–264.

Fiedler, F. E. (1978a). Recent developments in research on the contingency model. In L. Berkowitz (Ed.), *Group processes.* New York: Academic Press.

Fiedler, F. E. (1978b). The contingency model and the dynamics of the leadership process. In L. Berkowitz (Ed.), *Advances in experimental social psychology* (Vol. 11). New York: Academic Press.

Fiedler, F. E. (1981). Leadership effectiveness. *American Behavioral Scientist, 24,* 619–632.

Filardo, E. K. (1996). Gender patterns in African-American and white adolescents' social interactions in same-race, mixed gender groups. *Journal of Personality and Social Psychology, 71,* 71–82.

Finestone, H. (1964). Cats, kicks and color. In H. Becker (Ed.), *The other side* (pp. 281–297). New York: Free Press.

Fink, E. L., Kaplowitz, S. A., & Bauer, C. L. (1983). Positional discrepancy, psychological discrepancy, and attitude change: Experimental tests of some mathematical models. *Communication Monographs, 50,* 413–430.

Fischer, C. S. (1976). *The urban experience.* New York: Harcourt Brace Jovanovich.

Fischer, C. S. (1980). *Friendship, gender and the life-cycle.* Unpublished paper, Institute of Urban and Regional Development, University of California, Berkeley.

Fischer, C. S. (1984). *The urban experience* (2nd ed.). San Diego, CA: Harcourt Brace Jovanovich.

Fisek, M. H. (1974). A model for the evolution of status structures in task-oriented discussion groups. In J. Berger, T. L. Conner, & M. H. Fisek (Eds.), *Expectation states theory.* Cambridge, MA: Winthrop.

Fishbein, M. (1980). A theory of reasoned action: Some applications and implications. In H. Howe & M. Page (Eds.), *Nebraska symposium on motivation* (Vol. 27, pp. 65–116). Lincoln: University of Nebraska Press.

Fishbein, M., & Ajzen, I. (1975). *Belief, attitude, intention and behavior.* Reading, MA: Addison-Wesley.

Fisher, J., Nadler, D., & Whitcher-Alagna, S. (1982). Recipient reactions to aid. *Psychological Bulletin, 91,* 33–54.

Fishman, P. M. (1978). Interaction: The work women do. *Social Problems, 25,* 397–406.

Fishman, P. M. (1980). Conversational insecurity. In H. Giles, W. P. Robinson, & P. M. Smith (Eds.), *Language: Social psychological perspectives* (pp. 127–132). New York: Pergamon.

Fisicaro, S. A. (1988). A reexamination of the relation between halo error and accuracy. *Journal of Applied Psychology, 73,* 239–244.

Fiske, S. T., Kinder, D. R., & Larter, W. M. (1983). The novice and the expert: Knowledge-based strategies in political cognition. *Journal of Experimental Social Psychology, 19,* 381–400.

Fiske, S. T., & Linville, P. (1980). What does the schema concept buy us? *Personality and Social Psychology Bulletin, 6,* 543–557.

Fiske, S. T., & Taylor, S. E. (1991). *Social cognition* (2nd ed.). New York: McGraw-Hill.

Flavell, J., Shipstead, S., & Croft, K. (1978). *What young children think you see when their eyes are closed.* Unpublished report, Stanford University.

Fleming, J. H., Darley, J. M., Hilton, J. L., & Kojetin, B. A. (1991). Multiple audience problem: A strategic communication perspective on social perception. *Journal of Personality and Social Psychology, 58,* 593–609.

Flowers, M. L. (1977). A laboratory test of some implications of Janis' groupthink hypothesis. *Journal of Personality and Social Psychology, 35,* 888–896.

Fodor, E. M., & Smith, T. (1982). The power motive as an influence on group decision making. *Journal of Personality and Social Psychology, 42,* 178–185.

Ford, M. R., & Lowery, C. R. (1986). Gender differences in moral reasoning: A comparison of justice and care orientations. *Journal of Personality and Social Psychology, 50,* 777–783.

Forgas, J. P., & Bond, M. H. (1985). Cultural influences on the perception of interaction episodes. *Personality and Social Psychology Bulletin, 11,* 75–88.

Form, W. H., & Nosow, S. (1958). *Community in disaster.* New York: Harper.

Forsyth, D. R., Berger, R. E., & Mitchell, T. (1981). The effects of self-serving vs. other-serving claims of responsibility on attraction and attribution in groups. *Social Psychology Quarterly, 44,* 59–64.

Forthofer, M. S., Markman, H. J., Cox, M., Stanley, S., & Kessler, R. C. (1996). Associations between marital distress and work loss in a national sample. *Journal of Marriage and the Family, 58,* 597–605.

Foster, M. D., & Matheson, K. (1995). Double relative deprivation: Combining the personal and political. *Personality and Social Psychology Bulletin, 21,* 1167–1177.

Frank, F., & Anderson, L. R. (1971). Effects of task and group size upon group productivity and member satisfaction. *Sociometry, 34,* 135–149.

Franks, D., & Marolla, J. (1976). Efficacious action and social approval as interacting dimensions of self-esteem. *Sociometry, 39,* 324–341.

Fraser, C., Gouge, C., & Billig, M. (1971). Risky shifts, cautious shifts, and group polarization. *European Journal of Social Psychology, 1,* 7–30.

Fredricks, A., & Dossett, D. (1983). Attitude-behavior relations: A comparison of the Fishbein-Ajzen and the Bentler-Speckart models. *Journal of Personality and Social Psychology, 45,* 501–512.

Freedman, D. G. (1979). *Human sociobiology.* New York: Free Press.

Freese, L. (1976). The generalization of specific performance expectations. *Sociometry, 39,* 194–200.

Freese, L., & Cohen, B. P. (1973). Eliminating status generalization. *Sociometry, 36,* 177–193.

Freud, S. (1905). Fragment of an analysis of a case of hysteria. *Collected Papers* (Vol. 3). New York: Basic Books.

Freud, S. (1930). *Civilization and its discontents.* London: Hogarth Press.

Freud, S. (1950). Why war? In J. Strachey (Ed.), *Collected papers* (Vol. 5). London: Hogarth Press.

Frey, D. (1978). Reactions to success and failure in public and private conditions. *Journal of Experimental Social Psychology, 14,* 172–179.

Frey, D. L., & Gaertner, S. L. (1986). Helping and the avoidance of inappropriate interracial behavior: A strategy that perpetuates a nonprejudiced self-image. *Journal of Personality and Social Psychology, 50*(6), 1083–1090.

Frey, K. S., & Ruble, D. N. (1985). What children say when the teacher is not around: Conflicting goals in social comparison and performance assessment in the classroom. *Journal of Personality and Social Psychology, 48,* 550–562.

Friedrich-Cofer, L., & Huston, A. C. (1986). Television violence and aggression: The debate continues. *Psychological Bulletin, 100,* 364–371.

Frieze, I., & Weiner, B. (1971). Cue utilization and attributional judgments for success and failure. *Journal of Personality, 39,* 591–605.

Frieze, I. H., Parsons, J., Johnson, P., Ruble, D., & Zellman, G. (1978). *Women and sex roles: A social psychological perspective.* New York: Norton.

Fultz, J., Batson, C. D., Fortenbach, V. A., McCarthy, P. M., & Varney, L. L. (1986). Social evaluation and the empathy-altruism hypothesis. *Journal of Personality and Social Psychology, 50,* 761–769.

Gaertner, S., & Bickman, L. (1971). A nonreactive indicator of racial discrimination: The wrong number technique. *Journal of Personality and Social Psychology, 20,* 218–222.

Gaertner, S. L., Dovidio, J. F., Anastasio, P. A., Bachman, B. A., & Rust, M. C. (1993). The common ingroup identity model: Recategorization and the reduction of intergroup bias. In W. Stroebe & H. Hewstone (Eds.), *European Review of Social Psychology, 4,* 1–26.

Gaertner, S. L., Mann, J., Murrell, A., & Dovidio, J. F. (1989). Reducing intergroup bias: The benefits of recategorization. *Journal of Personality and Social Psychology, 57* (2), 239–249.

Gahagan, J. P., & Tedeschi, J. T. (1968). Strategy and the credibility of promises in the Prisoner's Dilemma game. *Journal of Conflict Resolution, 12,* 224–234.

Galton, M. (1987). An ORACLE chronicle: A decade of classroom research. *Teaching and Teacher Education, 3,* 299–313.

Gamson, W. A., & Modigliani, A. (1989). Media discourse and public opinion on nuclear power: A constructionist approach. *American Journal of Sociology, 95,* 1–37.

Gardner, L., & Shoemaker, D. J. (1989). Social bonding and delinquency: A comparative analysis. *Sociological Quarterly, 30,* 481–500.

Garfinkel, H. (1956). Conditions of successful degradation ceremonies. *American Sociological Review, 61,* 420–424.

Garfinkel, I., & McLanahan, S. S. (1986). *Single mothers and their children: A new American dilemma.* Washington, DC: The Urban Institute.

Gartner, R. (1990). The victims of homicide: A temporal and cross-national comparison. *American Sociological Review, 55,* 92–106.

Gaskell, G., & Smith, P. Group membership and social attitudes of youth: An investigation of some implications of social identity theory. *Social Behaviour, 1,* 67–77.

Gecas, V. (1972). Parental behavior and dimensions of adolescent self-evaluation. *Sociometry, 34,* 466–482.

Gecas, V. (1979). The influence of social class on socialization. In W. Burr, R. Hill, F. Nye, & I. Reiss (Eds.), *Contemporary theories about the family* (Vol. 1). New York: Free Press.

Gecas, V. (1990). Contexts of socialization. In M. Rosenberg & R. Turner (Eds.), *Social psychology: Sociological perspectives.* New Brunswick, NJ: Transaction.

Gecas, V., & Burke, P. J. (1995). Self and identity. In K. S. Cook, G. A. Fine, & J. S. House (Eds.), *Sociological perspectives on social psychology* (pp. 41–67). Needham Heights, MA: Allyn & Bacon.

Gecas, V., & Nye, F. (1974). Sex and class differences in parent-child interaction: A test of Kohn's hypothesis. *Journal of Marriage and the Family, 36,* 742–749.

Gecas, V., & Schwalbe, M. (1983). Beyond the looking-glass self: Social structure and efficacy-based self-esteem. *Social Psychology Quarterly, 46,* 77–88.

Geen, R. G. (1968). Effects of frustration, attack, and prior training in aggressiveness upon aggressive behavior. *Journal of Personality and Social Psychology, 9,* 316–321.

Geen, R. G. (1978). Some effects of observing violence upon the behavior of the observer. In B. A. Maher (Ed.), *Progress in experimental personality research* (Vol. 8). New York: Academic Press.

Geen, R. G., & Quanty, M. G. (1977). The catharsis of aggression: An analysis of a hypothesis. In L. Berkowitz (Ed.), *Advances in experimental social psychology* (Vol. 10, pp. 1–37). New York: Academic Press.

Geen, R. G., Stonner, L., & Shope, G. L. (1975). The facilitation of aggression by aggression: A study in response inhibition and disinhibition. *Journal of Personality and Social Psychology, 31,* 721–726.

Geis, M. L. (1995). *Speech acts and conversational interaction.* New York: Cambridge University Press.

Gelfand, D. M., & Hartmann, D. P. (1982). Response consequences and attributions: Two contributors to prosocial behavior. In N. Eisenberg (Ed.), *The development of prosocial behavior.* New York: Academic Press.

Geller, V. (1977). *The role of visual access in impression management and impression formation.* Unpublished doctoral dissertation, Columbia University, New York.

Gelles, R. J. (1980). A profile of violence toward children in the United States. In G. Gerbner, C. Ross, & E. Zigler (Eds.), *Child abuse.* New York: Oxford University Press.

Gelles, R. J., & Cornell, C. P. (1990). *Intimate violence in families* (2nd ed.). Newbury Park, CA: Sage.

Gerard, H. B. (1983). School desegregation: The social science role. *American Psychologist, 38,* 869–877.

Gerard, H. B., Wilhelmy, R. A., & Conolley, E. S. (1968). Conformity and group size. *Journal of Personality and Social Psychology, 8,* 79–82.

Gerbner, G., Ross, C. J., & Zigler, E. (Eds.) (1980). *Child abuse.* New York: Oxford University Press.

Gergen, K. (1971). *The self-concept.* New York: Holt, Rinehart and Winston.

Gergen, K. J., Ellsworth, P., Maslach, C., & Siepel, M. (1975). Obligation, donor resources and reactions to aid in three cultures. *Journal of Personality and Social Psychology, 31,* 390–400.

Gergen, K. J., & Gergen, M. M. (1983). Social construction of helping relationships. In J. D. Fisher, A. Nadler, & B. M. DePaulo (Eds.), *New directions in helping* (Vol. 1, pp. 144–163). New York: Academic Press.

Gergen, K. J., Gergen, M. M., & Meter, K. (1972). Individual orientation to prosocial behavior. *Journal of Social Issues, 28,* 105–130.

Gergen, K. J., Morse, S. J., & Bode, K. A. (1974). Overpaid or overworked? Cognitive and behavioral reactions to inequitable rewards. *Journal of Applied Social Psychology, 4,* 259–274.

Gerstal, N., Riessman, C., & Rosenfield, S. (1985). Explaining the symptomatology of separated and divorced women and men: The role of material conditions and social networks. *Social Forces, 64,* 84–101.

Gesell, A., & Ilg, F. (1943). *Infant and child in the culture of today.* New York: Harper & Row.

Gibbons, F. X., Smith, T. W., Ingram, R. E., Pearce, K., Brehm, S. S., & Schroeder, D. J. (1985). Self-awareness and self-confrontation: Effects of self-focussed attention on members of a clinical population. *Journal of Personality and Social Psychology, 48,* 662–675.

Gifford, R. (1982). Projected interpersonal distance and orientation choices: Personality, sex, and social situation. *Social Psychology Quarterly, 45,* 145–152.

Gilbert, T. F. (1978). *Human competence.* New York: McGraw-Hill.

Gilchrist, J. C., Shaw, M. E., & Walker, L. C. (1954). Some effects of unequal distribution of information in a wheel group structure. *Journal of Abnormal and Social Psychology, 49,* 554–556.

Giles, H. (1980). Accommodation theory: Some new directions. In S. deSilva (Ed.), *Aspects of linguistic behavior.* York, England: York University Press.

Giles, H., & Coupland, N. (1991). *Language: Contexts and consequences.* Pacific Grove, CA: Brooks/Cole.

Giles, H., Hewstone, M., & St. Clair, R. (1981). Speech as an independent and dependent variable of social situations: An introduction and new theoretical framework. In H. Giles & R. St. Clair (Eds.), *The social psychological significance of speech.* Hillsdale, NJ: Erlbaum.

Gill, V. T., & Maynard, D. W. (1995). On "labeling" in actual interactions: Delivering and receiving diagnoses of developmental disabilities. *Social Problems, 42,* 11–37.

Gilligan, C. (1982). *In a different voice.* Cambridge, MA: Harvard University Press.

Gladstein, D. (1984). Groups in context: A model of task group effectiveness. *Administrative Sciences Quarterly, 29,* 499–517.

Glaser, B. G., & Strauss, A. (1971). *Status passage: A formal theory.* Chicago: Aldine de Gruyter.

Glass, J., Bengston, V. L., & Dunham, C. (1986). Attitude similarity in three-generation families: Status inheritance or reciprocal influence? *American Sociological Review, 51,* 685–698.

Glick, P. C. (1997). Demographic pictures of African-American families. In H. P McAdoo (Ed.), *Black families* (3rd ed., pp. 118–137). Thousand Oaks, CA: Sage.

Glick, P., DeMorest, J. A., & Hotze, C. A. (1988). Keeping your distance: Group membership, personal space, and requests for small favors. *Journal of Applied Social Psychology, 18,* 315–330.

Goethals, G. R., & Zanna, M. P. (1979). The role of social comparison in choice shift. *Journal of Personality and Social Psychology, 37,* 1469–1476.

Goetting, A. (1986). The developmental tasks of siblingship over the life cycle. *Journal of Marriage and the Family, 48,* 703–714.

Goffman, E. (1952). Cooling the mark out: Some adaptations to failure. *Psychiatry, 15,* 451–463.

Goffman, E. (1959a). The moral career of the mental patient. *Psychiatry, 22,* 125–169.

Goffman, E. (1959b). *The presentation of self in everyday life.* Garden City, NY: Anchor/Doubleday.

Goffman, E. (1963a). *Behavior in public places.* New York: Free Press.

Goffman, E. (1963b). *Stigma: Notes on the management of spoiled identity.* Englewood Cliffs, NJ: Spectrum/Prentice-Hall.

Goffman, E. (1967). *Interaction ritual.* Chicago: Aldine de Gruyter.

Goffman, E. (1974). *Frame analysis.* New York: Harper & Row.

Goffman, E. (1983). Felicity's condition. *American Journal of Sociology, 89,* 1–53.

Goldberg, C. (1974). Sex roles, task competence, and conformity. *Journal of Psychology, 86,* 157–164.

Goldberg, C. (1975). Conformity to majority type as a function of task and acceptance of sex-related stereotypes. *Journal of Psychology, 89,* 25–37.

Goldberg, L. R. (1981). Language and individual differences: The search for universals in personality lexicons. In L. Wheeler

(Ed.), *Review of personality and social psychology* (Vol. 1, pp. 203–234). Hillsdale, NJ: Erlbaum.

Goldstein, B., & Oldham, J. (1979). *Children and work: A study of socialization.* New Brunswick, NJ: Transaction.

Gonos, G. (1977). "Situation" vs. "frame": The "interactionist" and the "structuralist" analysis of everyday life. *American Sociological Review, 42,* 854–867.

Gonzales, M. H., Davis, J. M., Loney, G. L., Lukens, C. K., & Junghans, C. M. (1983). Interactional approach to interpersonal attraction. *Journal of Personality and Social Psychology, 44,* 1192–1197.

Goodchilds, J. D., & Zellman, G. L. (1984). Sexual signaling and sexual aggression in adolescent relationships. In N. M. Malamuth & E. Donnerstein (Eds.), *Pornography and sexual aggression* (pp. 233–243). Orlando, FL: Academic Press.

Goode, E. (1978). *Deviant behavior: An interactionist approach.* Englewood Cliffs, NJ: Prentice-Hall.

Goodman, P. S., & Friedman, A. (1969). An examination of quantity and quality of performance under conditions of overpayment in piece rate. *Organizational Behavior and Human Performance, 4,* 365–374.

Goodwin, C. (1987). Forgetfulness as an interactive resource. *Social Psychology Quarterly, 50,* 115–131.

Goodwin, J. (1997). The libidinal constitution of a high-risk social movement: Affectual ties and the Huk Rebellion, 1946 to 1954. *American Sociological Review, 62,* 53–69.

Goodwin, R. (1991). A re-examination of Rusbult's responses to dissatisfaction typology. *Journal of Social and Personal Relationships, 8,* 569–574.

Gordon, C. (1968). Self-conceptions: Configurations of content. In C. Gordon & K. J. Gergen (Eds.), *The self in social interaction, I: Classic and contemporary perspectives* (pp. 115–136). New York: Wiley.

Gordon, R. A. (1996). Impact of ingratiation on judgments and evaluations: A meta-analytic investigation. *Journal of Personality and Social Psychology, 71*(1), 54–70.

Gordon, S. L. (1990). The sociology of sentiments and emotion. In M. Rosenberg & R. H. Turner (Eds.), *Social psychology: Sociological perspectives* (pp. 562–592). New Brunswick, NJ: Transaction.

Gorfein, D. S. (1964). The effects of a nonunanimous majority on attitude change. *Journal of Social Psychology, 63,* 333–338.

Gottlieb, J., & Carver, C. S. (1980). Anticipation of future interaction and the bystander effect. *Journal of Experimental Social Psychology, 16,* 253–260.

Gould, R. L. (1978). *Transformations.* New York: Simon & Schuster.

Gove, W., Hughes, M., & Geerken, M. R. (1980). Playing dumb: A form of impression management with undesirable effects. *Social Psychology Quarterly, 43,* 89–102.

Gowen, C. R. (1986). Managing work group performance by individual goals and group goals for an interdependent group task. *Journal of Organizational Behavior Management, 7* (3–4), 5–27.

Graen, G. B., Orris, J. B., & Alvares, K. M. (1971). Contingency model of leadership effectiveness: Some experimental results. *Journal of Applied Psychology, 55,* 196–201.

Graham, L., & Hogan, R. (1990). Social class and tactics: Neighborhood opposition to group homes. *Sociological Quarterly, 31,* 513–529.

Gramling, R., & Forsyth, C. J. (1987). Exploiting stigma. *Sociological Forum, 2,* 401–415.

Granberg, D. (1978). GRIT in the final quarter: Reversing the arms race through unilateral initiatives. *Bulletin of Peace Proposals, 9,* 210–221.

Granberg, D., & Holmberg, S. (1990). The intention-behavior relationship among U.S. and Swedish voters. *Social Psychology Quarterly, 53,* 44–54.

Granovetter, M. S. (1973). The strength of weak ties. *American Journal of Sociology, 78,* 1360–1380.

Grant, P. R., & Holmes, J. G. (1981). The integration of implicit personality schemas and stereotype images. *Social Psychology Quarterly, 44,* 107–115.

Grasmick, H. G., & Bryjak, G. J. (1980). The deterrent effect of perceived severity of punishment. *Social Forces, 59,* 471–491.

Grayshon, M. C. (1980). Social grammar, social psychology, and linguistics. In H. Giles, W. P. Robinson, & P. M. Smith (Eds.), *Language: Social psychological perspectives.* New York: Pergamon.

Green, J. A. (1972). Attitudinal and situational determinants of intended behavior toward blacks. *Journal of Personality and Social Psychology, 22,* 13–17.

Greenbaum, P., & Rosenfeld, H. (1978). Patterns of avoidance in response to interpersonal staring and proximity: Effects of bystanders on drivers at a traffic intersection. *Journal of Personality and Social Psychology, 36,* 575–587.

Greenberg, J., & Cohen, R. L. (1982). *Equity and justice in social behavior.* New York: Academic Press.

Greenberg, M. (1980). A theory of indebtedness. In K. J. Gergen, M. S. Greenberg, & R. H. Willis (Eds.), *Social exchange: Advances in theory and research.* New York: Plenum.

Greenberg, M., & Frisch, D. (1972). Effect of intentionality on willingness to reciprocate a favor. *Journal of Experimental Social Psychology, 8,* 99–111.

Greenberg, M. S., & Westcott, D. R. (1983). Indebtedness as a mediator of reactions to aid. In J. D. Fisher, A. Nadler, & B. M. DePaulo (Eds.), *New directions in helping: Vol 1. Recipient reactions to aid* (pp. 85–112). San Diego, CA: Academic Press.

Greenberger, E., & Steinberg, L. (1983). Sex differences in early labor force participation. *Social Forces, 62,* 467–486.

Greenfield, L. (1972). Situational measures of normative language views in relation to person, place and topic among Puerto Rican bilinguals. In J. A. Fishman (Ed.), *Advances in the sociology of language* (Vol. 2, pp. 17–35). The Hague: Mouton.

Greenley, J. (1979). Familial expectations, post-hospital adjustment and the societal reaction perspective on mental illness. *Journal of Health and Social Behavior, 20,* 217–227.

Greenwald, A. G., & Pratkanis, A. (1984). The self. In R. S. Wyer & T. K. Srull (Eds.), *Handbook of social cognition* (Vol. 3, pp. 129–178). Hillsdale, NJ: Erlbaum.

Gregory, S. W., Jr., & Webster, S. (1996). A nonverbal signal in voices of interview partners effectively predicts communication accommodation and social status perceptions. *Journal of Personality and Social Psychology, 70,* 1231–1240.

Grice, P. H. (1975). Logic and conversation. In P. Cole & J. L. Morgan (Eds.), *Syntax and semantics: Vol. 3. Speech acts* (pp. 41–58). New York: Academic Press.

Griffin, L., Wallace, M., & Rubin, B. (1986). Capitalist resistance to the organization of labor before the New Deal: Why? How? Success? *American Sociological Review, 51,* 147–167.

Grimshaw, A. D. (1973). On language in society. Part 1. *Contemporary Sociology, 2,* 575–585.

Grimshaw, A. D. (1990). Talk and social control. In M. Rosenberg & R. H. Turner (Eds.), *Social psychology: Sociological perspectives* (pp. 200–232). New Brunswick, NJ: Transaction.

Gross, A. E., & Latané, J. G. (1974). Receiving help, reciprocation, and interpersonal attraction. *Journal of Applied Social Psychology, 4,* 210–223.

Gross, A. E., & McMullen, P. A. (1983). Models of the help seeking process. In B. M. DePaulo, A. Nadler, & J. D. Fisher (Eds.), *New directions in helping: Vol. 2. Help seeking* (pp. 47–73). New York: Academic Press.

Gross, E., & Stone, G. P. (1970). Embarrassment and the analysis of role requirements. In G. P. Stone & H. A. Farberman (Eds.), *Social psychology through symbolic interaction* (pp. 174–190). Waltham, MA: Ginn-Blaisdell.

Guetzkow, H., & Dill, W. R. (1957). Factors in the organizational development of task-oriented groups. *Sociometry, 20,* 175–204.

Guimond, S., & Dubé-Simard, L. (1983). Relative deprivation theory and the Quebec nationalist movement: The cognition-emotion distinction and the personal-group deprivation issue. *Journal of Personality and Social Psychology, 44,* 526–535.

Gully, S. M., Devine, D. J., & Whitney, D. J. (1995). A meta-analysis of cohesion and performance: Effects of level of analysis and task interdependence. *Small Group Research, 26*(4), 497–520.

Gumperz, J. J. (1976). The sociolinguistic significance of conversational code-switching. In J. Cook-Gumperz & J. J. Gumperz (Eds.), *Papers on language and context.* Berkeley: University of California Language Behavior Research Laboratory.

Gurevitch, Z. D. (1990). The embrace: On the element of non-distance in human relations. *Sociological Quarterly, 31,* 187–201.

Gurin, G., Veroff, J., & Feld, S. (1960). *Americans view their mental health.* New York: Basic Books.

Gustafson, R. (1989). Frustration and successful vs. unsuccessful aggression: A test of Berkowitz's completion hypothesis. *Aggressive Behavior, 15*(1), 5–12.

Guttman, D. (1977). The cross-cultural perspective: Notes toward a comparative psychology of aging. In J. Birren & K. Schaie (Eds.), *Handbook of the psychology of aging.* New York: Van Nostrand Reinhold.

Haan, N. (1978). Two moralities in action contexts. Relationships to thought, ego regulation, and development. *Journal of Personality and Social Psychology, 36,* 286–305.

Haan, N. (1986). Systematic variability in the quality of moral action, as defined in two paradigms. *Journal of Personality and Social Psychology, 50,* 1271–1284.

Haan, N., Smith, M., & Block, J. (1968). Moral reasoning of young adults: Political-social behavior, family background, and personality correlates. *Journal of Personality and Social Psychology, 10,* 183–201.

Haas, R. G. (1981). Effects of source characteristics on cognitive responses and persuasion. In R. E. Petty, T. M. Ostrom, & T. C. Brock (Eds.), *Cognitive responses in persuasion.* Hillsdale, NJ: Erlbaum.

Hacker, H. M. (1981). Blabbermouths and clams: Sex differences in self-disclosure in same-sex and cross-sex friendship dyads. *Psychology of Women Quarterly, 5,* 385–401.

Hackman, J. R., & Morris, C. G. (1975). Group tasks, group interaction process, and group performance effectiveness: A review and proposed integration. In L. Berkowitz (Ed.), *Advances in experimental social psychology* (Vol. 8, pp. 45–99). New York: Academic Press.

Haines, H. H. (1984). Black radicalization and the funding of civil-rights: 1957–1970. *Social Problems, 32,* 31–43.

Halberstam, D. (1979). *The powers that be.* New York: Knopf.

Hall, E. T. (1966). *The hidden dimension.* Garden City, NJ: Doubleday.

Hallinan, M. T. (1981). Recent advances in sociometry. In S. R. Asher & J. M. Gottman (Eds.), *The development of children's friendships* (pp. 91–115). Cambridge, England: Cambridge University Press.

Hamilton, D. L. (1979). A cognitive-attributional analysis of stereotyping. In L. Berkowitz (Ed.), *Advances in experimental social psychology* (Vol. 12). New York: Academic Press.

Hamilton, D. L. (1981). Stereotyping and intergroup behavior: Some thoughts on the cognitive approach. In D. L. Hamilton (Ed.), *Cognitive processes in stereotyping and intergroup behavior* (pp. 333–353). Hillsdale, NJ: Erlbaum.

Hamilton, D. L., & Bishop, G. D. (1976). Attitudinal and behavioral effects of initial integration of white suburban neighborhoods. *Journal of Social Issues, 32,* 47–56.

Hamilton, D. L., & Zanna, M. P. (1972). Differential weighting of favorable and unfavorable attributes in impressions of personality. *Journal of Experimental Research in Personality, 6,* 204–212.

Han, G., & Lindskold, S. (1983). *A comparison of strategies with and without communication in a prisoner's dilemma game.* Unpublished manuscript, Ohio University.

Hanni, R. (1980). What is planned during speech pauses? In H. Giles, W. P. Robinson, & P. M. Smith (Eds.), *Language: Social psychological perspectives.* New York: Pergamon.

Hanson, R. (1986). Relational competence, relationships, and adjustment in old age. *Journal of Personality and Social Psychology, 50,* 1050–1058.

Harasty, A. (1997). The interpersonal nature of social stereotypes: Discussion patterns about in-groups and out-groups. *Personality and Social Psychology Bulletin, 23,* 270–284.

Hardy, C., & Latané, B. (1986). Social loafing on a cheering task. *Social Science, 71*(2–3), 165–172.

Hardy, R. C. (1975). A test of poor leader-member relations cells of the contingency model on elementary school children. *Child Development, 45,* 958–964.

Hardy, R. C., Sack, S., & Harpine, F. (1973). An experimental test of the contingency model on small classroom groups. *Journal of Psychology, 85,* 3–16.

Hare, A. P. (1992). *Groups, teams, and social interaction: Theories and applications.* New York: Praeger.

Harkins, S. G. (1987). Social loafing and social facilitation. *Journal of Experimental Social Psychology, 23,* 1–18.

Harkins, S. G., & Jackson, J. (1985). The role of evaluation in eliminating social loafing. *Personality and Social Psychology Bulletin, 11,* 457–465.

Harkins, S. G., & Petty, R. E. (1981a). The effects of source magnification of cognitive effort on attitudes: An information processing view. *Journal of Personality and Social Psychology, 40,* 401–413.

Harkins, S. G., & Petty, R. E. (1981b). The multiple source effect in persuasion: The effects of distraction. *Personality and Social Psychology Bulletin, 4,* 627–635.

Harkins, S. G., & Petty, R. E. (1982). Effects of task difficulty and task uniqueness on social loafing. *Journal of Personality and Social Psychology, 43,* 1214–1229.

Harkins, S. G., & Petty, R. E. (1983). Social context effects in persuasion: The effects of multiple sources and multiple targets. In P. B. Paulus (Ed.), *Basic group processes.* New York: Springer-Verlag.

Harkins, S. G., & Petty, R. E. (1987). Information utility and the multiple source effect. *Journal of Personality and Social Psychology, 52,* 260–268.

Harkins, S. G., & Szymanski, K. (1987). Social loafing and social facilitation. In C. Hendrick (Ed.), *Group processes and intergroup relations* (pp. 167–188). Newbury Park, CA: Sage.

Harkins, S. G., & Szymanski, K. (1989). Social loafing and group evaluation. *Journal of Personality and Social Psychology, 56,* 934–941.

Harlow, R. E., & Cantor, N. (1995). To whom do people turn when things go poorly? Task orientation and functional social contacts. *Journal of Personality and Social Psychology, 69,* 329–340.

Harlow, R. E., & Cantor, N. (1996). Still participating after all these years: A study of life task participation in later life. *Journal of Personality and Social Psychology, 71,* 1235–1249.

Harper, R. G., Wiens, A. N., & Matarazzo, J. D. (1978). *Nonverbal communication: The state of the art.* New York: Wiley.

Harris, L., et al. (1975). *The myth and the reality of aging in America.* Washington, DC: National Council on Aging.

Harris, L., et al. (1982). *Aging in the 80's.* Washington, DC: National Council on Aging.

Harris, M. B. (1974). Mediators between frustration and aggression in a field experiment. *Journal of Experimental Social Psychology, 10,* 561–571.

Harris, R., Ellicott, A., & Holmes, D. (1986). The timing of psychosocial transitions and changes in women's lives: An examination of women 45 to 60. *Journal of Personality and Social Psychology, 51,* 409–416.

Harris, R. J., Lee, D. J., Hensley, D. L., & Schoen, L. M. (1988). The effect of cultural script knowledge on memory for stories over time. *Discourse Processes, 11,* 413–431.

Harris, R. J., Messick, D. M., & Sentis, K. P. (1981). Proportionality, linearity, and parameter constancy: Messick and Sentis reconsidered. *Journal of Experimental Social Psychology, 17,* 210–225.

Harrison, A. (1977). Mere exposure. In L. Berkowitz (Ed.), *Advances in experimental social psychology* (Vol. 10, pp. 39–83). New York: Academic Press.

Harrison, A., & Saeed, L. (1977). Let's make a deal: An analysis of revelations and stipulations in lonely hearts advertisements. *Journal of Personality and Social Psychology, 35,* 257–264.

Hartman, R. L., & Johnson, J. D. (1990). Formal and informal group communication structures: An examination of their relationship to role ambiguity. *Social Networks, 12*(2), 127–151.

Harvey, J. H., Weber, A. L., & Orbuch, T. L. (1990). *Interpersonal accounts.* Oxford, England: Basil Blackwell.

Harvey, J. H., Yarkin, K. L., Lightner, J. M., & Tolin, J. P. (1980). Unsolicited interpretation and recall of interpersonal events. *Journal of Personality and Social Psychology, 38,* 551–568.

Hass, R. G. (1981). Effects of source characteristics on cognitive responses in persuasion. In R. E. Petty, T. M. Ostrom, & T. C. Brock (Eds.), *Cognitive responses in persuasion* (pp. 141–172). Hillsdale, NJ: Erlbaum.

Hastorf, A. H., Wildfogel, J., & Cassman, T. (1979). Acknowledgment of a handicap as a tactic in social interaction. *Journal of Personality and Social Psychology, 31,* 1790–1797.

Hater, J. J., & Bass, B. M. (1988). Supervisors' evaluations and subordinates' perceptions of transformational and transactional leadership. *Journal of Applied Psychology, 73,* 695–702.

Hatfield, E. (1982). What do women and men want from love and sex? In E. R. Allgeier & N. B. McCormick (Eds.), *Changing boundaries: Gender roles and sexual behavior.* Palo Alto, CA: Mayfield.

Hatfield, E., & Sprecher, S. (1986). Measuring passionate love in intimate relationships. *Journal of Adolescence, 9,* 383–410.

Hatfield, E., & Walster, G. W. (1978). *A new look at love.* Lanham, MD: University Press of America.

Hatfield, E., & Walster, G. W. (1983). *A new look at love* (2nd ed.). Reading, MA: Addison-Wesley.

Haveman, R., Wolfe, B., & Spaulding, J. (1991). Childhood events and circumstances influencing high school completion. *Demography, 28,* 133–157.

Hayduk, L. A. (1978). Personal space: An evaluation and orienting review. *Psychological Bulletin, 85,* 117–134.

Hedge, A., & Yousif, Y. H. (1992). Effects of urban size, urgency, and cost on helpfulness: A cross-cultural comparison between the United Kingdom and the Sudan. *Journal of Cross Cultural Psychology, 23*(1), 107–115.

Heider, E. R., & Olivier, D. (1972). The structure of the color space in naming and memory of two languages. *Cognitive Psychology, 3,* 337–354.

Heider, F. (1944). Social perception and phenomenal causality. *Psychological Review, 51,* 258–374.

Heider, F. (1958). *The psychology of interpersonal relations.* New York: Wiley.

Heimer, K., & Matsueda, R. (1994). Role-taking, role commitment and delinquency: A theory of differential social control. *American Sociological Review, 59,* 365–390.

Heirich, M. (1977). Change of heart: A test of some widely held theories about religious conversion. *American Journal of Sociology, 83,* 653–680.

Heiss, D. (1979). *Understanding events: Affect and the construction of social action.* New York: Cambridge University Press.

Heiss, D. A., & Calhan, C. (1995). Emotion norms in interpersonal events. *Social Psychology Quarterly, 53,* 223–240.

Heiss, J. (1981). Social roles. In M. Rosenberg & R. H. Turner (Eds.), *Social psychology: Sociological perspectives.* New York: Basic Books.

Heiss, J. (1990). Social roles. In M. Rosenberg & R. Turner (Eds.), *Social psychology: Sociological perspectives.* New Brunswick, NJ: Transaction.

Heiss, J., & Owens, S. (1972). Self-evaluations of blacks and whites. *American Journal of Sociology, 78,* 360–370.

Hendrick, C., & Hendrick, S. (1989). Research on love: Does it measure up? *Journal of Personality and Social Psychology, 56,* 784–794.

Hendrick, S., & Hendrick, C. (1992). *Liking, loving, and relating* (2nd ed.). Pacific Grove, CA: Brooks/Cole.

Henretta, J. C., Frazier, C., & Bishop, D. (1986). The effect of prior case outcome on juvenile justice decision-making. *Social Forces, 65,* 554–562.

Hensley, T. R., & Griffin, G. W. (1986). Victims of groupthink: The Kent State University board of trustees and the 1977 gymnasium controversy. *Journal of Conflict Resolution, 30,* 497–531.

Hepburn, C., & Locksley, A. (1983). Subjective awareness of stereotyping: Do we know when our judgments are prejudiced? *Social Psychology Quarterly, 45,* 311–318.

Herek, G. (1987). Can functions be measured? A new perspective on the functional approach to attitudes. *Social Psychology Quarterly, 50,* 285–303.

Heritage, J., & Greatbatch, D. (1986). Generating applause: A study of rhetoric and response at party political conferences. *American Journal of Sociology, 92,* 110–157.

Hess, T. M., & Slaughter, S. J. (1990). Schematic knowledge influences on memory for scene information in young and older adults. *Developmental Psychology, 26,* 855–865.

Hess, U., Banse, R., & Kappas, A. (1995). The intensity of facial affect is determined by underlying affective state and social situation. *Journal of Personality and Social Psychology, 69,* 280–288.

Hetherington, E. M. (1991). The role of individual differences and family relationships in children's coping with divorce and remarriage. In P. A. Cowan & E. M. Hetherington (Eds.), *Family transitions.* Hillsdale, NJ: Erlbaum.

Hetherington, E. M., Cox, M., & Cox, R. (1982). Effects of divorce on parents and children. In M. E. Lamb (Ed.), *Nontraditional families: Parenting and child development.* Hillsdale, NJ: Erlbaum.

Hewitt, J. P. (1997). *Self and society* (7th ed.). Boston: Allyn & Bacon.

Hewitt, J. P., & Stokes, R. (1975). Disclaimers. *American Sociological Review, 40,* 1–11.

Hewstone, M. (1990). The "ultimate attribution error"? A review of the literature on intergroup causal attribution. *European Journal of Social Psychology, 20,* 311–335.

Hewstone, M., & Brown, R. (1986). Contact is not enough: An intergroup perspective on the contact hypothesis. In M. Hewstone & R. Brown (Eds.), *Contact and conflict in intergroup encounters* (pp. 1–44). Oxford, England: Basil Blackwell.

Hewstone, M., & Giles, H. (1986). Social groups and social stereotypes in intergroup communication: A review and model of intergroup communication. In W. B. Gudykunst (Ed.), *Intergroup communication* (pp. 10–26). London: Edward Arnold.

Hewstone, M., & Jaspars, J. (1984). Social dimensions of attribution. In H. Tajfel (Ed.), *The social dimension* (Vol. 2). Cambridge, England: Cambridge University Press.

Hewstone, M., & Jaspars, J. (1987). Covariation and causal attribution: A logical model of the intuitive analysis of variance. *Journal of Personality and Social Psychology, 53,* 663–672.

Heyl, B. S. (1977). The madam as teacher: The training of house prostitutes. *Social Problems, 24,* 545–555.

Higbee, K. L. (1969). Fifteen years of fear arousal: Research on threat appeals, 1953–1968. *Psychological Bulletin, 72,* 426–444.

Higgins, E. T. (1989). Self-discrepancy theory: What patterns of self-beliefs cause people to suffer? In L. Berkowitz (Ed.), *Advances in experimental social psychology* (Vol. 22). New York: Academic Press.

Higgins, E. T., & Bargh, J. A. (1987). Social cognition and social perception. *Annual Review of Psychology, 38,* 369–425.

Higgins, E. T., & Bryant, S. L. (1982). Consensus information and the fundamental attribution error: The role of development and in-group versus out-group knowledge. *Journal of Personality and Social Psychology, 47,* 422–435.

Higgins, E. T., Klein, R., & Strauman, T. (1985). Self-concept discrepancy theory: A psychological model for distinguishing among different aspects of depression and anxiety. *Social Cognition, 3,* 51–76.

Hill, C., Rubin, Z., & Peplau, L. (1976). Breakups before marriage: The end of 103 affairs. *Journal of Social Issues, 32*(1), 147–168.

Hill, C., & Stull, D. (1981). Sex differences in effects of social and value similarity in same-sex friendship. *Social Psychology, 41,* 488–502.

Hiltrop, J. M. (1985). Mediator behavior and the settlement of collective bargaining disputes in Britain. *Journal of Social Issues, 41,* 83–99.

Hinkle, S., & Schopler, J. (1986). Bias in the evaluation of in-group and out-group performance. In S. Worchel & W. G. Austin (Eds.), *Psychology of intergroup relations* (2nd ed., pp. 196–212). Chicago: Nelson-Hall.

Hirsch, E. L. (1990). Sacrifice for the cause: Group processes, recruitment, and commitment in a student social movement. *American Sociological Review, 55,* 243–254.

Hirschi, T. (1969). *Causes of delinquency.* Berkeley: University of California Press.

Hochschild, A. R. (1975). Disengagement theory: A critique and a proposal. *American Sociological Review, 40,* 553–569.

Hochschild, A. R. (1983). *The managed heart: Commercialization of human feeling.* Berkeley: University of California Press.

Hodson, R. (1996). Dignity in the workplace under participative management. *American Sociological Review, 61,* 719–738.

Hoelter, J. W. (1983). The effects of role evaluation and commitment on identity salience. *Social Psychology Quarterly, 46,* 140–147.

Hoelter, J. W. (1984). Relative effects of significant others on self-evaluation. *Social Psychology Quarterly, 47,* 255–262.

Hoelter, J. W. (1986). The relationship between specific and global evaluations of self: A comparison of several models. *Social Psychology Quarterly, 49,* 129–141.

Hoffman, C., Lau, I., & Johnson, D. (1986). The linguistic relativity of person cognition: An English-Chinese comparison. *Journal of Personality and Social Psychology, 51,* 1097–1105.

Hoffman, M. L. (1977). Empathy, its development and prosocial implications. In C. B. Keasey (Ed.), *Nebraska symposium on motivation* (Vol. 25). University of Nebraska Press.

Hofling, C. K., Brotz, E., Dalrymple, S., Graves, N., & Pierce, C. M. (1966). An experimental study of nurse-physician relationships. *Journal of Nervous and Mental Disease, 143,* 171–180.

Hogan, D. P. (1981). *Transitions and social change.* New York: Academic Press.

Hogg, M. A., Terry, D. J., & White, K. M. (1995). A tale of two theories: A critical comparison of identity theory with social identity theory. *Social Psychology Quarterly, 58,* 255–269.

Holahan, C. J. (1977). Effects of urban size and heterogeneity on judged appropriateness of altruistic responses: Situational vs. subject variables. *Sociometry, 40,* 378–382.

Holahan, C., & Moos, R. (1987). Personal and contextual determinants of coping strategies. *Journal of Personality and Social Psychology, 52,* 946–955.

Holahan, C. J., & Moos, R. H. (1990). Life stressors, resistance factors, and improved psychological functioning: An extension of the stress resistance paradigm. *Journal of Personality and Social Psychology, 58,* 909–917.

Holden, R. T. (1986). The contagiousness of aircraft hijacking. *American Journal of Sociology, 91,* 874–904.

Holgartner, S., & Bosk, C. L. (1988). The rise and fall of social problems: A public arenas model. *American Journal of Sociology, 94,* 53–78.

Hollander, E. P. (1975). Independence, conformity and civil liberties: Some implications from social psychological research. *Journal of Social Issues, 31,* 55–67.

Hollander, E. P. (1985). Leadership and power. In G. Lindzey & E. Aronson (Eds.), *Handbook of social psychology* (3rd ed., Vol. 2, pp. 485–537). New York: Random House.

Hollander, E. P., Fallon, B. J., & Edwards, M. T. (1977). Some aspects of influence and acceptability for appointed and elected group leaders. *Journal of Psychology, 95,* 289–296.

Hollander, E. P., & Julian, J. W. (1970). Studies in leader legitimacy, influence, and innovation. In L. Berkowitz (Ed.), *Advances in experimental social psychology* (Vol. 5, pp. 33–69). New York: Academic Press.

Hollinger, R. C., & Clark, J. P. (1983). Deterrence in the workplace: Perceived certainty, perceived severity and employee theft. *Social Forces, 62,* 398–418.

Holmes, D. S. (1971). Compensation for ego threat: Two experiments. *Journal of Personality and Social Psychology, 18,* 234–237.

Holmes, J. G., & Grant, P. (1979). Ethnocentric reactions to social threats. In L. H. Strickland (Ed.), *Social psychology: East-west perspectives.* Oxford, England: Pergamon.

Homans, G. C. (1961). *Social behavior: Its elementary forms.* New York: Harcourt Brace Jovanovich.

Homans, G. C. (1974). *Social behavior: Its elementary forms* (2nd ed.). New York: Harcourt Brace Jovanovich.

Homer, P. M., & Kahle, L. R. (1988). A structural-equation test of the value-attitude-behavior hierarchy. *Journal of Personality and Social Psychology, 54,* 638–646.

Hood, W. R., & Sherif, M. (1962). Verbal report and judgment of an unstructured stimulus. *Journal of Psychology, 54,* 121–130.

Hopper, J. R., & Nielsen, J. M. (1991). Recycling as altruistic behavior: Normative and behavioral strategies to expand participation in a community recycling program. *Environment and Behavior, 23*(2), 195–220.

Horai, J., Naccari, N., & Fatoullah, E. (1974). The effects of expertise and physical attractiveness upon opinion agreement and liking. *Sociometry, 37,* 601–606.

Horai, J., & Tedeschi, J. T. (1969). The effects of threat credibility and magnitude of punishment upon compliance. *Journal of Personality and Social Psychology, 12,* 164–169.

Horne, W. C., & Long, G. (1972). Effect of group discussion on universalistic-particularistic orientation. *Journal of Experimental Social Psychology, 8,* 236–246.

Horney, J., Osgood, D. W., & Marshall, I. H. (1995). Criminal careers in the short-term: Intra-individual variability in crime and its relation to local life circumstances. *American Sociological Review, 60,* 655–673.

Hornstein, G. (1985). Intimacy in conversational style as a function of the degree of closeness between members of a dyad. *Journal of Personality and Social Psychology, 49,* 671–681.

Hornstein, H. A. (1965). The effects of different magnitudes of threat upon interpersonal bargaining. *Journal of Experimental Social Psychology, 1,* 282–293.

Hornstein, H. A. (1978). Promotive tension and prosocial behavior: A Lewinian analysis. In L. Wispe (Ed.), *Altruism, sympathy and helping.* New York: Academic Press.

Horowitz, R. (1987). Community tolerance of gang violence. *Social Problems, 34,* 437–450.

Horwitz, H. V., White, H. R., & Howell-White, S. (1996). Becoming married and mental health: A longitudinal study of a cohort of young adults. *Journal of Marriage and the Family, 58,* 895–907.

Hossain, Z., & Roopnarine, J. L. (1993). Division of household labor and child-care in dual-earner African-American families with infants. *Sex Roles, 29,* 571–583.

House, J. S. (1974). Occupational stress and coronary heart disease: A review and theoretical integration. *Journal of Health and Social Behavior, 15,* 17–21.

House, J. S. (1981). *Work stress and social support.* Reading, MA: Addison-Wesley.

House, J. S. (1990). Social structure and personality. In M. Rosenberg and R. Turner (Eds.), *Social psychology: Sociological perspectives* (pp. 525–561). New Brunswick, NJ: Transaction.

House, J. S., Lepkowski, J. M., Kinney, A. M., Mero, R. P., Kessler, R. C., & Herzog, A. R. (1994). The social stratification of aging and health. *Journal of Health and Social Behavior, 35,* 213–234.

House, J. S., & Wolf, S. (1978). Effects of urban residence on interpersonal trust and helping behavior. *Journal of Personality and Social Psychology, 36,* 1029–1043.

Hoyenga, K. B., & Hoyenga, K. (1979). *The question of sex differences: Psychological, cultural and biological issues.* Boston: Little, Brown.

Hoyert, D. (1991). *The formation and dissolution of multigenerational households.* Unpublished doctoral dissertation, University of Wisconsin, Madison.

Hrdy, S. B. (1977). Infanticide as a primate reproductive strategy. *American Scientist, 65,* 40–49.

Hrdy, S. B. (1982). Positivist thinking encounters field primatology, resulting in agonistic behavior. *Social Science Information, 21*(2), 245–250.

Hue, C., & Erickson, J. R. (1991). Normative studies of sequence strength and scene structure of 30 scripts. *American Journal of Psychology, 104,* 229–240.

Huesmann, L. R. (1986). Psychological processes promoting the relation between exposure to media violence and aggressive behavior by the viewer. *Journal of Social Issues, 42*(3), 125–139.

Huesmann, L. R., & Moise, J. (1996). Media violence: A demonstrated public health threat to children. *Harvard Mental Health Letter, 12*(12), 5–7.

Huffine, C., & Clausen, J. (1979). Madness and work: Short and long-term effects of mental illness on occupational careers. *Social Forces, 57,* 1049–1062.

Hughes, M., & Demo, D. H. (1989). Self-perceptions of black Americans: Self-esteem and personal efficacy. *American Journal of Sociology, 95,* 132–159.

Hull, C. L. (1943). *Principles of behavior.* New York: Appleton-Century-Crofts.

Hultsch, D., & Plemons, J. (1979). Life events and life span development. In P. Baltes & O. G. Brim, Jr. (Eds.), *Life-span development and behavior* (Vol. 2). New York: Academic Press.

Hundleby, J. D., & Mercer, G. W. (1987). Family and friends as social environments and their relationship to young adolescents' use of alcohol, tobacco and marijuana. *Journal of Marriage and the Family, 49,* 151–164.

Hunter, C. H. (1984). Aligning actions: Types and social distribution. *Symbolic Interaction, 7,* 155–174.

Hunter, J. A., Stringer, M., & Watson, R. P. (1991). Intergroup violence and intergroup attributions. *British Journal of Social Psychology, 30*(3), 261–266.

Huston, A. C., & Wright, J. C. (1982). Effects of communication media on children. In C. B. Kopp & J. B. Krakow (Eds.), *The child: Development in a social context.* Boston: Addison-Wesley.

Huston, T. L., Ruggiero, M., Conner, R., and Geis, G. (1981). Bystander intervention into crime: A study based on naturally-occurring episodes. *Social Psychology Quarterly, 44*(1), 14–23.

Huston-Stein, A., & Higgins-Trenk, A. (1978). Development of females from childhood through adulthood: Career and feminine role orientations. In P. Baltes (Ed.), *Life-span development and behavior* (Vol. 1). New York: Academic Press.

Hyde, J. S. (1984). How large are gender differences in aggression? A developmental meta-analysis. *Developmental Psychology, 20,* 722–736.

Hyde, J., & Linn, M. (Eds.). (1986). *The psychology of gender: Advances through meta-analysis.* Baltimore, MD: Johns Hopkins University Press.

Hymes, D. (1974). *Foundations in sociolinguistics.* London: Tavistock.

Ikle, F. C. (1971). *Every war must end.* New York: Columbia University Press.

Ingham, A. G., Levinger, G., Graves, J., & Peckham, V. (1974). The Ringelmann effect: Studies of group size and group performance. *Journal of Experimental Social Psychology, 10,* 371–384.

Insko, C. A., Arkoff, A., & Insko, V. M. (1965). Effects of high and low fear-arousing communications upon opinions toward smoking. *Journal of Experimental Social Psychology, 40,* 256–266.

Insko, C. A., Drenan, S., Solomon, M. R., Smith, R., & Wade, T. J. (1983). Conformity as a function of the consistency of positive self-evaluation with being liked and being right. *Journal of Experimental Social Psychology, 19,* 341–358.

Insko, C. A., Smith, R. H., Alicke, M. S., Wade, J., & Taylor, S. (1985). Conformity and group size: The concern with being right and the concern with being liked. *Personality and Social Psychology Bulletin, 11,* 41–50.

Isen, A. M. (1970). Success, failure, attention, and reaction to others: The warm glow of success. *Journal of Personality and Social Psychology, 15,* 294–301.

Isen, A. M., Clark, M., & Schwartz, M. F. (1976). Duration of the effect of good mood on helping: Footprints on the sands of time. *Journal of Personality and Social Psychology, 34,* 385–393.

Isen, A. M., & Levin, P. F. (1972). Effect of feeling good on helping: Cookies and kindness. *Journal of Personality and Social Psychology, 21,* 384–388.

Isen, A. M., & Simmons, S. F. (1978). The effect of feeling good on a helping task that is incompatible with good mood. *Social Psychology, 41,* 346–349.

Isenberg, D. J. (1986). Group polarization: A critical review and meta-analysis. *Journal of Personality and Social Psychology, 50,* 1141–1151.

Isozaki, M. (1984). The effect of discussion on polarization of judgments. *Japanese Psychological Research, 26,* 187–193.

Jaccard, J. (1981). Toward theories of persuasion and belief change. *Journal of Personality and Social Psychology, 40,* 260–269.

Jackson, J. (1954). The adjustment of the family to the crisis of alcoholism. *Quarterly Journal of Studies on Alcohol, 15,* 564–586.

Jackson, J. (1965). Structural characteristics of norms. In I. D. Steiner & M. Fishbein (Eds.), *Current studies in social psychology* (pp. 301–309). New York: Holt, Rinehart and Winston.

Jackson, J. M. (1986). In defense of social impact theory: Comment on Mullen. *Journal of Personality and Social Psychology, 50* (3), 511–513.

Jackson, J. M. (1987). Social impact theory: A social forces model of influence. In B. Mullen & G. R. Goethals (Eds.), *Theories of group behavior* (pp. 111–124). New York: Springer-Verlag.

Jackson, L. A., Hunter, J. E., & Hodge, C. N. (1995). Physical attractiveness and intellectual competence: A meta-analytic review. *Social Psychology Quarterly, 58,* 108–122.

Jackson, P. B. (1997). Role occupancy and minority mental health. *Journal of Health and Social Behavior, 38,* 237–255.

Jacques, J. M., & Chason, K. (1977). Self-esteem and low status groups: A changing scene? *Sociological Quarterly, 18,* 399–412.

Jahoda, M., Deutsch, M., & S. W. Cook (Eds.). (1951). *Research methods in social relations* (Vol. 2). New York: Holt, Rinehart and Winston.

James, W. (1890). *Principles of psychology.* New York: Holt, Rinehart and Winston.

Janis, I. L. (1982). *Groupthink* (2nd ed.). Boston: Houghton Mifflin.

Janis, I. L., & Mann, L. (1977). *Decision making: A psychological analysis of conflict, choice, and commitment.* New York: Free Press.

Jellison, J. M., & Riskind, J. (1970). A social comparison of abilities interpretation of risk-taking behavior. *Journal of Personality and Social Psychology, 15,* 375–390.

Jemmott, J. B., & Magloire, K. (1988). Academic stress, social support and secretory immunoglobulin A. *Journal of Personality and Social Psychology, 55,* 803–810.

Jenkins, C., Rosenman, R., & Zyzanski, S. (1974). Prediction of clinical coronary heart disease by a test for coronary prone behavior pattern. *New England Journal of Medicine, 290,* 1271–1275.

Jenkins, J. C. (1983). Resource mobilization theory and the study of social movements. In R. H. Turner & J. F. Short (Eds.), *Annual review of sociology* (Vol. 9). Palo Alto, CA: Annual Reviews.

Jenkins, J. C., & Eckert, C. (1986). Channeling black insurgency: Elite patronage and professional social movement organizations in the development of the black movement. *American Sociological Review, 51,* 812–829.

Jenkins, J. C., & Perrow, C. (1977). Insurgency of the powerless: Farm worker movements 1946–1972. *American Sociological Review, 42,* 249–268.

Jensen, G., Erickson, M., & Gibbs, J. (1978). Perceived risk of punishment and self-reported delinquency. *Social Forces, 57,* 57–78.

Jessor, R., Costa, F., Jessor, L., & Donovan, J. (1983). Time of first intercourse: A prospective study. *Journal of Personality and Social Psychology, 44,* 608–626.

Job, R. F. S. (1988). Effective and ineffective use of fear in health promotion campaigns. *American Journal of Public Health, 78,* 163–167.

Johnson, B. T. (1991). Insight about attitudes: Meta-analytic perspectives. *Personality and Social Psychology Bulletin, 17,* 289–299.

Johnson, B. T., & Eagly, A. H. (1989). Effects of involvement on persuasion: A meta-analysis. *Psychological Bulletin, 106,* 290–314.

Johnson, C. (1994). Gender, legitimate authority and leader-subordinate conversations. *American Sociological Review, 59,* 122–135.

Johnson, D. J., & Rusbult, C. E. (1989). Resisting temptation: Devaluation of alternative partners as a means of maintaining commitment in close relationships. *Journal of Personality and Social Psychology, 57,* 967–980.

Johnson, D. L., & Andrews, I. R. (1971). The risky-shift hypothesis tested with consumer products as stimuli. *Journal of Personality and Social Psychology, 20,* 382–385.

Johnson, K. J., Lund, D. A., & Dimond, M. F. (1986). Stress, self-esteem, and coping during bereavement among the elderly. *Social Psychology Quarterly, 49,* 273–279.

Johnson, M., & Leslie, L. (1982). Couple involvement and network structure: A test of the dyadic withdrawal hypothesis. *Social Psychology Quarterly, 45,* 34–43.

Johnson, M. M., Stockard, J., Acker, J., & Naffziger, G. (1975). Expressiveness reevaluated. *School Review, 83,* 617–644.

Johnson, N., & Feinberg, W. (1977). A computer simulation of the emergence of consensus in crowds. *American Sociological Review, 52,* 505–521.

Johnson, N. R. (1987). Panic at "the Who concert stampede": An empirical assessment. *Social Problems, 34,* 362–373.

Johnson, R. E. (1979). *Juvenile delinquency and its origins.* New York: Cambridge University Press.

Johnson-George, C., & Swap, W. (1982). Measurement of specific interpersonal trust: Construction and validation of a scale to assess trust in a specific other. *Journal of Personality and Social Psychology, 43,* 1306–1317.

Jones, E. E. (1964). *Ingratiation.* New York: Appleton-Century-Crofts.

Jones, E. E. (1979). The rocky road from acts to dispositions. *American Psychologist, 34,* 107–117.

Jones, E. E. (1985). Major developments in social psychology during the past five decades. In G. Lindzey & E. Aronson (Eds.), *Handbook of social psychology* (3rd ed., Vol. I, pp. 47–107). New York: Random House.

Jones, E. E., & Davis, K. E. (1965). From acts to dispositions. In L. Berkowitz (Ed.), *Advances in experimental social psychology* (Vol. 2). New York: Academic Press.

Jones, E. E., Davis, K. E., & Gergen, K. J. (1961). Role playing variations and their informational value for person perception. *Journal of Abnormal and Social Psychology, 63,* 302–310.

Jones, E. E., Farina, A., Hastorf, A. H., Markus, H., Miller, D. T., & Scott, R. A. (1984). *Social stigma.* New York: Freeman.

Jones, E. E., Gergen, K. J., Gumpert, P., & Thibaut, J. (1965). Some conditions affecting the use of ingratiation to influence performance evaluation. *Journal of Personality and Social Psychology, 1,* 613–626.

Jones, E. E., & Goethals, G. R. (1971). *Order effects in impression formation: Attribution context and the nature of the entity.* Morristown, NJ: General Learning Press.

Jones, E. E., & Harris, V. A. (1967). The attribution of attitudes. *Journal of Experimental Social Psychology, 3,* 1–24.

Jones, E. E., & McGillis, D. (1976). Correspondent inferences and the attribution cube: A comparative reappraisal. In J. H. Harvey, W. J. Ickes, & R. F. Kidd (Eds.), *New directions in attribution research* (Vol. 1). Hillsdale, NJ: Erlbaum.

Jones, E. E., & Nisbett, R. (1972). The actor and observer: Divergent perceptions of the causes of behavior. In E. E. Jones et al. (Eds.), *Attribution: Perceiving the causes of behavior.* Morristown, NJ: General Learning Press.

Jones, E. E., & Pittman, T. S. (1982). Toward a general theory of strategic self-presentation. In J. Suls (Ed.), *Psychological perspectives on the self* (Vol. 1). Hillsdale, NJ: Erlbaum.

Jones, E. E., Rock, L., Shaver, K. G., Goethals, G. R., & Ward, L. M. (1968). Pattern of performance and ability attribution: An unexpected primacy effect. *Journal of Personality and Social Psychology, 10,* 317–340.

Jones, E. E., & Wortman, C. (1973). *Ingratiation: An attributional approach.* Morristown, NJ: General Learning Press.

Jones, V. C. (1948). *The Hatfields and the McCoys.* Chapel Hill: University of North Carolina Press.

Jones, W. H., Briggs, S. R., & Smith, T. G. (1986). Shyness: Conceptualization and measurement. *Journal of Personality and Social Psychology, 51,* 629–639.

Joreskog, K. G., & D. Sorbom, D. (1979). *Advances in factor analysis and structural equation models.* Cambridge, MA: Abt Books.

Joseph, N., & Alex, N. (1972). The uniform: A sociological perspective. *American Journal of Sociology, 77,* 719–730.

Jourard, S. M. (1971). *Self-disclosure.* New York: Wiley.

Judd, C. M., Drake, R. A., Downing, J. W., & Krosnick, J. A. (1991). Some dynamic properties of attitude structures: Context-induced response facilitation and polarization. *Journal of Personality and Social Psychology, 60,* 193–202.

Jussim, L., Coleman, L., & Nassau, S. (1987). The influence of self-esteem on perceptions of performance and feedback. *Social Psychology Quarterly, 50,* 95–99.

Kahn, A. S., Mathie, V. A., & Torgler, C. (1994). Rape scripts and rape acknowledgment. *Psychology of Women Quarterly, 18,* 53–66.

Kahne, M., & Schwartz, C. (1978). Negotiating trouble: The social construction and management of trouble in a psychiatric context. *Social Problems, 25,* 461–475.

Kalven, H., Jr., & Zeisel, H. (1966). *The American jury.* Boston: Little, Brown.

Kandel, D. (1978). Similarity in real-life adolescent friendship pairs. *Journal of Personality and Social Psychology, 36,* 306–312.

Kanter, R. M. (1976). *Men and women of the corporation.* New York: Basic Books.

Kaplan, H. B., Johnson, R., & Bailey, C. A. (1987). Deviant peers and deviant behavior: Further elaboration of a model. *Social Psychology Quarterly, 50,* 277–284.

Kaplan, H. B., Martin, S. S., & Johnson, R. J. (1986). Self-rejection and the explanation of deviance: Specification of the structure among latent constructs. *American Journal of Sociology, 92,* 384–411.

Kaplan, M. F. (1977). Discussion polarization effects in a modified jury discussion paradigm: Informational influences. *Sociometry, 40,* 262–271.

Kaplan, M. F. (1987). The influencing process in group decision-making. In C. Hendrick (Ed.), *Group processes.* Newbury Park, CA: Sage.

Kaplan, M. F., & Miller, C. E. (1987). Group decision making and normative versus informational influence: Effects of type of issue and assigned decision rule. *Journal of Personality and Social Psychology, 53,* 306–313.

Karabenick, S. A. (1983). Sex-relevance of content and influenceability. *Personality and Social Psychology Bulletin, 9,* 243–252.

Karasek, R., Theorell, T., Schwartz, J., Schnall, P., Pieper, C., & Michela, J. (1988). Job characteristics in relation to the prevalence of myocardial infarction in the US Health Examination Survey (HES) and the Health and Nutrition Examination Survey (HANES). *American Journal of Public Health, 78,* 910–918.

Karau, S. J., & Williams, K. D. (1993). Social loafing: A meta-analytic review and theoretical integration. *Journal of Personality and Social Psychology, 65,* 681–706.

Karau, S. J., & Williams, K. D. (1995). Social loafing: Research findings, implications, and future directions. *Current Directions in Psychological Science, 4*(5), 134–140.

Karlins, M., & Abelson, H. I. (1970). *How opinions and attitudes are changed* (2nd ed.). New York: Springer.

Karlins, M., Coffman, T. L., & Walters, G. (1969). On the fading of social stereotypes: Studies on three generations of college students. *Journal of Personality and Social Psychology, 13,* 1–16.

Katz, D. (1960). The functional approach to the study of attitudes. *Public Opinion Quarterly, 24,* 163–204.

Katz, I. (1970). Experimental study in Negro-white relationships. In L. Berkowitz (Ed.), *Advances in experimental social psychology* (Vol. 5). New York: Academic Press.

Katz, I. (1981). *Stigma: A social psychological analysis.* Hillsdale, NJ: Erlbaum.

Katz, I., Wackenhut, J., & Glass, D. C. (1986). An ambivalence-amplification theory of behavior toward the stigmatized. In S. Worchel & W. G. Austin (Eds.), *Psychology of intergroup relations* (2nd ed., pp. 103–117). Chicago: Nelson-Hall.

Katz, J. (1988). *Seductions of crime: Moral and sensual attractions in doing evil.* New York: Basic Books.

Kauffman, D. R., & Steiner, I. D. (1968). Some variables affecting the use of conformity as an ingratiation technique. *Journal of Experimental Social Psychology, 4,* 400–414.

Keller, C. E., Hallahan, D. P., McShane, E. A., Crowley, E. P., & Blandford, B. J. (1990). The coverage of persons with disabilities in American newspapers. *Journal of Special Education, 24,* 271–282.

Kelley, H. H. (1950). The warm-cold variable in first impressions. *Journal of Personality, 18,* 431–439.

Kelley, H. H. (1967). Attribution theory in social psychology. In D. Levine (Ed.), *Nebraska symposium in motivation 1967.* Lincoln: University of Nebraska Press.

Kelley, H. H. (1973). The process of causal attribution. *American Psychologist, 28,* 107–128.

Kelley, H. H., Berscheid, E., Christensen, A., Harvey, H. H., Huston, T., Levinger, G., McClintock, E., Peplau, L. H., & Peterson, D. R. (1983). *Close relationships.* New York: Freeman.

Kelley, H. H., Condry, J. C., Jr., Dahlke, A. E., & Hill, A. H. (1965). Collective behavior in a simulated panic situation. *Journal of Experimental Social Psychology, 1,* 20–54.

Kelley, H. H., & Michela, J. L. (1980). Attribution theory and research. *Annual Review of Psychology, 31,* 457–501.

Kelley, H. H., & Thibaut, J. W. (1978). *Interpersonal relations: A theory of interdependence.* New York: Wiley.

Kelley, K., & Byrne, D. (1976). Attraction and altruism: With a little help from my friends. *Journal of Research in Personality, 10,* 59–68.

Kelman, H. C. (1974). Attitudes are alive and well and gainfully employed in the sphere of action. *American Psychologist, 29,* 310–324.

Kelman, H. C., & Hamilton, V. L. (1989). *Crimes of obedience: Toward a social psychology of authority and responsibility.* New Haven, CT: Yale University Press.

Kemper, T. D. (1973). The fundamental dimensions of social relationship: A theoretical statement. *Acta Sociologica, 16,* 41–57.

Kendon, A. (1970). Movement coordination in social interaction. Some examples described. *Acta Psychologica, 32,* 100–125.

Kendon, A., Harris, R. M., & Key, M. R. (1975). *Organization of behavior in face-to-face interaction.* The Hague: Mouton.

Kendzierski, D., & Whitaker, D. J. (1997). The role of self-schema in linking intention with behavior. *Personality and Social Psychology Bulletin, 23,* 139–147.

Kenney, D., & La Voie, L. (1982). Reciprocity of interpersonal attraction: A confirmed hypothesis. *Social Psychology Quarterly, 45,* 54–58.

Kenrick, D. T., McCreath, H. E., Govern, J., King, R., & Bordin, J. (1990). Person-environment intersections: Everyday settings and common trait dimensions. *Journal of Personality and Social Psychology, 58,* 685–698.

Kent, G., Davis, J., & Shapiro, D. (1978). Resources required in the construction and reconstruction of conversations. *Journal of Personality and Social Psychology, 36,* 13–22.

Kerber, K. W. (1984). The perception of nonemergency helping situations: Costs, rewards, and the altruistic personality. *Journal of Personality, 52,* 177–187.

Kerckhoff, A. C. (1974). The social context of interpersonal attraction. In T. Huston (Ed.), *Foundations of interpersonal attraction* (pp. 61–78). New York: Academic Press.

Kerckhoff, A. C., Back, K. W., & Miller, N. (1965). Sociometric patterns in hysterical contagion. *Sociometry, 28,* 2–15.

Kerr, N., & Bruun, S. (1981). Ringelmann revisited: Alternative explanations for the social loafing effect. *Personality and Social Psychology Bulletin, 7,* 224–231.

Kerr, N. L., & Bruun, S. (1983). The dispensability of member effort and group motivation losses: Free rider effects. *Journal of Personality and Social Psychology, 44,* 78–94.

Kessler, R. C. (1982). A disaggregation of the relationship between socioeconomic status and psychological distress. *American Sociological Review, 47,* 752–764.

Kessler, R. C., & Cleary, P. (1980). Social class and psychological distress. *American Sociological Review, 45,* 463–478.

Kessler, R. C., & McCrae, J., Jr. (1982). The effects of wives' employment on mental health of married men and women. *American Sociological Review, 47,* 216–226.

Kiesler, C. A., & Corbin, L. H. (1965). Commitment, attraction, and conformity. *Journal of Personality and Social Psychology, 2,* 890–895.

Kiesler, C. A., & Kiesler, S. B. (1969). *Conformity.* Reading, MA: Addison-Wesley.

Kiesler, C. A., Zanna, M., & DeSalvo, J. (1966). Deviation and conformity: Opinion change as a function of commitment, attraction, and the presence of a deviate. *Journal of Personality and Social Psychology, 3,* 458–467.

Kilham, W., & Mann, L. (1974). Level of destructive obedience as a function of transmitter and executant roles in the Milgram obedience paradigm. *Journal of Personality and Social Psychology, 29,* 696–702.

Killian, L. (1984). Organization, rationality and spontaneity in the civil rights movement. *American Sociological Review, 49,* 770–783.

Kimberly, J. C. (1984). Cognitive balance, inequality, and consensus: Interrelations among fundamental processes in groups. In E. J. Lawler (Ed.), *Advances in group processes* (Vol. 1). Greenwich, CT: JAI Press.

Kingston, P., & Nock, S. (1987). Time together among dual-earner couples. *American Sociological Review, 52,* 391–400.

Kiparsky, P. (1976). Historical linguistics and the origin of language. *Annals of the New York Academy of Sciences, 280,* 97–103.

Kirchmeyer, C. (1993). Nonwork-to-work spillover: A more balanced view of the experiences and coping of professional men and women. *Sex Roles, 28,* 531–552.

Kirk, R. E. (1982). *Experimental design* (2nd ed.). Belmont, CA: Brooks/Cole.

Kitayama, S., & Burnstein, E. (1988). Automaticity in conversations: A reexamination of the mindlessness hypothesis. *Journal of Personality and Social Psychology, 54,* 219–224.

Kitcher, P. (1985). *Vaulting ambition.* Cambridge, MA: MIT Press.

Kitsuse, J. (1964). Societal reaction to deviant behavior: Problems of theory and method. In H. Becker (Ed.), *The other side* (pp. 87–102). New York: Free Press.

Kleck, R. E. (1968). Physical stigma and nonverbal cues emitted in face-to-face interaction. *Human Relations, 21,* 19–28.

Kleck, R. E., & Strenta, A. (1980). Perceptions of the impact of negatively valued physical characteristics on social interaction. *Journal of Personality and Social Psychology, 39,* 861–873.

Kleidman, R. (1994). Volunteer activism and professionalism in social movement organizations. *Social Problems, 41,* 257–276.

Kleinke, C. L., & Kahn, M. L. (1980). Perceptions of self-disclosers: Effects of sex and physical attractiveness. *Journal of Personality, 48,* 190–205.

Klitzner, M., Gruenewald, P. J., & Bamberger, E. (1991). Cigarette advertising and adolescent experimentation with smoking. *British Journal of Addiction, 86,* 287–298.

Knapp, M. L., Stafford, L., & Daly, J. A. (1986). Regrettable messages: Things people wish they hadn't said. *Journal of Communication, 36,* 40–58.

Knottnerus, J. D. (1988). A critique of expectation states theory: Theoretical assumptions and models of social cognition. *Sociological Perspectives, 31,* 420–445.

Koestner, R., & Wheeler, L. (1988). Self-presentation in personal advertisements: The influence of implicit notions of attraction and role expectations. *Journal of Social and Personal Relationships, 5,* 149–160.

Kogan, N., & Wallach, M. (1964). *Risk taking: A study in cognition and personality.* New York: Holt, Rinehart and Winston.

Kohlberg, L. (1969). Stage and sequence: The cognitive-developmental approach to socialization. In D. Goslin (Ed.), *Handbook of socialization theory and research* (pp. 347–480). Chicago: Rand McNally.

Kohn, M. (1969). *Class and conformity: A study in values.* Homewood, IL: Dorsey.

Kohn, M. (1976). Occupational structure and alienation. *American Journal of Sociology, 82,* 111–130.

Kohn, M. (1977). Reassessment, 1977. *Class and conformity: A study in values* (2nd ed.). Chicago: University of Chicago Press.

Kohn, M., Naoi, A., Schoenbach, C., Schooler, C., & Slomczynski, K. M. (1990). Position in the class structure and psychological functioning in the United States, Japan and Poland. *American Journal of Sociology, 95,* 964–1008.

Kohn, M., & Schooler, C. (1973). Occupational experience and psychological functioning: An assessment of reciprocal effects. *American Sociological Review, 38,* 97–118.

Kohn, M., & Schooler, C. (1982). Job conditions and personality: A longitudinal assessment of their reciprocal effects. *American Journal of Sociology, 87,* 1257–1286.

Kohn, M., Schooler, C., with the collaboration of Miller, J., Miller, K., Schoenbach, S., & Schoenberg, R. (1983). *Work and personality: An inquiry into the impact of social stratification.* Norwood, NJ: Ablex.

Kollock, P., Blumstein, P., & Schwartz, P. (1985). Sex and power in interaction: Conversational privileges and duties. *American Sociological Review, 50,* 34–46.

Konecni, V. J. (1975). The mediation of aggressive behavior: Arousal level versus anger and cognitive labeling. *Journal of Personality and Social Psychology, 32,* 706–712.

Konecni, V. J. (1979). The role of aversive events in the development of intergroup conflict. In W. G. Austin & S. Worchel (Eds.), *The social psychology of intergroup relations.* Monterey, CA: Brooks/Cole.

Korte, C., Ypma, I., & Toppen, A. (1975). Helpfulness in Dutch society as a function of urbanization and environmental input level. *Journal of Personality and Social Psychology, 32,* 996–1003.

Korten, D. C. (1962). Situational determinants of leadership structure. *Journal of Conflict Resolution, 6,* 222–235.

Kortenhaus, C. M., & Demarest, J. (1993). Gender role stereotyping in children's literature: An update. *Sex Roles, 28,* 219–232.

Koskinen, S., & Martelin, T. (1994). Why are socioeconomic mortality differences smaller among women than among men? *Social Science and Medicine, 38,* 1385–1396.

Koss, M. P., & Leonard, K. E. (1984). Sexually aggressive men: Empirical findings and theoretical implications. In N. M. Malamuth & E. Donnerstein (Eds.), *Pornography and sexual aggression* (pp. 211–232). Orlando, FL: Academic Press.

Koss, M. P., & Oros, C. J. (1982). Sexual experiences survey: A research instrument investigating sexual aggression and victimization. *Journal of Consulting and Clinical Psychology, 50,* 455–457.

Kothandapani, V. (1971). Validation of feeling, belief and intention to act as three components of attitude and their contribution to prediction of contraceptive behavior. *Journal of Personality and Social Psychology, 19,* 321–333.

Kozielecki, J. (1984). Rodzaje samoprezentacji. ("Types of self-presentation.") *Psychologia Wychowawcza, 27,* 129–137. [Polish]

Kraus, R. M., & Fussell, S. K. (1996). Social psychological models of interpersonal communication. In E. T. Higgins & A. Kruglanski (Eds.), *Social psychology: Handbook of basic principles* (pp. 655–701). New York: Guilford.

Kraus, S. (Ed.). (1962). *The great debates.* Bloomington: Indiana University Press.

Krauss, R. M., Morrel-Samuels, P., & Colasante, C. (1991). Do conversational hand gestures communicate? *Journal of Personality and Social Psychology, 61,* 743–754.

Kraut, R. E. (1980). Humans as lie-detectors: Some second thoughts. *Journal of Communication, 30,* 209–216.

Kraut, R. E., Lewis, S. H., & Swezey, L. W. (1982). Listener responsiveness and the coordination of conversation. *Journal of Personality and Social Psychology, 43,* 718–731.

Kraut, R. E., & Poe, D. (1980). Behavior roots of person perception: The deception judgments of customs inspectors and laymen. *Journal of Personality and Social Psychology, 39,* 784–798.

Kravitz, D. A., & Martin, B. (1986). Ringelmann rediscovered: The original article. *Journal of Personality and Social Psychology, 50,* 936–941.

Krebs, D. L. (1975). Empathy and altruism. *Journal of Personality and Social Psychology, 32,* 1134–1146.

Krebs, D. L. (1982). Psychological approaches to altruism: An evaluation. *Ethics, 92,* 147–158.

Krebs, D. L., & Miller, D. T. (1985). Altruism and aggression. In G. Lindzey & E. Aronson (Eds.), *Handbook of social psychology* (3rd ed., Vol. 2, pp. 1–71). New York: Random House.

Krebs, R. (1967). *Some relations between moral judgement, attention and resistance to temptation.* Unpublished doctoral dissertation, University of Chicago.

Kressel, K., & Pruitt, D. G. (1985). Themes in the mediation of social conflict. *Journal of Social Issues, 41,* 179–198.

Kriesberg, L. (1973). *The sociology of social conflicts.* Englewood Cliffs, NJ: Prentice-Hall.

Kristiansen, C. M., & Harding, C. M. (1988). The comparison of coverage of health issues by Britain's quality and popular press. *Social Behaviour, 3,* 25–32.

Kritzer, H. (1977). Political protest and political violence: A non-recursive causal model. *Social Forces, 55,* 630–640.

Krohn, M. D. (1986). The web of conformity: A network approach to the explanation of delinquent behavior. *Social Problems, 33,* 581–593.

Krohn, M. D., Massey, J. L., & Zielinski, M. (1988). Role overlap, network multiplexity and adolescent deviant behavior. *Social Psychology Quarterly, 51,* 346–356.

Krosnick, J. A., & Alwin, D. F. (1989). Aging and susceptibility to attitude change. *Journal of Personality and Social Psychology, 57,* 416–425.

Krosnick, J. A., & Schuman, H. (1988). Attitude intensity, importance, and certainty and susceptibility to response effects. *Journal of Personality and Social Psychology, 54,* 940–952.

Kruglanski, A. W., & Mackie, D. M. (1990). Majority and minority influence: A judgmental process analysis. In W. Stroebe & M. Hewstone (Eds.), *European review of social psychology* (Vol. 1, pp. 229–261). Chichester, England: Wiley.

Kuhlman, C. E., Miller, M. H., & Gungor, E. (1973). Interpersonal conflict resolutions: The effects of language and meaning. In L. Rappoport & D. A. Summers (Eds.), *Human judgement and social interaction.* New York: Holt, Rinehart and Winston.

Kuhn, D., Langer, J., Kohlberg, L., & Haan, N. (1977). The development of formal operations in logical and moral judgement. *Genetic Psychology Monographs, 95,* 97–188.

Kuhn, M. H., & McPartland, T. (1954). An empirical investigation of self-attitudes. *American Sociological Review, 19,* 68–76.

Kulick, J. A., & Brown, R. (1979). Frustration, attribution of blame, and aggression. *Journal of Experimental Social Psychology, 15,* 183–194.

Kunkel, D. (1997). *Policy implications of the National Television Violence Study.* Presented at the American Psychological Association meetings, Chicago.

Kurdek, L. A. (1991). The relation between reported well-being and divorce history, availability of proximate adult, and gender. *Journal of Marriage and the Family, 53,* 71–78.

Kurmeyer, S. L., & Biggers, K. (1988). Environmental demand and demand engendering behavior: An observational analysis of the Type A pattern. *Journal of Personality and Social Psychology, 54,* 997–1005.

Kurtines, M. M. (1986). Moral behavior as rule governed behavior: Person and situation effects on moral decision-making. *Journal of Personality and Social Psychology, 50,* 784–791.

Labov, W. (1972). *Sociolinguistic patterns.* Philadelphia: University of Pennsylvania Press.

Lachman, S. J., & Bass, A. R. (1985). A direct study of halo effect. *Journal of Psychology, 119,* 535–540.

LaFrance, M., & Mayo, C. (1978). *Moving bodies: Nonverbal communication in social relationships.* Monterey, CA: Brooks/Cole.

LaFree, G., & Drass, K. A. (1996). The effect of changes in intra-racial income inequality and educational attainment on changes in arrest rates for African Americans and whites, 1957–1990. *American Sociological Review, 61,* 614–634.

Lakoff, R. T. (1979). Women's language. In O. Buturff & E. L. Epstein (Eds.), *Women's language and style.* Akron, OH: University of Akron.

Lamb, M. E. (1979). Paternal influence and the father's role: A personal perspective. *American Psychologist, 34,* 938–993.

Lamb, M. E. (1982). Maternal employment and child development: A review. In M. E. Lamb (Ed.), *Nontraditional families: Parenting and child development.* Hillsdale, NJ: Erlbaum.

Lamm, H. (1988). A review of our research on group polarization: Eleven experiments on the effects of group discussion on risk acceptance, probability estimation, and negotiation positions. *Psychological Reports, 62,* 807–813.

Lamm, H., & Sauer, C. (1974). Discussion-induced shift toward higher demands in negotiation. *European Journal of Social Psychology, 4,* 85–88.

Lamm, H., & Schwinger, T. (1980). Norms concerning distributive justice: Are needs taken into consideration in allocation decisions? *Social Psychology Quarterly, 43,* 425–429.

Lamm, H., & Trommsdorff, G. (1973). Group versus individual performance on tasks requiring ideational proficiency (brainstorming): A review. *European Journal of Social Psychology, 3,* 361–388.

Land, K. C., McCall, P. L., & Cohen, L. E. (1990). Structural correlates of homicide rates: Are there invariances across time and social space? *American Journal of Sociology, 95,* 922–963.

Landy, D., & Sigall, H. (1974). Beauty is talent: Task evaluation as a function of the performer's physical attractiveness. *Journal of Personality and Social Psychology, 29,* 299–304.

Langner, T. S. (1963). A twenty-two item screening score of psychiatric symptoms indicating impairment. *Journal of Health and Human Behavior, 3,* 269–226.

Langner, T. S., & Michael, S. (1963). *Life stress and mental health: The midtown Manhattan study.* New York: Free Press.

Lantz, D., & Stefflre, V. (1964). Language and cognition revisited. *Journal of Abnormal and Social Psychology, 69,* 472–481.

Lantz, H., Keyes, J., & Schultz, M. (1975). The American family in the preindustrial period: From base lines in history to change. *American Sociological Review, 40,* 21–36.

Lantz, H., Schultz, M., & O'Hara, M. (1977). The changing American family from the preindustrial to the industrial period: A final report. *American Sociological Review, 42,* 406–421.

LaPiere, R. (1934). Attitudes versus actions. *Social Forces 13,* 230–237.

Larson, L. L., & Rowland, K. (1973). Leadership style, stress, and behavior in task performance. *Organizational Behavior and Human Performance, 9,* 407–421.

Larzelere, R., & Huston, T. (1980). The dyadic trust scale: Toward understanding interpersonal trust in close relationships. *Journal of Marriage and the Family, 42,* 595–604.

Latané, B. (1981). The psychology of social impact. *American Psychologist, 36,* 343–356.

Latané, B., & Darley, J. M. (1970). *The unresponsive bystander: Why doesn't he help?* New York: Appleton-Century-Crofts.

Latané, B., & Nida, S. (1981). Ten years of research on group size and helping. *Psychological Bulletin, 89*(2), 308–324.

Latané, B., Nida, S. A., & Wilson, D. W. (1981). The effects of group size on helping behavior. In J. P. Rushton & R. M. Sorrentino (Eds.), *Altruism and helping behavior: Social, personality and developmental perspectives.* Hillsdale, NJ: Erlbaum.

Latané, B., & Rodin, J. (1969). A lady in distress: Inhibiting effects of friends and strangers on bystander intervention. *Journal of Experimental Social Psychology, 5,* 189–202.

Latané, B., Williams, K., & Harkins, S. (1979). Many hands make light the work: The causes and consequences of social loafing. *Journal of Personality and Social Psychology, 37,* 822–832.

Latané, B., & Wolf, S. (1981). The social impact of majorities and minorities. *Psychological Review, 88*, 438–453.

Lau, R. R. (1989). Individual and contextual influences on group identification. *Social Psychology Quarterly, 52*, 220–231.

Lau, R. R., & Russell, D. (1980). Attributions in the sports pages. *Journal of Personality and Social Psychology, 39*, 29–38.

Lauderdale, P. (1976). Deviance and moral boundaries. *American Sociological Review, 41*, 660–676.

Laufer, R., & Gallups, M. (1985). Life course effects of Vietnam combat and abusive violence: Marital patterns. *Journal of Marriage and the Family, 47*, 839–853.

Laughlin, P. R. (1980). Social combination processes of cooperative, problem-solving groups as verbal intellective tasks. In M. Fishbein (Ed.), *Progress in social psychology* (Vol. 1). Hillsdale, NJ: Erlbaum.

Laumann, E. O., Gagnon, J. H., Michael, R. T., & Michaels, S. (1994). *The social organization of sexuality.* Chicago: University of Chicago Press.

Lavine, H., Thomsen, C. J., & Gonzales, M. H. (1997). The development of interattitudinal consistency: The shared-consequences model. *Journal of Personality and Social Psychology, 72*, 735–749.

Lawler, E. E., & O'Gara, P. W. (1967). Effects of inequity produced by underpayment on work output, work quality, and attitudes toward work. *Journal of Applied Psychology, 51*, 403–410.

Lawler, E. J. (1975a). An experimental study of factors affecting the mobilization of revolutionary coalitions. *Sociometry, 38*, 163–179.

Lawler, E. J. (1975b). The impact of status differences on coalitional agreements: An experimental study. *Journal of Conflict Resolution, 19*, 271–285.

Lawler, E. J. (1983). Cooptation and threats as "divide and rule" tactics. *Social Psychology Quarterly, 46*, 89–98.

Lawler, E. J. (1986). Bilateral deterrence and conflict spiral: A theoretical analysis. In E. J. Lawler (Ed.), *Advances in group processes* (Vol. 3, pp. 107–130). Greenwich, CT: JAI Press.

Lawler, E. J., & Bacharach, S. B. (1987). Comparison of dependence and punitive forms of power. *Social Forces, 66*, 446–462.

Lawler, E. J., & Ford, R. (1995). Bargaining and influence in conflict situations. In K. S. Cook, G. A. Fine, & J. S. House (Eds.), *Sociological perspectives on social psychology* (pp. 236–256). Boston: Allyn & Bacon.

Lawler, E. J., Ford, R. S., & Blegen, M. A. (1988). Coercive capability in conflict: A test of bilateral deterrence versus conflict spiral theory. *Social Psychology Quarterly, 51*, 93–107.

Lawler, E. J., Youngs, J. A., Jr., & Lesh, M. D. (1978). Cooptation and coalition mobilization. *Journal of Applied Social Psychology, 8*, 199–214.

Lawson, E. B. (1964). Reinforced and non-reinforced four-man communication nets. *Psychological Reports, 14*, 287–296.

Leana, C. R. (1985). A partial test of Janis' groupthink model: Effects of group cohesiveness and leader behavior on defective decision making. *Journal of Management, 11*, 5–17.

Leary, M. R. (1995). *Self-presentation: Impression management and interpersonal behavior.* Madison, WI: Brown & Benchmark.

Leary, M. R., & Kowalski, R. M. (1990). Impression management: A literature review and two-component model. *Psychological Bulletin, 107*, 34–47.

Leary, M. R., Wheeler, D. S., & Jenkins, T. B. (1986). Aspects of identity and behavioral preference: Studies of occupational and recreational choice. *Social Psychology Quarterly, 49*, 11–18.

Leavitt, H. J. (1951). Some effects of certain communication patterns on group performance. *Journal of Abnormal and Social Psychology, 46*, 38–50.

Le Bon, G. (1895). *Psychologie des foules* (The Crowd). London: Unwin (1903).

Ledvinka, J. (1971). Race of interviewer and the language elaboration of black interviewees. *Journal of Social Issues, 27*, 185–197.

Lee, G. R., Seccombe, K., & Shehan, C. L. (1991). Marital status and personal happiness: An analysis of trend data. *Journal of Marriage and the Family, 53*, 839–844.

Lee, T. R., Mancini, J. A., & Maxwell, J. W. (1990). Sibling relationships in adulthood: Contact patterns and motivations. *Journal of Marriage and the Family, 52*, 431–440.

Leffler, A., Gillespie, D. L., & Conaty, J. C. (1982). The effects of status differentiation on nonverbal behavior. *Social Psychology Quarterly, 45*, 153–161.

Leigh, J. P., Markowitz, S. B., Fahs, M., Shin, C., & Landrigan, P. J. (1997). Occupational injury and illness in the United States: Estimates of costs, morbidity and mortality. *Archives of Internal Medicine, 157*, 1557–1568.

Leippe, M. R., & Elkin, R. A. (1987). When motives clash: Issue involvement and response involvement as determinants of persuasion. *Journal of Personality and Social Psychology, 52*, 269–278.

Lemert, E. (1951). *Social pathology.* New York: McGraw-Hill.

Lemert, E. (1962). Paranoia and the dynamics of exclusion. *Sociometry, 25*, 2–20.

Lemon, N. (1973). *Attitudes and their measurement.* New York: Wiley.

Lennon, M. C., Link, B. G., Marbach, J. J., & Dohrendwend, B. P. (1989). The stigma of chronic facial pain and its impact on social relationships. *Social Problems, 36*, 117–134.

Lepper, M., Greene, D., & Nisbett, R. (1973). Undermining children's intrinsic interest with extrinsic reward: A test of the "overjustification" hypothesis. *Journal of Personality and Social Psychology, 28*, 129–137.

Leventhal, G. S. (1979). Effects of external conflict on resource allocation and fairness within groups and organizations. In W. G. Austin & S. Worchel (Eds.), *The social psychology of intergroup relations.* Monterey, CA: Brooks/Cole.

Leventhal, G. S., & Lane, D. W. (1970). Sex, age, and equity behavior. *Journal of Personality and Social Psychology, 15*, 312–316.

Leventhal, G. S., Michaels, J. W., & Sanford, C. (1972). Inequity and interpersonal conflict: Reward allocation and secrecy about reward as methods of preventing conflict. *Journal of Personality and Social Psychology, 23*, 88–102.

Leventhal, G. S., Weiss, T. & Long, G. (1969). Equity, reciprocity, and reallocating the rewards in the dyad. *Journal of Personality and Social Psychology, 13*, 300–305.

Leventhal, H. (1970). Findings and theory in the study of fear communications. In L. Berkowitz (Ed.), *Advances in experimental social psychology* (Vol. 5). New York: Academic Press.

Leventhal, H. (1980). Toward a comprehensive theory of emotion. In L. Berkowitz (Ed.), *Advances in experimental social psychology* (Vol. 13). New York: Academic Press.

Leventhal, H. (1984). A perceptual motor theory of emotion. In K. R. Scherer & P. Ekman (Eds.), *Approaches to emotion.* Hillsdale, NJ: Erlbaum.

Leventhal, H., & Singer, R. P. (1966). Affect arousal and positioning of recommendations in persuasive communications. *Journal of Personality and Social Psychology, 4,* 137–146.

Levin, J., & Arluke, A. (1982). Embarrassment and helping behavior. *Psychological Reports, 51,* 999–1002.

Levin, P. E., & Isen, A. M. (1975). Something you can still get for a dime: Further studies on the effects of feeling good on helping. *Sociometry, 38,* 141–147.

Levine, J. M., Saxe, L., & Ranelli, C. J. (1975). Extreme dissent, conformity reduction, and the bases of social influence. *Social Behavior and Personality, 3,* 117–126.

LeVine, R. A., & Campbell, D. T. (1972). *Ethnocentrism: Theories of conflict, ethnic attitudes and group behavior.* New York: Wiley.

Levinger, G. (1974). A three-level approach to attraction: Toward an understanding of pair relatedness. In T. Huston (Ed.), *Foundations of interpersonal attraction.* New York: Academic Press.

Levinger, G. (1976). A social psychological perspective on marital dissolution. *Journal of Social Issues, 32*(1), 21–47.

Levinson, D. (1978). *The seasons of a man's life.* New York: Knopf.

Levinson, R., Powell, B., & Steelman, L. C. (1986). Social location, significant others and body image among adolescents. *Social Psychology Quarterly, 49,* 330–337.

Levitin, T. E. (1975). Deviants are active participants in the labelling process: The visibly handicapped. *Social Problems, 22,* 548–557.

Lewicki, P. A. (1982). Social psychology as viewed by its practitioners: Survey of SESP members' opinions. *Personality and Social Psychology Bulletin, 8,* 409–416.

Lewin, D., Feuille, P., & Kochan, T. (1977). *Public sector labor relations: Analysis and readings.* Glen Springs, NJ: Thomas Horton and Daughters.

Lewis, G. H. (1972). Role differentiation. *American Sociological Review, 37,* 424–434.

Lewis, M., & Brookes-Gunn, J. (1979). Toward a theory of social cognition: The development of self. In I. Uzgiris (Ed.), *Social interaction and communication during infancy. New directions for child development* (Vol. 4). San Francisco: Jossey Bass.

Lewis, S. A., Langan, C. J., & Hollander, E. P. (1972). Expectation of future interaction and the choice of less desirable alternatives in conformity. *Sociometry, 35,* 440–447.

Lex, B. W. (1986). Measurement of alcohol consumption in fieldwork settings. *Medical Anthropology Quarterly, 17,* 95–98.

Liberman, A., & Chaiken, S. (1992). Defensive processing of personally relevant health messages. *Personality and Social Psychology Bulletin, 18,* 669–679.

Lichterman, P. (1995). Piecing together multicultural community: Cultural differences in community building among grass-roots environmentalists. *Social Problems, 42,* 513–534.

Lieberman, A., & Chaiken, S. (1996). The direct effect of personal relevance on attitudes. *Personality and Social Psychology Bulletin, 22,* 269–279.

Lieberman, P. (1975). *On the origins of human language: An introduction to the evolution of human speech.* New York: Macmillan.

Lieberman, S. (1965). The effect of changes of roles on the attitudes of role occupants. In H. Proshansky & B. Seidenberg (Eds.), *Basic studies in social psychology* (pp. 485–494). New York: Holt, Rinehart and Winston.

Likert, R. (1932). A technique for the measurement of attitudes. *Archives of Psychology* (Whole no. 142).

Lin, N. (1974-1975). The McIntire march: A study of recruitment and commitment. *Public Opinion Quarterly, 38,* 562–573.

Lin, N., Ensel, W., & Vaughn, J. (1981). Social resources and strength of ties: Structural factors in occupational status attainment. *American Sociological Review, 46,* 393–405.

Lin, N., & Xie, W. (1988). Occupational prestige in urban China. *American Journal of Sociology, 93,* 793–832.

Lincoln, J., & Kalleberg, A. (1985). Work organization and workforce commitment: A study of plants and employees in the United States and Japan. *American Sociological Review, 50,* 738–760.

Linder, D. E., Cooper, J., & Jones, E. (1967). Decision freedom as a determinant of the role of incentive magnitude in attitude change. *Journal of Personality and Social Psychology, 6,* 245–254.

Lindskold, S. (1986). GRIT: Reducing distrust through carefully introduced conciliation. In S. Worchel & W. G. Austin (Eds.), *Psychology of intergroup relations* (2nd ed., pp. 305–323). Chicago: Nelson-Hall.

Lindskold, S., & Aronoff, J. R. (1980). Conciliatory strategies and relative power. *Journal of Experimental Social Psychology, 16,* 187–198.

Lindskold, S., Cullen, P., Gahagen, J., & Tedeschi, J. T. (1970). Developmental aspects of reaction to positive inducements. *Developmental Psychology, 3,* 277–284.

Lindskold, S., & Finch, M. L. (1981). Styles of announcing conciliation. *Journal of Conflict Resolution, 25,* 145–155.

Lindskold, S., Han, G., & Betz, B. (1986). The essential elements of communication in the GRIT strategy. *Personality and Social Psychology Bulletin, 12,* 179–186.

Lindskold, S., & Tedeschi, J. T. (1971). Reward power and attraction in interpersonal conflict. *Psychonomic Science, 22,* 211–213.

Link, B. G. (1982). Mental patient status, work and income: An examination of the effects of a psychiatric label. *American Sociological Review, 47,* 202–215.

Link, B. G. (1987). Understanding labelling effects in the area of mental disorders: An assessment of the effects of expectations of rejection. *American Sociological Review, 52,* 96–112.

Link, B. G., Cullen, F. T., Struening, E., Shrout, P. E., & Dohrenwend, B. (1989). A modified labeling theory approach to mental disorders: An empirical assessment. *American Sociological Review, 54,* 400–423.

Link, B. G., Struening, E. L., Rahav, M., Phelan, J. C., & Nuttbrock, L. (1997). On stigma and its consequences: Evidence from a longitudinal study of men with dual diagnosis of mental illness and substance abuse. *Journal of Health and Social Behavior, 38,* 177–190.

Linville, P. W. (1982). The complexity-extremity effect and age-based stereotyping. *Journal of Personality and Social Psychology, 42,* 193–211.

Linville, P. W., Fischer, G. W., & Salovey, P. (1989). Perceived distributions of the characteristics of in-group and out-group members: Empirical evidence and a computer simulation. *Journal of Personality and Social Psychology, 57*(2), 165–188.

Linville, P. W., & Jones, E. E. (1980). Polarized appraisals of out-group members. *Journal of Personality and Social Psychology, 38,* 689–703.

Lipe, M. G. (1991). Counterfactual reasoning as a framework for attribution theories. *Psychological Bulletin, 109,* 456–471.

Lipscomb, T. J., Larrieu, J. A., McAllister, H. A., & Bregman, N. J. (1982). Modeling and children's generosity: A developmental perspective. *Merrill-Palmer Quarterly, 28,* 275–282.

Lipscomb, T. J., McAllister, H. A., & Bregman, N. J. (1985). A developmental inquiry into the effects of multiple models on children's generosity. *Merrill-Palmer Quarterly, 31,* 335–344.

Lisak, D., & Roth, S. (1988). Motivational factors in noncarcerated sexually aggressive men. *Journal of Personality and Social Psychology, 55,* 795–802.

Liska, A. (1984). A critical examination of the causal structure of the Fishbein-Ajzen attitude-behavior model. *Social Psychology Quarterly, 47,* 61–74.

Littlepage, G. E. (1991). Effects of group size and task characteristics on group performance: A test of Steiner's model. *Personality and Social Psychology Bulletin, 17*(4), 449–456.

Littlepage, G., & Pineault, T. (1978). Verbal, facial, and paralinguistic cues to the detection of truth and lying. *Personality and Social Psychology Bulletin, 4,* 461–464.

Locke, E. A., & Latham, G. P. (1990). *A theory of goal setting and task performance.* Englewood Cliffs, NJ: Prentice-Hall.

Locke, K. D., & Horowitz, L. M. (1990). Satisfaction in interpersonal interactions as a function of similarity in level of dysphoria. *Journal of Personality and Social Psychology, 58,* 823–831.

Lofland, J. (1990). Collective behavior: The elementary forms. In M. Rosenberg & R. Turner (Eds.), *Social psychology: Sociological perspectives.* New Brunswick, NJ: Transaction.

Lohr, J. M., & Staats, A. (1973). Attitude conditioning in Sino-Tibetan languages. *Journal of Personality and Social Psychology, 26,* 196–200.

Lopata, H. (1988). Support systems of American urban widowhood. *Journal of Social Issues, 44,* 113–128.

Lord, C. G., Lepper, M. R., & Mackie, D. (1984). Attitude prototypes as determinants of attitude-behavior consistency. *Journal of Personality and Social Psychology, 46,* 1254–1266.

Lorence, J., & Mortimer, J. (1985). Job involvement through the life course: A panel study of three age groups. *American Sociological Review, 50,* 618–638.

Lorenz, K. (1966). *On aggression.* New York: Harcourt Brace Jovanovich.

Lorenz, K. (1974). *Civilized man's eight deadly sins.* New York: Harcourt Brace Jovanovich.

Lott, A., & Lott, B. (1965). Group cohesiveness as interpersonal attraction: A review of relationships with antecedent and consequent variables. *Psychological Bulletin, 64,* 259–309.

Lott, A., & Lott, B. (1974). The role of reward in the formation of positive interpersonal attitudes. In T. Huston (Ed.), *Foundations of interpersonal attraction.* New York: Academic Press.

Lottes, I. L. (1993). Nontraditional gender roles and the sexual experience of heterosexual college students. *Sex Roles, 29,* 645–669.

Lowenthal, M. F., Thurnher, M., & Chiriboga, D. (1975). *Four stages of life: A comparative study of women and men facing transitions.* San Francisco: Jossey-Bass.

Luchins, A. S. (1957). Experimental attempts to minimize the impact of first impressions. In C. I. Hovland (Ed.), *The order of presentation in persuasion.* New Haven, CT: Yale University Press.

Luckenbill, D. F. (1982). Compliance under threat of severe punishment. *Social Forces, 60,* 811–825.

Luhman, R. (1990). Appalachian English stereotypes: Language attitudes in Kentucky. *Language in Society, 19,* 331–348.

Lundman, R. J. (1974). Routine police arrest practices: A commonweal perspective. *Social Problems, 22,* 127–141.

Lundman, R. J., Sykes, R. E., & Clark, J. P. (1978). Police control of juveniles: A replication. *Journal of Research in Crime and Delinquency, 15,* 74–91.

Lurigis, A. J., & Carroll, J. S. (1985). Probation officers' schemata of offenders: Content, development and impact on treatment decisions. *Journal of Personality and Social Psychology, 48,* 1112–1126.

Lynch, J. C., Jr., & Cohen, J. L. (1978). The use of subjective expected utility theory as an aid to understanding variables that influence helping behavior. *Journal of Personality and Social Psychology, 36,* 1138–1151.

Maass, A., & Clark, R. D., III. (1984). Hidden impact of minorities: Fifteen years of minority influence research. *Psychological Bulletin, 95,* 428–450.

Maass, A., West, S. G., & Cialdini, R. B. (1987). Minority influence and conversion. In C. Hendrick (Ed.), *Group processes* (pp. 55–79). Newbury Park, CA: Sage.

Maccoby, E., & Jacklin, C. (1974). *The psychology of sex differences.* Stanford, CA: Stanford University Press.

MacEwen, K. E., & Barling, J. (1991). Effects of maternal employment experiences on children's behavior via mood, cognitive difficulties, and parenting behavior. *Journal of Marriage and the Family, 53,* 635–644.

Macintyre, S., Hunt, K., & Sweeting, H. (1996). Gender differences in health: Are things really as simple as they seem? *Social Science and Medicine, 42,* 617–624.

Mackie, D. M., & Goethals, G. R. (1987). Individual and group goals. In C. Hendrick (Ed.), *Group processes* (pp. 144–166). Newbury Park, CA: Sage.

Mackie, M. (1983). The domestication of self: Gender comparisons of self-imagery and self-esteem. *Social Psychology Quarterly, 46,* 343–350.

Maddi, S., Bartone, P., & Puccetti, M. (1987). Stressful events are indeed a factor in physical illness: Reply to Schroeder and Costa (1984). *Journal of Personality and Social Psychology, 52,* 833–843.

Maddux, J. E., & Rogers, R. W. (1980). Effects of source expertness, physical attractiveness, and supporting arguments on persuasion: A case of brains over beauty. *Journal of Personality and Social Psychology, 39,* 235–244.

Maddux, J. E., & Rogers, R. W. (1983). Protection motivation and self-efficacy: A revised theory of fear appeals and attitude change. *Journal of Experimental Social Psychology, 19,* 469–479.

Madon, S., Jussim, L., & Eccles, J. (1997). In search of the powerful self-fulfilling prophecy. *Journal of Personality and Social Psychology, 72,* 791–809.

Madsen, D. B. (1978). Issue importance and choice shifts: A persuasive arguments approach. *Journal of Personality and Social Psychology, 36,* 1118–1127.

Magdol, L. H. (1989). *Variations in social support from friends and kin.* Unpublished master's thesis, University of Wisconsin.

Mahon, N. E. (1982). The relationship of self-disclosure, interpersonal dependency, and life changes to loneliness in young adults. *Nursing Research, 31,* 343–347.

Maines, D., & Hardesty, M. (1987). Temporality and gender: Young adults' career and family plans. *Social Forces, 66,* 102–120.

Major, B., Schmidlin, A. M., & Williams, L. (1990). Gender patterns in social touch: The impact of setting and age. *Journal of Personality and Social Psychology, 58,* 634–643.

Major, B., Zubek, J. M., Cooper, M. L., Cozarelli, C., & Richards, C. (1997). Mixed messages: Implications of social conflict and social support within close relationships for adjustment to a stressful life event. *Journal of Personality and Social Psychology, 72,* 1349–1363.

Malamuth, N. M. (1983). Factors associated with rape as predictors of laboratory aggression against women. *Journal of Personality and Social Psychology, 45,* 432–442.

Malamuth, N. M. (1984). Aggression against women: Cultural and individual causes. In N. M. Malamuth & E. Donnerstein (Eds.), *Pornography and sexual aggression.* Orlando, FL: Academic Press.

Malamuth, N. M. (1989). Sexually violent media, thought patterns, and antisocial behavior. In G. Comstock (Ed.), *Public communication and behavior* (Vol. 2, pp. 159–204). Orlando, FL: Academic Press.

Malamuth, N. M., & Briere, J. (1986). Sexual violence in the media: Indirect effects on aggression against women. *Journal of Social Issues, 42*(3), 75–92.

Malamuth, N. M., & Check, J. V. P. (1981). The effects of mass media exposure on acceptance of violence against women: A field experiment. *Journal of Research in Personality, 15,* 436–446.

Malamuth, N. M., Heavey, C. L., & Linz, D. (1993). Predicting men's antisocial behavior against women: The interaction model of sexual aggression. In N. G. C. Hall & R. Hirschman (Eds.), *Sexual aggression: Issues in etiology and assessment, treatment and policy.* New York: Hemisphere.

Malamuth, N. M., Linz, D., Heavey, C. L., Barnes, G., & Acker, M. (1995). Using the confluence model of sexual aggression to predict men's conflict with women: A 10-year follow up study. *Journal of Personality and Social Psychology, 69,* 353–369.

Malinowski, C. I., & Smith, C. P. (1985). Moral reasoning and moral conduct: An investigation prompted by Kohlberg's theory. *Journal of Personality and Social Psychology, 49,* 1016–1027.

Mallozzi, J., McDermott, V., & Kayson, W. A. (1990). Effects of sex, type of dress, and location on altruistic behavior. *Psychological Reports, 67,* 1103–1106.

Mann, C. R. (1996). *When women kill.* Albany, NY: SUNY Press.

Mannheim, B. F. (1966). Reference groups, membership groups and the self-image. *Sociometry, 29,* 265–279.

Mannheimer, D., & Williams, R. M., Jr. (1949). A note on Negro troops in combat. In S. A. Stouffer, E. A. Suchman, L. C. DeVinney, S. A. Star, & R. M. Williams, Jr. (Eds.), *The American soldier* (Vol. 1). Princeton, NJ: Princeton University Press.

Manning, P. K., & Cullum-Swan, B. (1992). Semiotics and framing: Examples. *Semiotica, 92,* 239–257.

Mannix, E. A., Neale, M.A., & Northcraft, G. B. (1995). Equity, equality, or need? The effects of organizational culture on the allocation of benefits and burdens. *Organizational Behavior and Human Decision Processes, 63*(3), 276–286.

Manstead, A. S. R., Proffitt, C., & Smart, J. L. (1983). Predicting and understanding mother's infant-feeding intentions and behavior: Testing the theory of reasoned action. *Journal of Personality and Social Psychology, 44,* 657–671.

Manucia, G. K., Baumann, D. J., & Cialdini, R. B. (1984). Mood influences on helping: Direct effects or side effects? *Journal of Personality and Social Psychology, 46,* 357–364.

Manz, C. C., & Sims, H. P. (1982). The potential for "groupthink" in autonomous work groups. *Human Relations, 35,* 773–784.

Marangoni, C., & Ickes, W. (1989). Loneliness: A theoretical review with implications for measurement. *Journal of Social and Personal Relationships, 6,* 93–128.

Marchman, V. A. (1991). The acquisition of language in normally developing children: Some basic strategies and approaches. In I. Pavao-Martins, A. Castro-Caldas, H. Van Dongen, & A. Van Hout (Eds.), *Acquired aphasia in children.* NATO: Kleiwer Academic Press.

Marcus, D. K., Wilson, J. R., & Miller, R. S. (1996). Are perceptions of emotion in the eye of the beholder? A social relations analysis of judgments of embarrassment. *Personality and Social Psychology Bulletin, 22,* 1220–1228.

Marger, M. N. (1984). Social movement organizations and response to environmental change: The NAACP, 1960–1973. *Social Problems, 32,* 16–30.

Markovsky, B., Smith, L. F., & Berger, J. (1984). Do status interventions persist? *American Sociological Review, 49,* 373–382.

Markus, H. (1977). Self-schemas and processing information about the self. *Journal of Personality and Social Psychology, 35,* 63–78.

Markus, H., & Kitayama, S. (1991). Culture and the self: Implications for cognition, emotion and motivation. *Psychological Review, 98,* 224–253.

Markus, H., & Wurf, E. (1987). The dynamic self-concept: A social psychological perspective. *Annual Review of Psychology, 38,* 299–337.

Markus, H., & Zajonc, R. B. (1985). The cognitive perspective in social psychology. In G. Lindzey & E. Aronson (Eds.), *Handbook of social psychology* (3rd ed., Vol. 1, pp. 137–230). New York: Random House.

Marmot, M. G., Bosma, H., Hemingway, H., Brunner, E., & Stansfeld, S. (1997). Contribution of job control and other risk factors to social variations in coronary heart disease. *Lancet, 350,* 235–239.

Marsden, P., & Campbell, K. (1984). Measuring tie strength. *Social Forces, 63,* 482–501.

Marsh, H. W., Barnes, J., & Hocevar, D. (1985). Self-other agreement on multidimensional self-concept ratings: Factor analysis and multitrait-multimethod analysis. *Journal of Personality and Social Psychology, 49,* 1360–1377.

Marshall, S. E. (1985). Ladies against women: Mobilization dilemmas of antifeminist movements. *Social Problems, 32,* 348–362.

Marshall, S. E. (1986). In defense of separate spheres: Class and status politics in the antisuffrage movement. *Social Forces, 65,* 327–351.

Marsiglio, W. (1991). Paternal engagement activities with minor children. *Journal of Marriage and the Family, 53,* 973–986.

Martin, C. L. (1987). A ratio measure of sex stereotyping. *Journal of Personality and Social Psychology, 52,* 489–499.

Martin, J., Scully, M., & Levitt, B. (1990). Injustice and the legitimation of revolution: Damning the past, excusing the present, and neglecting the future. *Journal of Personality and Social Psychology, 58,* 281–290.

Martin, M. W., & Sell, J. (1985). The effect of equating status characteristics on the generalization process. *Social Psychology Quarterly, 48,* 178–182.

Martin, M. W., & Sell, J. (1986). Rejection of authority: The importance of type of distribution rule and extent of benefit. *Social Science Quarterly, 67,* 855–868.

Martin, R. (1988). Ingroup and outgroup minorities: Differential impact upon public and private responses. *European Journal of Social Psychology, 18,* 39–52.

Martinez, R., Jr. (1996). Latinos and lethal violence: The impact of poverty and inequality. *Social Problems, 43,* 131–145.

Marwell, G., Aiken, M. T., & Demerath, N. J., III. (1987). The persistence of political attitudes among 1960s civil rights activists. *Public Opinion Quarterly, 51,* 383–399.

Marwell, G., McKinney, K., Sprecher, S., Smith, S., & DeLamater, J. (1982). *Legitimizing factors in the initiation of heterosexual relationships.* Paper presented at the International Conference on Personal Relationships, Madison, WI.

Marx, G. T., & Wood, J. L.. (1975). Strands of theory and research in collective behavior. In A. Inkeles, J. Coleman, and N. Smelser (Eds.), *Annual review of sociology, 1,* 363–428.

Marx, K. (1964). *Early writings.* (Ed. and Trans. by T. B. Bottomore). New York: McGraw-Hill.

Mathews, K. E., Jr., & Canon, L. K. (1975). Environmental noise level as a determinant of helping behavior. *Journal of Personality and Social Psychology, 32,* 571–577.

Matsueda, R. (1982). Testing control theory and differential association: A causal modeling approach. *American Sociological Review, 47,* 489–504.

Matsueda, R. L. (1992). Reflected appraisals, parental labeling, and delinquency: Specifying a symbolic interactionist theory. *American Journal of Sociology, 97,* 1577–1611.

Matthews, L. S., Conger, R. D., & Wickrama, K. A. S. (1996). Work-family conflict and marital quality: Mediating processes. *Social Psychology Quarterly, 59,* 62–79.

Matthews, S., & Rosner, T. (1988). Shared filial responsibility: The family as primary care giver. *Journal of Marriage and the Family, 50,* 185–195.

Maynard, D. W. (1983). Social order and plea bargaining in the court. *Sociological Quarterly, 24,* 215–233.

Maynard, D. W., & Whalen, M. R. (1995). Language, action, and social interaction. In K. S. Cook, G. A. Fine, & J. S. House (Eds.), *Sociological perspectives on social psychology* (pp. 149–175). Needham Heights, MA: Allyn & Bacon.

Mazur, J. E. (1998). *Learning and behavior* (4th ed.). Upper Saddle River, NJ: Prentice-Hall.

McAdam, D. (1986). Recruitment to high-risk activism: The case of Freedom Summer. *American Journal of Sociology, 92,* 64–90.

McAdam, D. (1989). The biographical consequences of activism. *American Sociological Review, 54,* 744–760.

McAdoo, H. P. (1997). Upward mobility across generations of African-American families. In H. P. McAdoo (Ed.), *Black families* (3rd ed., pp. 139–162). Thousand Oaks, CA: Sage.

McArthur, L. Z. (1972). The how and what of why: Some determinants and consequences of causal attribution. *Journal of Personality and Social Psychology, 22,* 171–193.

McArthur, L. Z., & Post, D. L. (1977). Figural emphasis and person perception. *Journal of Experimental Social Psychology, 13,* 520–535.

McCall, G. J., & Simmons, J. L. (1978). *Identities and interactions.* New York: Free Press.

McCannell, K. (1988). Social networks and the transition to motherhood. In R. Milardo (Ed.), *Families and social networks.* Newbury Park, CA: Sage.

McCarthy, J. D., & Hoge, D. R. (1984). The dynamics of self-esteem and delinquency. *American Journal of Sociology, 90,* 396–410.

McCarthy, J. D., McPhail, C., & Smith, J. (1996). Images of protest: Dimensions of selection bias in media coverage of Washington demonstrations, 1982 and 1991. *American Sociological Review, 61,* 478–499.

McCarthy, J. D., & Wolfson, M. (1996). Resource mobilization by local social movement organizations: Agency, strategy, and organization in the movement against drinking and driving. *American Sociological Review, 61,* 1070–1088.

McCauley, C. (1989). The nature of social influence in groupthink: Compliance and internalization. *Journal of Personality and Social Psychology, 57,* 250–260.

McCauley, C., Stitt, C. L., & Segal, M. (1980). Stereotyping: From prejudice to prediction. *Psychological Bulletin, 87,* 195–208.

McClelland, D. (1958). Risk-taking in children with high and low need for achievement. In J. Atkinson (Ed.), *Motives in fantasy, action and society.* Princeton, NJ: Van Nostrand.

McClelland, D. (1961). *The achieving society.* Princeton, NJ: Van Nostrand.

McClelland, D., & Winter, D. (1969). *Motivating economic achievement.* New York: Free Press.

McCombs, M. E., & Eyal, C. H. (1980). Spending on mass media. *Journal of Communication, 30,* 153–158.

McCrae, R. R., & Costa, P. T., Jr. (1987). Validation of the five-factor model of personality across instruments and observers. *Journal of Personality and Social Psychology, 52,* 81–90.

McFarland, C., & Ross, M. (1982). The impact of causal attributions on affective reactions to success and failure. *Journal of Personality and Social Psychology, 43,* 937–946.

McGrath, J. E. (1984). *Groups: Interaction and performance.* Englewood Cliffs, NJ: Prentice-Hall.

McGuire, W. J. (1964). Inducing resistance to persuasion: Some contemporary approaches. In L. Berkowitz (Ed.), *Advances in experimental social psychology* (Vol. 1, pp. 191–229). New York: Academic Press.

McGuire, W. J. (1985). Attitude and attitude change. In G. Lindzey & E. Aronson (Eds.), *The handbook of social psychology* (3rd ed., Vol. 2). New York: Random House.

McGuire, W. J., & McGuire, C. (1982). Significant others in self-space: Sex differences and developmental trends in the social self. In J. Suls (Ed.), *Psychological perspectives on the self* (Vol. 1). Hillsdale, NJ: Erlbaum.

McGuire, W. J., & McGuire, C. (1986). Differences in conceptualizing self versus conceptualizing other people as manifested in contrasting verb types used in natural speech. *Journal of Personality and Social Psychology, 51,* 1135–1143.

McGuire, W. J., & Padawer-Singer, A. (1976). Trait salience in the spontaneous self-concept. *Journal of Personality and Social Psychology, 33,* 743–754.

McGuire, W. J., & Papageorgis, D. (1961). The relative efficacy of various types of prior belief-defense in producing immunity against persuasion. *Journal of Abnormal and Social Psychology, 62,* 327–337.

McLanahan, S., & Booth, K. (1989). Mother-only families: Problems, prospects and politics. *Journal of Marriage and the Family, 51,* 557–580.

McLanahan, S., & Bumpass, L. (1988). Intergenerational consequences of family disruption. *American Journal of Sociology, 94,* 130–152.

McLeod, J. M., Price, K. O., & Harburg, E. (1966). Socialization, liking and yielding of opinions in imbalanced situations. *Sociometry, 29,* 197–212.

McPhail, C. (1989). Blumer's theory of collective behavior: The development of a non-symbolic interactionist explanation. *Sociological Quarterly, 30,* 401–423.

McPhail, C. (1991). *The myth of the madding crowd.* Hawthorne, NY: Aldine de Gruyter.

McPhail, C. (1994). The dark side of purpose: Individual and collective violence in riots. *Sociological Quarterly, 35,* 1–32.

McPhail, C. (1997). Stereotypes of crowds and collective behavior: Looking backward, looking forward. *Studies in Symbolic Interaction, 3, Supplement,* 35–58.

McPhail, C., & Wohlstein R. T. (1983). Individual and collective behavior within gatherings, demonstrations and riots. In R. H. Turner & J. F. Short (Eds.), *Annual review of sociology* (Vol. 9). Palo Alto, CA: Annual Reviews.

McWorter, G. A., & Crain, R. L. (1967). Subcommunity gladiatorial competition: Civil rights leadership as a competitive process. *Social Forces, 46,* 8–21.

Mead, G. H. (1934). *Mind, self, and society.* Chicago: University of Chicago Press.

Mears, P. (1974). Structuring communication in a working group. *Journal of Communication, 24,* 71–79.

Meeker, B. F. (1981). Expectation states and interpersonal behavior. In M. Rosenberg & R. H. Turner (Eds.), *Social psychology: Sociological perspectives* (pp. 290–319). New York: Basic Books.

Mehrabian, A. (1972). *Nonverbal communication.* New York: Aldine-Atherton.

Mehrabian, A., & Ksionzky, S. (1970). Models for affiliative and conformity behavior. *Psychological Bulletin, 74,* 110–126.

Meier, R. F., & Johnson, W. T. (1977). Deterrence as social control: The legal and extralegal production of conformity. *American Sociological Review, 42,* 292–304.

Mendelsohn, H. (1973). Some reasons why information campaigns can succeed. *Public Opinion Quarterly, 37,* 50–61.

Merrens, M. R. (1973). Nonemergency helping behavior in various sized communities. *Journal of Social Psychology, 90,* 327–338.

Merton, R. (1957). *Social theory and social structure.* Glencoe, IL: Free Press.

Mesch, D. J., Farh, J. L., & Podsakoff, P. M. (1994). Effects of feedback sign on group goal setting, strategies, and performance. *Group and Organization Management, 19*(3), 309–333.

Messner, S. F., & Krohn, M. D. (1990). Class, compliance structure and delinquency: Assessing integrated structural-Marxist theory. *American Journal of Sociology, 96,* 300–328.

Metts, S., & Cupach, W. R. (1989). Situational influence on the use of remedial strategies in embarrassing predicaments. *Communication Monographs, 56,* 151–162.

Meyer, D. S., & Whittier, N. (1994). Social movement spillover. *Social Problems, 41,* 277–298.

Meyrowitz, J. (1985). *No sense of place: The impact of electronic media on social behavior.* New York: Oxford University Press.

Miall, C. E. (1986). The stigma of involuntary childlessness. *Social Problems, 33,* 268–282.

Michaels, J. W., Edwards, J. N., & Acock, A. C. (1984). Satisfaction in intimate relationships as a function of inequality, inequity and outcomes. *Social Psychology Quarterly, 47,* 347–357.

Michener, H. A., & Burt, M. R. (1974). Legitimacy as a base of social influence. In J. T. Tedeschi (Ed.), *Perspectives on social power.* Chicago: Aldine-Atherton.

Michener, H. A., & Burt, M. R. (1975a). Components of "authority" as determinants of compliance. *Journal of Personality and Social Psychology, 31,* 605–614.

Michener, H. A., & Burt, M. R. (1975b). Use of social influence under varying conditions of legitimacy. *Journal of Personality and Social Psychology, 32,* 398–407.

Michener, H. A., & Cohen, E. D. (1973). Effects of punishment magnitude in the bilateral threat situation: Evidence for the deterrence hypothesis. *Journal of Personality and Social Psychology, 26,* 427–438.

Michener, H. A., & Lawler, E. J. (1971). Revolutionary coalition strength and collective failure as determinants of status reallocation. *Journal of Experimental Social Psychology, 7,* 448–460.

Michener, H. A., & Lawler, E. J. (1975). Endorsement of formal leaders: An integrative model. *Journal of Personality and Social Psychology, 31,* 216–223.

Michener, H. A., & Lyons, M. (1972). Perceived support and upward mobility as determinants of revolutionary coalitional behavior. *Journal of Experimental Social Psychology, 8,* 180–195.

Michener, H. A., Plazewski, J. G., & Vaske, J. J. (1979). Ingratiation tactics channeled by target values and threat capability. *Journal of Personality, 47,* 36–56.

Michener, H. A., & Tausig, M. (1971). Usurpation and perceived support as determinants of the endorsement accorded formal leaders. *Journal of Personality and Social Psychology, 18,* 364–372.

Michener, H. A., Vaske, J. J., Schleifer, S. L., Plazewski, J. G., & Chapman, L. J. (1975). Factors affecting concession rate and threat usage in bilateral conflict. *Sociometry, 38,* 62–80.

Michener, H. A., & Wasserman, M. (1995). Group decision making. In K. S. Cook, G. A. Fine, & J. S. House (Eds.), *Sociological perspectives on social psychology* (pp. 336–361). Boston: Allyn & Bacon.

Milardo, R. M. (1982). Friendship networks in developing relationships: Converging and diverging social environments. *Social Psychology Quarterly, 45,* 162–172.

Milardo, R. (1988). Families and social networks: An overview of theory and methodology. In R. Milardo (Ed.), *Families and social networks.* Newbury Park, CA: Sage.

Milavsky, J. R., Kessler, R., Stipp, H., & Rubens, W. (1983). *Television and aggression: A panel study.* New York: Academic Press.

Milburn, M. A. (1987). Ideological self-schemata and schematically induced attitude consistency. *Journal of Experimental Social Psychology, 23,* 383–398.

Miles, R. H. (1977). Role-set configuration as a predictor of role conflict and ambiguity in complex organizations. *Sociometry, 40,* 21–34.

Milgram, S. (1963). Behavioral study of obedience. *Journal of Abnormal and Social Psychology, 67,* 371–378.

Milgram, S. (1965a). Some conditions of obedience and disobedience to authority. *Human Relations, 18,* 57–76.

Milgram, S. (1965b). Liberating effects of group pressure. *Journal of Personality and Social Psychology, 1,* 127–134.

Milgram, S. (1970). The experience of living in cities. *Science, 167,* 1461–1468.

Milgram, S. (1974). *Obedience to authority.* New York: Harper & Row.

Milgram, S. (1976). Obedience to criminal orders: The compulsion to do evil. In T. Blass (Ed.), *Contemporary social psychology: Representative readings* (pp. 175–184). Itasca, IL: Peacock.

Milgram, S., Liberty, H. J., Toledo, R., & Wackenhut, J. (1986). Response to intrusion into waiting lines. *Journal of Personality and Social Psychology, 51,* 683–689.

Milgram, S., & Toch, H. (1969). Collective behavior: Crowds and social movements. In G. Lindzey & E. Aronson (Eds.), *The handbook of social psychology* (2nd ed., Vol. 4). Reading, MA: Addison-Wesley.

Miller, A. G. (1976). Constraint and target effects on the attribution of attitudes. *Journal of Experimental Social Psychology, 12,* 325–339.

Miller, A. G. (Ed.). (1982). In the eye of the beholder: *Contemporary issues in stereotyping.* New York: Praeger.

Miller, A. G., Collins, B. E., & Brief, D. E. (1995). Perspectives on obedience to authority: The legacy of the Milgram experiments. *Journal of Social Issues, 51*(3), 1–19.

Miller, C. E., & Komorita, S. S. (1995). Reward allocation in task-performing groups. *Journal of Personality and Social Psychology, 69*(1), 80–90.

Miller, G. (1991). Family as excuse and extenuating circumstance: Social organization and use of family rhetoric in a work incentive program. *Journal of Marriage and the Family, 53,* 609–621.

Miller, J., Schooler, C., Kohn, M., & Miller, K. (1979). Women and work: The psychological effects of occupational conditions. *American Journal of Sociology, 85,* 66–94.

Miller, K., Kohn, M., & Schooler, C. (1986). Educational self-direction and personality. *American Sociological Review, 51,* 372–390.

Miller, L. K., & Hamblin, R. L. (1963). Interdependence, differential rewarding, and productivity. *American Sociological Review, 43,* 193–204.

Miller, N., & Carlson, M. (1990). Valid theory-testing meta-analyses further question the negative state relief model of helping. *Psychological Bulletin, 107*(2), 215–225.

Miller, R. S. (1987). Empathic embarrassment: Situational and personal determinants of reactions to the embarrassment of another. *Journal of Personality and Social Psychology, 53,* 1061–1069.

Miller, R. S. (1992). The nature and severity of self-reported embarrassing circumstances. *Personality and Social Psychology Bulletin, 18,* 190–198.

Miller-McPherson, J., & Smith-Lovin, L. (1982). Women and weak ties: Differences by sex in the size of voluntary organizations. *American Journal of Sociology, 87,* 883–904.

Minnigerode, F., & Lee, J. A. (1978). Young adults' perceptions of sex roles across the lifespan. *Sex Roles, 4,* 563–569.

Miranne, A. C., & Gray, L. N. (1987). Deterrence: A laboratory experiment. *Deviant Behavior, 8,* 191–203.

Mirowsky, J., & Ross, C. (1986). Social patterns of distress. In A. Inkeles, J. Coleman, & N. Smelser (Eds.), *Annual review of sociology* (Vol. 12). Palo Alto, CA: Annual Reviews.

Mirowsky, J., & Ross, C. E. (1995). Sex differences in distress: Real or artifact? *American Sociological Review, 60,* 449–468.

Mischel, W., & Liebert, R. (1966). Effects of discrepancies between deserved and imposed reward criteria on their acquisition and transmission. *Journal of Personality and Social Psychology, 3,* 45–53.

Miyamoto, S. F., & Dornbusch, S. (1956). A test of interactionist hypotheses of self-conception. *American Journal of Sociology, 61,* 399–403.

Mizrahi, T. (1984). Coping with patients: Subcultural adjustments to the conditions of work among internists-in-training. *Social Problems, 32,* 156–166.

Modigliani, A. (1971). Embarrassment, face-work, and eye contact: Testing a theory of embarrassment. *Journal of Personality and Social Psychology, 17,* 15–24.

Moede, W. (1927). Die Richtlinien der Leistungs-Psychologie. (Guidelines for a psychology of achievement). *Industrielle Psychotechnik, 4,* 193–209.

Moen, P., Erickson, M. A., & Dempster-McClain, D. (1997). Their mother's daughters? The intergenerational transmission of gender attitudes in a world of changing roles. *Journal of Marriage and the Family, 59,* 281–293.

Money, J., & Ehrhardt, A. (1972). *Man and woman. Boy and girl.* Baltimore, MD: Johns Hopkins University Press.

Monge, P. R., Edwards, J. A., & Kirste, K. K. (1983). Determinants of communication network involvement: Connectedness and integration. *Group and Organization Studies, 8*(1), 83–111.

Moore, J. C., Jr. (1968). Status and influence in small group interaction. *Sociometry, 31,* 47–63.

Moore, M. M. (1985). Nonverbal courtship patterns in women: Context and consequences. *Ethology and Sociobiology, 6,* 237–247.

Moore, M. M. (1995). Courtship signaling and adolescents: "Girls just wanna have fun"? *Journal of Sex Research, 32,* 319–328.

Moore, M. M., & Butler, D. L. (1989). Predictive aspects of nonverbal courtship behavior in women. *Semiotica, 76,* 205–215.

Moorhead, G., Ference, R., & Neck, C. P. (1991). Group decision fiascoes continue: Space shuttle *Challenger* and a revised groupthink framework. *Human Relations, 44*(6), 539–550.

Moorhead, G., & Montanari, J. R. (1986). An empirical investigation of the groupthink phenomenon. *Human Relations, 39,* 399–410.

Moran, G. (1966). Dyadic attraction and orientational consensus. *Journal of Personality and Social Psychology, 4,* 94–99.

Moreno, J. L. (Ed.). (1960). *The sociometry reader.* Glencoe, IL: Free Press.

Morgan, S., & Rindfuss, R. (1985). Marital disruption: Structural and temporal dimensions. *American Journal of Sociology, 90,* 1055–1077.

Morgan, W., Alwin, D., & Griffin, L. (1979). Social origins, parental values and the transmission of inequality. *American Journal of Sociology, 85,* 156–166.

Morris, W. N., & Miller, R. S. (1975). The effect of consensus-breaking and consensus-preempting partners on reduction of conformity. *Journal of Experimental Social Psychology, 11,* 215–223.

Morrissette, J. O. (1966). Group performance as a function of task difficulty and size and structure of groups, II. *Journal of Personality and Social Psychology, 3,* 357–359.

Morse, S., & Gergen, K. (1970). Social comparison, self-consistency, and the concept of self. *Journal of Personality and Social Psychology, 16,* 148–156.

Mortimer, J. T., Finch, M., & Kumka, D. (1982). Persistence and change in development: The multidimensional self-concept. In P. Baltes & O. Brim, Jr. (Eds.), *Life span development and behavior* (Vol. 4). New York: Academic Press.

Mortimer, J. T., & Lorence, J. (1995). Social psychology of work. In K. S. Cook, G. A. Fine, & J. S. House (Eds.), *Sociological perspectives on social psychology* (pp. 497–523). Needham Heights, MA: Allyn & Bacon.

Mortimer, J. T., & Simmons, R. (1978). Adult socialization. In R. Turner, J. Coleman, & R. Fox (Eds.), *Annual review of sociology* (Vol. 4). Palo Alto, CA: Annual Reviews.

Moscovici, S. (1980). Toward a theory of conversion behavior. In L. Berkowitz (Ed.), *Advances in experimental social psychology* (Vol. 13, pp. 209–239). New York: Academic Press.

Moscovici, S. (1985a). Innovation and minority influence. In S. Moscovici, G. Mugny, & E. Van Avermaet (Eds.), *Perspectives on minority influence* (pp. 9–52). Cambridge, England: Cambridge University Press.

Moscovici, S. (1985b). Social influence and conformity. In G. Lindzey & E. Aronson (Eds.), *Handbook of social psychology* (3rd ed., Vol. 2, pp. 347–412). New York: Random House.

Moscovici, S., & Lage, E. (1976). Studies in social influence III: Majority versus minority influence in a group. *European Journal of Social Psychology, 6,* 149–174.

Moscovici, S., Lage, E., & Naffrechoux, M. (1969). Influence of a consistent minority on the responses of a majority in a color perception task. *Sociometry, 32,* 365–379.

Moss, H., & Kagan, J. (1961). Stability of achievement and recognition seeking behavior from early childhood through adulthood. *Journal of Abnormal and Social Psychology, 62,* 504–513.

Muedeking, G. D. (1992). Authentic/inauthentic identities in the prison visiting room. *Symbolic Interaction, 15*(2), 227–236.

Mugny, G. (1982). *The power of minorities.* London: Academic Press.

Mugny, G. (1984). The influence of minorities: Ten years later. In H. Tajfel (Ed.), *The social dimension: European developments in social psychology* (Vol. 2, pp. 498–517). Cambridge, England: Cambridge University Press.

Mugny, G., & Papastamou, S. (1982). Minority influence and psycho-social identity. *European Journal of Social Psychology, 12,* 379–394.

Mullen, B. (1985). Strength and immediacy of sources: A meta-analytic evaluation of the forgotten elements of social impact theory. *Journal of Personality and Social Psychology, 48*(6), 1458–1466.

Mullen, B., Anthony, T., Salas, E., & Driskell, J. E. (1994). Group cohesiveness and quality of decision making: An integration of tests of the groupthink hypothesis. *Small Group Research, 25,* 189–204.

Mullen, B., & Copper, C. (1994). The relation between group cohesiveness and performance: An integration. *Psychological Bulletin, 115,* 210–227.

Mullen, B., & Hu, L. (1989). Perceptions of ingroup and outgroup variability: A meta-analytic integration. *Basic and Applied Social Psychology, 10,* 233–252.

Mullen, B., Johnson, C., & Salas, E. (1991a). Productivity loss in brainstorming groups: A meta-analytic integration. *Basic and Applied Social Psychology, 12,* 3–24.

Mullen, B., Johnson, C., & Salas, E. (1991b). Effects of communication network structure: Components of positional centrality. *Social Networks, 13*(2), 169–185.

Mullen, C. K., & Linz, D. (1995). Desensitization and resensitization to violence against women: Effects of exposure to sexually violent films on judgements of domestic violence victims. *Journal of Personality and Social Psychology, 69,* 449–459.

Muller, E. (1985). Income inequality, regime repressiveness and political violence. *American Sociological Review, 50,* 47–61.

Murray, J. P., and Kippax, S. (1979). From the early window to the late night show: International trends in the study of television's impact on children and adults. In L. Berkowitz (Ed.), *Advances in experimental social psychology* (Vol. 12). New York: Academic Press.

Murray, S. L., Holmes, J. G., & Griffin, D. W. (1996a). The benefits of positive illusions: Idealization and the construction of satisfaction in close relationships. *Journal of Personality and Social Psychology, 70,* 79–98.

Murray, S. L., Holmes, J. G., & Griffin, D. W. (1996b). The self-fulfilling nature of positive illusions in romantic relationships: Love is not blind, but prescient. *Journal of Personality and Social Psychology, 71,* 1155–1180.

Murstein, B. (1980). Mate selection in the 1970s. *Journal of Marriage and the Family, 42,* 777–792.

Myers, D. (1997). Racial rioting in the 1960s: An event history analysis of local conditions. *American Sociological Review, 62,* 94–112.

Myers, D. G. (1975). Discussion-induced attitude polarization. *Human Relations, 28,* 699–714.

Myers, D. G., Bruggink, J. B., Kersting, R. C., & Schlosser, B. A. (1980). Does learning others' opinions change one's opinion? *Personality and Social Psychology Bulletin, 6,* 253–260.

Myers, D. G., & Kaplan, M. F. (1976). Group-induced polarization in simulated juries. *Personality and Social Psychology Bulletin, 2,* 63–66.

Myers, D. G., & Lamm, H. (1976). The group polarization phenomenon. *Psychological Bulletin, 83,* 602–627.

Myers, F. E. (1971). Civil disobedience and organization change: The British Committee of 100. *Political Science Quarterly, 86,* 92–112.

Myers, M. A., & Hagan, J. (1979). Private and public trouble: Prosecutors and the allocation of court resources. *Social Problems, 26,* 439–451.

Myers, M. A., & Talarico, S. M. (1986). The social contexts of racial discrimination in sentencing. *Social Problems, 33,* 236–251.

Nadler, A. (1987). Determinants of help seeking behaviour: The effects of helper's similarity, task centrality and recipient's self esteem. *European Journal of Social Psychology, 17*(1), 57–67.

Nadler, A. (1991). Help-seeking behavior: Psychological costs and instrumental benefits. In M. S. Clark (Ed.), *Review of personality and social psychology: Vol. 12. Prosocial behavior* (pp. 290–311). Newbury Park, CA: Sage.

Nadler, A., & Fisher, J. D. (1984a). Effects of donor-recipient relationships on recipients' reactions to aid. In E. Staub, D. Bar-Tal, J. Karylowski, & J. Reykowski (Eds.), *Development and maintenance of prosocial behavior: International perspectives on positive morality.* New York: Plenum.

Nadler, A., & Fisher, J. D. (1984b). The role of threat to self-esteem and perceived control in recipient reaction to aid. In L. Berkowitz (Ed.), *Advances in experimental social psychology* (Vol. 17). New York: Academic Press.

Nadler, A., & Fisher, J. D. (1986). The role of threat to self-esteem and perceived control in recipient reaction to help: Theory development and empirical validation. In L. Berkowitz (Ed.), *Advances in experimental social psychology* (Vol. 19, pp. 81–122). San Diego, CA: Academic Press.

Nadler, A., Fisher, J. D., & Ben-Itzhak, S. (1983). With a little help from my friend: Effect of single or multiple act aid as a function of donor and task characteristics. *Journal of Personality and Social Psychology, 44*(2), 310–321.

Nadler, A., Mayseless, O., Peri, N., & Chemerinski, A. (1985). Effects of opportunity to reciprocate and self-esteem on help-seeking behavior. *Journal of Personality, 53,* 23–35.

Nadler, G. (1981). *The planning and design approach.* New York: Wiley.

Nagel, J. (1995). American Indian ethnic revival: Politics and the resurgence of identity. *American Sociological Review, 60,* 947–965.

Nakao, K., & Treas, J. (1994). Updating occupational prestige and socioeconomic scores: How the new measures measure up. *Sociological Methodology 1994, 24,* 1–72.

Neale, M., & Bazerman, M. (1991). *Cognition and rationality in bargaining.* New York: Free Press.

Nemeth, C. J. (1986). Differential contributions of majority and minority influence. *Psychological Review, 93,* 23–32.

Nemeth, C. J., & Kwan, J. L. (1985). Originality of word associations as a function of majority vs. minority influence processes. *Social Psychological Quarterly, 48,* 277–282.

Nemeth, C. J., & Kwan, J. L. (1987). Minority influence, divergent thinking and detection of correct solutions. *Journal of Applied Social Psychology, 17,* 788–799.

Nemeth, C. J., Swedlund, M., & Kanki, B. (1974). Patterning of the minority's responses and their influence on the majority. *European Journal of Social Psychology, 4,* 53–64.

Nemeth, C. J., Wachtler, J., & Endicott, J. (1977). Increasing the size of the minority: Some gains and some losses. *European Journal of Social Psychology, 7,* 15–27.

Netemeyer, R. G., Burton, S., & Johnston, M. (1991). A comparison of two models for the prediction of volitional and goal-directed behavior: A confirmatory analysis approach. *Social Psychology Quarterly, 54,* 87–100.

Neugarten, B. L., & Datan, N. (1973). Sociological perspectives on the life cycle. In P. Baltes & K. Schaie (Eds.), *Life-span developmental psychology: Personality and social processes.* New York: Academic Press.

Neugarten, B. L., Moore, J. W., & Lowe, J. C. (1965). Age norms, age constraints, and adult socialization. *American Journal of Sociology, 70,* 710–717.

New York Times. (1998, Feb. 15). Indonesians die as riots over price rises widen, p. A10.

Newcomb, T. M. (1943). *Personality and social change.* New York: Dryden.

Newcomb, T. M. (1961). *The acquaintance process.* New York: Holt, Rinehart and Winston.

Newcomb, T. M. (1968). Interpersonal balance. In R. P. Abelson et al. (Eds.), *Theories of cognitive consistency: A sourcebook.* Chicago: Rand McNally.

Newcomb, T. M. (1971). Dyadic balance as a source of clues about interpersonal attraction. In B. Murstein (Ed.), *Theories of attraction and love.* New York: Springer.

Newspaper Advertising Bureau. (1980). *Mass media in the family setting: Social patterns in media availability and use by parents.* New York: Author.

Newsweek. (1991, April 1). Violence in our culture, pp. 46 ff.

Newsweek. (1978, January 9). Gold in the streets, pp. 56–57.

Nisbett, R. E., Caputo, C., Legant, P., & Maracek, J. (1973). Behavior as seen by the actor and as seen by the observer. *Journal of Personality and Social Psychology, 27,* 154–164.

Nizer, L. (1973). *The implosion conspiracy.* New York: Doubleday.

Norman, R. (1975). Affective-cognitive consistency, attitudes, conformity, and behavior. *Journal of Personality and Social Psychology, 32,* 83–91.

Norstrom, T. (1995). The impact of alcohol, divorce, and unemployment on suicide. *Social Forces, 74,* 293–314.

Nyden, P. W. (1985). Democratizing organizations: A case study of a union reform movement. *American Journal of Sociology, 90,* 1179–1203.

Oakes, P. J., & Turner, J. C. (1980). Social categorization and intergroup behavior: Does minimal intergroup discrimination make social identity more positive? *European Journal of Social Psychology, 10,* 295–301.

O'Barr, W., & Atkins, B. (1980). Women's language or powerless language. In S. McConnell-Ginet, R. Borker, & N. Furman (Eds.), *Women and language in literature and society.* New York: Praeger.

Oberschall, A. (1973). *Social conflict and social movements.* Englewood Cliffs, NJ: Prentice-Hall.

Oberschall, A. (1978). Theories of social conflict. In R. Turner, J. Coleman, & R. Fox (Eds.), *Annual review of sociology* (Vol. 4). Palo Alto, CA: Annual Reviews.

O'Bryant, S. (1988). Sibling support and older widows' well-being. *Journal of Marriage and the Family, 50,* 173–183.

O'Connor, G. D. (1980). Small groups: A general system model. *Small Group Behavior, 11,* 145–174.

Oegema, D., & Klandermans, B. (1994). Why social movement sympathizers don't participate: Erosion and nonconversion of support. *American Sociological Review, 59,* 703–722.

Ohbuchi, K., Kameda, M., & Agarie, N. (1989). Apology as aggression control: Its role in mediating appraisal of and response to harm. *Journal of Personality and Social Psychology, 56,* 219–227.

O'Leary-Kelly, A. M., Martocchio, J. J., & Frink, D. D. (1994). A review of the influence of group goals on group performance. *Academy of Management Journal, 37*(5), 1285–1301.

Oliver, P. (1980). Rewards and punishments as selective incentives for collective action: Theoretical investigations. *American Journal of Sociology, 85,* 1356–1375.

Olver, R. (1961). *Developmental study of cognitive equivalence.* Unpublished doctoral dissertation, Radcliffe College.

Olzak, S. (1989). Labor unrest, immigration, and ethnic conflict in urban America, 1880–1914. *American Journal of Sociology, 94,* 1303–1333.

Olzak, S. (1992). *The dynamics of ethnic competition and conflict.* Palo Alto, CA: Stanford University Press.

Olzak, S., Shanahan, S., & McEneaney, E. H. (1996). Poverty, segregation, and race riots: 1990 to 1993. *American Sociological Review, 61,* 590–613.

Opp, K-D. (1988). Grievances and participation in social movements. *American Sociological Review, 53,* 853–864.

Oppenheimer, V. K. (1970). The female labor force in the United States. *Population Monograph Series* (No. 5). Berkeley, CA: Institute of International Studies.

Orbuch, T. L., & Custer, L. (1995). The social context of married women's work and its impact on black husbands and white husbands. *Journal of Marriage and the Family, 57,* 333–345.

Orbuch, T. L., House, J. S., Mero, R. P., & Webster, P. S. (1996). Marital quality over the life course. *Social Psychology Quarterly, 59,* 162–171.

Orcutt, J. (1975). Deviance as a situated phenomenon: Variations in the social interpretation of marijuana and alcohol use. *Social Problems, 22,* 346–356.

Orne, M. T. (1969). Demand characteristics and the concept of quasi-controls. In R. Rosenthal & R. Rosnow (Eds.), *Artifact in behavior research.* New York: Academic Press.

Orwell, G. (1949). *1984.* New York: Harcourt Brace Jovanovich.

Osborn, A. F. (1963). *Applied imagination* (3rd rev. ed.). New York: Scribner.

Osgood, C. E. (1962). *An alternative to war or surrender.* Urbana: University of Illinois Press.

Osgood, C. E. (1979). GRIT for MBFR: A proposal for unfreezing force-level postures in Europe. *Peace Research Review, 8,* 77–92.

Osgood, C. E. (1980, May). The GRIT strategy. *Bulletin of the Atomic Scientists,* pp. 58–60.

Osgood, C. E., Suci, G., & Tannenbaum, P. (1957). *The measurement of meaning.* Urbana: University of Illinois Press.

Osgood, D. W., Wilson, J. R., O'Malley, P. M., Bachman, J. G., & Johnston, L. D. (1996). Routine activities and deviant behavior. *American Sociological Review, 61,* 635–655.

Oskamp, S. (1971). Effects of programmed strategies on cooperation in the Prisoner's Dilemma and other mixed-motive games. *Journal of Conflict Resolution, 15,* 225–259.

Otten, C. A., Penner, L. A., & Waugh, G. (1988). That's what friends are for: The determinants of psychological helping. *Journal of Social and Clinical Psychology, 7*(1), 34–41.

Padden, S. L., & Buehler, C. (1995). Coping with the dual-income lifestyle. *Journal of Marriage and the Family, 57,* 101–110.

Page, A. L., & Clelland, D. A. (1978). The Kanawha County textbook controversy: A study of the politics of lifestyle concern. *Social Forces, 57,* 265–281.

Paicheler, G., & Bouchet, J. (1973). Attitude polarization, familiarization, and group process. *European Journal of Social Psychology, 3,* 83–90.

Palmore, E. (1981). *Social patterns in normal aging.* Durham, NC: Duke University Press.

Pandey, J. (1981). A note about social power through ingratiation among workers. *Journal of Occupational Psychology, 54*(1), 65–67.

Pantin, H. M., & Carver, C. S. (1982). Induced competence and the bystander effect. *Journal of Applied Social Psychology, 12,* 100–111.

Papastamou, S., & Mugny, G. (1985). Rigidity and minority influence: The influence of the social in social influence. In S. Moscovici, G. Mugny, & E. Van Avermaet (Eds.), *Perspectives on minority influence* (pp. 113–136). Cambridge, England: Cambridge University Press.

Park, B., & Rothbart, M. (1982). Perception of out-group homogeneity and levels of social categorization: Memory for the subordinate attributes of in-group and out-group members. *Journal of Personality and Social Psychology, 42*(6), 1051–1068.

Park, W-W. (1990). A review of research on groupthink. *Journal of Behavioral Decision Making, 3,* 229–245.

Parke, R. (1969). Effectiveness of punishment as an interaction of intensity, timing, agent nurturance and cognitive structuring. *Child Development, 40,* 213–235.

Parke, R. (1970). The role of punishment in the socialization process. In R. Hoppe, G. Milton, & E. Simmel (Eds.), *Early experiences and the processes of socialization.* New York: Academic Press.

Parke, R. D. (1996). *Fatherhood.* Cambridge, MA: Harvard University Press.

Parrott, W. G., & Smith, S. F. (1991). Embarrassment: Actual vs. typical cases, classical vs. prototypical representations. *Cognition and Emotion, 5,* 467–488.

Pascale, P. J., & Sylvester, J. (1988). Trend analyses of four large-scale surveys of high-school drug use 1977–1986. *Journal of Drug Education, 18,* 221–233.

Patrick, S. L., & Jackson, J. J. (1991). Further examination of the equity sensitivity construct. *Perceptual and Motor Skills, 73*(3, pt. 2), 1091–1106.

Patterson, M. L., Mullens, S., & Romano, J. (1971). Compensatory reactions to spatial intrusion. *Sociometry, 34,* 114–121.

Patterson, R. J., & Neufeld, R. W. J. (1987). Clear danger: Situational determinants of the appraisal of threat. *Psychological Bulletin, 101,* 404–416.

Patterson, T. E. (1980). *The mass media election: How Americans choose their president.* New York: Praeger.

Pearce, P. L. (1980). Strangers, travelers, and Greyhound terminals: A study of small-scale helping behaviors. *Journal of Personality and Social Psychology, 38*(6), 935–940.

Pearlin, L., & Johnson, J. (1977). Marital status, life-strains and depression. *American Sociological Review, 42,* 704–715.

Pearlin, L., & Radabaugh, C. (1976). Economic strains and the coping functions of alcohol. *American Journal of Sociology, 82,* 652–663.

Peirce, K. (1993). Socialization of teenage girls through teen-magazine fiction: The making of a new woman or an old lady? *Sex Roles, 29,* 59–68.

Pennebaker, J. W. (1980). Self-perception of emotion and internal sensation. In D. W. Wegner & R. R. Vallacher (Eds.), *The self in social psychology.* New York: Oxford University Press.

Peplau, L., Rubin, Z., & Hill, C. (1977). Sexual intimacy in dating relationships. *Journal of Social Issues, 33*(2), 86–109.

Perez, J. A., & Mugny, G. (1987). Paradoxical effects of categorization in minority influence: When being an outgroup is an advantage. *European Journal of Social Psychology, 17,* 157–169.

Perlman, D. (1988). Loneliness: A life-span family perspective. In R. Milardo (Ed.), *Families and social networks.* Newbury Park, CA: Sage.

Perry, J. B., & Pugh, M. D. (1978). *Collective behavior: Response to social stress.* St. Paul, MN: West.

Perry, L. S. (1993). Effects of inequity on job satisfaction and self-evaluation in a national sample of African-American workers. *Journal of Social Psychology, 133*(4), 565–573.

Personnaz, B. (1981). Study in social influence using the spectrometer method: Dynamics of the phenomena of conversion and covertness in perceptual responses. *European Journal of Social Psychology, 11,* 431–438.

Peters, L. H., Hartke, D. D., & Pohlmann, J. T. (1985). Fiedler's contingency theory of leadership: An application of the meta-analysis procedures of Schmidt and Hunter. *Psychological Bulletin, 97,* 274–285.

Petersen, T., & Morgan, C. A. (1995). Separate and unequal: Occupation-establishment sex segregation and the gender wage gap. *American Journal of Sociology, 101,* 329–365.

Petrunik, M., & Shearing, C. D. (1983). Fragile facades: Stuttering and the strategic manipulation of awareness. *Social Problems, 31,* 125–138.

Pettigrew, T. F. (1979). The ultimate attribution error: Extending Allport's cognitive analysis of prejudice. *Personality and Social Psychology Bulletin, 5,* 461–476.

Petty, R. E. (1995). Attitude change. In A. Tesser (Ed.), *Advanced social psychology* (pp. 195–255). New York: McGraw-Hill.

Petty, R. E., & Cacioppo, J. T. (1979). Issue involvement can increase or decrease persuasion by enhancing message-relevant cognitive responses. *Journal of Personality and Social Psychology, 37,* 1915–1926.

Petty, R. E., & Cacioppo, J. T. (1981). *Attitudes and persuasion: Classic and contemporary approaches.* Dubuque, IA: William C. Brown.

Petty, R. E., & Cacioppo, J. T. (1986a). *Communication and persuasion: Central and peripheral routes to attitude change.* New York: Springer-Verlag.

Petty, R. E., & Cacioppo, J. T. (1986b). The elaboration likelihood model of persuasion. In L. Berkowitz (Ed.), *Advances in experimental social psychology* (Vol. 19, pp. 207–249). New York: Academic Press.

Petty, R. E., & Cacioppo, J. T. (1990). Involvement and persuasion: Tradition versus integration. *Psychological Bulletin, 107,* 367–374.

Petty, R. E., Cacioppo, J. T., & Goldman, R. (1981). Personal involvement as a determinant of argument-based persuasion. *Journal of Personality and Social Psychology, 41,* 847–855.

Petty, R. E., Cacioppo, J. T., & Heesacker, M. (1981). Effects of rhetorical questions on persuasion: A cognitive response analysis. *Journal of Personality and Social Psychology, 40,* 432–440.

Petty, R. E., Cacioppo, J. T., Strathman, A. J., & Priester, J. R. (1994). To think or not to think: Exploring two routes to persuasion. In S. Shavitt & T. C. Brock (Eds.), *Persuasion: Psychological insights and perspectives* (pp. 113–147). Boston: Allyn & Bacon.

Phillips, D. P. (1974). The influence of suggestion on suicide: Substantive and theoretical implications of the Werther effect. *American Sociological Review, 39,* 340–354.

Phillips, D. P. (1979). Suicide, motor vehicle fatalities and the mass media: Evidence toward a theory of suggestion. *American Sociological Review, 84,* 1150–1174.

Piaget, J. (1954). *The construction of reality in the child.* New York: Basic Books.

Piaget, J. (1965). *The moral judgement of the child.* New York: Free Press.

Piliavin, I. M., & Briar, S. (1964). Police encounters with juveniles. *American Journal of Sociology, 70,* 206–214.

Piliavin, I. M., Piliavin, J. A., & Rodin, J. (1975). Cost, diffusion, and the stigmatized victim. *Journal of Personality and Social Psychology, 32,* 429–438.

Piliavin, I. M., Rodin, J., & Piliavin, J. A. (1969). Good Samaritanism: An underground phenomenon? *Journal of Personality and Social Psychology, 13,* 289–299.

Piliavin, I. M., Thornton, C., Gartner, R., & Matsueda, R. (1986). Crime, deterrence, and rational choice. *American Sociological Review, 51,* 101–119.

Piliavin, J. A., & Charng, H-W. (1990). Altruism: A review of recent theory and research. *Annual Review of Sociology, 16,* 27–65.

Piliavin, J. A., Dovidio, J. F., Gaertner, S. L., & Clark, R. D., III. (1981). *Emergency intervention.* New York: Academic Press.

Piliavin, J. A., & Unger, R. K. (1985). The helpful but helpless female: Myth or reality? In V. O'Leary, R. K. Unger, & B. S. Wallston (Eds.), *Women, gender, and social psychology* (pp. 149–186). Hillsdale, NJ: Erlbaum.

Pillemeer, K., & Suitor, J. J. (1991). "Will I ever escape my children's problems?" Effects of adult children's problems on elderly parents. *Journal of Marriage and the Family, 53,* 585–594.

Pittman, J., & Lloyd, S. (1988). Quality of family life, social support and stress. *Journal of Marriage and the Family, 50,* 53–67.

Pleck, J. H. (1976). The male sex role: Definitions, problems and sources of change. *Journal of Social Issues, 32,* 155–164.

Pleck, J. H. (1985). *Working wives/working husbands.* Beverly Hills, CA: Sage.

Plutzer, E. (1987). Determinants of leftist radical belief in the United States: A test of competing theories. *Social Forces, 65,* 1002–1017.

Podolny, J. M., & Baron, J. N. (1997). Resources and relationships: Social networks and mobility in the workplace. *American Sociological Review, 62,* 673–693.

Porter, J. R., & Washington, R. E. (1993). Minority identity and self-esteem. *Annual Review of Sociology, 19,* 139–161.

Powell, M. A., & Parcell, T. L. (1997). Effects of family structure on the earnings attainment process: Differences by gender. *Journal of Marriage and the Family, 59,* 419–433.

Powers, T. A., & Zuroff, D. C. (1988). Interpersonal consequences of overt self-criticism: A comparison with neutral and self-enhancing presentations of self. *Journal of Personality and Social Psychology, 54,* 1054–1062.

Poyatos, F. (1983). *New perspectives in nonverbal communication: Studies in cultural anthropology, social psychology, linguistics, literature and semantics.* Oxford, England: Pergamon.

Prager, I. G., & Cutler, B. L. (1990). Attributing traits to oneself and to others: The role of acquaintance level. *Personality and Social Psychology Bulletin, 16,* 309–319.

Pratkanis, A. R., & Greenwald, A. G. (1989). A sociocognitive model of attitude structure and function. In L. Berkowitz (Ed.), *Advances in experimental social psychology* (Vol. 22, pp. 245–285). New York: Academic Press.

Presser, H. (1988). Shift work and child care among young dual-earner American parents. *Journal of Marriage and the Family, 50,* 133–148.

Price, K. O., Harburg, E., & Newcomb, T. M. (1966). Psychological balance in situations of negative interpersonal attitudes. *Journal of Personality and Social Psychology, 3,* 265–270.

Priest, R. T., & Sawyer, J. (1967). Proximity and peership: Bases of balance in interpersonal attraction. *American Journal of Sociology, 72,* 633–649.

Pritchard, R. D., Jones, S. D., Roth, P. L., Stuebing, K. K., & Ekeberg, S. (1988). Effects of group feedback, goal setting, and incentives on organizational productivity. *Journal of Applied Psychology, 73,* 337–358.

Pritchard, R. D., & Watson, M. D. (1992). Understanding and measuring group productivity. In S. Worchel, W. Wood, & J. A. Simpson (Eds.), *Group process and productivity* (pp. 251–275). Newbury Park, CA: Sage.

Pruitt, D. G. (1983). Achieving integrative agreements. In M. Bazerman & R. Lewicki (Eds.), *Negotiating in organizations* (pp. 35–50). Beverly Hills, CA: Sage.

Pruitt, D. G., & Carnevale, P. J. (1993). *Negotiation in social conflict.* Buckingham, England: Open University Press.

Pruitt, D. G., & Insko, C. A. (1980). Extension of the Kelley attribution model: The role of comparison-object consensus, target-object consensus, distinctiveness, and consistency. *Journal of Personality and Social Psychology, 39,* 39–58.

Pryor, J. B., McDaniel, M. A., & Kott-Russo, T. (1986). The influence of the level of schema abstractness upon the processing of social information. *Journal of Experimental Social Psychology, 22,* 312–327.

Pugh, M. D., & Wahrman, R. (1983). Neutralizing sexism in mixed-sex groups: Do women have to be better than men? *American Journal of Sociology, 88,* 746–762.

Purdum, T. S. (1997, April 27). Legacy of riots in Los Angeles traces conflict. *New York Times,* pp. 1, 16.

Quarantelli, E. L., & Dynes, R. R. (1977). Response to social crisis and disaster. In A. Inkeles, J. Coleman, & N. Smelser (Eds.), *Annual review of sociology* (Vol. 3). Palo Alto, CA: Annual Reviews.

Quattrone, G. A. (1986). On the perceptions of a group's variability. In S. Worchel & W. G. Austin (Eds.), *Psychology of intergroup relations* (2nd ed., pp. 25–48). Chicago: Nelson-Hall.

Quattrone, G. A., & Jones, E. E. (1980). The perception of variability within in-groups and out-groups: Implications for the law of small numbers. *Journal of Personality and Social Psychology, 38,* 141–152.

Quinney, R. (1970). *The social reality of crime.* Boston: Little, Brown.

Rabbie, J. M., & Bekkers, F. (1978). Threatened leadership and intergroup competition. *European Journal of Social Psychology, 8,* 9–20.

Rabow, J., Neuman, C. A., & Hernandez, A. (1987). Cognitive consistency in attitudes, social support and consumption of alcohol: Additive and interactive effects. *Social Psychology Quarterly, 50,* 56–63.

Rahn, J., & Mason, W. (1987). Political alienation, cohort size, and the Easterlin hypothesis. *American Sociological Review, 52,* 155–169.

Rank, S. G., & Jacobson, C. K. (1977). Hospital nurses' compliance with medication overdose orders: A failure to replicate. *Journal of Health and Social Behavior, 18*(2), 188–193.

Ratzan, S. C. (1989). The real agenda setters: Pollsters in the 1988 presidential campaign. *American Behavioral Scientist, 32,* 451–463.

Raven, B. H. (1993). The bases of power: Origins and recent developments. *Journal of Social Issues, 49*(4), 227–251.

Raven, B. H., & Kruglanski, A. W. (1970). Conflict and power. In P. Swingle (Ed.), *The structure of conflict.* New York: Academic Press.

Raven, B. H., & Rietsema, J. (1957). The effects of varied clarity of group goal and group path upon the individual and his relation to the group. *Human Relations, 10,* 29–44.

Ray, M. (1973). Marketing communication and the hierarchy of effects. In P. Clarke (Ed.), *New models for communication research.* Beverly Hills, CA: Sage.

Rees, C. R., & Segal, M. W. (1984). Role differentiation in groups: The relations between instrumental and expressive leadership. *Small Group Behavior, 15,* 109–123.

Regan, D. T., & Fazio, R. (1977). On the consistency between attitudes and behavior: Look to the method of attitude formation. *Journal of Experimental Social Psychology, 35,* 21–30.

Reifenberg, R. J. (1986). The self-serving bias and the use of objective and subjective methods for measuring success and failure. *Journal of Social Psychology, 126,* 627–631.

Reis, H. T., Senchak, M., & Solomon, B. (1985). Sex differences in the intimacy of social interaction: Further examination of potential explanations. *Journal of Personality and Social Psychology, 48,* 1204–1217.

Rempel, J. K., Holmes, J. G., & Zanna, M. P. (1985). Trust in a close relationship. *Journal of Personality and Social Psychology, 49,* 95–112.

Repetti, R. (1987). Individual and common components of the social environment at work and psychological well-being. *Journal of Personality and Social Psychology, 52,* 710–720.

Repetti, R. (1989). Effects of daily workload on subsequent behavior during marital interaction: The roles of social withdrawal and spouse support. *Journal of Personality and Social Psychology, 57,* 651–659.

Report of the National Advisory Commission on Civil Disorders. (1968). New York: Bantam Books.

Reskin, B., & Hartmann, H. (Eds.). (1986). *Women's work, men's work: Sex segregation on the job.* Washington, DC: National Academy Press.

Reskin, B., & Padavic, I. (1994). *Women and men at work.* Thousand Oaks, CA: Pine Forge Press.

Rexroat, C., & Shehan, C. (1987). The family life cycle and spouses' time in housework. *Journal of Marriage and the Family, 49,* 737–750.

Reynolds, J. R. (1997). The effects of industrial employment conditions on job-related distress. *Journal of Health and Social Behavior, 38,* 105–116.

Reynolds, P. D. (1984). Leaders never quit: Talking, silence, and influence in interpersonal groups. *Small Group Behavior, 15,* 404–413.

Rhine, R. J., & Severance, L. J. (1970). Ego-involvement, discrepancy, source credibility, and attitude change. *Journal of Personality and Social Psychology, 16,* 175–190.

Rice, R. W., Marwick, N. J., Chemers, M. M., & Bentley, J. C. (1982). Task performance and satisfaction: Least preferred coworker (LPC) as a moderator. *Personality and Social Psychology Bulletin, 8,* 534–541.

Ridgeway, C. L. (1982). Status in groups: The importance of motivation. *American Sociological Review, 47,* 76–88.

Ridgeway, C. L. (1987). Nonverbal behavior, dominance, and the basis of status in task groups. *American Sociological Review, 52,* 683–694.

Ridley, M., & Dawkins, R. (1981). The natural selection of altruism. In J. P. Rushton & R. M. Sorrentino (Eds.), *Altruism and helping behavior: Social, personality, and developmental perspectives* (pp. 19–39). Hillsdale, NJ: Erlbaum.

Riggio, R. E., & Friedman, H. S. (1983). Individual differences and cues to deception. *Journal of Personality and Social Psychology, 45,* 899–915.

Rigney, J. (1962). *A developmental study of cognitive equivalence transformation and their use in the acquisition and processing of information.* Unpublished honors thesis, Radcliffe College, Cambridge, MA.

Riley, A., & Burke, P. J. (1995). Identities and self-verification in the small group. *Social Psychology Quarterly, 58,* 61–73.

Riley, M. (1987). On the significance of age in sociology. *American Sociological Review, 52,* 1–14.

Rindfuss, R., Swicegood, C. G., & Rosenfeld, R. (1987). Disorder in the life course: How common and does it matter? *American Sociological Review, 52,* 785–801.

Ring, K., & Kelley, H. H. (1963). A comparison of augmentation and reduction as modes of influence. *Journal of Abnormal and Social Psychology, 66,* 95–102.

Ringelmann, M. (1913). Recherches sur les moteurs animés: Travail de l'homme. [Research on animate sources of power: The work of man.] *Annales de l'Institut National Agronomique,* 2e série, tome XII, 1–40.

Riordan, C. (1978). Equal-status interracial contact: A review and revision of the concept. *International Journal of Intercultural Relations, 2,* 161–185.

Riordan, C., & Ruggiero, J. A. (1980). Producing equal-status interracial interaction: A replication. *Social Psychology Quarterly, 43,* 131–136.

Riordan, C. A., Marlin, N. A., & Kellogg, R. T. (1983). The effectiveness of accounts following transgression. *Social Psychology Quarterly, 46,* 213–219.

Robertson, J. F., & Simons, R. L. (1989). Family factors, self-esteem, and adolescent depression. *Journal of Marriage and the Family, 51,* 125–138.

Robinson, D. T., Smith-Lovin, L., & Tsoudis, O. (1994). Heinous crime or unfortunate accident: The effects of remorse on responses to mock criminal confessions. *Social Problems, 73,* 175–190.

Robinson, J. W., Jr., & Preston, J. D. (1976). Equal-status contact and the modification of racial prejudice: A re-examination of the contact hypothesis. *Social Forces, 54,* 911–924.

Robson, P. (1982). Patterns of mobility and activity among the elderly. In E. Warnes (Ed.), *Geographical perspectives on the elderly.* New York: Wiley.

Roethlisberger, F. J., & Dickson, W. J. (1939). *Management and the worker.* Cambridge, MA: Harvard University Press.

Rogers, M., Miller, N., Mayer, F. S., & Duvall, S. (1982). Personal responsibility and salience of the request for help: Determinants of the relation between negative affect and helping behavior. *Journal of Personality and Social Psychology, 43,* 956–970.

Rogers, R. G. (1995). Marriage, sex, and mortality. *Journal of Marriage and the Family, 57,* 515–526.

Rogers, R. W. (1980). Expressions of aggression: Aggression-inhibiting effects of anonymity to authority and threatened retaliation. *Personality and Social Psychology Bulletin, 6,* 315–320.

Rogers, T. B. (1977). Self-reference in memory: Recognition of personality items. *Journal of Research in Personality, 11,* 295–305.

Rohrer, J. H., Baron, S. H., Hoffman, E. L., & Swander, D. V. (1954). The stability of autokinetic judgments. *Journal of Abnormal and Social Psychology, 49,* 595–597.

Rokeach, M. (1973). *The nature of human values.* New York: Free Press.

Rommetveit, R. (1955). *Social norms and roles.* Minneapolis: University of Minnesota Press.

Ronis, D. L., & Lipinski, E. R. (1985). Value and uncertainty as weighting factors in impression formation. *Journal of Experimental Social Psychology, 21,* 47–60.

Rook, K., Dooley, D., & Catalano, R. (1991). Stress transmission: The effects of husbands' job stressors on the emotional health of their wives. *Journal of Marriage and the Family, 53,* 165–177.

Roopnarine, J. L. (1985). Changes in peer-directed behaviors following preschool experience. *Journal of Personality and Social Psychology, 48,* 740–745.

Rose, S., & Frieze, I. H. (1993). Young singles' contemporary dating scripts. *Sex Roles, 28,* 499–509.

Rosen, B., & D'Andrade, R. (1959). The psychological origins of achievement motivation. *Sociometry, 22,* 185–218.

Rosen, S. (1984). Some paradoxical status implications of helping and being helped. In E. Staub et al. (Eds.), *Development and maintenance of prosocial behavior: International perspectives on positive morality.* New York: Plenum.

Rosenbaum, M. E. (1986). The repulsion hypothesis: On the non-development of relationships. *Journal of Personality and Social Psychology, 51,* 1156–1166.

Rosenbaum, M. E., Moore, D. L., Cotton, J. L., Cook, M. S., Hieser, R. A., Shovar, M. N., & Gray, M. J. (1980). Group productivity and process: Pure and mixed reward structures and task interdependence. *Journal of Personality and Social Psychology, 39,* 626–642.

Rosenberg, L. A. (1961). Group size, prior experience, and conformity. *Journal of Abnormal and Social Psychology, 63,* 436–437.

Rosenberg, M. (1965). *Society and the adolescent self-image.* Princeton, NJ: Princeton University Press.

Rosenberg, M. (1973). Which significant others? *American Behavioral Scientist, 16,* 829–860.

Rosenberg, M. (1979). *Conceiving the self.* New York: Basic Books.

Rosenberg, M. (1990). The self-concept: Social product and social force. In M. Rosenberg & R. H. Turner (Eds.), *Social psychology: Sociological perspectives.* New Brunswick, NJ: Transaction.

Rosenberg, M., & Pearlin, L. (1978). Social class and self-esteem among children and adults. *American Journal of Sociology, 84,* 53–77.

Rosenberg, M., Schooler, C., & Schoenbach, C. (1989). Self-esteem and adolescent problems: Modeling reciprocal effects. *American Sociological Review, 54,* 1004–1018.

Rosenberg, M., Schooler, C., Schoenbach, C., & Rosenberg, F. (1995). Global self-esteem and specific self-esteem: Different concepts, different outcomes. *American Sociological Review, 60,* 141–156.

Rosenberg, M., & Simmons, R. (1972). *Black and white self-esteem: The urban school child.* Washington, DC: American Sociological Association.

Rosenberg, M. J., & Abelson, R. (1960). An analysis of cognitive balancing. In C. Hovland & M. Rosenberg (Eds.), *Attitude organization and change.* New Haven, CT: Yale University Press.

Rosenberg, S. V., Nelson, C., & Vivekananthan, P. S. (1968). A multidimensional approach to the structure of personality impressions. *Journal of Personality and Social Psychology, 9,* 283–294.

Rosenberg, S. V., & Sedlak, A. (1972). Structural representations in implicit personality theory. In L. Berkowitz (Ed.), *Advances in experimental social psychology* (Vol. 6). New York: Academic Press.

Rosenblatt, A., & Greenberg, J. (1988). Depression and interpersonal attraction: The role of perceived similarity. *Journal of Personality and Social Psychology, 55,* 112–119.

Rosenblatt, A., & Greenberg, J. (1991). Examining the world of the depressed: Do depressed people prefer others who are depressed? *Journal of Personality and Social Psychology, 60,* 620–629.

Rosenhan, D. L. (1973). On being sane in insane places. *Science, 179,* 250–258.

Rosenhan, D. L., Salovey, P., & Hargis, K. (1981). The joys of helping: Focus of attention mediates the impact of positive affect on altruism. *Journal of Personality and Social Psychology, 40,* 899–905.

Rosenthal, R. (1966). *Experimenter effects in behavioral research.* New York: Appleton-Century-Crofts.

Rosenthal, R. (1980). Replicability and experimenter influence: Experimenter effects in behavioral research. *Parapsychology, 11,* 5–11.

Ross, A. S. (1971). Effect of increased responsibility on bystander intervention: The presence of children. *Journal of Personality and Social Psychology, 19,* 306–310.

Ross, C. E., Mirowsky, J., & Goldsteen, K. (1990). The impact of the family on health: The decade in review. *Journal of Marriage and the Family, 52,* 1059–1078.

Ross, C. E., & Van Willigen, M. (1996). Gender, parenthood, and anger. *Journal of Marriage and the Family, 58,* 572–584.

Ross, C. E., & Van Willigen, M. (1997). Education and the subjective quality of life. *Journal of Health and Social Behavior, 38,* 275–297.

Ross, C. E., & Wu, C. (1995). The links between education and health. *American Sociological Review, 60,* 719–745.

Ross, L. (1977). The intuitive psychologist and his shortcomings: Distortion in the attribution process. In L. Berkowitz (Ed.), *Advances in experimental social psychology* (Vol. 10). New York: Academic Press.

Ross, L., & Ward, A. (1995). Psychological barriers to dispute resolution. In M. P. Zanna (Ed.), *Advances in experimental social psychology* (Vol. 27, pp. 255–304). San Diego, CA: Academic Press.

Ross, M., & Fletcher, G. (1985). Attribution and social perception. In G. Lindzey & E. Aronson (Eds.), *The handbook of social psychology* (3rd ed.). Reading, MA: Addison-Wesley.

Ross, M., & Lumsden, H. (1982). Attributions of responsibility in sports settings: It's not how you play the game but whether you win or lose. In H. Hiebsch, H. Brandstatter, & H. H. Kelley (Eds.), *Social psychology.* East Berlin: VEB Deutscher Verlag der Wissenschaften.

Ross, M., Thibaut, J., & Evenbeck, S. (1971). Some determinants of the intensity of social protests. *Journal of Experimental Social Psychology, 7,* 401–418.

Rossi, A. S. (1980). Parenthood in the middle years. In P. Baltes & O. Brim, Jr. (Eds.), *Life-span development and behavior* (Vol. 3). New York: Academic Press.

Rotenberg, K. J., & Mann, L. (1986). The development of the norm of the reciprocity of self-disclosure and its function in children's attraction to peers. *Child Development, 57*(6), 1349–1357.

Roth, D. L., Wiebe, D. J., Fillingian, R. B., & Shay, K. A. (1989). Life events, fitness, hardiness and health: A simultaneous analysis of proposed stress-resistance effects. *Journal of Personality and Social Psychology, 57,* 136–142.

Rothbart, M., Dawes, R., & Park, B. (1984). Stereotypes and sampling biases in intergroup perception. In J. R. Eiser (Ed.), *Attitudinal judgment* (pp. 109–134). New York: Springer.

Rothbart, M., Fulero, S., Jensen, C., Howard, J., & Birrell, B. (1978). From individual to group impressions: Availability heuristics in stereotype formation. *Journal of Experimental Social Psychology, 14,* 237–255.

Rothbart, M., & John, O. P. (1985). Social categorization and behavioral episodes: A cognitive analysis of the effects of intergroup contact. *Journal of Social Issues, 41,* 81–104.

Rotheram-Barus, M. J. (1990). Adolescents' reference group choices, self-esteem and adjustment. *Journal of Personality and Social Psychology, 59,* 1075–1081.

Rothstein, S. I., & Pierotti, R. (1988). Distinctions among reciprocal altruism, kin selection, and cooperation and a model for the initial evolution of beneficent behavior. *Ethology and Sociobiology, 9,* 189–209.

Rotton, J., & Frey, J. (1985). Air pollution, weather and violent crimes: Concomitant time-series analyses of archival data. *Journal of Personality and Social Psychology, 49,* 1207–1220.

Ruback, R. (1987). Deserted (and nondeserted) aisles: Territorial intrusion can produce persistence, not flight. *Social Psychology Quarterly, 50,* 270–276.

Ruback, R. B., Pape, K. D., & Doriot, P. (1989). Waiting for a phone: Intrusion on callers leads to territorial defense. *Social Psychology Quarterly, 52,* 232–241.

Rubin, B. A. (1986). Class struggle American style: Unions, strikes and wages. *American Sociological Review, 51,* 618–631.

Rubin, J. (1962). Bilingualism in Paraguay. *Anthropological Linguistics, 4,* 52–68.

Rubin, J. Z., & Brown, B. (1975). *The social psychology of bargaining and negotiation.* New York: Academic Press.

Rubin, J. Z., & Lewecki, R. J. (1973). A three-factor experimental analysis of promises and threats. *Journal of Applied Social Psychology, 3,* 240–257.

Rubin, L. (1979). *Women of a certain age: The midlife search for self.* New York: Harper & Row.

Rubin, L. (1983). *Intimate strangers: Men and women together.* New York: Harper & Row.

Rubin, Z. (1970). Measurement of romantic love. *Journal of Personality and Social Psychology, 16,* 265–273.

Rubin, Z. (1974). From liking to loving: Patterns of attraction in dating relationships. In T. Huston (Ed.), *Foundations of interpersonal attraction.* New York: Academic Press.

Rubin, Z., Hill, C., Peplau, L., & Dunkel-Scheker, C. (1980). Self-disclosure in dating couples: Sex roles and the ethic of openness. *Journal of Marriage and the Family, 42,* 305–317.

Rubinstein, E. A. (1983). Television and behavior: Research conclusions of the 1982 NIMH report and their policy implications. *American Psychologist, 38,* 820–825.

Rude, G. (1964). *The crowd in history.* New York: Wiley.

Rusbult, C. E. (1983). A longitudinal test of the investment model: The development (and deterioration) of satisfaction and commitment in heterosexual involvements. *Journal of Personality and Social Psychology, 45,* 101–117.

Rusbult, C. E., Johnson, D. J., & Morrow, G. D. (1986). Predicting satisfaction and commitment in adult romantic involvements: An assessment of the generalizability of the investment model. *Social Psychology Quarterly, 49,* 81–89.

Rusbult, C. E., Verette, J., Whitney, G. A., Slovik, L. A., & Lipkus, I. (1991). Accommodation processes in close relationships: Theory and preliminary empirical evidence. *Journal of Personality and Social Psychology, 60,* 53–78.

Rusbult, C. E., Zembrodt, I. M., & Gunn, L. K. (1982). Exit, voice, loyalty and neglect: Responses to dissatisfaction in romantic involvement. *Journal of Personality and Social Psychology, 43,* 1230–1242.

Rushton, J. (1978). Urban density and altruism: Helping strangers in a Canadian city, suburb, and small town. *Psychological Reports, 43*(3, pt. 1), 987–990.

Rushton, J. P., & Campbell, A. C. (1977). Modeling, vicarious reinforcement and extraversion on blood donating in adults: Immediate and long-term effects. *European Journal of Social Psychology, 7,* 297–306.

Ryder, N. B. (1965). The cohort as a concept in the study of social change. *American Sociological Review, 30,* 843–861.

Ryen, A. H., & Kahn, A. (1975). The effects of intergroup orientation on group attitudes and proxemic behavior. *Journal of Personality and Social Psychology, 31,* 302–310.

Sacks, H., Schegloff, E., & Jefferson, G. (1978). A simplest systematics for the organization of turn-taking in conversations. In J. Schenkein (Ed.), *Studies in the organization of conversational interaction.* New York: Academic Press.

Saegert, S. C., Swap, W., & Zajonc, R. B. (1973). Exposure, context and interpersonal attraction. *Journal of Personality and Social Psychology, 25,* 234–242.

Sagarin, E. (1975). *Deviants and deviance.* New York: Praeger.

Saito, Y. (1988). Situational characteristics as the determinants of adopting distributive justice principles: II. *Japanese Journal of Experimental Social Psychology, 27,* 131–138.

Sakurai, M. M. (1975). Small group cohesiveness and detrimental conformity. *Sociometry, 38*(3), 340–357.

Sales, S. (1969). Organizational roles as a risk factor in coronary heart disease. *Administrative Science Quarterly, 14,* 325–336.

Salovey, P., Mayer, J. D., & Rosenhan, D. L. (1991). Mood and helping: Mood as a motivator of helping and helping as a regulator of mood. In M. S. Clark (Ed.), *Review of personality and social psychology: Vol. 12. Prosocial behavior* (pp. 215–237). Newbury Park, CA: Sage.

Saltzer, E. B. (1981). Cognitive moderation of the relationship between behavioral intentions and behavior. *Journal of Personality and Social Psychology, 41,* 260–271.

Sammon, S., Reznikoff, M., & Geisinger, K. (1985). Psychosocial development and stressful life events among religious professionals. *Journal of Personality and Social Psychology, 48,* 676–687.

Sample, J., & Warland, R. (1973). Attitude and the prediction of behavior. *Social Forces, 51,* 292–304.

Sampson, H., Messinger, S., Towne, R., Russ, D., Livson, F., Bowers, M., Cohen, L., & Dorst, K. (1964). The mental hospital and marital family ties. In H. Becker (Ed.), *The other side*. New York: Free Press.

Sampson, R. J., & Laub, J. H. (1990). Crime and deviance over the life course: The salience of adult social bonds. *American Sociological Review, 55*, 609–627.

Samuels, F. (1970). The intra- and inter-competitive group. *Sociological Quarterly, 11*, 390–396.

Sanday, P. R. (1981). The socio-cultural context of rape: A cross-cultural study. *Journal of Social Issues, 37*(4), 5–27.

Sande, G. N., Goethals, G. R., & Radloff, C. E. (1988). Perceiving one's own traits and others': The multifaceted self. *Journal of Personality and Social Psychology, 54*, 13–20.

Sanders, C. (1988). Risk factors in bereavement outcome. *Journal of Social Issues, 44*, 97–111.

Sanitioso, R., Kunda, Z., & Fong, G. T. (1990). Motivated recruitment of autobiographical memories. *Journal of Personality and Social Psychology, 59*, 229–241.

Sarbin, T., & Rosenberg, B. (1955). Contributions to role-taking theory IV: A method for obtaining a qualitative estimate of the self. *Journal of Social Psychology, 42*, 71–81.

Sawyer, A. (1973). The effects of repetition of refutational and supportive advertising appeals. *Journal of Marketing Research, 10*, 23–33.

Schachter, S. (1964). The interaction of cognitive and physological determinants of emotional state. In L. Berkowitz (Ed.), *Advances in experimental social psychology* (Vol. 1). New York: Academic Press.

Schachter, S., & Singer, J. (1962). Cognitive, social and physiological determinants of emotional state. *Psychological Review, 69*, 379–399.

Schaller, M., & Cialdini, R. B. (1988). The economics of empathic helping: Support for a mood management motive. *Journal of Experimental Social Psychology, 24*, 163–181.

Schank, R. C., & Abelson, R. P. (1977). *Scripts, plans, goals and understanding*. Hillsdale, NJ: Erlbaum.

Scheff, T. (1966). *Being mentally ill*. Chicago: Aldine de Gruyter.

Schegloff, E. (1968). Sequencing in conversational openings. *American Anthropologist, 70*, 1075–1095.

Scheier, M. F., & Carver, C. (1981). Public and private aspects of the self. In L. Wheeler (Ed.), *Review of personality and social psychology* (Vol. 2). Beverly Hills, CA: Sage.

Scherer, K. R. (1979). Nonlinguistic indicators of emotion and psychopathology. In C. E. Izard (Ed.), *Emotions in personality and psychopathology*. New York: Plenum.

Scherer, S. E. (1974). Proxemic behavior of primary school children as a function of their socioeconomic class and subculture. *Journal of Personality and Social Psychology, 29*, 800–805.

Schiffenbauer, A., & Schiavo, R. S. (1976). Physical distance and attraction: An intensification effect. *Journal of Experimental Social Psychology, 12*, 274–282.

Schiffrin, D. (1977). Opening encounters. *American Sociological Review, 42*, 679–691.

Schifter, D. E., & Ajzen, I. (1985). Intention, perceived control and weight loss: An application of the theory of planned behavior. *Journal of Personality and Social Psychology, 45*, 843–851.

Schlenker, B. R. (1975). Self-presentation. Managing the impression of consistency when reality interferes with self-enhancement. *Journal of Personality and Social Psychology, 32*, 1030–1037.

Schlenker, B. R. (1980). *Impression management: The self-concept, social identity, and interpersonal relations*. Belmont, CA: Brooks/Cole.

Schlenker, B. R., Helm, B., & Tedeschi, J. T. (1973). The effects of personality and situational variables on behavioral trust. *Journal of Personality and Social Psychology, 25*, 419–427.

Schlenker, B. R., & Weigold, M. F. (1992). Interpersonal processes involving impression regulation and management. *Annual Review of Psychology, 43*, 133–168.

Schlenker, B. R., Weigold, M. E., & Hallam, J. K. (1990). Self-serving attributions in social context: Effects of self-esteem and social pressure. *Journal of Personality and Social Psychology, 58*, 855–863.

Schmitt, D. P., & Buss, D. M. (1996). Strategic self-promotion and competitor derogation: Sex and context effects in the perceived effectiveness of mate attraction tactics. *Journal of Personality and Social Psychology, 70*, 1185–1204.

Schmitt, D. R., & Marwell, G. (1972). Withdrawal and reward reallocation as responses to inequity. *Journal of Experimental Social Psychology, 8*, 207–221.

Schonbach, P. (1980). A category system for account phrases. *European Journal of Social Psychology, 10*, 195–200.

Schooler, C. (1996). Cultural and social-structural explanations of cross-national psychological differences. *Annual Review of Sociology, 22*, 323–349.

Schrauger, J. S., & Schoeneman, T. (1979). Symbolic interactionist view of self-concept: Through the looking glass darkly. *Psychological Bulletin, 86*, 549–573.

Schriesheim, C., & Kerr, S. (1974). Psychometric properties of the Ohio State leadership scales. *Psychological Bulletin, 81*, 756–765.

Schroeder, D. A., Penner, L. A., Dovidio, J. F., & Piliavin, J. A. (1995). *The psychology of helping and altruism: Problems and puzzles*. New York: McGraw-Hill.

Schrum, W., & Creek, N. A., Jr. (1987). Social structure during the school years: Onset of the degrouping process. *American Sociological Review, 52*, 218–223.

Schuman, H., & Johnson, M. (1976). Attitudes and behavior. In A. Inkeles, J. Coleman, & N. Smelser (Eds.), *Annual review of sociology* (Vol. 2). Palo Alto, CA: Annual Reviews.

Schuman, H., & Kalton, G. (1985). Survey methods. In G. Lindzey & E. Aronson (Eds.), *The handbook of social psychology* (3rd ed., Vol. 1, pp. 635–698). New York: Random House.

Schutte, J., & Light, J. (1978). The relative importance of proximity and status for friendship choices in social hierarchies. *Social Psychology, 41*, 260–264.

Schutte, N. S., Kendrick, D. T., & Sadalla, E. K. (1985). The search for predictable settings: Situational prototypes, constraint, and behavioral variation. *Journal of Personality and Social Psychology, 49*, 121–128.

Schwartz, R., & Skolnick, J. (1964). Two studies of legal stigma. In H. Becker (Ed.), *The other side*. New York: Free Press.

Schwartz, S., & Ames, R. (1977). Positive and negative referent others as sources of influence: A case of helping. *Sociometry, 40*, 12–20.

Schwartz, S. H. (1977). Normative influences on altruism. In L. Berkowitz (Ed.), *Advances in experimental social psychology* (Vol. 10). New York: Academic Press.

Schwartz, S. H. (1978). Temporal instability as a moderator of the attitude-behavior relationship. *Journal of Personality and Social Psychology, 36*, 715–724.

Schwartz, S. H., & Clausen, G. T. (1970). Responsibility, norms, and helping in an emergency. *Journal of Personality and Social Psychology, 16,* 299–310.

Schwartz, S. H., & Fleishman, J. (1978). Personal norms and the mediation of legitimacy effects on helping. *Social Psychology, 41,* 306–315.

Schwartz, S. H., & Gottlieb, A. (1976). Bystander reactions to a violent theft: Crime in Jerusalem. *Journal of Personality and Social Psychology, 34,* 1188–1199.

Schwartz, S. H., & Gottlieb, A. (1980). Bystander anonymity and reactions to emergencies. *Journal of Personality and Social Psychology, 39,* 418–430.

Schwartz, S. H., & Howard, J. A. (1980). Explanations of the moderating effect of responsibility denial on personal norm-behavior relationship. *Social Psychology Quarterly, 43,* 441–446.

Schwartz, S. H., & Howard, J. A. (1981). A normative decision-making model of altruism. In J. P. Rushton & R. M. Sorrentino (Eds.), *Altruism and helping behavior.* Hillsdale, NJ: Erlbaum.

Schwartz, S. H., & Howard, J. A. (1982). Helping and cooperation: A self-based motivational model. In V. J. Derlega & J. Grzelak (Eds.), *Cooperation and helping behavior: Theories and research* (pp. 327–353). New York: Academic Press.

Schwartz, S. H., & Howard, J. A. (1984). Internalized values as motivators of altruism. In E. Staub, E. Bar-Tal, J. Karylowski, & J. Reykowski (Eds.), *Development and maintenance of prosocial behavior: International perspectives on positive morality.* New York: Plenum.

Scott, M., & Lyman, S. (1968). Accounts. *American Sociological Review, 33,* 46–62.

Scott, R. (1976). Deviance, sanctions and social integration in small-scale societies. *Social Forces, 54,* 604–620.

Scott, W. R. (1981). *Rational, natural, and open systems.* Englewood Cliffs, NJ: Prentice-Hall.

Scotton, C. M. (1983). The negotiation of identities in conversation. *International Journal of the Sociology of Language, 44,* 115–136.

Searle, J. R. (1979). *Expression and meaning: Studies in the theory of speech acts.* New York: Cambridge University Press.

Sears, D. O., & Freedman, J. L. (1967). Selective exposure to information: A critical review. *Public Opinion Quarterly, 31,* 194–213.

Sears, D. O., & Whitney, R. E. (1973). Political persuasion. In I. deSola Pool, W. Schramm, N. Maccoby, & E. B. Parker (Eds.), *Handbook of communication* (pp. 253–289). Chicago: Rand McNally.

Sears, R. R., Maccoby, E., & Levin, H. (1957). *Patterns of child rearing.* New York: Harper & Row.

Seccombe, K. (1986). The effects of occupational conditions upon the division of household labor. *Journal of Marriage and the Family, 48,* 839–848.

Sedikides, C. & Jackson, J. M. (1990). Social impact theory: A field test of source strength, source immediacy and number of targets. *Basic and Applied Social Psychology, 11*(3), 273–281.

Seedman, A. A., & Hellman, P. (1975). *Chief.* New York: Avon Books.

Seeman, M. (1975). Alienation studies. In A. Inkeles, J. Coleman, & N. Smelser (Eds.), *Annual review of sociology* (Vol. 1). Palo Alto, CA: Annual Reviews.

Seeman, M., Seeman, M., & Sayles, M. (1985). Social networks and health status: A longitudinal analysis. *Social Psychology Quarterly, 48,* 237–248.

Segal, B. E. (1965). Contact, compliance and distance among Jewish and non-Jewish undergraduates. *Social Problems, 13,* 66–74.

Seltzer, J., & Bass, B. M. (1990). Transformational leadership: Beyond initiation and consideration. *Journal of Management, 16,* 693–703.

Semin, G. R., & Manstead, A. S. R. (1982). The social implications of embarrassment displays and restitution behaviour. *European Journal of Social Psychology, 12,* 367–377.

Semin, G. R., & Manstead, A. S. R. (1983). *The accountability of conduct: A social psychological analysis.* London: Academic Press.

Serbin, L., & O'Leary, K. (1975). How nursery schools teach girls to shut up. *Psychology Today, 9,* 56–58 ff.

Serpe, R. T. (1987). Stability and change in self: A structural symbolic interactionist explanation. *Social Psychology Quarterly, 50,* 44–55.

Sewell, W. H., & Hauser, R. M. (1975). *Education, occupation and earnings: Achievement in the early career.* New York: Academic Press.

Sewell, W. H., & Hauser, R. M. (1980). The Wisconsin longitudinal study of social and psychological factors in aspirations and achievements. *Research in Sociology of Education and Socialization, 1,* 59–99.

Sewell, W. H., Hauser, R. M., & Wolf, W. (1980). Sex, schooling, and occupational status. *American Journal of Sociology, 86,* 551–583.

Shanas, E. (1979). The family as a support system in old age. *Gerontologist, 19,* 169–174.

Shapiro, S. P. (1990). Collaring the crime, not the criminal: Considering the concept of white-collar crime. *American Sociological Review, 55,* 346–365.

Sharkey, W. F., & Stafford, L. (1990). Responses to embarrassment. *Human Communication Research, 17,* 315–342.

Shaver, P., & Freedman, J. (1976, August). Your pursuit of happiness. *Psychology Today,* pp. 26–32.

Shaw, M., & Costanzo, P. (1982). *Theories of social psychology* (2nd ed.). New York: McGraw-Hill.

Shaw, M. E. (1964). Communication networks. In L. Berkowitz (Ed.), *Advances in experimental social psychology* (Vol. 1, pp. 111–147). New York: Academic Press.

Shaw, M. E. (1978). Communication networks fourteen years later. In L. Berkowitz (Ed.), *Group processes* (pp. 351–361). New York: Academic Press.

Shaw, M. E., & Rothschild, G. H. (1956). Some effects of prolonged experience in communication nets. *Journal of Applied Psychology, 40,* 281–286.

Shaw, M. E., & Shaw, L. M. (1962). Some effects of sociometric grouping upon learning in a second grade classroom. *Journal of Social Psychology, 57,* 453–458.

Shell, R. M., & Eisenberg, N. (1992). A developmental model of recipients' reactions to aid. *Psychological Bulletin, 111*(3), 413–433.

Shelton, B. A., & John, D. (1996). The division of household labor. *Annual Review of Sociology, 22,* 299–322.

Sherif, M. (1935). A study of some social factors in perception. *Archives of Psychology, 27*(187).

Sherif, M. (1936). *The psychology of social norms.* New York: Harper & Row.

Sherif, M. (1966). *In common predicament.* Boston: Houghton Mifflin.

Sherif, M., Harvey, O. J., White, B. J., Hood, W. R., & Sherif, C. W. (1961). *Intergroup cooperation and competition: The Robbers Cave experiment.* Norman, OK: University Book Exchange.

Sherif, M., & Sherif, C. (1964). *Exploration into conformity and deviation of adolescents.* New York: Harper & Row.

Sherif, M., & Sherif, C. W. (1982). Production of intergroup conflict and its resolution—Robbers Cave experiment. In J. W. Reich (Ed.), *Experimenting in society: Issues and examples in applied social psychology.* Glenview, IL: Scott, Foresman.

Sherkat, D. E., & Blocker, T. J. (1994). The political development of sixties' activists: Identifying the influence of class, gender and socialization on protest participation. *Social Forces, 72*(3), 821–842.

Sherman, P. (1980). The limits of ground squirrel nepotism. In G. Barlow & J. Silverberg (Eds.), *Sociobiology: Beyond nature/nurture?* Boulder, CO: Westview.

Sherman, S. J. (1970). Effects of choice and incentive on attitude change in a discrepant behavior situation. *Journal of Personality and Social Psychology, 15*, 245–252.

Sherman, S. J., Judd, C. M., & Park, B. (1989). Social cognition. *Annual Review of Psychology, 40*, 281–326.

Sherrod, D. R., & Downs, R. (1974). Environmental determinants of altruism: The effects of stimulus overload and perceived control on helping. *Journal of Experimental Social Psychology, 10*, 468–479.

Sherwood, J. J. (1965). Self-identity and referent others. *Sociometry, 28*, 66–81.

Shibutani, T. (1961). *Society and personality.* Englewood Cliffs, NJ: Prentice-Hall.

Shiflett, S. (1979). Toward a general model of small group productivity. *Psychological Bulletin, 86*, 67–79.

Shihadeh, E. S. (1991). The prevalence of husband-centered migration: Employment consequences for married mothers. *Journal of Marriage and the Family, 53*, 431–444.

Shotland, R. L., & Stebbins, C. A. (1983). Emergency and cost as determinants of helping behavior and the slow accumulation of social psychological knowledge. *Social Psychology Quarterly, 46*, 36–46.

Shotland, R. L., & Straw, M. K. (1976). Bystander response to an assault: When a man attacks a woman. *Journal of Personality and Social Psychology, 34*, 990–999.

Shott, S. (1979). Emotion and social life: A symbolic interactionist analysis. *American Journal of Sociology, 84*, 1317–1334.

Shover, N., Novland, S., James, J., & Thornton, W. (1979). Gender roles and delinquency. *Social Forces, 58*, 162–175.

Shulman, N. (1975). Life-cycle variations in patterns of close relationships. *Journal of Marriage and the Family, 37*, 813–821.

Shweder, R. A. (1977). Likeness and likelihood in everyday thought: Magical thinking in judgments about personality. *Current Anthropology, 18*, 637–658.

Sigall, H., & Landy, D. (1973). Radiating beauty: The effects of having a physically attractive partner on person perception. *Journal of Personality and Social Psychology, 28*, 218–224.

Silberman, M. (1976). Toward a theory of criminal deterrence. *American Sociological Review, 41*, 442–461.

Simmons, R. G. (1991). Presidential address on altruism and sociology. *Sociological Quarterly, 32*, 1–22.

Simmons, R. G., Brown, L., Bush, D., & Blyth, D. (1978). Self-esteem and achievement of black and white adolescents. *Social Problems, 26*, 86–96.

Simon, R. (1997). The meanings individuals attach to role-identities and their implications for mental health. *Journal of Health and Social Behavior, 38*, 256–274.

Simpson, J. A. (1987). The dissolution of romantic relationships: Factors involved in emotional stability and emotional distress. *Journal of Personality and Social Psychology, 53*, 683–692.

Simpson, J. A., Gangestad, S. W., & Lerum, M. (1990). Perception of physical attractiveness: Mechanisms involved in maintenance of romantic relationships. *Journal of Personality and Social Psychology, 59*, 1192–1201.

Singer, J. L., & Singer, D. G. (1981). *Television, imagination and aggression: A study of preschoolers.* Hillsdale, NJ: Erlbaum.

Singer, J. L., & Singer, D. G. (1983). Psychologists look at television: Cognitive, developmental, personality, and social policy implications. *American Psychologist, 38*, 826–834.

Singh, D. (1995). Female judgment of male attractiveness and desirability for relationships: Role of waist-to-hip ratio and financial status. *Journal of Personality and Social Psychology, 69*, 1089–1101.

Sinnott, J. D. (1977). Sex-role inconstancy, biology, and successful aging. *Gerontologist, 17*, 459–463.

Sistrunk, F., & McDavid, J. W. (1971). Sex variable in conforming behavior. *Journal of Personality and Social Psychology, 17*, 200–207.

Sivacek, J., & Crano, W. (1982). Vested interest as a moderator of attitude-behavior consistency. *Journal of Personality and Social Psychology, 43*, 210–221.

Skager, R., & Fisher, D. G. (1989). Substance abuse among high school students in relation to school characteristics. *Addictive Behaviors, 14*, 129–138.

Skinner, B. F. (1953). *Science and human behavior.* New York: Macmillan.

Skinner, B. F. (1957). *Verbal behavior.* New York: Appleton-Century-Crofts.

Skinner, B. F. (1971). *Beyond freedom and dignity.* New York: Knopf.

Skowronski, J. J., & Carlston, D. E. (1989). Negativity and extremity biases in impression formation: A review of explanations. *Psychological Bulletin, 105*(1), 131–142.

Slater, P. (1963). On social regression. *American Sociological Review, 28*, 339–364.

Slater, P. E. (1955). Role differentiation in small groups. *American Sociological Review, 20*, 300–310.

Sloane, D., & Potrin, R. H. (1986). Religion and delinquency: Cutting through the maze. *Social Forces, 65*, 87–105.

Slomczynski, K. M., Miller, J., & Kohn, M. (1981). Stratification, work and values: A Polish-United States comparison. *American Sociological Review, 46*, 720–744.

Small, K. H., & Peterson, J. (1981). The divergent perceptions of actors and observers. *Journal of Social Psychology, 113*, 123–132.

Smith, D. A. (1987). Police response to interpersonal violence: Defining the parameters of legal control. *Social Forces, 65*, 767–782.

Smith, E. R., Fazio, R. H., & Cejka, M. A. (1996). Accessible attitudes influence categorization of multiply categorizable objects. *Journal of Personality and Social Psychology, 71*, 888–898.

Smith, E. R., & Henry, S. (1996). An in-group becomes part of the self: Response time evidence. *Personality and Social Psychology Bulletin, 22*, 635–642.

Smith, H. J., & Spears, R. (1996). Ability and outcome evaluations as a function of personal, and collective (dis)advantage: A group escape from individual bias. *Personality and Social Psychology Bulletin, 22*, 690–704.

Smith, W. P., & Anderson, A. (1975). Threats, communication, and bargaining. *Journal of Personality and Social Psychology, 32*, 76–82.

Smith, W. P., & Leginski, W. A. (1970). Magnitude and precision of punitive power in bargaining strategy. *Journal of Experimental Social Psychology, 6,* 57–76.

Smith-Lovin, L. (1990). Emotions as the confirmation and disconfirmation of identity: The affect control model. In T. D. Kemper (Ed.), *Research agendas in the sociology of emotion.* Albany, NY: SUNY Press.

Smith-Lovin, L., & Brody, C. (1989). Interruptions in group discussions: The effects of gender and group composition. *American Sociological Review, 54,* 424–435.

Sniezek, J. A., & May, D. R. (1990). Conflict of interests and commitment in groups. *Journal of Applied Social Psychology, 20,* 1150–1165.

Snow, D. A., & Oliver, P. E. (1995). Social movements and collective behavior: Social psychological dimensions and considerations. In K. S. Cook, G. A. Fine, & J. S. House (Eds.), *Sociological perspectives on social psychology* (pp. 571–599). Boston: Allyn & Bacon.

Snow, D. A., Rochford, E., Jr., Worden, S., & Benford, R. (1986). Frame alignment processes, micromobilization and movement participation. *American Sociological Review, 51,* 464–481.

Snyder, C. R., Higgins, R. L., & Stucky, R. J. (1984). *Excuses: Masquerades in search of grace.* New York: Wiley.

Snyder, C. R., Lassegard, M. A., & Ford, C. E. (1986). Distancing after group success and failure: Basking in reflected glory and cutting off reflected failure. *Journal of Personality and Social Psychology, 51,* 382–388.

Snyder, D. (1975). Institutional setting and industrial conflict: Comparative analysis of France, Italy and the United States. *American Sociological Review, 40,* 259–278.

Snyder, M. (1979). Self-monitoring. In L. Berkowitz (Ed.), *Advances in experimental social psychology* (Vol. 12). New York: Academic Press.

Snyder, M. (1981). On the self-perpetuating nature of social stereotypes. In D. L. Hamilton (Ed.), *Cognitive processes in stereotyping and intergroup behavior.* Hillsdale, NJ: Erlbaum.

Snyder, M., & Swann, W. B., Jr. (1978). Hypothesis-testing processes in social interaction. *Journal of Personality and Social Psychology, 36,* 1202–1212.

Snyder, M., Tanke, E. D., & Berscheid, E. (1977). Social perception and interpersonal behavior: On the self-fulfilling nature of social stereotypes. *Journal of Personality and Social Psychology, 35,* 656–666.

Solano, C. H., Batten, P. G., & Parish, E. A. (1982). Loneliness and patterns of self-disclosure. *Journal of Personality and Social Psychology, 43,* 524–531.

Solomon, D. S. (1982). Mass media campaigns for health promotion. *Prevention in Human Services, 2,* 115–123.

Sommer, R. (1969). *Personal space.* Englewood Cliffs, NJ: Prentice-Hall.

Sommers-Flanagan, R., Sommers-Flanagan, J., & Davis, B. (1993). What's happening on Music Television? A gender role content analysis. *Sex Roles, 28,* 745–753.

Sorrentino, R. M., & Field, N. (1986). Emergent leadership over time: The functional value of positive motivation. *Journal of Personality and Social Psychology, 50,* 1091–1099.

South, S., & Spitze, G. (1994). Housework in marital and nonmarital households. *American Sociological Review, 59,* 327–347.

Spencer, J. W. (1987). Self-work in social interaction: Negotiating role-identities. *Social Psychology Quarterly, 50,* 131–142.

Spencer, J. W., & Drass, K. (1989). The transformation of gender into conversational advantage: A symbolic interactionist approach. *Sociological Inquiry, 30,* 363–383.

Spiegel, J. P. (1969). Hostility, aggression and violence. In A. Grimshaw (Ed.), *Patterns in American racial violence.* Chicago: Aldine de Gruyter.

Spilerman, S. (1976). Structural characteristics of cities and severity of racial disorders. *American Sociological Review, 41,* 771–793.

Spitz, R. (1945). Hospitalism. *The Psychoanalytic Study of the Child, 1,* 53–72.

Spitz, R. (1946). Hospitalism: A follow-up report. *The Psychoanalytic Study of the Child, 2,* 113–117.

Spitze, G. D., & Huber, J. (1980). Changing attitudes toward women's nonfamily roles: 1938 to 1978. *Sociology of Work and Occupations, 7,* 317–335.

Sprecher, S., & Metts, S. (1989). Development of the "Romantic Beliefs Scale" and examination of the effects of gender and gender role orientation. *Journal of Social and Personal Relationships, 6,* 387–411.

Staats, A. W., & Staats, C. (1958). Attitudes established by classical conditioning. *Journal of Abnormal and Social Psychology, 57,* 37–40.

Stack, S. (1987). Celebrities and suicide: A taxonomy and analysis, 1948–1983. *American Sociological Review, 52,* 401–412.

Stafford, L., & Canary, D. J. (1991). Maintenance strategies by romantic relationship type, gender, and relational characteristics. *Journal of Social and Personal Relationships, 8,* 217–242.

Stafford, M. C., Gray, L. N., Menke, B. A., & Ward, D. A. (1986). Modelling the deterrent effects of punishment. *Social Psychology Quarterly, 49,* 338–347.

Staggenborg, S. (1986). Coalition work in the pro-choice movement: Organizational and environmental opportunities and obstacles. *Social Problems, 33,* 374–390.

Stang, D. J. (1972). Conformity, ability, and self-esteem. *Representative Research in Social Psychology, 3,* 97–103.

Stark, R., & Bainbridge, W. S. (1980). Networks of faith: Interpersonal bonds and recruitment in cults and sects. *American Journal of Sociology, 85,* 1376–1395.

Starrels, M. E. (1994). Husband's involvement in female gender-typed household chores. *Sex Roles, 31,* 473–491.

Stasser, G. (1992). Pooling of unshared information during group discussion. In S. Worchel, W. Wood, & J. Simpson (Eds.), *Group process and productivity* (pp. 48–67). Newbury Park, CA: Sage.

Stasser, G., & Titus, W. (1987). Effects of information load and percentage of shared information on the dissemination of unshared information during group discussion. *Journal of Personality and Social Psychology, 53,* 81–93.

Staub, E. (1974). Helping a distressed person: Social, personality, and stimulus determinants. In L. Berkowitz (Ed.), *Advances in experimental social psychology* (Vol. 7). New York: Academic Press.

Steblay, N. M. (1987). Helping behavior in rural and urban environments: A meta-analysis. *Psychological Bulletin, 102,* 346–356.

Steffensmeier, D. J., & Allan, E. (1996). Gender and crime: Toward a gendered theory of female offending. *Annual Review of Sociology, 22,* 459–487.

Steffensmeier, D. J., Allan, E. A., Haver, M. D., & Streifel, C. (1989). Age and the distribution of crime. *American Journal of Sociology, 94,* 803–831.

Steffensmeier, D. J., & Terry, R. M. (1973). Deviance and respectability: An observational study of reactions to shoplifting. *Social Forces, 51,* 417–426.

Stein, J. A., Newcomb, M. D., & Bentler, P. M. (1987). An eight-year study of multiple influences on drug use and drug use consequences. *Journal of Personality and Social Psychology, 53,* 1094–1105.

Steiner, D. D., & Rain, J. S. (1989). Immediate and delayed primacy and recency effects in performance evaluation. *Journal of Applied Psychology, 74,* 136–142.

Steiner, I. D. (1972). *Group process and productivity.* New York: Academic Press.

Steiner, I. D. (1974). *Task-performing groups.* Morristown, NJ: General Learning Press.

Steiner, I. D., & Rogers, E. (1963). Alternative responses to dissonance. *Journal of Abnormal and Social Psychology, 66,* 128–136.

Stemp, P., Turner, R., & Noh, S. (1986). Psychological distress in the postpartum period: The significance of social support. *Journal of Marriage and the Family, 48,* 271–277.

Stephan, W. G. (1987). The contact hypothesis in intergroup relations. In C. Hendrick (Ed.), *Group processes and intergroup relations* (pp. 13–40). Beverly Hills, CA: Sage.

Stephenson, W. (1953). *The study of behavior.* Chicago: University of Chicago Press.

Sternberg, R. J. (1985). Implicit theories of intelligence, creativity, and wisdom. *Journal of Personality and Social Psychology, 49,* 607–627.

Sternthal, B., Dholakia, R., & Leavitt, C. (1978). The persuasive effect of source credibility: A test of cognitive response analysis. *Journal of Consumer Research, 4,* 252–260.

Stevenson, M. B., Ver Hoeve, J. N., Roach, M. A., & Leavitt, L. A. (1986). The beginning of conversation: Early patterns of mother-infant vocal responsiveness. *Infant Behavior and Development, 9,* 423–440.

Stewart, R. H. (1965). Effect of continuous responding on the order effect in personality impression formation. *Journal of Personality and Social Psychology, 1,* 161–165.

Stiff, J. B. (1986). Cognitive processing of persuasive message cues: A meta-analytic review of the effects of supporting information on attitudes. *Communication Monographs, 53,* 75–89.

Stiles, W., Orth, J., Scherwitz, L., Hennrikus, D., & Vallbona, C. (1984). Role behaviors in routine medical interviews with hypertensive patients: A repertoire of verbal exchanges. *Social Psychology Quarterly, 47,* 244–254.

Stokes, J. P. (1985). The relation of social network and individual difference variables to loneliness. *Journal of Personality and Social Psychology, 48,* 981–990.

Stokes, J. P., & Levin, I. (1986). Gender differences in predicting loneliness from social network characteristics. *Journal of Personality and Social Psychology, 51,* 1069–1074.

Stokes, R., & Hewitt, J. P. (1976). Aligning actions. *American Sociological Review, 41*(5), 838–849.

Stoll, C. S. (1978). *Female and male: Socialization, social roles, and social structure.* Dubuque, IA: William C. Brown.

Stone, G. P. (1962). Appearances and the self. In A. Rose (Ed.), *Human behavior and social processes.* Boston: Houghton Mifflin.

Stoner, J. A. F. (1961). *A comparison of individual and group decisions involving risk.* Unpublished master's thesis, Massachusetts Institute of Technology.

Stoner, J. A. F. (1968). Risky and cautious shifts in group decisions: The influence of widely held values. *Journal of Experimental Social Psychology, 4,* 442–459.

Storms, M. D. (1973). Videotape and attribution process: Reversing actors' and observers' points of view. *Journal of Personality and Social Psychology, 27,* 165–175.

Strauman, T. J., Vookles, J., Berenstein, V., Chaiken, S., & Higgins, E. T. (1991). Self-discrepancies and vulnerability to body dissatisfaction and disordered eating. *Journal of Personality and Social Psychology, 61,* 946–956.

Straus, M., Gelles, R. J., & Steinmetz, S. K. (1980). *Behind closed doors: Violence in the American family.* Garden City, NJ: Anchor.

Street, W. R. (1974). Brainstorming by individuals, coacting and interacting groups. *Journal of Applied Psychology, 59*(4), 433–436.

Stricker, L. J., Jacobs, P. I., & Kogan, N. (1974). Trait interrelations in implicit personality theories and questionnaire data. *Journal of Personality and Social Psychology, 30,* 198–207.

Strodtbeck, F. L., Simon, R. J., & Hawkins, C. (1965). Social status in jury deliberations. In I. D. Steiner & M. Fishbein (Eds.), *Current studies in social psychology.* New York: Holt, Rinehart and Winston.

Stroebe, W., Thompson, V., Insko, C., & Reisman, S. R. (1970). Balance and differentiation in the evaluation of linked attitude objects. *Journal of Personality and Social Psychology, 16,* 38–47.

Strube, M. J., & Garcia, J. E. (1981). A meta-analytic investigation of Fiedler's contingency model of leadership effectiveness. *Psychological Bulletin, 90,* 307–321.

Struch, N., & Schwartz, S. H. (1989). Intergroup aggression: Its predictors and distinctness from in-group bias. *Journal of Personality and Social Psychology, 56*(3), 364–373.

Stryker, S. (1980). *Symbolic interactionism: A social structural version.* Menlo Park, CA: Benjamin/Cummings.

Stryker, S. (1987). The vitalization of symbolic interactionism. *Social Psychology Quarterly, 50,* 83–94.

Stryker, S., & Gottlieb, A. (1981). Attribution theory and symbolic interactionism: A comparison. In J. H. Hawes, W. Ickes, & R. F. Kidd (Eds.), *New directions in attribution theory* (Vol. 3). Hillsdale, NJ: Erlbaum.

Stryker, S., & Serpe, R. (1981). Commitment, identity salience and role behavior: Theory and research example. In W. Ickes & E. Knowles (Eds.), *Personality, roles and social behavior.* New York: Springer-Verlag.

Stryker, S., & Serpe, R. (1982). Towards a theory of family influence in the socialization of children. In A. Kerckhoff (Ed.), *Research in sociology of education and socialization* (Vol. 4). Greenwich, CT: JAI Press.

Suchner, R. W., & Jackson, D. (1976). Responsibility and status: A causal or only a spurious relationship? *Sociometry, 39,* 243–256.

Sudman, S., & Bradburn, N. M. (1974). *Response effects in surveys.* Chicago: Aldine de Gruyter.

Sueda, K., & Wiseman, R. L. (1992). Embarrassment remediation in Japan and the United States. *International Journal of Intercultural Relations, 16*(2), 159–173.

Suls, J. M., & Miller, R. (Eds.). (1977). *Social comparison processes: Theoretical and empirical perspectives.* New York: Wiley.

Sumner, W. G. (1906). *Folkways.* New York and Boston: Ginn.

Surra, C. A. (1985). Courtship types: Variations in interdependence between partners and social networks. *Journal of Personality and Social Psychology, 49,* 357–375.

Surra, C. A. (1990). Research and theory on mate selection and premarital relationships in the 1980s. *Journal of Marriage and the Family, 52,* 844–865.

Surra, C. A., & Longstreth, M. (1990). Similarity of outcomes, interdependence, and conflict in dating relationships. *Journal of Personality and Social Psychology, 59,* 501–516.

Sussman, N. M., & Rosenfeld, H. M. (1982). Influence of culture, language and sex on conversational distance. *Journal of Personality and Social Psychology, 42,* 66–74.

Sutherland, E., Cressey, D., & Luckenbill, D. (1992). *Principles of criminology* (11th ed.). Dix Hills, NY: General Hall.

Sutherland, E. H. (1937). *The professional thief.* Chicago: University of Chicago Press.

Suttles, G. (1968). *The social order of the slum.* Chicago: University of Chicago Press.

Sutton, J. R. (1990). Bureaucrats and entrepreneurs: Institutional responses to deviant children in the United States, 1890–1920s. *American Journal of Sociology, 95,* 1367–1400.

Swann, W. B., Jr. (1987). Identity negotiation: Where two roads meet. *Journal of Personality and Social Psychology, 53,* 1038–1051.

Swann, W. B., Jr., & Predmore, S. C. (1985). Intimates as agents of social support: Sources of consolation or despair? *Journal of Personality and Social Psychology, 49,* 1609–1617.

Swann, W. B., Jr., & Schroeder, D. G. (1995). The search for beauty and truth: A framework for understanding reactions to evaluations. *Personality and Social Psychology Bulletin, 21,* 1307–1318.

Swanson, D. L. (Ed.). (1979). The uses and gratifications approach to mass communications research. *Communication Research, 3,* 3–111.

Sweeney, P. D. (1990). Distributive justice and pay satisfaction: A field test of an equity theory prediction. *Journal of Business and Psychology, 4*(3), 329–341.

Sweet, J. A., & Bumpass, L. (1987). *American families and households.* New York: Russell Sage Foundation.

Swigert, V., & Farrell, R. (1977). Normal homicides and the law. *American Sociological Review, 42,* 16–32.

Syme, S. L., & Berkman, L. (1976). Social class, susceptibility and sickness. *American Journal of Epidemiology, 104,* 1–8.

Szymanski, K., & Harkins, S. G. (1987). Social loafing and self-evaluation with a social standard. *Journal of Personality and Social Psychology, 53,* 891–897.

Tajfel, H. (1981). *Human groups and social categories: Studies in social psychology.* Cambridge, England: Cambridge University Press.

Tajfel, H. (1982a). *Social identity and intergroup relations.* Cambridge, England: Cambridge University Press.

Tajfel, H. (1982b). Social psychology of intergroup relations. *Annual Review of Psychology, 33,* 1–39.

Tajfel, H., & Billig, M. (1974). Familiarity and categorization in intergroup behavior. *Journal of Experimental Social Psychology, 10,* 159–170.

Tajfel, H., Billig, M. G., Bundy, R. P., & Flament, C. (1971). Social categorization and intergroup behaviour. *European Journal of Social Psychology, 1,* 149–178.

Tajfel, H., and Turner, J. C. (1979). An integrative theory of intergroup conflict. In W. G. Austin and S. Worchel (Eds.), *The social psychology of intergroup relations* (pp. 33–47). Monterey, CA: Brooks/Cole.

Tajfel, H., & Turner, J. C. (1986). The social identity theory of intergroup behavior. In S. Worchel & W. G. Austin (Eds.), *Psychology of intergroup relations* (2nd ed., pp. 7–24). Chicago: Nelson-Hall.

Takooshian, H., Haber, S., & Lucido, D. J. (1977). Who wouldn't help a lost child? You, maybe. *Psychology Today, 10,* 67–68.

Tanford, S., & Penrod, S. (1984). Social influence model: A formal integration of research on majority and minority influence processes. *Psychological Bulletin, 95,* 189–225.

Tangney, J. P. (1992). Situational determinants of shame and guilt in young adulthood. *Personality and Social Psychology Bulletin, 18*(2), 199–206.

Tavris, C., & Offir, C. (1984). *The longest war: Sex differences in perspective* (2nd ed.). New York: Harcourt Brace Jovanovich.

Taylor, D. A., & Belgrave, F. Z. (1986). The effects of perceived intimacy and valence on self-disclosure reciprocity. *Personality and Social Psychology Bulletin, 12*(2), 247–255.

Taylor, D. M., & Jaggi, V. (1974). Ethnocentrism and causal attribution in a South Indian context. *Journal of Cross-Cultural Psychology, 5,* 162–171.

Taylor, D. M., & Moghaddam, F. M. (1987). *Theories of intergroup relations: International social psychological perspectives.* New York: Praeger.

Taylor, D. M., & Royer, L. (1980). Group processes affecting anticipated language choice in intergroup relations. In H. Giles, W. P. Robinson, & P. M. Smith (Eds.), *Language: Social psychological perspectives.* New York: Pergamon.

Taylor, D. W., Berry, P. C., & Block, C. (1958). Does group participation when using brainstorming facilitate creative thinking? *Administrative Science Quarterly, 3,* 23–47.

Taylor, R. J., Chatters, L. M., Tucker, M. B., & Lewis, E. (1990). Developments in research on black families: A decade review. *Journal of Marriage and the Family, 52,* 993–1014.

Taylor, S. E. (1981). A categorization approach to stereotyping. In D. L. Hamilton (Ed.), *Cognitive processes in stereotyping and intergroup behavior* (pp. 83–114). Hillsdale, NJ: Erlbaum.

Taylor, S. E., & Crocker, J. (1981). Schematic bases of social information processing. In E. T. Higgins, C. P. Herman, & M. P. Zanna (Eds.), *Social cognition: The Ontario symposium* (Vol. 1). Hillsdale, NJ: Erlbaum.

Taylor, S. E., & Fiske, S. T. (1978). Salience, attention and attribution: Top of the head phenomena. In L. Berkowitz (Ed.), *Advances in experimental social psychology* (Vol. 11). New York: Academic Press.

Taylor, S. E., Fiske, S. T., Etcoff, N. L., & Ruderman, A. J. (1978). The categorical and contextual bases of person memory and stereotyping. *Journal of Personality and Social Psychology, 36,* 778–793.

Taylor, S. P. (1967). Aggressive behavior and physiological arousal as a function of provocation and the tendency to inhibit aggression. *Journal of Personality, 35,* 297–310.

Teachman, J. (1987). Family background, educational resources and educational attainment. *American Sociological Review, 52,* 548–557.

Tedeschi, J. T. (Ed.). (1981). *Impression management theory and social psychological research.* New York: Academic Press.

Tedeschi, J. T., Bonoma, T. V., & Schlenker, B. R. (1972). Influence, decision, and compliance. In J. T. Tedeschi (Ed.), *The social influence processes.* Chicago: Aldine-Atherton.

Tedeschi, J. T., Schlenker, B. R., & Lindskold, S. (1972). The exercise of power and influence: The source of influence. In J. T. Tedeschi (Ed.), *The social influence processes.* Chicago: Aldine-Atherton.

Terry, D. J., & Hogg, M. A. (1996). Group norms and the attitude-behavior relationship: A role for group identification. *Personality and Social Psychology Bulletin, 22,* 776–793.

Tesser, A., & Campbell, J. (1983). Self-definition and self-evaluation maintenance. In J. Suls & A. G. Greenwald (Eds.), *Psychological perspectives on the self* (Vol. 2). Hillsdale, NJ: Erlbaum.

Teti, D. M. & Lamb, M. E. (1989.) Socioeconomic and marital outcomes of adolescent marriage, adolescent childbirth, and their co-occurrence. *Journal of Marriage and the Family, 51,* 203–212.

Teti, D., Lamb, M., & Elster, A. (1987). Long-range socioeconomic and marital consequences of adolescent marriage in three cohorts of adult males. *Journal of Marriage and the Family, 49,* 499–506.

Tetlock, P. E. (1980). Explaining teacher evaluations of pupil performance: A self-presentation interpretation. *Social Psychology Quarterly, 43,* 282–290.

Tetlock, P. E. (1981). The influence of self-presentational goals on attributional reports. *Social Psychology Quarterly, 44,* 300–311.

Tetlock, P. E. (1986). A value pluralism model of ideological reasoning. *Journal of Personality and Social Psychology, 50,* 819–827.

Tetlock, P. E., & Manstead, A. S. R. (1985). Impression management versus intrapsychic explanations in social psychology: A useful dichotomy? *Psychological Review, 92,* 59–77.

Thakerar, J. N., Giles, H., & Cheshire, J. (1982). Psychological and linguistic parameters of speech accommodation theory. In C. Fraser & K. R. Scherer (Eds.), *Advances in the social psychology of language.* Cambridge, England: Cambridge University Press.

Thayer, S., & Saarni, C. (1975). Demand characteristics are everywhere (anyway): A comment on the Stanford prison experiment. *American Psychologist, 30,* 1015–1016.

Thibaut, J., & Kelley, H. (1959). *The social psychology of groups.* New York: Wiley.

Thoennes, N. A., & Pearson, J. (1985). Predicting outcomes in divorce mediation: The influence of people and process. *Journal of Social Issues, 41,* 115–126.

Thoits, P. A. (1985). Self-labelling processes in mental illness: The role of emotional deviance. *American Journal of Sociology, 91,* 221–249.

Thomas, W. I., & Znaniecki, F. (1918). *The Polish peasant in Europe and America* (Vol. 1). Boston: Badger.

Thompson, L. (1990). Negotiation behavior and outcomes: Empirical evidence and theoretical issues. *Psychological Bulletin, 108,* 515–532.

Thompson, L. (1995). They saw a negotiation: Partisanship and involvement. *Journal of Personality and Social Psychology, 68,* 839–853.

Thompson, L., & Walker, A. J. (1989.) Gender in families: Women and men in marriage, work, and parenthood. *Journal of Marriage and the Family, 51,* 845–871.

Thompson, T. L., & Zerbinos, E. (1995). Gender roles in animated cartoons: Has the picture changed in 20 years? *Sex Roles, 32,* 651–673.

Thomson, E., McLanahan, S. S., & Curtin, R. B. (1992). Family structure, gender, and parental socialization. *Journal of Marriage and the Family, 54,* 368–378.

Thorlundsson, T. (1987). Bernstein's sociolinguistics: An empirical test in Iceland. *Social Forces, 65,* 695–718.

Thornberry, T. D., & Christenson, R. L. (1984). Juvenile justice decision-making as a longitudinal process. *Social Forces, 63,* 433–444.

Thornberry, T. D., & Farnsworth, M. (1982). Social correlates of criminal involvement: Further evidence on the relationship between social status and criminal behavior. *American Sociological Review, 47,* 505–518.

Thorndike, E. L. (1920). A constant error in psychological ratings. *Journal of Applied Psychology, 4,* 25–29.

Thorne, B. (1993) *Gender play: Girls and boys in school.* New Brunswick, NJ: Rutgers University Press.

Thorne, B., Kramerae, C., & Henley, H. (Eds.) (1983). *Language, gender and society.* Rowley, MA: Newbury.

Thornton, A. (1984). Changing attitudes toward separation and divorce: Causes and consequences. *American Journal of Sociology, 90,* 856–872.

Thornton, R., & Nardi, P. (1975). The dynamics of role acquisition. *American Journal of Sociology, 80,* 870–885.

Tilker, H. A. (1970). Socially responsible behavior as a function of observer responsibility and victim feedback. *Journal of Personality and Social Psychology, 14,* 95–100.

Tilly, C., Tilly, L., & Tilly, R. (1975). *The rebellious century, 1830–1930.* Cambridge, MA: Harvard University Press.

Toch, H. (1969). *Violent men: An inquiry into the psychology of violence.* Chicago: Aldine de Gruyter.

Toi, M., & Batson, C. D. (1982). More evidence that empathy is a source of altruism. *Journal of Personality and Social Psychology, 43,* 289–292.

Tolle, E. F. (1988). Management team building: Yes but! *Engineering Management International, 4,* 277–285.

Touhey, J. (1979). Sex-role stereotyping and individual differences in liking for the physically attractive. *Social Psychology Quarterly, 42,* 285–289.

Triandis, H. C. (1980). Values, attitudes and interpersonal behavior. In H. Howe & M. Page (Eds.), *Nebraska symposium on motivation* (Vol. 27). Lincoln: University of Nebraska Press.

Triandis, H. C. (1989). The self and social behavior in differing cultural contexts. *Psychological Review, 96,* 506–520.

Triandis, H. C. (1995). *Individualism and collectivism.* Boulder, CO: Westview.

Triandis, H. C., McCusker, C., & Hui, C. H. (1990). Multimethod probes of individualism and collectivism. *Journal of Personality and Social Psychology, 59,* 1006–1020.

Trivers, R. L. (1983). The evolution of cooperation. In D. L. Bridgeman (Ed.), *The nature of prosocial behavior.* New York: Academic Press.

Trope, Y., & Cohen, O. (1989). Perceptual and inferential determinants of behavior-correspondent attributions. *Journal of Experimental Social Psychology, 25,* 142–158.

Trope, Y., Cohen, O., & Maoz, Y. (1988). The perceptual and inferential effects of situational inducements on dispositional attribution. *Journal of Personality and Social Psychology, 55,* 165–177.

Turner, J. C. (1981). The experimental social psychology of intergroup behavior. In J. C. Turner & H. Giles (Eds.), *Intergroup behavior* (pp. 66–101). Oxford, England: Basil Blackwell.

Turner, J. C. (1982). Toward a redefinition of the social group. In H. Tajfel (Ed.), *Social identity and intergroup relations* (pp. 15–40). Cambridge, England: Cambridge University Press.

Turner, J. C. (1991). *Social influence.* Pacific Grove, CA: Brooks/Cole.

Turner, J. C., Wetherell, M. S., & Hogg, M. A. (1989). Referent informational influence and group polarization. *British Journal of Social Psychology, 28,* 135–147.

Turner, R. H. (1978). The role and the person. *American Journal of Sociology, 84,* 1–23.

Turner, R. H. (1990). Role change. In W. R. Scott & J. Blake (Eds.), *Annual review of sociology* (Vol. 16, pp. 87–110). Palo Alto, CA: Annual Reviews.

Turner, R. H., & Killian, L. M. (1972). *Collective behavior* (2nd ed.). Englewood Cliffs, NJ: Prentice-Hall.

Tyler, T. R., & Caine, A. (1981). The influence of outcomes and procedures on satisfaction with formal leaders. *Journal of Personality and Social Psychology, 41,* 642–655.

Tyler, T. R., & Folger, R. (1980). Distributional and procedural aspects of satisfaction with citizen-police encounters. *Basic and Applied Social Psychology, 1,* 281–292.

Tyler, T. R., & Lind, E. A. (1992). A relational model of authority in groups. In M. Zanna (Ed.), *Advances in experimental social psychology* (Vol. 25, pp. 115–191). San Diego, CA: Academic Press.

Tyler, T. R., & Sears, D. O. (1977). Coming to like obnoxious people when we must live with them. *Journal of Personality and Social Psychology, 35,* 200–211.

Tziner, A. (1982). Differential effects of group cohesiveness types: A clarifying overview. *Social Behavior and Personality, 10,* 227–239.

Udry, J. R., & Billy, J. O. G. (1987). Initiation of coitus in early adolescence. *American Sociological Review, 52,* 841–855.

Umberson, D., Chen, M. D., House, J. S., Hopkins, K., & Slaten, E. (1996). The effect of social relationships on psychological well-being: Are men and women really so different? *American Sociological Review, 61,* 837–857.

U.S. Bureau of Labor Statistics. (1989). *Handbook of labor statistics* (Bulletin No. 2340). Washington, DC: U.S. Government Printing Office.

U.S. Bureau of the Census. (1987). *School enrollment—social and economic characteristics of students: Oct. 1987* (Current Population Reports, Series P-20, No. 413). Washington, DC: U.S. Government Printing Office.

U.S. Bureau of the Census. (1992). *Statistical abstract of the United States: 1992* (112th ed.). Washington, DC: U.S. Government Printing Office.

U.S. Bureau of the Census. (1993). *My daddy takes care of me: Fathers as care providers.* Washington, DC: U.S. Government Printing Office.

U.S. Bureau of the Census. (1996). *Statistical abstract of the United States: 1996.* (116th ed.). Washington, DC: U.S. Government Printing Office.

Valle, V. A., & Frieze, I. H. (1976). Stability of causal attributions as a mediator in changing expectations for success. *Journal of Personality and Social Psychology, 35,* 579–589.

Vandewater, E. A., Ostrove, J. M., & Stewart, A. (1997). Predicting women's well-being in midlife: The importance of personality development and social role involvements. *Journal of Personality and Social Psychology, 72,* 1147–1160.

Van Gennep, A. (1960). *The rites of passage.* Chicago: University of Chicago Press. (Original work published 1908)

Van Maanen, J. (1976). Breaking in: Socialization to work. In R. Dubin (Ed.), *Handbook of work, organization and society.* Chicago: Rand McNally.

Van Yperen, N. W., Hagedoorn, M., & Geurts, S. A. E. (1996). Intent to leave and absenteeism as reactions to perceived inequity: The role of psychological and social constraints. *Journal of Occupational and Organizational Psychology, 69*(4), 367–372.

Vecchio, R. P. (1977). An empirical examination of the validity of Fiedler's model of leadership effectiveness. *Organizational Behavior and Human Performance, 19,* 180–206.

Vecchio, R. P. (1983). Assessing the validity of Fiedler's contingency model of leadership: A closer look at Strube and Garcia. *Psychological Bulletin, 93,* 600–603.

Vega, W. A., & Rumbaut, R. G. (1991). Ethnic minorities and mental health. *Annual Review of Sociology, 17,* 351–383.

Verbrugge, L. (1979). Marital status and health. *Journal of Marriage and the Family, 41,* 267–285.

Verplanck, W. S. (1955). The control of the content of conversation: Reinforcement of statements of opinion. *Journal of Abnormal and Social Psychology, 51,* 668–676.

Vidmar, N. (1974). Effects of group discussion on category width judgments. *Journal of Personality and Social Psychology, 29,* 187–195.

Vinokur, A., & Burnstein, E. (1978). Novel argumentation and attitude change: The case of polarization following group discussion. *European Journal of Social Psychology, 8,* 335–348.

Vinokur, A. D., Price, R. H., & Caplan, R. D. (1996). Hard times and hurtful partners: How financial strain affects depression and relationship satisfaction of unemployed persons and their spouses. *Journal of Personality and Social Psychology, 71,* 166–179.

Voissem, N. H., & Sistrunk, F. (1971). Communication schedule and cooperative game behavior. *Journal of Personality and Social Psychology, 19,* 160–167.

Volling, B. L., & Belsky, J. (1991). Multiple determinants of father involvement during infancy in dual career and single-earner families. *Journal of Marriage and the Family, 53,* 461–474.

von Baeyer, C. L., Sherk, D. L., & Zanna, M. P. (1981). Impression management in the job interview: When the female applicant meets the male (chauvinist) interviewer. *Personality and Social Psychology Bulletin, 7,* 45–51.

Voydanoff, P. (1990). Economic distress and family relations: A review of the eighties. *Journal of Marriage and the Family, 52,* 1099–1115.

Vygotsky, L. S. (1962). *Thought and language.* Cambridge, MA: MIT Press.

Waite, L., Haggstrom, G., & Kanouse, D. (1986). The effects of parenthood on the career orientations and job characteristics of young adults. *Social Forces, 65,* 43–73.

Waldron, I. (1976). Why do women live longer than men? *Social Science and Medicine, 10,* 349–362.

Walker, H. A., Thomas, G. M., & Zelditch, M., Jr. (1986). Legitimation, endorsement, and stability. *Social Forces, 64,* 620–643.

Walker, L. L., & Heyns, R. W. (1962). *An anatomy for conformity.* Englewood Cliffs, NJ: Prentice-Hall.

Wall, V. D., & Nolan, L. L. (1987). Small group conflict: A look at equity, satisfaction, and styles of conflict management. *Small Group Behavior, 18,* 188–211.

Wallach, L., & Wallach, M. A. (1991). Why altruism, even if it exists, cannot be demonstrated by social psychological experiments. *Psychological Inquiry, 2,* 153–155.

Wallin, P. (1950). Cultural contradictions and sex roles: A repeat study. *American Sociological Review, 15,* 288–293.

Walsh, E. J., & Taylor, M. (1982). Occupational correlates of multidimensional self-esteem: Comparisons among garbage collectors, bartenders, professors and other workers. *Sociology and Social Research, 66,* 252–258.

Walsh, E. J., & Warland, R. H. (1983). Social movement involvement in the wake of a nuclear accident: Activists and free riders in the TMI area. *American Sociological Review, 48,* 764–780.

Walster (Hatfield), E., Aronson, E., & Abrahams, D. (1966). On increasing the persuasiveness of a low prestige communicator. *Journal of Experimental Social Psychology, 2,* 325–342.

Walster (Hatfield), E., Aronson, V., Abrahams, D., & Rottman, L. (1966). The importance of physical attractiveness in dating behavior. *Journal of Personality and Social Psychology, 4,* 508–516.

Walster (Hatfield), E., Berschied, E., & Walster, G. W. (1973). New directions in equity research. *Journal of Personality and Social Psychology, 25,* 151–176.

Walster (Hatfield), E., Walster, G. W., & Berscheid, E. (1978). *Equity: Theory and research.* Boston: Allyn & Bacon.

Walters, J., & Walters, L. (1980). Parent-child relationships: A review, 1970–1979. *Journal of Marriage and the Family, 42,* 807–822.

Warner, L. C., & DeFleur, M. (1969). Attitude as an interactional concept: Social constraint and social distance as intervening variables between attitudes and action. *American Sociological Review, 34,* 153–169.

Wasserman, I. (1977). Southern violence and the political process. *American Sociological Review, 42,* 359–362.

Watson, D. (1982). The actor and the observer: How are their perceptions of causality divergent? *Psychological Bulletin, 92,* 682–700.

Watts, B. L., Messe, L. A., & Vallacher, R. R. (1982). Toward understanding sex differences in pay allocation: Agency, communion, and reward distribution behavior. *Sex Roles, 8*(12), 1175–1187.

Webb, E. J., Campbell, D. J., Schwartz, R. D., & Sechrest, L. (1981). *Unobtrusive measures: Nonreactive research in the social sciences* (2nd ed.). Chicago: Rand McNally.

Weber, R., & Crocker, J. (1983). Cognitive processes in the revision of stereotypic beliefs. *Journal of Personality and Social Psychology, 45*(5), 961–977.

Webster, M., Jr., & Driskell, J. E., Jr. (1978). Status generalization: A review and some new data. *American Sociological Review, 43,* 220–236.

Webster, M., Jr., & Foschi, M. (1988). *Status generalization: New theory and research.* Stanford, CA: Stanford University Press.

Webster, M., & Smith, L. F. (1978). Justice and revolutionary coalitions: A test of two theories. *American Journal of Sociology, 84,* 267–292.

Webster, P. S., Orbuch, T. L., & House, J. S. (1995). Effects of childhood family background on adult marital quality and perceived stability. *American Journal of Sociology, 101,* 404–432.

Weed, F. J. (1990). The victim-activist role in the anti-drunk driving movement. *Sociological Quarterly, 31,* 459–473.

Wegner, D. M., & Vallacher, R. R. (1977). *Implicit psychology: An introduction to social cognition.* New York: Oxford University Press.

Weigand, B. (1994). Black money in Belize: The ethnicity and social structure of black-market crime. *Social Forces, 73,* 135–154.

Weigel, R. H., & Newman, L. (1976). Increasing attitude-behavior correspondence by broadening the scope of the behavioral measure. *Journal of Personality and Social Psychology, 33,* 793–802.

Weinberg, M. (1976). The nudist management of respectability. In M. Weinberg (Ed.), *Sex research: Studies from the Kinsey Institute.* New York: Oxford University Press.

Weiner, B. (1985). An attributional theory of achievement motivation and emotion. *Psychological Review, 92,* 548–573.

Weiner, B. (1986). *An attributional theory of motivation and emotion.* New York: Springer-Verlag.

Weiner, B., Amirkhan, J., Folkes, V. S., & Verette, J. A. (1987). An attributional analysis of excuse giving: Studies of a naive theory of emotion. *Journal of Personality and Social Psychology, 52,* 316–324.

Weiner, B., Frieze, I., Kukla, A., Reed, L., Rest, B., & Rosenbaum, R. M. (1971). *Perceiving the causes of success and failure.* Morristown, NJ: General Learning Press.

Weiner, B., Heckhausen, H., Meyer, W. U., & Cook, R. E. (1972). Causal ascriptions and achievement behavior: A conceptual analysis of effort and reanalysis of locus of control. *Journal of Personality and Social Psychology, 21,* 239–248.

Weiner, B., Perry, R. P., & Magnusson, J. (1988). An attributional analysis of reactions to stigmas. *Journal of Personality and Social Psychology, 55*(5), 738–748.

Weinstein, E. A., & Deutschberger, P. (1963). Some dimensions of altercasting. *Sociometry, 26,* 454–466.

Weiss, R. S. (1973). *Loneliness: The experience of emotional and social isolation.* Cambridge, MA: MIT Press.

Weitzer, R. (1991). Prostitute's rights in the United States: The failure of a movement. *Sociological Quarterly, 32,* 23–41.

Weitzman, L. J., Eifler, D., Hokada, E., & Ross, K. (1972). Sex role socialization in picture books for pre-school children. *American Journal of Sociology, 77,* 1125–1150.

Weldon, E., & Weingart, L. R. (1993). Group goals and group performance. *British Journal of Social Psychology, 32,* 307–334.

Wellman, B. (1979). The community question: The intimate networks of east Yorkers. *American Journal of Sociology, 84,* 1201–1231.

Wellman, B., & Worley, S. (1990). Different strokes from different folks: Community ties and social support. *American Journal of Sociology, 96,* 558–588.

Werner, C., & Parmelee, P. (1979). Similarity of activity preferences among friends: Those who play together stay together. *Social Psychology Quarterly, 42,* 62–66.

Whalen, M. R., & Zimmerman, D. H. (1987). Sequential and institutional contexts in calls for help. *Social Psychology Quarterly, 50,* 172–185.

Wheeler, L. (1966). Toward a theory of behavioral contagion. *Psychological Review, 73,* 179–192.

White, C., Bushnell, N., & Regnemer, J. (1978). Moral development in Bahamian school children: A 3-year examination of Kohlberg's stages of moral development. *Developmental Psychology, 14,* 58–65.

White, G. (1980). Physical attractiveness and courtship progress. *Journal of Personality and Social Psychology, 39,* 660–668.

White, J. W., & Gruber, K. J. (1982). Instigative aggression as a function of past experience and target characteristics. *Journal of Personality and Social Psychology, 42,* 1069–1075.

White, L., & Edwards, J. N. (1990). Emptying the nest and parental well-being: An analysis of national panel data. *American Sociological Review, 55,* 235–242.

White, R. W. (1989). From peaceful protest to guerilla war: Micromobilization of the Provisional Irish Republican Army. *American Journal of Sociology, 94,* 1277–1302.

Whitt, H. P., & Meile, R. L. (1985). Alignment, magnification and snowballing: Processes in the definition of symptoms of mental illness. *Social Forces, 63,* 682–697.

Whittier, N. (1997). Political generations, micro-cohorts, and the transformation of social movements. *American Sociological Review, 62,* 760–778.

Whorf, B. L. (1956). *Language, thought and reality: Selected writings.* Cambridge, MA: MIT Press.

Wicker, A. W. (1969). Attitudes versus actions: The relationship of verbal and overt behavioral responses to attitude objects. *Journal of Social Issues, 25,* 41–78.

Wicklund, R. A. (1975). Objective self-awareness. In L. Berkowitz (Ed.), *Advances in experimental social psychology* (Vol. 8). New York: Academic Press.

Wicklund, R. A., & Brehm, J. (1976). *Perspectives on cognitive dissonance.* Hillsdale, NJ: Erlbaum.

Wicklund, R. A., & Frey, D. (1980). Self-awareness theory: When the self makes a difference. In D. M. Wegner & R. R. Vallacher (Eds.), *The self in social psychology.* New York: Oxford University Press.

Wickrama, K., Conger, R. D., Lorenz, F. O., & Matthews, L. (1995). Role identity, role satisfaction, and perceived physical health. *Social Psychology Quarterly, 58,* 270–283.

Wickrama, K. A. S., Lorenz, F. O., Conger, R. D., & Elder, G. H., Jr. (1997). Marital quality and physical illness: A latent growth curve analysis. *Journal of Marriage and the Family, 59,* 143–155.

Wiggins, J. S., Wiggins, N., & Conger, J. C. (1968). Correlates of heterosexual somatic preference. *Journal of Personality and Social Psychology, 10,* 82–90.

Wilcox, C., & Williams, L. (1990). Taking stock of schema theory. *Social Science Journal, 27,* 373–393.

Wilder, D. A. (1981). Perceiving persons as a group: Categorization and intergroup relations. In D. L. Hamilton (Ed.), *Cognitive processes in stereotyping and intergroup behavior* (pp. 213–257). Hillsdale, NJ: Erlbaum.

Wilder, D. A., & Thompson, J. E. (1980). Intergroup contact with independent manipulations on in-group and out-group interaction. *Journal of Personality and Social Psychology, 38*(4), 589–603.

Wiley, M. G. (1973). Sex roles in games. *Sociometry, 36,* 526–541.

Wilke, H., & Lanzetta, J. T. (1982). The obligation to help: Factors affecting response to help received. *European Journal of Social Psychology, 12,* 315–319.

Wilke, H. A. M. (1985). Coalition formation from a socio-psychological point of view. In H. A. M. Wilke (Ed.), *Coalition formation* (chapter 3). Amsterdam: Elsevier.

Wilke, H. A. M., Van Knippenberg, A. F. M., & Bruins, J. (1986). Conservative coalitions: An expectation states approach. *European Journal of Social Psychology, 16,* 51–63.

Williams, D. R. (1990). Socioeconomic differentials in health: A review and redirection. *Social Psychology Quarterly, 53,* 81–99.

Williams, K. D., Harkins, S. G., & Latané, B. (1981). Identifiability as a deterrent to social loafing: Two cheering experiments. *Journal of Personality and Social Psychology, 40,* 303–311.

Williams, K. D., Nida, S. A., Baca, L. D., & Latané, B. (1989). Social loafing and swimming: Effects of identifiability on individual and relay performance of intercollegiate swimmers. *Basic and Applied Social Psychology, 10,* 73–81.

Williams, R. H. (1995). Constructing the public good: Social movements and cultural resources. *Social Problems, 42,* 124–144.

Williams, R. M., Jr. (1977). *Mutual accommodation: Ethnic conflict and cooperation.* Minneapolis: University of Minnesota Press.

Willis, F. N., & Carlson, R. A. (1993). Singles ads: Gender, social class, and time. *Sex Roles, 29,* 387–404.

Wills, T. A. (1992). The helping process in the context of personal relationships: In S. Spacapan & S. Oskamp (Eds.), *Helping and being helped: Naturalistic studies* (pp. 17–48). Newbury Park, CA: Sage.

Wilmoth, J. K., & Ball, P. (1995). Arguments and action in the life of a social problem: A case study of "overpopulation," 1946–1990. *Social Problems, 42,* 318–343.

Wilson, E. O. (1971). *The insect societies.* Cambridge, MA: Belknap.

Wilson, E. O. (1975). *Sociobiology: The new synthesis.* Cambridge, MA: Harvard University Press.

Wilson, E. O. (1978). *On human nature.* Cambridge, MA: Harvard University Press.

Windle, M., & Dumenci, L. (1997). Parental and occupational stress as predictors of depressive symptoms among dual-income couples: A multilevel modeling approach. *Journal of Marriage and the Family, 59,* 625–634.

Winterbottom, M. (1958). The relation of need for achievement to learning experiences in independence and mastery. In J. Atkinson (Ed.), *Motives in fantasy, action and society.* Princeton, NJ: Van Nostrand.

Wisconsin State Journal (Madison, WI). (1997, January 26). Body rebuilding, p. 1G.

Wit, A. P., Wilke, H. A. M., & Van Dijk, E. (1989). Attribution of leadership in a resource management situation. *European Journal of Social Psychology, 19,* 327–338.

Wofford, J. (1970). Factor analysis of managerial behavior. *Journal of Applied Psychology, 54,* 169–173.

Wolf, S. (1985). Manifest and latent influence of majorities and minorities. *Journal of Personality and Social Psychology, 48,* 899–908.

Wolf, S., & Bugaj, A. M. (1990). The social impact of courtroom witnesses. *Social Behaviour, 5*(1), 1–13.

Wolf, S., & Latané, B. (1983). Majority and minority influences on restaurant preferences. *Journal of Personality and Social Psychology, 45*(2), 282–292.

Woll, S. B., & Young, P. (1989). Looking for Mr. or Ms. Right: Self-presentation in videodating. *Journal of Marriage and the Family, 51,* 483–488.

Won-Doornink, M. J. (1979). On getting to know you. The association between the stage of a relationship and the reciprocity of self-disclosure. *Journal of Experimental Social Psychology, 15,* 229–241.

Won-Doornink, M. J. (1985). Self-disclosure and reciprocity in conversation: A cross-national study. *Social Psychology Quarterly, 48,* 97–107.

Wood, W., Lundgren, S., Ouellette, J. A., Busceme, S., & Blackstone, T. (1994). Minority influence: A meta-analytic review of social influence processes. *Psychological Bulletin, 115,* 323–345.

Worchel, S. (1986). The role of cooperation in reducing intergroup conflict. In S. Worchel & W. G. Austin (Eds.), *Psychology of intergroup relations* (2nd ed., pp. 288–304). Chicago: Nelson-Hall.

Worchel, S., Lind, E., & Kaufman, K. (1975). Evaluations of group products as a function of expectations of group longevity, outcome of competition, and publicity of evaluations. *Journal of Personality and Social Psychology, 31,* 1089–1097.

Worchel, S., & Norvell, N. (1980). Effect of perceived environmental conditions during cooperation on intergroup attraction. *Journal of Personality and Social Psychology, 38*(5), 764–772.

Wright, E. O., Costello, C., Hachen, D., & Sprague, J. (1982). The American class structure. *American Sociological Review, 47,* 709–726.

Wrightsman, L. S. (1969). Wallace supporters and adherence to law and order. *Journal of Personality and Social Psychology, 13,* 17–22.

Wu, X., & DeMaris, A. (1996). Gender and marital status differences in effects of chronic strain. *Sex Roles, 34,* 299–319.

Wulfert, E., & Wan, C. K. (1995). Safer sex intentions and condom use viewed from a health belief, reasoned action, and social cognitive perspective. *Journal of Sex Research, 32,* 299–311.

Wyer, R. S., Jr. (1966). Effects of incentive to perform well, group attraction, and group acceptance on conformity in a judgmental task. *Journal of Personality and Social Psychology, 4,* 21–26.

Wyer, R. S., Jr., & Srull, T. K. (Eds.). (1984). *Handbook of social cognition* (Vols. 1–3). Hillsdale, NJ: Erlbaum.

Wylie, R. C. (1979). *The self-concept: Theory and research on selected topics.* (rev. ed., Vol. 2). Lincoln: University of Nebraska Press.

Yarrow, M., Schwartz, C., Murphy, H., & Deasy, L. (1955). The psychological meaning of mental illness in the family. *Journal of Social Issues, 11,* 12–24.

Young, R. A. (1984). Vocational choice and values in adolescent women. *Sex Roles, 10,* 485–492.

Youngs, G. A., Jr. (1986). Patterns of threat and punishment reciprocity in a conflict setting. *Journal of Personality and Social Psychology, 51,* 541–546.

Yukl, G. (1981). *Leadership in organizations.* Englewood Cliffs, NJ: Prentice-Hall.

Zaccaro, S. J. (1984). Social loafing: The role of task attractiveness. *Personality and Social Psychology Bulletin, 10,* 99–106.

Zajonc, R. B. (1968). The attitudinal effects of mere exposure. *Journal of Personality and Social Psychology, 9* (monograph supplement no. 2), Part 2, 1–27.

Zander, A. (1985). *The purposes of groups and organizations.* San Francisco: Jossey-Bass.

Zanna, M., & Fazio, R. (1982). The attitude-behavior relation: Moving toward a third generation of research. In M. Zanna, E. Higgins, & C. Herman (Eds.), *Consistency in social behavior: The Ontario symposium* (Vol. 2). Hillsdale, NJ: Erlbaum.

Zanna, M. P., & Hamilton, D. L. (1977). Further evidence for meaning change in impression formation. *Journal of Experimental Social Psychology, 13,* 224–238.

Zantra, A. J., Reich, J. W., & Guarnaccia, C. A. (1990.) Some everyday life consequences of disability and bereavement for older adults. *Journal of Personality and Social Psychology, 59,* 550–561.

Zartman, I. W., & Touval, S. (1985). International mediation: Conflict resolution and power politics. *Journal of Social Issues, 41,* 27–45.

Zebrowitz, L., Voinescu, L., & Collins, M. A. (1996). "Wide-eyed" and "crooked-faced": Determinants of perceived and real honesty across the life span. *Personality and Social Psychology Bulletin, 22,* 1258–1269.

Zelditch, M., Jr. (1972). Authority and performance expectations in bureaucratic organizations. In C. G. McClintock (Ed.), *Experimental social psychology.* New York: Holt, Rinehart and Winston.

Zelditch, M., Jr., Lauderdale, P., & Stublarec, S. (1980). How are inconsistencies between status and disability resolved? *Social Forces, 58,* 1025–1043.

Zelditch, M., Jr., & Walker, H. A. (1984). Legitimacy and the stability of authority. In E. J. Lawler (Ed.), *Advances in group processes* (Vol. 1, pp. 1–26). Greenwich, CT: JAI Press.

Zeller, R. A., Neal, A., & Groat, H. (1980). On the reliability and stability of alienation measures: A longitudinal analysis. *Social Forces, 58,* 1195–1204.

Zelnik, M., & Kantner, J. (1981). Sexual and contraceptive experience of young unmarried women in the United States, 1976 and 1971. In F. F. Furstenberg, Jr., R. Lincoln, & J. Mencken (Eds.), *Teenage sexuality, pregnancy and childbearing.* Philadelphia: University of Pennsylvania Press.

Zey, M. (1993). *Banking on fraud: Junk bonds and buyouts.* New York: Aldine de Gruyter.

Zillman, D. (1978). Attribution and misattribution of excitatory reactions. In J. Harvey, W. Ickes, & R. F. Kidd (Eds.), *New directions in attribution research* (Vol. 2). Hillsdale, NJ: Erlbaum.

Zillman, D. (1979). *Hostility and aggression.* Hillsdale, NJ: Erlbaum.

Zimbardo, P. G. (1969). The human choice: Individuation, reason and order versus de-individuation, impulse and chaos. In W. J. Arnold & D. Levine (Eds.), *Nebraska symposium on motivation, 1969.* Lincoln: University of Nebraska Press.

Zimmerman, D. H., & West, C. (1975). Sex roles, interruptions and silences in conversations. In B. Thorne & N. Henley (Eds.), *Language and sex: Difference and dominance.* Rowley, MA: Newbury.

Zimmerman, R., & DeLamater, J. (1983). *Threat of topic, social desirability, self-awareness and accuracy of self-report.* Unpublished manuscript.

Zipf, S. G. (1960). Resistance and conformity under reward and punishment. *Journal of Abnormal and Social Psychology, 61,* 102–109.

Zollar, A., & Williams, J. (1987). The contribution of marriage to the life satisfaction of black adults. *Journal of Marriage and the Family, 49,* 87–92.

Zuckerman, M., DePaulo, B. M., & Rosenthal, R. (1981). Verbal and nonverbal communication of deception. In L. Berkowitz (Ed.), *Advances in experimental social psychology* (Vol. 14). New York: Academic Press.

Zuckerman, M., Koestner, R., & Alton, A. O. (1984). Learning to detect deception. *Journal of Personality and Social Psychology, 46,* 519–528.

Zurcher, L. A., & Snow, D. A. (1990). Collective behavior and social movements. In M. Rosenberg & R. H. Turner (Eds.), *Social psychology: Sociological perspectives.* New Brunswick, NJ: Transaction.

Zvonkovic, A. M., Greaves, K. M., Schmiege, C. J., & Hull, C. D. (1996). The martial construction of gender through work and family decisions: A qualitative analysis. *Journal of Marriage and the Family, 58,* 91–100.

CREDITS

Photographs

Chapter 1 p. 7: Frank Siteman/Jeroboam; p. 11: Bob Daemmrich/Stock, Boston; p. 13: Mark Tetrault/Picture Cube.

Chapter 2 p. 28: David Frazier; p. 34: Stock, Boston; p. 37: Mike Kagan/Monkmeyer; p. 39: Michael Newman/PhotoEdit.

Chapter 3 p. 47: Michael Newman/PhotoEdit; p. 57: David Shaefer/Jeroboam; p. 59: David Sams/Stock, Boston; p. 61: David Woo/Stock, Boston; p. 65: Paul Damien/Tony Stone; p. 69: Paul Conklin/Monkmeyer.

Chapter 4 p. 77: Elizabeth Crews; p. 79: Robert Hollister Davis/Gamma Liaison; p. 87: Geoffrey Hiller/Black Star; p. 94: Reuters/Laszlo Balogh/Archive; p. 98: Hector Amezcua—The Fresno Bee/AP/Wide World.

Chapter 5 p. 109: Judy S. Gelles/Stock, Boston; p. 115: Don Smetzer/Tony Stone; p. 124: Dean Abramson/Stock, Boston; p. 127: Ajit Kumar/AP/Wide World.

Chapter 6 p. 134: Farley Andrews/Picture Cube; p. 144 (left): courtesy American Cancer Society; p. 144 (right): Brown & Williamson Tobacco Company; p. 147: Sylvia Johnson/Woodfin Camp; p. 156: Peter Menzel/Stock, Boston.

Chapter 7 p. 160: Bob Daemmrich/Stock, Boston; p. 167: Jane Scherr/Jeroboam; p. 169: George Bellerose/Stock, Boston; p. 170: courtesy of Dr. Paul Ekman; p. 173: R. Lord/Image Works; p. 181: Bob Daemmrich/Stock, Boston.

Chapter 8 p. 190: Sepp Seitz/Woodfin Camp; p. 193: Rhoda Sidney/Monkmeyer; p. 198: courtesy Revlon, Inc.; p. 204: AP/Wide World; p. 208: John Nordell/Picture Cube.

Chapter 9 p. 217: Sybil Shelton/Monkmeyer; p. 218: Oscar Palmquist/Lightwave; p. 220: Steven Baratz/Picture Cube; p. 221: Richard Younker/Tony Stone; p. 230: Union Tribune Publishing Company.

Chapter 10 p. 239: N. R. Rowan/Stock, Boston; p. 248: Keith Horan/Stock, Boston; p. 253: Dave Bowman/Daily Press/Gamma Liaison; p. 256: Helayne Seidman/Sipa Press.

Chapter 11 p. 266: Richard Hutchings/PhotoEdit; p. 271: Lara Regan/SABA; p. 274: Spencer Grant/Stock, Boston; p. 278: Mark Burnett/Stock, Boston; p. 281: Rick Kopstein/Monkmeyer.

Chapter 12 p. 287: Spencer Grant/Monkmeyer; p. 289: Eric Kroll; p. 301: James Carroll/Stock, Boston.

Chapter 13 p. 316: Bob Daemmrich; p. 320: Tony Freeman/PhotoEdit; p. 324 (top): R. Rotolo/Gamma Liaison; p. 324 (bottom): Dan Bryant Photos.

Chapter 14 p. 343: Tony Freeman/PhotoEdit; p. 346: Billy E. Barnes/Stock, Boston; p. 350: Rose Skytta/Jeroboam; p. 354: John Lawlor/Tony Stone.

Chapter 15 p. 361: Duomo; p. 364: Jean-Claude Lejeune/Stock, Boston; p. 368: Hugh Rogers/Monkmeyer; p. 374 (top): Phyllis Graber Jensen/Stock, Boston; p. 374 (bottom): courtesy of Catherine Comet, Conductor of the Grand Rapids, Michigan, Symphony and Shaw Concerts, Inc.

Chapter 16 p. 389: AP/Wide World; p. 391: Spencer Grant/Picture Cube; p. 396: Michael Siluk; p. 401: Mark Richards; p. 406: Mark Pichards/PhotoEdit.

Chapter 17 p. 416: Myleen Ferguson/PhotoEdit; p. 420: Frank Siteman; p. 431: Spencer Grant/Picture Cube; p. 435: Peter Freed/Newsweek.

Chapter 18 p. 442: Mary Kate Denny/PhotoEdit; p. 447 (left): Tim Barnwell/Stock, Boston; p. 447 (right): Sidney/Image Works; p. 450: Bob Rashid/Monkmeyer; p. 455: Bill Leisner/TexaStock; p. 465: Judy Gelles/Stock, Boston.

Chapter 19 p. 470: Chris Pizzolorusso/Monkmeyer; p. 474: Catherine Karnow/Woodfin Camp; p. 480: Bohdan Hrynwice/Stock, Boston; p. 484: Frank Fournier/Woodfin Camp; p. 486: The Boston Globe; p. 489: Bob Daemmrich/Tony Stone; p. 492: Spencer Grant./Monkmeyer.

Chapter 20 p. 499: Dave Cannon/Allsport; p. 501: AP/Wide World; p. 506: Lester Sloan/Woodfin Camp; p. 513: Tom Carter/Monkmeyer; p. 521 (left): Brown Brothers; p. 521 (right): Grant LeDuc/Stock, Boston.

Literary Acknowledgments

Chapter 3 Excerpt from "A Change Has Come Over Me" by Richard Cohen. © 1982 Washington Post Writers Group. Reprinted by permission. Table 3-1: THE INFANT AND CHILD IN THE CULTURE OF TODAY by Arnold Gesell and Frances L. Ilg. © 1943 by Arnold Gesell and Frances L. Ilg. Copyright renewed 1971 by Frances L. Ilg, Gerhard Gesell, and Katherine Gesell Walden. Reprinted by permission of Harper-Collins Publishers, Inc. Table 3-2: "Social Structure During the School Years: Onset of the De-Grouping Process" by W. Shrum and N. H. Cheek. © *American Sociological Review*, 52, 1987.

Reprinted with permission. Table 3-3: Adapted from "Stage and sequence: The cognitive developmental approach to socialization" by Lawrence Kohlberg. © 1969 Lawrence Kohlberg. Used with permission.

Chapter 4   Box 4.1: SELF-CONCEPTIONS: CONFIGURA-TIONS OF CONTENT by Gordon and Gergen. Copyright © 1968 by John Wiley & Sons, Inc. Reprinted by permission of the publisher. Table 4-1: "Psychological Perspectives on the Self" by J. Suls. Copyright ©1982 by Lawrence Erlbaum Associates, Inc. Reprinted by permission. Figure 4-1: Adapted from "Trait Salience in the Spontaneous Self-Concept" by W. J. McGuire and A. Padawer-Singer. Copyright © 1976 by the American Psychological Association. Adapted with permission. Figure 4-2: Adapted from "Threat of Own Prejudice and Reverse Discrimination in Interracial Situations" by D. Dutton and R. Lake. Copyright © 1973 by the Americal Psychological Association. Adapted with permission.

Chapter 5   Figure 5-1: Adapted from "Relationships Among Attributes: A Mental Map" by Rosenberg and Vivekonanthon. Copyright © 1968 by the American Psychological Association. Adapted with permission. Figure 5-2: Adapted from "The Focus of Attention Bias" by S. E. Taylor and S. T. Fiske. Copyright © 1978 by Academic Press, Inc. All rights of reproduction in any form reserved. Reprinted by permission of the author.

Chapter 6   Figure 6-5: Adapted from "Attitude as an Interactional Concept: Social Constraints and Social Distance as Intervening Variables between Attitudes and Action" by L. C. Warner and M. DeFleur. Copyright 1969 *American Sociological Review*, 34, 1969. Adapted by permission. Figure 6-6: Adapted from UNDERSTANDING ATTITUDES AND PREDICTING SOCIAL BEHAVIOR by Ajzen and Fishbein, © 1980. Adapted by permission of Prentice-Hall, Inc., Upper Saddle River, NJ.

Chapter 7   Table 7-1: Adapted from "Nonverbal Courtship Patterns in Women: Context and consequences: by M. M. Moore. Copyright © 1985 M. M. Moore. Table 7-3: Adapted from "Single Emotion Judgment Task" by P. Ekman and M. Friesan. Copyright © 1987 by the American Psychological Association. Adapted with permission. Figure 7-2: Adapted from "Invasions of Personal Slpace" By N. J. Felipe and R. Sommer. © 1966 by The Society for the Study of Social Problems. Adapted from *Social Problems*, Vol. 14, No. 2, by permission.

Chapter 8   Figure 8-3: Adapted from "Effects of Personal Involvement on Persuasion" by Petty, Cacioppo and Golman. Copyright © 1981 by the American Psychological Association. Adapted with permission. Box 8.2: Reprinted from *Social Forces*, Vol. 60, March 1982. "Compliance Under Threat of Severe Punishment" by D. F. Luckenbill. Reprinted with permission. Figure 8-4: Adapted from "Compliance With Threats" by Tedeschi and Horai. Copyright © 1969 by the American Psychological Association. Adapted by permission of the author.

Chapter 9   Figure 9-1: Adapted from "Perceptions of the Impact of Negatively Valued Characteristics on Social Interaction" by R. E. Kleck and A. Strenta. Copyright © 1980 by the Americal Psychological Association. Adapted with permission.

Chapter 10   Figure 10-1: Adapted from "Explanations of the Moderating Effect of Responsibility Denial on the Personal Norm-Behavior Relationship" by S. H. Schwartz and J. A. Howard. Copyright by the *American Sociological Review*, Vol. 43, 1980. Figure 10-2: Reprinted from "Decisions Leading to Intervention in an Emergency" by B. Latané and J. M. Darley. Copyright © 1970 and reprinted by permission of the author. Figure 10-3: Reprinted from "The Bystander Effect" by B. Latané and J. M. Darley. Copyright © 1968 by the American Psychological Association. Reprinted with permission.

Chapter 11   Figure 11-1: Adapted from "Effects of Legitimacy of Frustration on Aggression" by Kulick and Brown. Copyright © 1979 by the American Sociological Association. Table 11-1: Adapted from "Marital Violence by Husbands and Wives" by R. J. Gelles and C. P. Cornell. Copyright 1990. Adapted by permission of Sage Publications. Table 11-2: Adapted from "Control of Aggression in a Nursery School Class" by P. Brown and R. Elliott. Copyright 1988, *Journal of Experimental Child Psychology*, 2. Adapted by permission of Academic Press. Figure 11-2: Adapted from "Aggression, Catharsis and Subsequent Aggression" by R. G. Green, L. Stonner and G. I. Shope. Copyright © 1975 by the American Psychological Association. Adapted with permission. Box 11.1: Reprinted from "Rape Myth Acceptance Scale—Selections" by M. R. Burt. Copyright © 1980 by the Americal Psychological Association. Reprinted with permission. Table 11-3: Adapted from "A Typology of Sexually Assaultive Men" by Koss and Leonard. Copyright 1984 by Academic Press, Inc. Reprinted by permission of the author.

Chapter 12   Figure 12-1: Reprinted from "Levels of Pair Relatedness" by G. Levinger and J. D. Snock. Copyright 1972 by General Learning Press. Reprinted by permission of the author. Figure 12-2: Reprinted from "Self-Disclosure and Reciprocity in Conversation: A Cross-National Study" by M. J. Won-Doormick. Copyright 1985 *Social Psychology Quarterly*, Vol. 48. Reprinted by permission. Table 12-1: Adapted from "The Dyadic Trust Scale: Toward Understanding Interpersonal Trust in Close Relationships" by R. E. Larzelere and T. L. Huston. Copyright 1980 by the National Council on Family Relations, 3989 Central Ave., NE, Suite 550, Minneapolis, MN 55421. Adapted by permission. Figure 12-3: Adapted from "The Dyadic Trust Scale: Toward Understanding Interpersonal Trust in Close Relationships" by R. E. Larzelere and T. L. Huston. Copyright 1980 by the National Council on Family Relations, 3989 Central Ave., NE, Suite 550, Minneapolis, MN 55421. Adapted by permission. Box 12.2: Reprinted from "Scale for Determining Passionate Love" by Hatfield and Sprecher. Copyright 1986 by the *Journal of Adolescence*. Reprinted by permission of Academic Press, Inc. Figure 12-4: Adapted from "Occurrence of the Romantic-Love Ideal in Major American Magazines, 1741–1865" by J. Lantz, J. Keyes and M. Schultz and "The American Family in the Preindustrial Period: From Base Lines in History to Change" by H. Lantz, M. Schultz and M. O'Hara. *American Sociological Review*, Vol. 40 and Vol. 42. Copyright 1975 and 1979 by the American Sociological Association. Adapted with permission. Table 12-2: Adapted from "Breakups Before Marriage: The End of 103 Affairs" by C. Hill, Z. Rubin and L. Peplau. Copyright © 1976 *Journal of Social Issues*, 32(1). Adapted with permission.

NAME INDEX

SUBJECT INDEX